MODERN MARINE ENGINEER'S MANUAL

Volume II

CORNELL MARITIME PRESS
A Division of Schiffer Publishing, Ltd.

4880 Lower Valley Road, Atglen, PA 19310

MODERN MARINE ENGINEER'S MANUAL

Volume II

THIRD EDITION

EVERETT C. HUNT, *Editor-in-Chief*
Consulting Engineer

JAMES A. HARBACH, *Associate Editor*
U.S. Merchant Marine Academy

ALAN L. ROWEN, *Associate Editor*
Webb Institute of Naval Architecture

CONTRIBUTING EDITORS

Robert S. Behr
U.S. Department of Transportation, Maritime Administration

Davis Brown
New Sulzer Diesel, Ltd.

R. D. Jacobs
Consultant

Aaron R. Kramer
State University of New York Maritime College

Thomas P. Mackey
Hyde Marine, Inc.

James Mercanti
Camar Corporation, Inc.

William B. Morgan
Retired, Formerly David Taylor Model Basin

Keith Wilson
Consultant

Conrad C. Youngren
State University of New York Maritime College

CORNELL MARITIME PRESS
A Division of Schiffer Publishing, Ltd.
4880 Lower Valley Road, Atglen, PA 19310

Library of Congress Cataloging-in-Publication Data

Modern marine engineer's manual. — 3rd ed. / Everett C. Hunt, editor-in-chief ;
contributing editors, Gus Bourneuf, Jr. ... [et al.]
p. cm.
Includes bibliographical references and index.
ISBN 0-87033-537-5 (v.2)
1. Marine engineering. I. Hunt, Everett C., 1928–
II. Bourneuf, Gus.
VM600.M65 1999
623.8'7—dc21 99-14987
CIP

ISBN: 978-0-87033-537-2
Printed in China
First edition, 1943. Third edition, 2002;
second printing, 2010. third printing 2017

Published by Schiffer Publishing, Ltd.
4880 Lower Valley Road
Atglen, PA 19310
Phone: (610) 593-1777; Fax: (610) 593-2002
E-mail: Info@schifferbooks.com
Web: www.schifferbooks.com

For the seamen of the U.S. Merchant Marine,
who in times of national emergency
have never been found wanting

Contents

CHAPTER 18

Marine Refrigeration Systems

James A. Harbach

CHAPTER 19

Heating, Ventilation, and Air-Conditioning

James A. Harbach

CHAPTER 20

Cryogenic Cargo Systems

James A. Harbach

CHAPTER 21

Hull Machinery

Thomas P. Mackey and Alan L. Rowen

CHAPTER 22

Marine Electrical Systems

Conrad C. Youngren

CHAPTER 23

Electrical Machinery

Conrad C. Youngren

CHAPTER 24

Shipboard Central Operations Systems

Aaron R. Kramer

CHAPTER 25

Shipboard Vibration Analysis

Everett C. Hunt

CHAPTER 26

Inert Gas Systems and Crude Oil Washing Machinery

Everett C. Hunt and James Mercanti

CHAPTER 27

Coal Burning Technology

Everett C. Hunt

CHAPTER 28

Waste Disposal Systems

Everett C. Hunt

CHAPTER 29

Propellers and Propulsion

William B. Morgan

CHAPTER 30

Machinery Tests and Trials

Robert S. Behr and Everett C. Hunt

Preface

At the beginning of the twenty-first century, there are two marine engineering texts published in the United States: *Marine Engineering* and *Modern Marine Engineer's Manual,* Volumes I and II. *Marine Engineering*, published by the Society of Naval Architects and Marine Engineers, is a design-oriented text that is frequently referenced herein. *Modern Marine Engineer's Manual*, Volumes I and II, is prepared for students and engineers engaged in ship operations, ship maintenance and repair, and shipbuilding.

The first edition of this volume was edited by Alan Osbourne of the U.S. Maritime Commission in 1943 as the United States rapidly expanded the Merchant Marine to meet the waterborne transportation needs of World War II. The second edition, prepared fifty years later, was an entirely new text that reflected the vast changes in ship technology. This third edition is an updated text designed to provide information concerning new and changing technologies that have developed during the past ten years. The contributors to this new edition are qualified and experienced marine engineers most of whom were involved in the preparation of the prior edition.

This edition of Volume II begins with Chapters 16 and 17, covering principles of diesel engines and descriptions of engines frequently selected. Since diesel engines are now the most popular selection for ship propulsion, these may be the most important chapters in the volume. Chapter 16 now includes a section on exhaust emissions in recognition of international environmental concerns. Chapter 17 has been revised to incorporate a second detailed description of a popular medium speed engine.

The chapters covering refrigeration and HVAC have been revised to incorporate new refrigerants that replace the fluorocarbons being phased out by the Clean Air Act. Other international protocols governing operation and maintenance of refrigeration systems are discussed.

The propulsion systems compatible with the unique cargo of LNG vessels are described in Chapter 20. Chapter 21, covering hull machinery, has been completely rewritten to describe machinery in current use.

The computer-based integrated and centralized control system of alarms, condition monitoring, management, record keeping, and communications found on modern passenger vessels and other specialized vessels are described in Chapter 24. On such vessels the engineer officer of the watch is located in an officelike environment surrounded by computer consoles and display monitors.

Technical advances in monitoring and analysis of machinery and hull vibrations have been added to Chapter 25. New equipment and systems for handling trash, sewage, bilge water, and other waste materials have been added to Chapter 28. Chapter 30 has been rewritten to provide current and more detailed information on ship trial procedures.

In the cases where the material from the second edition has been used in this edition, the chapters have been carefully reviewed to assure that all material is correct for current usage.

Consistent with Volume I, both American customary units and SIU are used throughout the text. SIU units are the most common measurement units used in the international maritime community. In particular recognition of the needs of students, each chapter concludes with a set of review questions and references for further study.

This volume represents the total experience of the editors in the various fields of marine engineering covered. However, preparation of a book of this type requires help and cooperation from many sources. The editors are indebted to the many companies and individuals who contributed materials presented in this edition. These contributions are acknowledged at the end of each chapter.

MODERN MARINE ENGINEER'S MANUAL

Volume II

CHAPTER 16

Marine Diesel Engines

ALAN L. ROWEN AND R. D. JACOBS

INTRODUCTION

Current Status

IN 2001 diesel engines are by far the dominant choice for propulsion of merchant ships and naval auxiliary vessels. The radical increases in fuel oil prices which followed the Middle East war of 1973 elevated the fuel component of ship operating cost to the point of dwarfing most of the other factors, including machinery maintenance. The higher efficiency of diesel engines relative to steam and gas turbine plants made them the obvious choice for new construction and many major conversions. In the years since, evolutionary developments in diesel engine design, which have not been matched in steam or gas turbine plants, have emphasized these differences.

Classification

Diesel engines are probably best defined as reciprocating, *compression-ignition* engines, in which the fuel is ignited on injection by the hot, compressed charge of air in the cylinder. Beyond this they may be classified as follows:

Speed. Traditionally, diesel engines are grouped into categories of *low*, *medium,* and *high speed*, depending on crankshaft RPM and/or mean piston speed. Engine design appears to have overtaken the traditional definitions of the boundaries among these categories, however, especially when one attempts to distinguish between the medium and high speed groups, and a case can be made for additional categories. Low speed engines might best be defined as those whose crankshaft speeds are a suitable match for direct connection to a ship's propeller without reduction gearing, and so tend to have rated crankshaft speeds below 250 to 300 RPM. Most engineers would place the upper limit of the medium speed group, and the start of the high speed group, in the range of 900 to 1,200 RPM. With reference to the discussions which follow, low speed engines are usually two-stroke, in-line, crosshead engines with high stroke-to-bore ratios, while medium and high speed engines may be two- or four-stroke, in-line or V, and, with few exceptions, are trunk piston types with low stroke-to-bore ratios.

Thermodynamic cycle. Theoretical thermodynamic cycles for internal combustion engines include the *Otto* cycle, the *diesel* cycle, and a combination of the two called the *dual combustion,* mixed, or Sabathé cycle. While these are theoretical cycles that are only approached in reality, it is the dual combustion cycle that most accurately represents the operation of most diesel engines of current design.

Operating cycle. This can be *two-stroke,* in which the entire sequence of events takes place in one revolution, or *four-stroke,* in which the sequence requires two revolutions.

Cylinder grouping. Most engines of current design are vertical. There may be up to 12 cylinders in-line, or as many as 24 in a V configuration.

Air supply. This can be provided in one of three ways: (1) *Turbocharged,* in which air is supplied to the engine at a pressure above atmospheric by a compressor driven by the exhaust gases. Most engines of current design are turbocharged. (2) Turbocharged and *aftercooled,* in which the air leaving the turbocharger, at high temperature as a result of compression, is cooled before entering the cylinders. Most engines of current design, especially the larger ones, are not only turbocharged but also aftercooled. (3) *Naturally* (or normally) *aspirated,* in which the engine draws its air directly from its surroundings at atmospheric pressure. Two-stroke cycle engines that are not turbocharged are incapable of drawing in air on their own, and so must be provided with some means of supplying air to the cylinders, such as under-piston scavenging or an engine-driven low pressure blower.

Running gear can include a *trunk piston,* in which the cylinder wall must carry the side thrust of the connecting rod, or a *crosshead,* in which the side thrust is transmitted directly to the engine structure by a crosshead and crosshead guide.

Power pulses. Engines may be *single acting,* in which combustion produces one power thrust toward the crankshaft, or *double acting,* in which combustion occurs alternately on both sides of the piston, producing power thrusts alternating toward and away from the crankshaft. All major engines of current design are single acting, although some double acting engines remain in service. Another type is the *opposed piston engine,* in which combustion takes place between two pistons in each cylinder, each of which is single acting. Doxford opposed piston, low speed engines remained in production in Britain until 1981, while the Fairbanks Morse medium speed engine remains in production in 1990.

Method of fuel injection. With the *solid injection method,* fuel is injected at very high pressure developed mechanically by an engine-driven fuel pump. Solid injection is the normal method of fuel injection on engines of current design. *Air injection* uses an engine-driven high pressure air compressor to inject the fuel, and is now generally obsolete.

Combustion chamber design. In a *direct* or open chamber, the fuel is injected directly into the cylinder. Most engines of current design are of this type. In a *pre-combustion chamber* design, a portion of the cylinder volume is partially isolated to receive the fuel injection. Some higher speed engines are so designed.

Cylinder proportions. Cylinder proportions may be expressed as the *stroke-to-bore ratio.* Low speed engines may have very high ratios of 3:1 or more, but medium and high speed engines are usually constrained by air flow considerations to ratios close to one.

Cooling. An engine may be *water cooled,* in which case water is circulated through cooling passages around the combustion chamber, or *air cooled,* in which air is circulated over the external surfaces of the engine. Most marine engines are water cooled in a closed circuit by treated fresh water, which is then cooled in a closed heat exchanger by seawater, although for some applications, such as emergency generator engines, the heat exchanger may be an air-cooled radiator as in automotive applications. In any event, the lubricating oil serves as an intermediate coolant of the bearings and, in most cases, of the piston as well.

OPERATING PRINCIPLES

Thermodynamic Cycles

Theoretical thermodynamic cycles for internal combustion engines include the Otto cycle, the diesel cycle, and a combination of the two, called the dual combustion, mixed or Sabathé cycle. While these are theoretical cycles that are only approached in reality, it is the dual combustion cycle that most accurately represents the operation of most diesel engines of current design.

In the Otto cycle, a charge of fuel and air is ignited by a spark and burns explosively, so rapidly that combustion is completed before the piston begins to move down, and therefore takes place at constant volume. Otto cycle engines usually operate on gasoline and are classified as spark-ignition engines.

In the diesel and dual combustion cycles, ignition occurs when fuel is injected into the hot, compressed charge, and combustion continues after injection ceases until the fuel is consumed. Engines operating on these cycles are categorized as compression-ignition engines. The diesel cycle, in which the rate of combustion is so matched to the descent of the piston that pressure during the combustion period is constant, is difficult to achieve in practical engines. The dual combustion cycle assumes the initial combustion process to be explosive and the rest to occur at constant pressure, which more closely approximates conditions in diesel engines of current design. Most oil-burning diesel engines of current design are compression-ignition types whose thermodynamic cycle is approximated by the dual combustion cycle.

Otto and dual combustion cycles are related by the manner in which combustion takes place; if all other factors were equal, the theoretical thermal efficiency would be higher for an Otto cycle. Practical considerations prevent all these other factors from being equal, but the fact remains that the closer the dual combustion cycle can be made to approach an Otto cycle, with a large fraction of the fuel burning rapidly before the piston commences its downward stroke, the higher will be its theoretical thermal efficiency.

A modification of the dual combustion cycle known as the Miller cycle or the modified Atkinson cycle has been used with diesel engines fitted with two-stage turbochargers.

Basic Terminology

Refer to Figure 16-1. The *piston* operates in the *cylinder block,* which, in all but the smallest engines, is fitted with a replaceable cast iron *cylinder liner* as well as a separate *cylinder head.* The liner and the head are usually water cooled, while the piston is usually oil cooled except in some of the large, low speed engines where it is water cooled. The reciprocating motion of the piston is converted to rotary motion of the crankshaft by the *connecting rod,* which swivels about the *wrist* (piston or gudgeon) *pin* at the top, and at its bottom end rotates about the *crank pin.*

The inside diameter of the cylinder is the *bore.* The uppermost position of the piston (and therefore of the crank) is *top dead center,* or TDC, while the lowest is *bottom dead center,* or BDC. The distance travelled by the piston between TDC and BDC is the *stroke,* which, when multiplied by the

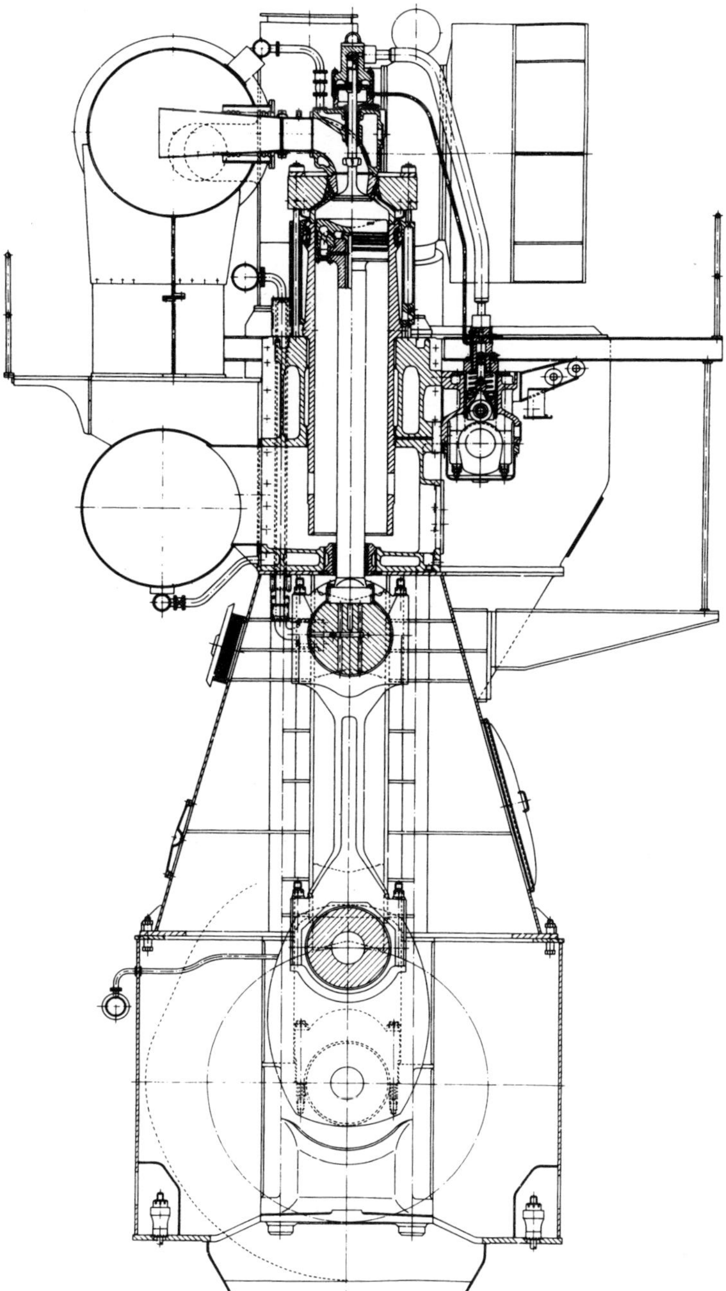

Figure 16-1. Two-stroke crosshead engine

cross-sectional area of the cylinder bore, yields the volume displaced by the piston during its stroke, called the *displacement.*

Four-Stroke Cycle Events

Cycle events are most easily understood in relation to the four-stroke cycle, shown diagrammatically in Figures 16-2 and 16-3. The latter figure represents the pressure in the cylinder plotted against the piston position, which in turn is directly proportional to the cylinder volume displaced by the piston at that point in its travel.

In four-stroke cycle engines the head contains passages connecting to air supply and exhaust manifolds, and also carries the air and exhaust valves as well as the fuel injector. Air and exhaust valves are opened into the cylinder mechanically by *push rods* and rocker arms operated by the *camshaft,* and are closed by the combination of pressure within the cylinder and the force of the valve springs. The camshaft is gear- or chain-driven from the crankshaft at one-half of crankshaft RPM, in order to complete one cycle of events in two revolutions. (Each revolution causes two strokes of the piston: one up, one down.) Starting with the piston at top dead center at the start of the charging stroke, the events are as follows:

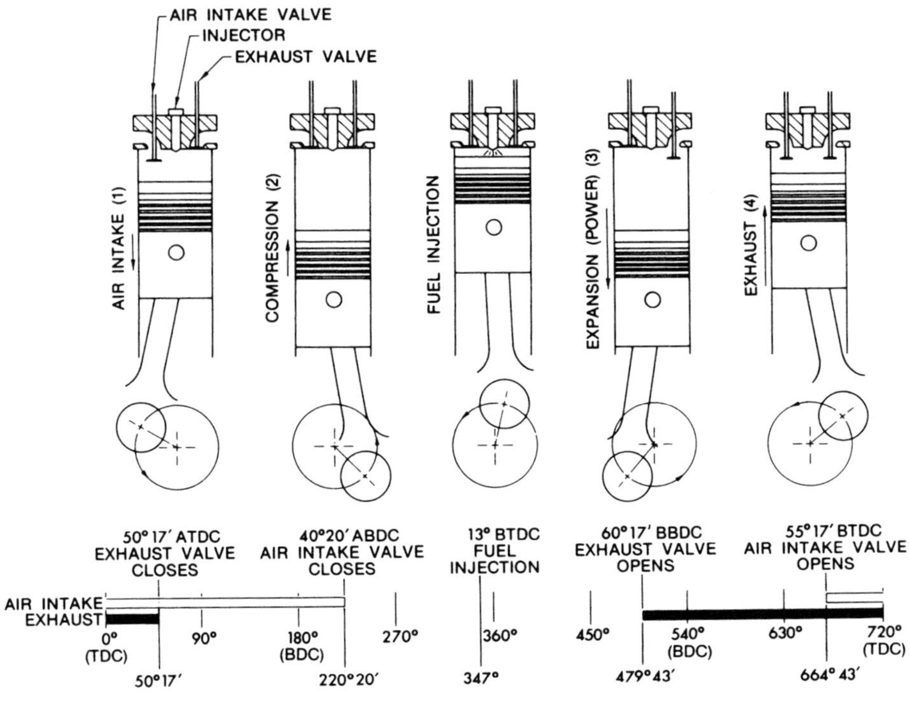

Figure 16-2. Four-stroke cycle events

1. The charging stroke (in naturally aspirated engines, this is the intake or suction stroke). The air valve is open but the exhaust valve is closed. The piston has passed the top dead center position and is being moved down by the connecting rod as the crankshaft rotates. As the piston descends, air flows into the cylinder because the pressure in the cylinder is slightly less than that in the air manifold. Power to turn the crankshaft is provided by the other cylinders in a multiple-cylinder engine, or by energy stored in the flywheel.
2. The compression stroke. The air valve closes as the piston passes through bottom dead center, trapping the charge of air in the cylinder. The piston is driven up as the crankshaft rotates, compressing the charge to one-tenth to one-twentieth of its initial volume (the actual value, called the compression ratio, is at the lower end of this range in turbocharged engines). As the charge is compressed, its temperature rises until, toward the end of the stroke, it is well above the ignition temperature of the fuel.
3. Fuel injection. Fuel injection begins during the compression stroke, before the piston reaches top dead center. Ignition will occur as soon as the first droplets of fuel are heated to ignition temperature by the hot charge. The brief time between the beginning of injection and ignition is the *ignition delay* period. (The fuel which accumulates during the ignition delay period accounts for the initial explosive combustion phase of the dual combustion cycle.)
4. The power stroke. After the piston passes through TDC, the pressure developed by the combustion of the fuel begins to force the piston down. As

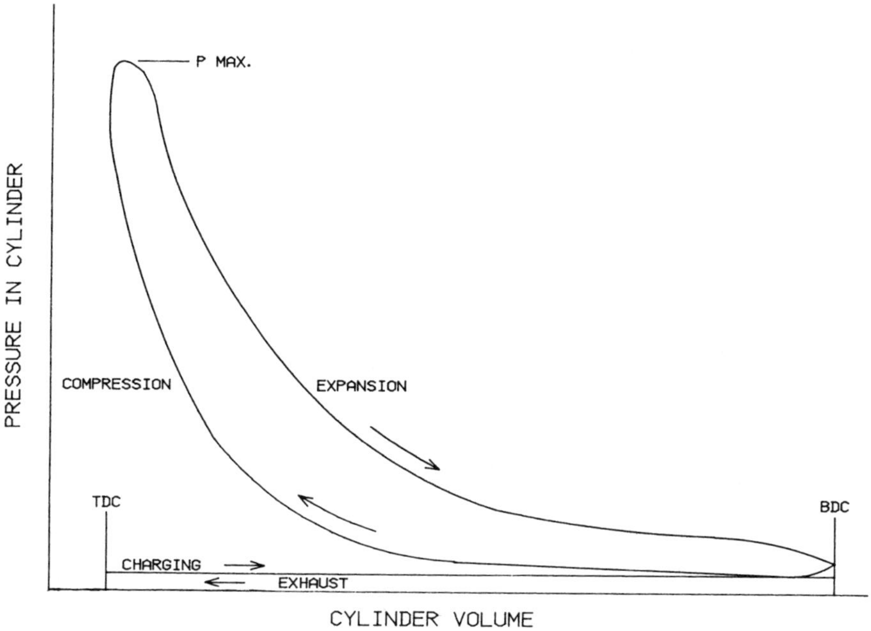

Figure 16-3. Pressure-volume diagram

the cylinder volume increases, however, the continued combustion maintains the pressure in the cylinder until injection and then combustion cease (points that are called, respectively, *cutoff* and *burnout*). After burnout, the piston continues to be forced down by the expanding gas.

The power developed is related to the quantity of fuel burned in the cylinder. The quantity is in turn proportional to the length of the injection period, if the fuel is injected at a constant rate (as it is in most engines). If the beginning of the injection period is fixed (the more common case) then at light loads cutoff occurs early, followed by a long expansion period.

5. The exhaust stroke. The exhaust stroke actually begins just before the piston reaches bottom dead center, when the exhaust valve opens and the residual high pressure in the cylinder is relieved into the exhaust manifold as the gases *blow down.* As the crankshaft pushes the connecting rod and piston up, most of the gas remaining in the cylinder is forced out. At top dead center only a fraction of the gas remains. In turbocharged engines this will be swept out as the air valve opens, just before the exhaust valve closes. This brief period when both valves are open is the *overlap* period, and the process in which incoming air sweeps the cylinder clear of exhaust gas is called *scavenging.*

As the piston passes through top dead center, with the exhaust valve closing and the air valve opening, the cycle repeats.

Two-Stroke Cycle Events

Engines operating on the two-stroke cycle may be *loop-scavenged* or *uniflow-scavenged,* as illustrated diagrammatically in Figures 16-4 and 16-5.

Figure 16-4. Loop scavenging with under-piston boost

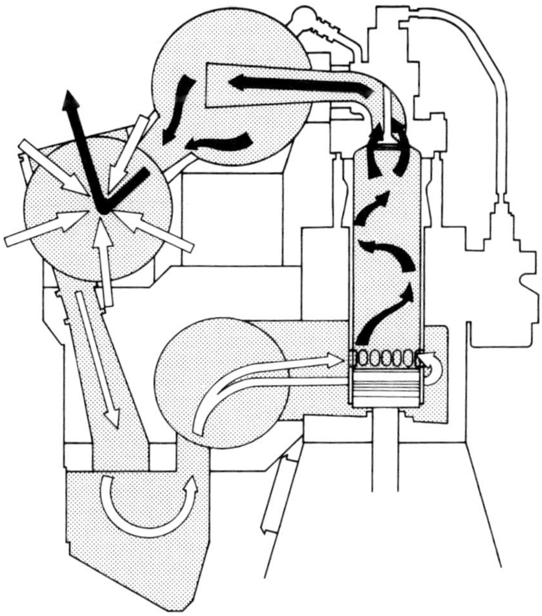

Figure 16-5. Uniflow scavenging

In general, in two-stroke cycle engines, air is supplied to the cylinder through a row of ports arranged around the circumference of the cylinder liner just above the bottom dead center position of the piston crown, the piston and ports therefore serving the same function as the air valves of the four-stroke cycle engine. In loop-scavenged engines, exhaust also takes place through a row of ports in the cylinder, these being arranged just above the air ports. Uniflow-scavenged engines (except opposed piston engines) exhaust through a valve (or two valves) in the cylinder head, which is operated by the camshaft. Since, in the two-stroke cycle, one cycle of events is completed in each revolution of the crankshaft, the camshaft speed is the same as that of the crankshaft.

In discussing events in the two-stroke cycle, it is important to bear in mind that air is always supplied to the cylinders under pressure, either the higher discharge pressure of a turbocharger or the lower pressure of a boost blower (or, in two-stroke engines without supercharging, of a scavenge air blower).

With the piston at bottom dead center at the start of a cycle, events are as follows:

1. Scavenging and charging. As the piston passes through bottom dead center, the air ports are open, as are the exhaust ports (or valves). Scavenging occurs

as the incoming air sweeps out the exhaust gases, a process which is likely to be more effective in a uniflow engine, especially in cylinders of high stroke-to-bore ratios. As the piston rises it closes off the air ports, then the exhaust ports in the loop-scavenged engine. In uniflow engines the exhaust valve is closed at this time. With the charge trapped in the cylinder, compression begins.

2. The compression stroke. As in the four-stroke cycle engine, as the piston rises, it compresses the charge to perhaps one-tenth to one-twentieth of its initial volume (the actual value, called the compression ratio, is at the lower end of the range in turbocharged engines). As the charge is compressed, its temperature rises until, toward the end of the stroke, it is well above the ignition temperature of the fuel.
3. Fuel injection. Fuel injection begins during the compression stroke, before the piston reaches top dead center. Ignition will occur as soon as the first droplets of fuel are heated to ignition temperature by the hot charge. The brief time between the beginning of injection and ignition is the *ignition delay* period. (The fuel which accumulates during the ignition delay period accounts for the initial explosive combustion phase in the dual combustion cycle.)
4. The power stroke. After the piston passes TDC the pressure developed by the combustion of the fuel begins to force the piston down. As the cylinder volume increases, however, the continued combustion will maintain the pressure in the cylinder until injection and then combustion cease (points which are called, respectively, cutoff and burnout). Subsequently, the piston continues to be forced down by the expanding gas.

The power developed is related to the quantity of fuel burned in the cylinder. The quantity is in turn proportional to the length of the injection period, if the fuel is injected at a constant rate (as it is in most engines). If the beginning of the injection period is fixed (the more common case) then at light loads cutoff occurs early, followed by a long expansion period.

5. Exhaust. Exhaust begins in the loop-scavenged engine as soon as the descending piston exposes the exhaust ports, and the residual high pressure in the cylinder is relieved into the exhaust manifold as the gases *blow down*. In the uniflow engine the exhaust valves are opened at about this time and the resulting action is similar. As the piston continues its descent, the air ports are exposed and incoming air begins to sweep the cylinder clear of exhaust gas.

As the piston passes through bottom dead center, with air ports as well as exhaust ports (or valves) open, the cycle repeats.

Deviations from the Norm

Opposed piston engines. Opposed piston engines operate on the two-stroke cycle, generally as described above, but achieve uniflow scavenging without exhaust valves. See Figure 16-6. The cylinder liner has ports arranged at each end: one set is for air, the other for exhaust, and each set is controlled by one of the pistons.

Loop-scavenged engines with exhaust valves. Some older loop-scavenged engines are fitted with rotating valves in the passage from the exhaust ports to the manifold so that, even when the exhaust ports remain exposed by the piston on the upstroke, compression can begin earlier.

Glow plug and hot bulb engines. In these engines, generally considered obsolete, the compression ratio is insufficient to raise the temperature of the charge air above the ignition temperature of the fuel. The glow plug or hot bulb is an ignition source in the cylinder that will ignite the fuel as it is injected. *Glow plugs* are heated electrically, while a *hot bulb* is simply an uncooled portion of the cylinder head that can be heated initially by a blowtorch to start the engine, after which it will be kept hot by the combustion process. Some small engines use glow plugs as a cold-starting aid, but run on a normal dual combustion cycle once warmed.

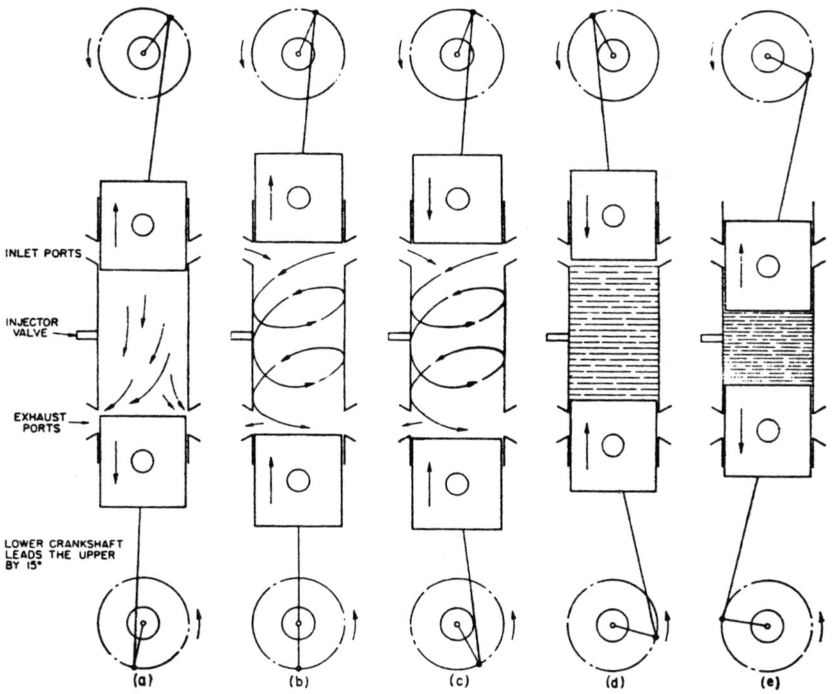

Figure 16-6. Opposed piston engine: cycle events

Indicator Cards, IHP, BHP, Pressures

Indicator cards. Figure 16-3 represents the pressure in an engine cylinder plotted against the piston position, which in turn is directly proportional to cylinder volume, and is therefore called a pressure-volume, or P-V, diagram. When the P-V diagram is obtained from the engine itself, using an engine indicator for low speed engines or electronic means for higher speed engines, it is called an *indicator card.*

IHP. In thermodynamic terms, the work done during a cycle is the product of the pressure at any point in the cycle times the volume displaced by the piston at that point. It is therefore proportional to the area enclosed by the curve on the P-V diagram. The area enclosed can be determined by measurement with a planimeter, or by graphical or mathematical integration. Once multiplied by the appropriate constants, this area is the *net work* (W_{net}) done by the piston during the cycle; i.e., it is all the work delivered by the piston to the crankshaft during the power stroke, plus or including the work to overcome friction and to drive engine accessories, less the work obtained from the crankshaft to drive the piston on the other strokes. The mathematical expression is:

$$W_{net} = C\oint PdV$$

where C is the constant of integration, P is cylinder pressure, and V is cylinder volume. When the net work is multiplied by the RPM, the result is the *indicated power* developed by the cylinder, expressed in horsepower as its IHP. The IHP of the engine is the sum of the IHP of the cylinders. Stated mathematically:

$$IHP = \Sigma\, W_{net} \times RPM$$

If all the cylinders were assumed to be performing equally, the expression would be:

$$IHP = N \times W_{net} \times RPM$$

where N is the number of cylinders.

It is important to stress that the IHP includes the power consumed by friction and by engine-driven accessories. However, since the IHP can be accurately measured by operators of low speed engines, it remains a primary tool for assessing the performance of these engines in service.

BHP. The power that can actually be obtained from the engine is called the *brake power,* or BHP when expressed in horsepower. It can be measured during shop tests of an engine, when the engine is connected to drive a dynamometer. (One antiquated form of dynamometer is the prony brake, hence the terminology. The prony brake was named after the French engineer G. C. F. M. Riche, Baron de Prony.)

The ratio of BHP to IHP, expressed as a percentage, is the *mechanical efficiency* of the engine. The mechanical efficiency can be accurately measured only when the BHP can be measured. However, once determined, values of mechanical efficiency provide a means of estimating BHP from measured values of IHP. Expressed mathematically:

$$\text{BHP} = \text{mechanical efficiency} \times \text{IHP}$$

Maximum, boost, and mean effective pressures. The highest pressure reached in the combustion chamber during the cycle is the *maximum pressure,* also called the maximum firing pressure or the peak pressure. It can be readily measured in service with a special pressure gauge, and is therefore a useful diagnostic tool, especially for medium and high speed engines for which conventional indicator cards cannot easily be taken. The maximum pressure is usually reached shortly after injection begins, just beyond TDC. It is the maximum pressure developed when the engine is running at full load or rated output, which, with margin applied, the cylinder components must be designed to withstand.

The *boost pressure* is the pressure in the charge air manifold of engines with turbochargers or blowers.

The *mean effective pressure* (MEP) and the *mean indicated pressure* (MIP) are the average pressures during the complete cycle. These values are calculated from measured data: When calculated from the indicated power the resulting value is the MIP, while a calculation from the BHP will yield the MEP. The two differ because of mechanical efficiency. The appropriate expressions are as follows:

$$\text{MIP} = C\,\frac{W_{net}}{V_{dis}}$$

$$\text{MEP} = C\,\frac{\text{BHP}}{\text{RPM} \times V_{dis}}$$

$$\text{MEP} = \text{mechanical efficiency} \times \text{MIP}$$

where C represents the appropriate unit conversion factors and V_{dis} is the displacement of the cylinder(s).

Relationship of Power, MEP, MIP, and RPM. The expressions below relate power to MEP, MIP, and RPM:

$$IHP = C \times MIP \frac{\pi L}{4} B^2N$$

$$BHP = C \times MEP \frac{\pi L}{4} B^2N$$

where C = the appropriate unit conversion factor
L = stroke
B = bore
N = RPM for two–stroke cylinders, or RPM/2 for four–stroke cylinders

These relations can easily be remembered if one uses P for pressure and notes that the cylinder cross-sectional area, A, is $(\pi/4)B^2$. Then the expressions become:

$$\text{Power} = \text{PLAN}$$

Among the important conclusions which can be drawn from these relations are the following:

1. If all other things were equal, a two-stroke engine could deliver twice the power of a similar size four-stroke engine. In actual fact, however, attainable levels of work per cycle or MEP in two-stroke engines are about half those of four-stroke engines, mainly because of the improved cooling possible between power strokes in the four-stroke cycle. The doubled number of power strokes per revolution of two-stroke engines therefore tends to compensate for their lower MEP.
2. For any given engine, power is proportional to the product of MEP and RPM. The implication of this relation is that, should the RPM of the engine be reduced to a lower-than-rated value, the power output will have to be reduced in accord with the limiting values of MEP. For example, a main propulsion engine driving a fixed-pitch propeller will be forced to a lower RPM as the hull fouls over a period of time; an attempt to maintain engine output under these conditions results in high values of MEP, usually reflected in high exhaust temperatures.
3. To the extent that limiting values of MEP are indicative of limiting maximum pressure (therefore approaching maximum permissible component stress levels), it should be obvious from the relations that, if component strength were roughly equal for all engines, then a particular power output could be achieved by a cylinder of large dimensions at low RPM, or by a smaller cylinder at higher RPM.
4. Conversely, if component strength were increased to permit higher values of MEP, then higher engine output could be achieved from an engine of given size and RPM. This has been, in fact, the path of design evolution of most

engines, as they have been matched to turbochargers of increasing efficiency, permitting the attainment of higher MEP.

5. Just as there is a limiting value of MEP for a particular engine, so will there be a limiting value of torque. For this reason diesel engines are considered torque-limited machines; i.e., they are prevented, at reduced RPM, from reaching their rated BHP by a torque limit.

Specific Fuel Consumption

The amount of fuel consumed by an engine over a period of time, divided by the power output of the engine, is the specific fuel consumption (SFC). It will usually be measured on a test-bed at constant RPM and load, in accord with an established standard test code, for a fuel of given quality, and will be expressed as grams (or pounds) of fuel per metric (or British) brake horsepower (or kilowatt) per hour. Among the standards used in this country are those of ASME, SAE, ISO, and DEMA. Even for the same engine, the SFC will vary with ambient conditions, with load, with RPM, and with fuel quality. It is most important, in comparing values of SFC, to ascertain that these factors are all the same, and to determine whether or not there are parasitic loads being imposed by such auxiliaries as engine-driven cooling or oil pumps.

FUELS FOR DIESEL ENGINES

Introduction

Fuels are discussed in Chapter 8 of Volume I; the discussion below is limited to fuels for marine diesel engines. It should be noted that in the years since Volume I was published there have been substantial changes in sources of crude oil, in refining techniques, and in distribution and marketing procedures, changes that have had generally harmful effects on the characteristics of fuels used aboard motorships. Even distillate fuels are often at the limits of the specification. It remains true that a balance must be struck between the lower cost of the heavier fuel oils, and the inconvenience and greater cost of the fuel treatment combined with the increased engine maintenance associated with their use. At this point the great majority of low speed engines and a good number of medium speed engines are operated on heavy fuels, while an increasing number of high speed engines are proving capable of operation on at least lighter blends.

Terminology: Heavy Fuels versus Light Fuels

Refining separates crude oil into a number of hydrocarbon products in a process based on their boiling points, with the lightest products having the lowest boiling points. At the light end of this spectrum are the distilled products, including the light distillate fuels known as gas oil or number 2.

These fuels are suitable for combustion in diesel engines without preheating (except in the coldest climates), so that fuel treatment can be limited to settling and filtration, although it is good practice to centrifuge even the distillate fuels.

Present refining techniques are aimed at extracting the largest quantity of distilled products feasible from the crude. The resulting residual tends to be of very high viscosity, with most of the undesirable constituents of the crude, and it is frequently contaminated with the highly abrasive particles from catalytic converters called *catalytic fines*. This is the residual fuel used in most steamship boilers without any further treatment other than heating, settling, and rather coarse filtering.

Residual fuel is rarely used alone as a fuel for diesel engines; far more frequently it is blended with a distilled product (the *cutting stock*) to produce a less viscous intermediate fuel, which, depending on the proportions used, can itself be described as light or heavy. Even the lighter blends will require preheating before pumping, settling, centrifuging, and combustion, so it is reasonable to refer to any intermediate fuel as heavy fuel. A blending chart can be used to determine the proportions necessary to produce a blend of selected viscosity, as shown, for example, in Table 16-1. It can be seen that relatively small fractions of distilled product can reduce the viscosity substantially, so that even the lighter blends will contain significant amounts of undesirable constituents.

TABLE 16-1

Intermediate Fuels Produced by Blending

	Cutting Stock (distillate)	*Blends*			*Residual*
		IF80	*IF180*	*IF280*	
Viscosity, cST at 50°C	40	80	180	280	600
Percent cutting stock	100	27	14	10	0
Percent residual	0	73	86	90	100
Vanadium, ppm	nil	290	345	360	400
Sulfur, percent by weight	0.5	3.1	3.5	3.7	4.0

Fuel Properties and Constituents and Their Consequences

Fuel properties are defined in Chapter 8 of Volume I. The following discussion is an amplification of that material as it applies to diesel engines of current design.

Viscosity. Because fuel is usually sold according to its viscosity, viscosity is often considered an index of fuel quality. This can be misleading since full consideration must be given to undesirable constituents and properties. Viscosity of fuel alone may present no problem as long as the fuel can be heated sufficiently at each point in the system to permit pumping, settling, filtration, centrifuging, and atomization. Reasons for incorrect fuel temperature (and therefore higher viscosity) include inadequate steam supply, inadequate or fouled heating surfaces, damaged or missing insulation, and poorly calibrated or malfunctioning thermometers or viscosimeters. At the very high end of the viscosity spectrum problems may arise if the fuel must be heated to the point where it is subject to thermal cracking, or where thermal expansion of the injection pump components is sufficient to move their clearances outside intended limits.

It is essential when burning heavy fuel in a diesel engine that the viscosity at the injection pumps and injectors be within design limits at all times. The volume of fuel consumed by an engine will be small in relation to the volume available in the piping; therefore, in installations intended for operation on heavy fuels, the residence time between the heaters and the injectors can be sufficient, especially at low loads, for the fuel to cool. To prevent this cooling, a much larger flow rate is maintained, two or three times engine consumption at maximum continuous rating (MCR), with the unconsumed excess leaving the spill valves of the injection pumps and recirculating back to the booster pump suction (see "Fuel forwarding system" near the end of this chapter).

Specific gravity. The ability to separate water and solids from a fuel by settling and centrifuging is dependent primarily on their differences in weight from the fuel (and is also affected by the fuel viscosity). These differences increase as the fuel is heated. Conventional centrifuges can achieve adequate separation of water from suitably heated fuel with a specific gravity as high as 0.995 at ambient temperature. More sophisticated centrifuges with water-sensing controls can separate even heavier fuels.

It should be borne in mind that an injection pump is a volume-measuring device: at constant engine output rack settings will vary depending on both the specific gravity and the heating value of the fuel.

Heating value. The heating value (per unit mass or weight) of residual fuels is typically some 6 percent lower than that of distillates, a difference which carries over in proportion to the blended fuels. There is an inverse relation between the heating value and the specific gravity, as both properties are determined by the chemical composition of the fuel, i.e., the ratio of carbon to hydrogen, and the presence of other combustible elements, especially sulfur.

Engine builders' published data for specific fuel consumption, as well as most shop test data, are usually based on the use of distillate fuel of a standard heating value. The specific fuel consumption determined for an engine in service must therefore be corrected for the difference in heating value of the fuel actually used if comparison to such data is intended.

Ignition quality. The ignition quality is an indication of the time necessary for the fuel to ignite after it has been injected into the cylinder of an engine: fuel of low ignition quality will take longer to ignite, thus the ignition delay will be longer. The ignition quality of distillate fuels can be measured, and is usually presented as the *cetane number*. For heavy fuels the ignition quality is calculated and presented as an approximate *cetane index*. More recently, a Calculated Carbon Aromaticity Index (CCAI) has been introduced.

The long ignition delay associated with fuels of low ignition quality can result in a late and therefore more explosive start to the combustion period, with higher peak pressures, manifested as rough, noisy operation that, if sustained, can result in damage to cylinder heads, liners, pistons, and rings. The end of the combustion period can also be delayed, resulting in rough and incomplete combustion and, therefore, high fuel consumption and fouling of the combustion space. Because the ignition quality is related to time, slower turning engines are less affected by fuels of low ignition quality, and to some extent the injection timing can be advanced to compensate for the long ignition delay. Conversely, higher speed engines require fuels of higher ignition quality.

Ignition delay is reduced at higher temperatures, and some manufacturers recommend that, for operation on low ignition quality fuel at low loads, the temperature of the jacket and piston coolants be maintained at high levels, and that the temperature of the charge air leaving the charge air cooler be increased.

Carbon residue. The standard carbon residue tests are meant to provide an indication of carbon formation at high temperatures. Fuel with a high carbon residue index can be expected to leave more deposits after combustion, and fouling and wearing of cylinder liners, rings, ring grooves, exhaust valves, and turbocharger turbine nozzles. Effects on cylinder components can be reduced by the use of detergent cylinder oils. Turbocharger fouling is countered by frequent water washing.

Carbon can also accumulate on the nozzle tips, interfering with the spray pattern, an effect best limited by frequent withdrawals of the injectors for cleaning.

Solids and ash. Solid particles carried into the engine with the fuel can cause abrasive wear of fuel injection pumps, injectors, cylinder liners and rings, exhaust valve seats, and turbochargers. The larger solid particles will be removed in settling, filtration, and centrifugal purification.

The solids that have proven particularly difficult to remove are the highly abrasive particles carried over into the residual from the silica-alumina–based catalyst used in catalytic cracking processes at many refineries and called *catalytic fines.* The most effective procedure for reducing the presence of catalytic fines aboard ship includes the full-time use of multiple centrifuges arranged to process the fuel in series, with the first set up as a purifier (water and solids removal) and subsequent units as clarifiers (solids only).

Fine filters, used alone, can provide adequate protection only for engines burning the cleanest of fuels. Fine filters are usually fitted to fuel systems that handle lower quality fuels only as a final backup in the event of purifier malfunction. When fitted as the sole means of protection, fine filters may clog at inconveniently frequent intervals.

Sulfur. Sulfur is carried through to residual fuels from the crude, and consequently into blended fuels. In the combustion process the sulfur is reduced to sulfur dioxide, which can subsequently convert, in the presence of unused oxygen, to sulfur trioxide, which can then combine with water vapor to form gaseous sulfuric acid. At temperatures below about 150° C, condensation of the sulfuric acid begins. The presence of sulfur in the fuel therefore indicates a potential for *cold end corrosion,* i.e., acidic attack of surfaces exposed to the exhaust gas when they are at or below about 150° C. It can also cause contamination of the lubricating oil.

Engine components that are most vulnerable to cold end corrosion can include the lower ends of the cylinder liners and pistons, especially those of engines operated at low power for sustained periods. This problem can be countered by maintaining the temperature of the jacket and piston coolants at high levels, and by increasing the temperature of the charge air leaving the charge air cooler. In addition, the oil used for cylinder lubrication should have a high alkaline content (high *total base number,* TBN) in order to neutralize the acid.

A corollary problem exists in the use of lubricating oils of high TBN: if there is insufficient sulfur present in the fuel to neutralize the alkaline ingredients of the lubricating oil, the resulting deposits can cause scoring of the liner and wear on the rings. This problem can arise when an engine that is normally operated on high sulfur fuel is later supplied with low sulfur fuel for an extended period, without a change of the oil lubricating the cylinders.

In crosshead engines the purely vertical movement of the piston rod permits a packing gland to be fitted to separate the combustion space from the crankcase, preventing combustion blowby and prohibiting excess cylinder oil from reaching the crankcase. Crosshead engines do not, therefore, require a crankcase oil of high TBN, but use a high TBN cylinder oil in a separate cylinder oil system. On the other hand, contamination of the crankcase oil can be a problem with trunk piston engines, which are usually supplied with crankcase oils having a high TBN.

Vanadium. Vanadium is carried through to the residual and blended fuels from the crude. During the combustion process, and especially in combination with sodium (see below), gaseous oxides will form, some of which will begin to change phase and form adhering deposits on combustion space surfaces whose temperatures exceed about 500°C. The surfaces most susceptible to such deposits are piston crowns, exhaust valves, and turbocharger turbine nozzles and blades.

A more minor problem with vanadium deposits is corrosive attack and its ultimate effect on piston crowns and the bottom faces of exhaust valves. When the deposits occur on the seating surfaces of the exhaust valves, however, the results are more immediate, as the valve can overheat and burn, through the following mechanism of failure:

1. Most valves are cooled intermittently through contact with their seating surfaces when closed, and the deposits interfere with the good contact required.
2. Further, if the deposits prevent the valve from closing tightly, the hot gases will find passages between the valve and its seat during the combustion and expansion periods, eroding a groove in the valve (*wire-drawing*).

Vanadium problems will be minimized in engines where the surfaces in question can be kept below about 500°C. Valve cooling is usually not a problem in the large low speed engines, but can be difficult in high speed engines and some of the medium speed engines. The two-stroke, loop-scavenged engines are all but immune to vanadium attack. Manufacturers' limits specified for vanadium and sodium take these factors into account.

Sodium. Most of the sodium in fuels is introduced through seawater contamination, and most will be removed with the water if settling and centrifuging procedures are adequate. The principal problem with sodium is in its combination with vanadium, described above. A rough rule of thumb limits sodium content to one-third of the vanadium content. It should be borne in mind that since sodium, unlike vanadium, can enter the fuel during transport to the ship or later, while stored aboard, analyses of samples from the fuel supplier or of samples taken during bunkering may give a false impression of the sodium content of the fuel reaching the engine.

Flash point. The minimum permitted flash point, usually 60°C, is indicative of the maximum safe storage temperature for fuel oil. A problem can arise in a diesel plant burning heavy fuel if the fuel leaving the centrifuge, where it might be heated to 98°C, in turn raises the day tank temperature above the flash point. The problem can be solved if a cooler is fitted after the purifier in the line to the day tank.

In some cases, particularly when plants have been converted from distillate to heavy oil, the heated returns from the engine are returned to the day tank, heating it above the flash point. The correct arrangement includes a mixing tank of limited capacity, so that its high temperature represents less of a hazard.

Crude oil, because it contains the light fractions, may have a flash point below the legal minimum, and may therefore be unacceptable for direct use as a fuel.

Pour point. The pour point indicates the temperature to which fuel must be heated to permit pumping. The temperature of fuel can fall below the pour point not only in storage tanks and transfer lines, but also in the service system of an idle plant. Most plants burning heavy oil have fuel lines that are extensively steam-traced beneath the insulation.

Incompatibility. Not all fuel constituents will mix compatibly with each other, so there is the possibility of constituents separating in tanks, often precipitating a heavy sludge, and leading to fluctuations in flow as the separated constituents reach key points in the system unevenly. It is the responsibility of the fuel supplier to ensure that fuels blended ashore do not contain incompatible constituents. Aboard ship, it is important to avoid mixing fuels from different deliveries, or blending fuels, without first undertaking a spot test for compatibility.

Incompatibility in a fuel can reveal itself by increased sludge accumulations in tanks and at filters and centrifuges, by fluctuating pump discharge pressures, and by frequent viscosimeter excursions. Other than discharging the fuel ashore at the next opportunity, the only cure for the operator is to cope with the incompatibility as best he can until the fuel is consumed.

Fuel Oil Analysis

In order to treat the fuel properly, and because of the potential for damage to the engine, it is important that complete analyses of all fuel coming on board be available. A complete analysis includes all of the properties cited above. Many operators have found that analyses provided by fuel suppliers are incomplete or otherwise unreliable, and have resorted to taking their own samples from the bunkering line during delivery and sending them ashore for independent analysis. The principal problem with such arrangements is in getting results of the analysis back to the ship before the newly bunkered fuel is needed.

Care must be taken to ensure that a sample is truly representative. One recommended method is to drip-feed a reservoir from the bunkering line during the entire bunkering period, with the sample being extracted from the reservoir.

TURBOCHARGING

Introduction

Although some applications are best served by naturally aspirated or mechanically blown engines, the vast majority of main propulsion engines and generator drive engines are turbocharged and aftercooled.

An engine and its turbocharger(s) are interdependent in their performance: a defective or mismatched turbocharger will preclude proper engine performance.

Description and Classification

Figure 16-7 shows a typical turbocharger. The important characteristic to note is that the rotor is freewheeling, driven only by the engine exhaust gases as they expand through the turbine. Turbochargers may be classified as follows:

Number of stages. In general, turbochargers use a single stage compressor, driven by a single stage turbine. (Where engines have been fitted with two-stage turbocharging, two single stage units are fitted, with turbines and compressors in series.)

Compressor type. Turbochargers almost always have centrifugal compressors.

Turbine type. Turbines of large turbochargers are usually *axial flow,* as in Figure 16-7, while those of smaller units are usually *radial flow,* as in Figure 16-8.

Discharge pressure. Compressor discharge pressure is usually described by its ratio to intake pressure, called the *pressure ratio.* Currently, turbochargers are suitable for pressure ratios as high as 4.0.

Turbine cooling. Traditionally, large turbocharger turbines are cooled by circulating engine jacket water through passages in the casing, as in Figure 16-7. Figure 16-9 shows an *uncooled* turbocharger in which, while water may still be used to cool the turbine bearing, the turbine casing is not cooled, improving waste heat recovery from the exhaust gases downstream of the turbocharger.

Bearing location. When the bearings are located at the extreme ends of the shaft, as in Figure 16-7, they are *outboard bearings.* Because the ends

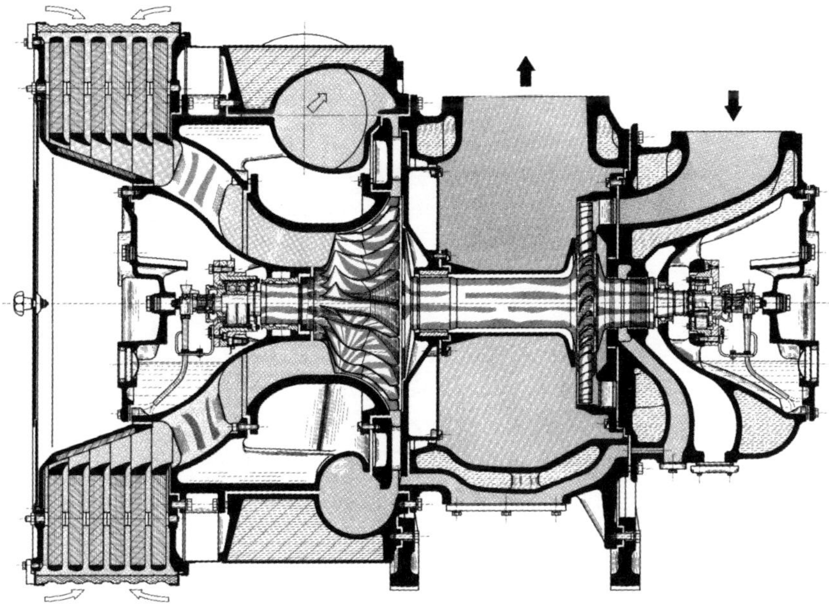

Figure 16-7. Turbocharger with axial flow turbine and water-cooled casing

of the shaft can have reduced diameters, outboard bearings will usually have lower friction losses. *Inboard bearings* are located between the turbine disk and the compressor impeller as in Figure 16-8.

Principles of Turbocharging

Reasons for turbocharging and aftercooling. The principal reason for turbocharging is to increase the power output of an engine of given size and speed, by enabling the cylinders to be charged with air at high pressure, hence at higher density than atmospheric. Since the greater mass of air then present will permit a correspondingly greater mass of fuel to be burned, the engine output will be higher.

The effect on the cycle is an increase in the intensity or duration of the combustion period and an increase in the work per cycle and, therefore, the MEP. From the relations cited previously, it can be seen that an increase in MEP will result in an increase in power output. The power increase will be directly proportional to the increase in MEP if other factors, including cylinder dimensions and RPM, are unchanged.

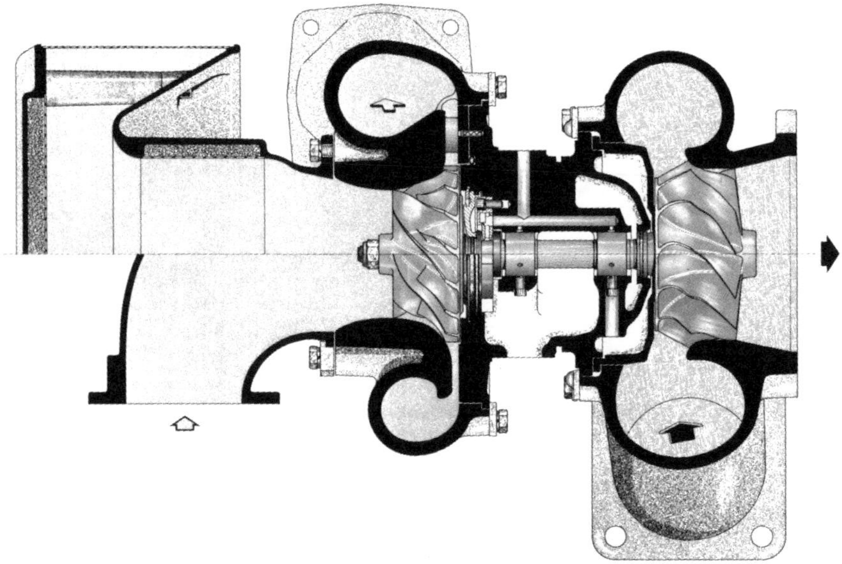

Figure 16-8. Turbocharger with radial flow turbine

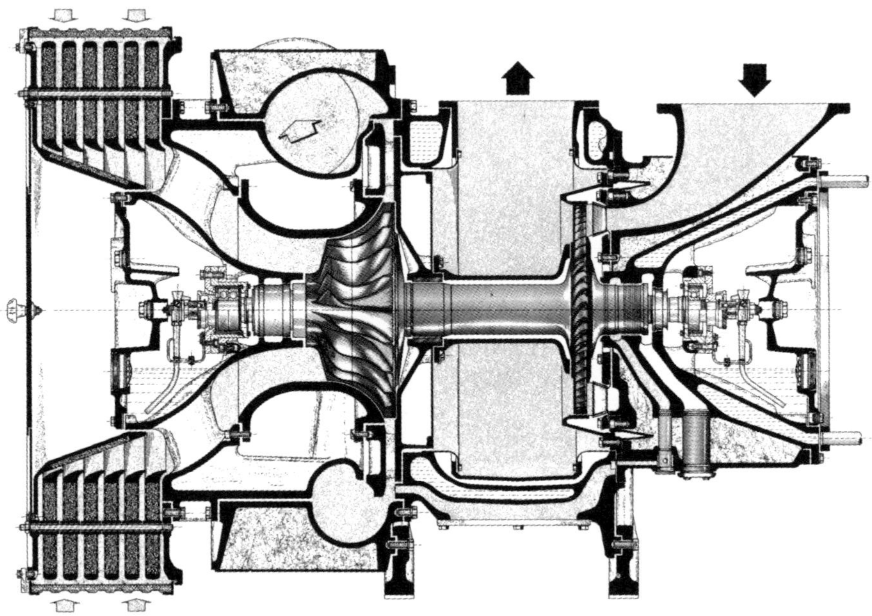

Figure 16-9. Turbocharger with “uncooled” casing

Some of the aforementioned effect will be lost, however, if the air leaving the compressor is not cooled, because the temperature rise of the air during the compression process has the opposite effect, decreasing the density. Consequently, in most applications, charge air coolers are fitted after the compressor and are therefore often called aftercoolers.

Turbocharging tends to reduce fuel consumption, in part because the friction losses of the turbocharged engine do not increase as rapidly as the power output, and in part because the improved charging results in better combustion conditions.

Boost ratio, pressure ratio, and compression ratio. The *boost ratio* compares the pressure in the air manifold to atmospheric pressure. The pressure ratio compares the pressure at the compressor discharge to intake pressure. The two ratios differ because of the pressure drops across the charge air cooler and intake air system.

The *compression ratio,* while actually a ratio of cylinder volume at BDC to cylinder volume at TDC, is also indicative of the pressure rise during the compression stroke.

Turbocharged engines, which by definition have elevated boost ratios, tend to have relatively low compression ratios in order to avoid excessive maximum pressures.

Turbocharger efficiency. A turbocharger is an energy recovery device: a more efficient turbine will recover more energy from the exhaust gas stream; low friction rotor shaft bearings will absorb less of the turbine output; and a more efficient compressor will better utilize the remaining energy to compress more air to a higher pressure. Expressed mathematically:

$$\eta_{tc} = \eta_t \times \eta_c \times \eta_m$$

where η_{tc} = overall turbocharger efficiency
η_t = turbine efficiency
η_c = compressor efficiency
η_m = mechanical efficiency

Thus, small but simultaneous improvements in the efficiencies of components, through improved component design and manufacturing, have compounded effects on overall turbocharger efficiency. While these improvements tend to result in higher turbocharger cost, the environment of high fuel costs that has prevailed since the mid-Seventies makes the cost increase acceptable because of the resulting improvements in engine fuel consumption and power output.

Improvements in turbocharger efficiency lead to the attainability of higher boost pressures; if engine components are redesigned appropriately, the new generation engine that results can have higher MEP, higher power output, and lower SFC.

Compressor characteristics and the surge limit. Centrifugal compressor characteristics are similar to those of centrifugal pumps. Most compressors used for turbocharging have essentially radial vanes, though slight backward curvature is increasingly used. In either event a plotted characteristic at constant RPM would appear similar to Figure 16-10. At constant speed the discharge pressure first rises as volumetric flow increases, then drops off rather sharply. The compressor efficiency curve also rises to a peak, although at any constant speed this peak is slightly to the right of the pressure peak.

The power consumed by the compressor is related to the product of discharge pressure and flow rate. Thus, in the region to the right of the peak in the pressure curve, operation will be stable: in this region a momentary drop in volumetric flow rate, for example, perhaps brought on

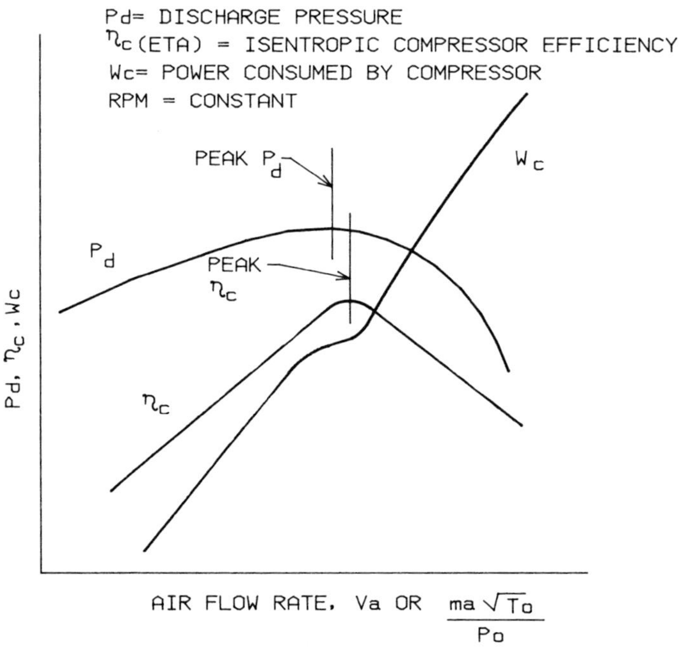

Figure 16-10. Centrifugal compressor performance characteristics at constant RPM

by a momentary reduction in engine speed, will be countered by a rise in pressure, with little or no effect on the turbine. In the region to the left of the pressure peak, a momentary drop in volumetric flow rate will be accompanied by a drop in discharge pressure and a reduction in compressor power consumption. Operation in the unstable area to the left of the pressure peak may result in *compressor surge.* As the pressure at the compressor discharge falls below that downstream, the flow can reverse. The result can simply be a pulsation if the situation is not severe or of long duration, or the reversed flow can continue to the air intake and become audible, ranging in volume from a soft sneezing to a very loud backfiring sound.

Obviously, operation in the surge region should be avoided; consequently, turbocharger designers establish a line, called the surge limit, through the pressure characteristic slightly to the right of the peak.

Figure 16-10 represents compressor characteristics at only one speed. In order to completely define the characteristics of a particular compressor, similar data must be obtained at several constant speeds covering the range of its operation, and plotted together on the same axes. The resulting diagram, of which Figure 16-11 is an example, is called a *compressor performance map.*

Effects of wheel diameter and diffuser vane height. A map such as Figure 16-11 describes the performance of a particular compressor, comprising a wheel of given design and diameter, and a diffuser with vanes of given height. In practice, turbocharger manufacturers design a series or "family" of geometrically similar compressors with a range of compressor wheel diameters to cover a range of flow rates. When the compressor performance maps for the whole family are plotted together, the result will be similar to Figure 16-12.

For each compressor wheel, a narrow range of performance variation is possible by exchanging the diffuser for one with a different vane height: higher vanes will shift compressor performance slightly to the right, while lower vanes will move the performance slightly to the left. In general, these adjustments of vane height away from the optimum are accompanied by a small penalty in compressor efficiency.

Turbocharger frame size and turbine characteristics. In general, turbine characteristics are more straightforward than compressor characteristics. Usually, for any given turbocharger series, selection of a compressor wheel diameter specifies the turbocharger; i.e., for each compressor wheel diameter there is a given compressor casing, turbine casing, and turbine disk. Adjustment of turbine performance is then obtained by selection of nozzle and blade characteristics. Figure 16-13 is an example of a selection curve for turbine characteristics, from which, given the exhaust gas flow

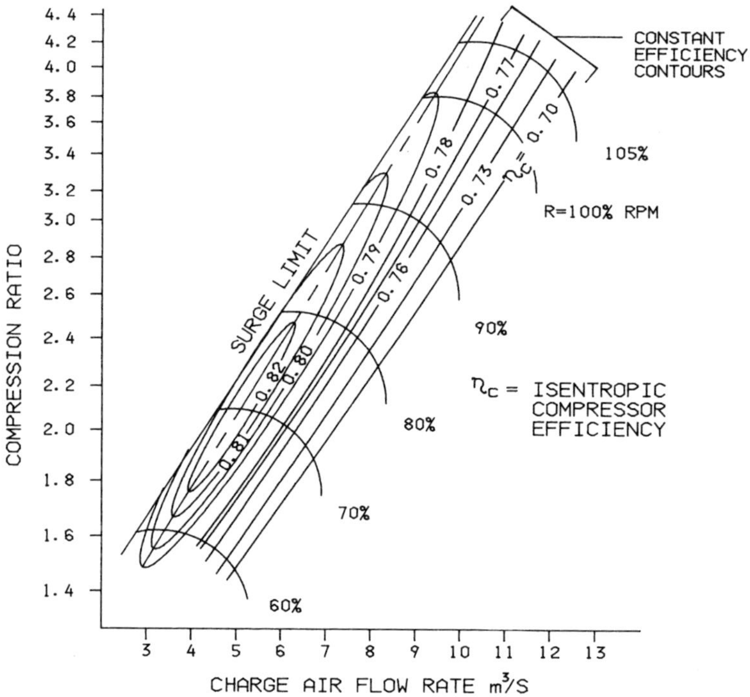

Figure 16-11. Compressor performance map

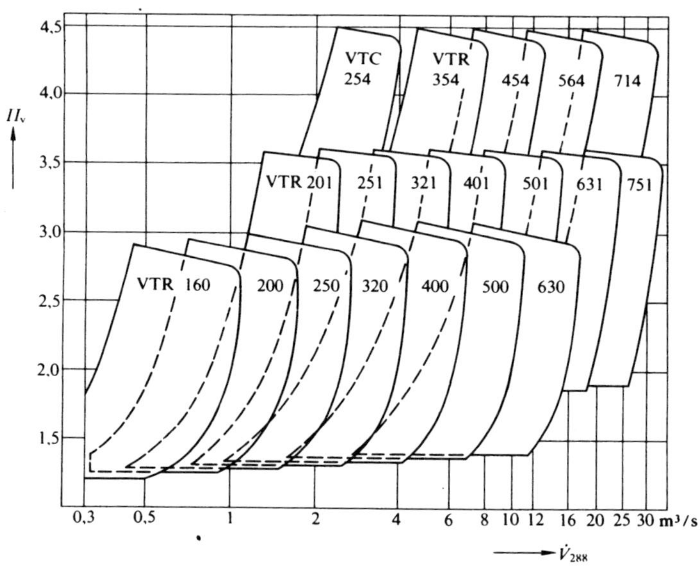

Figure 16-12. "Families" of compressor performance maps

and the expansion ratio across the turbine, the appropriate combination of nozzle plate and blade angle can be obtained. Then, from a curve such as Figure 16-14, the resulting turbine efficiency is estimated.

Turbocharger Matching

Description of matching procedure. The procedure by which a turbocharger is selected to mate with an engine is called turbocharger matching. Usually this is done by the engine designer in the course of development of an engine design, or in upgrading an engine design to keep pace with advances in turbocharger or engine technology. On occasion, an existing engine will be rematched with a new or modified turbocharger, perhaps to suit new operating conditions.

While the operating engineer will not normally be involved in turbocharger matching, a familiarity with the procedure will lead to a better understanding of the interdependent relationship between engines and their turbochargers, and of the effects, in service, of operation off the design point.

Figure 16-15 is provided to identify the terminology used in the simplified procedure outlined below.

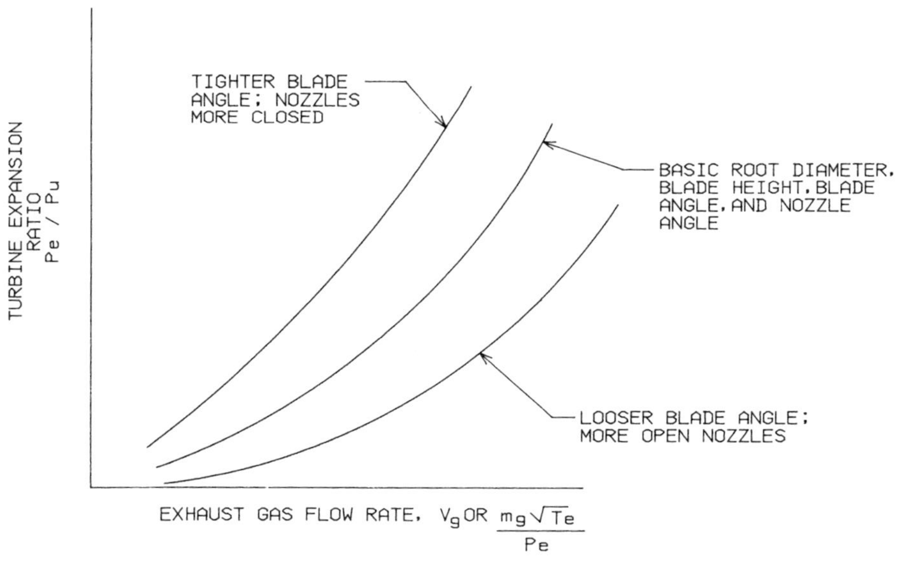

Figure 16-13. Selection curve for turbine blade and nozzle angles

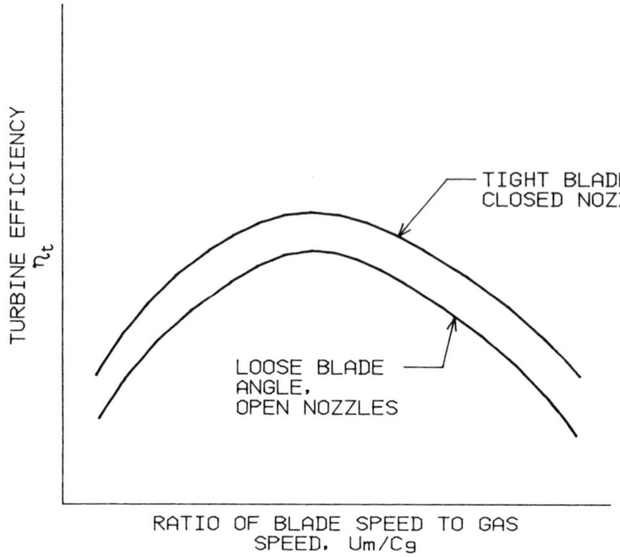

Figure 16-14. Turbine characteristics: efficiency versus blade speed/gas speed ratio

1. An estimate is made of the anticipated BHP that the engine will develop at a particular engine RPM. Normally the rated RPM will be selected, but under some conditions the turbocharger may best be matched to the engine at a different RPM.
2. Inlet conditions for the turbocharger, P_o and T_o, are selected. Normally these will be standard atmospheric conditions, with P_o corrected for a pressure drop across the intake filter. However, if operation in an abnormally hot or cold environment is expected, or if a long run of intake ducting is to be fitted, then conditions should be selected to suit.
3. An estimate might now be made of the amount of air, m_a, that the engine will require at this condition. This can be obtained from a combination of basic principles and empirical data, including previous engine performance. Once m_a is determined, the volumetric air flow, V_a, at standard conditions of pressure and temperature, can be calculated.
4. The engineer must now determine the air manifold pressure, P_1. As with the estimate of air flow, he will have a good idea of the approximate value to use as a first estimate. (Turbocharger matching is an iterative procedure in which the results of the first series of calculations become the assumptions for the next series; the calculations are repeated until the results equal the assumed values.)
5. Using the air cooler manufacturer's data for pressure drop versus air flow rate, the air cooler pressure drop can be added to P_1 to yield the compressor discharge pressure, P_d, required. (The air cooler data will be appropriate for

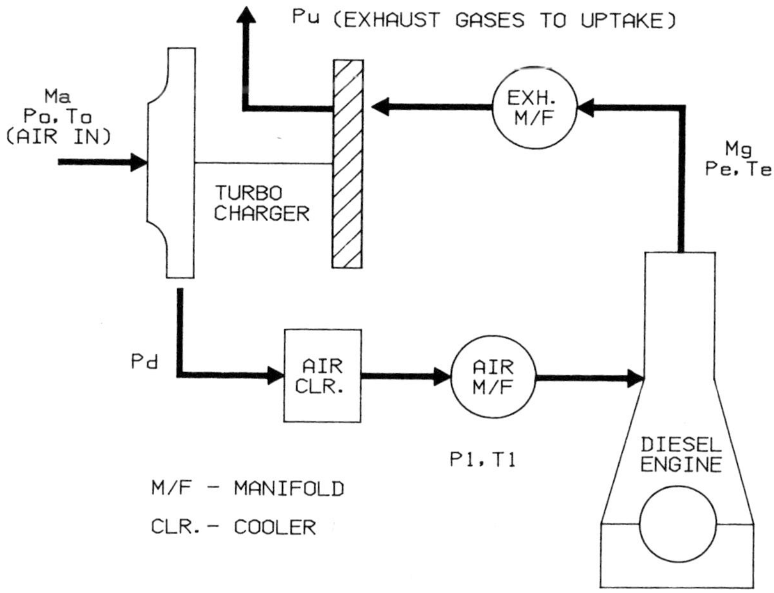

Figure 16-15. Schematic for turbocharger matching

a clean air cooler only: obviously, if the air cooler is fouled, a higher P_d would be needed to achieve the same value of P_1). The compressor pressure ratio, P_d/P_o, can be calculated.

6. The compressor frame size can now be selected by entering the family of compressor performance maps (of which Figure 16-11 was an example) with the values of P_d/P_o and V_a just determined. These same data are now used to pinpoint the first estimate of operating point on the compressor performance map for the selected compressor, as in Figure 16-16. The operating point must have adequate margin from the surge limit; i.e., it must be 15 percent to 20 percent to the right of the surge limit at the value of P_d/P_o. Inadequate margin will invite turbocharger surge under service conditions, while excessive margin will place the compressor in a region of low efficiency. If the surge margin is outside the recommended range, then maps for the same frame size but with higher or lower diffuser vanes, should be checked; if the margin against surge is still inadequate the next compressor frame size will be required.

Once this is done, a preliminary selection of compressor, and therefore of turbocharger, has been made, based on the initial assumptions. What must be done next is to confirm that the power produced by the turbine will be sufficient to drive the compressor; i.e., that the initial assumptions were correct.

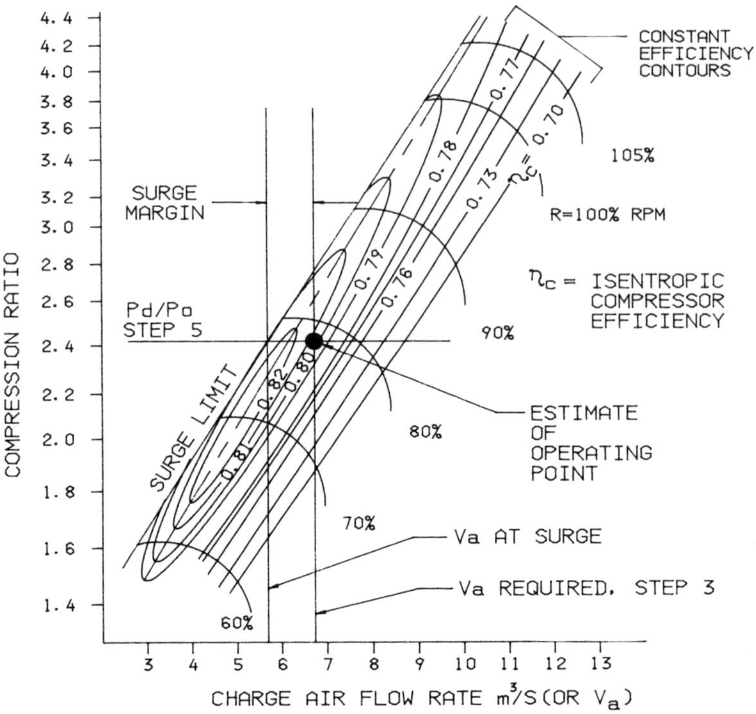

Figure 16-16. Compressor performance map: estimate of operating point, showing adequate surge margin

7. From the performance map for the selected compressor and diffuser, compressor efficiency (η_c) and turbocharger RPM are read at the operating point.
8. The power needed to drive the compressor can now be calculated from the following equation (derived from basic thermodynamic principles):

$$W_c = \frac{m_a\, c_{pa}\, T_o}{\eta_c}\left[1 - \left(\frac{P_d}{P_o}\right)^{\frac{k-1}{k}}\right]$$

where W_c = compressor power consumption
m_a = air flow, from step 3
c_{pa} = specific heat capacity of air, from property tables or similar source
T_o = absolute air inlet temperature, step 2
k = specific heat ratio for ideal air, 1.4

9. The power absorbed in overcoming bearing friction can be determined from

the turbocharger RPM and the known characteristics of the particular bearings. The sum of bearing power absorption and compressor power is the amount of power that must be developed by the turbine:

$$W_{t,\ req} = W_{brg} + W_c$$

where $W_{t,req}$ = *required* turbine output
W_{brg} = power absorbed by bearings
W_c = compressor power consumption

At this point, a first estimate of the power *required* from the turbine has been made. The next steps will determine, from the turbine characteristics for the turbocharger under consideration, whether this is available.

10. Estimates must be made of the gas conditions at the turbine inlet. For the simpler case of engines with constant pressure turbocharger systems it suffices to know the turbine inlet pressure and temperature, P_e and T_e. (For engines with pulse charging systems the procedure is more complex, though similar in principle.) For any particular engine, P_e and T_e can be estimated from basic principles and empirical data, including previous engine performance.
11. The exhaust pressure from the turbine, P_u, must also be estimated. Generally this can be done by adding an amount to the standard atmospheric pressure sufficient to allow for typical uptake losses. If, however, a waste heat recovery boiler or other device will be fitted in the uptake, or if the standard atmospheric pressure is not representative of anticipated operating conditions for the engine, more appropriate data should be used. Once these two pressure estimates are made, the expansion ratio, P_e/P_u, can be calculated.
12. The mass flow rate of the gas, m_g, can be calculated by adding the mass flow rate of the fuel, m_f, to the air mass flow, m_a, previously estimated:

$$m_g = m_f + m_a$$

    The volumetric flow rate of the gases, V_g, can also be calculated.
13. In general, the selection of a compressor wheel diameter predetermines turbine characteristics, which may include wheel mean diameter and blade length. With the values of P_e/P_u and V_g obtained in the previous steps, a turbine blade and nozzle angle selection curve, such as Figure 16-17, can be entered for the frame size under consideration, to select nozzle opening and blade angle.
14. The turbine efficiency can then be obtained from a curve such as Figure 16-18. However it will first be necessary to calculate the ratio of blade speed to ideal gas speed from the following relations:

$$U_m = \pi \times D_m \times RPM$$

$$C_g = \left(2c_{pg}\, T_e \left[1 - \left(\frac{P_u}{P_e} \right)^{\frac{k-1}{k}} \right] \right)^{1/2}$$

where U_m = mean tangential velocity of blade
D_m = mean diameter of turbine wheel
C_g = ideal gas speed at nozzle exit, sometimes called the ideal spouting velocity
c_{pg} = specific heat capacity of the exhaust gas
T_e = absolute temperature of exhaust gas at turbine nozzle inlet
k = specific heat ratio for exhaust gas, taken as equal to that of air, 1.4

Entering Figure 16-18 with the ratio U_m/C_g, and the nozzle and blade data of step 14, a turbine efficiency η_t, is determined.

15. The turbine power output can now be calculated from the following equation, derived from basic principles of thermodynamics:

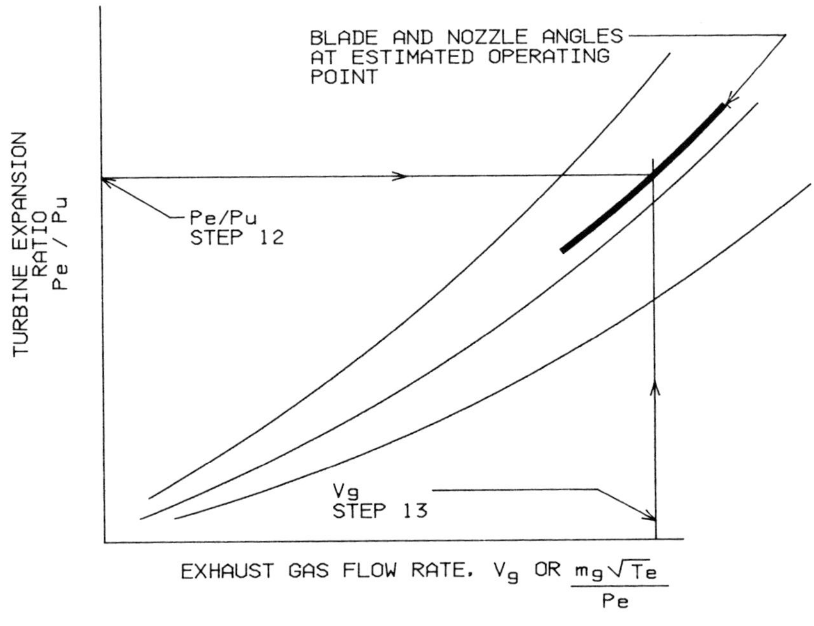

Figure 16-17. Selection of turbine blade and nozzle angles

$$W_{t,\,ach} = m_g\, c_{pg}\, \eta_t\, T_e \left[1 - \left(\frac{P_u}{P_e}\right)^{\frac{k-1}{k}}\right]$$

where $W_{t,\,ach}$ = *achieved* turbine power output

16. It is now necessary to compare the turbine output achieved, $W_{t,\,ach}$, with the turbine output required, $W_{t,\,req}$, to drive the compressor and to overcome bearing friction. If $W_{t,\,ach}$ is greater than $W_{t,\,req}$ then the estimate of air manifold pressure, P_1, made in step 4, can be raised, and the procedure repeated with this new assumption. If $W_{t,\,ach}$ is lower, a lower value of P_1 should be assumed.

This iterative process is continued until the required turbine output matches the achieved value within a percent or two, with the final matching to be done by actually running the engine and the turbocharger together on a test-bed, and making final adjustments by changing compressor diffuser vanes and turbine nozzle rings.

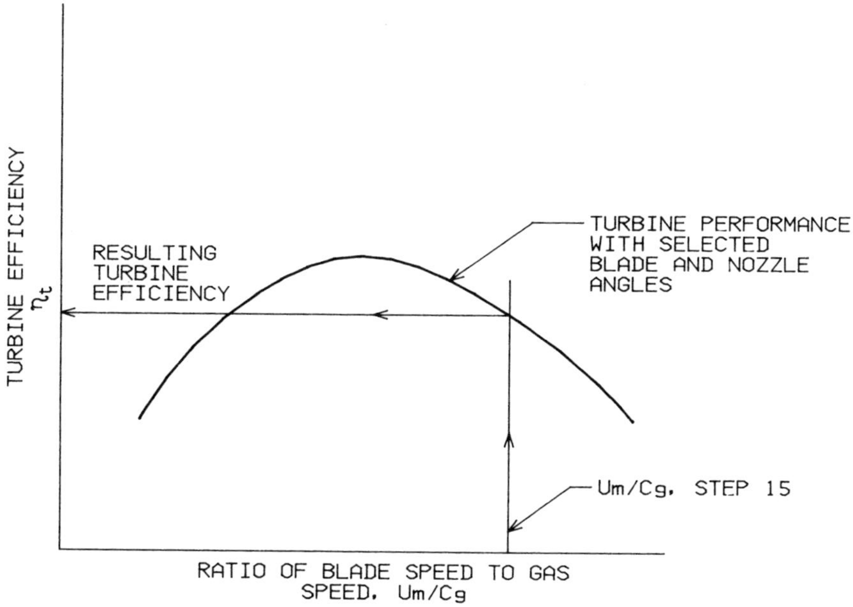

Figure 16-18. Determination of turbine efficiency

Turbocharger matching: conclusions. The most important conclusion to be reached in reviewing the turbocharger matching procedure is that a turbocharger is matched to an engine for a particular set of conditions. Operation at different conditions will be less than optimum and may, in extreme cases, be so unsatisfactory as to justify the retrofit of a new turbocharger, matched to the new conditions. Off-design-point operation may be permanent or temporary, intentional or inadvertent. A few examples follow.

An over-pitched propeller, a heavily fouled hull, single screw operation of a twin screw ship, or single engine operation of a pair of engines geared together are among those conditions that may require an engine to deliver higher than anticipated power at reduced RPM. Turbocharger surge would not be surprising under such circumstances.

Long or complex runs of intake ducting, elevated intake air temperatures, fouled intake air filters, or a dirty or damaged compressor are all likely to result in lower than expected air manifold pressures.

A fouled air cooler can force the compressor to operate at high RPM, close to the surge margin.

Fouled turbine nozzles can sometimes force a turbocharger into surge; under other circumstances the result would be reflected in reduced air manifold pressures.

Fouled turbine blades, a heavily fouled waste heat boiler, or a constricted uptake can prevent the turbine from reaching projected performance, and might first be reflected in low air manifold pressures.

The interdependent relation of the engine and turbocharger, and of the turbine and compressor, means that the system is prone to chain reactions. As an example, low air manifold pressure, which indicates lower air flow, can lead to dirty exhaust, resulting in turbine fouling, which can further aggravate the situation.

Effect of improved turbocharger efficiency. With reference to step 16 of the matching procedure, if a turbocharger of improved efficiency became available, the balance between turbine output required and turbine output achieved would occur at a higher boost pressure or greater air flow rate. There are three possibilities:

By changing exhaust timing of the engine or the configuration of its exhaust system, an engine's SFC can be improved without altering its rating.

The greater mass of air trapped in the cylinder could be used for the combustion of more fuel; i.e., engine output could be increased. Furthermore, the improved turbocharger efficiency would be reflected in a reduced SFC for the engine. However, the engine would have to be capable of this greater output, as cycle pressures and therefore component stress levels would rise. In most cases, in fact, as more efficient turbochargers have become available, engine components have had to be upgraded in order to permit the potential for higher ratings to be realized.

The potential for excess turbine output can be realized in the provision of an exhaust gas turbine driving a mechanical load. Three of the many possible configurations are (a) an exhaust gas turbine-driven generator; (b) an exhaust gas turbine geared to the engine output shaft, forming, in effect, a combined cycle or turbo-compound arrangement (the power contributed by the turbine might be up to 5 percent); or (c) a combination of these, in which the exhaust gas turbine is connected at the power take-off gear of an engine fitted with a shaft-driven generator.

Boost Blowers

As engine output is reduced, boost pressure falls. While four-stroke engines, by virtue of piston movement on charging and exhaust strokes, will continue to draw in their own charge air and expel most of the exhaust gases, two-stroke engines rely on elevated charge air pressure to scavenge and charge the cylinder. Below approximately half power, therefore, two-stroke engines must be provided with an auxiliary means of pressurizing the air manifold. In most engines this takes the form of an electric motor-driven *boost blower,* which is switched on automatically in response to air manifold pressure. In some smaller engines boost pressure is provided by mechanical drive of the turbocharger through an overrunning clutch and gear train from the crankshaft. Other methods, such as the use of the piston undersides as reciprocating pumps, or the provision of reciprocating pumps driven by links from the crossheads, may be considered obsolete, at least for the larger engines.

Turbocharger Water Washing Systems

In order to avoid the decline in performance that is caused by fouled turbines and compressors, many engines, including most intended for operation on heavier fuels, are fitted with water washing systems. Most commonly, these systems take the form of small tanks piped to the compressor inlet and the turbine inlet, fitted with water-filling and compressed air connections. In use, the engine load is reduced, and the charge of water, limited by the size of the tank, is injected over a period of about one minute. Solvents are usually not recommended: it is the impact of the water which does the cleaning.

Frequency of use will depend on the rate of fouling, determined from experience. Water wash of the compressor will most likely be required infrequently. On the other hand it is not uncommon, in the case of engines run on the heaviest fuels, for the turbine to be washed daily.

Two-stage Turbocharging

Two-stage turbocharging, in which two turbochargers are connected in series on both compressor and turbine ends (usually with charge air coolers at each compressor discharge), has been used to attain higher boost ratios and therefore higher MEP than even the most efficient turbochargers could make possible in a single stage. In most of the applications to date, it has been a matter of attaining very high power output on infrequent occasions from engines installed in compartments where there were space and weight restrictions. Maximum pressures were limited by reducing the compression ratio. In part because of the rather recent availability of more efficient turbochargers, interest in two-stage turbocharging for merchant ships has diminished.

INTAKE AND EXHAUST SYSTEMS

Intake Systems

The function of the intake system is to ensure a supply of clean air to the engine, within reasonable limits of temperature and pressure. The components of the system mounted on the engine may not be alone in achieving this: an engine installed in a clean, warm engine room may appear to have the most rudimentary intake system but in this case the engine room and its air supply system must be considered part of the engine air intake system as well. Not every installation will necessarily include all of the components described below.

Direct versus external air intake. Typically, main propulsion engines and ship's service generator engines are installed in well-ventilated engine rooms, from which they draw their intake air. Care must be taken in laying out and operating the ventilating system to ensure that fresh air is supplied to the vicinity of the engine intakes. Location of the main engine intake in a poorly ventilated area of the engine room can result in air intake temperatures that are sufficiently in excess of conditions used in matching the turbochargers to bring on surge, air starvation, poor combustion, and high exhaust temperatures.

Some engines are provided with external air intakes. As long as outside ambient temperatures are near the conditions for which the turbocharger was matched, and the location and configuration of the intakes are such as to avoid water ingestion, this is usually beneficial. However, very low intake air temperatures can cause the turbochargers to surge. Surge can also be caused by low intake pressures at the compressor brought about by excessive pressure drop in the intake system because of its length, or tight turns, or restricted air flow areas.

Intake filter and silencer. Engines with direct air intake have the filter mounted locally at the engine. Small, naturally aspirated engines will have intake filters of either the oil bath or the disposable dry media type mounted on the air intake manifold. In the most common configuration for large, turbocharged engines, washable dry media panels are mounted in an array surrounding the circumference of the compressor inlet. The design of these filters usually provides adequate silencing, but in some installations a plenum may be installed for further silencing.

On engines fitted with external air intakes, a filter box may be mounted behind a set of louvers that will provide a level of salt spray protection. The design of the filter box and the ducting to the engine must take silencing into account.

Charge air cooler. Most marine charge air coolers are configured as a bank of finned water tubes over which the air flows, but sometimes compact heat exchangers of proprietary design are fitted. In either case, the air side will be prone to fouling and, because of the impact that this has on engine performance, maintaining cleanliness of the surface is of paramount importance, even though some compact cooler designs may be particularly difficult to clean. (Frequency of cleaning is best determined by observation of the air pressure drop across the cooler.)

Proper cleaning of the oily residue that accumulates on charge air coolers requires the use of a solvent and time for the solvent to soak into the residue. Therefore, cleaning with the engine in operation is impractical. Ideally, charge air coolers would be arranged to allow cleaning with a minimum of dismantling, but this is not always the case and ad hoc arrangements are common. Often a blind flange is inserted to close off the lowest end of the cooler, allowing the entire external heat exchange surface to be immersed in solvent while the cooler remains in place. In other installations the cooler is broken out of its location and lowered into a solvent-filled tank, an arrangement which, after soaking, permits more thorough cleaning by use of a compressed-air hose.

Most charge air coolers for large engines are cooled by seawater, but increasingly ships are being fitted with central seawater-to-freshwater coolers which then allow the charge air coolers to be circulated with fresh water. The advantage of the more complex central system is in its reducing the potential for fouling and corrosion of the charge air cooler water sides and water piping. Most smaller engines have charge air coolers that are freshwater cooled by the engine jacket water.

The air entering the charge air cooler can exceed 200°C in the case of engines with high boost ratios and, in some more recent plants fitted with extensive waste heat recovery systems for turbogenerator drive, this source is used to preheat feedwater to the boiler (see Figures 16-53 and

16-55). For this purpose the first rows of charge air cooler tubes are separated from the remainder of the bank in order to allow circulation by boiler feedwater.

Charge air heating. The normal function of the charge air cooler is to reduce the temperature of the air leaving the turbocharger in order to increase its density. However, the incoming temperature of the air will vary with engine load, and consequently the cooling water flow to the charge air cooler must be regulated. This is usually done automatically in response to the temperature of the air leaving the cooler. At low engine loads the air leaving the charge air cooler may be too cool for optimal combustion conditions and may become saturated if it cools to below ambient temperature. In those engines where the cooler is normally circulated with jacket water, limited low load air heating is inherent in the design, but this can be increased using cooling water crossovers at low loads to circulate the charge air cooler with jacket water leaving the jackets.

Air manifold. The air manifold serves to distribute the air uniformly to the cylinders. In turbocharged engines, it is typically located below both the turbochargers and the charge air coolers. It is important to note, therefore, that water leakage from the charge air cooler or turbocharger will accumulate in the intake manifold while the engine is stopped. If the engine is subsequently rotated without draining the manifold, the water will be drawn into the cylinders, where, because it is relatively incompressible, it can cause cracking of the piston crowns, skirts, liners, or heads, bending of the connecting rods, or damage to the bearings or crankshaft.

Exhaust Systems

The typical exhaust system of a turbocharged engine comprises ducting and manifolds, the turbocharger, often a waste heat boiler or other heat recovery device, and a silencer, and usually terminates with a spark arrestor at the top of the ship's stack. Exhaust systems of multiple engine installations are usually independent of each other.

Pulse versus constant pressure turbocharging. A pressure probe located at the exhaust port of a cylinder will indicate a sharp pressure peak as the port first opens, called the *blowdown pulse.* A second, lesser, scavenging pulse occurs when the air ports first open, and charge air sweeps through the cylinder to the exhaust port. When the exhaust piping is designed to maintain these pulses all the way to the turbine inlet nozzles so that their energy can be utilized in the turbine, the engine is said to be *pulse turbocharged.* In its simplest form, pulse charging would require that separate exhaust ducts be led from each cylinder to separate groups of turbine nozzles in order to avoid the interference of pulses from different cylinders, but in fact it is possible to combine exhaust branches from groups of two or more cylinders that are sufficiently far apart in their firing order (see Figure 16-19 for an example). If the selected cylinders are too close in firing order, or if the valve timing is incorrect, the exhaust pulse of one cylinder can interfere with the exhaust of another. Pulse charging uses small diameter piping from the cylinders to the turbine to prevent the pulses from dissipating en route.

Constant pressure turbocharging is characterized by a large diameter exhaust manifold running the length of the engine, into which the cylinders exhaust through short branches. Refer to Figure 16-19. The energy of the pulses can be partly recovered as the gas enters the manifold if these entrances are carefully designed as diffusers (sometimes called *pulse converters*), which will elevate manifold pressure above what it would otherwise be.

Generally, pulse charging permits energy to be recovered at lower engine output than constant pressure charging, and enables a somewhat more compact installation. These advantages must be weighed against the more efficient operation of the turbine in a constant pressure system, where the turbine benefits not only from the nearly constant inlet pressure but from full peripheral admission as well.

Exhaust gas heat recovery. The energy in the gas leaving the turbocharger turbine, at temperatures ranging from a low of about 250°C for some large two-stroke engines to a high of about 500°C for some higher speed, four-stroke engines, is often recovered in waste heat boilers or other heat exchangers. The extent of waste heat recovery and the use to which the

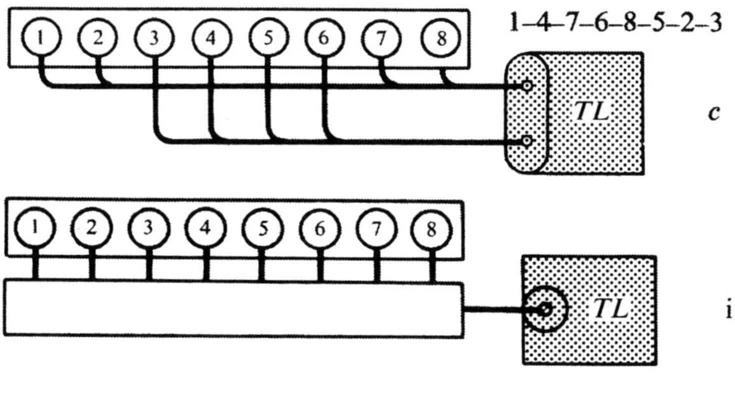

Figure 16-19. Pulse turbocharging versus constant pressure turbocharging

recovered heat is put are matters which must be determined by examining the economic trade-offs involved; these, in turn, are affected by different operating patterns of the ship as well as fluctuations in costs, principally of fuel. A few of the many possibilities are described below.

Almost all ships burning heavy fuel are fitted with waste heat boilers in the main engine uptakes sufficient to meet fuel oil heating requirements, plus domestic needs. It is usually not feasible to recover sufficient heat to meet more than a small portion of a tanker's cargo-heating requirements, for which an oil-fired boiler is necessary.

In the case of ships with minimal electrical load (bulk carriers and tankers), sufficient heat can be recovered from main engines operated at as little as about 10,000 to 15,000 BHP to supply the ship's normal electrical and steam requirements from a waste heat boiler/steam turbogenerator plant (see Figures 16-53 and 16-55). These systems are common on high-powered ships. The BHP threshold will be lower in the case of four-stroke engines with their higher exhaust temperatures than for two-stroke engines, but in either case it can be reduced further by using charge air coolers for feedwater heating, using multiple pressure boilers and turbines, reducing the electrical load by using engine-driven auxiliaries, supplementing steam production with the oil-fired boiler, or supplementing electrical supply with shaft-driven generators. Supplementing the turbogenerator with diesel generators may be done only when the diesel generator can be kept sufficiently loaded for trouble-free operation, typically above about 35 percent load.

Generally, waste heat boilers are fitted to the main engines, but under some circumstances—for example, passenger cruise ships with high electrical loads and relatively low utilization of the main engines—waste heat boilers are sometimes fitted to the auxiliary engines. In these cases much of the steam produced is used for fresh water production.

Usually the waste heat recovery fluid is water but, in some special cases, other fluids, usually proprietary in composition, are more suitable.

It should be noted that, in addition to the exhaust gases, the engine cooling water also contains recoverable heat. The use of the air cooler as a boiler feed heater is mentioned above. Most oceangoing motor ships utilize the jacket water as the heat source for the fresh water generators (i.e., the evaporators or distilling plant).

FUEL INJECTION AND COMBUSTION

Introduction

The fuel injection system must accurately meter the fuel in response to required output, then inject it into the cylinder as a finely atomized spray in order to enable complete combustion. Without exception, modern oil-burning diesel engines achieve these goals with solid injection systems. Of the three types of solid injection systems, the most commonly applied is the jerk pump system. Common rail systems and distributor pump systems are confined in their application to the smaller, higher speed engines, although the large Doxford opposed piston engines, which remained in production until 1981, had common rail systems. Only the jerk pump system will be described.

The fuel injection system is also the fuel metering system. Therefore, the first requirement of the system is:

1. The fuel injection system must accurately meter the fuel in response to required output.

In addition, the following points are of absolute importance in obtaining good combustion:

2. The fuel must enter the cylinder at a precise moment during the compression stroke.
3. The fuel must enter as a finely atomized spray. This condition must obtain from the very beginning of the injection period through to the end.
4. The droplets must penetrate far enough into the combustion space to ensure that they are evenly distributed.
5. The fuel droplets must not penetrate so far that they impinge on the surrounding surfaces.
6. The fuel must be supplied to the cylinder at a predetermined rate (a constant rate is usually required).
7. At the end of the injection period the cutoff must be sharp and complete.

Jerk Pump Injection System

The jerk pump system comprises one *injection pump* and up to four *injectors* for each cylinder. Fuel is delivered at nominal pressure to the injection pump—a reciprocating, positive displacement, plunger pump—with the reciprocation provided by connecting the plunger directly to a cam follower (the term *jerk pump* derives from the short, sharp strokes which result). The principal types of injection pumps have a constant stroke, with metering provided by closing, then opening, spill ports during the stroke. Discharge of the injection pump is led directly to the injector, which comprises a spring-loaded *fuel valve* surmounting the *fuel nozzle.* The injector is enslaved to the pump, in that it is the discharge pressure of the fuel alone that forces the fuel valve open.

In the jerk pump system the requirements for good combustion are obtained as follows:

Timing the start and end of injection is dependent, first, on proper cam timing, and then on the correct internal calibration of the injection pump to ensure that spill port operation occurs correctly relative to plunger movement.

Atomization and penetration are obtained by forcing the fuel through the holes of the fuel nozzle at very high pressure, typically on the order of 300 to 1,000 or more atmospheres. Obtaining good atomization from the beginning of the injection period through to the end is dependent on the sharp rise and then fall of pump discharge pressure, as well as the rapid opening and closing of the injector.

The spray pattern is a function of the configuration of the nozzle, which must be selected to avoid droplet impingement on the liner or the piston crown.

Distribution of the fuel droplets will be assisted by air turbulence, which can be obtained by suitably shaping the piston crown and the cylinder head, and by orienting the air inlet ports to induce a swirling motion to the air.

Injection pumps. Figure 16-20 shows a typical example of a helix-controlled injection pump, which is the most common type (a valve-controlled pump is described in the next chapter). Note the helical recess in the periphery of the plunger. As the plunger rises, the spill port will close as the top of the plunger passes it. This traps the fuel above the plunger and initiates the effective portion of the stroke. The rise in fuel pressure as the plunger continues its stroke will be very sharp, since the fuel is almost incompressible. When the edge of the recess in the plunger exposes the spill port, the effective stroke terminates with a sharp pressure drop. Most injection pumps are fitted with a spring-loaded discharge check valve which will then close. Because of the helical shape of the recess, rotation of the plunger will alter the length of the effective stroke and therefore meter the amount of fuel injected: when the vertical edge of the recess is aligned with the spill port, no fuel is injected. Rotation of the plunger is achieved by lateral movement of the *fuel rack,* which is in mesh with a pinion on the plunger shaft.

The discharge check valve ensures that a residual pressure is maintained in the high pressure fuel line between injections. This residual pressure aids in ensuring a prompt beginning of each injection and also helps to avoid the cavitation that would be likely if line pressure dropped too low. The residual pressure will vary with engine speed and output, however, and many injection pumps are fitted with a relief valve that bypasses the check valve, enabling a constant residual pressure to be maintained over the whole load range, while also helping to prevent secondary injections.

In the injection pump of Figure 16-20 the top of the plunger closes the spill port at the same point regardless of its angular position, so that the injection always begins at the same time in the cycle regardless of engine output. It is increasingly common for the top of the injection pump plunger to be shaped to vary the beginning of the effective stroke, hence the timing

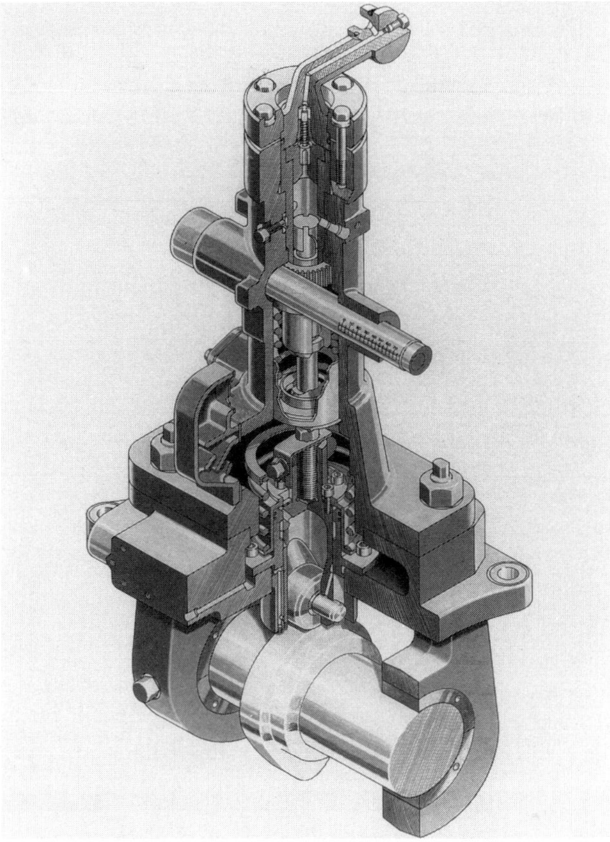

Figure 16-20. Fuel injection pump

of the start of injection. A common pattern advances the injection furthest at settings corresponding to about 80 percent of engine output in order to maintain maximum cylinder pressure throughout the upper end of the engine output range. This favors specified fuel consumption by improving the thermodynamic cycle (injecting a larger fraction of the fuel before the piston reaches its TDC position; see "Thermodynamic cycle" at the beginning of this chapter) and also enables the high temperatures conducive to complete combustion to be maintained.

Injectors. Figure 16-21 shows a typical fuel injector. The shoulder on the plunger provides the lifting area on which the pressure of the fuel initially acts against the spring to start to open the injector. As soon as the plunger begins to rise, the additional lifting area at the bottom is exposed and the injector snaps open sharply.

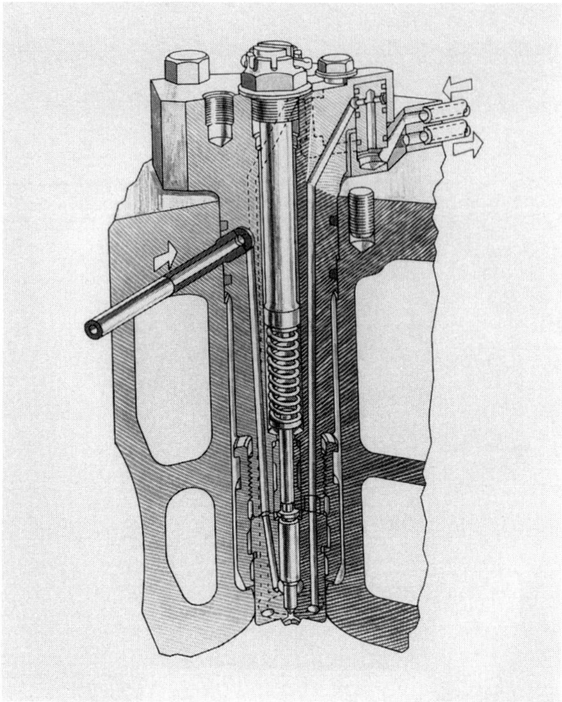

Figure 16-21. Fuel injector

The injector of Figure 16-21 is water cooled. The injector cooling circuit may be separate from the other engine cooling circuits to facilitate temperature control. In many engines sufficient cooling can be obtained by conduction to the cylinder head cooling circuit, simplifying injector manufacture and renewal.

High pressure fuel line. The third component of the jerk pump system is the high pressure fuel line connecting the injection pump to the injector. In most engines the injection pumps are fitted on the side of the engine, convenient to the camshaft, thus necessitating piping of some length. Several possible problem areas must be addressed as a result:

Even in the largest engines, the quantity of fuel injected per stroke is small relative to the volume of the high pressure line: the fuel discharged by the pump displaces fuel already in the injector, which opens in response to the pressure pulse traveling the length of the line. The time needed for this pulse to travel the distance accounts for *injection lag,* the delay between spill port closure at the pump and the beginning of injection.

Any irregularity in the interior of the high pressure passage can be sufficient to set up a reflected pressure pulse which, on reaching the injector after it closes, can cause it to reopen, resulting in a *secondary injection.*

High pressure fuel line leaks were one of the prevalent causes of fire in motorships, as the fuel lines were usually in the vicinity of hot exhaust surfaces. These lines are usually required to be fitted with shielding to reduce the fire risk.

On some engines these problems are minimized or eliminated by uniting the injection pump and the injector in a single *unit injector.* Usually camshaft motion is then brought to the pump by push rods and rocker arms.

Jerk pump injection system problems. Problems in jerk pump systems on diesel engines of mature design are more likely related to component wear or to improper settings than to initial design. The cams, for example, must be correctly timed: it may not be sufficient to set the camshaft timing alone, since the cams are not always integral with the shaft. Because of the high loadings on the cam face, surface damage sufficient to affect the injection timing can occur.

Injection pumps and injectors operate with close tolerances and are subject to wear from abrasive particles in the fuel. Poor quality fuels or weaknesses in the fuel treatment system will aggravate the situation. Items most subject to fuel abrasion are the injection pump plunger, barrel and valve seats, and, at the injector, the plunger, seat, and orifices.

Cavitation erosion can be a problem in the high pressure parts of the system, affecting pump plunger, barrel, valve seating surfaces, high pressure line, and passages in the injector body. The erosion may have an obvious cause, such as cavitation induced at an irregularity or a change in the shape of a passage. Where there is no such obvious cause, the cavitation may be occurring in the wake of the pressure waves induced by the sharp closure of the injector, perhaps as the result of a failure by a discharge check valve to maintain an adequate residual pressure between injections.

In engines operated on heavy fuels, the injector is likely to be the engine component most frequently removed for cleaning, testing, and parts renewal. On the test stand, the injector is checked for correct opening pressure, for leakage, and for spray pattern. Spring compression can be adjusted or a weak spring renewed; if the plunger seat or the orifices are worn, the nozzle must be replaced.

Correct calibration of the injection pump is essential: spill port operation must be correct relative to plunger movement, the calibration between rack and plunger position must be correct, and the right number of shims must be inserted at adjustment points.

Combustion

Fuel combustion in a diesel cylinder may be considered to occur in four phases (see Figure 16-22):

1. Ignition delay period
2. Rapid combustion period
3. Steady burning period
4. Afterburning period

Combustion in a diesel cylinder does not take place at the tip of the injector, but rather at a distance away from it, as the individual fuel droplets will have to travel (*diffuse*) through the hot cylinder contents for sufficient time (the *preparation time*) to heat, begin to vaporize and mix with air, and finally ignite.

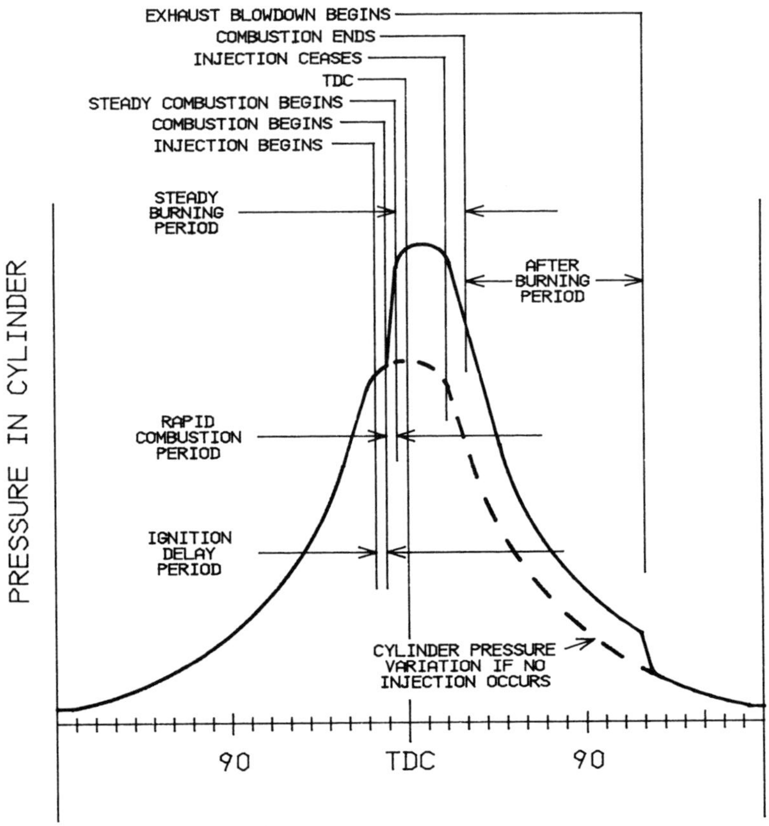

Figure 16-22. Four stages of combustion

Phases of combustion

1. The ignition delay period. The ignition delay period is the interval between injector opening and the start of ignition. During this period the first droplets to enter the cylinder are heated by the surrounding charge of compressed air, begin to vaporize, and finally ignite. Until ignition occurs, there is no noticeable increase in the pressure in the cylinder above what it would be had no injection occurred.

   The ignition delay period is primarily a function of the *ignition quality* of the fuel, hence of its chemical composition. Fuels of low ignition quality (i.e., of low *cetane number*) will require more preparation time, and the delay period will therefore be longer. It is important to note that in a high speed engine the crankshaft rotates farther in a given period of time than in a low speed engine, which explains the generally lower tolerance of high speed engines for fuel of low ignition quality.
2. The rapid combustion period. During this period, the fuel that has accumulated in the cylinder during the delay period before ignition burns rapidly. Because the fuel has already mixed with the charge air and begun the process of preparation for combustion, this is sometimes called the *premixed combustion phase.* The rapid combustion is accompanied by a sharp rise in cylinder pressure. If the pressure rises too sharply the combustion becomes audible, a phenomenon known as *diesel knock.*
3. The steady combustion period. Once combustion has been established in the cylinder, further fuel droplets entering the cylinder will burn as soon as they have penetrated, heated, vaporized, and mixed, so that the combustion rate lags behind the injection rate by the preparation time. Because the droplets burn as they diffuse into the cylinder, this is sometimes called the *diffusion combustion phase.* This period ends shortly after the injector closes (*cutoff*), when the last of the fuel has burned.

   Cylinder pressure usually peaks just after TDC, near the middle of the steady combustion period, and then falls off smoothly after cutoff as the expansion stroke begins.
4. The afterburning period. If all the fuel has burned cleanly and completely by the end of the steady combustion period, the pressure trace will be smooth through the expansion stroke, and the afterburning period could be neglected. Typically, however, there will be some irregularities reflecting combustion of incompletely burned fuel or of intermediate combustion products, and some delayed *chemical end reactions.* It is during this period that soot and other pollutants are produced.

Combustion problems. Difficulties in the combustion process are usually symptomatic of other problems, often related to the quality of the fuel and its preparation and injection, to air supply, or to maloperation; these are discussed under the appropriate headings. In engines of mature design such causes of combustion difficulties as component configuration are likely to have been eliminated.

MOMENTS, FORCES, AND VIBRATION

Introduction

The marine engineer is usually most concerned about the forces and moments generated by a diesel engine because of their potential for causing (*exciting*) vibration of hull structure and such connected equipment as reduction gearing and propeller shafting. The engine is only one of several possible sources of vibration, however; propeller excitation is the most common cause of problems. The forces and moments (*disturbances*) developed by an engine are entirely predictable in both magnitude (or *amplitude*) and frequency. Whether they will cause problems depends on the response of the ship's structure or connected equipment. This response will lie between two extremes:

1. If the frequency of the disturbance is even slightly different from the natural frequency of the structure or connected equipment, then, if they are sufficiently robust, the structure or connected equipment may absorb disturbances of large magnitude.
2. On the other hand, if the frequency of the disturbance (or an integral multiple of the frequency) is a sufficiently close match to the natural frequency of the structure or connected equipment, then even small disturbances can excite *resonant* responses in the structure or connected equipment that are much larger in magnitude than the exciting disturbance.

It is useful to consider two categories of diesel engine–induced disturbance:

1. External forces and moments that arise from the reciprocating motion of the pistons and running gear, and could cause an unrestrained engine to pitch, roll, or yaw. (With the engine installed in the ship, these disturbances can excite a response from the hull.)
2. Torsional vibration in the propulsion drive train that arises from the discrete power strokes of the engine and the resulting periodic application of torque, and generally affects only shaft-connected equipment, including reduction gearing and propeller shafting.

Engine bearings and structure are designed to absorb internal forces and moments; thus, they are rarely transmitted to the hull.

External Forces and Moments

Overview. Because the source of external forces and moments generated by a diesel engine is in the reciprocating motion of the pistons and running gear, it can be noted that:

The magnitude of the individual forces and moments will be proportional to the masses involved and also to the square of the engine RPM.

In multiple cylinder engines, it is possible to arrange the cylinders so that some of the external forces and moments generated by one cylinder are cancelled by other cylinders; in certain configurations the effect is complete.

The lower the RPM of the engine, the lower will be the frequency of the disturbances it generates. Therefore, given the typically low natural frequencies of most hull structures, there is great likelihood that a response will be excited.

Consequently:

From the standpoint of exciting hull vibration, a worst case would be presented by a large bore, low speed engine with four, five, or six cylinders. A simple solution is available, however, in the form of balancers, which are frequently fitted to these engines.

Smaller, high speed engines may generate disturbances which, while unlikely to excite hull vibration, may cause local vibration. Balancers may be used, and the engine may be installed on resilient, vibration-absorbing mountings.

Piston motion and resulting forces. A simple analysis of the geometry of piston and crank motion (see Figure 16-23) will yield the following relationship between piston position and crank position:

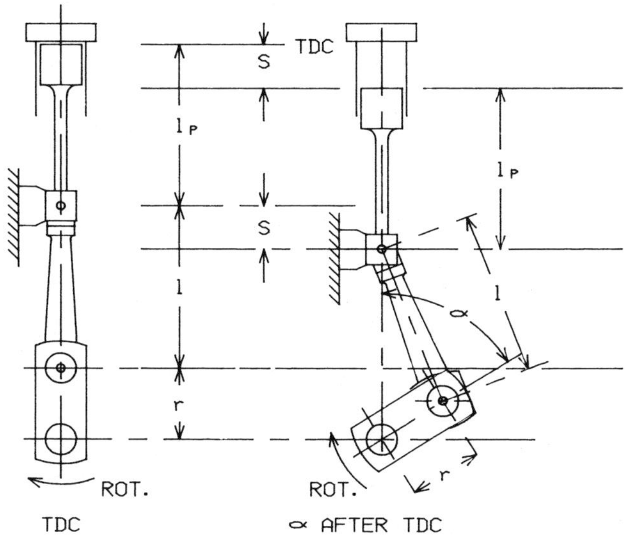

Figure 16-23. Geometry of piston and crank motion

$$s = r\left[1 + \frac{l}{r} - \cos\alpha - \left\{\left(\frac{l}{r}\right)^2 - \sin^2\alpha\right\}^{1/2}\right]$$

where s = piston position measured from TDC
l = length of the connecting rod
r = throw of the rock
α = angular travel of the crank measured from TDC

From this relation the approximate expression shown below can be derived, which relates the acceleration of the reciprocating parts, including the piston, to the angular velocity of the crank (i.e., its RPM):

$$a = \omega^2 r\,[\cos\alpha + \frac{\cos 2\alpha}{l/r} + \text{higher order terms}]$$

where α = acceleration of piston, its contents, and all connected reciprocating parts
ω = angular velocity of the crank

The higher order terms in the relationship contain, successively, cos3α, cos4α, etc., but are divided by successively higher powers of l/r and are therefore of rapidly decreasing magnitude.

This relation can be inserted in Newton's expression for the force opposing acceleration, F = ma, to obtain the vertical force (in a vertical, in-line engine) that will act upward at the beginning of the downstroke to oppose the acceleration of the piston (and connected parts), then downward after the middle of the downstroke as the piston decelerates towards BDC. Because this force opposes the acceleration of the piston, it is generally called the vertical inertia force, but may be more accurately termed the reciprocating force, F_{rec}.

$$F_{rec} = \frac{m_{rec}}{g_c}\,\omega^2 r\,[\cos\alpha + \frac{\cos 2\alpha}{l/r} + \text{higher order terms}]$$

where m_{rec} = mass of the piston, its contents, and all connected reciprocating parts
g_c = the gravitational constant relating mass to weight

The connecting rod motion is complex: at the wrist pin it reciprocates but at the crankpin it rotates. A reasonable approximation is to take a portion of its mass as reciprocating, with the balance considered to rotate.

It is useful to break the reciprocating force into components according to their order, i.e., their frequency as a multiple of RPM:

$$\text{first order: } \frac{m_{rec}}{g_c}\,\omega^2 r\cos\alpha$$

$$\text{second order: } \frac{m_{rec}}{g_c}\,\omega^2 r\,\frac{\cos 2\alpha}{l/r}$$

$$\text{higher order: } \frac{m_{rec}}{g_c}\,\omega^2 r\ \text{(higher order terms)}$$

The first and second order terms are plotted in Figure 16-24 for one revolution of the crank. Because l/r is usually in the range of 2.5 to 4.5, the *amplitude* of the second order component is correspondingly smaller than that of the first order component. The higher order components will be of successively lower magnitude. To summarize, for a single cylinder:

The strongest component of the vertical reciprocating force fluctuates with a frequency equal to engine RPM.
The second order component, fluctuating at twice engine RPM, is smaller.
There are smaller, higher order components.

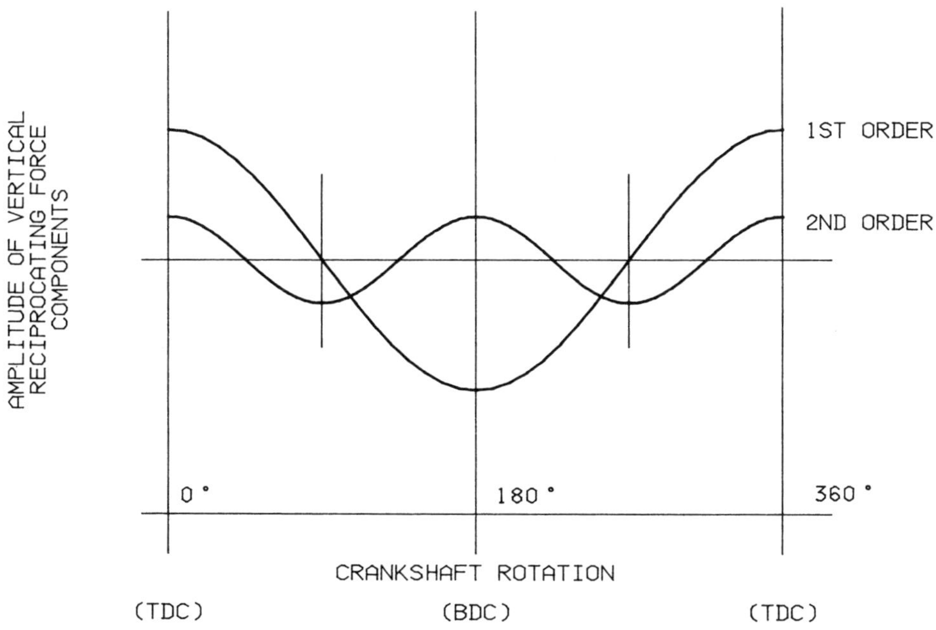

Figure 16-24. Amplitude of vertical reciprocating force components for one cylinder

External forces and moments generated by a single cylinder. The reciprocating force is only one of a number of forces acting on a cylinder. The others are illustrated in Figure 16-25 and described below:

F_g is the gas force exerted by cylinder contents on the piston and head. As long as the cylinder contents are above ambient pressure (as they always are in supercharged engines under load) F_g will act up on the head and down on the piston. F_g can be determined from the indicator card at any point in the stroke, by multiplying the indicated pressure by the piston area.

W_{recip} is the weight of the reciprocating components, including a share of the connecting rod weight considered to reciprocate.

F_n is the normal or guide force acting between crosshead and guide of a crosshead engine, or between piston skirt and cylinder wall of a trunk piston engine. This force acts in a vertical plane perpendicular to the crankshaft centerline. As the crank passes through BDC, the normal force will reverse its direction.

F_{cr} is the force on the connecting rod, placing it in compression as long as the gas force is above ambient.

F_t is the torque force, the tangential component of F_{cr} as it acts on the crankpin, which, on the power stroke, turns the crankshaft to drive the load. During the compression stroke, F_t will oppose the turning of the crankshaft as torque is absorbed to compress the charge.

F_{rot} is the rotating inertial force, or centrifugal force, caused by the eccentric mass of the crankpin, webs, and the portion of the connecting rod considered to rotate. This force is usually balanced by fitting a counterweight to the crankshaft.

W_{rot} is the weight of the rotating components, including the share of the connecting rod weight considered to rotate.

F_b is the bearing force.

By performing vector sums of the forces on successive components, and letting W represent the total weight of piston and running gear, the force acting on the main bearing can be determined. Resolved into vertical (F_{bv}) and horizontal (F_{bh}) components, and letting W be the weight of all the parts supported by the bearing, the bearing force is:

$$F_{bv} = F_g + W + F_{recip}$$

$$F_{bh} = F_n$$

Figure 16-26 shows the forces imposed on the stationary parts of the engine, which can be added vertically and horizontally to determine the forces transmitted to the foundation:

$$\Sigma F_V = W + F_{recip}$$

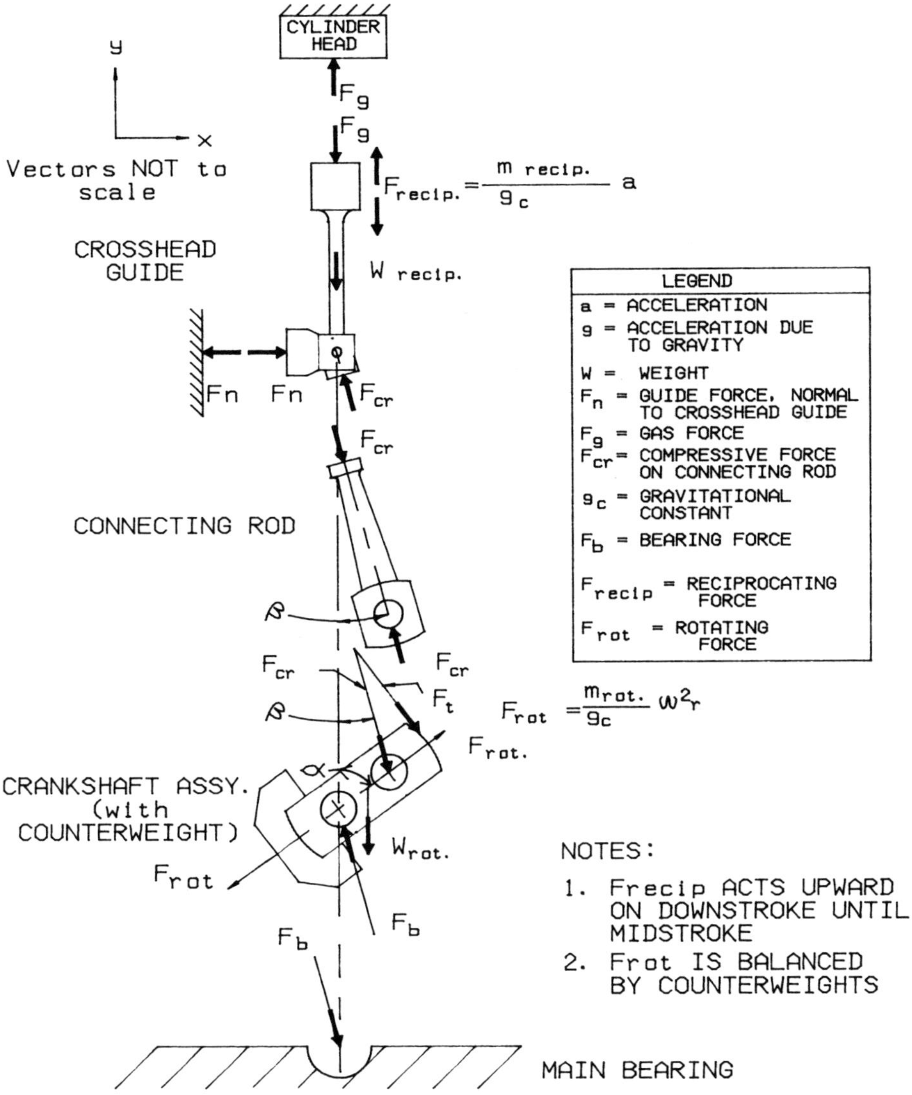

Figure 16-25. Forces generated by a single cylinder unit

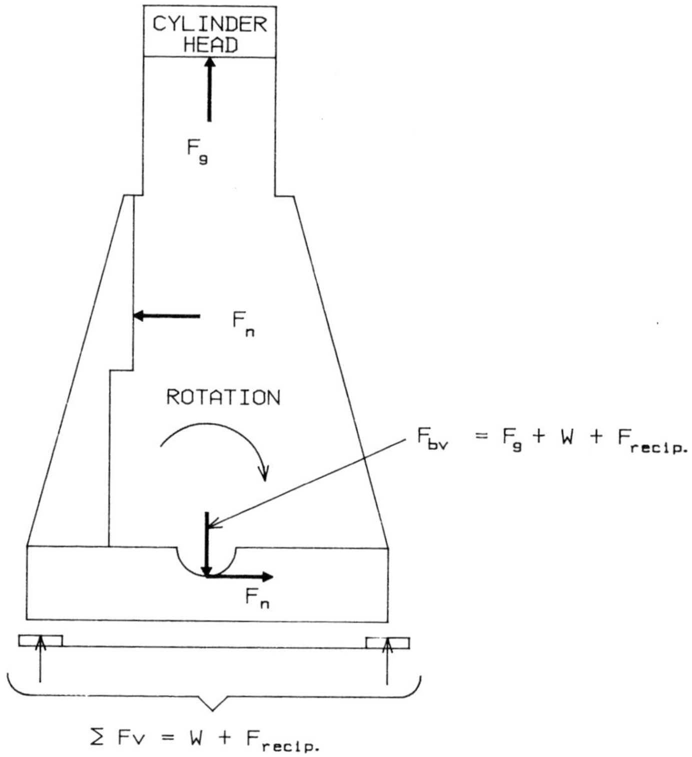

Figure 16-26. Forces on stationary components of a reciprocating engine

$$\Sigma F_h = 0$$

Two conclusions can be drawn:

A single cylinder engine would impose a vertical force on its foundation that is made up of the engine weight plus the vertical reciprocating force, and that fluctuates in magnitude and direction at single and higher multiples of crankshaft RPM.

Horizontally, although the *forces* are balanced, a *torque reaction* couple (also called a guide force moment) is generated because the guide force and the horizontal component of bearing reaction are vertically displaced. This couple would tend to roll, capsize, or tip the single cylinder engine of Figure 16-26 counterclockwise on the downstroke and clockwise on the upstroke, more severely when F_g is high, as in power and compression strokes. The torque reaction couple, therefore, has a major fluctuation at a frequency equal to crankshaft RPM in a single-cylinder, two-stroke engine, and equal to half crankshaft RPM in a single-cylinder, four-stroke engine, but because the guide force is affected by the vertical reciprocating force, the moment will have higher order components of low magnitude.

In addition, because the crankshaft is not absolutely rigid, the cranks deflect longitudinally under load. This will produce a pulsating axial force containing first and higher order components, all of low magnitude.

External forces and moments generated by multiple cylinders. It may be obvious from Figure 16-24 that if a second cylinder were arranged on the crankshaft, but with its crank 180 degrees out of phase with the first, as in Figure 16-27, then the first order vertical reciprocating force of the second cylinder would always balance that of the first cylinder, as shown in Figure 16-28. The second order vertical reciprocating forces, however, would always reinforce each other, creating a severe second order vertical reciprocating force imbalance. It is worth noting for this crank arrangement that, with the cylinders 180 degrees out of phase, equally spaced power strokes would occur if the cylinders operated on a two-stroke cycle: a two-cylinder, four-stroke cycle engine, with power strokes of each cylinder at 720-degree intervals, would require the cylinders to be 360 degrees out of phase, resulting in reinforced first and second order vertical reciprocating forces. While two-cylinder engines have practical application, it is confined to the very low output range. It is, therefore, worth moving on to examine more practical configurations.

Case study of a four-cylinder, two-stroke cycle engine

Timing. Since each cylinder of a two-stroke engine will complete its cycle in 360 degrees, even application of torque to the shaft requires that a four-cylinder, two-stroke engine have the cylinders arranged 90 degrees apart, as in Figure 16-29. The firing order of the cylinders, in this case

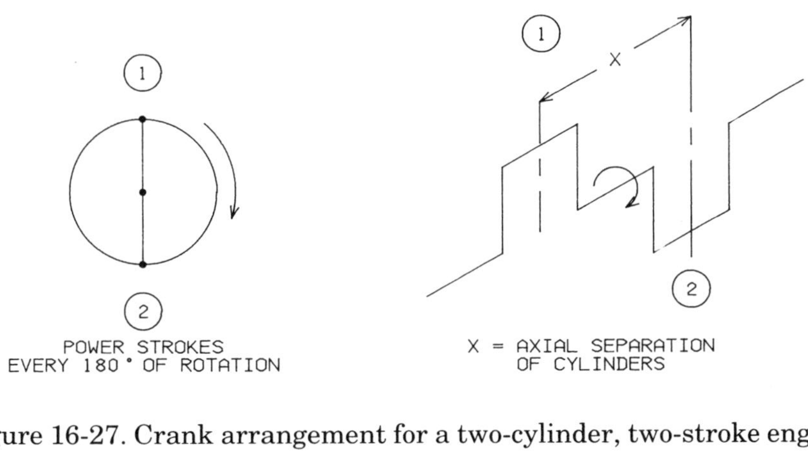

Figure 16-27. Crank arrangement for a two-cylinder, two-stroke engine

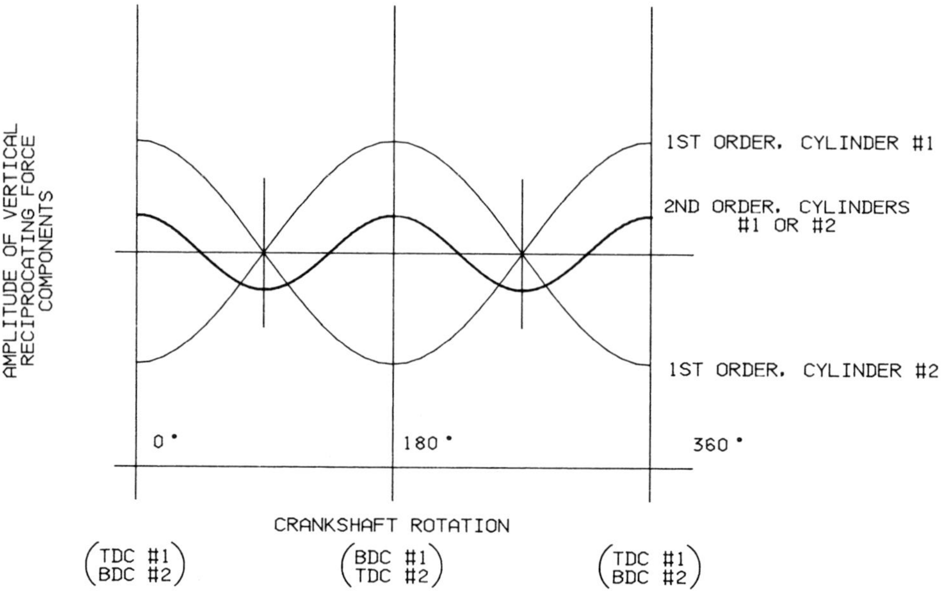

Figure 16-28. Amplitude of vertical reciprocating force components for a two-cylinder, two-stroke engine

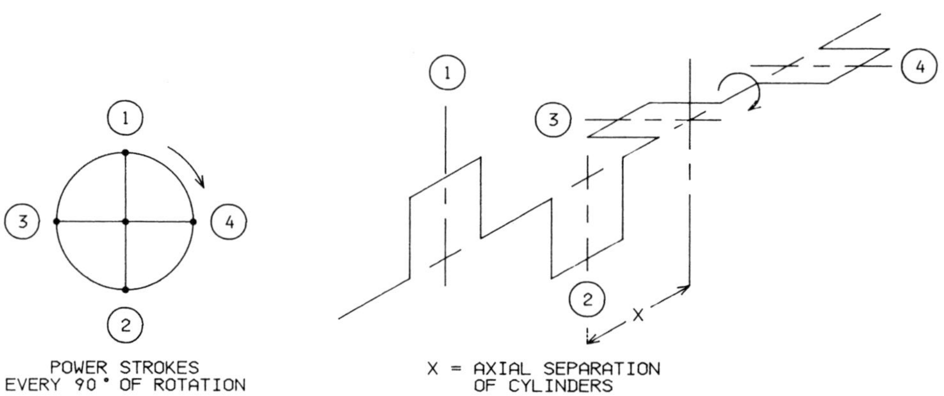

Figure 16-29. Crank arrangement for a four-cylinder, two-stroke engine

1-3-2-4, is selected to limit the longitudinal distance between opposing cylinders.

Rotating imbalance. The longitudinal displacement of the eccentric masses of the crankpins and webs would produce a rotating couple, best eliminated by fitting counterweights to the crankshaft.

Vertical forces and their effects. With the cylinders 90 degrees out of phase, both first and second order vertical reciprocating forces are completely balanced, as shown in Figure 16-30. However, if the crankshaft is examined from the side (as in Figure 16-31, where it is positioned with cylinder 1 at TDC), it can be seen that, while the vertical reciprocating forces of cylinders 1 and 2 cancel each other, then because the forces act through the centerlines of the cylinders and are therefore longitudinally (or axially) displaced, a couple is generated tending to pitch the engine about a transverse axis. An analogous couple will be generated by cylinders 3 and 4. Thus, first, second, and higher order pitching couples will arise from the longitudinal displacements of first, second, and higher order vertical reciprocating forces.

The first order pitching couple will be the largest in magnitude, as it arises from the largest force. Pairs of cylinders whose first order forces balance each other should be adjacent: to do otherwise would increase the pitching moment arm. It is this consideration that produces a firing order of 1-3-2-4 (although 1-4-2-3 would be equally suitable).

The first order pitching couple can be countered by fitting additional counterweights to the crankshaft, i.e., in addition to the counterweights

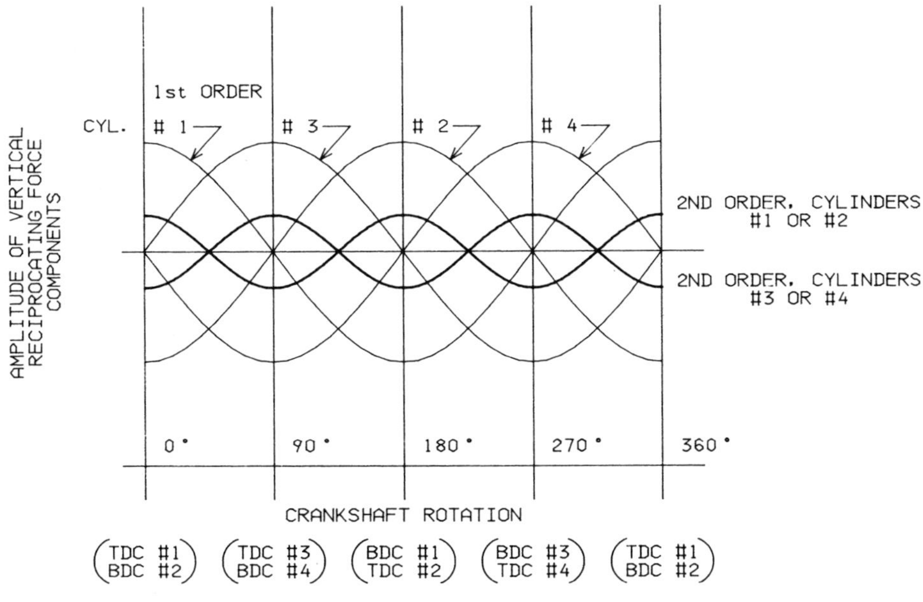

Figure 16-30. Amplitude of vertical reciprocating force components for a four-cylinder, two-stroke engine

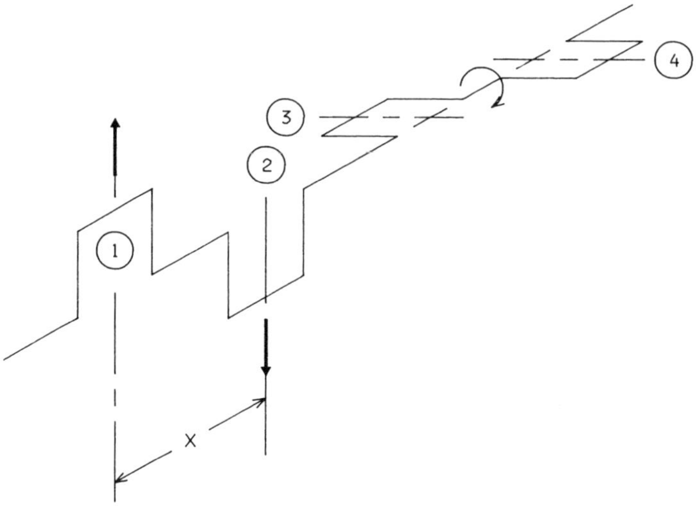

Figure 16-31. Axial displacement of vertical forces in a four-cylinder, two-stroke engine: cylinder 1 at TDC

fitted to balance the crankpins and webs. However, while the pitching couples are exclusively in the vertical plane, the counterweights would, as they rotate, generate a horizontal first order yawing couple. In fact, this is often an acceptable situation for main propulsion engines, as ships' hulls tend to be more rigid in the transverse direction. It is, therefore, the general practice of the engine designers to fit additional counterweights sufficient to cancel half of the first order pitching couple, thereby imposing a first order yawing couple of equal magnitude, i.e., half the magnitude of the original, vertical pitching couple.

Alternatively, the first order pitching couple could be completely cancelled by two pairs of counterweights rotating in opposite directions at crankshaft RPM. Figure 16-32 illustrates an arrangement where one weight of such a pair is on the crankshaft, and the other is on a balance shaft geared to the crankshaft. Since the weights rotate in opposite directions (and in the same transverse plane), the horizontal components of the forces they generate will always cancel, leaving only a vertical force fluctuating at first order frequency. By fitting two pairs of such weights, longitudinally separated along the crankshaft, the first order pitching couple can be cancelled without generating a yawing couple.

The same principle can be used to balance the second order pitching couples, if the pairs of opposing counterweights are driven at twice crankshaft RPM. This arrangement is called a Lanchester balancer, and a chain-driven example is shown in Figure 16-33.

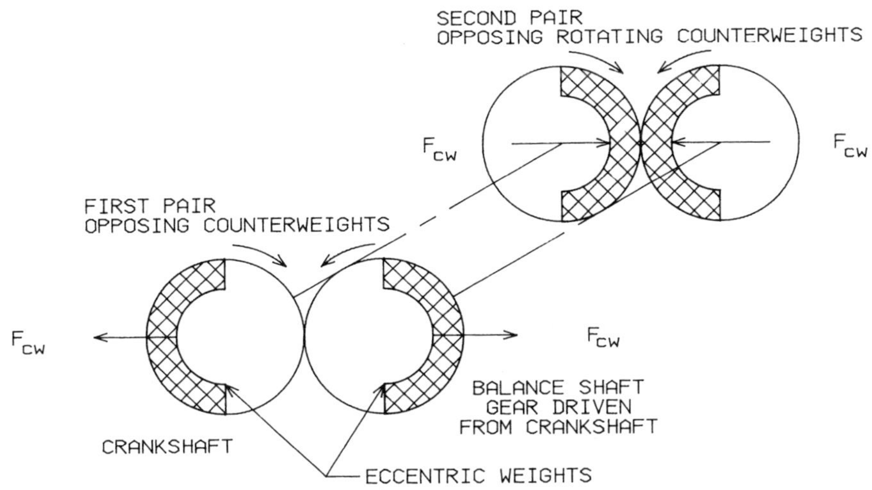

Figure 16-32. Principle of opposing counterweights to balance first order vertical couples

In principle, the higher order pitching couples could be countered by similar means; in fact they are usually not of sufficient magnitude to cause problems.

A pulsating axial force will be produced by the deflection of the crank webs under load and will contain first and higher order components, all of low magnitude and therefore not normally a source of trouble. Occasionally, however, a higher order of this axial force may coincide with the natural frequency of the crankshaft itself. The usual solution in this case is to fit a damper, consisting of a dummy piston under engine oil pressure, at the free end of the crankshaft.

Torque reaction couples. Each of the cylinders will develop a torque reaction roll couple that will tend to rock the engine about a longitudinal axis because of the vertical displacement of the guide force from the horizontal bearing reaction. Because the frequency of the largest component of this disturbance is at crankshaft RPM for each cylinder of a multicylinder two-stroke engine, the engine will have a torque reaction roll couple whose largest component is at a frequency equal to the RPM multiplied by the number of cylinders.

In addition to the torque reaction roll couple, the longitudinal displacement of the cylinders will cause the guide forces and the horizontal bearing reaction forces to generate equal but opposite yaw moments, one moment acting at the height of the wrist pins and the other moment acting at the height of the main bearings. The resulting torque reaction yawing couple

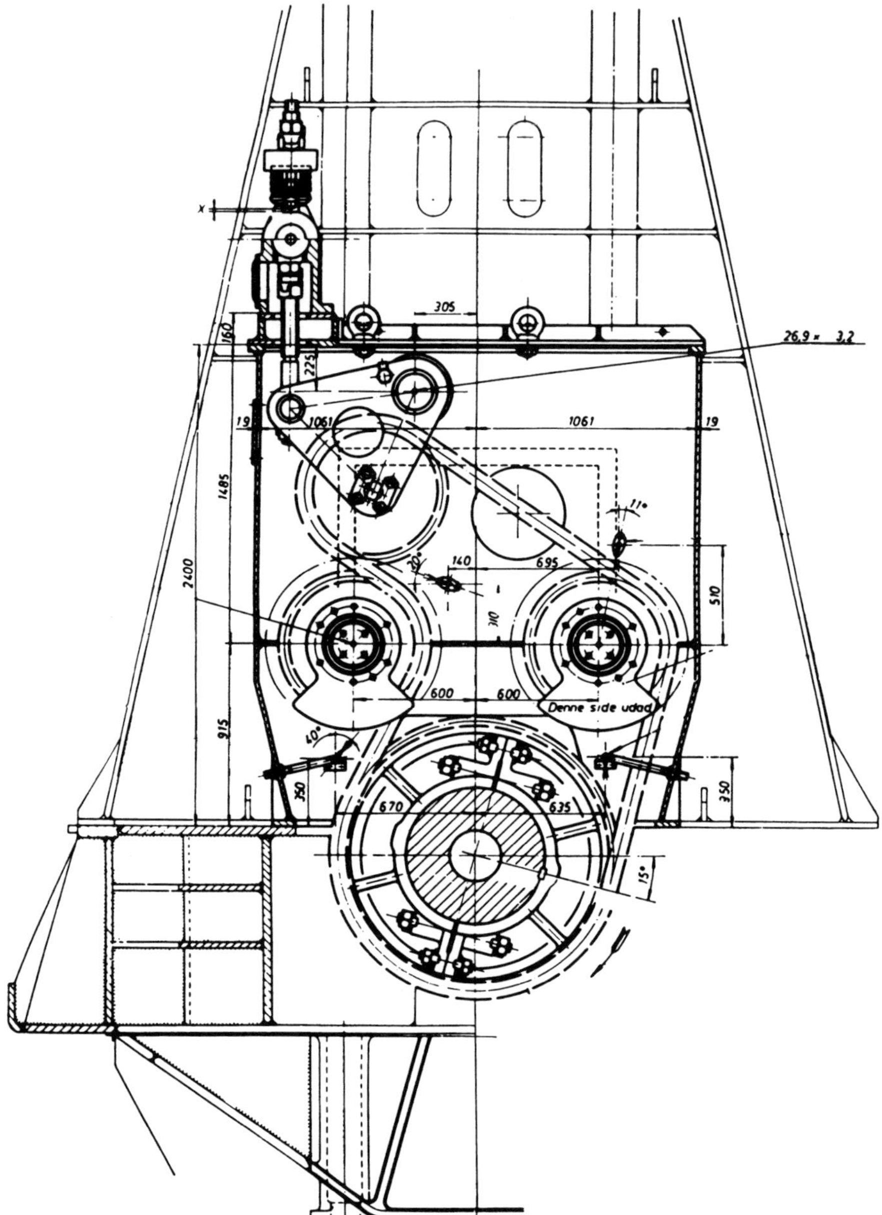

Figure 16-33. Lanchester second order balancer

will tend to force the top of a tall engine into a racking movement (i.e., an X pattern when viewed from above) since the base of the engine is constrained. A four-cylinder, two-stroke engine may have a significant first order component of this racking couple, but for engines with more cylinders there are usually higher magnitude racking couples at higher order frequencies.

Torque reaction couples of both types can be countered when necessary by bracing the top of the engine to an adjacent hull structure with transverse struts.

Summary. Depending on the timing, the four-cylinder, two-stroke engine will have no external unbalanced forces of significance; unbalanced first, second, and higher order pitching couples, correctable by pairs of counter-rotating counterweights; an unbalanced first order yawing couple if additional counterweights have been fitted to reduce the first order vertical pitching couple; and unbalanced torque reaction couples, the most significant being a roll couple at four times engine RPM, countered when necessary by braces to hull structure.

Case study of a four-cylinder, four-stroke engine

Timing. Since each cylinder of a four-stroke engine will complete its cycle in two revolutions or 720 degrees, even application of torque requires that a four-cylinder, four-stroke engine have the cylinders arranged 180 degrees apart, as in Figure 16-34.

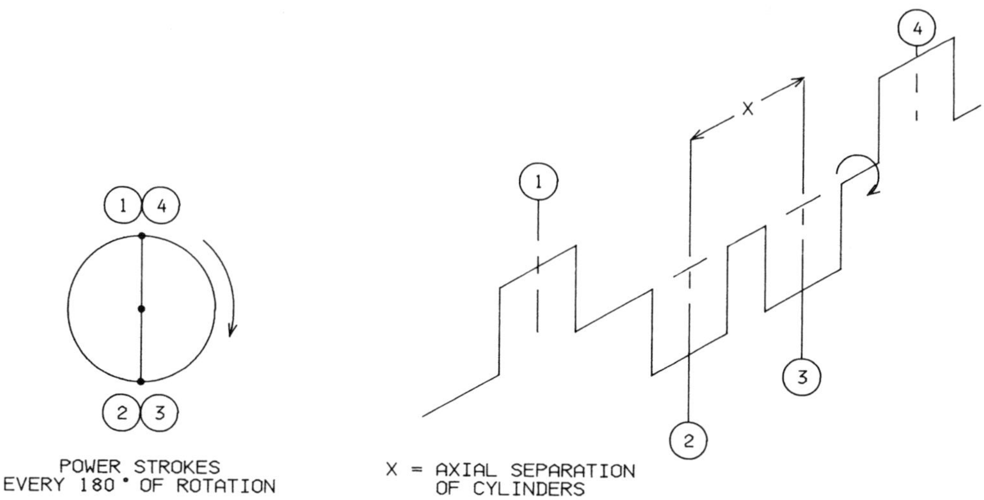

Figure 16-34. Crank arrangement for a four-cylinder, four-stroke engine

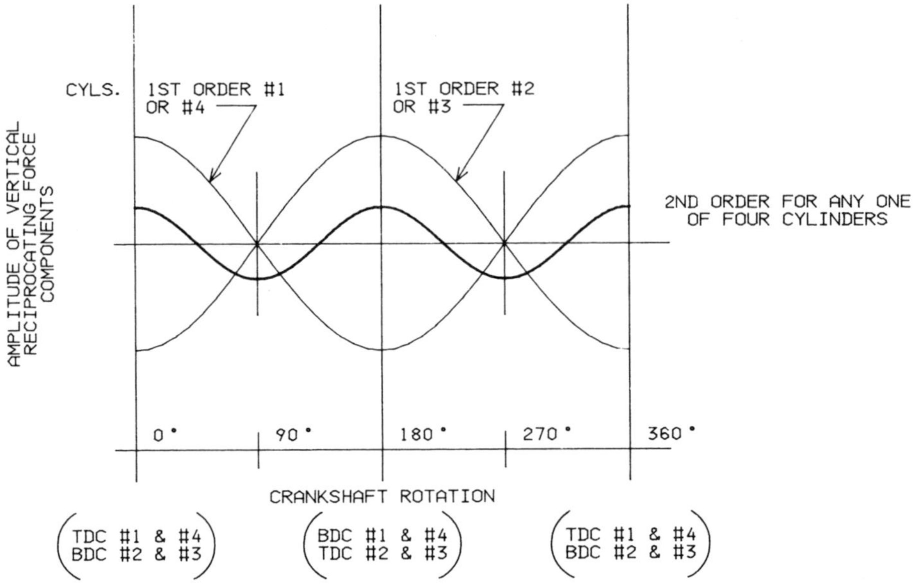

Figure 16-35. Amplitude of vertical reciprocating force components for a four-cylinder, four-stroke engine

Rotating imbalance. The longitudinal displacement of the eccentric masses of the crankpins and crank webs would produce a rotating couple, best eliminated by fitting counterweights to the crankshaft.

Vertical forces and their effects. As can be seen in Figure 16-35, with cylinders spaced at 180 degrees, only the first order vertical reciprocating forces balance, while the second order forces actually reinforce each other. With reference to the equation for second order terms, the resulting combined, unbalanced, second order force will have a magnitude four times that of a single cylinder, i.e.:

$$4\left(\frac{m_{rec}}{g_c}\,\frac{\omega^2 r}{l/r}\right)$$

Since values of l/r range from about 2.5 to 4.5, the magnitude will rival or exceed that of the first order vertical reciprocating force of any single cylinder.

The unbalanced second order force can be cancelled by paired, horizontally opposed counterweights driven at twice crankshaft RPM, i.e., a Lanchester balancer. As a practical matter, many of the higher output designs of four-stroke engines are simply not manufactured in the four-cylinder configuration because of the unbalanced second order force.

If the crankshaft is viewed from the side, it can be seen that the couple generated by the longitudinal displacement of first order vertical reciprocating forces of cylinders 1 and 2 as they oppose each other is itself cancelled by an analogous couple from cylinders 3 and 4, so that no external first order pitching couple remains. With secondary reciprocating forces acting in phase for all cylinders, there will be no second order pitching couple either.

As in the two-stroke engine, the pulsating axial force produced by the deflection of the crank webs may contain a higher order component whose frequency could coincide with the natural frequency of the crankshaft itself, necessitating the installation of an axial damper.

Torque reaction couples. Because the frequency of the largest component of the torque reaction couple is at half crankshaft RPM for each cylinder of a four-stroke engine, the four-cylinder in-line engine will have a torque reaction disturbance whose largest component is at a frequency equal to half of the RPM multiplied by the number of cylinders.

Summary. Depending on the timing, the four-cylinder four-stroke engine will have a severe unbalanced second order external force, correctable by pairs of horizontally opposed counterweights rotating at twice crankshaft RPM. There will be no unbalanced pitching couples of significance, but there will be unbalanced torque reaction couples, the most significant being a roll couple at twice engine RPM.

Summary of External Forces and Couples of In-line Engines

It should be noted that pitching couples are independent of gas force and are influenced only by the weight and inertia of the components. If a cylinder must be cut out, balance will usually be less affected if piston and running gear remain in place.

For in-line vertical engines with evenly spaced power strokes and no balancing gear or additional counterweights, the most serious external moments and forces generally encountered are described in the following paragraph. It should be noted that most of these disturbances can be corrected as described above. Higher order disturbances may be present in all planes including the longitudinal, but these only occasionally cause problems.

Two-stroke in-line engines. Four-cylinder engines will usually have severe first and second order pitching couples and a severe fourth order roll couple. Five-cylinder engines will usually have a moderate first order pitching couple, but severe second order pitching and fifth order roll couples. Six-cylinder engines will usually have no first order pitching couple, but a severe second order pitching couple and a severe sixth order roll couple. Engines with seven or more cylinders will usually have moderate or negligible first and second order pitching couples and moderate roll couples at an order equal to the number of cylinders. Eight- and twelve-cylinder engines may have racking moments sufficient to require countermeasures, typically at third, fourth, and fifth orders.

Four-stroke in-line engines. Four-cylinder engines will usually have a severe second order vertical force and a severe second order roll couple. Five-cylinder engines will usually have a moderate first order pitching couple, but a severe second order pitching couple and a severe 2.5 order roll couple. Engines with six or more cylinders will usually have moderate or negligible first and second order couples and moderate roll couples at an order equal to half the number of cylinders.

Summary of External Forces and Couples of V Engines

V engines are arranged with opposite cylinders in each bank acting on the same crank and therefore firing in succession. For even distribution of power strokes, the angle between the banks would have to be equal to the firing interval but, to simplify manufacturing, a constant V angle (typically 45 or 50 degrees) is usually used regardless of the number of cylinders. Where an uneven distribution of power strokes results, the effect is mitigated by the large number of cylinders present.

Several configurations will be perfectly balanced in regard to first and second order pitch and yaw, both forces and couples, regardless of the bank angle. Engines in this category include four-stroke V12, V16, and V24 engines.

V8, V12, V16, and V20 engines with a 90-degree bank angle can, depending on firing order, have a rotating first order couple (i.e., equal yaw and pitch couples) and second order pitch and yaw balance. Since a rotating couple can be readily cancelled by a pair of opposing counterweights on the crankshaft, these engines can have perfect first and second order balance when so equipped. The 90-degree bank angle will also provide evenly distributed power strokes in a four-stroke V8, which accounts for the popularity of this configuration in automotive applications.

With smaller bank angles the first order yaw couple diminishes while the first order pitch couple increases, but the opposing counterweights on the crankshaft can still be used to eliminate the yaw couple and reduce the pitch couple, or to convert the remaining pitch couple to a yaw couple. If a counter-rotating shaft, geared to run at crankshaft RPM, is fitted, the remaining first order imbalance can be completely eliminated.

Torque reaction couples will be present in V engines, most significantly as a roll couple at an order equal to half the number of cranks for four-stroke engines, and at an order equal to the number of cranks for two-stroke engines. Because of the inherently more rigid structure of these engines compared with the tall crosshead engines, and also because of the higher frequencies resulting from the higher RPM and greater number of cylinders, these couples are not normally a source of concern.

Described in the following paragraph are the most serious external moments and forces generally encountered in four-stroke V engines without balancing gear or additional counterweights beyond those necessary to reduce first order rotating couples, and with a typical bank angle of 45 to 50 degrees. Most first and second order disturbances can be corrected using combinations of opposite-rotating counterweights. Higher order components may be present in all planes but only occasionally cause problems. The uneven firing that results when the bank angle differs from the firing interval is not usually an important factor.

Four-stroke V engines. Eight-cylinder engines can be perfectly balanced in regard to first and second order pitching and yawing couples, but only if the bank angle is equal to the firing interval of 90 degrees; smaller angles can yield first order disturbances sufficient to require correction. Ten cylinder engines will usually have moderate first and second order pitching and yawing couples and a moderate 2.5 order roll couple. Engines with 12, 16, or 24 cylinders will generally be perfectly balanced in regard to low order pitching and yawing couples, although they will usually have low to moderate roll couples at an order equal to half the number of cranks (i.e., at a quarter the number of cylinders). Engines with 14, 18, or 20 cylinders will generally have moderate or negligible first and second order pitching and yawing couples as well as low to moderate roll couples.

Two-stroke V engines. There are far fewer designs of two-stroke V engines than four-stroke, but inasmuch as the Electro-Motive Division of General Motors has been building its engines exclusively to this configuration since the 1930s, there are vast numbers of them to be found in marine service. These engines have a firing order selected to eliminate secondary forces and couples, but with a 45-degree bank angle they would have a rather large primary pitching couple, even after being fitted with crankshaft counterweights, were it not for the use of the camshaft counterweights to provide first order moment compensation. This is possible because the camshafts run at engine RPM and are geared to run in opposite directions. The 45-degree bank angle and the selected firing order yield evenly distributed power strokes in the V8 and V16 engines, but not in the V12 and V20 engines, where the large number of cylinders provides adequate compensation.

Torsional Vibration

Nature and effects. Torsional vibration arises from periodically varying torque superimposed upon the steady torque being transmitted to the load. The sources of torque variation in propulsion systems are:

The discrete power strokes of the engine, which generate torque pulsations once per crank throw per cycle, and at higher orders of this frequency, with a high magnitude at first order. In ships with direct-connected low speed diesels, this is usually the dominant source of torque variation. (A lower order torque variation of significant magnitude can be imposed if one or more cylinders are unbalanced or not firing.)

The discrete number of blades on the propeller, which imposes a torque variation at a frequency equal to the number of blades times the propeller RPM, with higher order components of lower magnitude.

Gear tooth irregularities, with the frequency of the torque disturbance dependent on the particular irregularity. Given modern gear-manufacturing techniques, only damage is likely to cause the gearing to generate a torque variation of significant magnitude.

In direct drive electrical generating sets, the source of torque variation is normally the engine, but under certain circumstances the generator can produce a torque variation as well. However the range of operating RPM of a generator set is normally predetermined and torsional vibration problems are usually avoided.

When a propulsion system is operated in its normal range of RPM, torsional vibration becomes a problem if a source of excitation has a first or higher order frequency close to one of the natural frequencies for torsional vibration of the rotating system. Under these circumstances, the torque variation of the source will force the rotating system into resonance, causing it to oscillate torsionally at high angular amplitude, with corresponding high stresses that could result in shaft line component fatigue failure. In any event, the torsional oscillation can produce transverse components or can excite axial vibration of the shafting, either of which can in turn be transmitted to adjacent structures. In geared installations at low power levels, the oscillations can cause the gear tooth loaded faces to periodically separate and chatter.

Determination of resonant speeds; barred speed ranges. Torsional vibration calculations are required by classification societies at an early stage in the design process, as soon as the engine has been selected and the configuration of the rotating system—including shafting, couplings, clutches, gearing, bearings, and propeller—is known. The natural frequencies for the rotating system are estimated for at least the three most likely modes of torsional vibration. Traditionally, a method called a *Holzer tabulation* is used, but more sophisticated techniques are available. The resulting natural frequencies can each be divided by the orders of torque variation of each source in order to determine values of RPM at which resonances will occur. These are called *critical speeds*. For example, for a direct-drive propulsion system driven by a five-cylinder, two-stroke engine, where the first mode natural frequency has been determined to be 40.1 radians/second or 383 RPM, there would be a major, or first order, critical speed at 76.6 RPM (383 divided by 5), and a second order critical speed at 38.3 RPM. If the stress levels arising at these speeds at any point in the shaft system—taking into account the energy absorbed in friction and by the entrained water surrounding the propeller (*damping*)—are in excess of permitted values, sustained operation near these speeds would be *barred*. Barred speed ranges may be an acceptable solution provided they are well below normal operating RPM.

Avoiding torsional vibration. If the imposition of a barred speed range is not acceptable, usually because the barred range is too close to operating speed, the following solutions are available at the design stage:

With direct connected engines, the intermediate shaft diameter can be increased. This lowers the operating stress on what is usually the weakest part of the rotating system, while raising the natural frequency of the system and moving the major critical speed above the operating RPM. This approach may be impractical for installations with long shaft lines.

With direct connected engines and a long shaft line, a heavy flywheel (at the aft end of the crankshaft) can be fitted to reduce the natural frequency of the system, thereby shifting critical speeds to a lower RPM.

A torsional vibration damper (or detuner) can be fitted at the forward end of the crankshaft to alter the natural frequency of the system and reduce the amplitude of the torque variation. The most frequently used dampers are of the spring-loaded or viscous fluid types. These dampers are occasionally used with direct connected engines and are commonly fitted to medium and high speed engines.

Torsionally flexible couplings are almost always fitted to geared installations to isolate the gearing, and therefore the rest of the system aft of the gearing, from engine excitation. The most commonly fitted couplings are of the spring-loaded or rubber element types. These couplings are in addition to the quill shaft, which almost always connects the input flange of the reduction gear to the pinion, and is itself an effective torsionally flexible coupling.

Flexible couplings and spring-loaded torsional dampers are both susceptible to low frequency excitation, such as that produced when one cylinder is cut out. Consequently, torsional vibration calculations are usually required for this mode of operation as well, and additional barred speed ranges may be imposed.

ENGINE PERFORMANCE: MATCHING ENGINES TO THEIR LOADS

Introduction

Diesel engines are typically furnished with a *rating* or a maximum continuous rating (*MCR*), i.e., a stated value of power output at a corresponding RPM. This is an inadequate description of the performance characteristics of the engine, however, which will rarely be matched to its load, or operated continuously, at the rating. More complete information can take the form of an engine performance map, of which Figure 16-36 is a hypothetical example. The map defines an envelope within which operation is recommended by the engine designers. When an engine is initially selected for a particular installation, it must be correctly matched to its load so that conditions encountered in service do not force the engine to be operated outside of this envelope. Unfortunately, not all future service conditions can be foreseen, and operating engineers frequently find themselves struggling with engines that must be operated near or even beyond the boundaries of the envelope. An understanding of the implications of

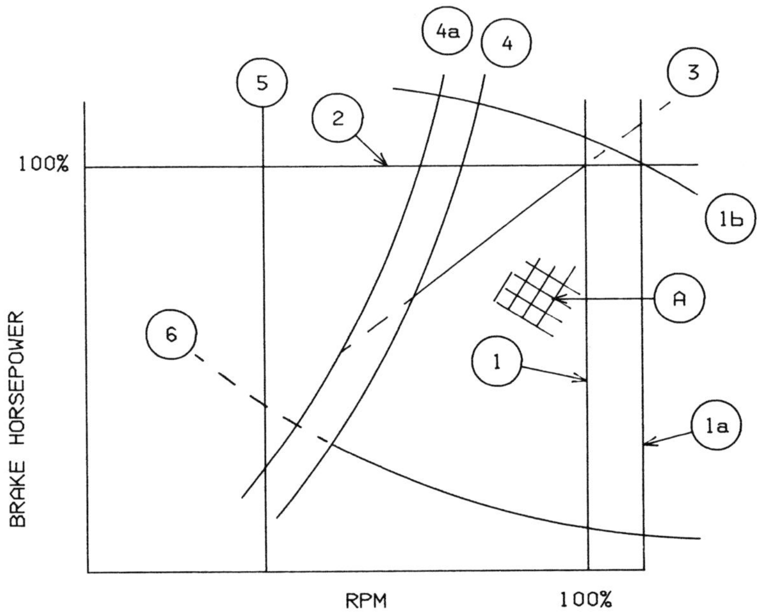

Figure 16-36. Typical engine performance map

these boundaries, and a familiarity with the load-matching procedure, will lead to a better understanding of what can be expected from a given installation.

Engine Performance Maps

Figure 16-36 defines the envelope in terms of power output, in this case BHP, versus RPM: some designers prefer to define the output in terms of MEP. The lines identified as 100 percent coincide with the MCR, and lines 1 through 6 with recommended limits for continuous operation under what the engine designer would define as normal conditions, with tolerable maintenance requirements. Normal conditions, including ambient pressures, temperatures, humidity, and fuel quality, must be accurately defined by the engine builder as they may or may not comply with international standards.

A point outside the limiting envelope at which an engine can be safely operated will, more than anything else, be a function of the particular engine design, reflecting design margins and the extent to which these have been proven by tests and by service experience, and the extent to which the design has evolved, been up-rated, or improved.

Most designers have established their ratings and recommendations by probing the areas beyond the boundaries in prototype testing. Such testing, however, is usually confined to one or a few configurations of a particular model. For some models, therefore, margins may have been proven by, for example, several thousand hours of operation with an overload of 20 or 30 percent, while for other models the margins may represent extrapolations of test data.

Engine performance limits. The boundaries of the operating envelope are as follows:

Rated engine RPM, Line 1 in Figure 16-36, is usually set by applying a design margin to the mechanical strength of those engine components which are subjected to the inertia forces. The turbocharger overspeed limit, Line 1b, is usually set by turbine blade root strength. Line 1a represents the setting of the overspeed governor or trip, usually 110 to 120 percent of rated RPM. Except for occasions when an engine driving a fixed pitch propeller is run at a modest overspeed on trials in order to maximize the load, operation beyond rated RPM is usually accidental.

Lines 2 and 3 are, respectively, rated power (MCR) and rated MEP (or torque), and operation beyond these limits constitutes an overload. Overload operation results in increased stress and higher temperatures for combustion chamber surroundings (i.e., cylinder heads, liners, and piston crowns), a condition often referred to as high thermal load, as well as increased mechanical loadings on other engine components. Catastrophic failure in a properly maintained engine is considered unlikely because of the design margins, and because inspections would reveal such effects as burning, cracking, or distortion in time for component renewal.

In general, MEP and cylinder exhaust temperature are better parameters of thermal load than power output, with temperature being the preferred choice of the operator because of the ease with which it can be measured.

Line 4 represents marginal combustion air supply, and in turbocharged engines is set by applying a margin to the surge limit of the turbocharger, represented by Line 4a. Operation to the left of Line 4 incurs increased likelihood of combustion chamber and turbine fouling, and smoke emission.

Line 5 is the engine idle speed, typically 25 to 40 percent of rated RPM. At lower engine speeds piston ring leakage during the compression stroke can so deprive the cylinders of air that combustion becomes unreliable.

Line 6 is a limit for the sustained use of heavy or sulfur-bearing fuels, below which operation on such fuels is not recommended. Combustion chamber temperatures at low loads are likely to be too low to effectively burn out some constituents of heavy fuels, and fouling of the combustion areas and the turbine can result. In addition, the low temperatures toward the bottom of the cylinder liners at such low loads can lead to sulfuric acid attack.

Normal operation, fuel rate, and service recommendations. Generally, diesel engines are so matched to their loads that the normal operation in service is at some high fraction of rated output, normally under 90 percent of MCR, and at, or somewhat below, rated RPM, i.e., in region A of Figure 16-36. This region usually coincides with the best range of specified fuel consumption. In addition, anticipated component lives, and therefore service recommendations for inspection, maintenance, renewal, and overhaul intervals, assume operation in this region.

Nevertheless, most designers consider a certain amount of operation beyond Lines 2, 3, and 4 inevitable and take this into account in their service recommendations. Statements permitting time-limited operation in these regions, limited to perhaps one hour in ten or twelve, or a cumulative total of perhaps 500 to 2,000 hours per year, are typical.

Variations in ratings. Apart from evolutionary effects, which usually result in an increase in rating as an engine design matures, identical or nearly identical engines are often offered with different ratings depending on the application. Some of the possibilities are listed below, where for purposes of this discussion the point corresponding to 100 percent RPM and 100 percent output in Figure 16-36 is referred to as the *basic* rating. These variations from the basic rating may involve differences in the engine components or outfit.

Economy ratings. By derating an engine, i.e., by setting a new MCR lower than the basic rating, perhaps at a lower RPM as well, and by fine-tuning the turbocharger, fuel nozzles, and timing for this point, it is usually possible to obtain a lower fuel consumption, typically by 2 to 4 percent over the whole operating range. The basic rating may not then be achievable because of these changes. Since engine acquisition cost, weight, and volume will be essentially unchanged, the *specific* cost, weight, and volume (per horsepower or kilowatt) are all higher, but the long-term reduced fuel and maintenance often yield an economic gain. This is particularly true for direct-drive propulsion applications, where a reduced engine RPM be translated by the designer into a more efficient propeller.

High performance ratings. High performance operation is equivalent to sustained overload operation of the base engine, usually at the expense of increased maintenance requirements. The table below, for a particular line of four-stroke engines, is illustrative, and assumes no upgrading of engine components:

Percent of basic MCR	*Inspection / renewal interval as percent of interval at 80 to 90 percent MCR*
80 to 90	100
100	75
110	30
120	3

Fuel consumption will also suffer. The trade-off might be justified where low specific weight or volume is the overriding concern, for example, in naval combatants, or where engine use is limited, as in emergency generators.

Upgrading of components—improving the materials of pistons and exhaust valves, for example—can improve the maintenance requirements but with the penalty of increased acquisition cost. In some cases the component changes are sufficient to justify a different model designation for the high performance engine.

Heavy fuel ratings. Low speed engines and the vast majority of medium speed engines of recent design, as well as some high speed engines, were intended from the beginning for heavy fuel operation. Other engines, however, may carry a lower rating for use with heavy fuels. The heavy fuel version of the engine may differ in components and outfit, with fuel nozzles, exhaust valves, and piston material being common changes.

Generator RPM ratings. Engines driving alternating current generators are usually directly connected and must therefore run at a synchronous RPM. While the basic rated RPM may be a synchronous RPM in some cases, in other cases there will be a rated RPM for generator service, usually the closest synchronous RPM below the basic rated RPM.

Engine Matching for Ship Propulsion

The case of engine matching most likely to be troublesome for the marine engineer, and the one which is least understood and most abused in service, is the selection of engines for ship propulsion.

Hull power requirements. Figure 16-37 is a hypothetical but typical example of speed-power curves for a ship. The shape of a speed-power curve can be projected at the design stage from model tests or from a data base that includes standard series as well as trial run data from previously built ships of similar hull form.

Line 1 in Figure 16-37 may represent the power required at loaded draft and trim, with the hull clean and smooth, and in good weather and calm seas (i.e., the expected performance on loaded trials). In service, the hull and propeller roughen through fouling by marine growth as well as general degradation (flaking paint, physical damage, corrosion, etc.); as this occurs, more power will be required for a given ship speed, so that the curve will shift to the left over time. Line 2 therefore represents loaded performance at a later time. An analogous curve for the ship in ballast would always lie lower and to the right of the corresponding curve for the loaded condition.

Apart from weather and sea conditions, the rate at which performance deteriorates is a function of many factors, not all of which are predictable, but which include the quality of the antifouling paint and the time spent

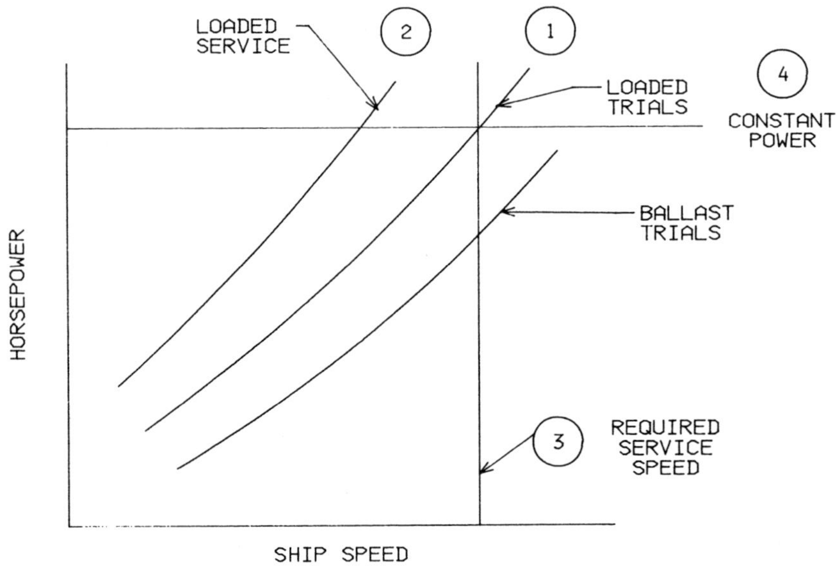

Figure 16-37. Typical speed-power curves

in warm water ports where fouling is most likely. For ships coated with conventional antifouling paints, the power required to maintain a constant ship speed might rise 30 to 50 percent or, if constant power is maintained, the ship speed may fall by some 5 to 15 percent over the interval before the next drydocking. If more sophisticated antifouling coatings with self-polishing properties are used, the rise in power or drop in speed will be less. The typical interval between drydockings is 12 to 24 months, although the interval can be extended by partial hull cleaning and even repainting with the ship afloat. When the ship is drydocked the hull is cleaned and repainted, and the propeller cleaned and perhaps polished. Not all of the general degradation is normally recoverable, however, so that even after cleaning and repainting the hull will normally be rougher than at the previous drydocking. As a consequence, the loaded performance curve after each drydocking will lie progressively to the left and above Line 1. After several drydocking cycles it may become economically attractive to attempt a more complete restoration to an "as-new" condition by blasting the hull to bare metal before repainting.

Line 3 in Figure 16-37 represents a required speed to be achieved in service at loaded draft. If the engine were matched to the intersection of Line 3 and Line 1, where it would deliver MCR at the required service speed with the hull clean and smooth, it should be obvious that this speed could not be maintained as the hull and propeller roughen, even with good weather and in calm seas, except by overloading the engine. Consequently a propulsion engine must be matched so that only a portion of its rated power is required to meet the desired speed with the hull and propeller clean and smooth, leaving a power reserve, in addition to the engine margin, to offset the performance deterioration. Hull performance will also be affected by weather and sea conditions. Depending on the size of the total power reserve, as well as the trade route and the consequences of speed loss, an additional margin may or may not be deemed necessary. The total margin, the difference between installed power or MCR and power required with hull and propeller clean and smooth, is the *service margin.* It may be divided into constituent parts, as engine margin and *sea margin.*

Propeller RPM. The efficiency of any propeller will peak at an optimal value of RPM, and propellers are usually designed so that this RPM coincides with the normal range of operation. That peak efficiency will be higher, however, for lower optimal values of RPM. A rough rule of thumb is that peak propeller efficiency rises 2 to 3 percent for every 10 RPM by which the optimal RPM can be reduced. It is largely this fact that has driven designers of low speed engines intended for direct connection to propellers to design for lower RPM. A limit on propeller RPM, and therefore on achievable efficiency, is set by the fact that the lower the RPM, the larger the propeller diameter. Propeller diameter is in turn limited by the necessity for adequate propeller tip-to-hull clearance and by the need to keep the propeller immersed even in light ballast conditions, together with the desirability, at least for most merchant ships, of keeping the propeller above the baseline of the hull. (The achievable efficiency of a controllable pitch propeller is usually considered to be slightly lower than that of a fixed pitch propeller designed for the same application, because of the larger hub required.)

The relation of the RPM of a fixed pitch propeller to the ship speed is shown in Figure 16-38 where Line 1 represents this relation at loaded draft and trim, with the hull clean and smooth, and in good weather and calm seas, i.e., the expected performance on loaded trials. In service, as the hull and propeller roughen, the RPM at any given ship speed will rise slightly; or, stated differently, the *slip* would increase. The curve thus shifts upward with time, and Line 2 represents loaded performance at a later time. An analogous curve for a ballast condition would always lie lower than the curve for the loaded condition.

Line 3 in Figure 16-38 represents a required ship speed to be achieved in service on loaded draft. If the engine were matched to the intersection of Line 3 and Line 1, where it would run at 100 percent RPM at the required service speed with the hull clean and smooth, it should be obvious that the engine would exceed its rated RPM increasingly thereafter whenever the ship was at its service speed, as the hull and propeller roughen, even with good weather and in calm seas. Consequently a propulsion engine must be matched so that the ship speed achieved at 100 percent RPM, with hull and propeller clean and smooth, exceeds the required service speed.

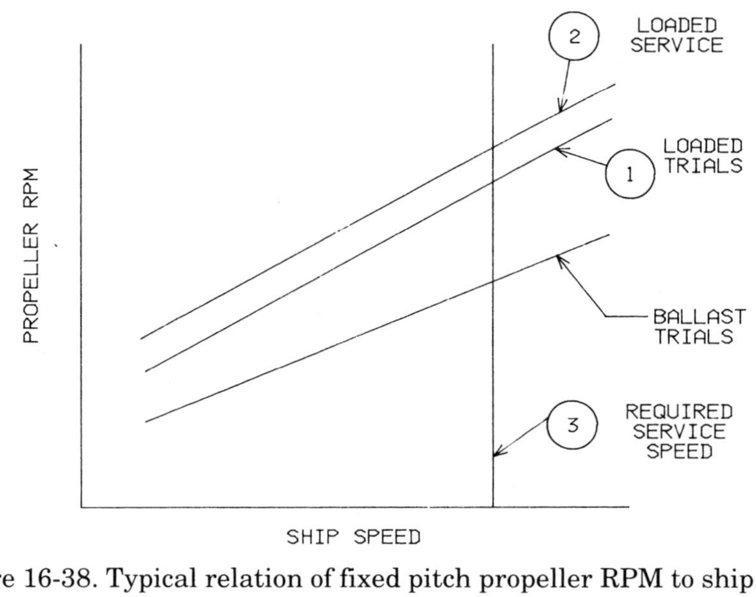

Figure 16-38. Typical relation of fixed pitch propeller RPM to ship speed

It is important to note that the rise in RPM for a given ship speed is much less than the rise in power, perhaps only 6 to 8 percent over an interval between drydockings.

Effect of propeller RPM on engine performance. With reference to Figures 16-37 and 16-38, for installations with fixed pitch propellers, if constant power is maintained as the hull and propeller roughen, then the RPM drops as the ship speed declines. Therefore, the MEP rises as the hull and propeller roughen in service:

$$\text{MEP} \propto \frac{\text{BHP}}{\text{RPM}}$$

Thus an engine can be forced to operate at the edges of its envelope even without exceeding rated power. As a numerical example, consider the case of a ship operated at constant power for which, towards the end of an interval between drydockings, the loaded hull speed has fallen by some 15 percent, to 85 percent of its clean hull value. The propeller RPM at this ship speed, which might have been 85 percent if the hull were clean, has risen by perhaps 6 percent, to 90 percent RPM. If the engine is run at a constant 90 percent power output, the MEP will be as follows:

$$\text{MEP} = .90\ \text{BHP}/.90\ \text{RPM} = 1.0, \text{ or } 100\%$$

This shows that the engine is operating at its MEP limit even though BHP and RPM are well below rated values; attempting to increase ship speed by increasing engine output would force the engine into its overload range. If the hull condition were allowed to degrade further the power would have to be decreased, resulting in an even more severe drop in ship speed.

With reference to Figure 16-36 it can be seen that the point can be reached where the engine envelope is no longer bounded by the MEP, but by the air limit, Line 4, invoking the possibility of combustion chamber fouling, smoke, and, ultimately, turbocharger surge, if the power is not reduced even further.

These situations could be alleviated if a controllable pitch propeller had been fitted, enabling the engine to operate at or near its rated RPM, independently of ship speed. Any comparison would have to take into account the slightly lower design point efficiency and the higher costs of the controllable pitch propeller.

Matching engine performance to hull requirements. The matching procedure for a fixed pitch propeller superimposes a curve for hull power requirement on an engine performance map, as in Figure 16-39. A curve for displacement hull power requirements at any particular condition of the hull can be approximated by the *cube law,* also called the propeller law:

$$\text{power} \propto \text{RPM}^3$$

Figure 16-39 illustrates the following rule:

Match the engine, propeller, and hull so that no more than 90 percent of rated output is absorbed at 100 percent rated RPM, with the hull clean and smooth, at loaded draft and trim, and in calm seas and fair weather.

The speed that the ship might then achieve in service can be estimated from the following rule, illustrated in Figure 16-40:

Estimate service speed by applying a 15- to 25-percent power margin to the clean hull performance curve, and determine the service speed achievable in calm seas and fair weather along this curve at a power corresponding to no more than 90 percent of engine rated output.

Lower match points are frequently used:

Eighty to 85 percent for bulk carriers and tankers that will be operated at constant power, and therefore declining speed, where the larger power-reserve will allow extended intervals between drydockings.

Seventy-five percent if the owner is attempting to allow for the long-term effects of hull and propeller roughness, i.e., those not recoverable in normal drydockings.

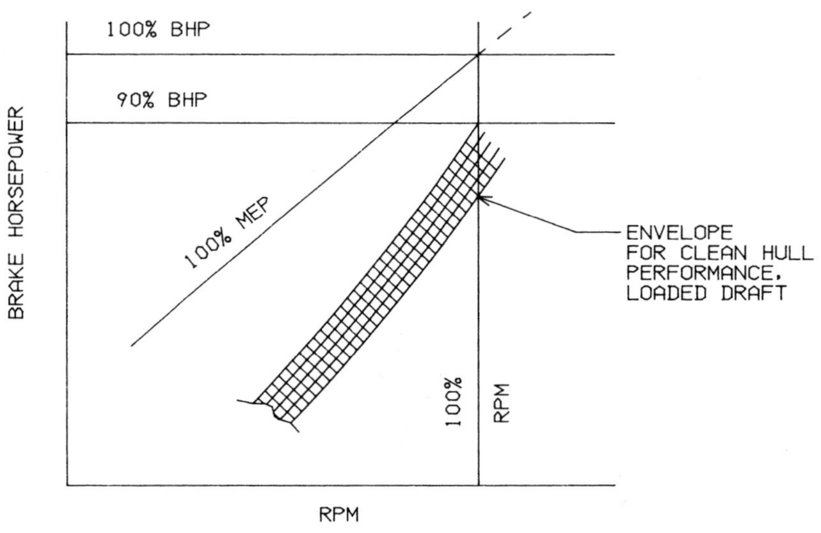

Figure 16-39. Matching region for fixed pitch propeller

Seventy percent or less if the ship is to operate at constant speed.

Under these circumstances the propeller is described as a *light propeller* because it absorbs less than 100 percent of engine output at 100 percent RPM.

The consequences of a low match point are that more power must be installed, so acquisition cost will be higher and cargo capacity lower; and, even if the ship can be ballasted down to loaded draft on trials, it will not

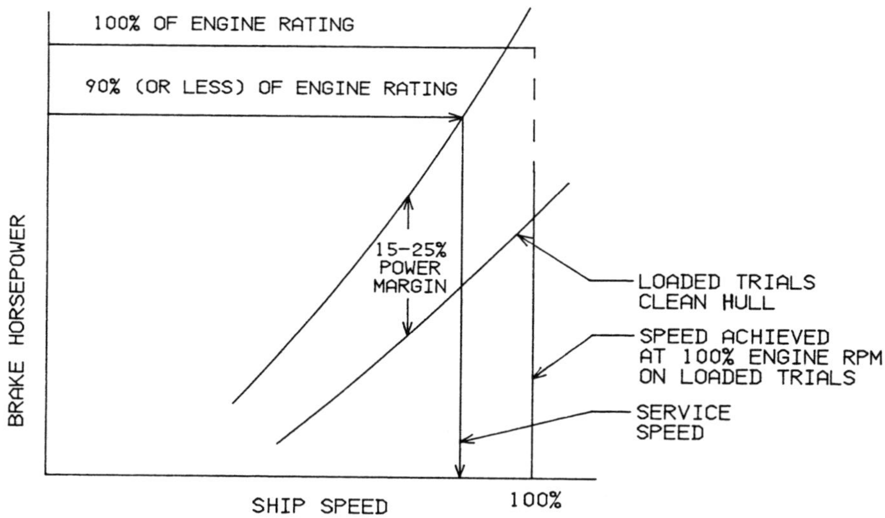

Figure 16-40. Estimate of service speed

be possible to achieve MCR without overspeeding the engine, perhaps beyond the limit imposed by the engine designers. However, the result will be that higher average power output can be utilized in service as the hull and propeller roughen, without excessive stress (as reflected in high MEP and cylinder exhaust temperatures), enabling higher ship speeds to be achieved.

Constant speed versus constant power. Examining Figure 16-37 it can be seen that if a constant service speed is to be maintained from one drydocking to the next, the power required will advance along Line 3, increasing as hull and propeller roughen. Even without separate allowance for weather and sea conditions, the engine would have to be matched to a Line 2 corresponding to conditions on the last loaded voyage before drydock; therefore, a large service margin would be required. Bearing the comments of the preceding section in mind, a service margin of 50 percent or more might be needed with conventional antifouling protection. Nevertheless, this would be the course adopted if constant speed were required, for example, in the case of a container ship operated on a rigid schedule.

On the other hand, if the power level were maintained at a constant value (Line 4 in the figure) the ship speed would decline. Cargo-carrying capability would be assessed on the basis of an average service speed, but service margins are typically lower than for constant speed applications. A representative range for motorships with conventional antifouling protection may be taken as 10 to 20 percent.

Where the trade permits, constant power operation is more economical; a high service margin means a higher installed power, hence higher acquisition cost and greater machinery weight and volume. In addition the fuel use will be greater in constant speed operation: a 50 percent rise in power to maintain constant speed might correspond to a 10 to 15 percent drop in speed at constant power.

Case study: two engines geared to a single fixed pitch propeller. Figure 16-41 represents the combined performance envelope for two turbocharged engines with a single propeller matched to absorb 85 percent of the total

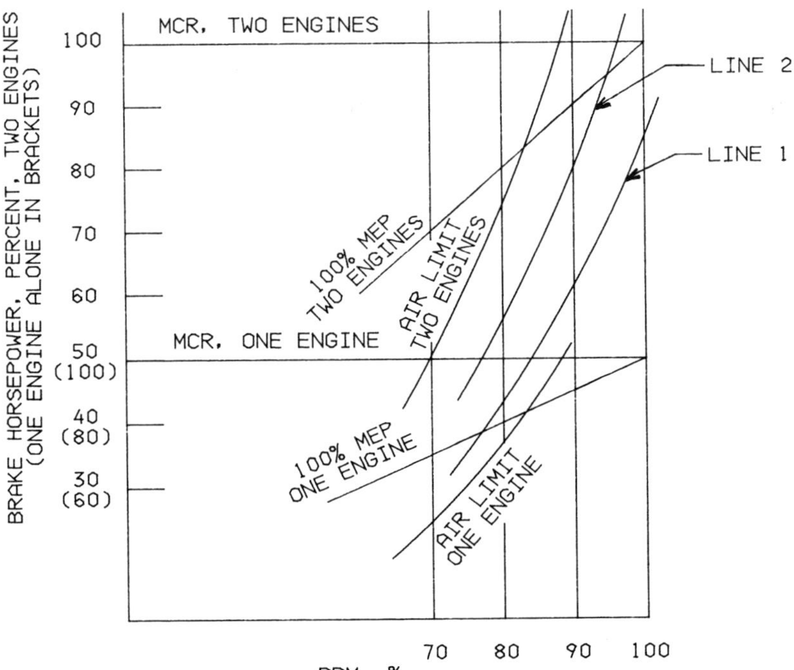

Figure 16-41. Two engines geared to a single fixed pitch propeller

BHP at 100 percent RPM in the clean hull, loaded condition, in good weather, as indicated by the cube law, Line 1. Line 2 represents loaded hull performance at a later date. The maximum power output of one engine (50 percent of the total) intersects Line 1 at $(.50/.85)^{1/3}$, or 84 percent RPM. The MEP would then be as follows:

$$\text{MEP} = 1.0\ \text{BHP}/.84\ \text{RPM} = 1.19, \text{ or } 119\%$$

The excessive MEP indicates a high overload and will be accompanied by excessive exhaust temperatures, forcing a reduction in output until Line 1 intersects the 100 percent MEP line for one engine. In this case the intersection is at about 76 percent RPM where the engine performance envelope is bounded by air supply rather than MEP. Here, it appears that sustained single engine use is precluded by the air limit, even with the hull clean and smooth, at least at loaded draft.

If a ship is to be fitted with two engines geared to a single propeller, with the intention of single engine operation at reduced ship speed, there are the following possibilities:

Fit a controllable pitch propeller, thereby enabling the engine to operate at or near its rated RPM, independent of ship speed.

Use a lower match point; i.e, fit larger engines. In the example above, however, even a match point at 70 percent of combined MCR would not permit single engine operation at loaded draft with the hull clean.

As an alternative to the lower match point, a “father-son” configuration is possible, with the larger engine meeting requirements for single engine operation.

PROPULSION ENGINE SUPPORT SYSTEMS

Introduction

The arrangement of support systems will vary from ship to ship. These variations depend on many factors, among them the type of main and auxiliary engines, the trade, the arrangement of the ship, and the shipowner’s preferences. The systems illustrated and discussed here are meant as typical examples only and may therefore fail to comply with systems recommended by a manufacturer, for example, or in place in a given plant.

Starting Air/Compressed Air System

A typical compressed air system is illustrated in Figure 16-42. The three segments of the system provide air for engine starting, for instrumentation and control, and for miscellaneous ship’s services.

Some engines, mostly smaller auxiliary engines, are started by cranking motors, which may be battery-, hydraulically, pneumatically, or

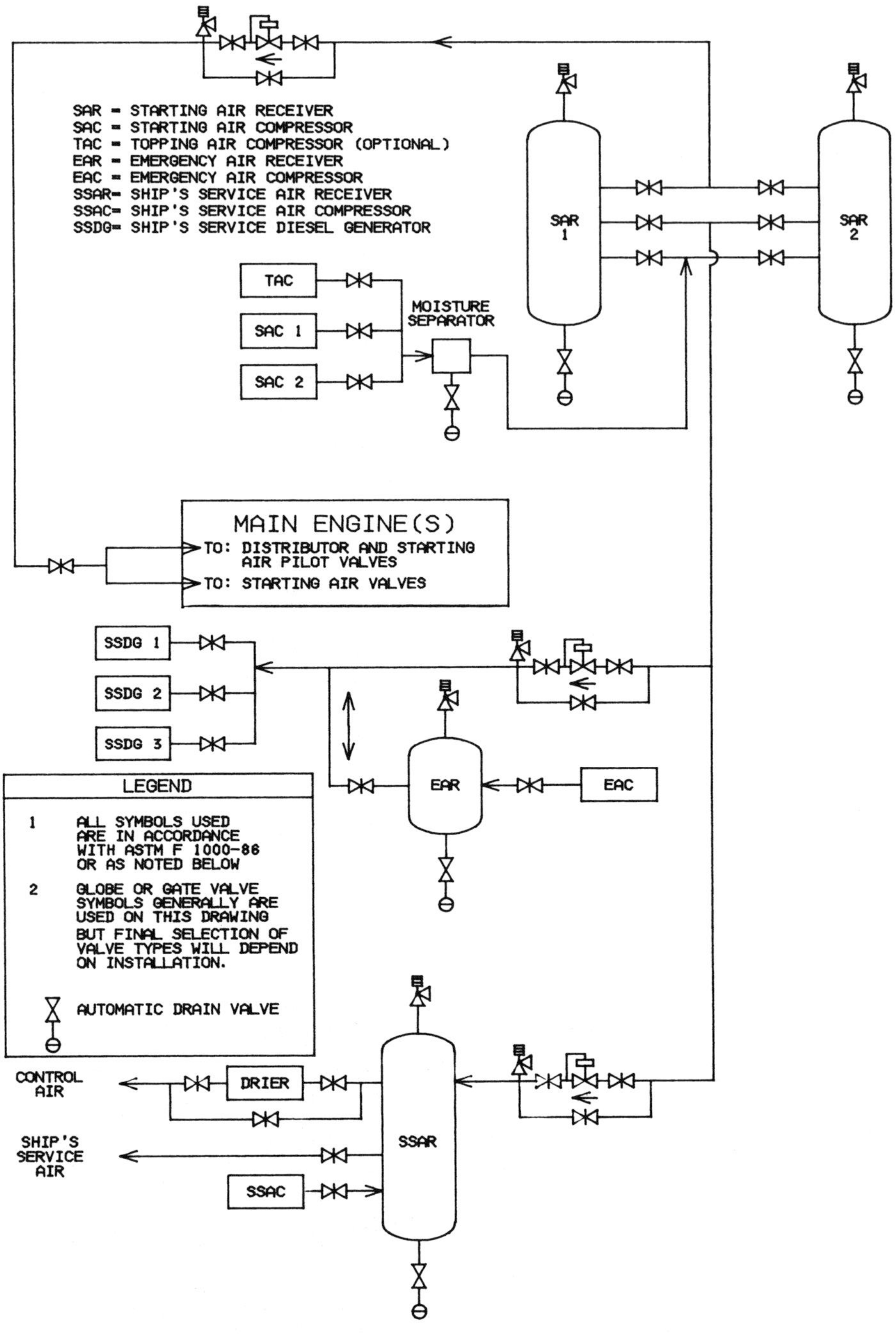

Figure 16-42. Typical compressed air system

mechanically driven. Most larger auxiliary engines, however, as well as most propulsion engines, are started by the timed introduction of compressed air directly into those cylinders that were stopped in positions corresponding to the beginning of their power strokes in the selected direction of rotation. The compressed air drives those pistons down, in firing order sequence, thereby compressing air trapped in other cylinders. As one or more revolutions are completed, fuel is introduced in the normal manner into those cylinders whose pistons are completing a compression stroke, which then fire. The starting air is cut off and the engine accelerates to its idle speed, under control of the governor. Typically each cylinder of an in-line engine is fitted with a starting air valve, but commonly only the cylinders of one bank of a V engine are so fitted. The valves are usually opened by pilot air supplied via a camshaft-driven starting air distributor. In direct-reversing engines the distributor timing is shifted for reverse rotation (together with the timing of the inlet and exhaust valves and the fuel injection pumps), directing pilot air to those cylinders whose pistons have stopped just short of TDC on the upstrokes, so that the engine is rotated in the opposite direction for starting.

Because the maneuverability of a ship is tied to the availability of starting air, the minimum number and size of starting air receivers must comply with regulatory body requirements. Typically, sufficient air must be stored to enable at least six consecutive starts of a nonreversing engine, or twelve of a direct-reversing engine, without recharging, in at least two receivers. Although the pressure may be reduced for admission to the cylinders, the pressure at which the air is stored will be twenty-five to thirty bars or more, with this higher pressure allowing smaller receivers.

It is usually necessary to provide an independent means of starting the ship's service diesel generators. In the system shown, a small (and therefore rapidly recharged) receiver floats on the starting air main. In the absence of another source of compressed air, this emergency receiver can in turn be charged by an emergency air compressor, which may be driven by, for example, a hand-started diesel engine or a motor supplied power from the emergency switchboard.

Air for instrumentation and control should be dry and free from oil. It may be supplied from an independent, oil-free control air compressor with separate receiver; or, as shown, it may be bled from the starting air system and passed through a drier. In the latter case, a topping air compressor is advisable, matched to control air requirements, thereby relieving the large starting air compressors of this duty.

In many geared diesel plants, the engines drive the pinions through air-actuated clutches, which are usually supplied with operating air from the starting air system through a reducing valve, and control and instrumentation air from the control air system.

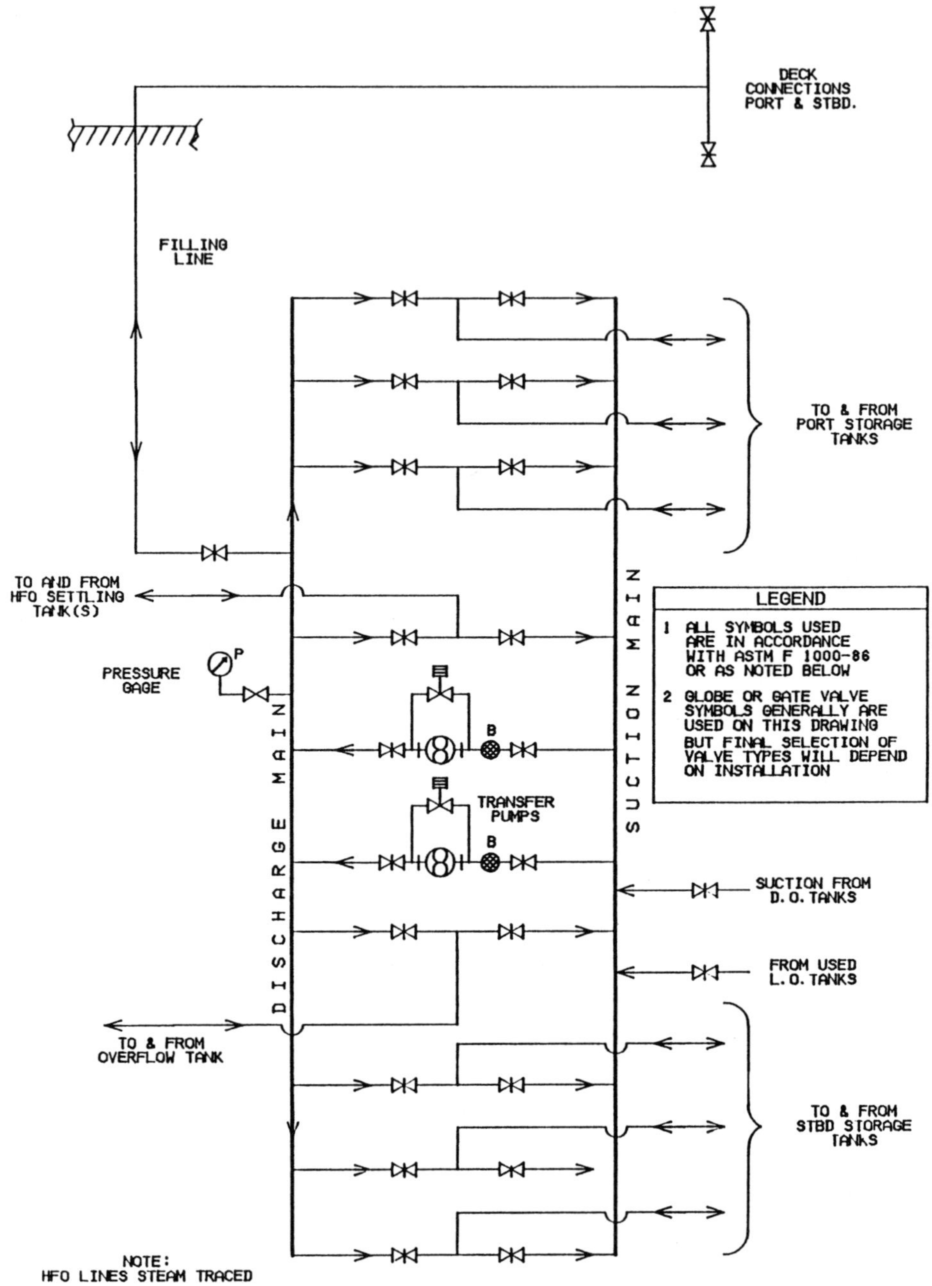

Figure 16-43. Typical HFO filling and transfer system

Ship's service air supply may be taken from the starting air receivers as long as demand is limited. Where demand is high, a separate, low pressure compressor should be used in order to limit running hours on the large starting air compressors. The system of Figure 16-42 provides for either alternative.

Compressed air systems on modern ships tend to be completely automated even in plants with attended machinery. Automatic drain traps are fitted at receivers and moisture separators, and compressors may be started and stopped automatically by pressure switches on the receivers, or they may start automatically, after which they run continuously until manually secured, being unloaded and loaded automatically as receiver pressure rises or falls. The pressure switches are set to start compressors successively, rather than simultaneously, as pressure falls in response to demand.

Fuel Systems

A typical fuel system can be subdivided into separate filling and transfer systems for heavy fuel oil (Figure 16-43) and diesel oil, separate treatment systems (Figure 16-44), and a fuel oil forwarding system (Figure 16-45).

Filling and transfer systems. The heavy fuel oil (HFO) filling and transfer system enables all HFO bunker tanks to be filled under pressure from pumps ashore or aboard a bunker barge. Good design practice calls for all the valves to be concentrated in one location to facilitate one-man operation. By using the valve at the foot of the filling line from deck to hold the pressure in the filling main below the static head of the overflow/vent pipes of the tanks, the possibility of an overflow to deck is reduced: overflow will be to the designated overflow tank instead, and this should be the last tank to be filled.

The transfer pumps are normally used to transfer fuel from bunker tanks to the settling tanks, but can also serve between tanks or back up the filling line if it becomes necessary to discharge the contents of a tank ashore or to a barge. The suction from the distillate fuel oil (DO) transfer system enables the main engine to be run on DO for extended periods when necessary. Usually the suction main and the branches to the tanks will be steam traced and insulated.

The transfer pumps are generally positive displacement rotary pumps, with coarse suction strainers for their own protection. The capacity of the transfer pumps is dictated by operational considerations: it may be reasonable to size each pump to fill the settling tank within an eight-hour workday. If the machinery arrangement permits, the duplicate transfer pump may be deleted, with standby provided by cross connections to one of the HFO booster pumps. The (DO) filling and transfer system is a

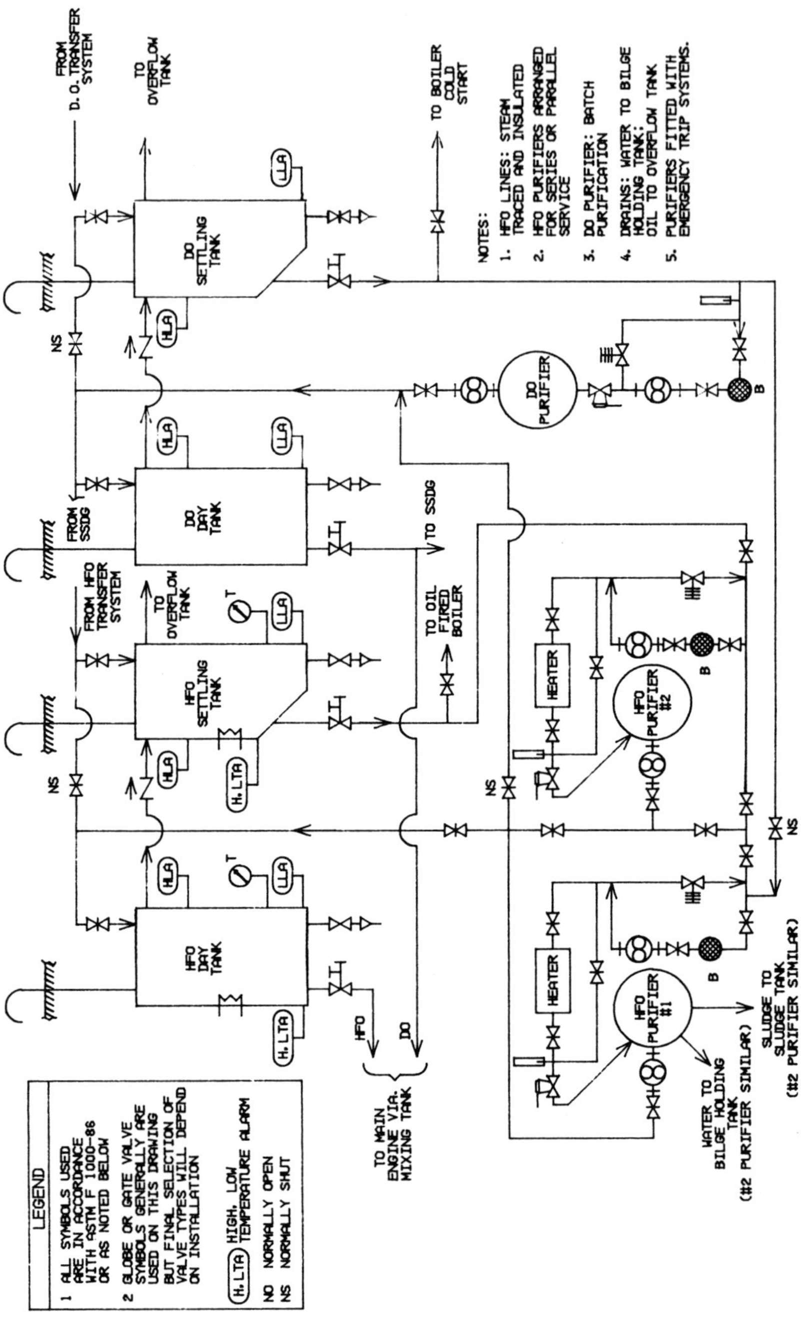

Figure 16-44. Typical fuel treatment system

simplified version of the HFO system, with fewer tanks and no need for steam tracing or insulation.

Fuel treatment systems. Fuel treatment systems include the settling tanks and purifiers, which enable most of the water and solids in the fuels to be removed. While clean distillate fuels are sometimes considered suitable for combustion in diesel engines without any treatment other than settling and filtration, given current refining practices it is advisable to centrifuge even the distillate fuel. In normal operation, fuel is transferred directly into the settling tanks from the bunker tanks, but passes to the day tanks only via the purifiers.

Figure 16-44 shows a single HFO settler but two are preferable, each of 24-hour capacity, so that fuel can settle undisturbed for an extended period. To avoid drawing settled water and sediment into the purifier, the settling tank should be fitted with a sloping bottom, with the suction connection at the upper end, rising about 50 mm into the tank. HFO settling tank temperature will normally be 40°C to 50°C but should be kept well below the flash point.

Most plants are fitted with centrifugal purifiers, with at least two units intended for full-time HFO operation either in series or in parallel, as described earlier. The rated capacity of each of the HFO purifiers should, at the very least, meet the main engine consumption at MCR with a 10 percent margin to allow for cleaning and other maintenance. The benefit of this apparent oversizing is more effective purification. (It should be noted that rated throughput of a given purifier when handling HFO may be only a fifth or less of its rating when handling DO.) Where existing piping precludes the flexibility of series or parallel operation of the HFO purifiers, a rearrangement of the piping should be undertaken. If existing purifiers are of low capacity or are otherwise inadequate, installation of at least one new purifier should be considered.

Modern purifiers tend to be self-cleaning, i.e., sludge-ejecting, and fully automated, with each HFO unit equipped with its own heater. Frequently, one of the HFO purifiers is arranged to stand by for a single DO purifier, as shown.

Generally, purifier feed pumps are attached, positive displacement units, with throughput controlled by dumping back to the suction, in preference to throttling a suction valve. An alternative is to fit independent, motor-driven, variable displacement feed pumps. In most installations, oil leaves the purifiers under sufficient head to reach the day tanks, but where this is not the case discharge pumps must be fitted. HFO must be heated close to the boiling point of water to facilitate purification both by reducing the viscosity and by enhancing the difference in specific gravity between the fuel and the water; this hot oil input to the day tank can alone

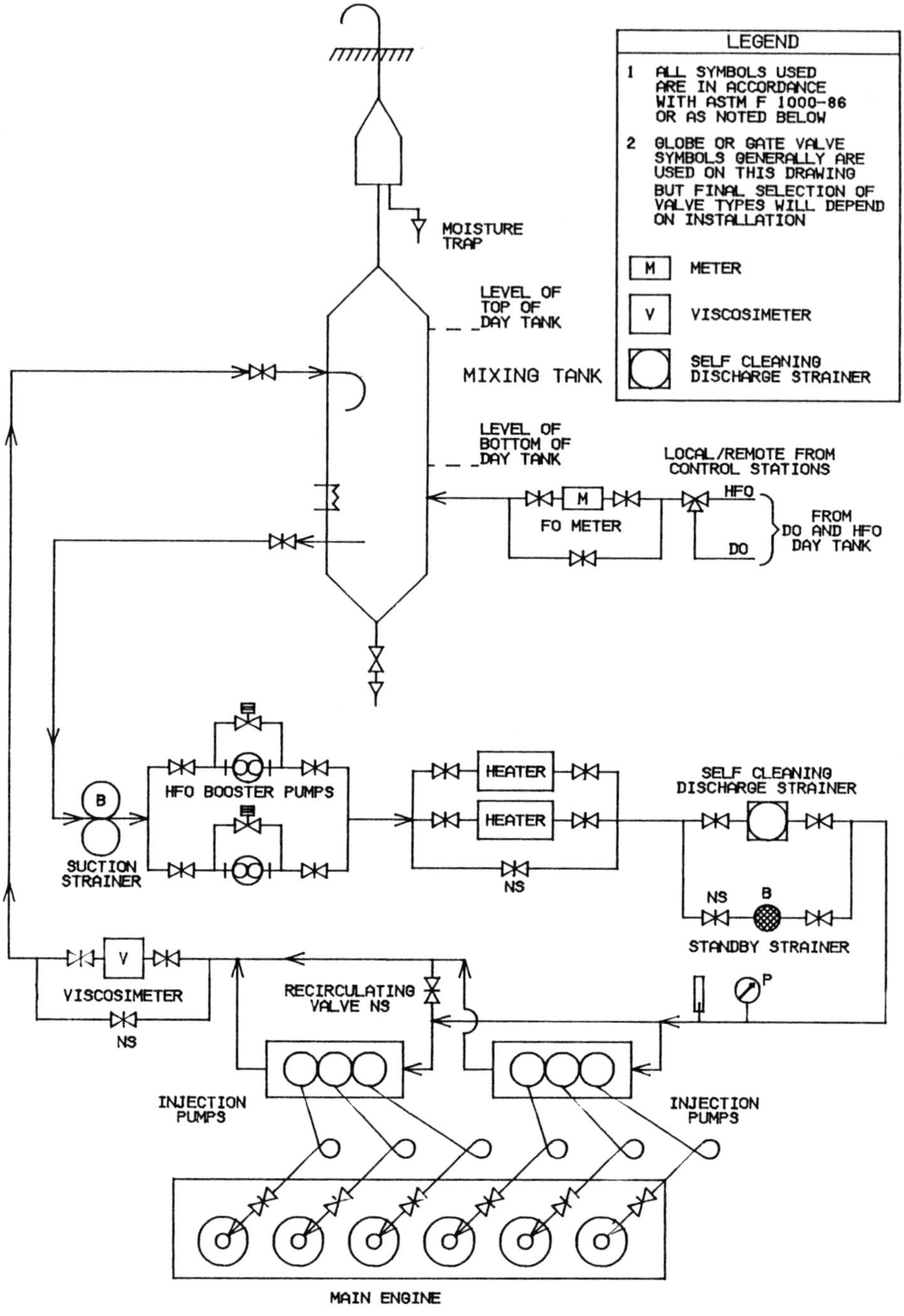

Figure 16-45. Typical fuel forwarding system

force the day tank temperature above the flash point. Consequently, some installations are fitted with a fuel cooler in the HFO purifier discharge line.

Day tanks and settlers are likely to be fitted with level alarms, as well as local and remote level indicators. Settling tanks may be arranged so that a high level will stop the transfer pumps automatically. HFO tanks are fitted with steam heating coils, temperature indicators, and, often, alarms for low as well as high temperature. (DO will not normally require heating.) Remote operated valves at the tank outlets, that are capable of operation from outside the space containing the tank in the event of a fire, are a requirement of most regulatory bodies. Where the configuration permits, the day tank in each system may be arranged to overflow into the settlers, with the settlers in turn overflowing into the overflow tank in the double bottom.

Fuel forwarding system. The fuel forwarding system of Figure 16-45 serves to supply HFO to engine injection pumps at the correct viscosity. Because the volume of fuel consumed will be small in relation to the volume available in the piping, the residence time between the heaters and the injectors would be sufficient, especially at low loads, for the fuel to cool. To prevent this cooling, a much larger flow rate is maintained, two or three times engine consumption at MCR, with the unconsumed excess leaving the spill valves of the injection pumps and recirculating back to the booster pump suction via the mixing tank. The amount drawn into the mixing tank from the day tank therefore corresponds to the amount consumed, while the oil temperature entering the heaters will reflect the amount of hot oil recirculated.

The mixing tank is sometimes called the buffer tank since it enables the transition from hot HFO to cold DO (or vice versa) to occur gradually; it is also called a deaerating or degassing tank, since lighter fractions in the HFO can vent. In the typical configuration shown, the mixing tank is supplied by gravity from the day tank; its bottom, therefore, must be lower than the bottom of the day tank, while the top must be higher than the top of the day tank to enable degassing to occur. Nevertheless, to facilitate a reasonably rapid changeover from HFO to DO and vice versa, the volume of the mixing tank must be limited; consequently the mixing tank is most often configured as a pipe column.

Because of the high temperatures to which the heaviest fuels must be heated and the increased amount of gas formation which would occur at atmospheric pressures, some engine manufacturers now recommend that the hot circuit of the forwarding system be designed as a closed and pressurized loop. This requires an additional set of low pressure booster pumps to feed the mixing tank from the day tank, but the mixing tank can then be more conveniently proportioned and located. Venting of gas that forms in the pressurized loop must be done through a relief valve.

Booster pumps are fitted in duplicate and are usually of the positive displacement, rotary type. While the booster pumps will normally handle partly preheated oil, they must also be capable of handling DO on occasion, as well as cold HFO being recirculated while the system is brought up to operating conditions.

At least two fuel heaters are fitted, each with sufficient capacity to heat the heaviest fuels likely to be encountered, and each conservatively rated in regard to fouling margins. The steam supply to the heaters is controlled by the viscosimeter, which must be sited in close proximity to the injection pumps, either just ahead of them or in the fuel return line as shown.

When a ship's auxiliary engines are fitted for operation on HFO, the forwarding system may be duplicated on an appropriate scale, and may incorporate an in-line blending unit, comprising metering pumps and mixing devices. To permit operation of the auxiliary engines on HFO when steam is unavailable, an electric fuel heater will be provided in parallel with the steam heaters.

A final fuel filter is fitted after the heaters, where filtration is facilitated by the reduced viscosity. This filter serves principally as a backup to the purifiers. Modern installations are most likely to have a fine mesh, self-cleaning unit as shown, or a disposable-element type of even finer mesh. On older ships, fitted with duplex basket strainers, an upgrade may be advisable.

All of the HFO piping in the forwarding system, sometimes including the filters and pumps, is likely to be steam traced and insulated. The extent of automation of the system will vary but will typically include automatic changeover for the pumps, low pressure alarms at pump discharge points, temperature alarms after the heaters, and a differential pressure alarm at the final filter.

Lubrication Systems

The complexity of motorship lubricating oil (LO) systems is the result of the number of different grades of oil required. A geared, medium speed diesel plant will usually require different grades of oil for the main and auxiliary engines, and other grades for gearing and miscellaneous uses. A LO system for a low speed diesel plant will be equally complex, involving at least two grades of oil for the main engine (one for the circulating system and the other for the cylinders) and a third for the auxiliary engines. It is important to note that most trunk piston engines do not use a separate grade of oil for cylinder lubrication.

Main engine LO circulating system. A typical circulating oil system appears in Figure 16-46. Oil draining from bearings and cooling passages to the bottom of the crankcase passes into a drain tank (or remote sump) built into the double bottom below the engine, from which it is drawn by

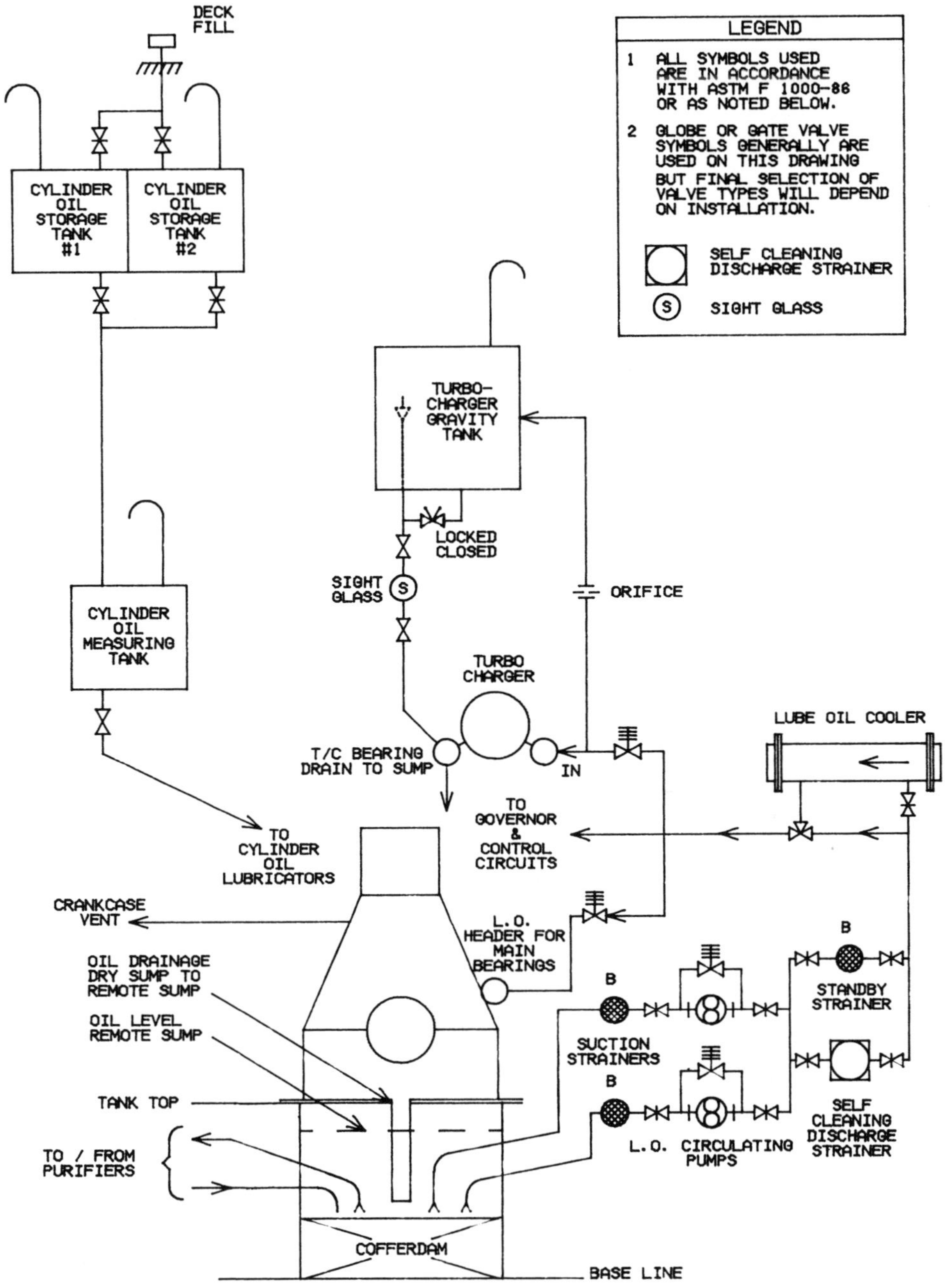

Figure 16-46. Typical main engine LO system

the LO circulating pump for redistribution via a cooler and a filter. Although the drain tank is shown as an integral part of the double bottom structure, in some cases independent tanks have been installed. In smaller engines the additional cost and complexity of a drain tank are often eliminated in favor of a wet sump formed by the crankcase bottom. The suction bell mouths clear the bottom by about 100 mm to avoid ingesting water or sediment, and may incorporate a foot valve to keep the suction line full when the pump is stopped. The drain tank or wet sump must be so designed that suction will be maintained at the most extreme conditions of list, trim, and sloshing.

The LO circulating pumps are most often positive displacement rotary pumps, and in larger plants are fitted in duplicate. Both pumps are motor driven in installations with low speed diesels, but higher speed engines are often fitted with an engine-driven pump, relying on the motor-driven pump for standby service. Each pump will have a coarse suction strainer for its own protection. Providing each pump with its own suction line is an advisable precaution.

A full flow filter is provided in the pump discharge line before or after the cooler. It may be of the duplex, basket type, but better filtration will be provided by a disposable-element or self-cleaning simplex unit, with a standby filter in a bypass. Filtered oil is distributed to engine bearings, for governing and control service, to valve gear, and, on trunk piston engines as well as some crosshead engines, to pistons for cooling. Some of these services may require higher pressure oil, which can be obtained by fitting booster pumps in the line. In other cases some equipment is served by independent circulating systems that are scaled-down versions of the main system.

Usually oil is also supplied to turbocharger bearings from the circulating oil pumps, but some turbochargers are provided with their own self-contained LO system. In either case the turbochargers must be provided with an emergency supply of LO, as they will continue to spin at high speed following a LO failure. The system of Figure 16-46 includes a gravity tank for this purpose.

In crosshead engines, the crankcase, and therefore the circulating oil, is protected by the piston rod packing from contamination by combustion products blowing by the piston rings, whereas this kind of contamination in trunk piston engines is usually inevitable. Consequently, while a straight mineral oil with corrosion and oxidation inhibitors is usually recommended for the circulating oil of most crosshead engines regardless of the fuel in use, manufacturers of trunk piston engines generally recommend a detergent oil with alkaline additives (measured as total base number or TBN) matched to the likely sulfur content of the fuel to be burned. While the circulating LO in a crosshead engine rarely requires replacement in the normal course of events, this is not usually the case with trunk piston engines, where, in most cases, the LO must be renewed periodically. The life of trunk piston engine LO, already extended by the regular addition of fresh makeup oil to compensate for oil burned in cylinder lubrication, can be further extended by taking such extra measures as the fitting of additional extra-fine filtration loops and, if necessary, the occasional addition of chemical additive packages to the oil.

There is the possibility of a crankcase overpressure condition or explosion in any diesel engine. A common pattern begins when an overheating bearing vaporizes oil, creating an explosive mixture which can then be ignited by further overheating of the bearing or, in trunk piston engines, by blowby. To give warning of overheating bearings, alarm systems of large engines include temperature probes at major bearings within the crankcase, while photoelectric oil mist detectors continuously sample the air drawn from likely accumulation points within the crankcase. Should an ignition occur, an initial rise in crankcase pressure will be limited by air available within the crankcase, but if this is followed by an influx of air, a much larger secondary explosion can occur. To minimize the possibility of the crankcase being breached by a primary pressure rise and of air entering to cause a secondary explosion, large engines are fitted with spring-loaded pressure relief valves (usually mounted on the crankcase doors) which open to relieve an overpressure and then snap shut. In addition, the crankcase vent pipe is restricted in size to limit the ingress of air.

The extent of additional automation of the system will vary, but will typically include automatic changeover for the pumps, low pressure alarms, a low LO pressure trip, and a differential pressure alarm at the filter. LO temperature is usually controlled by a three-way, thermostatically controlled valve on the oil side of the cooler.

Main engine cylinder oil system. Crosshead engines are fitted with an independent cylinder oil system for lubrication of the piston rings. A typical system is included in Figure 16-46. The cylinder oil is stored in one or, preferably, two tanks and is transferred daily to a small capacity measuring tank, from which it passes by gravity to the cylinder lubricators on the engine. The lubricators are precisely calibrated injectors, mechanically driven by the engine and timed to inject a metered quantity of the oil into the cylinder as the piston ring pack rises past the injection points. The oil is ultimately consumed. In crosshead engines in good condition the cylinder oil consumption may range from below 0.5 g/hp-hr to below 1.0 g/hp-hr. Because the quantities of oil injected per stroke are small, the measuring tank provides for consumption to be determined accurately as a drop in level over an elapsed time period.

Cylinder oil is a high viscosity mineral oil, with a TBN matched to the anticipated sulfur content of the fuel. Two cylinder oil storage tanks provide flexibility in this regard by enabling cylinder oil of different TBN

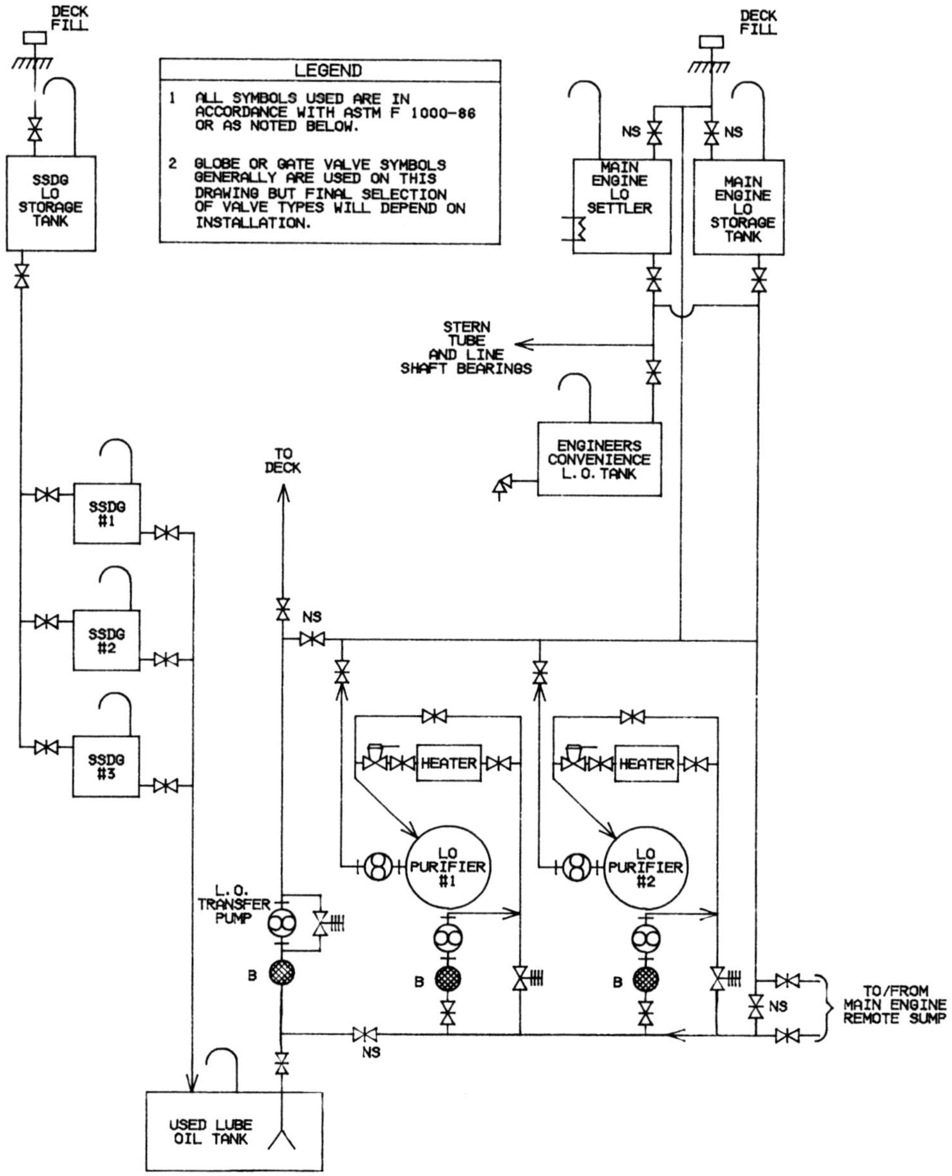

Figure 16-47. Typical LO storage, transfer, and purifcation system

to be carried. Cylinder oil storage tanks are filled from deck by gravity, a fact which may preclude filling the measuring tank from the storage tanks by gravity as well, necessitating a small hand- or motor-driven transfer pump.

In trunk piston engines in good condition, cylinder lubrication consumes up to 1 g/hp-hr or more of circulating oil, which usually reaches the ring pack and cylinder liner walls by a controlled leakage from the wrist pin bearing. In some of the larger medium speed engines, circulating oil is injected for cylinder lubrication in the same manner as described above for cylinder oil in low speed engines. In these engines the oil is usually taken from the circulating system, but separate oil tanks and piping can be arranged to bring only clean, unused detergent oil of high TBN to the injectors. The arrangement will still fall short of what can be achieved in a crosshead engine, since an unburned portion of the injected oil, carrying entrained contaminants, will drain to the crankcase.

LO storage, transfer, and purification system. While smaller high speed main engines may rely solely on filtration and occasional oil changes to maintain the quality of the circulating oil, most larger main engines are arranged for continuous bypass purification using centrifugal purifiers, as shown in Figure 16-47.

Two purifiers are shown, although one is a standby unit, and it is often possible, by means of blanked cross connections or shared components, to rely on a reserve FO purifier for standby. Normally, the purifier draws from a bell mouth in the suction well of the main engine drain tank and returns the oil to the forward end of the same tank, thereby avoiding short-circuiting. The purifier suction will generally be located aft of the circulating pump suction, and lower—about 50 mm above the tank bottom—to draw water before it reaches the circulating pump suctions.

The main LO storage tank, with a capacity equal to at least one charge for each engine it serves plus sufficient margin to meet miscellaneous needs, is filled from deck by gravity, while the settling tank is normally empty. Should an engine's circulating oil be massively contaminated, for example by water, it can be transferred to the settler using the transfer pump, and fresh oil brought down from the storage tank. If the oil cannot then be redeemed by a combination of settling and purification, it can be discharged ashore or to a barge for reclamation.

Modern purifiers tend to be self cleaning and fully automated. Generally purifier feed pumps are attached, positive displacement units, requiring the LO purifiers to be located on the floor plates to minimize the suction lift required from the main engine drain tank. An alternative arrangement would place a motor-driven purifier feed pump on the floor plates, to provide more flexibility in purifier location. In most installations, oil leaves the purifiers under sufficient head to reach the settling tank, but where

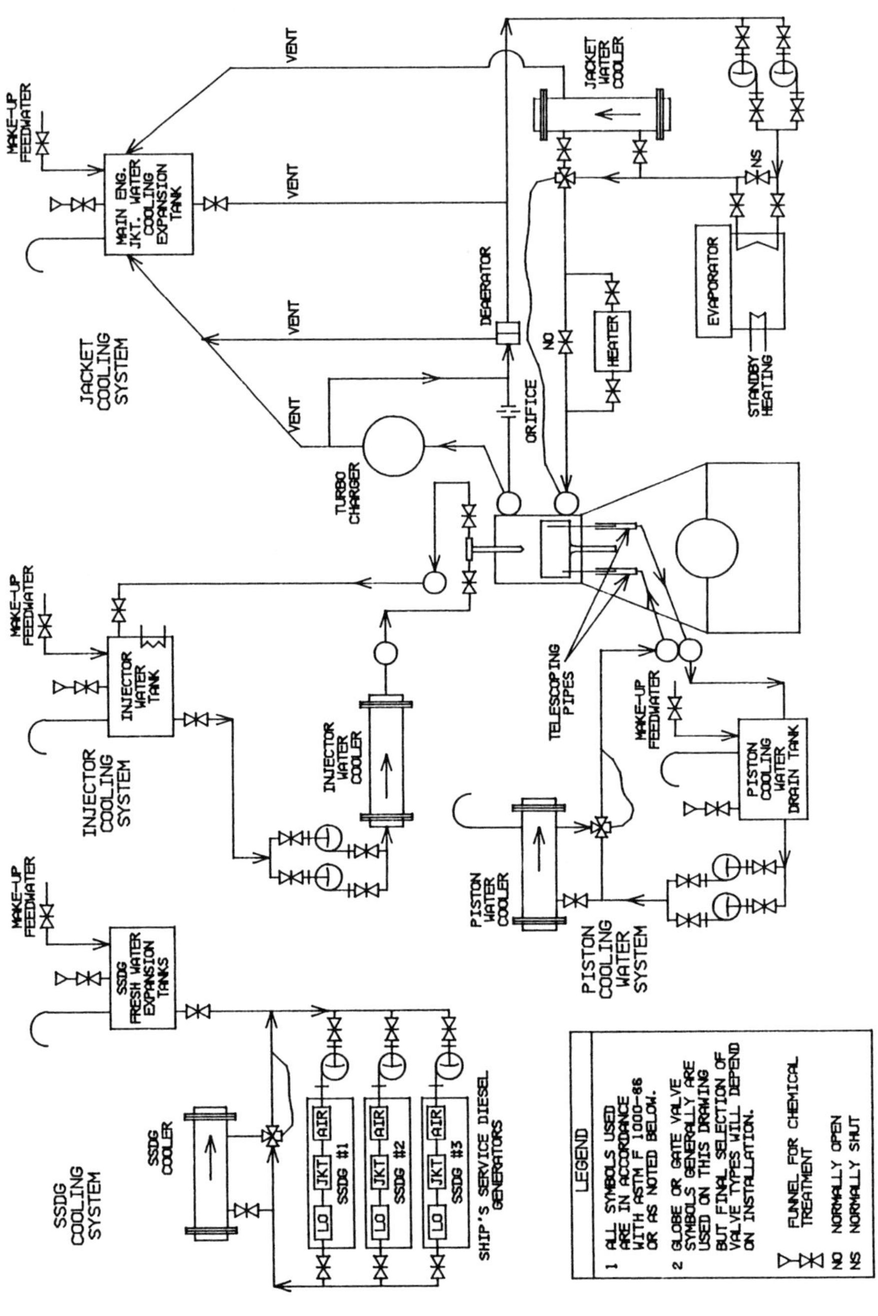

Figure 16-48. Typical freshwater cooling system

this is not the case, a discharge pump must be fitted. Each purifier is normally sized to circulate the main engine drain tank three to five times per day.

Purifier connections can be provided for batch purification of auxiliary engine LO but are unlikely to be used, since relatively small quantities are involved, since the alkaline-additive detergent oil will require renewal at intervals in any event, and since there is potential for contamination of the main engine oil. When auxiliary engines of substantial size are involved, however, a separate purifier may be justified. The fixed piping shown for filling and draining the auxiliary engine sumps may not be provided for smaller installations.

An analogous problem exists in medium speed diesel plants, where the reduction gear contains a straight mineral oil that is subject to water contamination from condensation within the gear case, but which would otherwise last indefinitely. Again because of the danger of contamination, in this case of the gear oil by the alkaline main engine oil, even when purifier connections are fitted they are unlikely to be used. Some operators fit water-absorbing filters in the gear lubrication system, while others rely on occasional batch purification using a portable purifier or filter.

Freshwater Cooling Systems

Figure 16-48 illustrates a typical freshwater cooling system for a plant with a crosshead engine with water-cooled pistons and three diesel generators. The system can be divided into separate subsystems for cooling main engine cylinder jackets, main engine pistons, main engine injectors, and the generator engines. (Charge air coolers of main engines are usually cooled by seawater; see below.)

Jacket water cooling system. The jacket cooling system shown differs from that used on trunk piston engines only in the fact that the LO cooler and the charge air cooler, which are seawater cooled here, may be included in the same circuit.

An elevated expansion tank maintains a static head on the suction side of the system and provides a convenient point for collecting vents, adding makeup feed, and adding chemicals to inhibit corrosion and formation of scale. Good venting is important: air carried with the coolant will enhance the potential for corrosion and can also accumulate at points to block coolant flow.

The turbocharger supply and return lines are shown, since even turbochargers with uncooled casings usually require cooling water for the turbine-end bearing. In either event, water flow must be forced by an orifice in the bypass line, as shown.

The jacket water circulating pumps are usually centrifugal pumps, and in larger plants are fitted in duplicate. Both pumps are motor driven in installations with low speed diesels, but medium and high speed engines are often fitted with an engine-driven pump, relying on the motor-driven pump for standby service.

Most seagoing ships recover heat from the main engine jacket water for fresh water generation. The fresh water generator is usually located ahead of the jacket cooler, and may be fitted with a supplemental steam or hot water heating coil for use when insufficient jacket water heat is available.

The jacket water heater is used when the engine is idle. Maintaining the engine in a warm condition assists in minimizing corrosion.

Automation is likely to include alarms for low pressure, high temperature, and low level in the expansion tank; automatic changeover of the pumps; and thermostatic control of the three-way valve at the cooler.

Piston cooling water system. While all trunk piston engines, as well as some crosshead engines, use oil to cool the pistons, a number of crosshead engines use a cooling water system separate from the jacket water system. The water reaches and leaves the pistons through telescoping tubes enclosed within compartments inside the crankcase in order to avoid contamination of the LO should a gland fail.

Because of the high temperature of water draining from the pistons and the resulting potential for flashing at the pump suction, some manufacturers recommend that deep well pumps immersed in the tank be used. Both pumps will be motor driven. Automation and other features will be similar to those described for the main engine jacket water system.

Injector cooling system. Some engines are fitted with fuel injector cooling systems and others are not. In some cases injectors are cooled only through conduction to the heads; in other cases they are circulated with water as part of the jacket cooling system; and sometimes they are circulated with diesel oil in a closed loop. When a separate injector water circuit is fitted, it is a scaled-down version of the jacket cooling system, often without a cooler, and with a steam coil in the expansion tank to maintain temperature at low engine output when use of HFO is intended.

Diesel generator cooling system. Auxiliary engines tend to have self-contained cooling circuits, with charge air cooler, cylinder jackets, and LO cooler circulated by a single cooling pump on each engine. The system of Figure 16-48 combines these circuits into a common system, with a central generator engine cooler and expansion tank, and with motor-driven pumps. This offers the advantage of circulating the idle generator engines with warm cooling water as protection against corrosion, and enables the standby engine to start and pick up load more rapidly. For reasons of reliability, however, many operators prefer that the auxiliary engines have

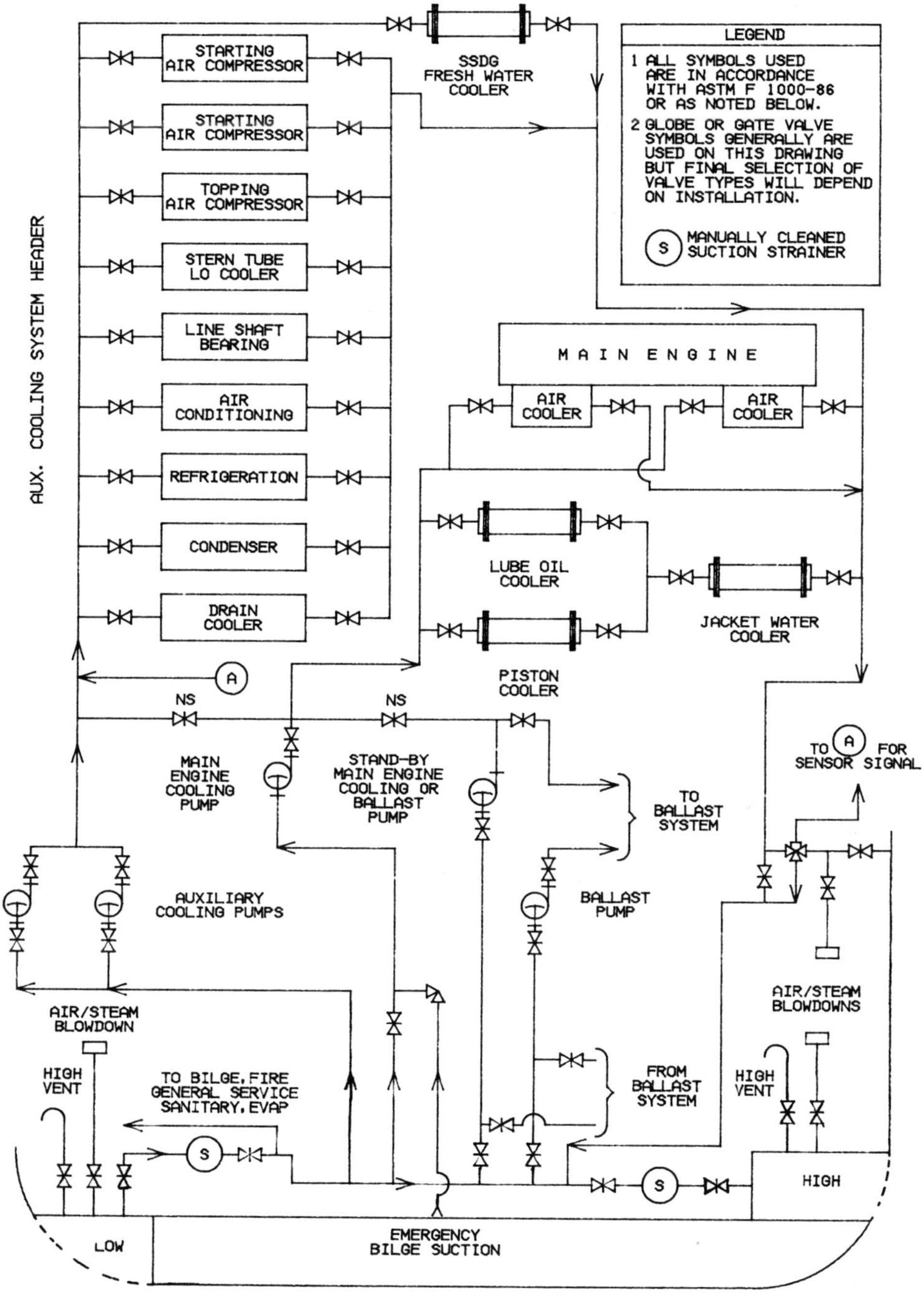

Figure 16-49. Typical seawater cooling system

separate cooling systems. Automation and other features will be similar to those described for the main engine jacket water system.

Central freshwater cooling systems. There is trend toward increased use of freshwater cooling for equipment previously cooled by seawater, with the logical extrapolation being a complete central freshwater cooling system in which fresh water is circulated from seawater-cooled central heat exchangers to main engine and auxiliary machinery coolers. The central system may be divided into separate high and low temperature subsystems. The benefit of a central freshwater cooling system is in reduced maintenance, since only the central coolers are exposed to seawater.

Seawater Systems

Included under this heading are not only the seawater cooling circuits but also such ship's service systems as ballast and fire main. Figure 16-49 is a typical example.

In the example, seawater suction is taken from a crossover main connecting a low sea chest (suction box) on one side of the ship to a high sea chest on the other side. The low sea suction is used at sea where it is more likely to remain immersed as the ship rolls and pitches, while the high sea suction is used in port to reduce ingestion of sand and bottom debris. Both sea chests are vented by standpipes rising well above the highest waterline, and both are fitted with hose connections for blowing them out with compressed air or steam. Where operation in freezing water is frequent, the sea chests may be fitted with piped connections to introduce hot water from the overboard discharge or from separate steam heaters. The sea chests are covered with grids mounted flush with the hull; a reasonable precaution is to secure these with stainless steel bolts removable by divers. Further protection is provided by coarse basket strainers at each end of the crossover main.

Seawater pumps in saltwater service tend to be high maintenance items and good design practice calls for pumps of similar capacity to be identical, simplifying spare parts requirements in service. In the illustrated example, three identical centrifugal pumps are provided, one for main engine cooling and two for ballast, with one ballast pump piped to stand by for engine cooling. This arrangement is suited to dry cargo ships and gas carriers, as most tankers with conventional cargo systems tend to locate the ballast pumps (which are most likely to be turbine driven) in the cargo pump room. In a tanker, however, the standby main engine cooling pump might serve to circulate the cargo and ballast pump condenser. The emergency bilge suction on the main engine cooling pump meets a regulatory body requirement.

In the illustrated example, two auxiliary cooling pumps are provided, drawing from the crossover main and discharging to meet all miscellaneous seawater cooling needs. The heated seawater leaving most of the auxiliaries is then combined with seawater leaving the main engine coolers; downstream is a three-way thermostatically controlled recirculating valve that can return some of the heated water to the crossover main and discharge the rest overboard. The recirculating system enables the seawater used for cooling to be maintained above 20°C, even in very cold ambient conditions. For the system to operate in port with the main engine secured, the thermostat should sense the seawater temperature in the auxiliary cooling system.

It is usually good practice to limit the number of hull penetrations; to the extent practical, additional connections to the crossover main can provide sea suctions for fire, bilge, general service, after ballast, evaporator, and sanitary pumps. Where all these pumps cannot be clustered in the proximity of the crossover, however, an additional sea chest may be necessary. (In any event, one fire pump must have an independent suction.) Similar logic is applied to overboard discharge valves. Each overboard discharge is fitted with a hose connection for blowing out.

Automation is likely to include, in addition to the recirculation valve, automatic starting of the standby pumps and low pressure alarms in both subsystems.

Steam Systems

Most diesel plants of any size are fitted with exhaust heat recovery systems to meet the heating requirements of the fuel and LO systems as well as hotel needs. Occasionally a thermal fluid is used but most often the working fluid is water. Where sufficient waste heat is available, a waste heat turbogenerator may be fitted.

The design pressure of the steam system may be dependent on its use for fuel heating: the heaviest fuels require heating to about 170°C, in turn requiring saturated steam at the fuel heaters at a saturation pressure of about eight bars.

Figure 16-50 illustrates a simple steam system frequently used on ships where steam requirements are limited, in which the oil-fired package boiler serves as the steam drum for a forced-circulation water tube waste heat boiler. This kind of system is easily automated: excess steam produced is dumped through a pressure-regulating valve to the seawater-circulated condenser, while a shortage of steam triggers a pressure switch to supplementally fire the oil-fired boiler. The fact that the oil-fired boiler is filled with saturated steam and water at all times results in a rapid response. Sometimes gas bypasses are fitted to the waste heat boiler to enable control of steam production.

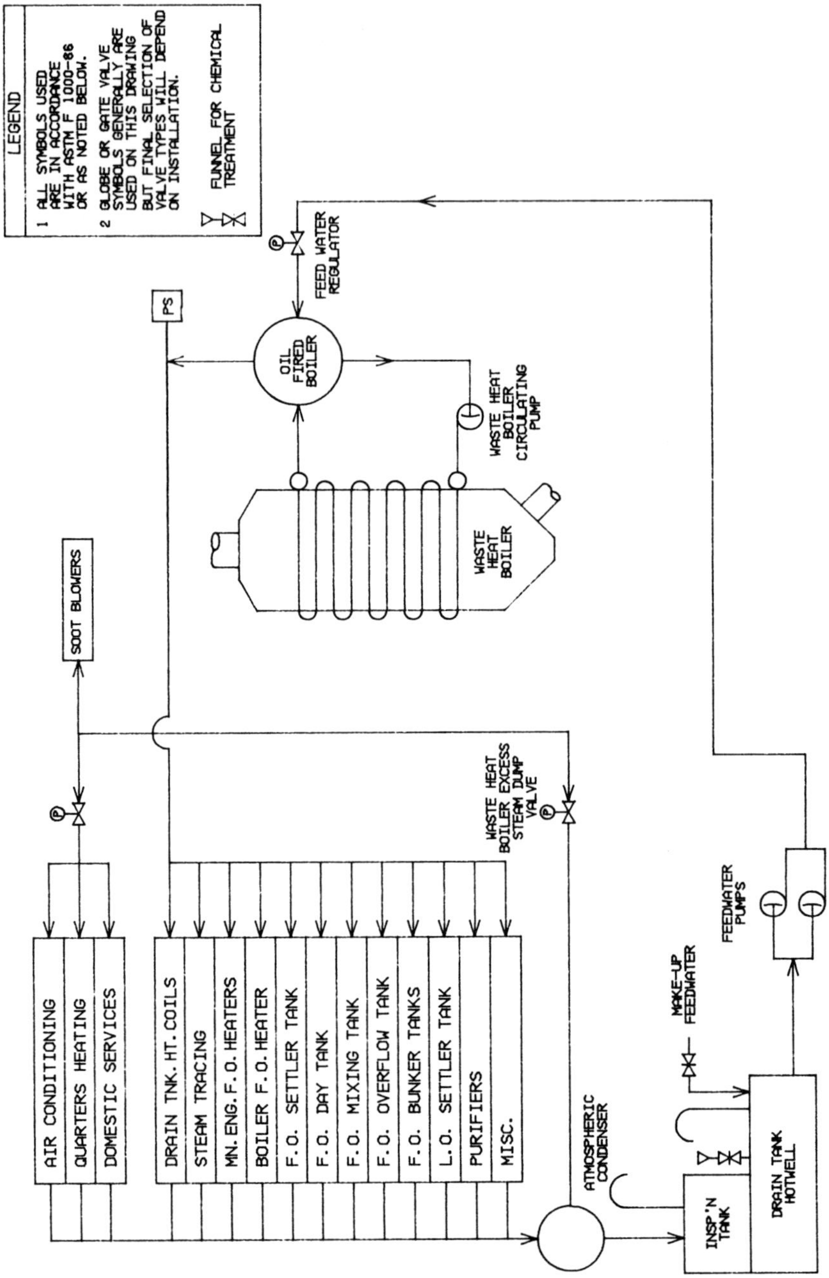

Figure 16-50. Typical steam system

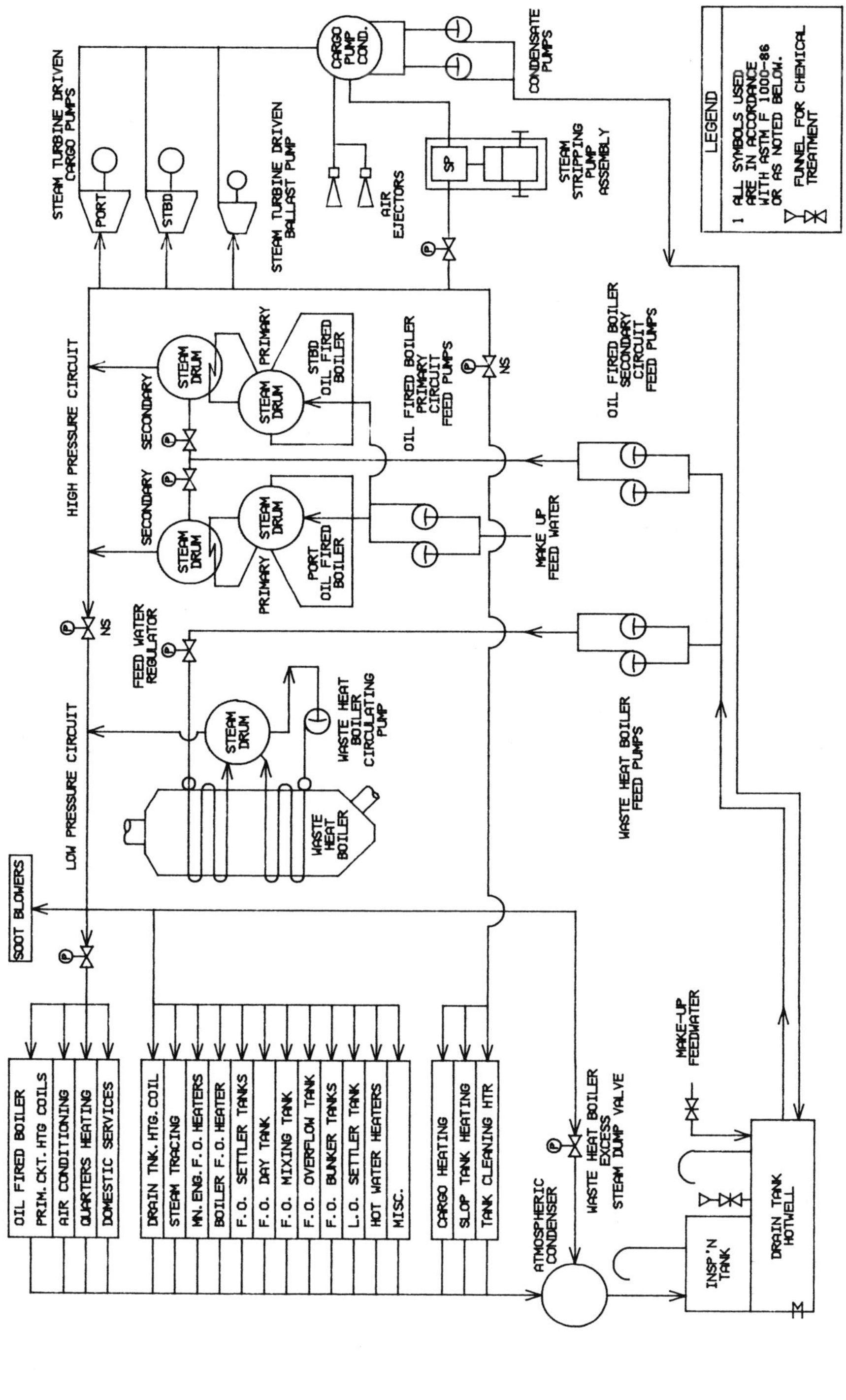

Figure 16-51. Typical tanker steam system

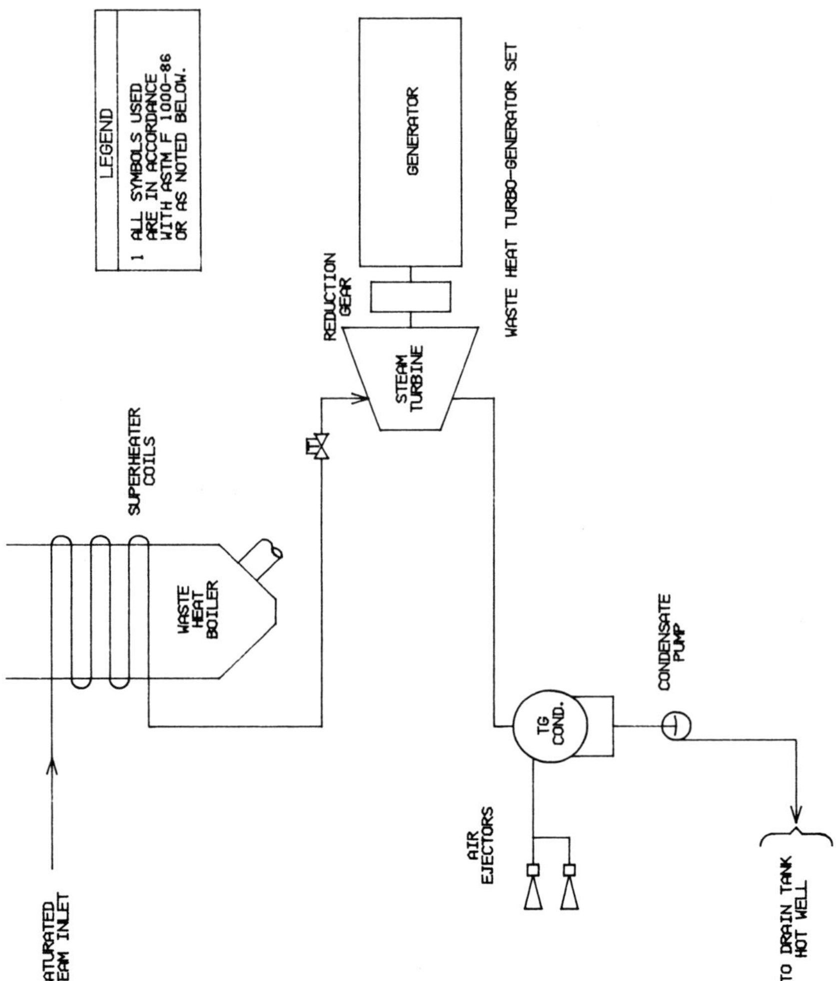

Figure 16-52. Superheater, turbogenerator, and condenser

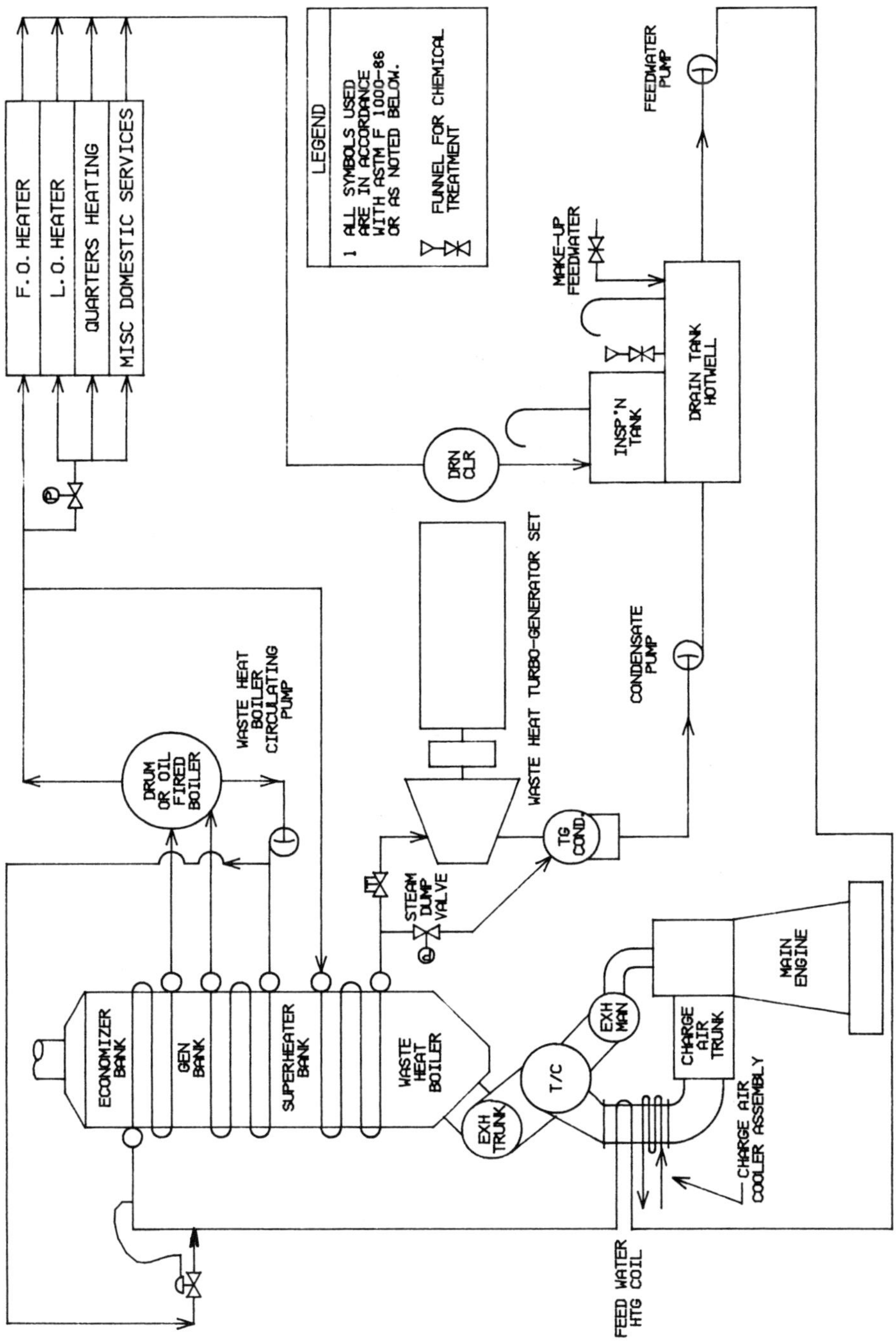

Figure 16-53. Steam plant with turbogenerator and feed heater

Some ships are fitted with waste heat boilers of the gas tube type in which steam production can be controlled by varying the water level. The oil-fired boiler would not necessarily be in the circuit, but it can be kept warm by a steam heating coil.

Where the steam demand is in the range of about 2,000 kg/hr or less (the usual case on dry cargo vessels), the oil-fired boiler is most often of the fire tube type.

On tankers, where steam may be required for cargo heating, cargo and ballast pumping, and tank cleaning, the steam plant becomes more complicated. Figure 16-51 is a typical example, in which two oil-fired boilers supply steam at a pressure sufficient for cargo and ballast pump turbines, and also serve to supplement the output of the waste heat boiler. Because the waste heat boiler is operating at lower pressure it is provided with its own steam drum, and the oil-fired boilers are fitted with heating coils. The oil-fired boilers in this example are of the double circuit type: an oil-fired, closed, primary steam circuit generates steam in an attached secondary drum, precluding contamination of the primary circuit via a leaking

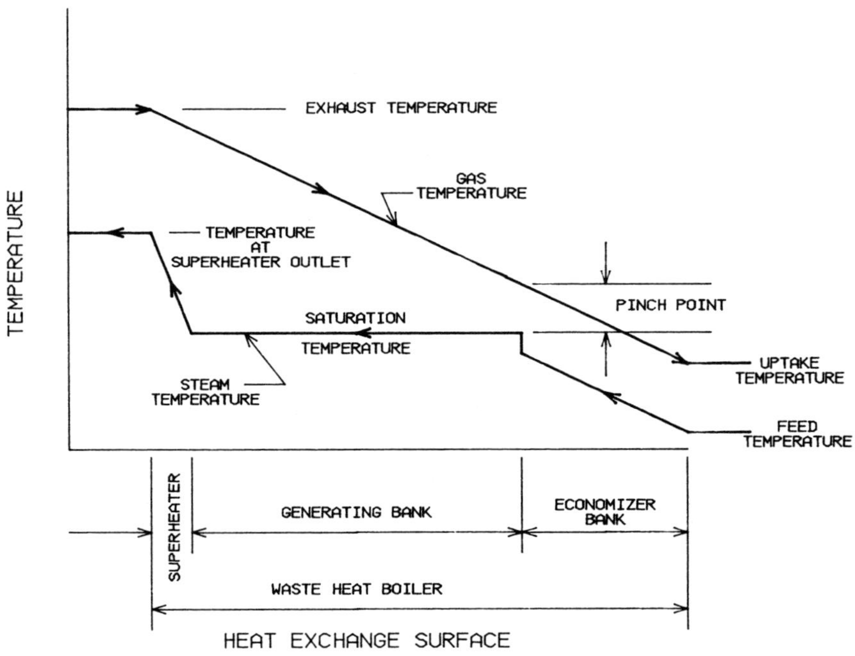

Figure 16-54. Pinch point diagram for single pressure boiler

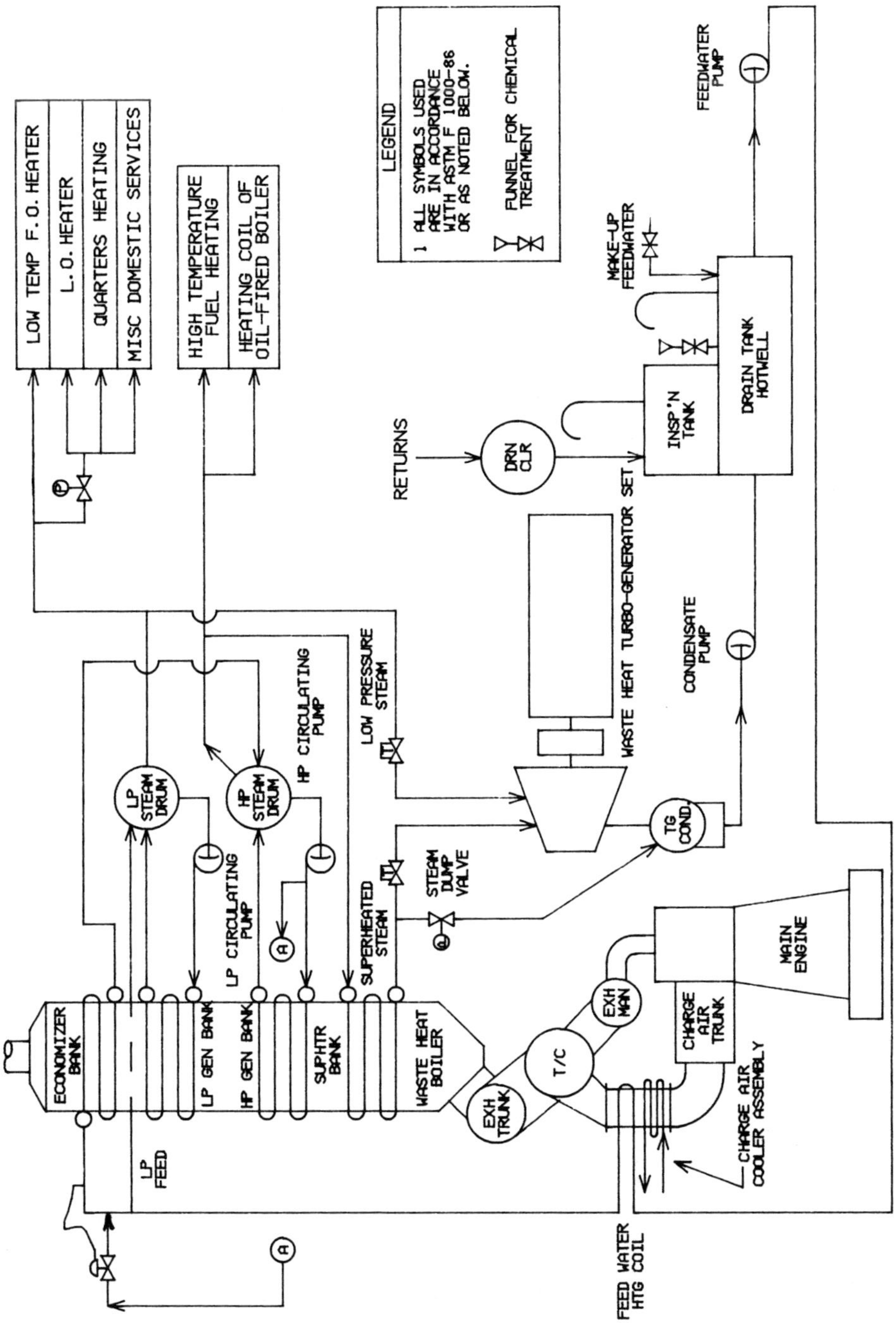

Figure 16-55. Steam system for maximum heat recovery

cargo-heating coil. The same results can be achieved using a contaminated evaporator (low pressure steam generator) to supply cargo-heating steam.

Where sufficient waste heat is available for a turbogenerator to be fitted, the system can be as simple as in Figure 16-52, intended as a fly sheet to Figures 16-50 and 16-51. A problem with this system arises from the fact that a high condenser vacuum and its corresponding low condensate temperature will depress feedwater temperature and increase the heat required from the waste heat boiler, unless a feed heater is fitted. In Figure 16-53, a section of the charge air cooler is arranged as a feed heater.

Because of the danger of sulfuric acid attack on cold economizer tubes, the systems of Figures 16-53 and 16-55 incorporate injection points upstream of the economizer for saturated water, thereby enabling the feed temperature to be maintained even at low engine output. This is sometimes called *economizer recirculation.*

The quantity of steam generated in a waste heat recovery cycle is governed by the available heat exchange surface in the waste heat boiler, and by the gas temperature, which must exceed the steam temperature at every point in the boiler, as illustrated in Figure 16-54. The point at which the temperatures come closest is the *pinch point,* for which a practical minimum may be about 15°C. The situation can be improved by lowering the design steam pressure, but an alternative means might then be

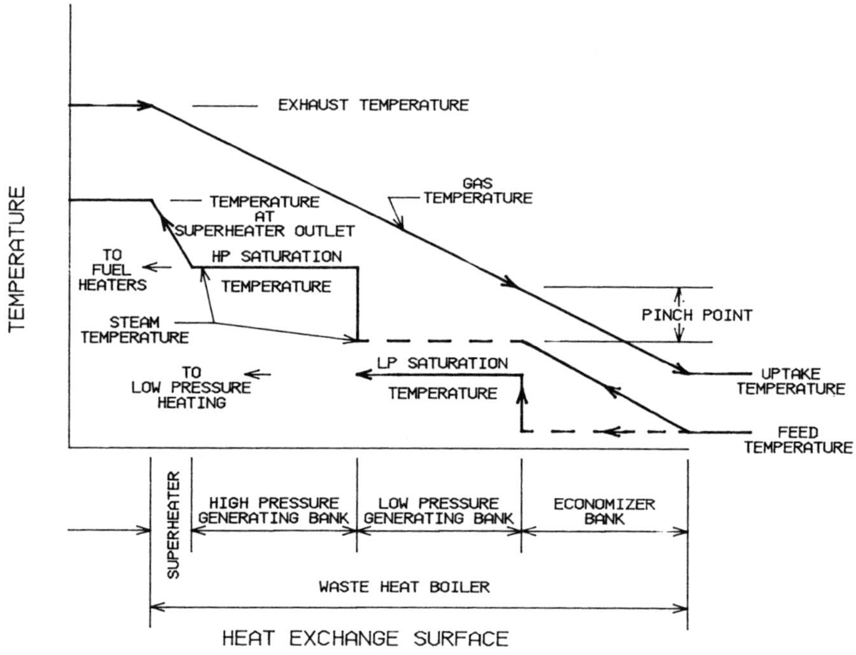

Figure 16-56. Pinch point diagram for dual pressure boiler

required for final fuel heating. Figure 16-55 illustrates a steam system for maximum heat recovery that incorporates a dual pressure waste heat boiler, enabling the steam temperatures to more closely match the gas temperature, as shown in Figure 16-56. There may be a trade-off involved when a more sophisticated system is adopted, as the increased complication may result in increased maintenance.

A properly designed waste heat boiler may include the following features:

- Sootblowers.
- A vestibule design, enabling the U-bends to be outside the gas path (and thus not subject to external corrosion in addition to the internal corrosion and erosion), while improving access for isolating and plugging leaking tubes.
- Provision for gas side washdown when the main engine is secured, usually encompassing access hatches above and below the tube banks, a drain fitting at the bottom, and a means for blanking the gas inlet to protect the turbochargers.
- The ability to pass engine exhaust with the water side dry without damage (dry-steamed) in an emergency.

Steam plants on modern motorships are likely to be fully automated for normal operations even if the machinery spaces are normally manned. All the normal procedures and precautions regarding the maintenance of water quality are necessary, including purity of makeup feed, regular sampling and testing, blowdown, and chemical treatment to preclude corrosion and scale formation.

OPERATING AND MAINTENANCE PROCEDURES

Introduction

The procedures described here are general and meant to provide background information only. They are not intended to be comprehensive or all-inclusive. Recommended procedures for specific equipment should be sought out and followed.

Normal Operations

Starting. Prior to starting an engine that has been shut down for an extended period or has been overhauled, tools, eyebolts and other lifting gear, and rags and debris must be cleared from the crankcase, from the air and exhaust manifolds, from the air intakes, and from the general vicinity of the engine.

Fluid levels and quality must be verified. If there is any doubt about the condition of the coolant, the LO, or the fuel, the system should be drained and refilled. On main engines, the LO is circulated through the purifier and brought close to its normal temperature.

Where motor-driven cooling pumps are fitted, one of these is started (in each circuit) and, on large engines, the cooling water is heated toward the operating temperature. High points and other air traps in the cooling system are vented.

A seawater cooling pump is started and coolers are vented on their seawater sides.

Where independent LO circulating pumps are fitted, one is started, and a flow to all bearings and service points assured. On engines with prelubrication pumps, these are operated until all bearings receive oil. Crosshead engine cylinder lubricators are operated manually until all cylinders receive oil.

Drains in air and exhaust manifolds are opened, as are indicator cocks on each cylinder. Where a compression release device is fitted, it should be used. If the engine is direct connected, permission to rotate the engine must be obtained from the bridge. After checking to see that all hands are clear, the engine is rotated through several revolutions with the turning gear or a barring device to ensure freedom from interference, to help establish oil films at bearings, and to ensure that cylinders and manifolds are free of water. The turning gear or barring device must then be disengaged.

If access permits, turbochargers should be rotated by hand to ensure freedom of rotation.

Fuel systems are primed and vented. Most engines that run on HFO are started on distillate fuel and shifted over to HFO only after starting; however, where the engine will be directly started on HFO, a booster pump is started with the recirculating line open. Steam is lined up to heaters, to steam-traced fuel lines, and, where appropriate, to the injector cooling circuit. The fuel is recirculated through the heaters (and injectors, on some engines) until hot fuel is present at the injectors, or as close to them as the system permits.

On air-started engines the starting air receivers are charged and drained.

An inspection is made to ensure that all gauge valves are open, that all sensors are in place and connected, and that standby pumps not in use are lined up and operable. Safety devices, including interlocks and alarms, are tested.

Before starting propulsion engines that are directly connected to the propeller, permission must be obtained from the bridge. Further, if the propeller is fixed pitch the engine will have to be stopped immediately to avoid putting way on the ship and straining docking lines and bollards.

On reversing engines, the camshaft and starting air distributor are set in the desired direction. The fuel rack is set to a starting position, usually about 25 percent of maximum. Starting air is lined up to the starting valve.

When the starting valve is opened, the engine will crank through one or two revolutions before firing. As the engine accelerates, the fuel rack is repositioned to the idle speed setting. If a decompression device is fitted and used, it is returned to its normal setting.

When possible, the engine should be allowed to warm up at light load. When a higher load must be applied immediately, the lowest initial load should be applied and increases should be gradual. The engine manufacturer's suggestions should be followed. Immediately after starting, all systems are checked for proper operation.

On engines fitted with attached pumps, but started using independent motor-driven pumps, the motor-driven pumps can be stopped once the engine stabilizes and "full away" is rung up on the telegraph. If the motor-driven pumps are fitted for automatic standby they must be left correctly lined up for this function.

Maneuvering. Propulsion engine control may be effected by direct mechanical linkage or by pneumatic, hydraulic, or electrical servomotors, and may be applied from a local control stand at the engine or from a remote console in an engine control room or on the bridge. Once the engine has been prepared for starting, with support systems in operation, control can be turned over to a remote console. Automation interposed between the remote console and the engine may be such that the operator at the console need not understand, when he makes a change in speed or direction, the procedural details necessary to accomplish the change. Sufficient instrumentation is provided, however, to confirm his actions.

Speed and load control of a diesel engine are achieved by positioning the fuel racks of the injection pumps, which are mechanically linked to operate simultaneously. In some installations, the racks may be positioned directly by the operator, but in most cases they are positioned by a governor, whose set point is adjusted by the operator.

Engines that directly drive fixed pitch propellers are necessarily reversing engines. To comply with a stop order, the engine itself must be stopped by moving the fuel racks to the zero position. Then, when ordered to run ahead or astern, the camshaft and starting air distributor are positioned in the appropriate direction, the fuel racks are set to the starting position, and air is admitted to restart the engine. If there is way on the ship when the fuel is cut off in response to a stop order, the engine will continue to be rotated by the propeller until starting air is applied in the reverse direction, or until the shaft brake, where fitted, is applied.

In fixed pitch propeller installations where reversing engines are clutched to the propeller shaft (usually at gearing) the engines can be warmed up in advance of the maneuvering period. Normal maneuvering may then be as described above, with the clutches engaged throughout; or, alternatively, a stop may be achieved by declutching the engines and allowing them to idle, while a shaft brake holds the propeller shaft stationary. To reverse propeller rotation, however, the engines must be stopped, the camshafts and starting air distributors shifted, the engines started again in the reverse direction, and the clutches engaged.

Where nonreversing engines drive a fixed pitch propeller through reversing gearing, a shaft brake must be fitted. Reversing is achieved by reducing engine speed to idle, declutching the ahead gear train, applying the shaft brake to stop the shaft, engaging the clutch on the astern gear train, and then raising engine speed to the desired setting. In this arrangement the clutch movements and shaft brake application would normally be programmed to operate in sequence from a single lever.

After each use of starting air, the compressors are started, usually automatically, to recharge the receivers. During intensive maneuvering periods the compressors may be running continuously, cycling on their unloaders. In most cases the receivers and moisture separators will be fitted with automatic drain traps, but if this is not the case they must be drained at frequent intervals.

The minimum engine speed will typically be 25 to 40 percent of rated RPM and, unless a slip clutch is included in the drive train, or electric drive is used, this minimum RPM will determine minimum shaft speed. To maintain extremely low ship speeds in most ships with fixed pitch propellers requires that the engines be stopped and started repeatedly or, where clutches are fitted, that they be clutched in and out repeatedly.

If there is a barred speed range, it is passed through quickly.

Ships with controllable pitch propellers may be maneuvered entirely by pitch control, with the engine speed left constant throughout the maneuvering period, at about 75 to 90 percent of rated RPM.

During maneuvering periods, all support systems and engine instrumentation are monitored. Depending on the extent and condition of automation present, it may be necessary to make adjustments to keep temperatures and pressures within bounds. Boost blowers on two-stroke engines should be operating at low engine output.

If an engine is idled or run at low loads for an extended period, the charge air cooling water is regulated (automatically or manually) to maintain the charge air temperature above the intake temperature to avoid condensation in the manifold. Where charge air coolers are cooled by jacket water, it is frequently possible under these conditions to elevate the charge air temperature well above ambient, to help ensure prompt ignition, to aid combustion, and to help avoid sulfuric acid attack in the cylinders. For engines normally maneuvered on HFO it may be advisable, during a long idle period, to switch to distillate fuel, but this must be done gradually, usually through a mixing tank, since a sudden fuel temperature reduction can cause injection pump plungers to bind.

At the close of a departure maneuvering period, engines should be loaded in increments that extend over at least the first hour at sea.

Running-in after an overhaul. Running-in is necessary to enable new piston rings to wear to conformity with the cylinder liner. (Modern bearing materials, on the other hand, usually need no running-in period.) An engine with one or more pistons fitted with new rings should be started and run on DO, with load applied gradually over a period of at least six hours. On engines with separate cylinder oil lubricators, a straight mineral cylinder oil should be fed to the re-ringed cylinders at an increased rate during the running-in period.

Steady running. During steady running, all engine indicators and all support systems are monitored regularly, and adjustments made as required. Even in plants with extensive automation, engineers should regularly walk around the engines, listening for unusual noises, looking for leaks and sources of unusual vibration, and checking sight glasses and fluid levels. Trends are noted for their value in planning maintenance and in avoiding unexpected failure.

Every day at sea, fuel is transferred to the settler, and, where two settlers are provided, purifier suction is changed over. Where HFO is used, tank heating and line tracing steam is supplied to the tank and lines that will be used in the following day's fuel transfer. Water and sediment are drained from settling tanks daily, and other tanks are checked for accumulations at least weekly.

In crosshead engine installations, the cylinder oil measuring tank is refilled daily, and the levels before and after refilling noted in order to calculate the consumption. The cylinder oil injectors are then adjusted as required.

Moisture is drained daily from charge air coolers and from the air manifold.

On engines burning HFO, the turbocharger turbines are water washed, often every day, in accordance with manufacturer's instructions. The compressor will require cleaning less frequently.

At least weekly, or once per passage for vessels on shorter voyages, indicator cards are taken on low speed engines, and maximum pressure readings on higher speed engines, in order to confirm that the load is balanced among the cylinders.

Fuel oil consumption is calculated daily and lubricating oil drain tank (i.e., sump) levels checked at least as often.

Self-cleaning purifiers will have to be opened for manual cleaning at intervals determined by experience. Until that experience is gained it may be reasonable to open HFO purifiers weekly, and LO and DO purifiers monthly. Older purifiers that do not eject accumulated sludge will require more frequent cleaning—at least daily for the HFO purifiers and perhaps weekly for the others.

Stopping. Before stopping an engine normally run on HFO but started on DO, it will be necessary to change over to DO, although up to half an hour of operation at high output may be needed for the DO to fill the system through to the injectors. Even for engines normally started on HFO this may be advisable, as it will simplify any maintenance requiring the opening of the fuel system.

When steam for fuel heaters and for fuel line tracing is no longer needed, it is secured.

On engines with independent cooling water and LO circulating pumps, these are kept in operation for about a half hour after the engine has been stopped.

Main propulsion engines cannot be secured until the bridge has given and confirmed the order “finished with engines.”

Once an engine is secured, fuel pumps are stopped, the starting air stop valve is shut, and the turning gear is engaged to prevent accidental rotation of the engine. Cylinder indicator cocks, drains at air coolers, and air and exhaust manifolds are opened and left open.

LO purifiers on propulsion engines are kept in operation for at least 12 to 24 hours after the engine is stopped.

Crankcase doors must not be opened until after the crankcase has cooled down because of the danger of a crankcase explosion.

Emergency Operation

It is often necessary to put an engine into temporary service with one or more cylinders cut out, or with a turbocharger secured. Detailed procedures and precautions differ from one engine to another, and instruction manuals should be consulted, but some general comments are in order:

- Depending on the nature of the problem, it is usually better from the standpoint of engine balance if the piston and running gear of the cut-out cylinder remain in place. In this case, care must be taken to ensure adequate lubrication.
- If the piston must be hung up or removed, the starting air valve must be isolated.
- Load should be reduced to avoid overloading the remaining cylinders, and reduced further if limits in operating parameters are approached.
- Where machinery is normally unattended, watches should be set.
- It must be ascertained if barred speed ranges exist for operation with a reduced number of cylinders, and operations in these ranges must be avoided. In any event, RPM and/or load must be changed if vibration intensifies.

Care must be taken on stopping to see that the engine is in a position to be restarted; if not, a swing on air in the opposite direction will usually bring the engine to a starting position in the correct direction.

The master must be advised of anticipated maneuvering constraints.

Repairs should be undertaken as soon as possible.

Maintenance

Maintenance is necessary to keep machinery in sound condition, thereby reducing the likelihood of unexpected failure. *Preventive maintenance* is performed according to a schedule recommended by the engine and auxiliary manufacturers as amended by the operator's experience and, in addition, as called for by an analysis of performance data (*condition monitoring* or *trend analysis*).

Performance data analysis. Performance data analysis may be restricted to observing trends in pressures and temperatures read from thermometers and pressure gauges, and recorded in the logbook, but more sophisticated methods are being increasingly applied. These methods often enable recommended maintenance intervals to be exceeded with confidence, or can warn of early failure. As an example, some new engines are fitted with electronic pressure sensors in cylinders and high pressure fuel lines, together with position sensors on injectors and valve stems; these are keyed to crank angle sensors that enable the production of indicator cards, crank angle diagrams, fuel line pressure diagrams, and needle and valve lift diagrams accurately and instantly.

It must be noted that for data analysis to be of any use, the data must be reasonably accurate. Periodically, sensors and instruments must be checked and calibrated, gauge lines blown clear, and circuits tested. Where readings are questionable and time permits, similar sensors or instruments may be interchanged between cylinders or locations to confirm a reading.

One of the most important parameters providing an indication of engine condition is the exhaust gas temperature. Exhaust temperature will be indicated at the turbocharger for small engines, and, on larger engines, at individual cylinder exhaust branches as well. Rising exhaust temperatures overall are indicative of increasing load and should correlate with the fuel rack position. Where exhaust temperatures are rising overall at constant or declining rack setting, and without appreciable change in air temperatures, the cause might be air starvation because of fouled air coolers or fouled or damaged turbochargers. When individual cylinder exhaust temperatures are rising, the cause may be fouled or otherwise blocked ports, a maladjusted fuel rack, a defective injection pump or injector, a worn fuel cam, or a leaking or burned exhaust valve. When individual cylinder exhaust temperatures fall, a maladjusted fuel rack or a defective injection pump may be indicated, or a compression failure resulting from broken or stuck piston rings, a stuck or badly damaged exhaust valve, a stuck intake valve, or a failed gasket.

At least weekly, indicator cards or maximum pressure readings should be taken. Indicator cards or their electronic equivalent can be used to detect problems with valves and combustion, and a fault-free card can provide sufficient information to balance the cylinder outputs. More discretion is required when relying on maximum pressure readings, which should be taken in conjunction with individual cylinder exhaust temperatures: a low maximum pressure accompanied by a high exhaust temperature usually reflects a leaking exhaust valve rather than an unbalanced cylinder.

Cylinder compression readings can indicate that rings and valves are sealing tightly, independent of fuel rack position. The engine manual should be consulted for the recommended method and values: readings may be taken cold by barring or jacking the engine; or closely approximated using a maximum pressure indicator or electronic pressure sensors when the engine is idling without load; or they may be taken hot, with the engine near normal speed and the fuel cut off each cylinder in turn. With this confirmation that the cylinders are tight, maximum pressures or indicator cards taken shortly thereafter, with the engine loaded, can be used to balance the cylinders.

Lubricating oil drain tank levels should be observed at least daily, and trends noted. A rising level may be because of leakage of LO from a storage tank, leakage of water from a heating coil or heat exchanger, or contamination by distillate fuel oil. A falling level may reflect a leak from the system or, in trunk piston engines, a rise in LO consumption, perhaps indicating, in turn, worn or stuck piston rings or worn valve guides.

Lubricating oil samples provide useful information on engine condition, and may indicate a need for oil renewal:

The presence of tin, copper, lead, or other bearing materials may be a warning of bearing wear.

Traces of iron or chromium may indicate liner or ring wear; of aluminum, in engines with aluminum pistons, piston wear.

Trace quantities of water treatment chemicals may indicate water leakage from cooling passages through cracks or seals.

Traces of fuel constituents, a decrease in viscosity, or a drop in the flash point can be the result of fuel dilution, perhaps through leaking seals at fuel pumps.

An abnormal drop in the TBN is usually the result of excessive blowby, indicating stuck or worn piston rings.

A gradually rising pressure in the crankcase of trunk piston engines usually indicates blowby.

Fuel oil consumption should be calculated daily but, because of the many factors that can contribute to a rising trend in fuel consumption, it will not serve, on its own, to pinpoint problem areas adequately.

Smoke observations provide a coarse indication of problems that, in large, reasonably well instrumented engines, are more likely to have been detected earlier. Generally, dark smoke is composed of incompletely burned fuel, and so indicates overload, air starvation, defective or maladjusted injection equipment, or low fuel temperature. When there is a blue tinge to the smoke, the cause is usually excess lubricating oil in the combustion space, indicating worn or stuck piston rings, worn valve guides, or maladjusted cylinder lubricators. White smoke usually indicates low combustion temperature, caused by overcooling the cylinder or by an advanced or retarded injection, or it could be steam from a leaking jacket, cylinder head, turbocharger, or exhaust gas boiler. A smoke problem can be traced to a particular cylinder by successively cutting off fuel to each cylinder in turn.

Falling pressures in pumped systems may be indicative of pump wear, leakage (including internal leakage through a bypass or relief valve), improperly set stop valves, or blockage at a strainer, heat exchanger, or elsewhere in the system. Falling pressures in lubricating oil systems could also be the result of excessive bearing wear, overheating of the oil, or fuel dilution.

Discharge filters in LO and fuel systems, and charge air intake filters and coolers are normally fitted with pressure gauges or manometers indicating pressure drop. When the pressure drop rises to the prescribed limit, the unit must be cleaned.

Vibration sensing and analysis have proven to be a very reliable monitoring technique for rotating machinery, indicating bearing deterioration or imbalances resulting from damage or deposits. The technique is widely applied to auxiliary machinery, but its usefulness on engine-driven or otherwise attached equipment is limited by the extent to which the inherent vibration of the engine's many reciprocating components can be filtered out. Nevertheless, vibration-monitoring techniques have successfully detected bearing wear and piston skirt wear in high speed engines on an experimental basis, and wider application to large engines for these purposes is likely.

As more sophisticated measuring techniques and instruments become available and are proven to be effective and reliable in service, the number of parameters monitored will increase and therefore the effectiveness of condition monitoring will grow. At present some engines are fitted with proximity detectors that locate piston rings and measure their wear, and others that measure valve and fuel plunger positions and accelerations. This information, taken together with data from temperature and pressure probes embedded at strategic points, can provide sufficient data to enable a computer to track performance, detect ominous trends, diagnose likely causes, and suggest possible solutions.

Maintenance schedules. Engine manufacturers provide schedules of routine maintenance tasks and their recommended frequencies. These schedules are usually predicated on the basis of a fuel that meets a certain specification and operation at 80 to 90 percent of rating. When fuels of poor quality are being used or the engine is operating under such adverse conditions as sustained overload or low load, more frequent maintenance may be necessary (see table in section on "Variations in ratings" earlier in this chapter) and would generally be indicated by an analysis of the performance data. The intervals cited below are for background information only and may be contradicted by manufacturers' requirements for specific engines, to which strict attention should be paid. In general, when run on fuel of poor quality, low speed engines will require less maintenance than higher speed engines.

Major machinery maintenance tasks aboard ship are generally staggered to provide a tolerable amount of work during each port visit. Classification societies offer continuous machinery survey provisions to suit this practice. Operators of ships in seasonal trades usually attempt to restrict all planned maintenance to the lay-up period, when complete overhauls might be undertaken.

Spares are kept in an overhauled, subassembled condition, ready for use. For example, a fully outfitted cylinder head, immediately available for installation in place of one in service, will expedite both emergency repairs and staggered maintenance schemes. A used component, withdrawn from the engine, is either reconditioned to become the next spare or replaced.

Each time bunkers are received, a sample should be obtained and analyzed, and the fuel thereby proven to be acceptable, before the fuel is used.

Several times daily, temperatures, pressures, levels, and flows are monitored, and inspections made for noises, vibrations, and leaks.

Daily, water is drained from the charge air cooler and the air manifold, and from points in the compressed air system not fitted with automatic traps; sludge is drained from the HFO settler; and other fuel tanks are checked for accumulated water or sediment.

On crosshead engines, the cylinder oil measuring tank is refilled daily, the consumption calculated, and the lubricators inspected and adjusted as required.

Weekly or more often, turbocharger turbines of engines run on HFO might be water washed. Compressors will require water washing less frequently.

At intervals, typically of 500 to 1,000 hours, samples of LO are withdrawn from the main engine and auxiliary engines for analysis; oil must be renewed if the analysis shows this is required (renewal will be more commonly necessary for trunk piston engines than for crosshead engines: see discussion, "Main engine LO circulating system," earlier in this chapter). At similar intervals, fresh water in each cooling circuit is sampled, tested, and treated or renewed as indicated.

Every 1,000 hours or so, control levers and linkages are inspected, cleaned, and lubricated.

At intervals of 1,000 to 2,000 hours on engines burning HFO, or as experience dictates, the fuel injectors must be removed, tested, cleaned, reset, and reconditioned or replaced as necessary.

On medium and higher speed engines burning HFO and fitted with exhaust valves, exhaust valve performance may become suspect at intervals as short as 1,000 hours or less, but more usually by the time 5,000 or 6,000 hours have passed, depending on factors that include fuel quality, valve materials, valve cooling, and operating practices. (This problem is less often encountered in low speed engines, where valve cooling can generally be more effective.) On engines fitted with caged exhaust valves, the valves are withdrawn for inspection, cleaning, and lapping or grinding of the seats. Where the heads must be withdrawn to expose the valves, however, an effort should be made to confirm that the exhaust valve is indeed the problem: the air start and relief valves should be pulled and checked; the exhaust valve should be examined to see that it moves freely in its guide, and has the proper spring and sufficient tappet clearance to ensure closure; if a borescope is available it might be used to inspect the exhaust valve seats through the opening for the air start valve, the injector, or the relief valve. If the exhaust valve remains suspect, there is no alternative to pulling the head to inspect, clean, lap, grind, or replace the valve and/or the seats.

At intervals of 1,000 to 3,000 hours air manifolds and, on crosshead engines, under-piston spaces are inspected and cleaned. On large two-stroke engines the air ports usually permit a limited visual inspection of the cylinder liner, piston crown, rings, and skirt from the air manifold.

At intervals of 1,000 to 3,000 hours, cam surfaces are inspected for pitting; timing gears or sprockets are inspected for tooth wear and cracks; timing chain tension is checked and adjusted; and tappet clearances are checked and adjusted. On engines with valves, the valve gear is inspected for free and proper operation; valves should be shown to move easily but not loosely in their guides; exhaust valve rotators must be checked for proper functioning; and valve and injector timing should be checked. Improper operation must be rectified and defective components replaced as necessary.

At intervals of 3,000 to 6,000 hours, major nuts and bolts, including foundation bolts and tie-rods, are checked to ensure that they remain properly tightened; crankshaft deflection readings are taken and suspect main bearing clearances measured; telescopic tubes or swivel tubes used for oil and water delivery to pistons and crossheads of crosshead engines are examined and glands are tightened or packed as required.

At intervals of 6,000 to 12,000 hours in engines running on HFO, or up to 25,000 hours in distillate fuel engines, cylinder heads are removed, inspected for cracks and leaks, and, if acceptable, rebuilt with new parts. Pistons are pulled; inspected for scuffing, cracking, burning, and fuel impingement; measured for circularity and ring groove wear; and wrist pin condition and clearances in trunk pistons are determined. Acceptable pistons will be returned to service, usually with new rings. Cylinder liners are inspected for scuffing, fuel impingement, cracking, and ridges or shoulders; measured to determine wear; and dressed, honed, or replaced as required. The cylinders are reassembled with new gaskets and seals.

At 6,000- to 12,000-hour intervals, usually in coordination with the cylinder overhaul described above, connecting rod bearing condition and clearances are checked; in crosshead engines, crosshead guides are also examined.

At intervals of 8,000 to 12,000 hours in engines running on HFO, or up to 25,000 hours in engines burning distillate fuel, the turbochargers are disassembled, cleaned, inspected, and, if acceptable, balanced and reassembled with new bearings and seals.

At intervals of 12,000 to 25,000 hours the condition of the main bearings, including the thrust bearing, is checked and their clearances measured. At similar intervals attached oil and water pumps are inspected and overhauled as required; injection pumps and the governor are overhauled and recalibrated.

Special tools and maintenance aids. The premier maintenance aid in a diesel plant is an overhead crane. While a simple overhead beam fitted with a trolley for a chain hoist may suffice for small engines, a permanently installed gantry crane, electrically or pneumatically powered, with mobile remote control, capable of longitudinal and transverse positioning, is the norm in large plants. In plants with eight or more main engine cylinders and short port stays, consideration might be given to a second gantry carriage on reinforced rails, to enable two different cylinders to be worked simultaneously. By similar reasoning, auxiliary engines should have their own lifting gear, even when the arrangement might permit the main engine gantry crane to plumb the auxiliary engines.

Maintenance tasks are simplified by the use of special tools and access gear, usually supplied by the engine builder, sometimes as an option but often as standard equipment. Access gear is usually intended for use with the overhead crane and may include built-in rails, lifting points, and jacking pads within the crankcase to facilitate removal of large items. Equipment for a large engine might include the following:

Lifting beams or brackets for heavy components, including heads, valve cages, pistons, liners, connecting rods, bearing caps, turbocharger rotor.
Tensioning devices (hydraulic or electric) for all major studs and bolts, including cylinder head studs, tie rods, foundation bolts, bearing bolts.
Strongbacks for removing components whose withdrawal might be difficult, such as valve cages, fuel injectors, air starting valves, bushings and guides, cylinder liners.
Motor driven grinders for exhaust valves and seats.
An injector testing unit.
Mandrels for lapping fuel and starting air valve seats.
Piston ring spreading and compression devices.
Measuring devices, including trammels, bridge gauges, brackets for jacks, dial indicators used in measuring bearing clearances, extension rods for inside micrometer for cylinder liner, go/no-go gauges for valve guides and small bearings.
Support devices, including such equipment as brackets to hold the crosshead in place while removing the connecting rod, brackets to support the connecting rod while removing bearing caps, hydraulic jacks and cradle for crankshaft to enable removal of main bearing bottom half.
Blanks and caps, for openings and exposed threads, for example, to protect sensitive equipment from damage when adjacent parts are being serviced.

Stud tensioning devices are common outfit. Smaller bolts and studs can be tensioned using a properly calibrated torque wrench, but larger bolts and studs are most often tensioned using a hydraulic jack to stretch the bolt or stud by an amount that coincides with the predetermined pressure developed by a hand pump. Figure 16-57 shows a typical unit. At the specified tension the nut is easily turned and may be backed off for removal, or, when assembling, the nut is run up, the tension is released, and the bolt or stud remains prestressed. In an increasingly common application, a group of units is mounted in a ring to tension all cylinder head studs simultaneously, as in Figure 16-57. In some applications the tensioning device is an electric heating rod inserted in a central boring of the bolt or stud to elongate it.

Related to the use of tensioning devices is the use of bolts and studs of calibrated length: when properly made up, the bolt or stud will have elongated by a predetermined amount, which can be measured with a micrometer (calibrated studs have an internal boring and are measured with a depth micrometer). Obviously, when a calibrated bolt retains its elongation with the nut slack, an overstress has occurred.

Maintenance and repair procedures. Maintenance and repair procedures will differ from one engine to another, and instruction manuals should be consulted. Nevertheless, some general comments are in order:

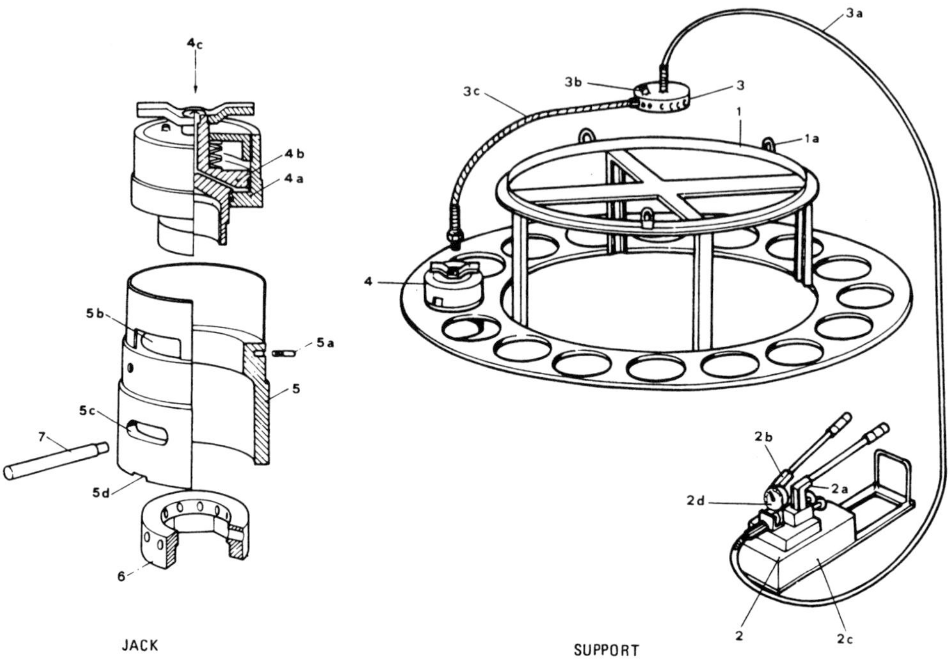

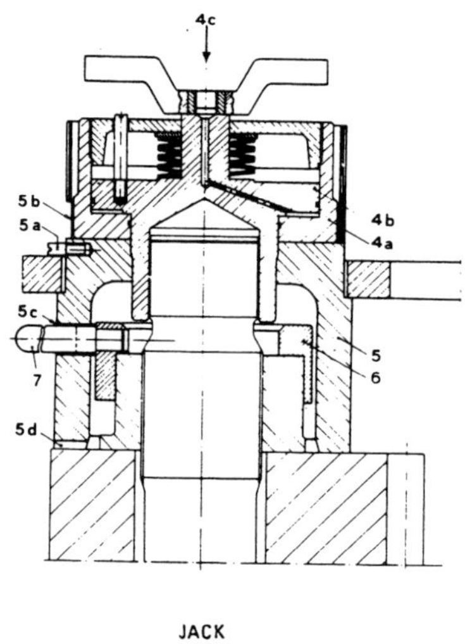

PART LIST

1 – Supporting frame
1a – Eyelets for lifting and transport

2 – Pump unit
2a – Low pressure pump
2b – High pressure pump
2c – Lube oil tank
2d – Manometer
3 – Multiple union
3a – Delivery flexible pipe from the pumps
3b – Air valve
3c – Flexible pipe for jack
4 – Complete jack
4a – Jack body
4b – Jack piston
4c – Oil inlet to jack from the pumps
5 – Spacer
5a – Ring-grub screw
5b – Inspection slit
5c – Tightening slit
5d – Control slit
6 – Box spanner
7 – Lever to actuate spanner

Figure 16-57. Hydraulic bolt tensioning device; arrangement for simultaneously tensioning cylinder head bolts

After running the engine, crankcase doors must not be opened until after the crankcase has cooled down, because of the danger of a crankcase explosion.

Manufacturers' instruction manuals should be studied prior to undertaking a task.

Proper tools and materials must be on hand, including spares for any components that might, in the course of the job, be discovered to require replacement.

Particularly in regard to such high speed precision equipment as turbochargers and purifiers, extreme care must be taken to avoid causing even minor damage that might lead to catastrophic consequences when the units are returned to service.

Consumable components such as gaskets and lock washers should be renewed on reassembly. Consideration should be given to including piston rings in this category.

Engines must be safely secured against rotation while work is underway. This is usually best done by engaging the jacking gear. On engines without jacking gear, brackets or blocks intended for this purpose must be installed. Individual components should be fastened in place.

Before turning the engine on the jacking gear it must be ascertained that personnel and equipment are clear of moving parts.

The temptation to use the gantry crane or a chain fall to jerk free an item which is difficult to withdraw must be resisted.

Care must be taken to avoid contaminating LO or cooling water, by using only appropriate methods, such as draining it into clean tanks, and by shielding or blanking connections.

Fire precautions must be observed.

Electronic Monitoring and Control

Electronic devices are increasingly used in monitoring and control of diesel engines. Although no engines are yet in service with complete electronic control, prototype engines are already running in which the crankshaft position, determined by proximity indicators, is used to signal electrohydraulic solenoids for operation of fuel injectors and air and exhaust valves, thereby eliminating the need for timing gears, camshafts, pushrods, and rocker arms. Fuel is supplied from a high pressure common rail, making precision-controlled jerk pumps unnecessary. Advantages of electronic controls include reduced manufacturing cost, enhanced reliability, and simplified maintenance, but an even more important advantage is the ability to adjust the timing of the key cycle events of fuel injection and valve operation while the engine is running. This flexibility has immediate ramifications in improved fuel consumption and in reducing exhaust pollutants.

Electronic sensors also enable continuous monitoring of engine parameters that were previously difficult to measure without disassembly. Embedded sensors for cylinder-liner and bearing wear, and for vibration, pressure, and temperature, are commonly fitted. The output of these devices not only can be continuously monitored, but also can be continuously integrated with other data to permit computerized trend monitoring and a degree of automatic fault prediction and maintenance supervision.

Exhaust Emissions

Diesel engine exhaust gas is composed primarily of nitrogen, oxygen, water vapor, and carbon dioxide, but smaller quantities of gases that are harmful to the environment, including nitrogen oxides, sulfur dioxide, carbon monoxide, and hydrocarbons, as well as particulates, are also present. With the possible exception of the sulfur dioxide, all of these constituents will be present whether the engine is running on heavy fuel or distillate fuel.

To reduce the quantities of harmful gas emitted, high compression ratios, long stroke-to-bore ratios, higher fuel-injection pressures, and short but intensive fuel-injection periods are used, which, with the high air-to-fuel ratios possible with modern turbochargers, generally bring benefits in emission control as well as in fuel consumption. However, some adjustments to injection timing and to the rate of fuel injection that reduce harmful emissions can increase fuel consumption. Electronic control of fuel injection is therefore especially advantageous in emission-controlled engines, not only because it facilitates the varied timing needed, but also because it is possible to adjust the timing in service to optimize engine performance for the prevailing circumstance, for example, for low fuel consumption at sea but for low exhaust emissions in restricted areas. The precision and response of electronic control of fuel injection is also helpful in reducing hydrocarbon emissions during maneuvering.

However, these methods are not sufficient by themselves to meet all of the exhaust-emission criteria that are coming into force. Further methods of emission reduction now in use include exhaust-gas recirculation, the introduction of freshwater to the cylinder before or during the combustion period, and the treatment of the exhaust gas by catalytic converters.

Water is introduced into the combustion process by emulsifying it into the fuel or by directly injecting it into the cylinder. The water acts primarily to depress the temperatures during combustion, which contributes to reduced generation of nitrogen oxides and suppression of visible hydrocarbons. The rate of water use depends on the degree of emission control needed, but for maximum effect the rate of water use can be up to 50 percent or more of the fuel rate. Water injection does not affect engine component life, but in addition to the freshwater consumption and the added water-handling equipment, it generally causes an increase in specific fuel consumption.

A catalytic converter, also called a selective catalytic reactor, is a large steel housing similar in dimensions to a large exhaust-gas boiler, fitted internally with racks to support the renewable catalyst elements. Exhaust gas from the engine is prepared for the catalyst by the injection of a urea solution upstream of the converter. The urea decays into ammonia and carbon dioxide as it vaporizes into the hot gas, and in the converter, the ammonia combines with the nitrogen oxides in the exhaust gas to form gaseous nitrogen and water vapor. An oxidizing section in the converter will also reduce emissions of carbon monoxide. The gas entering the converter must be at a temperature of about 300°C or more, which, in low speed engines, may force the converter to be located before the turbocharger turbines. The cost of operation includes the urea solution and the renewal of the catalyst elements at intervals of three-to-five years, but there is little or no effect on the specific fuel consumption.

The matter of sulfur dioxide emissions is principally an economic issue, as the sulfur is introduced in the fuel. High-sulfur fuels, less marketable ashore, have had a ready market among oceangoing ships. When these ships enter restricted areas, the fuel supply is switched to a tank containing low-sulfur fuel.

REVIEW

1. What is meant by "compression ignition"?
2. List distinguishing features of low speed engines and of medium speed engines.
3. Explain the difference between two-stroke engines and four-stroke engines.
4. What is the difference between a trunk piston engine and a crosshead engine?
5. Draw and label pressure-volume diagrams for two-stroke and four-stroke cycles.
6. What accounts for the rapid initial rise in pressure that follows the end of the compression stroke in a dual-combustion cycle?
7. How does brake power differ from indicated power?
8. What is meant by mean effective pressure? How is it related to work and power?
9. How are heavy fuels treated differently from light fuels in preparing them for use in a diesel engine?
10. For each of the fuel properties or constituents mentioned in the text, cite the effect on a diesel engine and the usual methods used to counter these effects.
11. Why are engines turbocharged? Why are turbocharged engines usually fitted with charge-air coolers?
12. What is meant by the term "blowdown pulse"?
13. Describe the operation of a fuel-injection pump.
14. Describe the operation of a fuel-injector.
15. Explain the connection between the fuel rack and the steady-burning combustion period.
16. Why are the gas forces acting on a piston not transmitted to the ship's structure?
17. What is the source of pitching couples in a diesel engine?

18. What is meant by the term “barred speed range”?
19. Why are four-cylinder, four-stroke engines notorious for their second-order vertical forces?
20. Describe a typical engine performance map. Identify each of the boundaries.
21. Why are engines normally matched to their loads at power levels below their maximum continuous rating (MCR)?
22. How does the sea margin differ from the service margin?
23. Estimate the power that might be saved if a propeller were designed for 75 RPM in lieu of 106 RPM.
24. Show how a speed-power curve can be used to estimate the average service speed of a ship.
25. Describe each of the engine-support systems discussed in the text.
26. Why is the crankcase oil of a crosshead engine less subject to contamination than that of a trunk piston engine?
27. What is meant by “pinch point”? Why might it limit waste heat recovery?
28. List the steps necessary to start a main engine and prepare it for maneuvering.
29. Why are individual cylinder exhaust-gas temperatures so closely watched?
30. How often are cylinder heads removed? How is this frequency affected by fuel quality?

REFERENCES

Benson, R. S., and N. D. Whitehouse. 1979. *Internal Combustion Engines.* New York: Pergamon Press.

Harrington, R. L., ed. 1992. *Marine Engineering.* Jersey City, N.J.: Society of Naval Architects and Marine Engineers (SNAME).

Henshall, S. H. 1972. *Medium and High Speed Diesel Engines for Marine Use.* London: Institute of Marine Engineers.

MAN B&W Diesel A/S. 1999. *The Intelligent Engine: Development Status and Prospects.* Copenhagen: MAN B&W Diesel A/S.

Rowen, A. L. 1999. *Notes on Marine Diesel Engines.* Glen Cove, N.Y.: Webb Institute of Naval Architecture.

Society of Naval Architects and Marine Engineers. 1990. *Marine Diesel Power Plant Practices,* T&R Bulletin No. 3-49. Jersey City, N.J.: SNAME.

Stinson, K. W. 1983. *Diesel Engineering Handbook,* 12th ed. Norwalk, Ct.: Business Journals.

Wärtsilä NSD Corporation. 1999. “Direct Water Injection” and “Compact Selective Catalytic Reduction.” Vaasa, Finland: Wärtsilä NSD Corporation.

ACKNOWLEDGMENTS

The authors are indebted to Lawrence R. Wiersum (Computer Graphics of Northern Illinois) for the line drawings appearing in this chapter, and to the following manufacturers, who provided the remaining figures.

Asea Brown Boveri Ltd.
Colt Industries, Fairbanks Morse Engine Division
MAN-B&W Diesel A/S
Wärtsilä NSD Corp. (previously Sulzer Diesel, Ltd.)

CHAPTER 17

Engine Descriptions

KEITH WILSON, DAVID BROWN, AND ALAN L. ROWEN

INTRODUCTION

TWO-STROKE, low speed, crosshead engines likely to be found in service today were manufactured to the designs of Burmeister & Wain, Sulzer, Mitsubishi, MAN, or GMT. In 2001 low speed, crosshead engine design is left to only three firms: MAN-B&W (Burmeister & Wain engines), Wärtsilä (Sulzer engines), and Mitsubishi. A greater number of designs of medium speed, trunk piston engines are in manufacture, including those by Wärtsilä, MAN-B&W, SEMT-Pielstick (now a division of MAN-B&W), Sulzer (now part of Wärtsilä), MaK (now part of Caterpillar), Fairbanks Morse, EMD (General Motors), GMT (now part of Wärtsilä), Daihatsu, Deutz, Ruston (now a division of MAN-B&W), and others. Different designs of high speed engines are in even greater number, including Wärtsilä, Cummins, Caterpillar, MTU, Paxman, Lister, Volvo, and many others. No single engine design can be said to be typical of its type except in the most shallow sense, although low speed designs do tend to share more of the same characteristics than the medium speed and high speed engines, where differences extend even to the fact that both two-stroke and four-stroke cycles are represented. Space precludes descriptions of more than a few examples: The Sulzer RTA program of two-stroke, low speed, crosshead engines, and the SEMT-Pielstick PC2.5 and Wärtsilä 20 four-stroke, medium speed, trunk piston engines.

SULZER RTA TWO-STROKE DIESEL ENGINES

Introduction and Evolution

Figures 17-1, 17-2, and 17-3 show Sulzer RTA engines. Table 17-1 lists current characteristics of the RTA program, and Figure 17-4 illustrates the

Figure 17-1. Sulzer 6RTA62

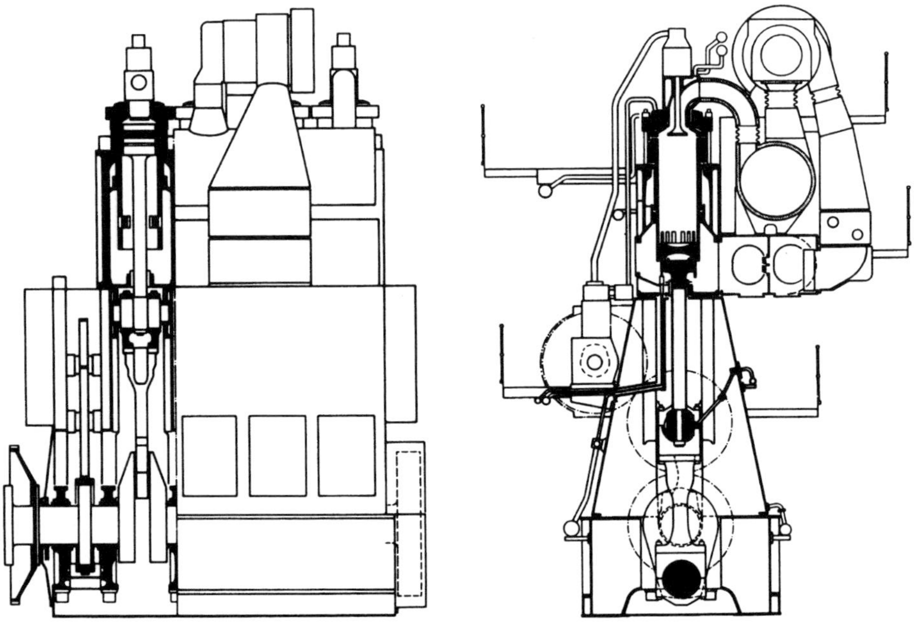

Figure 17-2. Sections of RTA58 to RTA84 engines

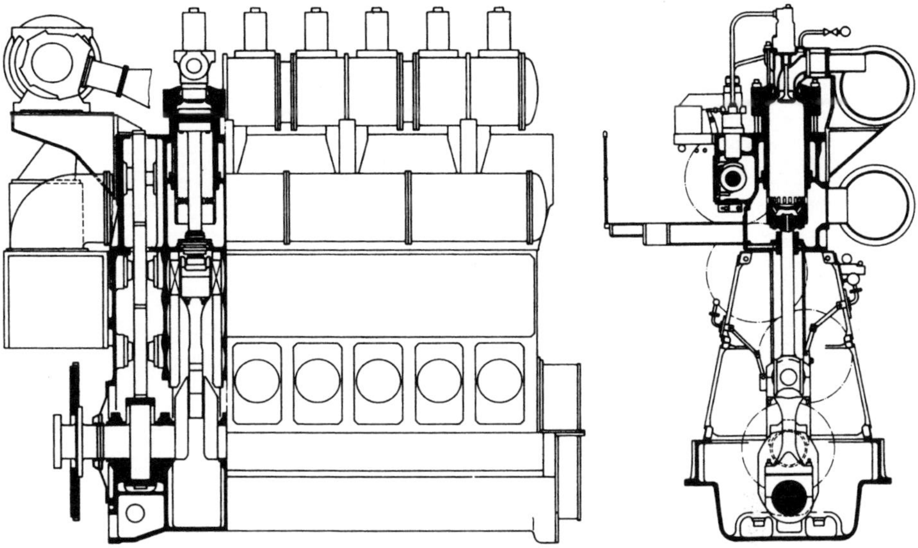

Figure 17-3. Sections of RTA38

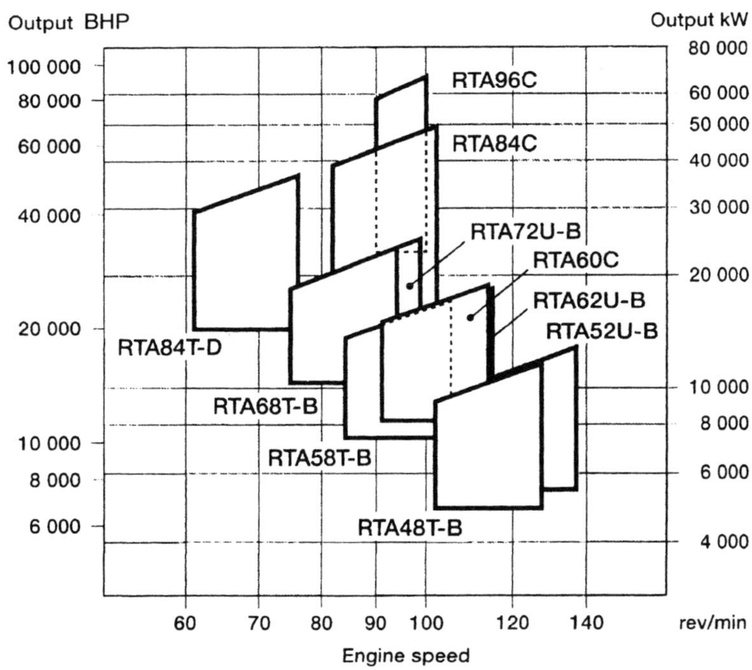

Figure 17-4. Power and speed ranges of the RTA family

range of output and RPM covered, from under 7,000 BHP for a 5RTA48T derated to 102 RPM, to almost 90,000 BHP for a 12RTA96C at its maximum rating. The RTA, introduced in 1981, represents the latest evolution in a two-stroke, low speed, crosshead engines, which began with the introduction of the SD in the 1940s.

A comparison of the RTA84 with its immediate predecessor, the RLB90 (Figure 17-5), shows both basic similarities (attesting to the evolutionary development typical of diesel engines) and distinct differences: a much higher stroke-to-bore ratio and a uniflow scavenging arrangement. These differences result from market conditions prevailing toward the end of the 1970s when shipowners' fuel costs tended to be at least as important a consideration as acquisition costs. Consequently, while rising power density had previously been the main parameter characterizing two-stroke engine development, improved specific fuel consumption (SFC) and lower RPM (to match the more efficient, slower turning propellers) became dominant development aims. The longer stroke made the lower RPM possible but necessitated the switch to uniflow scavenging; however, both of these features tended to improve SFC. These trends are illustrated by Figures

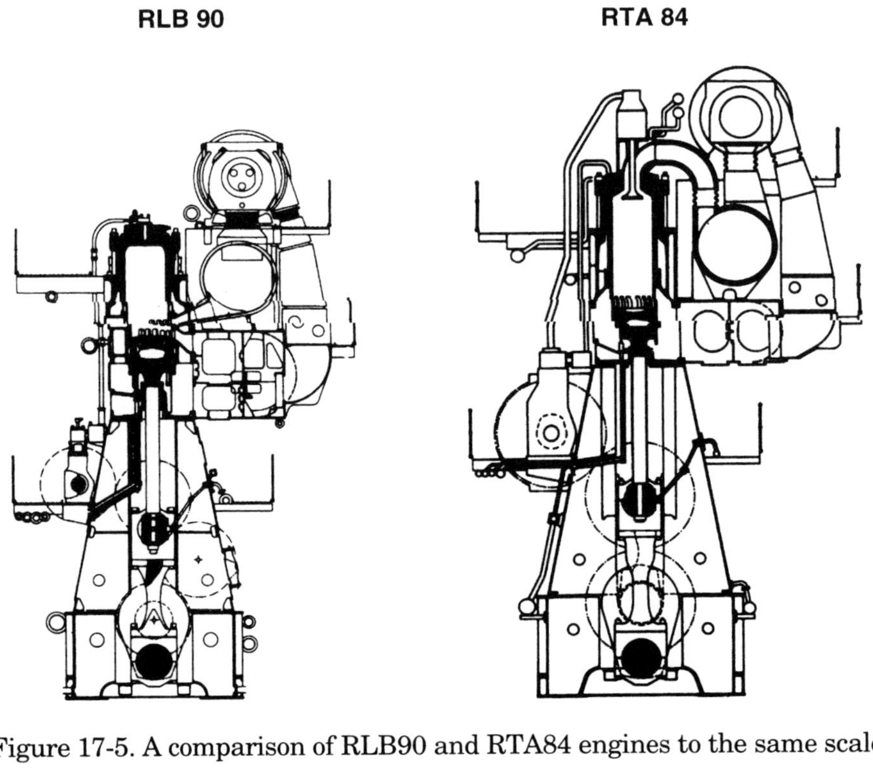

Figure 17-5. A comparison of RLB90 and RTA84 engines to the same scale

17-6 through 17-8, which trace Sulzer low speed engine evolution since 1960. The evolution of the competing Mitsubishi and B&W engines shows similar trends.

There is a further point to be made in regard to evolutionary design of the RTA program. When initially introduced in 1981, it consisted of six cylinder sizes, designated by their bore in centimeters as RTA38, 48, 58, 68, 76, and 84, all with stroke-to-bore ratios of about 2.9. The larger engines continued Sulzer practice in many details of construction and in having water-cooled pistons, while the smaller engines introduced several concepts not previously applied by Sulzer for low speed engines, including oil-cooling of the pistons. When additional engines were added to the program—the RTA52, 62, 72, and 84M, all with stroke-to-bore ratios of almost 3.5—sufficient satisfactory experience had been gained with oil-cooled pistons and some of the other new concepts to incorporate these features in these new engine designs.

Flexible Ratings

A further reflection of the trade-off between acquisition cost and operating cost in the RTA program (as well as in competing B&W and Mitsubishi designs) is shown in the variety of different rating points available for each engine design. Referring to Table 17-1 and Figure 17-9, the R1 rating represents maximum continuous rating (MCR); the design is such that MEP can be maintained at the maximum value for selected RPM values down to 72 percent of maximum RPM, the R3 rating, to suit slower turning, more efficient propellers. Along the line from R1 to R3, therefore, the engine is derated in proportion to the RPM; the SFC will fall only slightly, but total fuel costs will be reduced because of the potential for improvements in propeller efficiency. Further reduction in fuel costs through reduced SFC can be had if the engine is derated to lower output, in the direction of the R2 and R4 ratings, as illustrated in Figure 17-10.

Fuel Oil Quality

RTA engines are designed for operation on heavy fuels, with the following quality limits:

Maximum viscosity at 50°C	700 cSt
Maximum density at 15°C	0.991 kg/l
Maximum Conradson carbon test residue, by weight	22%
Maximum sulfur content, by weight	5%
Maximum ash content, by weight	0.2%
Maximum vanadium content, by weight	600 ppm
Maximum aluminum content, by weight	30 ppm
Maximum water content, by volume	1%

The density limit is dependent upon the capability of the fuel oil purifiers: if the latest generation of flow-controlled purifiers is used the limit is raised to 1.010 kg/l at 15°C. The aluminum content is used as an indication of the presence of catalytic fines.

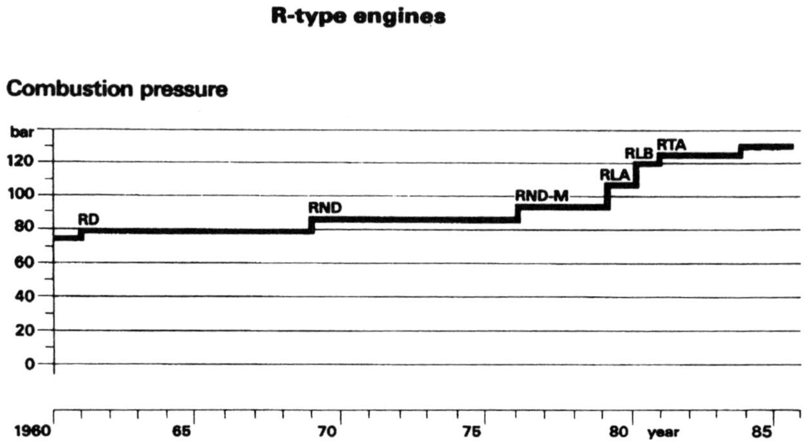

Figure 17-6. Trend in combustion pressure of Sulzer low speed engines

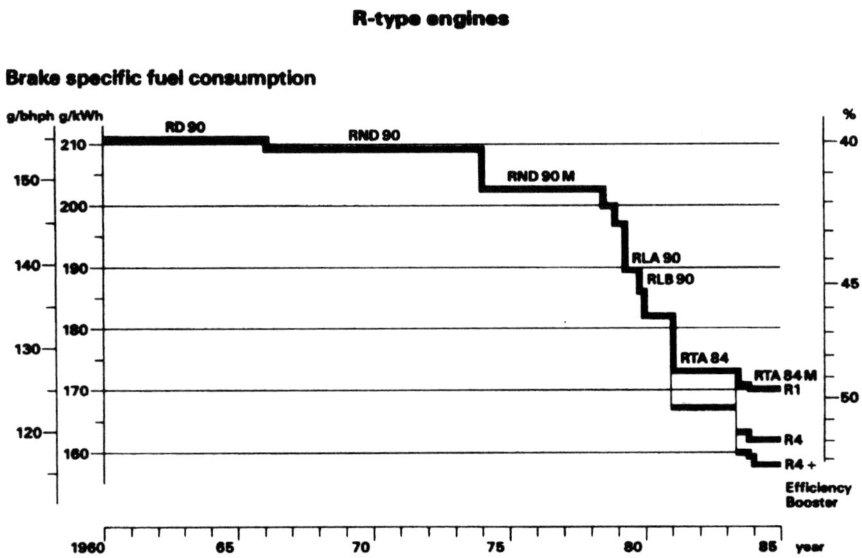

Figure 17-7. Trend in fuel consumption of Sulzer low speed engines

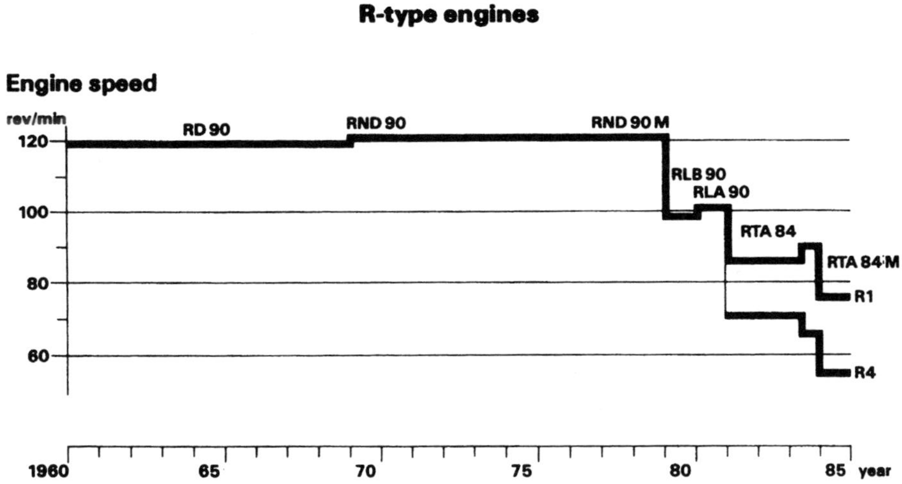

Figure 17-8. Trend in RPM at rating of Sulzer low speed engines

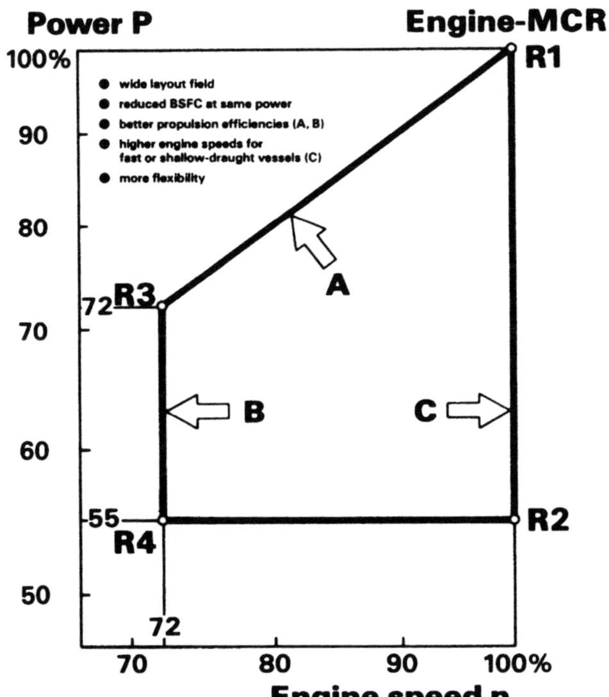

Figure 17-9. RTA layout field (not applicable to all RTA engines)

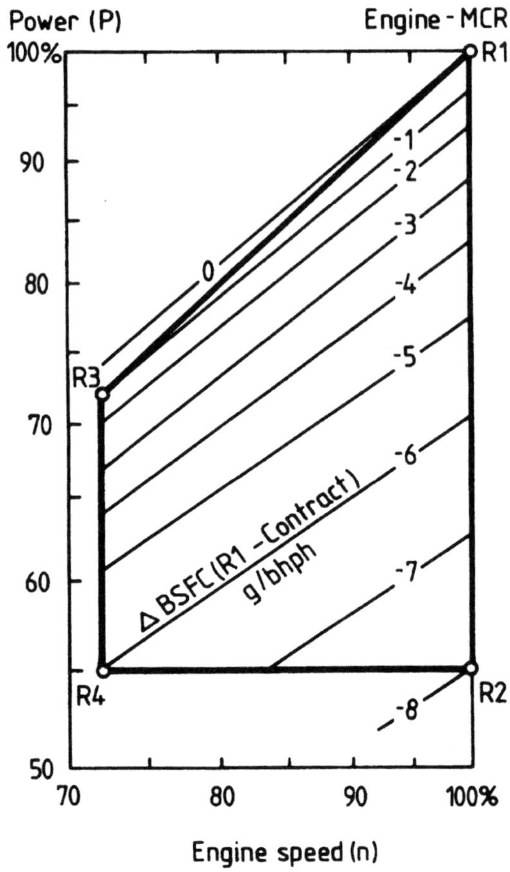

Figure 17-10. Reduction in fuel consumption achieved across the layout field

Engine Structure

For the larger bore engines the bedplate is of welded steel construction; for the RTA38 and 48 cast iron is used. An integral thrust bearing is incorporated at the aft end. Cross members incorporate the main bearing saddles and tie bolt housings. Hold down bolts are accessible in a single line on either side of the bedplate, outside of the bedplate walls. The engines can be seated on either conventional cast iron chocks or on epoxy resin. Thrust transfer to the tank top can be through fitted hold down bolts or through welded thrust brackets (Figure 17-11), although if epoxy resin is used the welded brackets are mandatory.

TABLE 17-1

Sulzer RTA Series Characteristics
R1 designates the highest rating of the engine

	RTA 96C	*RTA 84C*	*RTA 84T*	*RTA 72U*
bore × stroke mm	960 × 2500	840 × 2400	840 × 3150	720 × 2500
mep, bars	18.2	17.9	18.5	18.3
RPM at R1	100	102	76	99
kW/BHP per cylinder at R1	5490/7470	4050/5510	4100/5580	3080/4190
bsfc, g/kWh at 100% R1	171	171	168	171
number of cylinders	6–12	4–12	5–9	5–8

	RTA 68T	*RTA 62U*	*RTA 62*	*RTA 60C*
bore × stroke mm	680 × 2150	620 × 2150	620 × 2150	600 × 2250
mep, bars	19.0	18.2	17.2	19.5
RPM at R1	94	115	109	114
kW/BHP per cylinder at R1	2940/4000	2285/3110	2030/2760	2360/3210
bsfc, g/kWh at 100% R1	169	173	173	167
number of cylinders	5–8	5–8	4–8	5–8

	RTA 58T	*RTA 52U*	*RTA 52*	*RTA 48T*
bore × stroke mm	580 × 2416	520 × 1800	520 × 1800	480 × 2000
mep, bars	18.3	18.3	17.1	19.0
RPM at R1	105	137	130	127
kW/BHP per cylinder at R1	2125/2890	1600/2175	1420/1930	1455/1980
bsfc, g/kWh at 100% R1	170	174	174	171
number of cylinders	5–8	4–8	4–8	4–8

In the smaller bore engines the bedplate seating is arranged above the bottom of the bedplate to suit installation on smaller ships where the propeller centerline may be relatively closer to the baseline, while still allowing sufficient depth of double bottom. See Figure 17-12.

The crankcase is assembled from A-frame columns—fabricated steel in the larger bore engines and steel-reinforced cast iron for the smaller engines—connected by steel plate and girder to form a box girder. The A-frames provide mountings for the crosshead guides. In way of the thrust block the A-frames form a double walled enclosure for the camshaft drive gears.

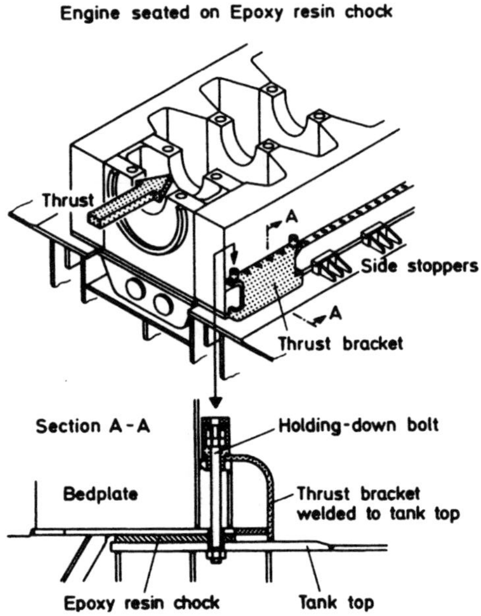

Figure 17-11. Arrangement of thrust bracket for epoxy resin chocks

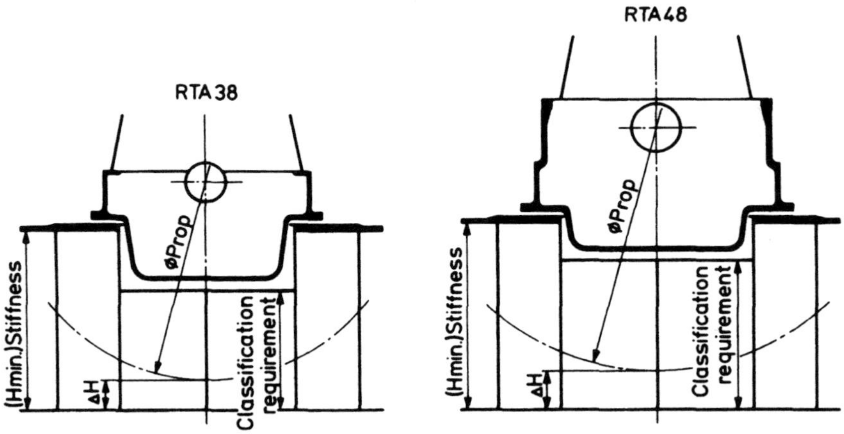

Figure 17-12. Seating arrangements for the small bore RTA engines

Depending on the number of cylinders, crankcase and bedplate may each be a single fabrication, or they may be divided to meet installation limitation.

The cylinder blocks are individual cast iron units bolted together, and further secured to the bedplate by tie bolts extending from the top of the blocks to the underside of the bearing saddles. The tie-rods are prestressed to ensure the rigidity of the structure and to avoid fretting and fatigue that would result from relative motion between the components. Cylinder blocks for the RTA38 and 48 are cast with an integral camshaft housing (see Figure 17-3).

Running Gear

Crankshafts are steel forgings, machined all over to avoid stress-raising surface irregularities. The larger engines use semibuilt crankshafts, with crankpin and webs forged from single ingots, shrink-fitted to separately forged journals, while the RTA38 crankshaft is a single forging. Depending on the number of cylinders, the crankshaft may be a single fabrication or may be divided by a flanged joint.

Figure 17-13, a longitudinal section, shows the large diameters of journal and crankpin needed to handle the high torque, as well as the long webs resulting from the long stroke. Also evident is the close approach of the crosshead to the webs in the bottom dead center (BDC) position, a result of the short connecting rod used to limit engine height, and the close longitudinal spacing of the cylinders to limit engine length.

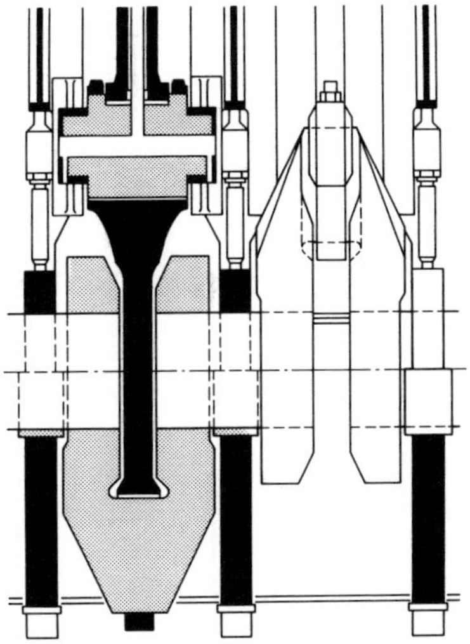

Figure 17-13. Longitudinal section in way of running gear

Main and bottom end bearings are thin-walled shells lined with white metal. Connecting rods are forged steel, with marine-type bottom ends. In engines with water-cooled pistons the crosshead bearing housing is divided, as illustrated in Figure 17-14, where the division permits the crosshead to clear the telescopic pipes carrying piston cooling water (see Figure 17-15). In engines with oil-cooled pistons the division is unneces-

Figure 17-14. Crosshead assembly for an RTA with water-cooled pistons

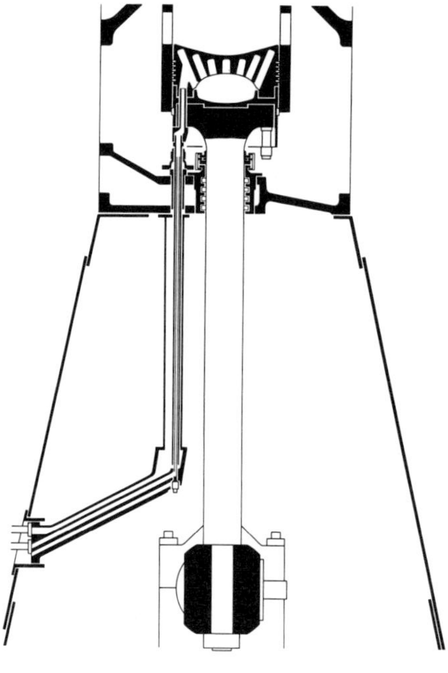

Figure 17-15. Water supply for water-cooled pistons in the RTA engine series

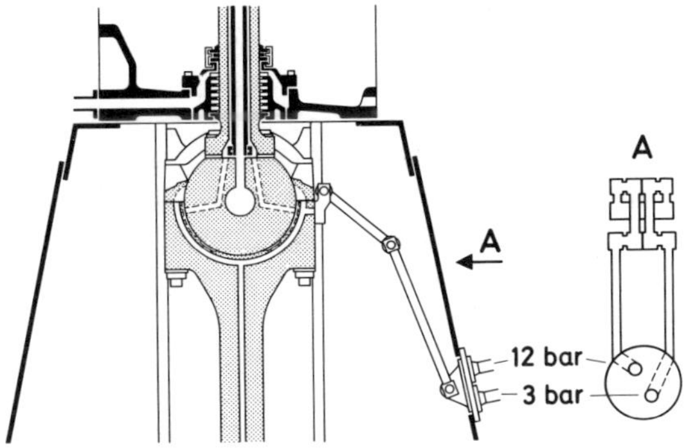

Figure 17-16. Cross section through crosshead of RTA engines with oil-cooled pistons

sary and the piston rod is bolted into the top of the crosshead pin, permitting the full lower half of the pin to be supported in a full-width bearing, as shown in Figures 17-16, 17-17, and 17-18. The divided bearings are thin-walled steel shells with a tin-aluminum lining, a design evolved from succesful experience as illustrated in Figure 17-19, which also shows the increased bearing loadings and their extended duration on the later model engines. In the engines with a full-width lower half to the crosshead bearing, the specific loads are low enough to permit the use of white metal.

All of the RTA engines incorporate hydrostatic lubrication for the crosshead bearings, using high pressure oil (12 to 16 bars). This has been a Sulzer practice since the early 1970s.

Bore Cooling

The practice of bore cooling was originally patented by Sulzer in 1936, when it was applied to cooling the ribs between exhaust gas ports of the loop- scavenged engines. Since 1968 bore cooling has been applied progressively in the combustion chamber area, beginning with the liner of the RND engine, as illustrated by Figure 17-20; on the RTA the liner, the piston, and the head are all bore cooled. Bore cooling is a successful compromise between the conflicting requirements for, on the one hand, thick sections to withstand the gas pressure and, on the other, the thin sections needed for cooling. This success is illustrated by Figure 17-21, showing the reduced stress levels in cylinder heads of successive engine designs, even as gas pressures rose, and by Figure 17-22, which shows the modest temperatures of an RTA piston crown. A further advantage of bore cooling is in the simplification of castings and forgings and the improved quality control that results.

Figure 17-17. Crosshead bearing for RTA62 engine showing full-width lower half

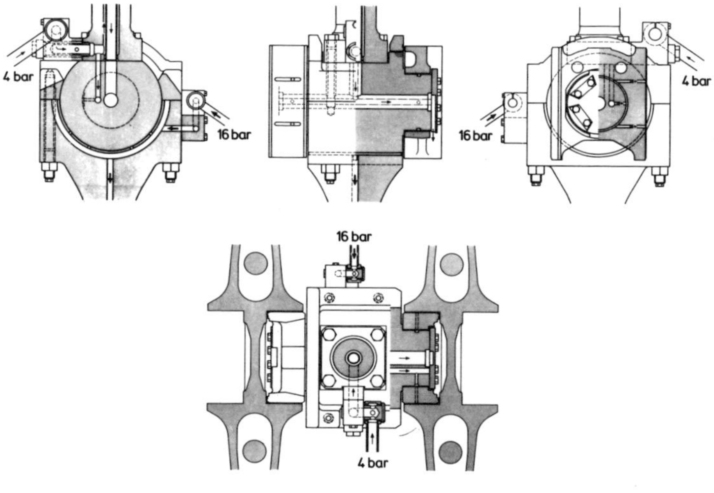

Figure 17-18. Crosshead arrangement for RTA38 and 48 engines

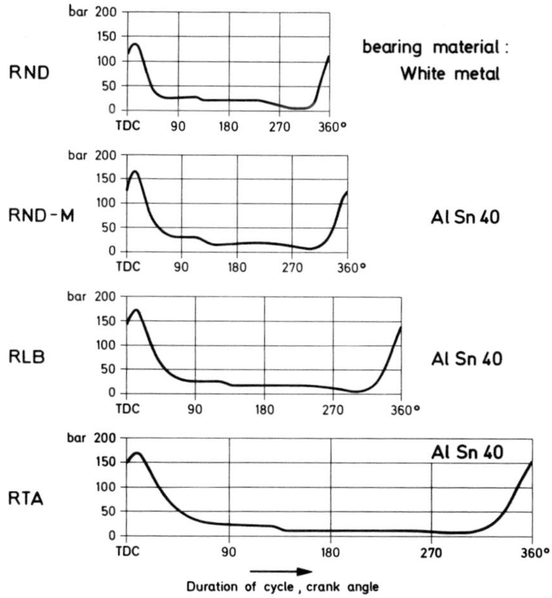

Figure 17-19. Maximum bearing loads for R-type engine crossheads

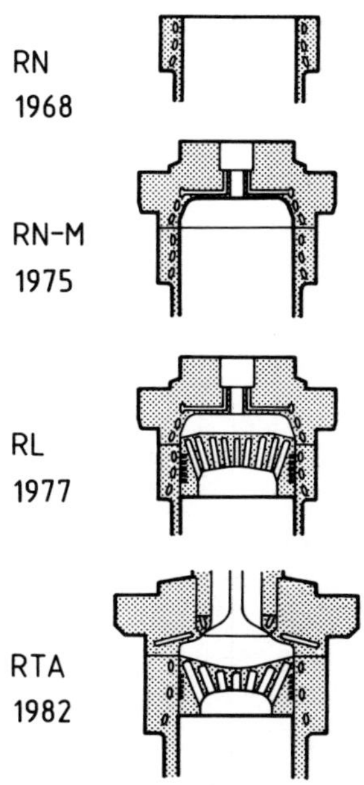

Figure 17-20. Evolution of the use of bore cooling in Sulzer low speed engines

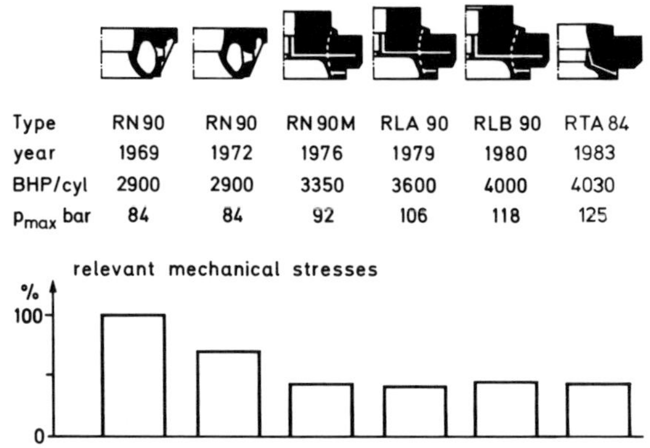

Figure 17-21. Trend in cylinder head stress of Sulzer low speed engines

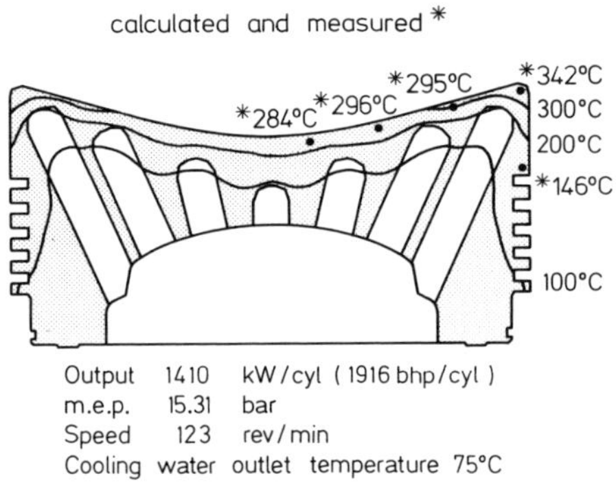

Figure 17-22. Temperatures in an RTA58 water-cooled piston

Cylinder Liner and Piston

The cylinder liner is of alloy cast iron with a thick, bore-cooled collar at the top to provide the strength to withstand the gas pressures.

Cylinder lubrication is by eight quills, each with its own spring-loaded accumulator, arranged around the liner about a fifth of the stroke below the TDC position of the top ring. The cylinder oil is metered by pumps, one for each cylinder, driven by hydraulic motors. The hydraulic motors are driven by bearing oil, with their speed regulated by an adjustable linkage from the fuel rack, giving adjustable but load-dependent oil feed rates. The discharge of each metering pump is separately adjustable, enabling a cylinder with new liner or rings to be fed at a greater rate.

Piston crowns are forged alloy steel, carrying five cast iron rings free to float in their grooves.

The arrangement of the water-cooled pistons is shown in Figure 17-15. Stationary supply and return pipes are housed in a watertight enclosure in the crankcase, while telescopic tubes attached to the piston reciprocate over them. The watertight enclosure precludes the possibility of lubricating oil being contaminated by water leakage, a problem easily revealed, since the enclosure is open to the exterior at the bottom. Circulation of the water within the piston relies on the "cocktail shaker" effect.

Figures 17-23, 17-24, 17-25, and 17-26 show the arrangement of the oil-cooled pistons. Oil is supplied to a connection on the crosshead through hinged pipes (see Figure 17-16) and then rises through a bored passage in the piston rod, which also carries the return pipe. Circulation of the oil to the ends of the bores is forced by jets at the piston rod top, shown in Figure 17-24.

Piston Rod Packing Gland

Figures 17-15 and 17-27 show the packing gland and diaphragm separating the scavenge space from the crankcase. In way of the gland and extended out to one side of the engine, the diaphragm is doubled to form a neutral space, with separate packing rings at the scavenge space bottom and at the crankcase top. The neutral space permits the collection of dirty oil scraped from the rod.

Cylinder Head and Exhaust Valve

The bore-cooled, forged steel cylinder head used on the larger bore engines is illustrated in Figure 17-25. The major cooling passages are radial bores from the outer periphery, continued with nearly vertical bores into an

Figure 17-23. Underside of an oil-cooled piston crown for an RTA62

Figure 17-24. Top flange of an RTA62 piston rod showing jets for cooling oil

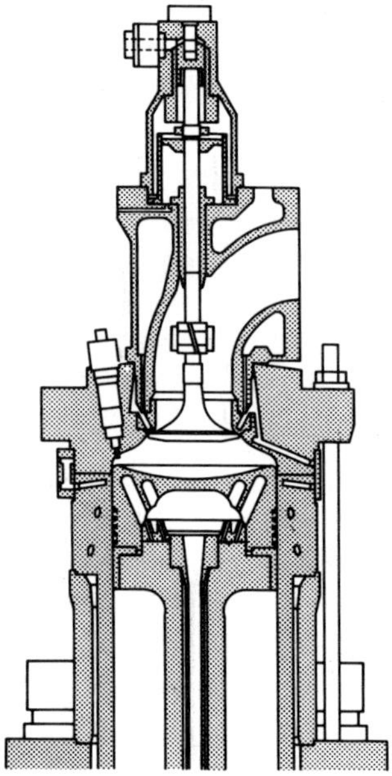

Figure 17-25. Combustion chamber with exhaust valve assembly for larger bore RTA engines

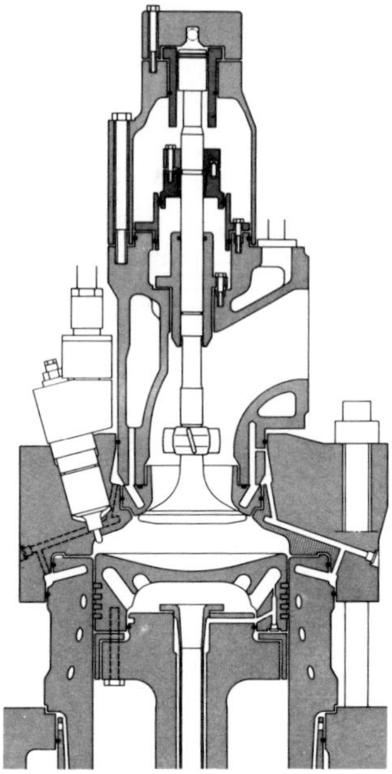

Figure 17-26. Combustion chamber with exhaust valve assembly for RTA38 and 48

annular space at the base of the exhaust valve pocket, as shown in detail in Figure 17-28. From the annular space blind bores in the exhaust valve cage carry the water close to the valve seat, while vertical bores lead up to the valve cage cooling jacket.

The head used on the RTA38 and 48 is shown in Figure 17-26. It differs from the head of the larger bore engines at the junction with the liner, where it overlaps to provide added support to the liner, and to form an integral water guide.

Bored pockets in the head carry the air starting valve, the fuel injectors, the relief valve, and the indicator cock. Depending on the bore, there will be two, three, or four injectors per cylinder.

The exhaust valve and its cage and hydraulic actuator are shown in Figures 17-29 and 17-30. The concept is similar to that used on the latest B&W and Mitsubishi low speed engines: the valve is opened by hydraulic pressure from the cam-driven actuator and closed by compressed air (an "air spring"). The valve is rotated by vanes on the stem. The valve cage is cast iron with a hardened steel valve seat ring, while the valve itself is of nimonic alloy. The lower portion of the cage, the seat ring, and the valve itself have full rotational symmetry to avoid thermal or mechanical distor-

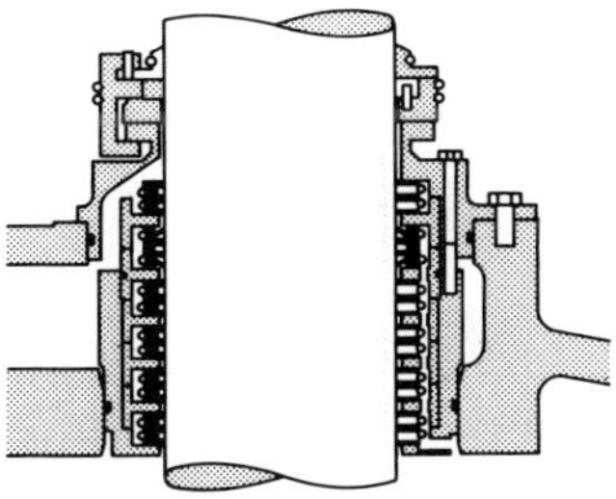

Figure 17-27. Packing gland for larger bore RTA engines

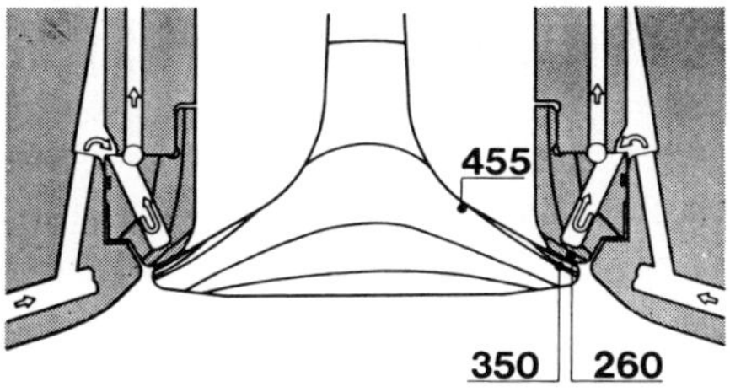

Figure 17-28. Exhaust valve and seat temperatures for a 7RTA58 at R1 rating

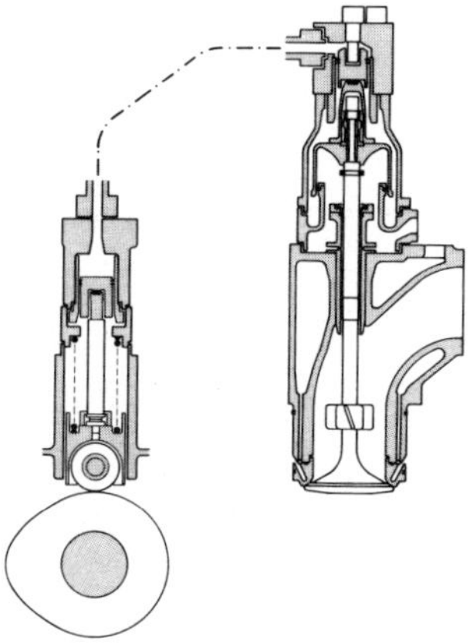

Figure 17-29. Hydraulic push rod system used on RTA engines

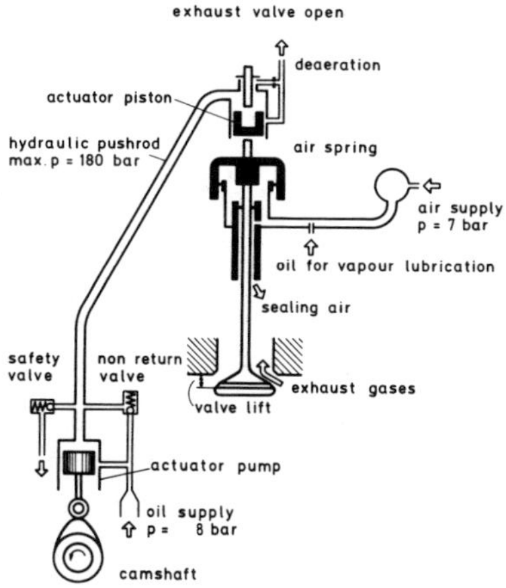

Figure 17-30. Schematic of push rod system and air spring arrangement

tion. This symmetry, the rotation of the valve, and the intensive cooling of the seat ring ensure that seating surface temperatures are evenly distributed and sufficiently low to avoid vanadium deposition, as indicated in Figure 17-28. The exhaust valve stems are lubricated with cylinder oil metered by an additional pump on the cylinder oil pump shaft.

Camshaft

The camshaft is gear-driven from the crankshaft, using three gears on the larger bore engines where the camshaft is at midheight on the engine, and four gears on the RTA38 and 48 (see Figure 17-31). For engines with up to eight or nine cylinders the drive is taken from the rim of the thrust collar; for large-bore engines with more than eight cylinders the drive is taken from the rim of the crankshaft coupling flange at midlength of the engine. To isolate the camshaft from the effect of torsional vibration, a Holset damper is fitted at the camshaft or, where engine-driven balancing gear is fitted, a Geislinger coupling is inserted in the drive train.

The camshaft extends the full length of the engine and is assembled in segments, each of which carries cams for exhaust valve actuators and fuel injectors serving a pair of cylinders, and each of which is mounted in a separate cam box. The arrangement is shown in Figures 17-32 and -33.

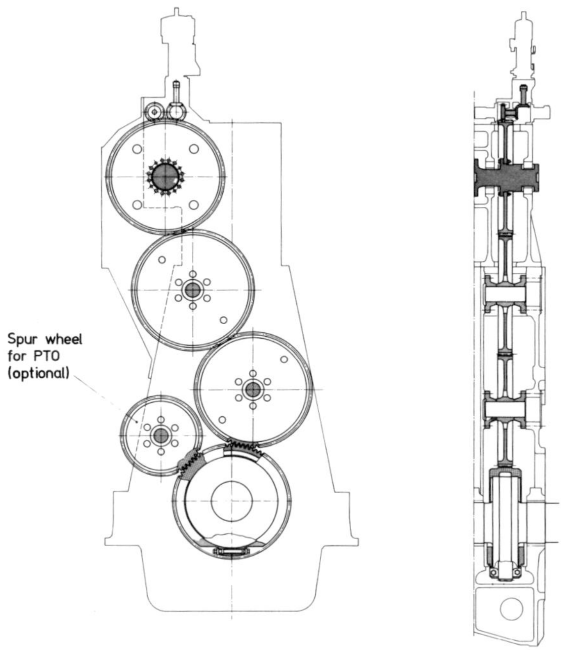

Figure 17-31. Camshaft drive gearing for the smaller bore engines

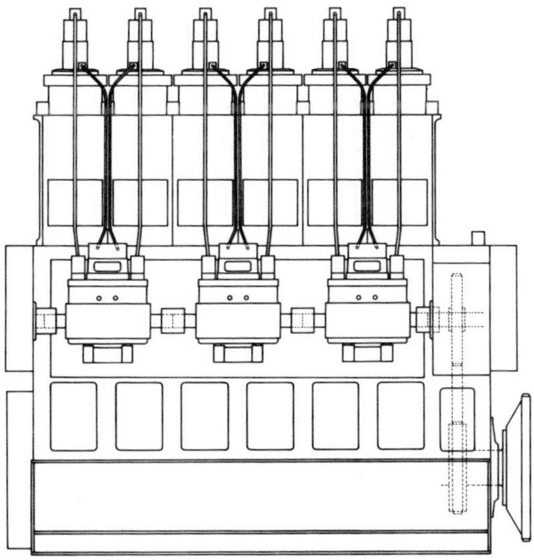

Figure 17-32. The fuel pump side of the large RTA engines

Figure 17-33. Camshaft unit with fuel pumps and exhaust valve actuator

The exhaust valve timing is sufficiently symmetrical about BDC to remain unchanged when the engine is reversed, but this is not true for the injection timing. Consequently the fuel cams are rotated relative to the camshaft for reversing. To accomplish this the exhaust cams are arranged as the outer pair in each cam box, as shown in Figure 17-33. The fuel cams for each pair of cylinders are combined with a hydraulic servomotor which turns with the shaft (see Figure 17-34). Oil is supplied to the servomotor to rotate the fuel cams relative to the camshaft, through a stationary sleeve on the larger engines (Figure 17-34), or through drillings in the camshaft on the RTA38 and 48 (Figure 17-35). The oil supply is from the same high pressure lubrication circuit that serves the crosshead bearings.

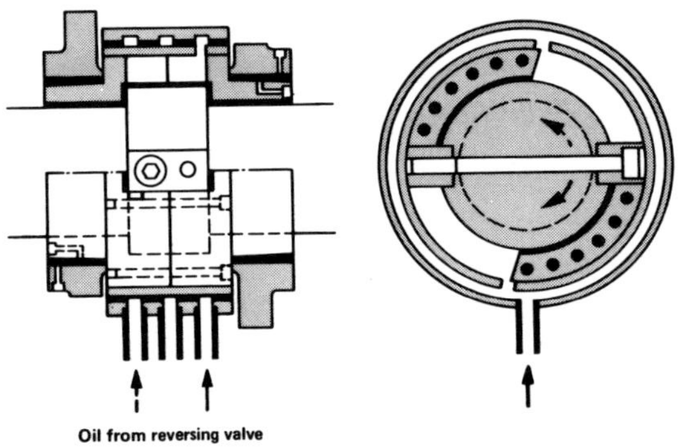

Figure 17-34. Fuel cams mounted on hydraulic servomotor for reversing

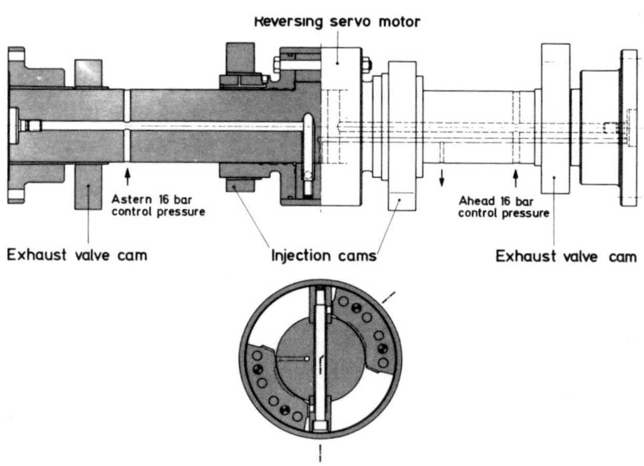

Figure 17-35. RTA38 camshaft showing drilled passages for reversing oil supply

The camshaft drives the governor, the pilot air distributor, the cylinder oil hydraulic pump, remote and local tachometers, and the rotation direction safeguard.

Fuel Injection

Unlike most diesel engines, which are fitted with injection pumps of the helix-controlled type, but continuing Sulzer's own long-standing practice, the larger bore RTA engines are fitted with valve-controlled pumps (see Figure 17-36). Injection begins when the suction valve is closed and continues until the spill valve is opened. In a system first applied to the RLA engines and shown in Figure 17-37, the control shafts of the two valves are linked together, enabling the timing of the beginning of fuel injection, as well as its duration, to be varied according to the engine output. In the Sulzer system, this variable injection timing (VIT) is done automatically by the governor, and enables the maximum combustion pressure to be held constant at the MCR value down to 85 percent output, and at higher values than would otherwise be obtained throughout the load range (see the upper curves of Figure 17-38), thereby achieving reduced SFC at all partial loads.

Incorporated in the VIT linkage is a manually operated lever (the fuel quality setting lever in Figure 17-37) that will advance the fuel injection timing to compensate for the longer ignition delay period of fuels with poor ignition quality. The effects are shown in the lower curves of Figure 17-38.

The RTA38 and 48 are fitted with helix-controlled injection pumps. The recess in the plunger is shaped at the top as well as at the bottom to provide load-dependent VIT, while the spill helix is stepped to reduce injection pressure toward the end of the injection period and so avoid cavitation.

For all engines, the tappet space below the pump plunger is sealed to avoid lubricating oil contamination from fuel leakage. Each pump is provided with a manually operated lever to lift the tappet clear of the camshaft, enabling individual cylinders to be isolated from service.

Two to four fuel injectors are fitted to each cylinder, depending on the bore. The injectors are cooled by conduction through the cylinder head to cooling bores adjacent to the injector pocket: there is no separate injector cooling circuit. Figure 17-39 shows an injector equipped with a fuel circulation valve that is shown in detail in Figure 17-40. The circulation valve functions automatically to maintain flow under booster pump pressure through the high pressure piping and the injector, both when the engine is stopped and between injection strokes. This circulation ensures the availability of hot fuel right up to the nozzle at all times, facilitating pier-to-pier operation on heavy fuel.

Air and Exhaust System and Turbochargers

The RTA engines are constant-pressure turbocharged, using either Brown Boveri or Mitsubishi turbochargers with uncooled casings. One or two

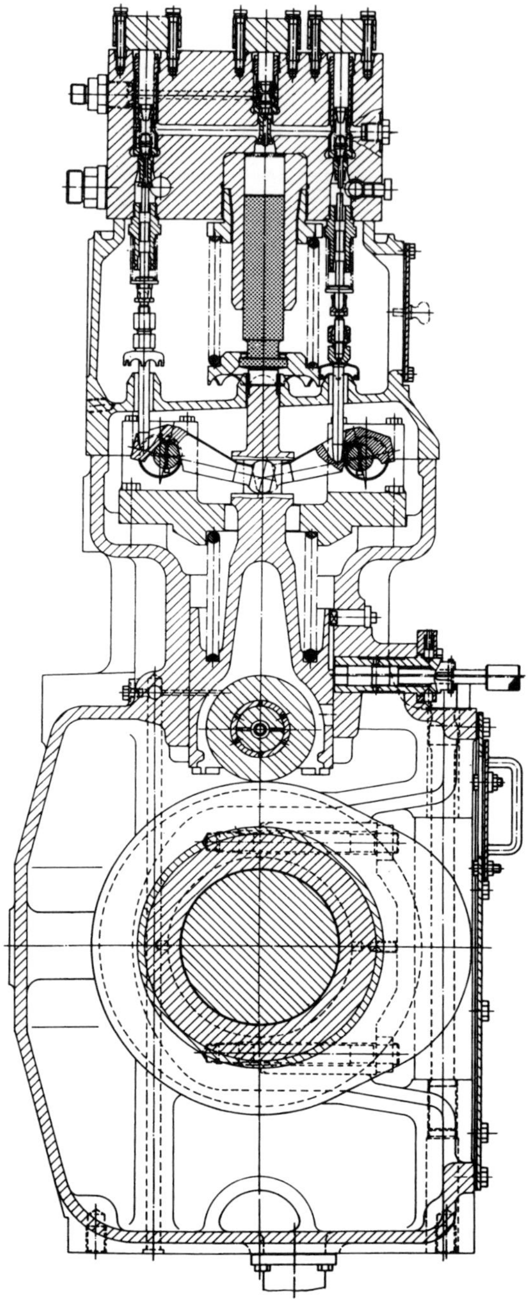

Figure 17-36. Valve-controlled injection pump used on the larger RTA engines

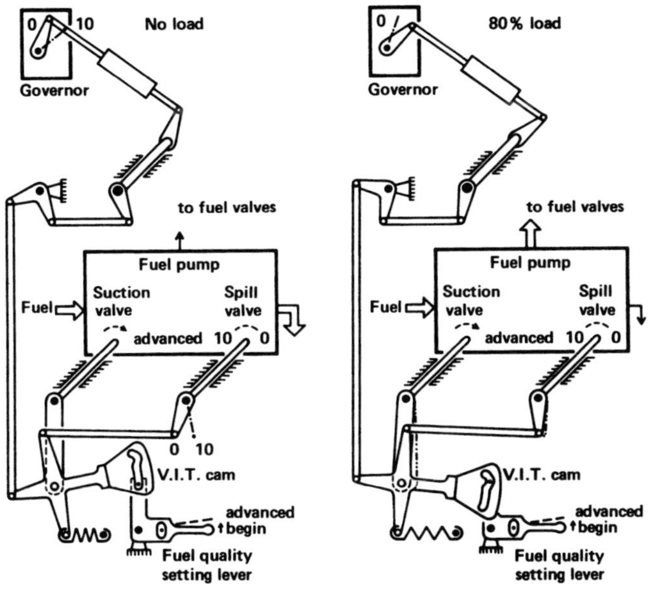

Figure 17-37. Variable injection timing mechanism

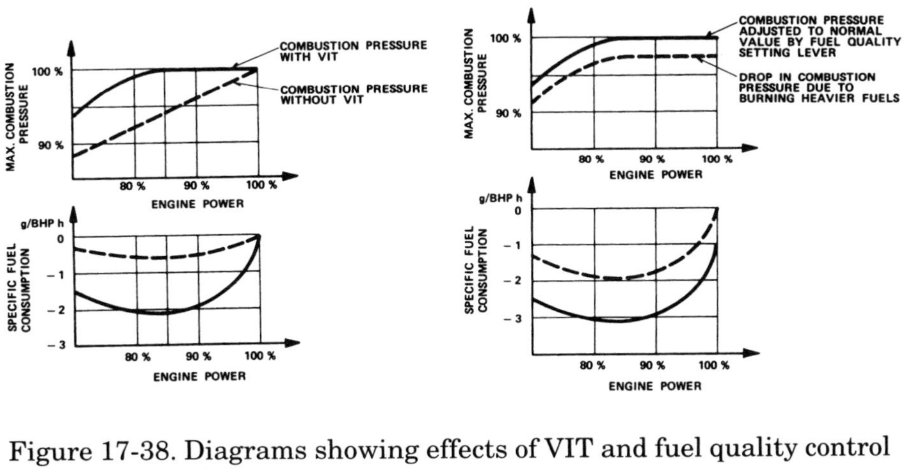

Figure 17-38. Diagrams showing effects of VIT and fuel quality control

turbochargers are fitted where possible but the 12-cylinder RTA84 and 84M engines require four. The uncooled casings result in a higher exhaust gas temperature after the turbine, which improves the potential for waste heat recovery. On the larger bore engines the turbochargers are mounted alongside the cylinders, above the exhaust manifold, as in Figure 17-2. For the RTA38 and 48, a single turbocharger is used in all cases and is mounted

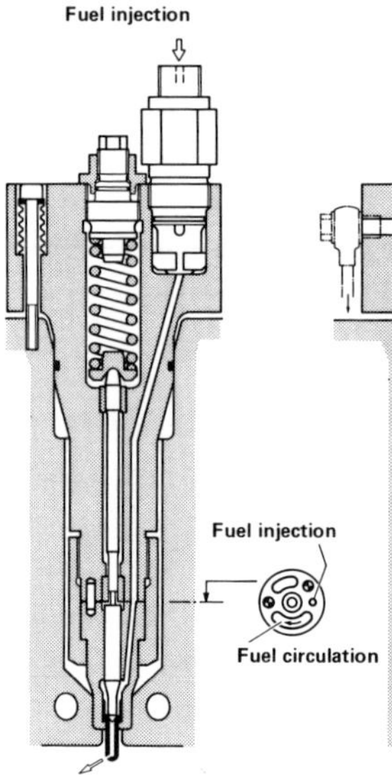

Figure 17-39. RTA fuel injector

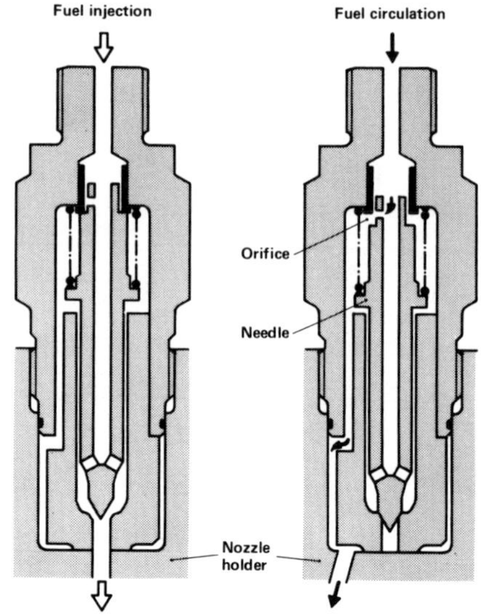

Figure 17-40. Detail of fuel circulation valve

at one end of the engine or the other, as in Figure 17-3, to suit the confines of engine rooms on smaller ships.

The air cooler is a finned tube unit, fitted one per turbocharger. The water side can be divided for circulation from up to three different sources to facilitate waste heat recovery. A water separator is fitted at the air manifold inlet from each air cooler.

The air manifold is bolted to the cylinder blocks and, on the larger bore engines, provides structural support for the exhaust manifold. On these engines the manifold is divided longitudinally by a bulkhead carrying reed-type check valves (see Figure 17-2). A pair of motor-driven boost blowers, mounted one at each end of the air manifold, cuts in automatically as engine output is reduced below about 40 percent of MCR. The boost blowers draw from the upstream side of the bulkhead, discharging to the downstream side, and are therefore in series with the turbochargers. On the RTA38 and 48 the air cooler, water separator, boost blowers, and check valves are grouped into a compact module at the end of the engine, below the turbocharger, as shown in Figure 17-41.

Balancing and Vibration

In common with competing B&W and Mitsubishi engines, the RTA engines have no unbalanced free forces, and unbalanced external couples of first and second order are minimized. In the standard configuration, additional counterweights are fitted to the crankshaft to correct the first order pitching moment, thereby imposing a first order yawing moment. For engines with seven or more cylinders external couples will not normally excite hull vibration, but external, first order couples of four-cylinder engines, and second order couples of four-, five-, and six-cylinder engines can; consequently, these engines may be fitted with balancers: second order balancers will almost certainly be fitted to the largest bore engines with six or fewer cylinders and may be fitted to smaller bore engines; first order balancers may be required for four cylinder engines. Figure 17-42 shows the arrangement of combined first and second order balancing gear used on the larger bore four-cylinder engines, where it is built onto the engine and driven through Geislinger couplings and chains from the timing gears and camshaft; when balancing gear is required for the smaller bore engines, independent motor-driven balancers are installed.

For installations where torque reaction moments excite hull or engine vibration, transverse stays, usually incorporating hydraulic dampers, are fitted at cylinder block height to connect the engine to adjacent ship's structure.

Where torsional vibration calculations indicate the likelihood of dangerous resonance levels at critical speeds, shafting diameters or flywheel mass may be modified, a barred speed range may be imposed, or a damper may be fitted at the forward end of the crankshaft. A damper may also be

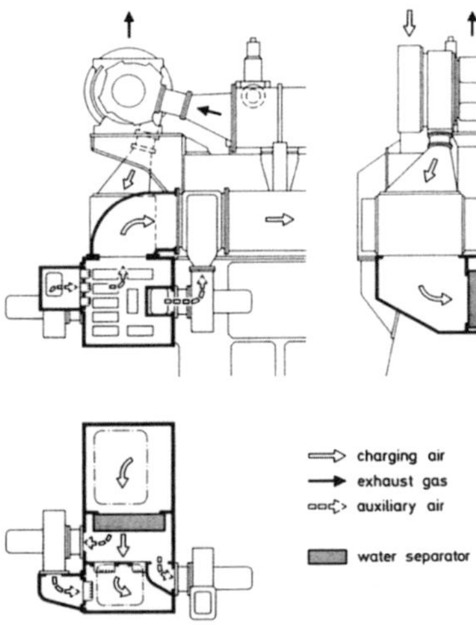

Figure 17-41. Modular arrangement of turbocharger and air cooler used on the RTA38 and 48

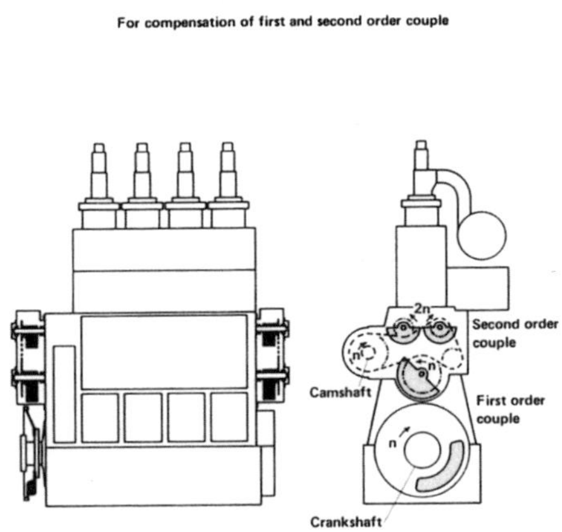

Figure 17-42. Combined balancing gear to compensate for first and second order couples

fitted at the forward end of the crankshaft if axial vibration is likely to pose a problem.

Support Systems

Support systems recommended for the RTA engines differ in detail from the typical systems described in the previous chapter. The following features may be worthy of note:

The remote-controlled starting air master valve incorporates a slow turning valve (item 17 in Figure 17-43) that will pass sufficient starting air to crank the engine over at a speed of five to ten RPM with the indicator cocks open.

The lubricating oil system, shown schematically in Figure 17-44 for engines with water-cooled pistons, includes a high pressure circuit for crosshead bearing hydrostatic lubrication, for reversing servomotors, and for exhaust valve actuators. The following recommendations are made regarding the lubricating oil systems:

1. For engines with water-cooled pistons, the system oil should be an SAE 30 oil, rust and oxidation inhibiting; a light alkaline oil can be used. For engines with oil-cooled pistons where the oil is subject to higher temperatures, the system oil should be a mildly alkaline SAE 30 detergent oil.
2. For normal service the cylinder oil should be an SAE 50, with alkalinity matched to the sulfur content of the fuel, supplied at a feed rate of 0.4 to 0.6 g/BHP-hr. When running in new liners or rings a straight mineral oil should be used, at a feed rate of about 1.3 g/BHP-hr.

The RTA engines use a pressurized fuel system (Figure 17-45) to prevent gas formation at the high temperatures to which the heaviest fuels must be heated for atomization. Fuel from the injection pump spill valves and from the injector circulation valves returns to the buffer tank via the venting unit.

The seawater system, for engines with water-cooled pistons, is arranged so that seawater flows to the lubricating oil cooler first, then the charge air cooler; after this, it divides to flow through the jacket water cooler and the piston water cooler, which are arranged in parallel with each other. For engines with oil-cooled pistons the flow of seawater is first through the lubricating oil cooler, after which it divides to flow through the jacket water cooler and the charge air cooler, which are arranged in parallel with each other.

The long-standing practices of waste heat recovery using jacket water for the evaporator and exhaust gas for steam generation may be supplemented at the charge air cooler. The water side of the cooler may be divided for circulation from up to three different sources for this purpose: in one of many possible schemes, the hottest section may be used to heat water for domestic needs and other low temperature heating services, while the second section can serve as a feed heater for the exhaust gas boiler; the third section is reserved for seawater or central freshwater cooling of the charge air.

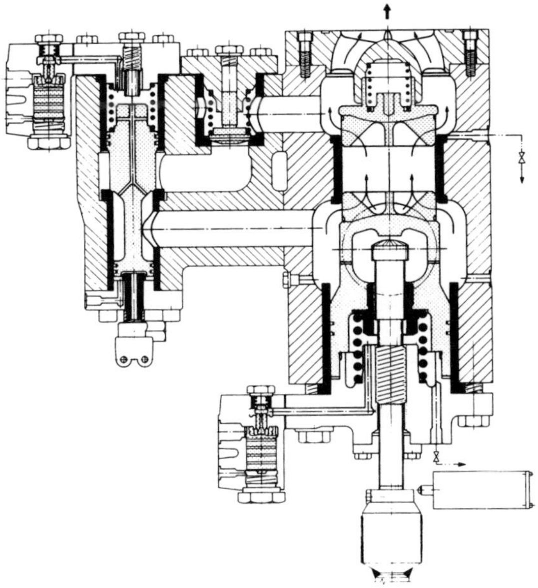

Figure 17-43. Starting air master valve

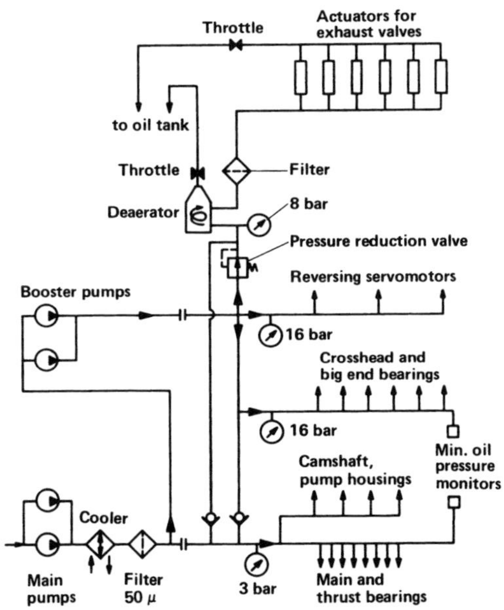

Figure 17-44. Lubricating oil system for RTA engine with water-cooled pistons

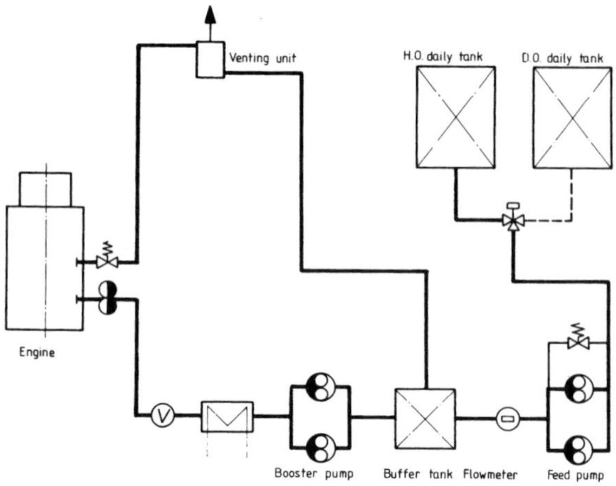

Figure 17-45. Pressurized fuel oil system

Control and Monitoring Systems

The standard control system supplied by Sulzer incorporates a remote pneumatic maneuvering unit, usually installed in the engine control room console, as well as in a local control stand at the engine. The system may be augmented with an electro-pneumatic bridge control system, with automatic programs for control of engine acceleration, deceleration, and crash maneuvers, or with special features to suit particular installations, such as those with controllable pitch propellers.

The pneumatic maneuvering unit in the control room includes a reversing lever, a start button, a speed setting lever, and a control transfer button. In its ahead or astern position, the reversing lever sends an air signal to the reversing control valve, which controls the oil flow to the servomotors at the fuel cams and the starting air distributor; the servomotors set these into their ahead or astern positions. The reversing lever also has a stop position, in which an air signal is sent to release a pneumatic shutdown servomotor; this servomotor shifts the injection pump suction and spill valve shafts to the zero injection setting. The speed setting lever adjusts the set point of the governor through a pneumatic link; the governor then operates a mechanical linkage that adjusts the injection pump suction and spill valve settings. Pushing the start button sends an air signal to open the starting air master valve, permitting starting air to flow to the distributor and the appropriate starting air valves. A rotation direction safeguard, driven by the governor drive gear, will release the shutdown servomotor if the direction of engine rotation is not the same as the direction called for by the reversing lever and reversing control valve.

For a normal start, the reversing lever is set to the correct direction and the speed setting lever is set to a low speed; when the start button is pushed, the engine begins to turn as starting air is admitted, and the rotation direction safeguard activates the shut-down servomotor, permitting fuel flow to the injectors: the engine accelerates until the governor takes control. To reverse the engine, the reversing lever is changed over to the opposite direction; since the rotation direction safeguard then senses the conflict between engine direction and lever setting, it releases the shutdown servomotor and fuel injection ceases. The operator can use starting air to brake the engine against the rotation of the propeller by pushing the start button, and can subsequently start the engine turning on air in the new direction; when the rotation direction indicator senses this, it activates the shutdown servomotor to permit injection, and the engine will accelerate until the governor takes control.

The local control stand can be activated using the control transfer switch. At this stand, the reversing lever, start button, speed setting lever, and control transfer button are again present and have the same functions, but the pneumatic linkages are reduced or eliminated: the reverse lever is mechanically linked to the reverse control valve, the start button actuates the starting air solenoid mechanically, and the speed control lever can either operate via the governor or be latched to a direct mechanical linkage to the injection pump suction and spill valve shafts. In addition, the pneumatic control of the starting air master valve can be bypassed using an extended valve stem adjacent to the local control stand.

When bridge control is fitted, the bridge console will have a single telegraph-style lever serving for starting, reversing, stopping, and speed control, and a control transfer switch.

In addition to the shutdown servomotor, each pair of injection pumps is fitted with a spring-loaded safety cut-out that, unless restrained by air pressure, holds the suction valves open to prevent fuel injection. These are usually arranged to release if any of the following events occur:

Overspeed
Emergency stop button pressed
Loss of air pressure in exhaust valve air springs
Bearing lubricating oil failure
Cylinder lubricating oil failure
Cylinder cooling water failure
Piston cooling water or oil failure

Air is blocked from the starting air solenoid when the turning gear is engaged. Sensors on the charge air manifold act at high pressure to limit fuel injection, thereby preventing excess torque and smoke, and, at low pressure in consort with a sensor on the fuel linkage, to start the motor-driven boost blowers.

The slow turning facility, which uses starting air supplied through a bypass on the starting air master valve to turn the engine (Figure 17-43), is operated from the control room console by push buttons.

Instrumentation and alarm points vary, depending on requirements of regulatory bodies and owner's preferences, but the standard engine design permits most parameters to be monitored locally or remotely, or to be fitted with alarms or other safeguards as appropriate.

Power Take-Off

An RTA engine may be fitted with an integral power take-off from the crankshaft gear wheel of the camshaft drive train (Figure 17-46) to drive a generator, as shown in Figure 17-47. In this case the generator is driven at constant speed by a Sulzer-SLM drive unit.

The constant speed drive unit, shown schematically in Figure 17-48, utilizes two epicyclic gear sets in tandem, and a superimposed hydraulic speed control system. The speed control comprises two hydraulic units that act as either pumps or motors, a constant displacement unit geared to an annulus of the first epicyclic, and a variable displacement unit on the planet carrier of the first set, which is also the annulus of the second set, together with a speed sensor and an electromechanical control unit. When the input speed is 100 percent of synchronous speed the annulus and planet carrier rotate at equal speed. If the main engine RPM is decreased so that input speed falls below synchronous, the displacement of the variable displacement unit is increased so that it acts as a pump, driving the other unit to maintain synchronous output speed; on an increase in input speed the variable displacement unit becomes a motor. For the RTA application,

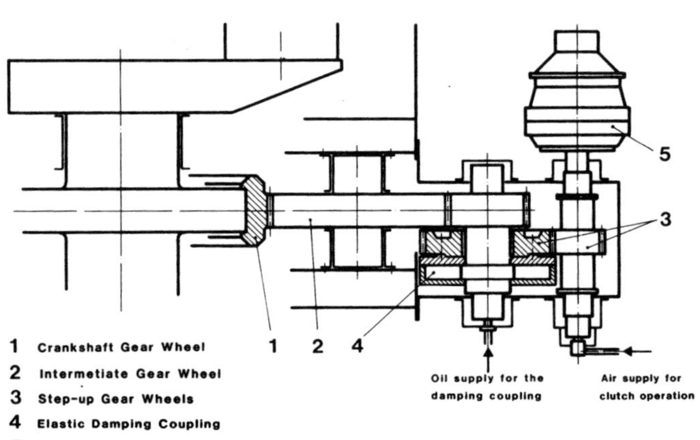

Figure 17-46. Integral PTO drive in horizontal section

Figure 17-47. Engine-driven generator on an RTA58

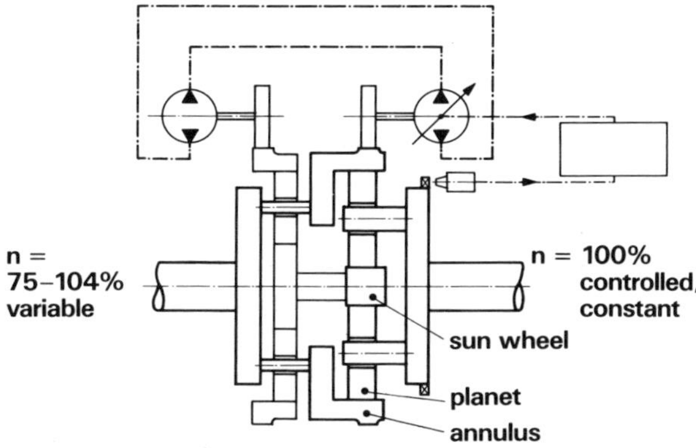

Figure 17-48. Schematic arrangement of the constant speed gear

these units are self-contained and intended for mounting, together with coupling and generator, on a base plate supported by brackets on the main engine bedplate.

Exhaust Gas Turbine

The introduction in 1984 of turbochargers that were even more efficient than the BBC series-4 units, which the RTA program was originally designed to use, opened the possibilities of increasing engine ratings to match the increased air flow or, alternatively, of diverting some of the exhaust flow to a power turbine. It was the latter course that was adopted for the already high-rated RTA engines, under the name "efficiency booster system." The standard configuration is shown schematically in Figure 17-49: the turbine is a single stage axial flow unit derived from turbocharger practice (Figure 17-50) that drives into the integral power take-off gear (Figure 17-51) via an epicyclic gear set and a fluid coupling. As can be seen in Figure 17-52, the reduction in SFC increases with the percent of gas diverted from the turbocharger, but so does the valve seat temperature. Diverted gas flow is limited to 10 percent for this reason.

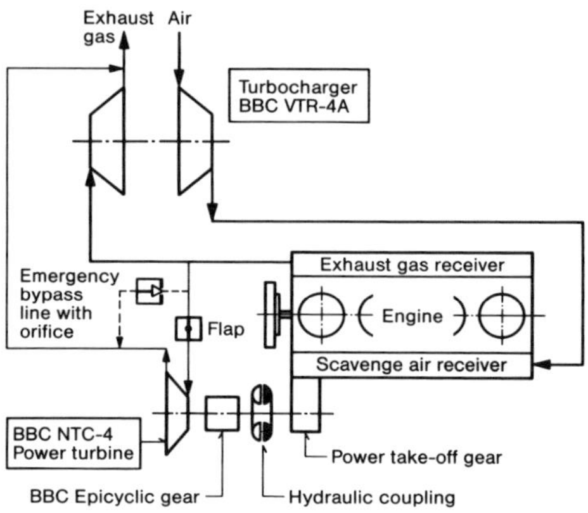

Figure 17-49. Schematic arrangement of efficiency booster system

Figure 17-50. Efficiency booster power turbine fitted to a 6RTA62

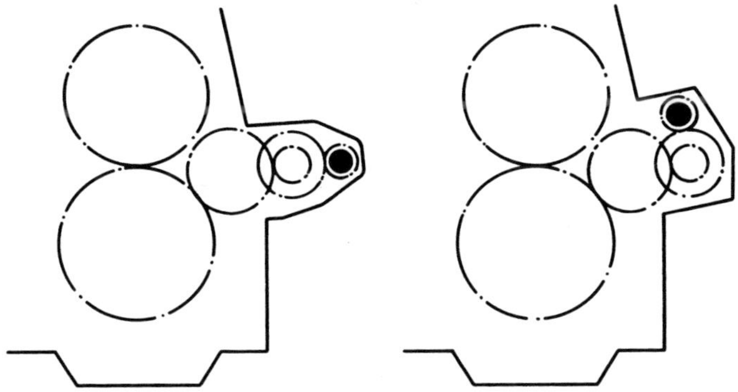

Figure 17-51. Integral PTO drive of the RTA engines

For partial output operation, at about 50 percent or less, the gas flow to the turbine is closed so that all of the exhaust passes to the turbocharger. The step improvement in partial output fuel rates and airflow is evident in Figure 17-53.

In Figure 17-54 the case is made that the additional power and improvement in efficiency resulting from the combination of a more efficient turbocharger with an exhaust gas turbine can render an engine so equipped competitive with a larger but derated engine: in this example a 6RTA62, derated to 75 percent output, can be replaced with a 5RTA62 matched at 90 percent output.

Maintenance

As with most other large diesel engines, the RTA engines are normally outfitted with an extensive array of hydraulic tools to facilitate maintenance tasks (see Figures 17-55 to 17-57). Such other specialized tools as valve seat grinders may be provided as well.

Guideline maintenance intervals for an RTA with water-cooled pistons are as follows:

Daily	Check and top up governor oil level. Check drain outlet on air manifold. Wash turbocharger compressor.
500 hours	Undertake spot test of system lubricating oil. Wash turbocharger turbine. Retension newly installed balancer drive chain (thereafter at 1,500-hour intervals).
1,500 hours	Check hold-down bolts for the first time (thereafter, at 6,000- to 8,000-hour intervals).

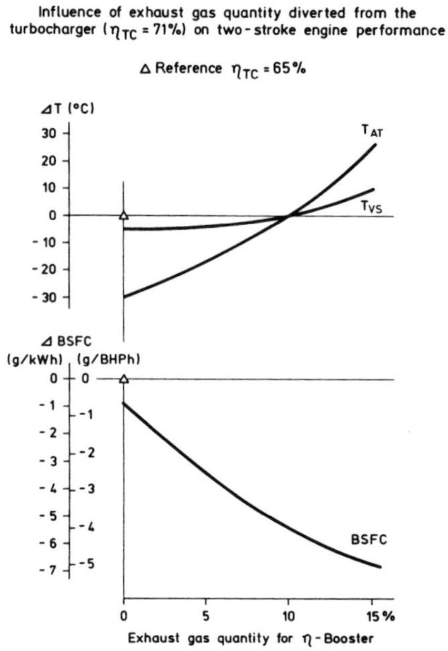

Figure 17-52. Influence of exhaust gas quantity diverted from the turbocharger

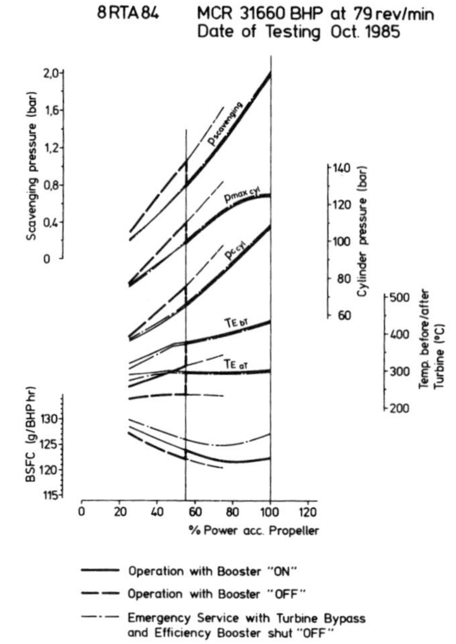

Figure 17-53. Test results according to the propeller law of an 8RTA84 with an efficiency booster system

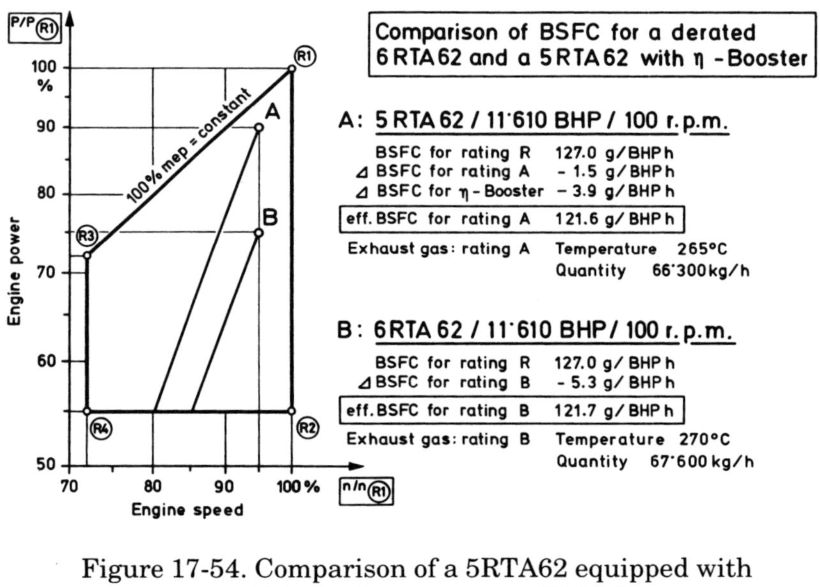

Figure 17-54. Comparison of a 5RTA62 equipped with an efficiency booster and a 6RTA62

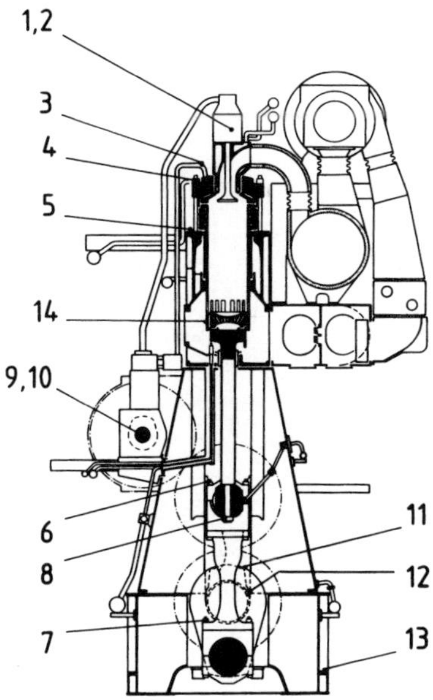

Figure 17-55. Hydraulic tools for the RTA engines

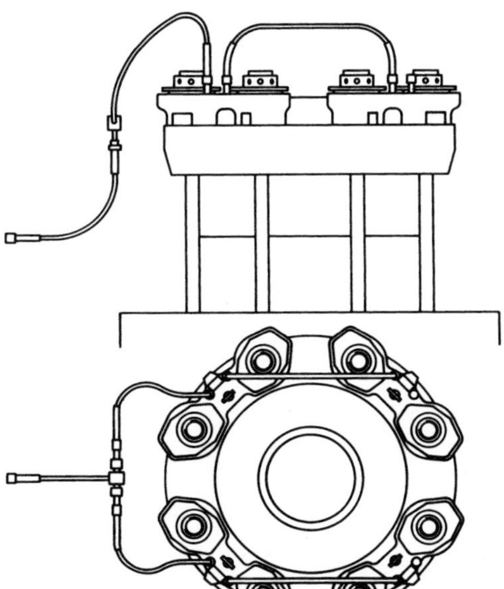

Figure 17-56. Tensioning jacks for the cylinder head studs

Figure 17-57. Cylinder head and exhaust valve cage for an RTA62

	Inspect fuel and exhaust cam surfaces. Check performance of starting air master valve. Check PTO gearbox externally for oil leakage and proper venting. Retension balancer drive chain.
3,000 hours	Sample system lubricating oil for laboratory analysis. Test fuel injector spray pattern. Inspect piston visually through the air ports. Check condition of camshaft drive gear teeth. Change governor oil after flushing casing. Inspect, clean, or renew check valves in air manifold. Check condition of PTO gear teeth and determine bearing clearances.
6,000-8,000 hours	Check hold-down bolt tension. Check main bearing clearance; adjust if necessary. Check tie-rod tension for the first time; adjust if necessary (subsequently at four-year intervals). Overhaul and regrind exhaust valve seats as needed. Measure crankshaft deflections. Check connecting rod bearing clearances. Check crosshead guide shoe clearances. Remove cylinder heads; withdraw and clean pistons; measure ring grooves; clean or renew piston rings; disassemble and inspect cooling water standpipes, telescoping tubes, and glands; measure liner wear; remove wear ridge from liner; check cylinder oil quills; inspect combustion side of head. Check backlash and clearance of intermediate camshaft drive gear. Check thrust bearing radial and axial clearances; check that bottom drain hole is clear. Dismantle, clean, and inspect valves of fuel injection pumps; check fuel pump timing and adjust if necessary; lubricate injection pump linkage. Clean air manifold. Open one piston annually to clean internal cooling spaces. Clean boost blower impeller and casing. Drain and wash out cylinder oil pumps. Check exhaust pipe bolts; clean sliding feet.

	Check or recalibrate pressure gauges and pyrometers.
15,000-20,000 hours	Check thrust bolt tension; adjust if necessary. Inspect connecting rod bearings and renew if necessary. Dismantle, clean, and inspect starting air distributor; check pilot valves for easy movement. Dismantle, clean, and inspect starting air master valve. Check performance of rotation direction safeguard. Dismantle and clean shutdown servomotor. Dismantle, clean, and inspect safety cut-out devices. Dismantle, clean, and inspect Geislinger coupling of PTO gear.
35,000-40,000 hours	Remove sample of silicon fluid from viscous vibration damper on camshaft.
40,000-60,000 hours	Renew spring packs of Geislinger coupling of PTO gear.
As necessary or when required	Remove and inspect main bearings and thrust bearing. Remove and replace cylinder liners; renew rubber rings. Chamfer edges of air ports. Recut cylinder oil grooves. Clean or remachine fuel injector seat. Clean and overhaul fuel injectors. Check, top-up, or renew turning gear lubricating oil. Check and clean undersides of pistons. Check and clean water separators in air manifold. Clean turbocharger air filter. Clean air and water sides of charge air cooler. Measure stretch of balancer drive chain; renew if necessary. Clean or renew oil filter inserts. Renew starting air valves.

SEMT-PIELSTICK PC2.5 FOUR-STROKE DIESEL ENGINES

Introduction and Evolution

The Pielstick-type PC medium speed diesel engine designed by the Société d'Etudes de Machines Thermiques (SEMT) has been manufactured by licensees around the world since 1951. The PC2.5-400 series, rated at 650 BHP per cylinder at 514 to 520 RPM, is the successor to the previous PC2.3 series, and is built in V configurations of 12, 14, 16, and 18 cylinders by the Fairbanks Morse Division of Colt Industries, and in both V and in-line configurations by other licensees. The PC2.5 has a bore and stroke of 400 mm and 460 mm, respectively. Table 2 shows external dimensions and weights of the Fairbanks Morse engines.

TABLE 17-2

External Dimensions and Weights of Fairbanks Morse PC2.5V Marine Engines

Number of Cylinders	*Length Overall*	*Width Overall*	*Height Overall*	*Approximate Weight, lb**
12	22′7½″	11′5¾″	11′5″	144,200
14	25′0″	11′5¾″	11′5″	161,200
16	28′0″	12′1¾″	12′8¼″	178,200
18	30′5¼″	12′1¾″	12′8¼″	195,200

*Dry weight for non-reversing engines without pumps

A cross section through a PC2.5 engine appears as Figure 17-58. Comparison with a section of a PC2.3, with the higher-rated PC2.6, or with the much larger PC4.2 reveals the direct line of evolution of these engines.

Frame

The engine frame is a welded fabrication of steel plate, castings, and forgings intended to provide great strength with minimum weight. The lower portion forms the crankcase, and is divided from the upper portion by a cast steel lower deck, clearly visible in Figure 17-58. The upper portion of the frame houses the bottom halves of the cylinders, which are inserted through the steel plate upper deck and secured by studs from the lower deck. Precision bored holes in the two decks, with a counterbore in the lower deck, ensure the 45-degree alignment of the cylinders. Upper and lower decks extend out to house the two camshafts.

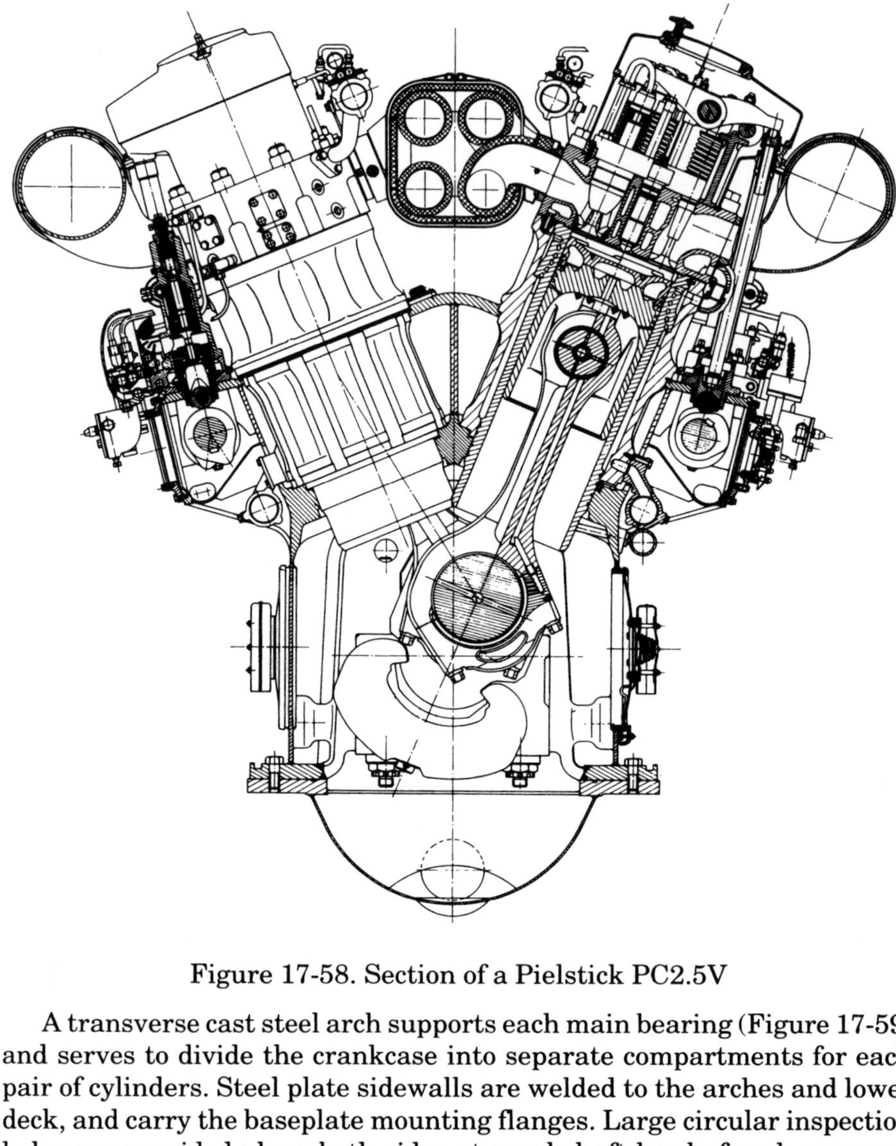

Figure 17-58. Section of a Pielstick PC2.5V

A transverse cast steel arch supports each main bearing (Figure 17-59) and serves to divide the crankcase into separate compartments for each pair of cylinders. Steel plate sidewalls are welded to the arches and lower deck, and carry the baseplate mounting flanges. Large circular inspection holes are provided along both sides at crankshaft level of each compartment to facilitate inspection and service.

The frame is extended at the drive end of the engine to house the timing gears and pump drives.

Main Bearings

A main bearing is shown in detail in Figure 17-59. Each saddle is inserted upward into its arch, and each is fastened by two vertical studs to the top

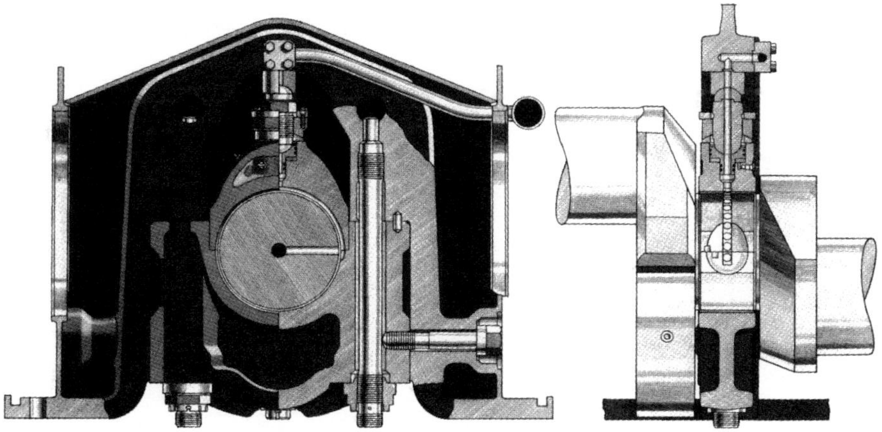

Figure 17-59. PC2.5 main bearings

of the arch and secured transversely by two lateral crossbolts. The bearing caps are each held in place by a single jack screw bearing down from the top of the arch.

Thin wall precision bearings are used; the mild steel shell is lined with a copper-lead alloy and a lead-tin overlay; backs and sides of the shells are lead-tin plated. The bearings are positioned within the housings by tongues that fit into grooves in the saddle and cap.

The lubrication arrangement can be traced in Figure 17-59. Oil from the external header passes down through the jack screw and cap to a circumferential groove extended over the upper 55 percent of the bearing, then through holes in the shell to a distribution groove in the lining. A radial bore in the crankshaft in way of the circumferential groove carries oil into an axial bore for distribution to connecting rod bearings, then through the connecting rods to the pistons.

In most applications an engine is fitted with an external bearing on the drive end to provide additional support for such external loads as a flywheel, a resilient coupling, shafting, or a generator rotor. The bearing is of the same materials as the internal main bearings, and is carried in a cast steel housing bolted to the frame. The sides of the shells are lined to serve as thrust faces operating against two collars machined onto the crankshaft, to maintain the longitudinal position of the crankshaft.

Crankshaft

The alloy steel crankshaft is forged in a single piece and machined all over. Coupling flanges are provided at both ends, with the flange at the free end available for barring gear, a torsional vibration damper, or power take-off.

At the drive end the crankshaft carries a shoulder flange for mounting the timing gear and a journal with two collars for the external bearing.

Passages are bored through the journals, webs, and crankpins to carry lubricating oil from the main bearings to the connecting rod bearings and the piston (see Figures 17-59 and 17-60). Dovetail slots for the balance weights are machined in the crank webs (visible in Figures 17-58 and 17-60). Each weight is secured against its slots by a single set screw in compression.

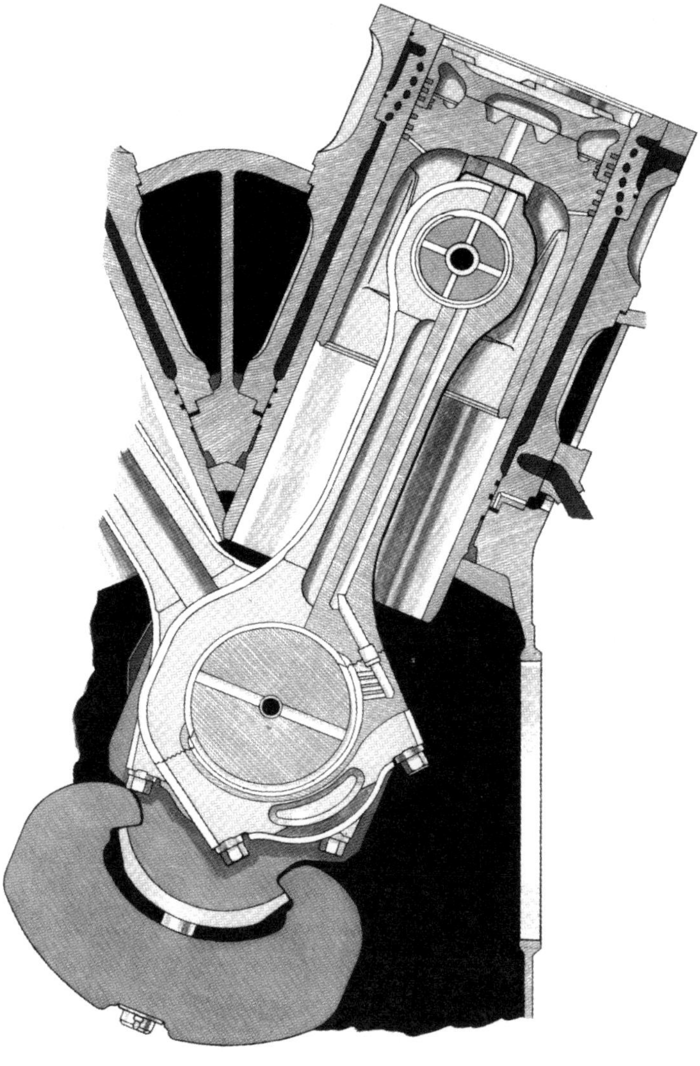

Figure 17-60. Connecting rod

Torsional Vibration Damper

Selection of the torsional vibration damper is made on the basis of a torsional vibration analysis of the engine and connected loads, including gearing, shafting propeller, and attached auxiliaries. Where a damper is required, either of the two types shown in Figure 17-61 might be fitted. In both types a heavy rim floats around a hub to which it is connected only by multiple spring packs; the rim is restrained axially by bolted side plates that overlap the hub. The mass of the rim and the stiffness of the springs are selected to provide the amount of damping needed.

Camshafts, Reversing System, Timing Gears, and Auxiliary Drive

Camshafts. Air and exhaust valves and fuel injection pumps are actuated by camshafts located outboard of each cylinder bank. The camshafts also drive the starting air distributor and a lubricating oil pump for the rocker arms. Individual cams for each service are keyed and shrunk onto the camshafts. The camshafts are underslung below the side extensions at the top of the frame (Figure 17-58) and are supported by sleeve bearings. At each cylinder, two sleeve bearings are fastened to the injection pump mounting, so that injection pump operating stresses are isolated from the engine frame. Bearings at the drive end aft of the timing gears include thrust faces for nonreversing engines; thrust faces for reversing engines are in the reversing lever bearings. The bearings are mounted to permit removal of a camshaft within its own length.

Reversing System. For direct-reversing engines, each cam has two profiles, one for each direction, connected by a sloping surface to support the cam roller when the reversing lever shifts the camshaft axially. The camshaft drive gear is axially restrained on the shaft and moves with it when reversing, its teeth sliding on pinions whose face width is extended for this purpose.

Timing, Timing Gears, and Auxiliary Drive. The camshafts are driven by two reduction gear trains, one for each camshaft, as shown in Figure 17-62. The large gear on the crankshaft drives two intermediate gears that turn at 1.6 times crankshaft speed. Each intermediate gear has an integral pinion that drives the larger camshaft gear at half crankshaft speed to achieve the cycle of events in Figure 17-63. The camshaft gears are resiliently mounted to isolate the camshaft from torsional vibration.

The governor, the overspeed trip, and a tachometer are driven by one of the intermediate gears through a step-up pinion; attached pumps for lubricating oil, jacket water, and raw water, and the fuel oil booster pump may be driven through step-up pinions from either intermediate gear.

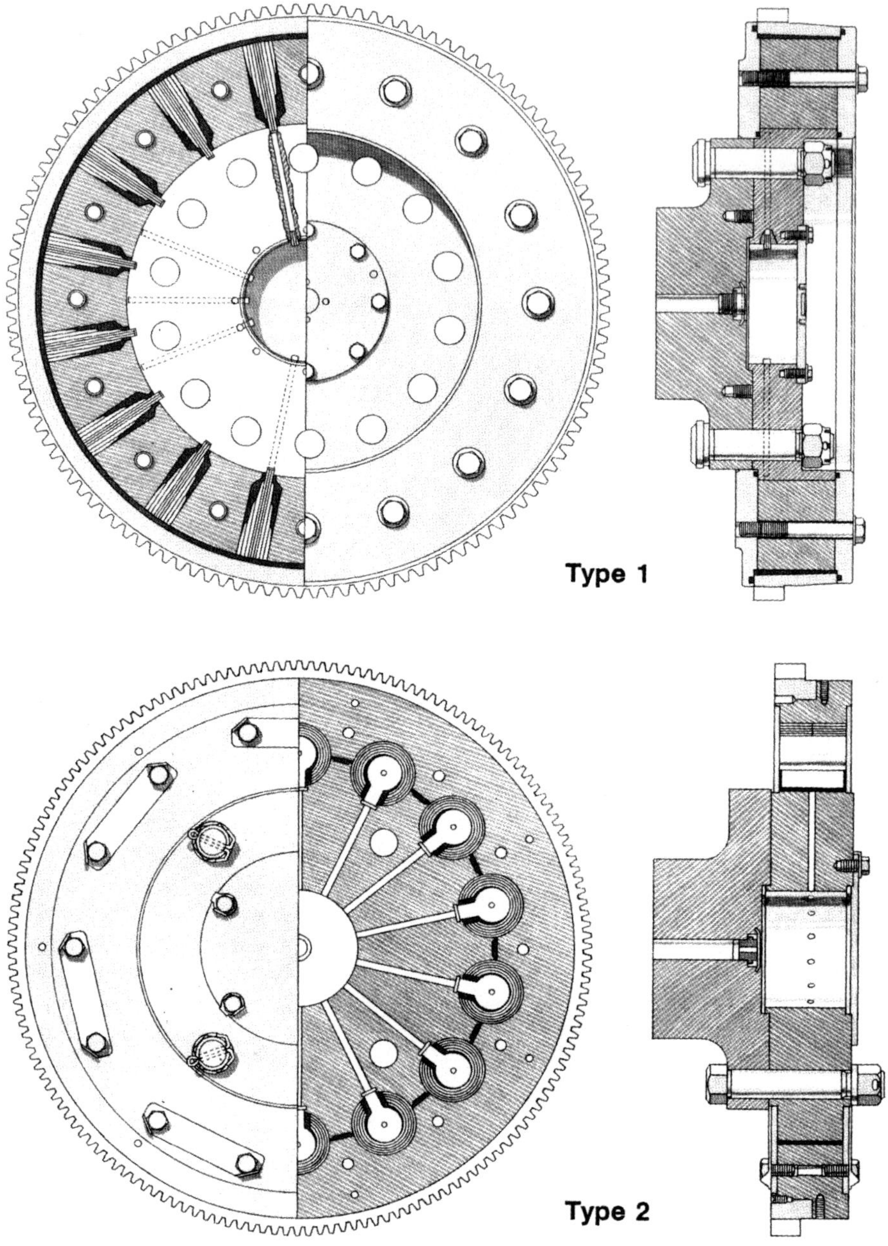

Figure 17-61. PC2.5 torsional vibration dampers

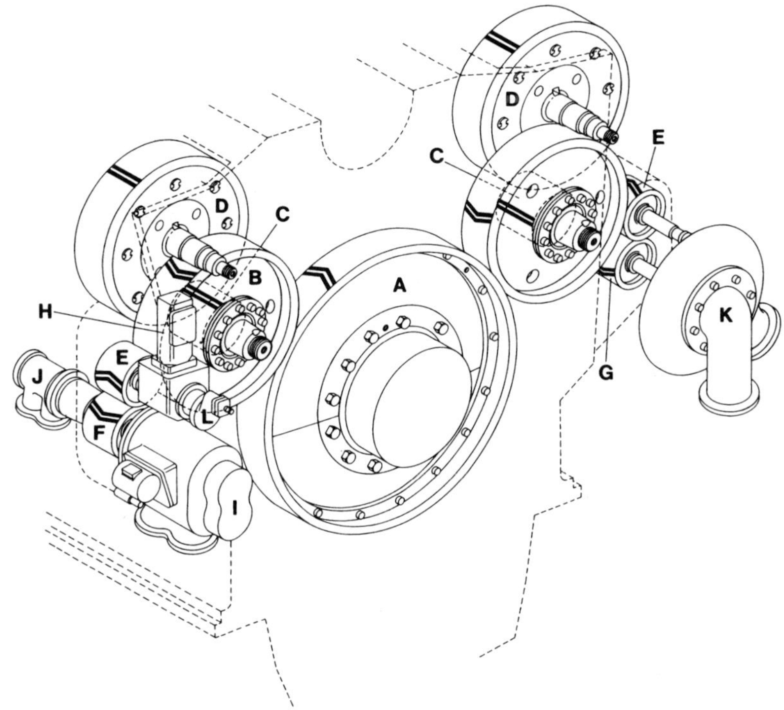

Timing Gear and Auxiliary Drives

A. — 168 tooth crank gear
B. — 105 tooth wheels—intermediate train
C. — 30 tooth pinions—intermediate train
D. — 96 tooth wheels on camshafts
E. — 36 tooth pinions
F. — 40 tooth pinion
G. — 35 tooth pinion
H. — Speed governor
I. — Lube-oil pump
J. — Fuel booster pump
K. — Water pump
L. — Overspeed trip

Figure 17-62. Timing gear and auxiliary drives

Cylinders

Figure 17-64 is a section through a complete cylinder assembly. Each cylinder consists of a centrifugally cast gray iron liner and a gray cast iron water jacket. The liner-to-jacket joint is made watertight at both ends by synthetic rubber O-rings that allow sufficient flexibility to accommodate relative movement due to thermal expansion and strain. The cylinder assembly fits into the frame to about half its length, and is clamped against the frame upper deck by the cylinder head, which is secured by eight studs

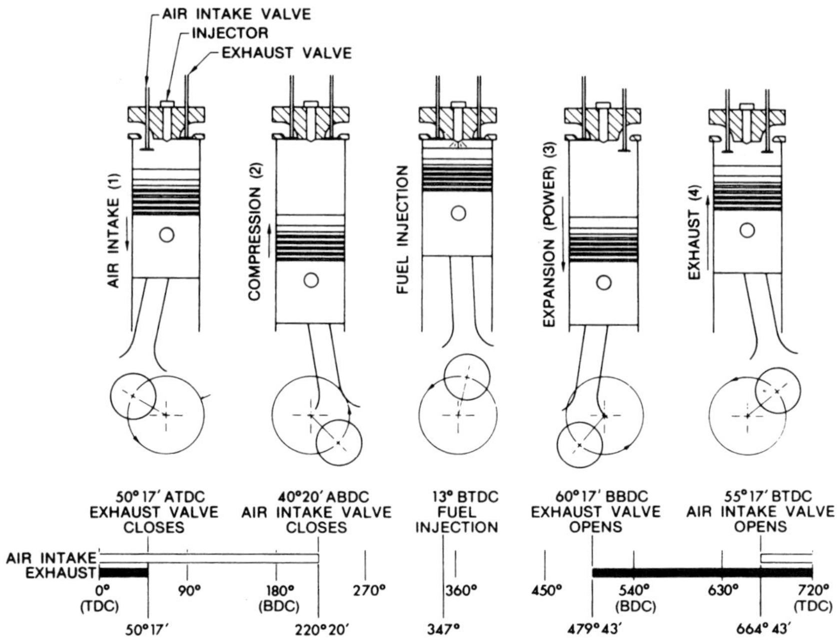

Figure 17-63. Timing diagram

threaded into the lower deck of the frame. The head bears on the liner, with a collar on the liner transmitting the clamping force to the jacket, which in turn has a shoulder bearing on the frame upper deck. A pilot shoulder on the liner fits a counterbore in the frame upper deck to maintain alignment. The liner extends below the jacket, through a bore in the lower deck of the frame. An O-ring seals the liner-to-lower-deck joint against water leakage from the cylinder jacket, which would be revealed by a drain from the frame above the lower deck.

Cooling water from the headers, exterior to the frame on each side of the engine, passes to the lower end of each jacket; it circulates around the liner to the top of the jacket, then through the bore-cooled collar of the liner, from which it passes into the cylinder head through an external duct. An air vent passage is provided between the jacket and the head at the high point, nearest the engine centerline.

Pistons

Two types of pistons are used in the engine: a two-piece piston with a forged steel crown and an aluminum skirt, and a one-piece cast iron piston (see Figure 17-65). All pistons in an engine are of the same type.

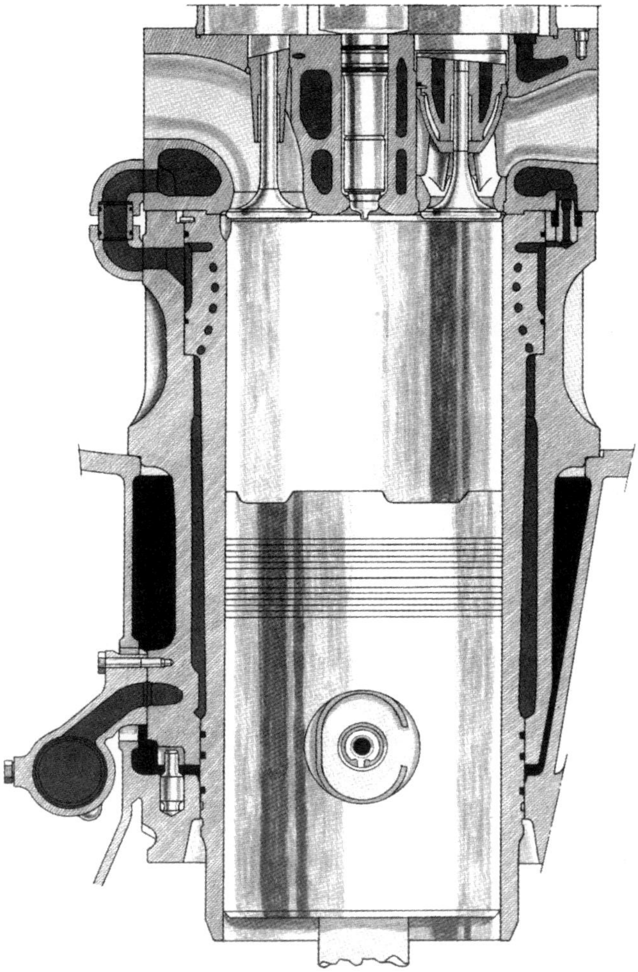

Figure 17-64. Section through cylinder

Regardless of piston type, six piston rings are fitted, all located above the piston pin to ensure good lubrication of the skirt. The upper four rings are compression rings. The top ring is faced with chrome, the second with molybdenum, and the next two with copper. The lowest two rings are oil scraper rings, loaded by garter springs.

The piston pin is hollow and mounts directly in the piston bosses with a slight interference fit. Drilled passages in the pin and the bosses carry oil from the connecting rod to cavities in the crown. Oil flow from these

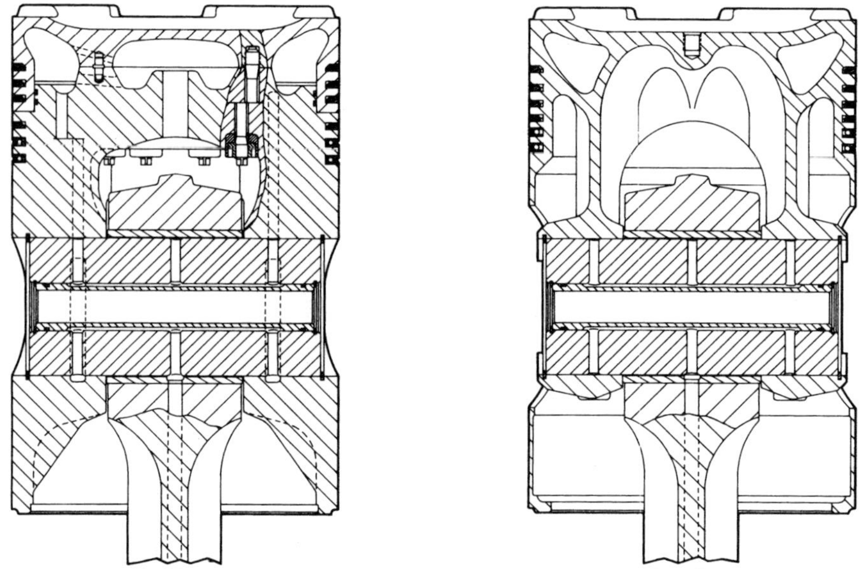

Figure 17-65. Pistons

cavities is controlled to ensure that sufficient oil is retained to cool the crown by "cocktail shaker" action.

Connecting Rods

Each pair of cylinders in a V shares a crankpin (see Figure 17-60) and, to accommodate two connecting rods side by side on a crankpin, cylinder transverse centerlines are staggered by a distance equal to the width of the connecting rod bottom end.

Connecting rods are made of drop-forged steel with an I-section shank. The crankpin bearing is split diagonally and is serrated, with the bearing cap secured by four cap screws. The bearing itself is of the thin-walled type, with steel shells lined with copper-lead, faced with a lead-tin flashing to reduce friction while running in. The backs of the shells are tin-plated to improve heat transfer. The piston pin bearing is a phosphor bronze bushing shrunk into the connecting rod.

The diagonally split crankpin bearing on the PC2.5 is an evolutionary development from the horizontally split bearing used on the 2.3, permitting an enlarged bottom end that accommodates larger cap screws to absorb the increased stress, while still allowing the connecting rod to be withdrawn through the cylinder liner.

Oil for the piston pin bearing and for piston cooling and lubrication passes from the crankpin through holes in the bearing shells to a circumferential groove in the cap, then through drilled passages in the connecting rod, as shown in Figure 17-60.

Cylinder Heads and Valves

Each cylinder head assembly includes two air valves, two exhaust valves, an air start/check valve, a relief valve, and an indicator cock, as well as a fuel injector.

Cylinder heads. Cylinder heads are hollow iron castings divided internally by intermediate decks, walls, baffles, and passages (see Figure 17-66). Cooling water enters the lower chamber of the head via an external jumper from the cylinder jacket. In the lower chamber the water is forced to flow at high velocity over the hottest surfaces. It then passes to the upper chamber through two holes at opposite sides of the intermediate deck, cooling this area of lower heat emission at lower velocity. Most of the water then flows through the exhaust valve cages before rejoining the rest and flowing to the outlet header via an orifice. The orifice ensures that water enters all cavities in the head and valve cages.

Air and exhaust valves. Two air valves and two exhaust valves are fitted in each head. Each valve operates in an alloy iron bushing carried in a cast iron body or cage secured to the top of the head by three studs. The air valve seats are machined directly into the bottom of the head while each exhaust valve is carried in a cage that includes the valve seat (Figure 17-67). Each valve is closed by two helical springs of opposite pitch, acting on the stem via a collar and a conical, split lock. The air valves and bodies are identical, and the two exhaust valves in each cylinder head are also identical except for the cages, which are mirror images.

Three types of exhaust valve assemblies are available for the PC2.5 engines:

1. The standard assembly, with water-cooled valve guide and Stellite-faced valve seat, suitable for engines burning light distillate fuel.
2. An upgraded assembly, with the standard cage but with a valve with a nimonic valve head and a Rotocap, suitable with fuel containing up to about 150 ppm vanadium.
3. The further upgraded assembly shown in Figure 17-67, where the valve seat is intensively water cooled and the valve with nimonic head and Rotocap is used. This arrangement is used with fuel of higher vanadium content.

The Rotocap device provides positive rotation of the valve to help keep seating surfaces free of deposits, to prevent hot spots, and to prevent

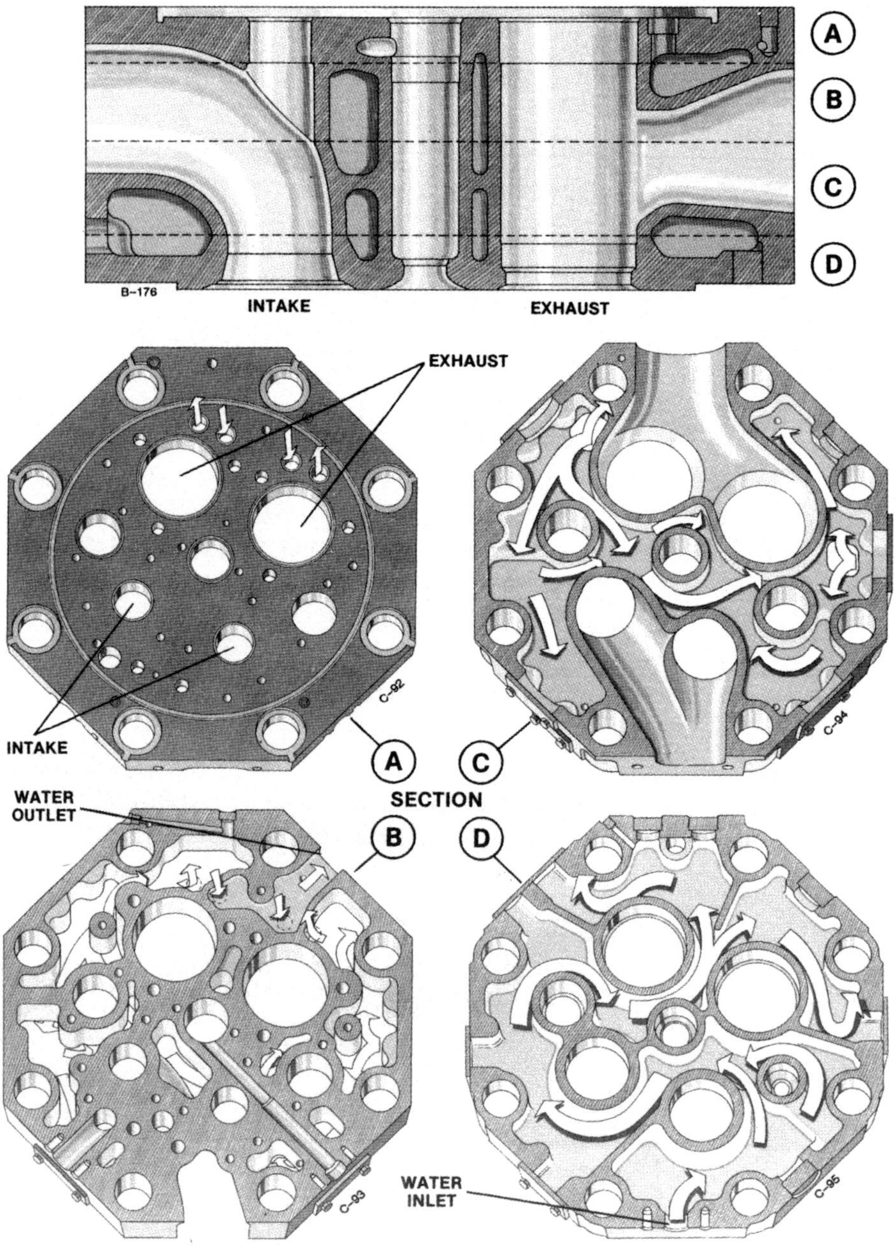

Figure 17-66. Cylinder head horizontal sections, showing cooling water flow

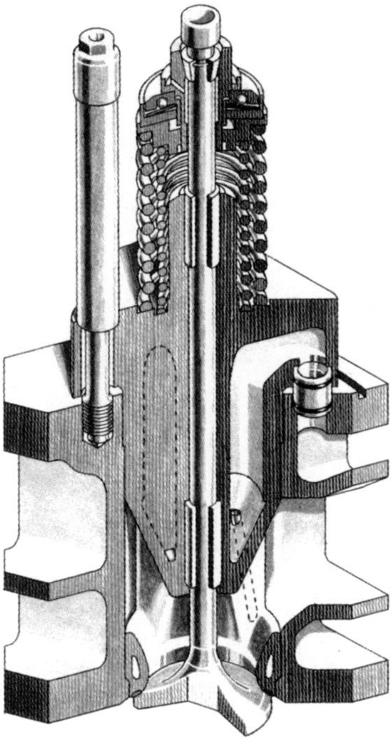

Figure 17-67. Caged exhaust valve

localized erosion. The nimonic valve head provides high strength and resistance to oxidation, corrosion, and creep at high temperatures. The water-cooled valve seat is designed to keep the temperature of the seating surfaces below that at which serious vanadium attack is likely.

Air and exhaust valves are operated by cams through push rods and rocker arms. Each exhaust valve rocker arm is forked to operate both exhaust valves from a single cam and push rod (see Figure 17-68). Both air valves on each cylinder are operated simultaneously by one rocker arm, which is keyed to one end of a pivot shaft (visible in the figure), which carries a half arm at the other end for the second valve. Tappet adjustment is provided at the tips of the rocker arms for each valve.

Relief valves. Each cylinder head on marine propulsion engines is fitted with a relief valve that will open at cylinder pressures of 2,100 to 2,200 psi. These valves can be removed for inspection and maintenance without disturbing the cylinder heads.

Air start/check valves. Air start valves are normally fitted to only one bank of cylinders. These are pilot-activated valves, and when opened by air from

Figure 17-68. PC2.5 cylinder head complete with rocker arms

the starting air distributor they admit high pressure air from the main air header to the cylinders in correct sequence to rotate the engine. They also function as check valves, closing against pilot air pressure as soon as cylinder pressure exceeds starting air pressure. Air start/check valves can be removed for inspection and maintenance without disturbing the cylinder heads.

Indicator cocks. Each cylinder head is fitted with an indicator cock for checking compression and firing pressures. These cocks can be removed for inspection and maintenance without disturbing the cylinder heads.

Fuel Injection System

A separate fuel injection system is fitted for each cylinder and consists of a cam-operated injection pump, a single injector centered in each cylinder head, and the interconnecting high pressure pipe. About twice the quantity of fuel as is burned at full load is supplied at low pressure to the injection pump, which meters the fuel according to the load and returns the excess to the day tank or mixing tank. The injection pump discharges the metered fuel into the high pressure pipe, at pressures that reach about 14,000 psi at maximum RPM, at the proper time to initiate the injection process. When the pressure upstream of the injector builds up to the proper pressure for atomization (about 3,500 psi) the injector opens, allowing the fuel to spray into the cylinder.

Injection pumps. An injection pump is shown in Figure 17-69. The barrel has two ports on opposite sides for fuel inlet at low pressure from the booster pump. The plunger has two helical recesses on opposite sides to control the amount of fuel discharged at high pressure. When the cam forces the plunger up, the top of the plunger closes the inlet ports in the barrel and fuel is discharged at high pressure through the check valve at the top of the barrel, until the helical recesses rise into alignment with the inlet ports. The duration of the discharge stroke, and therefore the amount of fuel discharged to the high pressure line, is established by the angular position of the plunger relative to the inlet ports in the barrel. This position is in turn established by rotating the plunger in the barrel by means of the pinion, engaged by the fuel rack, which is in turn positioned by the governor. The fuel rack is spring loaded to the zero position at each pump.

Individual plungers and barrels are lapped together for a precision fit and operate without rings or seals. They are interchanged only as matched sets. Small amounts of fuel leaking past the plungers are collected via gravity drains. The opposing inlet ports and plunger recesses prevent lateral forces between the plunger and barrel that would cause excessive wear.

The discharge check valve prevents the fuel in the high pressure line from returning to the pump during the down stroke. Timing of the start of the injection is adjustable at the tappet.

Injectors. An injector is shown in Figure 17-70. Each injector consists of a body, a spring-loaded, fuel pressure-actuated needle valve, and a nozzle. Spring loading is factory set to permit opening at 3,500 psi. Two studs secure each injector to the head through a gasket at a shoulder in the cylinder head bore close to the bottom of the head. The injector body is sealed to the bore in the head at the top by O-rings to prevent contamination of rocker arm lubricating oil. Needles are precision fitted to nozzle center bores and are interchanged only in matched sets. Small amounts of fuel leaking past the needles are collected via gravity drains.

The nozzle has ten orifices symmetrically oriented at a 70-degree angle to the vertical centerline. Orifice diameters of 0.60 mm and 0.65 mm are in use, with the smaller ones fitted where lower viscosity fuels will be burned.

Injectors are cooled by water passages in the body and nozzle that connect to supply and return tubes above the top of the head. On heavy fuel engines the injector circuit is separate to enable the injector to be maintained at the correct temperature for injection, or to be heated in case the engine has been stopped without first changing over to distillate fuel.

High pressure pipes. Each injection pump carries a forged steel delivery fitting that has bored fuel passages and connects, at a fitting on the side

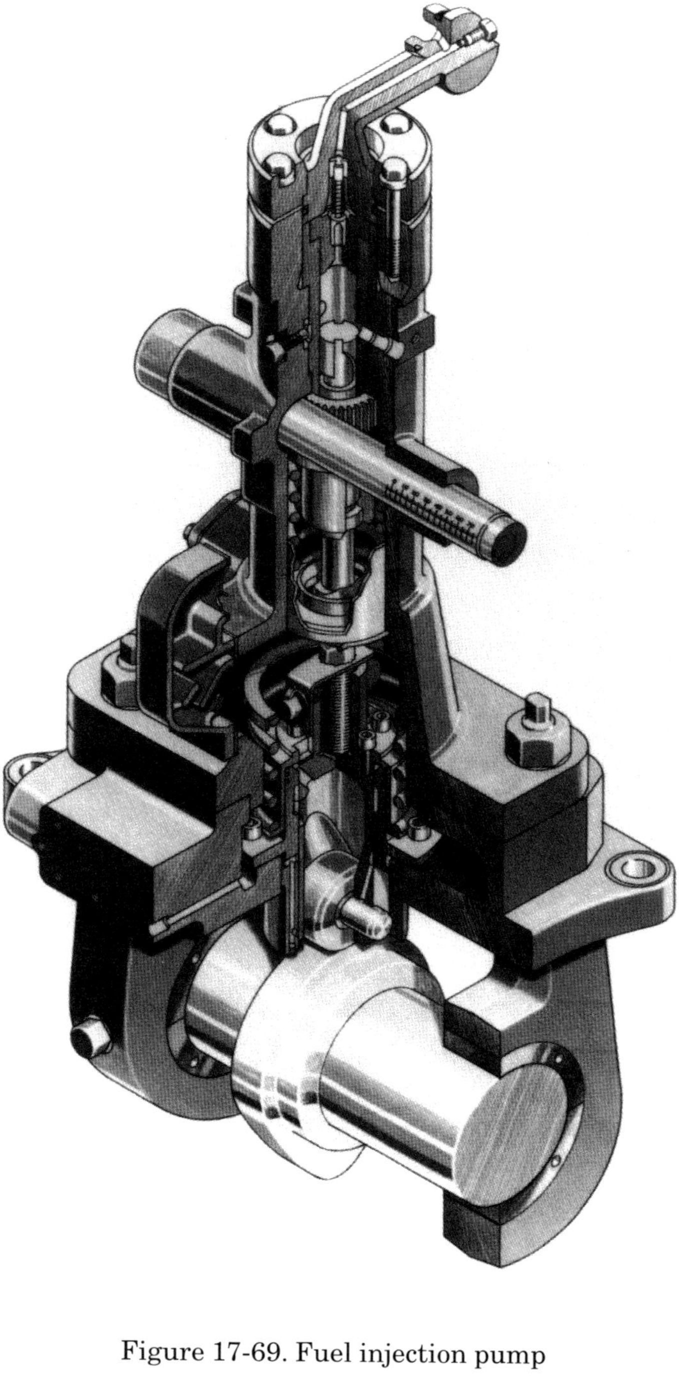

Figure 17-69. Fuel injection pump

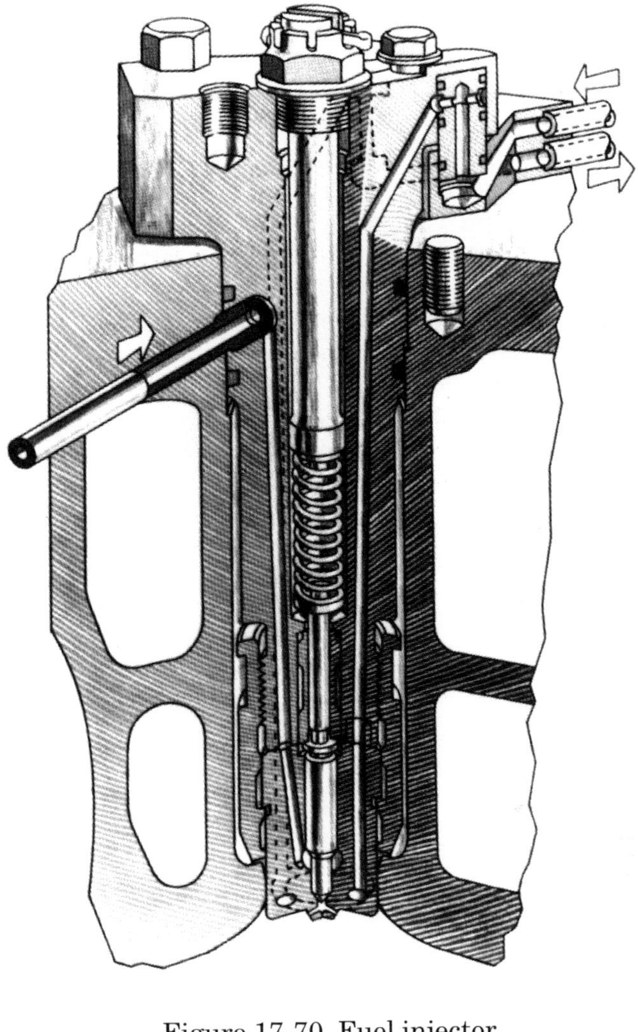

Figure 17-70. Fuel injector

of the cylinder head, with a heavy steel tube run in a bore in the head to the injector. The annular space between the bore in the head and the heavy tube provides a passage for fuel leaking past the injector needle to the gravity drain system.

Turbocharging

Each bank of cylinders is fitted with a turbocharger, an aftercooler, a moisture separator, and an air manifold, all independent of the other bank. Exhaust gas may be led from selected groups of cylinders in each bank to separate groups of turbine nozzles in a pulse-charged configuration. Alternatively, each cylinder may exhaust through a pulse converter to a common manifold for each bank, which is then led to an undivided nozzle block on the turbine. Turbochargers and aftercoolers may be mounted at either end of the engine but are usually at the drive end of marine propulsion engines. With reference to Figure 17-58, the air manifolds are outboard of each bank, while the exhaust lines are run in the insulated housing between the banks.

Turbochargers may be fitted for water washing of both the turbine and the compressor. In the recommended system, a one-liter tank is installed for compressor washing and a 200-liter tank for turbine washing. The tanks are filled with pure, untreated fresh water injected by compressed air. The compressor is washed with the engine at full power, the full charge being injected over a period of four to ten seconds; washing must be followed by at least another hour of operation under load. The turbine wash is done with engine load adjusted until the turbine is running at 3,200 RPM, the water being injected over a period of ten to twenty minutes, with the turbine casing drain open. After a turbine wash the engine should be run under load for at least a further thirty minutes.

Control and Monitoring Systems

Propulsion engines are usually arranged for remote control from consoles in an engine control room and on the ship's bridge, with a local control fitted for maintenance and emergency use. Engine controls are part of a larger system that may include controls for clutches, shaft brakes, controllable propeller pitch, and attached auxiliaries, as appropriate. The remote control system may or may not be supplied by the engine builder.

The engine functions that are operated by the control system are starting, stopping, speed, and, on reversing engines, direction. Starting, stopping, and reversing controls are pneumatic. In some cases the pneumatic controls interface with solenoid-actuated pilot valves for electric control. Speed control may be pneumatic or electric.

Starting system. An engine is started by admitting compressed air to one bank of cylinders only, in the timed sequence of their firing order. The starting system of a nonreversing engine is shown schematically in Figure 17-71. Prior to starting, valves between the air tanks and the engine are opened, directing air under pressure to the main air control valve and the three-way solenoid start valve. Starting is initiated by energizing the three-way solenoid start valve, which directs air through the barring gear interlock to the pilot piston of the main air control valve, opening the valve to pressurize the main air header. Pilot air also passes to the air cylinder of the pneumatic/hydraulic actuator, which positions the fuel racks for starting. When the main air control valve is opened, a second pilot air circuit is pressurized, bringing air to the air start distributor. The air start

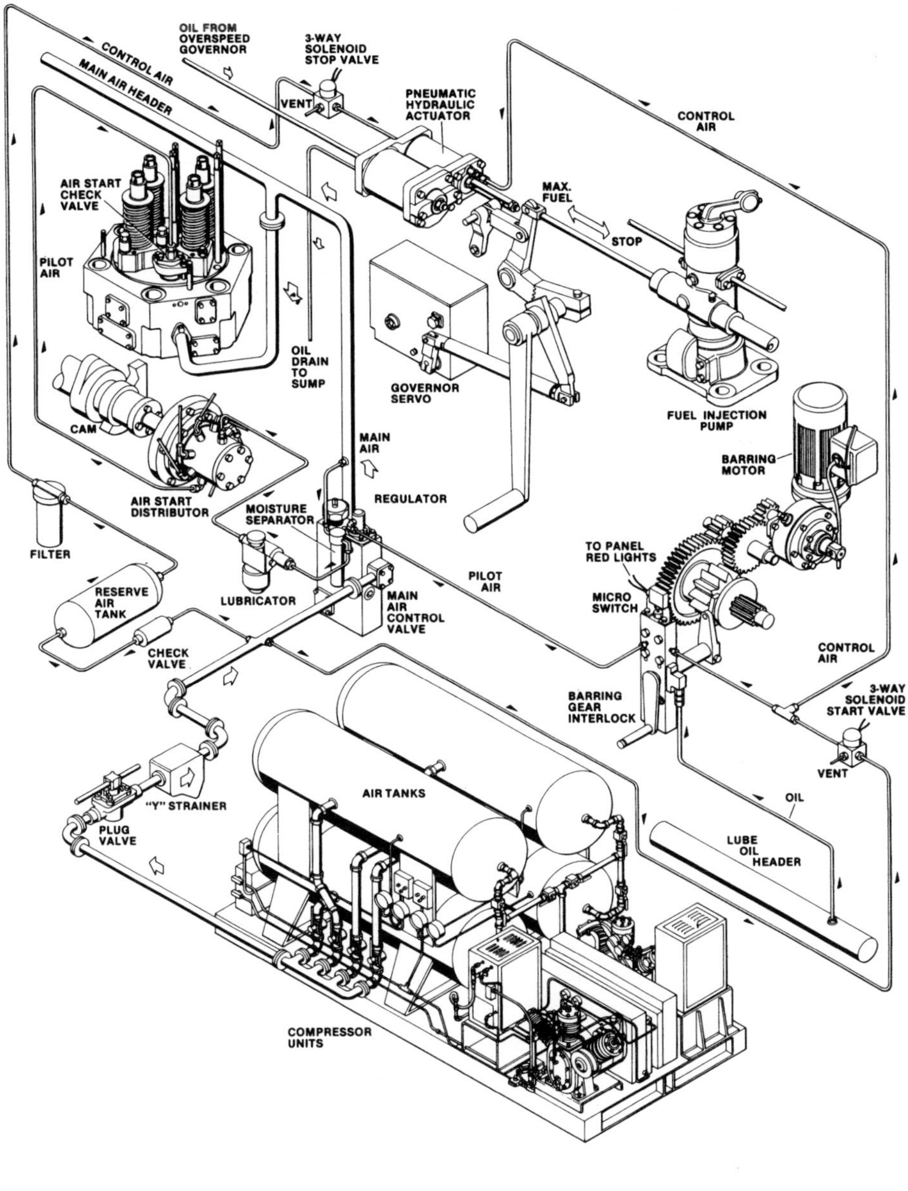

Figure 17-71. Air start system

distributor is driven by the camshaft of the bank fitted with air start/check valves, and determine the sequence by which pilot air is sent to open air start/check valves; these admit air from the main air header to each cylinder in the bank during its power stroke, turning the engine over. Each air start/check valve functions as a pilot-actuated air admission valve until the cylinder begins to fire, then acts as a check valve to prevent the discharge of combustion gas into the main air header. The engine will accelerate until the governor takes control of the fuel rack setting.

An engine can also be started manually by positioning the fuel racks using the lever visible in Figure 17-71, then using another lever to open the main air control valve. Manual operation bypasses the barring gear interlock.

For reversing engines the starting system is similar, but incorporates additional interlocks operated by a multiported pilot air valve called a reversing gear selector, which is mechanically linked to the crosshead of the reversing gear. In addition there is a rotation detector which mechanically senses the direction of rotation of the engine. One set of ports on the reversing gear selector is in series with the barring gear interlock, and ensures that air can reach the main air control valve pilot only if a camshaft shift is complete. The rotation detector is in series with the pneumatic/hydraulic actuator, and will prevent the fuel racks from being moved from the zero injection position unless the direction of rotation coincides with the direction for which the camshafts are set.

Stopping. An engine is normally stopped by energizing the three-way solenoid stop valve (see Figure 17-71), which admits air to the air cylinder of the pneumatic/hydraulic actuator, forcing the fuel racks to the zero injection position. The reserve air tank and check valve ensure a supply of shutdown air independent of the starting air receivers. An engine can be manually stopped using the lever on the fuel rack operating linkage.

Reversing system. The reversing levers for both camshafts are operated by a single hydraulic reversing cylinder mounted at the drive end of the engine. In the reversing cylinder the piston separates two closed oil supplies, one for each direction, each with its own small reservoir: control air pressure applied to either reservoir displaces oil into the cylinder, moving the piston from one end of the cylinder to the other, in turn shifting the reversing levers and the camshafts. Once its stroke is completed, the reversing piston, and therefore the levers and camshafts, is locked in position by an oil-actuated detent. Pneumatic interlocks prevent engine starting during camshaft shift and prevent camshaft shift while the engine is running.

Speed control. In normal operation engine speed is controlled by adjusting the set point of the governor, which acts on the fuel racks through a hydraulic servo. A knob on the governor permits manual adjustment of the set point. The connection between the servo and the fuel rack operating linkage is made by a spring-loaded link that is rigid in normal operation but can be overpowered to move the racks to the zero injection position.

Monitoring and automatic shutdown. Instrumentation and alarm points vary, depending on requirements of regulatory bodies and owner's preferences, but the standard engine design permits most parameters to be monitored locally or remotely, or to be fitted with alarms or other safeguards as appropriate.

Each engine is fitted with an overspeed trip, a mechanical device comprising spring-loaded flyweights driven by the timing gear train (see Figure 17-62). On overspeed, the flyweights overpower their springs and trip a pawl, releasing a spring-loaded valve that directs lubricating oil to the oil cylinder of the pneumatic/hydraulic actuator. The actuator then forces the fuel racks to the zero injection position. An electric switch on the valve senses the trip, and energizes the three-way solenoid stop valve to simultaneously shift the air cylinder of the pneumatic/hydraulic actuator to the zero injection position, providing a redundant shutdown system. The overspeed trip must be manually reset after tripping.

In addition to the overspeed trip, engines may be arranged to automatically shut down (depending on the requirements of the regulatory bodies involved) if any of the following events occur:

Emergency stop button pressed
Low lubricating oil pressure at engine inlet
Low rocker arm lubricating oil pressure
Low jacket water pressure at engine inlet
High jacket water temperature at engine outlet
Low injector cooling pump discharge pressure
High injector cooling water temperature at engine outlet
Low reduction gear lubricating oil pressure

Usually an automatic shutdown will occur only after alarms have sounded and, in the case of low pressure events, a standby pump has been started. Where appropriate the shutdown may occur only after a time delay. Also where appropriate, a manual override is provided to enable engine operation to continue, in situations that are critical to the ship, until the engine fails.

Support Systems

Support systems recommended for the PC2.5 engines differ only in detail from the typical systems described in the previous chapter. The following features may be worthy of note:

Some of the pumps in support systems may be engine driven from the timing gear train (Figure 17-62). Typically these would include the lubricating oil pump, the jacket water pump, and the seawater pump. In each case an electric motor-driven standby pump is provided, although in twin engine plants a single standby for each service may serve either engine. Engines run on distillate fuels only may have engine-driven fuel booster pumps, but heavy fuel engines are fitted with motor-driven booster pumps.

Engines burning heavy fuels are fitted with separate closed-circuit injector cooling systems that allow the injector to be maintained at the correct temperature for injection, or to be heated in case the engine has been stopped without first having been changed over to distillate fuel. The cooling water from each injector is individually piped to an overhead expansion tank that is fitted with a steam coil, from which a motor-driven pump takes suction discharging to the injectors via a seawater-circulated cooler. On distillate fuel engines, cooling water reaches the injectors from the jacket water system via a separate seawater-circulated cooler.

The seawater system is usually arranged with both of the air coolers and the injector water cooler in parallel, followed by the lubricating oil cooler and then the jacket water cooler in series.

An independent lubricating oil system is provided for rocker arm lubrication to prevent contamination of the main system by water or fuel leakage in the cylinder head area. The system includes a tank, camshaft- and motor-driven pumps, filter, and pressure-regulating valve. Oil is fed to the rocker arm pivot bushings and then passes through drilled passages in the pivot pins and rocker arms to reach the push rods and tappets, returning to the tank by gravity.

The recommended lubricating oil is an SAE 40 detergent, dispersant oil, with a principally paraffinic base. For heavy fuel engines the TBN should be in the 20 to 40 range, matched to neutralize the sulfur in the fuel oil.

Waste heat can be recovered from the exhaust gases, from the jacket water, and from the charge air coolers.

The engine crankcase is maintained at a very slight vacuum by a motor-driven fan that takes suction through an oil separator and discharges to a weather deck.

Fuel Oil Specification

PC2.5 engines may be operated on distillate fuel or, when the engines and supporting systems are correctly designed and outfitted, on heavy fuels with the following quality limits:

Maximum viscosity at 50°C	380 cSt
Maximum Conradson carbon test residue, by weight	12.5%
Maximum sulfur content, by weight	3.5%
Maximum ash content, by weight	0.1%
Maximum bottom sediment and water, by volume	2.0%
Minimum cetane number	30
Minimum gravity, degrees API	12.5
Maximum sodium content, by weight	100 ppm
Maximum vanadium content, by weight	400 ppm

Water and sediment must be reduced to 0.10 percent by purification before use. The specifications for sodium and vanadium are based on the presence of one or the other; if both are present lower limits may be applicable.

Maintenance

Man-hour requirements. Guideline man-hour requirements for maintenance tasks, beginning with the engine in a fully assembled condition (except as otherwise noted), are as follows:

Dismantling and refitting a cylinder head	2 men, 3 hours
Dismantling and refitting a connecting rod bearing (piston and connecting rod in place)	2 men, 1 hour
Dismantling and refitting a main bearing (piston, connecting rod, and counterweights in place)	2 men, 3 hours
Dismantling and refitting two exhaust valves (cylinder head in place)	2 men, 1 hour
Dismantling and refitting a fuel injector (cylinder head in place)	1 man, 45 minutes
Dismantling and refitting an injection pump	1 man, 1 hour
Dismantling and refitting two air valves (cylinder head removed)	2 men, 30 minutes
Dismantling and refitting a piston and connecting rod (cylinder head already removed)	2 men, 45 minutes

Changing cylinder liner (cylinder head and piston already removed)	1 man, 1 hour
Cleaning and measuring cylinder liner (head, piston, and connecting rod already removed)	2 men, 1 hour

Special tools. Tools normally furnished with the engines include hydraulic tighteners for main bearings and cylinder heads, extension gauges, component lifting attachments, alignment tools, valve refacing and grinding tools, pullers and extractors, fuel nozzle tester, maximum cylinder pressure indicator, special wrenches and pliers, special turbocharger tools, an engine barring lever, and conventional wrenches.

Maintenance program. The following program is recommended for engines burning heavy fuel and operating at a high load factor. It is conservative and should be modified as experience is gained.

Daily	Check engine lubricating oil level.
Biweekly	Analyze jacket and injector cooling water and treat as necessary.
As required	Clean or replace lubricating oil and fuel oil filters and strainers, and air filters when differential pressures reach limiting values.
1,250-1,500 hours	Check freedom of air and exhaust valves by observation while barring engine. Check valve-to-rocker clearance. Change turbocharger lubricating oil. Inspect turbocharger cooling spaces for deposits and clean if necessary. Clean rocker arm lubricating oil tank. Take lubricating oil sample for spot test or analysis. Check exhaust valve Belleville washers. Examine gear train for tooth wear. Examine camshafts and camshaft bearings, cam surfaces, and push rods.
2,500-3,000 hours	Check that main bearing jackscrews and crossbolts are tight. Take crank web deflections. Drain and refill governor oil. Clean air cooler water side if necessary.

	Test alarms and automatic shutdown systems. Check movement of air starting valves and lubricate. Check exhaust valve Rotocaps. Remove fuel injectors and test for popping pressure and spray pattern; clean without disassembly. If no distress is noted, subsequent inspection may be at 5,000- to 6,000-hour intervals.
5,000-6,000 hours	Select two cylinders for disassembly: check crankpin bearing clearances with feeler gauge; pull both piston assemblies for inspection and determination of ring life; inspect crankpin bearing shells. Remove and inspect all exhaust valves; clean and reface if necessary; check wear of stems and guide bushings. Disassemble and clean main air control valve. Disassemble and clean starting air distributor. Remove air start valves; disassemble and clean. Clean turbocharger wheels and diffusers. Clean air and water sides of air coolers. Inspect relief valve on lubricating oil pump.
10,000-12,000 hours	Inspect gear train; measure backlash. Check crankpin bearing clearances with feeler gauge. Pull piston assemblies (see Note 2 below); measure compression ring grooves; renew compression and scraper rings; renew piston pin seals and lock plates; check that piston pin tubes are tight; measure liner wear; deglaze liner; inspect crankpin bearing shells; inspect and measure piston pin bearing bushings; check air valve stem and bushing wear; reface air valves and seats if necessary; clean head seating surfaces; reassemble with new gaskets. Disassemble turbochargers for complete inspection; renew ball bearings and lubricating oil pumps. Disassemble engine-driven pumps for cleaning and complete inspection; renew ball bearings and seals if necessary. Inspect and clean overspeed trip.

20,000-24,000 hours — Remove cylinder heads, piston and connecting rod assemblies, liners, and jackets; separate liners from jackets for inspection and cleaning, then reassemble with new sealing rings; check crankpin bearing bolts by magnetic particle detection.
Replace main bearing shells.
Disassemble push rods and followers; check clearances at camshaft bearings and push rod guides.
Inspect governor and governor drive.
Install replacement injection pumps and return used pumps to manufacturer for overhaul.

Notes:

1. Log readings taken with accurate instruments are an important part of the maintenance program. Trends should be noted, for they serve as an early warning of problems developing.
2. Experience should be used to modify the program. While the program calls for piston overhauls at 10,000-12,000 hours, overhaul might not be required until 18,000-20,000 hours. The inspection of two pistons at 5,000-6,000 hours should provide adequate indication of when an overhaul will be necessary.

WÄRTSILÄ 20 FOUR-STROKE DIESEL ENGINES

Introduction and Design Philosophy

The Wärtsilä 20 engine is representative of a new type of medium speed engine intended for merchant ship auxiliary generator drive and for small-ship propulsion, designed from the beginning for heavy fuels, and featuring low acquisition cost, reduced maintenance requirements, and minimized NOx emissions. These engines tend towards modularized construction, using unitized components and cast-in passages, and extensive use of hydraulically tightened bolts. As a group, these are turbocharged, charge-air cooled, four-stroke cycle, direct-injected, nonreversing engines with attached auxiliaries. The Wärtsilä 20 has a bore and stroke of 200 mm and 280 mm respectively (about 8" and 11"), and is rated from 130 bkW (about 175 hp) per cylinder at 720 RPM to 180 bkW (about 240 hp) per cylinder at 1000 RPM (except for the five-cylinder engine, which is only rated at 900 to 1000 RPM). The limiting mean effective pressure is about 25 bars (about 360 psi). The engine is available only as an in-line engine. Table 17-3 shows external dimensions and weights of the engine and approximate dimensions and weights of typical generator sets. The dimensions do not include overhaul clearances.

A cross section of the Wärtsilä 20 engine appears as Figure 17-72.

TABLE 17-3

External Dimensions and Weights of Wärtsilä 20 Engines and Generator Sets

	Engine alone, in mm				*Generator Set, in mm*			
Number of Cylinders	*Length Overall*	*Width Overall*	*Height Overall*	*Weight, metric tons*	*Length Overall*	*Width Overall*	*Height Overall*	*Weight, metric tons*
4	2430	1426	2093	6.8	5100	1630	2500	14
5	2730	1360	2093	7.6	5600	1630	2500	16
6	3123	1360	2242	8.4	6000	1630	2500	17
8	3731	1360	2350	10.5	6550	1730	2500	22
9	4031	1360	2350	12.5	6820	1730	2500	23

Fuel Oil Specification

The Wärtsilä 20 is designed for operation on heavy fuel as well as distillate fuel. The following heavy fuel quality limits apply:

Maximum viscosity at 50°C *or* maximum viscosity at 100°F	730cSt 7200 sR1
Maximum density at 15°C *or* maximum specific gravity at 60 °F	0.991kg/l (see note below) 0.991
Maximum Conradson carbon test residue, by weight	22%
Maximum sulfur content, by weight	5%
Maximum vanadium content	600 ppm
Maximum sodium content	50 ppm
Maximum ash content, by weight	0.2%
Maximum water content, by volume	1% (0.3% at the engine)
Maximum pour point	30°C (86°F)
Maximum asphaltene content, by weight	14%
Maximum aluminum and silicon content	80 ppm

Heavy fuel is expected to be cleaned in centrifugal purifiers. With modern purifiers capable of treating high density fuels, the upper limit on density can be raised to 1.010 kg/l at 15°C (specific gravity of 1.010 at 60 °F). Heavy fuel must be heated to yield a viscosity at the engine of 12 to 24 cSt (55 to 100 sR1), with a recommended viscosity of 16 cSt (75 sR1).

Block

The block is a single, very rigid, nodular cast-iron unit that incorporates integral housings for the main bearings and camshaft bearings, as well as the charge-air manifold, cooling-water manifold, and lubricating oil

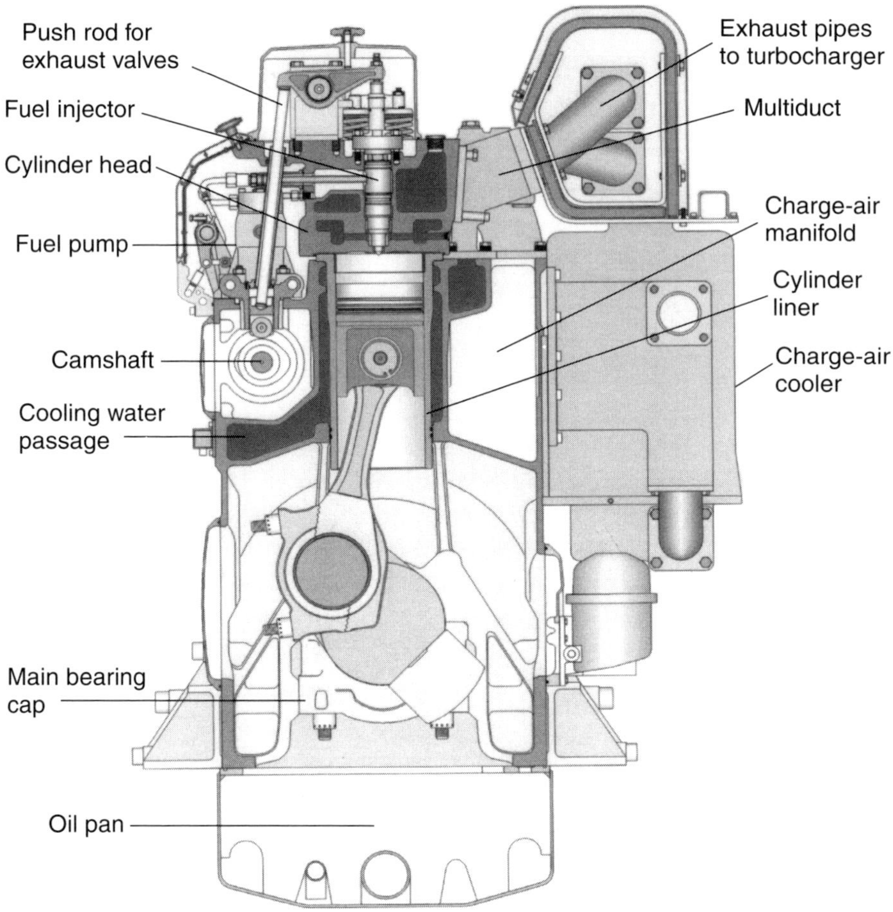

Figure 17-72. Sectional view of a Wärtsilä 20 engine

passages. In Figure 17-72, the air manifold is the large space to the right of the cylinder, the water-supply manifold is to the left, below the camshaft housing, and the water-return manifold is in the upper left corner of the air manifold. Gear housings are cast into the block at both ends, at the drive end for the timing gears and the lubricating oil pump, and at the free end for water-pump drive gears. The gear housings are closed by cast-iron covers. Mountings cast into the block accept bolted brackets for mounting the engine to the foundation or bedplate. Large access openings on both sides of the block at each cylinder are closed by sheet-metal covers. Some of these covers carry explosion-relief doors. The oil pan is a nonstructural sheet-metal or cast-iron assembly that includes lubricating oil outlet connections for the lubricating pumps and the purifier, and the oil-supply manifold for the main bearings.

Main Bearings

The main bearings are inverted, with the caps bolted in place from below with two bolts at each cap, and further secured by horizontal bolts, one from each side. The bearings are horizontally split, steel-backed precision bearings, lined with a corrosion-resistant, bimetallic alloy. There is an oil groove in the upper shell. The last bearing at the drive end of the engine has thrust faces on both sides.

Crankshaft

The crankshaft is a one-piece forging with bolted-on counterweights (except for the four-cylinder engine, which has integral counterweights). At the drive end is an integral thrust collar and a hub for the shrunk-on timing-gear ring. The flywheel is bolted to the drive flange. At the free end of the shaft is a press-fitted gear ring for driving the attached water pumps, and, on most engines, a torsional-vibration damper. An extension shaft can be fitted at the free end for power takeoff. Main journals are drilled radially for oil distribution to passages drilled in the crank webs and crankpins.

Pistons

Two-piece pistons are used, each comprising a nodular cast-iron skirt with a bolted-on forged-steel crown. Three piston rings are fitted. The top two rings are compression rings, and the bottom ring is a spring-loaded oil-control ring. All three rings are chrome plated. The top two grooves in the piston are hardened to reduce wear. The skirt has a phosphate-graphite overlay to assist in lubrication.

Piston pins are of the floating type, free in both the piston bosses and in the connecting rod, held in place axially by snap rings. Drilled passages in the piston pins and the skirts carry lubricating oil from the connecting rods to the cavities below the crowns where it cools the underside of the crowns on the "cocktail shaker" principle.

Connecting Rods

A connecting rod is shown in Figure 17-73. It is made of drop-forged steel with an H-section shank. The crankpin bearing is split vertically in the offset pattern visible in the figure, and serrated, with the bearing cap secured by two nuts. This arrangement allows a very large crankpin diameter, commensurate with the high firing pressure, while still permitting the connecting rod to be removed with the piston by lifting it through the cylinder liner.

The crankpin bearing is a split, steel-backed precision bearing, lined with a corrosion-resistant, bimetallic alloy. There is an oil distribution groove in the bottom shell. The small end of the connecting rod, together with the inserted lead-bronze bushing, is stepped to maximize the bottom

Figure 17-73. Connecting rod of Wärtsilä 20 engine

bearing surface to carry the high gas forces. Lubrication is provided by oil passing through a drilled passage in the shank of the rod.

Cylinder Liners

A cylinder liner is shown in Figure 17-74 and in sectional view in Figure 17-72. It is of centrifugally cast iron, flanged at the top to fit a recess in the top of the block. The upper part is thicker and, as can be seen in Figure 17-72, has additional support from the block to better withstand the high firing pressures. The liner is of the wet type, and the liner-to-block joint is made watertight at both ends with a gasket under the flange at the top and O-rings at the bottom. The dimensions of the water passage around the liner are carefully selected to control the water velocity, in order to preclude excessive cooling at the bottom and mid-height of the liner (which can cause cold-end corrosion traceable to the high sulfur content of the heavy fuel) while providing better cooling at the upper end.

A special feature is the anti-polishing ring inserted into the top of the cylinder liner as shown in the figure. The ring prevents excessive accumu-

Figure 17-74. Cylinder liner and anti-polishing ring, Wärtsilä 20 engine

lation of carbon deposits on the outer circumference of the piston crown above the uppermost piston ring, which could then be deposited lower on the cylinder liner, leading to "polishing" of the bore, and, in turn, to poor lubrication and increased wear and lubricating oil consumption.

Cylinder Heads

Cylinder heads are deep, rigid cast-iron units carrying two air valves, two exhaust valves, rocker arms, a centrally positioned fuel injector, as well as an indicator cock. A head is shown in the Figure 17-75 and, in sectional view, in Figure 17-72.

Figure 17-75. Cylinder head, Wärtsilä 20 engine

The rigidity of the head facilitates maintenance by allowing it to be secured to the block by only the four studs at the corners. Contributing to the rigidity is the use of bored passages to supplement the cast-in cooling cavities, enabling thinner sections in the lower part of the head, and therefore better cooling of the fire deck, the fuel nozzle, and the exhaust valve seats.

Connections for the head to the air manifold and the exhaust manifold, and for cooling water are simplified. The air inlet from the air manifold cast into the block and the water-cooled exhaust gas outlet are contained in a single iron casting shown in Figure 17-72, immediately to the right of the head. This unit, called the multiduct by the engine builder, also carries cooling water outlet and vent connections, and it normally remains on the block when the head is lifted. A single water inlet from the block is cast into the head and is therefore made up when the head is attached. A single water outlet to the jacket surrounding the exhaust gas passage of the multiduct is similarly made up when the head is secured to the multiduct. Separate small connections are necessary for lubricating oil to and from the rocker arms, for the high pressure fuel from the injection pump, and for the return of leak-off fuel from the injector.

Visible in Figure 17-75 are the yokes that enable each rocker arm to open two valves. Each of the valves, air as well as exhaust, is fitted with a rotating device, has a chrome-plated stem and a stellite-plated seating surface, and has a pressed-in valve-stem guide bushing and valve seat that can be renewed when the head is removed from the engine. The exhaust valve seats are directly water cooled.

Camshaft

The camshaft is driven by timing gears from the drive-end of the crankshaft; it carries a fuel-pump cam, an air-valve cam, and an exhaust-valve cam for each cylinder. There is a thrust bearing at the driven end of the camshaft and a journal bearing at each end, with one more journal bearing between each pair of cylinders. The single-piece journal-bearing bushings are pressed into the block, and to enable axial withdrawal of the camshaft, the journals are sufficiently large in diameter to permit the cams to pass through. The cams and journals are integral with the camshaft, which is a single drop-forged unit for the four-cylinder engines, but is assembled from two or three drop-forged segments for the larger engines. Cam surfaces are case hardened. The camshaft is drilled axially and radially for lubricating oil distribution.

Balancing Gear

In-line four-cylinder four-stroke engines are subject to severe second-order vertical forces, and in-line five-cylinder engines to severe second-order couples, as described in the previous chapter. The Wärtsilä 20 four- and five-cylinder engines are therefore fitted with second-order balancers of the Lanchester type, in the form of paired eccentric weights carried on parallel balance shafts in the cast-iron sumps below the crankshaft. The balance shafts are driven from the water-pump drive gear at twice crankshaft speed.

Fuel-Injection Pump and Valve Tappet Assembly

A unique feature of the Wärtsilä 20 is the modularized unit shown in Figure 17-76, which integrates the injection-pump housing with the air and exhaust valve tappet guides for each cylinder, along with cast-in low pressure fuel supply and return passages. When the units are bolted down to the top of the camshaft housing the fuel passages form continuous headers running the length of the engine.

The injection pump itself is of the common helix-controlled type already described in the previous chapter. The pump element, including the barrel and plunger, can be replaced without removing the module from the engine.

The high pressure fuel line from the pump is led directly to the injector through the short line visible in Figure 17-72. Note from the figure that the

Figure 17-76. Fuel injection pump with integral tappet housing, Wärtsilä 20 engine

bank of fuel pump modules, together with the high pressure fuel line and low pressure passages, is enclosed under an insulated sheet-metal cover that will contain any leakage or spray.

Turbocharger, Charge Air, and Exhaust Systems

The turbocharger is mounted transversely at the free end of the engine, on a bracket bolted to the block, as shown in Figure 17-77. Pulse charging is used in order to improve low-load operation and response. Cast-iron exhaust elbows or tees from each cylinder are joined in groups of two or three by steel bellows sections, led together to the turbine. The exhaust piping is enclosed in an insulated sheet metal housing, visible at the upper right side of Figure 17-72.

Air is normally drawn by the turbocharger compressor directly from the engine room and discharged through piping to the charge-air cooler on the side of the engine. See Figure 17-77. From the cooler the air flows directly into the air manifold cast into the block. The air cooler is also visible in Figure 17-72 as the large rectangle below the exhaust manifold. The air cooler is of the shell-and-finned-tube type, with engine cooling water circulating through the tubes. The tube nest is designed for easy withdrawal to facilitate cleaning of the air side.

The turbochargers are fitted with water-wash connections for the turbine and for the compressor.

Figure 17-77. Free end of Wärtsilä 20 engine showing turbocharger at top with compressor discharge line to charge-air cooler at left, and cooling water pumps

Lubricating Oil System

All engines may be fitted with wet sumps, but the six-, eight-, and nine-cylinder engines may have dry sumps and an independent drain tank. The dry sump arrangement is typical when the engines are used for propulsion. Oil is drawn from the sump by a gear pump driven from the timing-gear train at the drive end of the engine, discharged through the tubes of a shell-and-tube cooler, and then passed to a bank of filters before distribution. A motor-driven gear pump is fitted in parallel with the engine-driven pump and is normally used for prelubrication service before the engine is started, or when the engine is on standby. Both pumps are fitted with pressure-regulating bypass valves, and the cooler is fitted with a three-way, thermostatically controlled bypass valve. These components, pumps, cooler, regulating valves, and filters, are assembled into two adjacent modules bolted to brackets on the block at the drive end of the engine. The filters are disposable paper-element units, arranged horizontally to facilitate changing, which can be done one by one when the engine is running. To extend the life of the disposable filter elements, one or two centrifugal filters may be fitted upstream of the disposable filters in a bypass branch at the pump discharge, returning cleaned oil to the sump.

For engines operated on heavy fuel, the lubricating oil is separately purified, usually in centrifugal purifiers, either continuously or in batches, in a bypass loop, drawing from one end of the sump and returning treated oil to the other end.

The lubricating oil distribution system includes three parallel paths. Most of the oil flows to a header in the sump for the main bearings and for the turbocharger bearings. Oil from the main bearings enters drilled passages in the crankshaft to flow to the crankpins, then through the connecting rods to the piston pins, where the flow divides for cylinder and piston-ring lubrication, and for piston cooling. Another path carries the oil to drilled passages in the camshaft for distribution to the camshaft bearings, cams, and tappets, and through separate lines to the valve gear. A third distribution path carries oil to nozzles for lubrication of the timing-gear train and the pump-drive gears.

Cooling Water System

The cooling system of the engine is divided into separate high temperature and low temperature circuits, each with its own engine-driven pump carried on the cast-iron cover at the free end of the block. See Figure 17-76. The pumps are identical and replaceable as modularized units. Most all of the water passages are cast into the block or built into the attached modules.

The high temperature pump discharges to all of the cylinders in parallel, via the cooling water manifold in the block. Depending on the particular turbocharger fitted, water may be supplied from the high temperature circuit to the turbine casing or bearing housing as well. At each cylinder the water rises through the jacket, then passes through the head, then to the exhaust elbow in the multiduct module, which also carries the vent to the elevated expansion tank, and then drains to the return manifold in the block. In the low temperature circuit the pump discharges first through the tubes of the charge air cooler, and then to the lubricating oil cooler.

From both circuits the water flows to the separately mounted seawater cooler, from which the pumps draw. An expansion tank, elevated to at least seven meters, is connected to the pump suction line to maintain a positive head. The temperature in each circuit is maintained by three-way valves that permit some of the return flow to bypass the cooler. In the high temperature circuit the three-way valve is thermostatically controlled to keep the water at about 95°C (about 200°F) at all times. In the low temperature circuit, however, the three-way valve is electronically controlled in accord with the engine power output, so that the cooling water heats the charge air at low output in order to improve combustion and to avoid cold-end corrosion of the cylinder liners.

In all cases the high temperature cooling water should be preheated prior to starting the engine, but for engines running on heavy fuel, preheating is required. A separate preheater and motor-driven circulating pump are usually fitted.

The cooling water is treated with scale, corrosion, and cavitation inhibiting chemicals.

Starting Air System

The engine is fitted with a pneumatic starting motor at the flywheel. Normal air supply to the motor is limited to 10 bars (about 150 psi) by a regulating valve at the engine, but the usual storage pressure will be about 30 bars (about 450 psi).

Controls

The standard outfit of the engine includes the following local instrumentation:

Lubricating oil pressure
Fuel oil supply pressure
High temperature cooling water pressure
Low temperature cooling water pressure
Charge air pressure
Starting air pressure
Fuel oil temperature
Lubricating oil temperatures before and after the cooler
High temperature cooling water temperature before and after the engine

Low temperature cooling water temperature before and after the charge air cooler and after the lubricating oil cooler
Charge air temperature after the charge air cooler
Engine RPM
Turbocharger RPM
Running hours
Exhaust gas temperatures at each cylinder and after the turbine

The engine is also fitted with sensors sufficient for remote monitoring and alarm in most applications. These may include the following alarm sensors:

Low fuel oil supply pressure
Low lubricating oil pressure
Low starting air pressure
Low high temperature cooling water pressure
Low low temperature cooling water pressure
High exhaust gas temperatures at each cylinder and after the turbine
High lubricating oil temperature
High high temperature cooling water temperature
High charge air temperature
Leakage at high pressure fuel injection line
Low level in wet oil sump
High lubricating oil filter pressure drop
Low pneumatic overspeed trip air pressure
Overload

The following trip sensors may also be included:

Low lubricating oil pressure
High cooling water temperature
Overspeed

The following starting interlock sensors may also be included:

Turning gear engaged
Low prelubricating oil pressure
Manual block

The sensors plug into rails on either side of the engine that terminate in a common junction box mounted at the drive end of the engine for integration with a ship's central monitoring and control system, or for installation of alternative monitoring and control equipment.

Maintenance

The following maintenance schedule conforms to the guidance offered by the engine manufacturer. The schedule applies to engines run on heavy

fuel, and the suggestion is made that when running on distillate fuel or on intermediate fuel of good quality, intervals may be lengthened depending on engine load.

Every second day when engine is not in service	Check operation of prelubrication system. Rotate crankshaft to a new position.
Weekly when engine is not in service	Test start the engine.
50 running hours after starting from new or after overhaul	Check tension of connecting rod bolts, main bearing bolts, and valve clearances.
Every 50 running hours	Check for proper drainage at charge air cooler. Check water level in expansion tank. Check fuel and lubricating oil pressure drops; change filters as necessary. Read and record temperatures, pressures, and power output. Check oil level in governor. Check oil level in sump. Check oil level at turbocharger bearings. Water wash turbocharger compressor.
Every 250 running hours	Clean centrifugal oil filter. Check, clean, and lubricate fuel-rack linkage. Take lubricating oil sample for analysis. Water wash turbocharger turbine.
Every 500 running hours	Check and correct cooling water chemistry. Take and record maximum pressures of all cylinders. Change oil in turbochargers with separate lubricating oil systems.
Every 1,000 running hours	Clean turbocharger air filter. Test alarms and trips. Replace fuel filter cartridges. Replace lubricating oil filter cartridges. With engine stopped, drain lubricating oil, clean and inspect sump, and refill. Send sample for analysis. Interval can be extended with experience. With engine stopped, check that air and exhaust valves are free in guides; check valve clearances. With engine stopped, apply pneumatic test to cylinders to check valve tightness and piston ring effectiveness.

Interval	Task
Every 2,000 running hours	In new engine, inspect waterside of charge air cooler; clean if necessary. If satisfactory, raise interval to 4,000 hours. Remove injectors, test opening pressure, clean, check plunger lift, reassemble with new O-rings. Change governor lubricating oil. Test all pressure gauges and thermometers. Test overspeed trip.
Every 4,000 running hours	Inspect and clean water and air sides of charge air cooler and apply pressure test. Inspect cam faces and cam followers. Confirm rotation of follower rollers. Check fuel-rack linkage for wear. Check crankshaft alignment. Pull one cylinder liner to allow inspection of cooling water spaces for fouling. (Note that this task entails lifting the cylinder head and pulling piston and connecting rod.) Check and tighten bolts in exhaust connections and manifold. If lubricating oil cooler outlet temperature is high, open, clean, and inspect cooler. After inspection and test of fuel nozzles, consider replacement. Inspect and clean water passages at water-cooled turbocharger.
Every 8,000 to 12,000 running hours	Lift all cylinder heads; disassemble, clean, inspect; lap or grind valve seats; check rotators; replace valve guide O-rings. Pull all pistons and connecting rods; clean and inspect; remove piston rings, measure grooves, replace with new rings; check piston-pin snap rings. Remove several connecting rods from pistons to spot-check small-end bushes for wear. Inspect and measure all cylinder liners; replace as needed; hone those retained. Inspect anti-polishing rings, inverting or, if necessary, replacing, any showing wear. Inspect all crankpin journals; inspect and measure bearing-shell thickness, renewing as necessary. Open and inspect main bearings and journals; measure bearing-shell thickness; renew as necessary; measure thrust-bearing clearance. In four- and five-cylinder engines, remove one balance-shaft bearing bush to spot-check for wear.

	Inspect camshaft drive gear; measure clearance and backlash; replace parts as necessary. Inspect governor drive gear; replace parts as necessary. Inspect water and lubricating oil pump drive gear; measure backlash; replace parts as necessary. Open and inspect water and lubricating oil pumps; replace parts as necessary. Clean and inspect temperature control valves for cooling water and lubricating oil systems Replace turbocharger bearings
Every 16,000 running hours	Inspect camshaft bearings; replace as necessary. With pistons withdrawn, remove all connecting rods to inspect small-end bushes; replace as necessary. With main and crankpin bearings open, measure all journal diameters. Inspect fuel-feed pump drive; replace parts as necessary. Overhaul all fuel-injection pumps: remove elements, disassemble, clean, inspect; replace parts or element as necessary. Inspect valve-gear bearings. Inspect governor-drive bearing. Open and inspect prelubrication pump; replace parts as necessary. Dismantle and inspect vibration damper; send oil sample for analysis.
Every 24,000 running hours	Disassemble drive-shaft flexible coupling for inspection. Check tightness of foundation bolts and hold-down bolts and/or condition of flexible mounts. Inspect governor: check settings and adjustments; replace parts as necessary. Overhaul and test overspeed trip.
Every 48,000 running hours	Replace turbocharger rotor.
Every 64,000 running hours	Completely overhaul engine.

REVIEW

1. Sulzer RTA engines are fitted with valve-controlled fuel-injection pumps, in contrast to the more common helix-controlled pumps

described and illustrated in the previous chapter. Contrast the different methods used in each type to achieve the same goals.

2. Explain the advantages of variable injection timing as used in the Sulzer RTA engines.
3. The RTA engines use constant-pressure turbocharging, while the Wärtsilä 20 uses pulse charging. Draw on the material in the previous chapter to cite the differences in these methods, and use it to justify the designers' selections.
4. Compare the balancing equipment described for the RTA engines and for the Wärtsilä 20 with the balancers described in the previous chapter. Does the information correlate? Explain.
5. Why is there almost no discussion of balancing gear for the Fairbanks Morse PC2.5V engines? Refer to the previous chapter in your explanation.
6. Compare the maintenance schedule offered for the RTA engines with those for the PC2.5V and the Wärtsilä 20 engines. Cite and explain similarities and differences.
7. Working from the bore and stroke for the RTA68, use the cross-section drawing of the engine to estimate its overall height. Then, noting that maintenance requires that pistons be lifted with the piston rod attached, clear of the top of the cylinder liner (after the head is removed, of course) estimate the total overhaul height required from the tank top to the hook of the gantry crane. From the cross-section of PC2.5V, consider if piston withdrawal adds similarly to the engine height. Explain your conclusion.
8. Following the guidance given in the previous chapter, if 85 to 90 percent of the rated output of an engine is continuously available, how many cylinders of an RTA 68, at its R1 rating, would be needed for a single-screw ship which requires 15,000 BHP to achieve its average service speed? If PC2.5V engines are to be selected as an alternative, and allowing a gear efficiency of 98 percent, how many cylinders would be required, bearing in mind that each V-engine requires an even number of cylinders?
9. It can be assumed that the man-hour requirements cited for maintenance tasks for the PC2.5V engines are similar to those required for the RTA engines. On this basis, use the maintenance schedules to quantify the maintenance man-hours required for the two power plants selected in the previous question, over a five-year period, if the ship operates at service speed about 275 days per year. Draw conclusions about the ratio of labor costs involved.
10. For the two power plants selected in Question 8 above, and using the information on heights derived in answering Question 7, compare the necessary heights of the engine rooms for the two power plants. How would this height comparison affect the choice of machinery if the ship

were a large tanker? A containership? A cruise ship? A ro-ro with open cargo decks and a stern ramp?

REFERENCES

In addition to manuals, project guides, and other technical publications of the engine manufacturers, current versions of which are readily available to serious users of the information on request from the manufacturers, the following books are cited for their engine and component descriptions and illustrations.

Griffiths, D. 1999. *Marine Medium Speed Diesel Engines*. MEP Series, Vol 1, Pt 3. London: Institute of Marine Engineers.

Henshall, S. E. 1978. *Slow Speed Diesel Engines*. MEP Series, vol. 2, Pt. 17. London: Institute of Marine Engineers.

Knak, C. 1990. *Diesel Motor Ships' Engines and Machinery*. London: Institute of Marine Engineers.

Stinson, K. W. 1983. *Diesel Engineering Handbook*, 12th ed. Norwalk, Ct.: Business Journals.

Woodyard, D. F. 1998. *Pounder's Marine Diesel Engines*. England: Butterworth-Heinemann.

ACKNOWLEDGMENTS

All of the material in this chapter was drawn from manuals, project guides, and other technical publications of the engine manufacturers, Wärtsilä NSD Corporation; its predecessor, Sulzer Diesel, Ltd.; and Colt Industries, Fairbanks Morse Engine Division. The authors are indebted to these firms for their assistance in preparing the text of this chapter and for their provision of all of the figures.

CHAPTER 18

Marine Refrigeration Systems

JAMES A. HARBACH

REFRIGERATION PRINCIPLES

Refrigeration

REFRIGERATION is defined as the branch of science and engineering that deals with the process of reducing and maintaining the temperature of a space or material below that of the surroundings. Since heat flows from a region of higher temperature to one of lower temperature, there is always a flow of heat from the surroundings into the refrigerated area. The amount of this flow of heat can be controlled by proper insulation. To maintain a constant temperature in the refrigerated space, the refrigeration system must remove heat from the space at the same rate as the heat is entering from the surroundings.

Ice Refrigeration

Melting ice was not many years ago the only refrigerant available for use in domestic and small commercial refrigerators. Ice refrigeration has certain disadvantages: it cannot be used to maintain low temperatures; the ice must constantly be replenished; and the water resulting from the melting of the ice must be disposed of. However, even today, the capacity of a refrigeration system in tons is based on the cooling effect of melting ice. One ton of ice has a cooling effect of 288,000 Btu (144 Btu per pound times 2000 pounds = 288,000 Btu). One ton of refrigeration is the cooling effect of one ton of ice melting over 24 hours, i.e., 12,000 Btu/hour or 200 Btu/minute (3.52 kw). Figure 18-1 shows the temperature-enthalpy characteristics of water.

Refrigerants

A refrigerant is the working fluid which picks up the heat from the enclosed refrigerated space and transfers it to the surroundings. It should be understood that there is no such thing as an ideal refrigerant. Factors that enter into the selection of a refrigerant for a particular application include thermodynamic properties, cost, safety aspects, and environmental considerations. In the early part of the twentieth century, substances such as carbon dioxide (CO_2) and ammonia (NH_3) were commonly used. For most of the last half of the twentieth century, the fluorocarbon refrigerants R-12 (dichlordifluoromethane—CCl_2F_2), R-11 (trichloromonofluoromethane—CCl_3F), and R-22 (monochlorodifluoromethane—$CHClF_2$) have been the most popular for industrial and marine applications. Fluorocarbons are based on the substances methane (CH_4) or ethane (C_2H_6) in which one or more of the hydrogen atoms has been replaced by fluorine or chlorine atoms. If all the hydrogen atoms have been replaced by fluorines and chlorines, the refrigerant is referred to as a chlorofluorocarbon (CFC). If all of the hydrogen molecules have not been replaced by fluorines and chlorines, the refrigerant is referred to as a hydrochlorofluorocarbon (HCFC). R-12 and R-11 are examples of CFCs, and R-22 is an example of a HCFC. A substance with no chlorine atoms is referred to as a hydrofluorocarbon (HFC). Table 18-1 summarizes the characteristics of a number of different refrigerants.

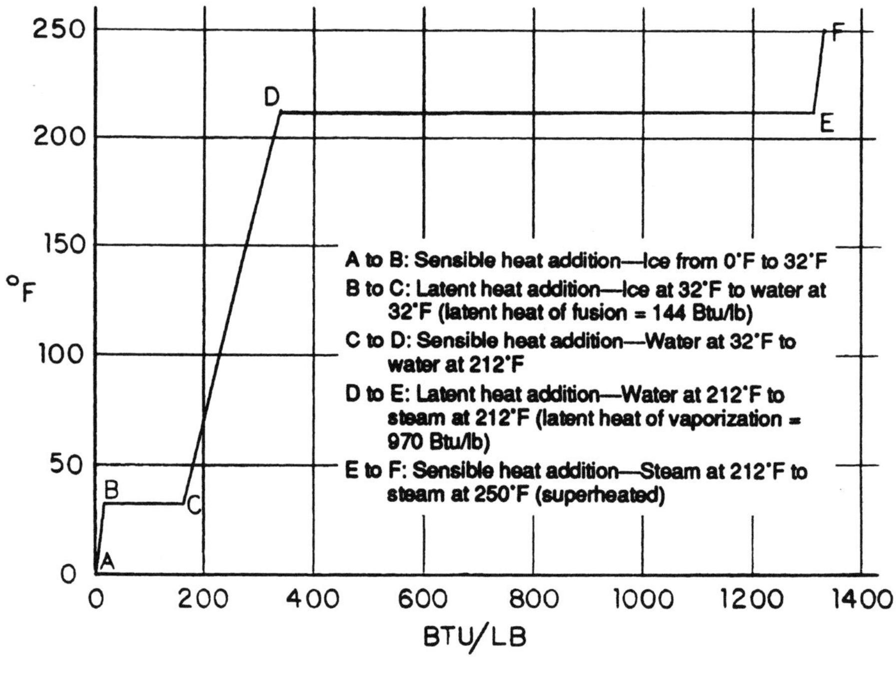

Figure 18-1. Temperature-enthalpy characteristic of water at 14.7 psia

TABLE 18-1

Comparative Refrigerant Characteristics*

Refrigerant	*Boiling Point @14.7 psia*	*Evaporator Pressure psig*	*Condensor Pressure psig*	*Refrigerant Flow lb/min*	*Compressor Displacement, CFM*
R-12 (CCl_2F_2)	–21.6	11.8	93.3	4.00	5.83
R-22 ($CHClF_2$)	–41.4	28.2	158.2	2.86	3.55
R-134a (CH_2FCF_3)	–15.3	9.1	97.0	3.17	6.10
R-11 (CCl_3F)	–74.7	23.9†	3.5	2.99	36.54
R-744 (CO_2)	–109	317.5	1031.0	3.62	0.96
R-717 (NH_3)	–28.0	19.6	154.5	0.422	3.44

*Based on 5°F evaporation and 86°F condensation for a 1-ton system
†Inches of Hg vacuum

Refrigerant Numbering System

In the 1950s DuPont developed a standard numbering system for refrigerants based on their chemical composition. This system has become an industry standard. The system for numbering single component refrigerants is explained as follows:

A refrigerant is given the number R-XYZ

where

X = the number of carbon atoms minus one (omitted if zero)
Y = the number of hydrogen atoms plus one
Z = the number of fluorine atoms

Thus for dichlorodifluoromethane, CCl_2F_2:

$X = 1 - 1 = 0$ (omit)
$Y = 0 + 1 = 1$
$Z = 2$

Therefore the refrigerant is R-12.

Current practice is to use the designation CFC, HCFC or HFC along with the refrigerant number, rather than just "R." Thus R-12 becomes CFC-12, R-22 becomes HCFC-22, and R-134a becomes HFC-134a. Note that a refrigerant with a two-digit number will be based on methane with one carbon at the center of the molecule, while a one-hundred series refrigerant will be based on ethane with two carbons at the center. The letter at the end of the number of some refrigerants indicates that it is an isomer. Isomers have the same chemical composition but differ in the arrangement of the atoms that make up the molecule.

Blends are mixtures of two or more different substances. Blends can be classified as azeotropes or zeotropes. An azeotrope is a mixture of two or more liquids of precise proportions which have a constant boiling point at a constant temperature and maintain a nearly constant chemical composition throughout the refrigeration cycle. A zeotrope will experience a change in mixture ratio during phase changes and show a variation in temperature during phase changes at constant pressure referred to as temperature glide. A blend that approaches that of a azeotrope is referred to as near-azeotropic. Zeotropes are given four-hundred-series numbers and azeotropes are given five-hundred-series numbers.

Ozone Depletion and Alternative Refrigerants

In the 1970s, concerns were raised that some chlorinated compounds were destroying ozone in the upper atmosphere. Ozone (O_3) is formed when high energy ultraviolet light breaks apart an oxygen (O_2) molecule, liberating two oxygen molecules, which may then bind with an (O_2) molecule to form ozone. Ozone in the stratosphere scatters and absorbs UV radiation, reducing the levels that reach the earth's surface. Reduction of ozone levels in the upper atmosphere leads to increased ultraviolet radiation at the earth's surface that can result in increased rates of skin cancer and cataracts, and also possible suppression of the body's immune system. Increased levels of ultraviolet light can also affect the growth of photoplankton, disrupting the marine food chain. The Rowland-Molina theory states that certain chlorinated compounds will remain intact in the atmosphere for years, eventually reaching the stratosphere where the higher UV radiation levels will break down the molecule, releasing the chorine atoms. While the process of ozone depletion is complex, it can be explained by the following simplified process. The free chlorine atoms react with ozone (O_3), forming chlorine monoxide (ClO) and oxygen (O_2). A free oxygen atom will then react with the chlorine monoxide, forming oxygen and releasing the chlorine atom to repeat the process. The chlorine acts as a catalyst in the process, and a single chlorine atom can destroy thousands of ozone molecules.

These concerns led to the banning of CFCs for propellants in aerosol sprays in 1978. In 1987, the United States and many other countries signed the "Montreal Protocol on Substances That Deplete the Ozone" which regulates the production and use of CFCs. Subsequent agreements included HCFCs in the substances under control, and in 1990 President

Bush signed amendments to the Clean Air Act which put in place schedules for the phaseout of the production of CFCs and HCFCs, regulations on their handling and use, and requirements for the licensing of technicians involved in the repair and maintenance of refrigeration and air-conditioning systems.

In response to the phaseout of CFCs, refrigerant manufacturers have introduced a number of new compounds for use in new systems and also as replacements in existing systems after the necessary conversion process. Some of these new refrigerants are chlorine free (HFCs) and are expected to be used far into the future. Others contain HCFCs and are considered interim refrigerants, designed for use as retrofits to existing equipment or in new systems where a chlorine-free substance is not yet available. Some of these new refrigerants are single component types containing only one type of molecule, while others are blends of two or three different substances. Table 18-2 shows the saturation pressure-temperature relationships of some older refrigerants and several of the new replacements.

HFC-134a is the most widely used of the new single component refrigerants. It has a similar pressure-temperature relationship to CFC-12 and has been used in many new applications where CFC-12 was used in the past. It is not recommended for applications with evaporator temperatures of –10°F or below. Like most HFCs, it is not compatible with traditional mineral lubricating oil, requiring the use of synthetic oils such as polyolester (POE). This refrigerant may be used in retrofit applications, but requires complete flushing of the old mineral oil and replacement of certain system components.

HCFC-123 has been proposed as an interim replacement for CFC-11 in centrifugal chillers because of its similar temperature-pressure characteristics. Caution must be taken when using this refrigerant as it has been shown to have toxic properties. Refrigerant leakage alarms must be installed and technicians must be trained in the use of breathing apparatus and go through annual fit testing. Retrofit in an existing CFC-11 system is an engineered conversion requiring significant parts replacement.

As many CFC-12 compressors are also designed for use with HCFC-22, it should be considered a possible retrofit for such systems. One possible difficulty is that operating pressures are higher than with CFC-12 and compressor speed may have to be reduced to avoid motor overloads.

A wide variety of blends are being marketed for retrofit and new applications. Blends not yet in wide use will have manufacturer's designations such as HP-80, AZ-50, and MP-66. Blends that achieve acceptance will receive a four- or five-hundred-series designation. One such blend is 404a, a blend of HFC-125, HFC-143a, and HFC-134a sold as Forane FX 70 and Suva HP-62. As an HFC blend, it has potential as a long-term solution and has found use in new low and medium temperature applications.

TABLE 18-2

Saturation Pressure-Temperature Relationships of Some Older Refrigerants and Several New Replacements

Temperature		*Pressure in Psig (Bold in Inches HG Vacuum)*					
°F	*°C*	*R-12*	*R-134a*	*R-22*	*R-11*	*R-123*	*R-404a*
–40	–40	**11.0**	**14.7**	0.5			5.2
–35	–37.22	**8.4**	**12.4**	2.6			7.1
–30	–34.44	**5.5**	**9.8**	4.9		**28.5**	10.5
–25	–31.67	**2.3**	**7.0**	7.4		**28.2**	13.6
–20	–28.89	0.6	**3.8**	10.1	**27.0**	**27.8**	16.8
–15	–26.11	2.4	**0.3**	13.2	**26.5**	**27.4**	20.7
–10	–23.33	4.5	1.8	16.5	**26.0**	**27.0**	24.2
–5	–20.56	6.7	3.9	20.1	**25.4**	**26.5**	28.4
0	–17.78	9.2	6.3	24.0	**24.7**	**26.0**	33.0
5	–15.00	11.8	8.8	28.2	**23.9**	**25.4**	38.5
10	–12.22	14.6	11.6	32.8	**23.1**	**24.7**	43.3
15	–9.44	17.7	14.7	37.7	**22.1**	**23.9**	50.0
20	–6.67	21.0	18.0	43.0	**21.1**	**23.0**	55.2
25	–3.89	24.6	21.6	48.8	**19.9**	**22.0**	63.4
30	–1.11	28.4	25.6	54.9	**18.6**	**20.8**	68.9
35	1.67	32.6	29.9	61.5	**17.2**	**19.6**	78.7
40	4.44	37.0	34.5	68.5	**15.6**	**18.2**	84.6
45	7.22	41.7	39.5	76.0	**13.9**	**16.7**	96.2
50	10.00	46.7	44.9	84.0	**12.0**	**15.0**	102.4
55	12.78	52.0	50.7	92.6	**10.0**	**13.2**	114.6
60	15.56	57.7	56.9	101.6	**7.8**	**11.2**	125.1
65	18.33	63.8	63.6	111.2	**5.4**	**9.0**	136.3
70	21.11	70.2	70.7	121.4	**2.8**	**6.6**	148.2
75	23.89	77.0	78.3	132.2	0	**4.0**	160.1
80	26.67	84.2	86.4	143.6	1.5	**1.1**	174.0
85	29.44	91.8	95.0	155.7	3.2	1.0	188.1
90	32.22	99.8	104.2	168.4	4.9	2.6	202.9
95	35.00	108.3	113.9	181.8	6.8	4.4	218.6
100	37.78	117.2	124.3	195.9	8.8	6.3	235.1
105	40.56	126.6	135.2	210.8	10.9	8.3	252.5
110	43.33	136.4	146.8	226.4	13.2	10.5	270.8
115	46.11	146.8	159.0	242.7	15.6	12.9	290.0
120	48.89	157.7	171.9	259.9	18.2	15.4	310.3
125	51.67	169.1	185.5	277.9	21.0	18.1	331.3
130	54.44	181.0	199.8	296.8	24.0	21.0	353.9
135	57.22	193.5	214.8	316.6	27.1	24.1	377.3
140	60.00	206.6	230.5	337.3	30.4	27.3	401.9

THE VAPOR-COMPRESSION CYCLE AND REFRIGERATION SYSTEMS

The Vapor-Compression Cycle

While other cycles such as the absorption are sometimes used for refrigeration systems, almost all marine refrigeration systems operate on the vapor-compression cycle. Any cycle consists of a repetitive series of thermodynamic processes. The operating fluid starts at a particular state or condition, passes through the series of processes, and returns to the initial condition. The vapor-compression cycle consists of the following processes: (1) expansion, (2) vaporization, (3) compression, and (4) condensation.

A simple vapor-compression cycle is shown in Figure 18-2. Starting at the receiver, high temperature, high pressure liquid refrigerant flows from the receiver to the expansion valve. The pressure of the refrigerant is reduced by the expansion valve so that the evaporator temperature will be below the temperature of the refrigerated space. Some of the liquid refrigerant flashes to a vapor as the pressure is reduced. In the evaporator, the liquid vaporizes at a constant temperature and pressure as heat is picked up through the walls of the cooling coils. The compressor draws the vapor from the evaporator through the suction line into the compressor inlet. In the compressor, the refrigerant vapor pressure and temperature are increased and the high temperature, high pressure vapor is discharged into the hot gas line. The vapor then flows to the condenser where it comes in contact with the relatively cool condenser tubes. The refrigerant vapor gives up heat to the condenser cooling medium, condenses to a liquid, and drains from the condenser into the receiver, ready to be recirculated.

High Side/Low Side

A refrigerating system can be divided into two parts according to the pressure exerted by the refrigerant. The low pressure portion of the system (the "low side") consists of the expansion valve, the evaporator, and the suction line. This is the pressure at which the refrigerant is vaporized in the evaporator. The high pressure portion of the system (the "high side") consists of the compressor, the hot gas line, the condenser, the receiver, and the liquid line. This is the pressure at which the refrigerant is condensed in the condenser. The dividing points between the high side and the low side are the expansion valve and the compressor.

Evaporator Temperature

In order for refrigeration to take place, the evaporator coil temperature must be below that of the refrigerated space. The evaporator temperature is controlled by varying the pressure in the evaporator since the vaporiza-

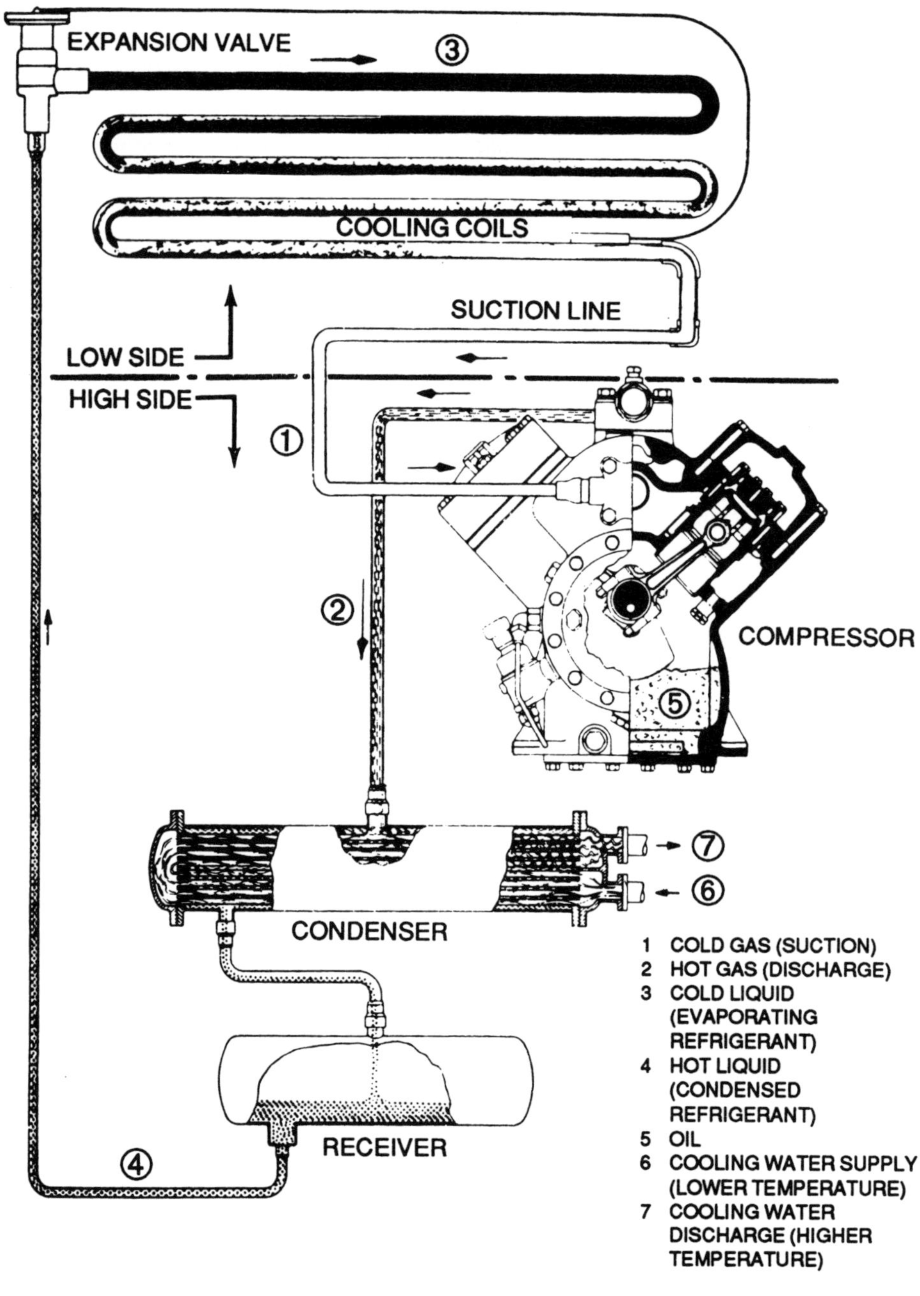

Figure 18-2. Basic vapor-compression refrigeration cycle

tion of the refrigerant occurs at the saturation temperature corresponding to the evaporator pressure. Raising the evaporator pressure raises the evaporator temperature, and lowering the pressure lowers the temperature.

Condensing Temperature

In order for the refrigerant gas to condense to a liquid in the condenser, its saturation temperature must be above that of the condenser cooling medium. Raising the condenser pressure raises the condensing temperature, and lowering the pressure lowers the temperature. To provide continuous refrigeration, the refrigerant vapor must be condensed at the same rate as the refrigerant liquid is vaporized in the evaporator. Obviously, any increase in the rate of vaporization will increase the required rate of heat transfer in the condenser. Heat transfer in the condenser is a function of (1) the condenser surface area, (2) the condenser heat transfer coefficient, and (3) the temperature difference between the condensing refrigerant vapor and condensing medium. Since the first two items are normally fixed, it follows that the condenser heat transfer varies with the temperature difference. The condensing temperature thus varies directly with the cooling medium temperature and the rate of refrigerant vaporization in the evaporator.

Deviations from the Simple Cycle

Actual refrigeration cycles deviate somewhat from the simple cycle discussed above. For example, pressure drops occur in the piping, across the condenser and evaporator, and across other components due to the flow of refrigerant through them. Subcooling of liquid refrigerant can occur in the receiver and liquid line. Superheating of the refrigerant gas can occur in the evaporator and suction line. Figure 18-3 compares an actual cycle with the simple cycle on a pressure-enthalpy diagram. In general, pressure drops are kept to a minimum consistent with economic pipe sizes. The effect of superheating of the refrigerant gas on cycle performance depends on where the superheating occurs. In general, superheating is kept to a minimum consistent with complete vaporization and avoiding liquid refrigerant flooding back to the compressor. Subcooling of the liquid refrigerant generally has a desirable effect on cycle performance. It is common on modern systems to install a liquid-suction heat exchanger (heat interchanger). The relatively cool refrigerant gas coming from the evaporator is used to subcool the liquid refrigerant flowing to the expansion valve. This increases the efficiency and capacity of the system.

Marine Applications of Refrigeration Systems

Mechanical refrigeration is used aboard ship for many purposes, including (1) refrigerated ship's stores, (2) air-conditioning, (3) refrigerated cargo,

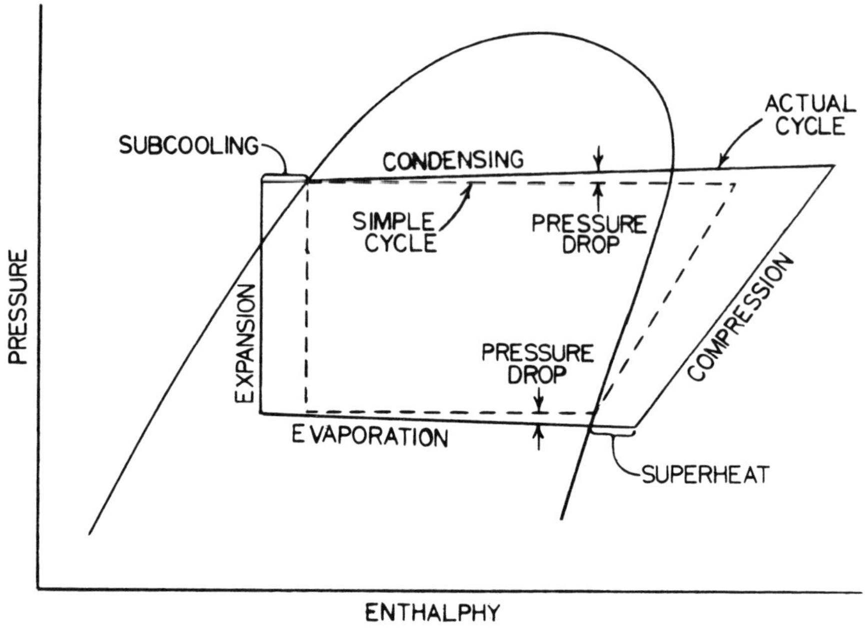

Figure 18-3. Comparison of simple cycle and actual cycle

and (4) drinking water cooling. Most marine refrigeration systems use reciprocating compressors; however, systems using rotary and centrifugal compressors are becoming more common. The reciprocating and rotary types are positive displacement while the centrifugal type uses the centrifugal force created by a high speed impeller to provide the compression. Reciprocating compressors are especially flexible. They are used in high temperature (air-conditioning) as well as low temperature (cryogenic) applications, and in sizes from less than 1 ton to 250 tons or more. This flexibility when considered along with the reciprocating compressor's reliability and efficiency accounts for its widespread popularity. Rotary compressors are replacing reciprocating compressors in certain applications, while centrifugal compressors are used primarily in large tonnage air-conditioning or refrigerated cargo applications.

Automatic Operation of Reciprocating Compressor Systems

Most marine refrigeration systems operate on a method of control known as the "pump-down cycle." The refrigerated space temperature is monitored by a thermostat which acts to open and close the solenoid valve in the liquid line. The routine starting and stopping of the compressor is controlled by the low pressure cutout switch in the compressor suction line. As the space temperature is reduced to the set point of the thermostat, the thermostat contacts open, deenergizing the solenoid valve and stopping the flow of refrigerant to the evaporator. Continued compressor operation reduces the suction pressure to the set point of the low pressure switch, stopping the compressor. Due to the stopping of refrigeration, the space temperature will slowly rise. When the temperature reaches the thermostat set point, the contacts close, energizing the solenoid valve, thus permitting liquid refrigerant to enter the evaporator. The refrigerant vaporizes, raising the suction pressure to the cut-in point of the low pressure switch, thus starting the compressor.

The pump-down cycle reduces the possibility of liquid flooding back to the compressor during start-up and substantially reduces the dilution of crankcase oil by the refrigerant. An understanding of the pump-down cycle control sequence is essential in troubleshooting refrigeration systems.

Refrigerated Ship's Stores

Ship's stores refrigeration equipment is installed to preserve food required for consumption by the crew and passengers. Insulated walk-in type storage compartments are usually installed. A typical installation consists of a freeze room, dairy room, fruit and vegetable room, and two condensing units. Figure 18-4 is a diagram of a typical system. The system is designed to maintain the freeze room at 0°F (–18°C) and chill rooms at 33°F (0.5°C). Unit coolers or natural convection bare tube evaporators are commonly used. Each condensing unit consists of a compressor, condenser, receiver,

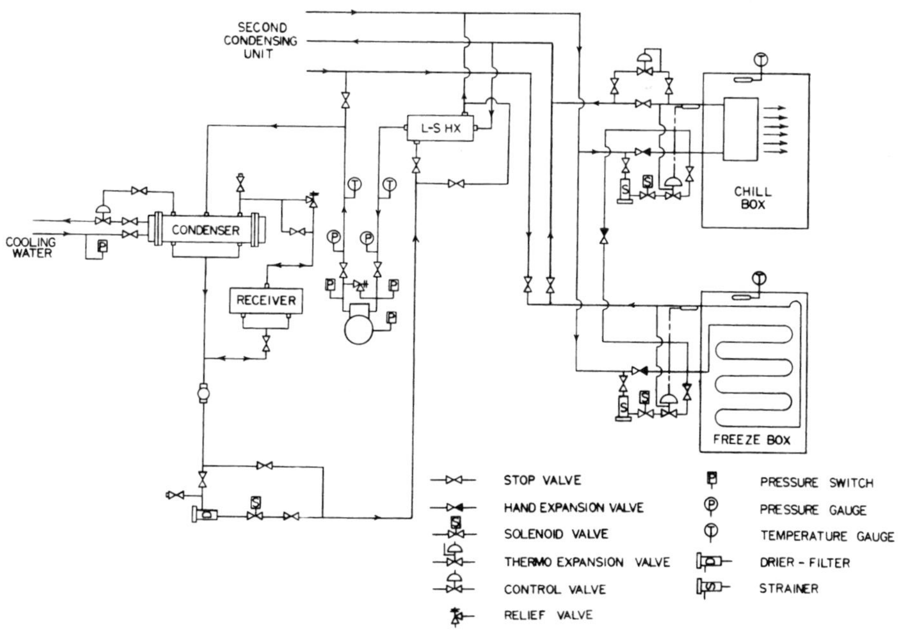

Figure 18-4. Ship's stores refrigeration system diagram

heat exchanger, controls, valves, and associated piping. One condensing unit can handle the entire plant load during holding operation while both units can be used for pulldown operation. It is estimated that lowering the product temperatures after loading should not take more than two days under extreme design conditions. Provisions are made for the defrosting of freeze box evaporator coils to permit removal of accumulated ice to maintain efficiency. Defrosting is accomplished by electric heaters, or by use of hot refrigerant gas from the compressor.

Air-Conditioning

Air-conditioning is the control of temperature and humidity of enclosed spaces to make the environment more comfortable for the people living and working there. While technically it includes winter heating, air-conditioning is normally taken to mean the cooling and dehumidifying of air in warm weather. A typical air-conditioning system is similar to that used for

Figure 18-5. Condensing unit for 60-ton air-conditioning system, front view. Courtesy Carrier Corp.

Figure 18-6. Condensing unit for 60-ton air-conditioning system, back view. Courtesy Carrier Corp.

refrigerated ship's stores except that higher temperatures are involved and system tonnage is larger. The evaporators operate above 32°F (0°C), and therefore no defrosting provisions are required. Evaporators of smaller systems are usually of the direct expansion type while larger systems typically are of the chilled water type. The evaporator in a chilled water system is located in the refrigeration machinery room, and it cools water to 45°F to 50°F (7°C to 10°C). The chilled water is then circulated to the remotely located cooling coils. Air-conditioning applications are different from refrigerated storage applications in that standby condensing capacity is normally not furnished. At peak loads, all condensing units are usually required. The plant is typically arranged to permit cross connection of condensing units thus allowing securing unneeded units.

Refrigerated Cargo

Refrigerated cargo spaces are installed to permit the shipment of perishable cargo. These systems vary in size from the small self-contained unit on a refrigerated container to a complex brine system on a refrigerated cargo vessel. The systems are usually designed for maximum flexibility to permit the carriage of different cargoes at different temperatures. Defrosting provisions are required for all evaporators designed for below 32°F (0°C) operation. Where the installation is extensive and uses forced air cooling coils, hot seawater defrosting is common. Hot seawater is heated and sprayed over the coils to melt the frost and carry it away down the drains. The fans are shut down during defrosting to minimize the carryover of heat into the storage area.

Some cargoes such as fruit and vegetables give off carbon dioxide (CO_2) during storage. To prevent dangerous concentrations, ventilation systems are commonly provided to force fresh air into the refrigerated space and exhaust stale compartment air to the outside. Since the introduction of warm outside air imposes a significant load on the refrigeration system, the fan is sized for one air change per hour and is operated for 20 minutes per hour during normal operation (not pulldown) only.

In large systems, it is frequently economical to employ brine as a secondary refrigerant in an indirect system. The primary refrigerant evaporator is employed to chill the secondary refrigerant, the brine, which is then circulated to the refrigerated spaces. Calcium chloride and sodium chloride are the most common types of brine used. Calcium chloride brine can achieve temperatures of 67°F (–55°C) while sodium chloride brine can achieve temperatures of –6°F (–21°C). Tables 18-3 and 18-4 give the freezing points of calcium chloride and sodium chloride brine solutions. The density of the brine should be such that the freezing point is 10°F to 15°F (5°C to 8°C) below the required brine temperature. Note that the

TABLE 18-3

Properties of Calcium Chloride Brine

Percent CaCl (by weight)	*Specific Gravity (60°F/60°F)*	*lb CaCl2/gal*	*Crystallization Temperature (°F)*
0	1.000	0.000	32.0
5	1.044	0.436	27.7
10	1.087	0.908	22.3
15	1.133	1.418	13.5
20	1.182	1.970	–0.4
25	1.233	2.574	–21.0
29.87	1.290	3.16	–67.0
30	1.295	3.22	–50.8
32	1.317	3.49	–19.5
34	1.340	3.77	4.3

TABLE 18-4

Properties of Sodium Chloride Brine

Percent NaCl (by weight)	*Specific Gravity (59°F/39°F)*	*lb NaCl/gal*	*Crystallization Temperature (°F)*
0	1.000	0.000	32.0
5	1.035	0.432	27.0
10	1.072	0.895	20.4
15	1.111	1.392	12.0
20	1.150	1.920	1.8
23	1.175	2.256	–6.0
25	1.191	2.488	16.1

crystallization point of a brine solution decreases with increasing salt content, then starts to increase again. The concentration at which the crystallization temperature is at a minimum is called the eutectic point. At concentration below eutectic, ice crystals form on lowering temperature. At concentrations above eutectic, salt crystals form on lowering temperatures. The brine density should be checked at least once a month and additional salt added as required.

Refrigerated containers are fitted with self-contained electric heating and cooling units. The typical unit is mounted flush with the front face of a standard-sized container. Cooling is provided by an R-12 refrigeration system with a semihermetic reciprocating compressor and an air-cooled condenser. Figure 18-7 is a typical diagram of the refrigeration system. Axial flow fans are installed to provide ventilation across the evaporator and condenser. Figure 18-8 shows the air flow through the unit. Heating and defrosting are accomplished with electric resistance heating elements located in the evaporator section. The heating and cooling cycles are controlled automatically by a thermostat, while the defrost cycle is initiated by a timer or a differential pressure switch monitoring the air pressure across the evaporator. The evaporator fan is secured automatically during the defrost cycle. Electric power to run the units can be supplied from the ship's electrical system, from deck-mounted packaged diesel generators, or from individual diesel generators mounted in the container front. Some units are built with a diesel as the primary drive and an electric motor backup. The units with diesel engines or diesel generators can be carried over land by trucks without generating capacity.

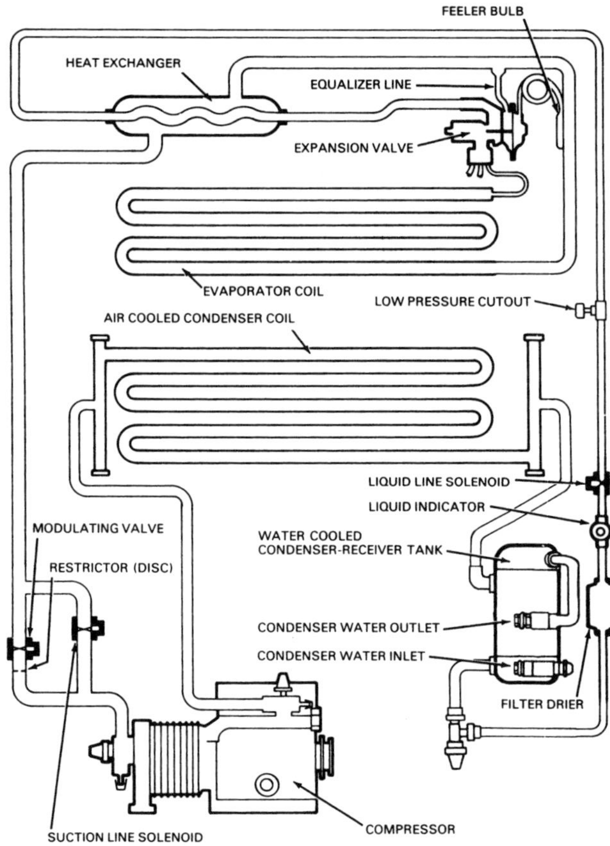

Figure 18-7. Refrigerated container system diagram.
Courtesy Thermo King Corp.

REFRIGERATION SYSTEM COMPONENTS

Refrigeration Compressor Types

Three types of compressors are used in refrigeration systems: (1) reciprocating, (2) rotary, and (3) centrifugal. The reciprocating and rotary types are positive displacement machines. In the reciprocating compressor, the gas is compressed by a piston, while in a rotary compressor a roller, vane, or screw accomplishes it. The centrifugal compressor utilizes a high speed impeller to generate the compression force by centrifugal action. Each compressor type has certain advantages and disadvantages. The type selected depends on such factors as the size and type of the installation and on the refrigerant used.

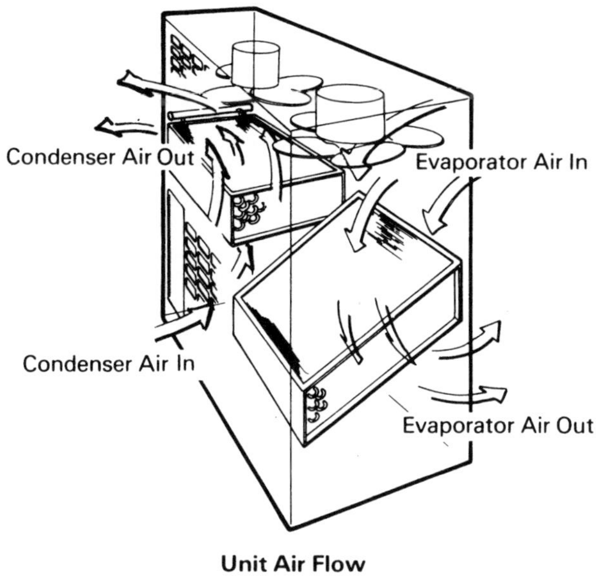

Figure 18-8. Refrigerated container air flow.
Courtesy Thermo King Corp.

Reciprocating Compressors

The reciprocating compressor is the most common type found in marine refrigeration systems. It is found in systems rated from less than one ton to several hundred tons and in high, medium, and low temperature applications. It is durable and efficient and can be manufactured economically.

Marine reciprocating compressors are typically of the single-acting enclosed type. In single-acting compressors, vapor compression occurs on only one side of the piston, while in double-acting compressors, vapor compression occurs on both sides of the piston. Double-acting compressors are only found on older large industrial installations. Enclosed-type compressors drive the piston by a connecting rod driven by the crankshaft. The crankcase is airtight and exposed to the system refrigerant. Reciprocating compressors can be further classified as open, hermetic, or semihermetic. An open type unit has a separate motor and compressor. A hermetic unit is a sealed motor-compressor assembly and is common on smaller units. A semihermetic or "accessible hermetic" is a unit in which the shell of the assembly is bolted rather than welded. This allows field servicing.

The number of cylinders varies from one to as many as twelve or more. In multicylinder units, the cylinders may be arranged in-line, radially, or in a V or W pattern. Figure 18-9 shows a modern open-type W pattern

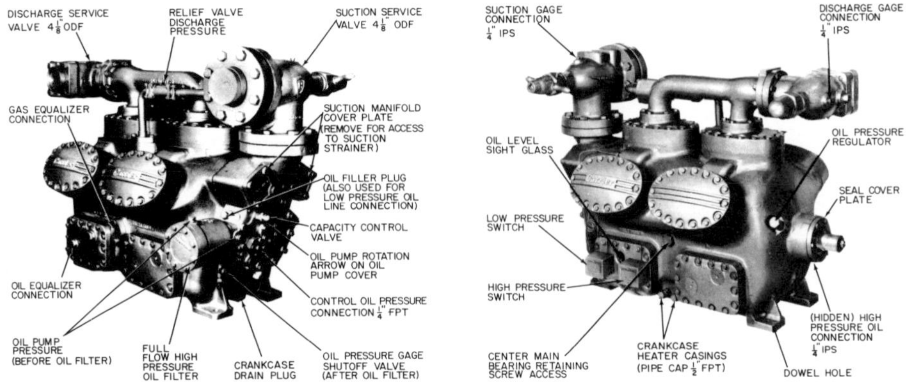

Figure 18-9. Open-type W pattern compressor. Courtesy Carrier Corp.

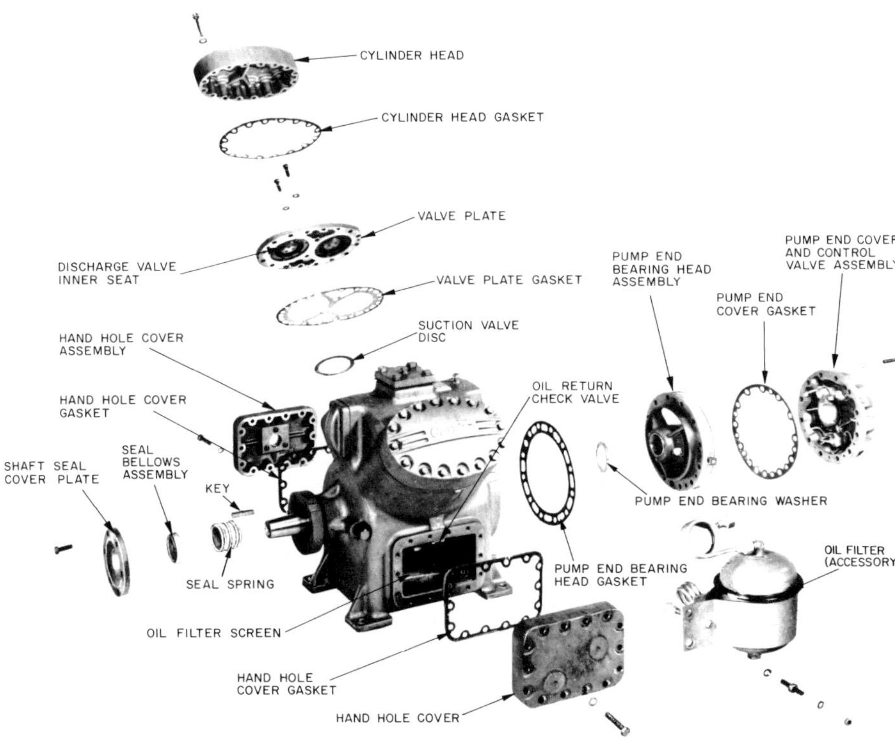

Figure 18-10. Compressor assembly. Courtesy Carrier Corp.

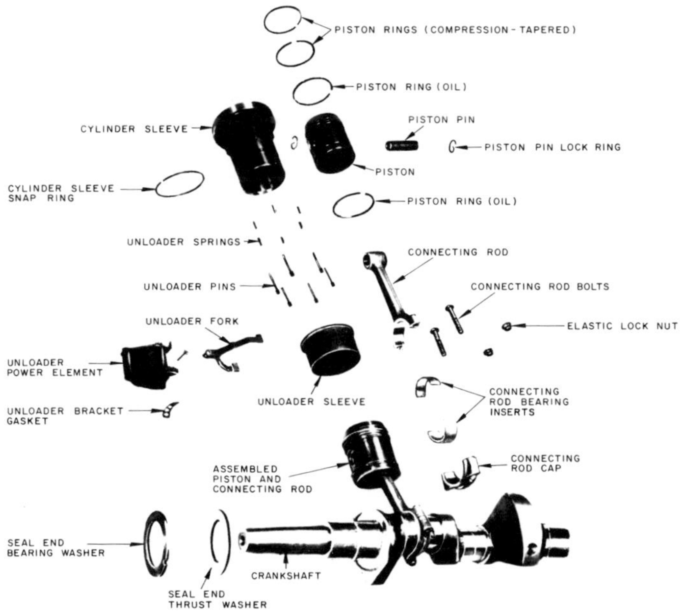

Figure 18-11. Compressor internal parts. Courtesy Carrier Corp.

compressor. Figures 18-10 and 18-11 show the construction details of a typical compressor.

Pistons are of the automotive or the double-trunk types. When automotive type pistons are used, the gas enters and exits through valves located in the cylinder head. In double-trunk piston units, the gas enters through ports in the cylinder wall and passes into the cylinder through suction valves located in the top of the piston. Figure 18-12 shows a compressor with double-trunk pistons.

The valve types used in refrigeration compressors are the reed, the ring, and the poppet. All three types open and close automatically based on the pressure differential across the valve. Reed valves are flexible metal strips that fit over slots in the valve seat. Ring valves consist of one or more circular rings, valve springs, and a retainer. The poppet valve is similar in construction to an automobile valve that is spring loaded. Poppet valves are used only in low speed compressors while reed and ring valves can be used in low and high speed machines. Figure 18-13 shows the gas flow through a typical compressor. Figure 18-14 shows a typical suction and discharge valve assembly using ring valves.

Most larger compressors are fitted with a spring loaded safety head. Under normal operating conditions, the head is held in place by heavy springs. In the event that a slug of liquid refrigerant should enter the cylinder, the safety head will rise and prevent damage to the compressor.

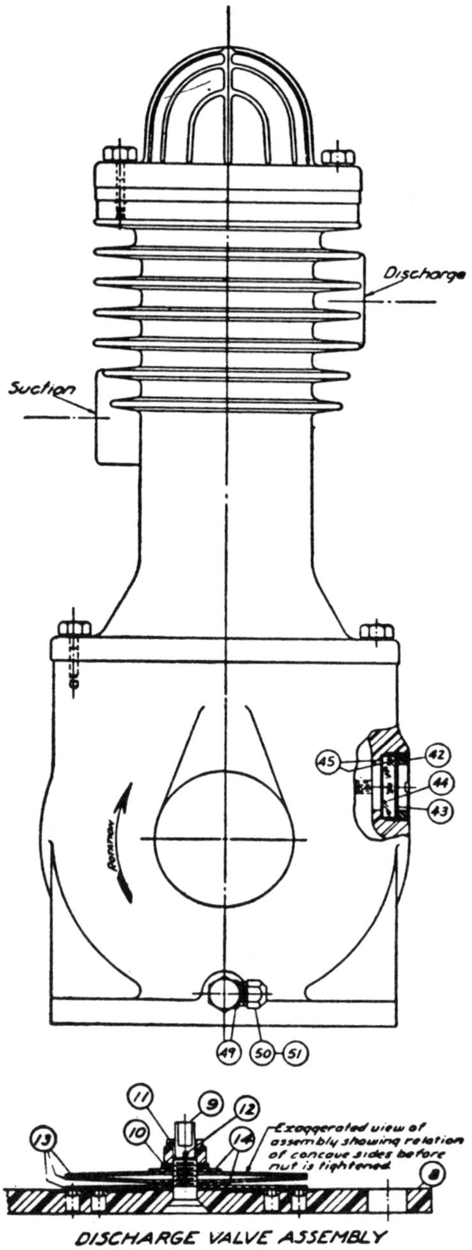

1 Crankcase
2 Bearing head
3 Crankshaft bearing
4 Shaft seal ring cover plate
5 Shaft seal ring cover plate drain pipe
6 Cylinder
7 Top head
8 Discharge valve plate
9 Discharge valve bolt
10 Discharge valve washer
11 3/8″ castellated nut (24th'ds)
12 3/32″ × 1cotter pin
13 3 3/4″ diaphragm valve
14 Diaphragm valve spacer
15 Connecting rod
16 Connecting rod bolt
17 9/16″ castellated nut (18-thread)
18 1/8″ × 1 7/8″ wire keeper
19 Taper pin #2 × 3/4″ long
20 Piston pin bushing
21 Piston
22 4″ × 3/16″ piston ring, plain
23 4″ × 3/16″ piston ring, ventilated
24 Piston pin
25 Suction valve screw
26 Crankshaft
27 1 1/4″ nut, 12 threads
28 5/16″ × 2 1/8″ Woodruff key #28
29 1/4″ steel ball
30 Shaft seal collar
31 Shaft seal assembly
32 Felt washer
33 Felt washer retainer
34 Flywheel 23″ diameter
37 Gasket (shaft seal diaphragm)
38 Gasket (cylinder to crankcase)
39 Gasket (bearing hd. to crankcase)
40 Gasket (discharge valve plate)
41 Gasket (top head)
42 Oil sight nut
43 Oil sight washer
44 Oil sight glass
45 Oil sight gasket
46 Hex head cap screw 9/16″ × 2″
47 Hex head cap screw 9/16″ × 1 1/2″
48 Hex head cap screw 9/16″ × 1″
49 Angle valve, 3/8″ mp × 3/8″ flare
50 Seal cap 3/8″
51 Flare gasket 3/8″

Figure 18-12. Two-cylinder in-line compressor.
Courtesy York Ice Machinery Corp.

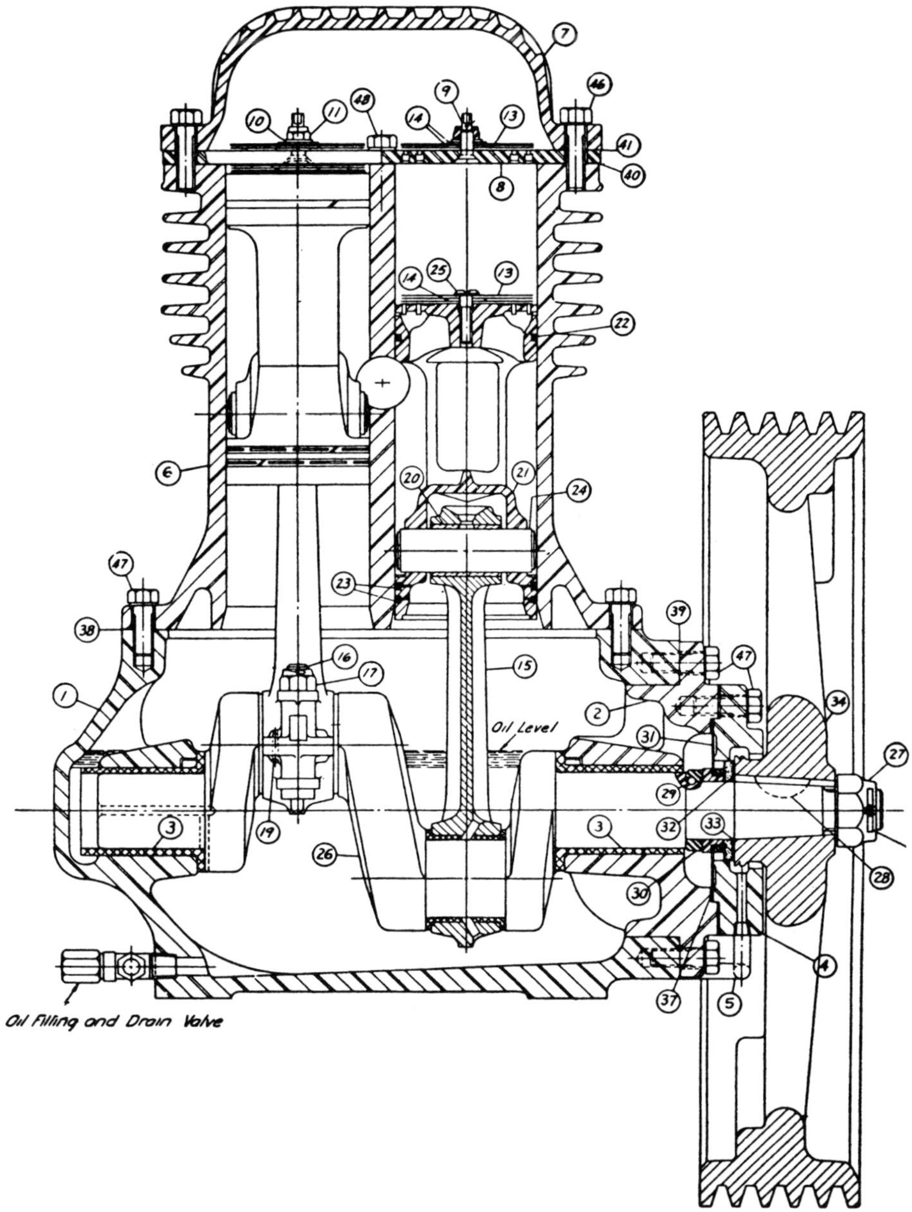

Figure 18-12 (continued).

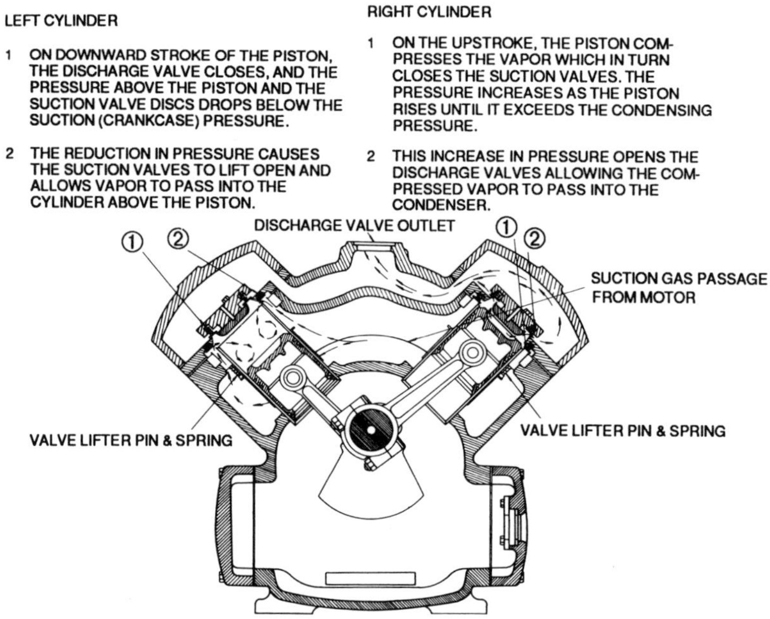

Figure 18-13. Compressor gas flow

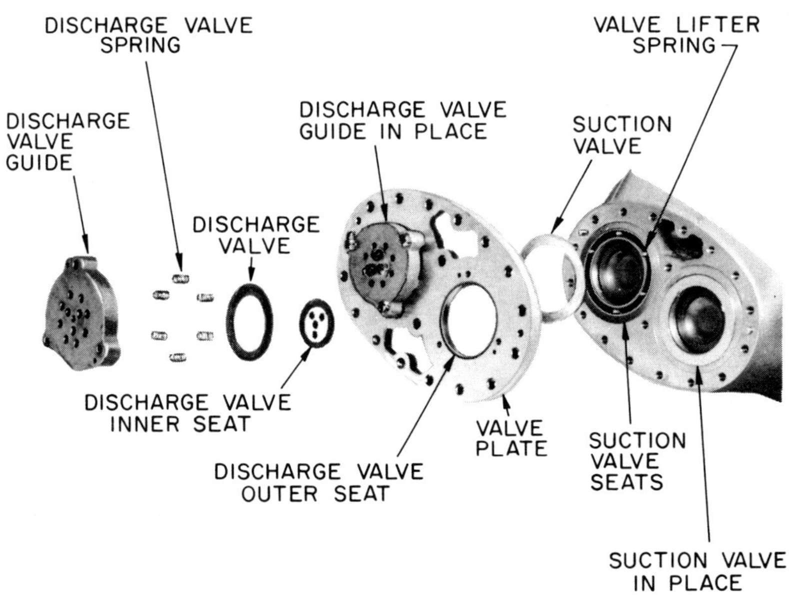

Figure 18-14. Suction and discharge valve assemblies. Courtesy Carrier Corp.

Open type compressors require a crankshaft seal where the shaft exits the casing. The most popular type of crankshaft seal is the mechanical seal, Figures 18-15 and 18-16. It consists of a spring-loaded carbon or bronze rotating seal which runs against a stationary seal plate with a highly polished face. A rubber gasket seals the crankshaft and the rotating seal. The spring maintains the rotating seal in firm contact with the stationary seal plate. An oil film is required to lubricate the two smooth surfaces and form a vapor-tight seal.

There are two basic types of lubrication systems found on refrigeration compressors: (1) splash and (2) forced feed. Splash lubrication is common on smaller compressors (10 HP and less) while larger compressors typically have some form of forced lubrication. A combination of splash and forced lubrication can be found on a single compressor.

In splash lubrication, the crankcase is filled with oil to approximately the centerline of the crankshaft. As the crankshaft turns, oil is splashed onto the bearings, the cylinder walls, and other surfaces. Oil scoops are sometimes used to force oil through drilled passages to lubricate the connecting rod bearings. Oil cavities usually are located above the main bearings to collect oil which then feeds the bearings by gravity.

In forced feed lubrication, a pump is used to deliver the oil to the various parts requiring lubrication. The pump is typically a small positive displace-

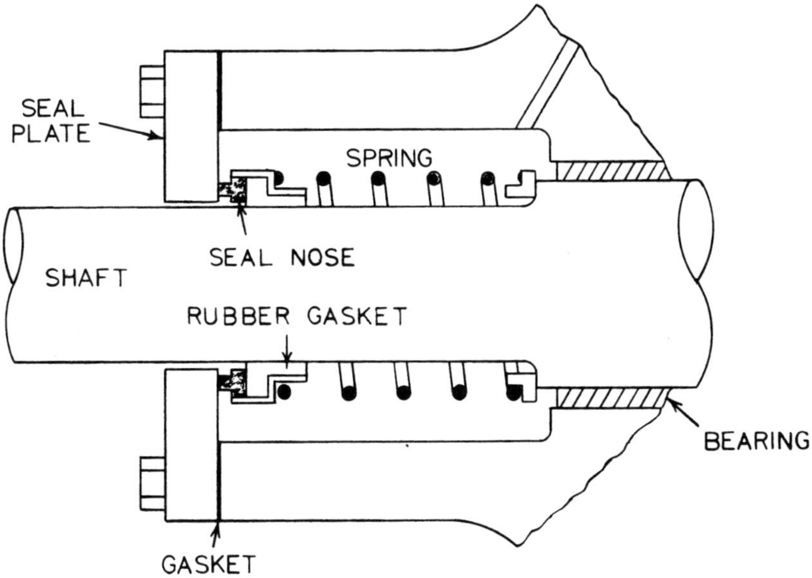

Figure 18-15. Mechanical shaft seal

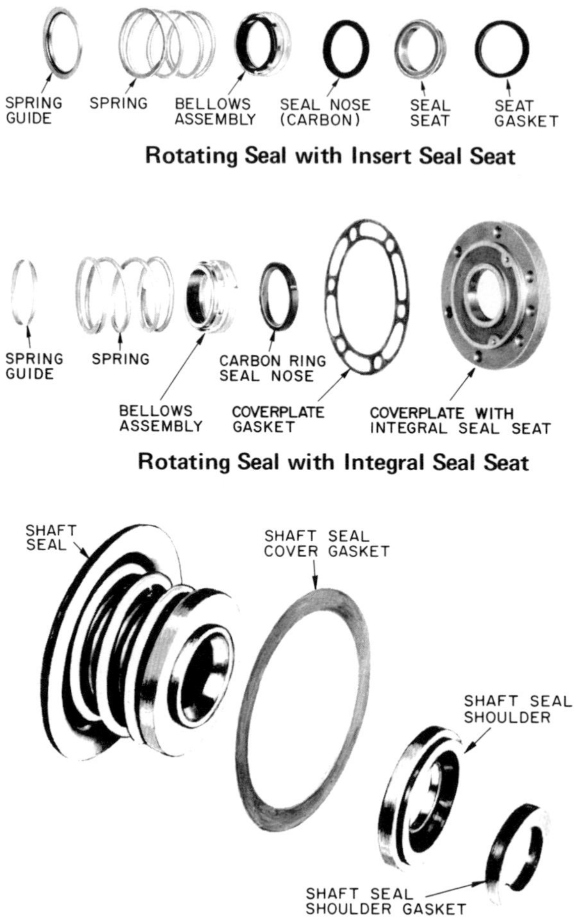

Figure 18-16. Compressor shaft seals. Courtesy Carrier Corp.

ment unit driven by gears from the crankshaft. The oil is delivered under pressure through oil tubes and/or drilled passages in the crankshaft and connecting rods. Oil strainers are located at the pump suction, and oil filters are sometimes installed at the pump discharge. An oil pressure failure switch is installed to warn of loss of proper lubrication. Figure 18-17 shows a typical compressor lubrication system.

The oil selected for lubrication of refrigeration compressors must be chemically stable, have low pour and cloud points, and have the proper viscosity for the operating temperature range. In addition the oil used in hermetic units must have a high dielectric strength to avoid grounding or

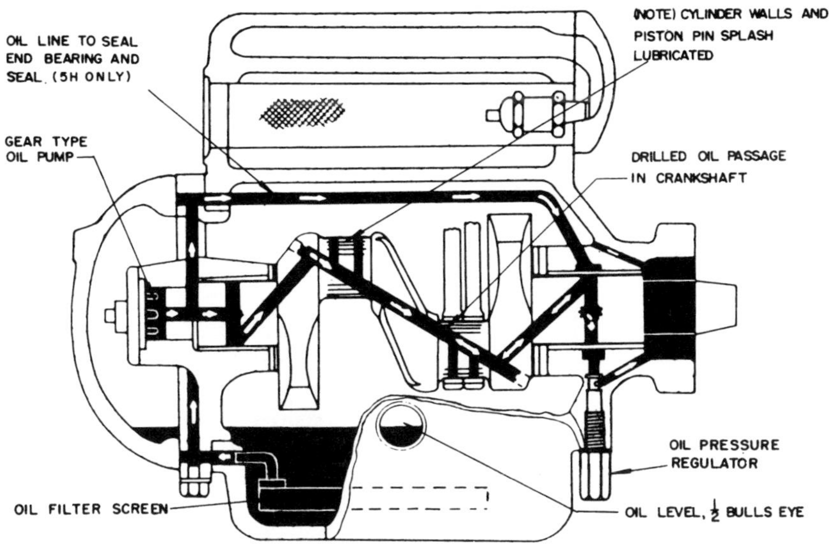

Figure 18-17. Compressor lubrication system

shorting of the motor windings. Consult the manufacturer's technical manual for the recommended oils for each compressor.

A certain amount of refrigerant will always be dissolved in the lubricating oil. However, large amounts of refrigerant in the oil are undesirable. Excessive dilution can result in inadequate lubrication. In addition, during compressor start-up, the lowering of the crankcase pressure will cause oil foaming due to the vaporization of the refrigerant. In severe cases, this can disrupt lubrication and can cause carryover of the liquid refrigerant and oil into the cylinder. Since marine systems typically operate on the pump-down cycle, the low crankcase pressure at shutdown limits refrigerant absorption by the oil. Crankcase heaters which come on automatically during the compressor off cycle can be used to keep the oil warm and reduce refrigerant absorption. A common cause of excessive oil dilution by liquid refrigerant is an improperly adjusted thermostatic expansion valve which allows overfeeding of the evaporator. The liquid refrigerant then carries over into the compressor crankcase.

There are a number of methods of controlling the capacity of reciprocating compressors. One method is to vary the speed of the compressor. Multispeed motors or steam turbines can be used to drive the unit. The most common method of varying the capacity of multicylinder compressors is to vary the number of active cylinders by holding the suction valves open. The capacity control system unloads cylinders (i.e., cuts cylinders out of operation) in response to decreases in refrigeration load based on suction pressure. Under high loads (high suction pressures) none of the suction valves are held open, and all the cylinders are in operation. As the load decreases (and the suction pressure falls off), the cylinders are cut out in sequence. If the suction pressure continues to fall off, the compressor will stop on the low pressure switch. Figure 18-18 shows a typical capacity control system. Compressor lubricating oil is used to operate the valve lifting mechanism. Since oil pressure is required to load the cylinder, the compressor will start with all controlled cylinders unloaded, thus reducing the starting load on the compressor motor.

Rotary Compressors

Rotary compressors are positive displacement machines. They have had limited usage in marine systems, but are becoming more common in industrial installations. The two basic types are vane and screw compressors. Vane compressors are commonly used in small domestic refrigerators, freezers, and room air conditioners. Vane compressors are similar in construction to a vane pump. Screw compressors first became popular for refrigeration service in Europe during the 1950s and 1960s. They are now popular in the United States for use in systems of 200 to 500 tons capacity. Construction is similar to a double screw pump. While some early compressors used timing gears to avoid lubrication of the rotors, common practice today is to drive one rotor and then inject oil between the rotors for lubrication and sealing. An oil separator is located in the system

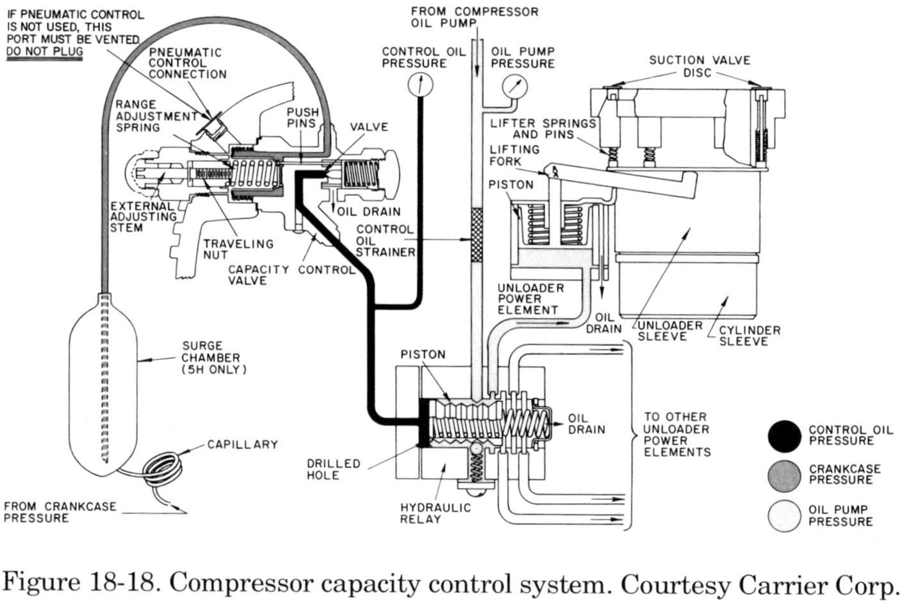

Figure 18-18. Compressor capacity control system. Courtesy Carrier Corp.

after the compressor discharge. Many screw compressors are fitted with a sliding valve for capacity control. As the valve is opened, it delays the position at which compression begins along the rotor.

Centrifugal Compressors

Centrifugal compressors are commonly used on systems of 300 tons and larger. They are especially suitable for large chilled water air-conditioning systems. While not common on commercial vessels, centrifugal compressor systems are used on many larger naval vessels.

Centrifugal compressors are similar in construction to centrifugal pumps. Figure 18-19 shows the construction of a typical centrifugal compressor. Due to the low density of the typical refrigerant gas, the pressure developed per stage is typically fairly low. Multistage compressors are common, even when relatively modest discharge pressures are required. Refrigerants such as R-11 which require a relatively large displacement per ton are especially suitable for these systems. Figure 18-20 is a diagram of a multistage centrifugal compressor system. Like most, it is close-coupled to minimize pressure drops, and employs a secondary refrigerant (chilled water or brine). It shows the use of flash gas removal during the expansion process which improves the cycle efficiency. Figure 18-21 shows a typical centrifugal compressor chiller unit.

Centrifugal compressors are pressure lubricated either by a shaft-driven or external motor-driven oil pump. The system is arranged to avoid contact of the oil by the system refrigerant, thus simplifying lubrication. The system also includes oil coolers to maintain oil temperature during operation and oil filters to remove contaminants. Since many systems operate with low side pressures below atmospheric, a purge recovery unit is commonly installed to continuously remove noncondensable gases and to recover and return refrigerant mixed with the purged gases. The system consists of a small reciprocating compressor, a condenser, and a separating tank. The suction of the compressor is connected to the main system condenser. The gases flow to the purge condenser where the refrigerant condenses. In the separating tank, a float valve controls the return of the liquid refrigerant to the main system, and a relief valve vents the noncondensable gases to the atmosphere.

Centrifugal compressor systems are controlled by sensing (or monitoring) the secondary refrigerant temperature. The system capacity is varied as required to maintain the chiller outlet temperature constant. The methods that can be employed to vary compressor capacity are: (1) prerotation inlet vanes, (2) variable speed operation, (3) varying condenser pressure, and (4) bypassing discharge gas. The first two methods are the most popular, but the latter two are sometimes used in combination with one of the other methods.

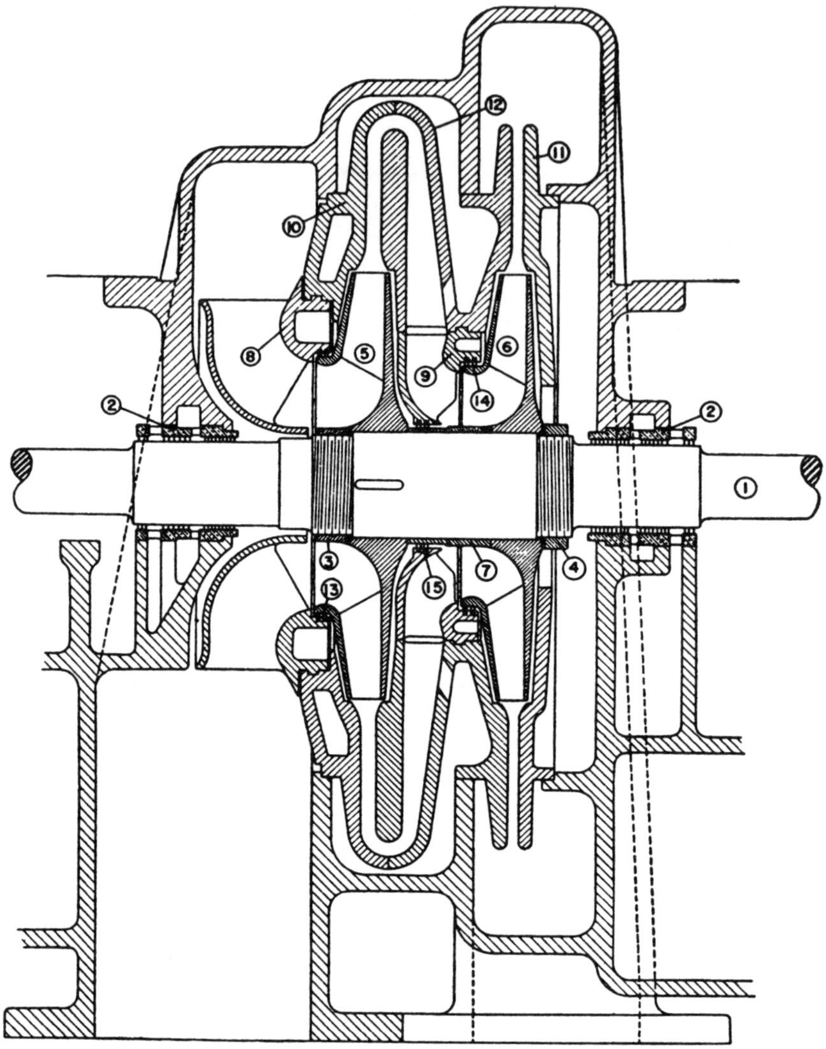

1 Shaft
2 Shaft labyrinth, either end
3 Impeller lock nut, suction end
4 Impeller lock nut, discharge end
5 Impeller, first stage
6 Impeller, second stage
7 Impeller spacer
8 Inlet guide vane, first stage
9 Inlet guide vane, second stage
10 Intake wall
11 Discharge wall
12 Diaphragm, first stage
13 Inlet labyrinth, first stage
14 Inlet labyrinth, second stage
15 Spacer labyrinth

Figure 18-19. Centrifugal compressor

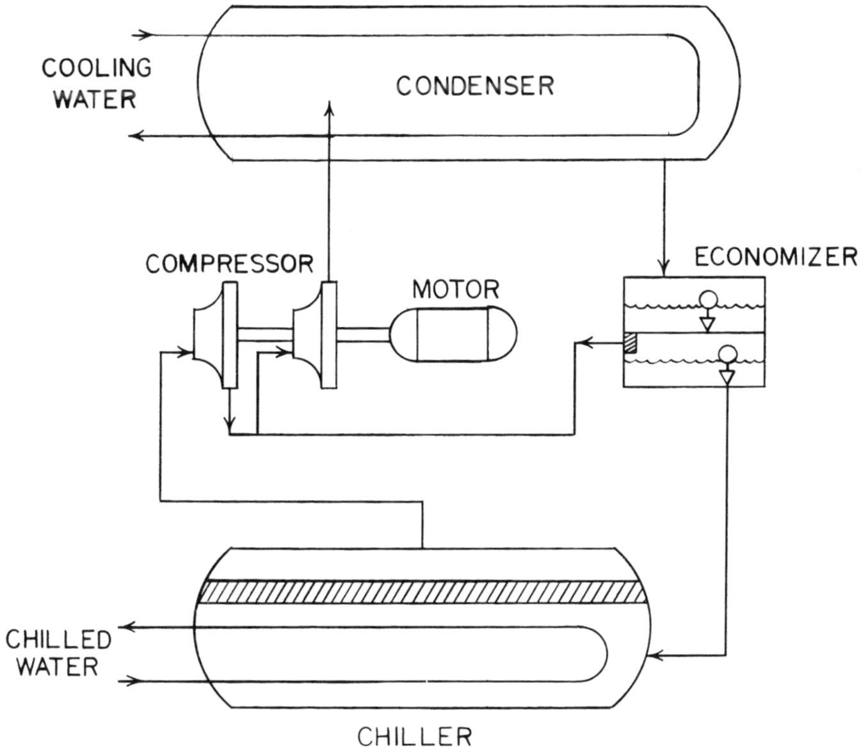

Figure 18-20. Centrifugal compressor system

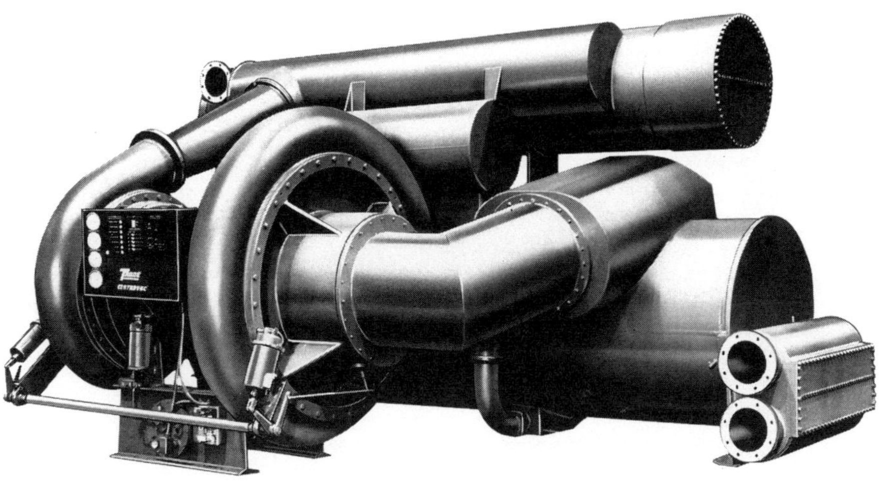

Figure 18-21. Centrifugal chiller. Courtesy The Trane Company

Condenser

Most marine refrigeration condensers are of the water-cooled, multipass, shell-and-tube type as shown in Figure 18-22. Seawater is circulated through the tubes, and the hot gas from the compressor discharge is admitted to the shell and condenses on the outer surfaces of the tubes. The condenser is typically constructed of a steel shell, copper-nickel tubes and tube sheets, and bronze waterheads. Gas inlet, liquid outlet, purge, and water regulating valve control connections are provided. Small systems such as found in drinking water coolers are usually air-cooled, with finned tubes and a cooling fan.

Receiver

The receiver collects the liquid refrigerant draining from the condenser. It consists of a steel shell with steel dished heads welded on each end. Sight glasses or a liquid level indicator is installed to permit determination of the amount of liquid refrigerant in the receiver. The receiver will typically have sufficient capacity to hold the entire system refrigerant charge and will retain a small liquid level during full load operation. High levels indicate overcharge and low levels indicate undercharge.

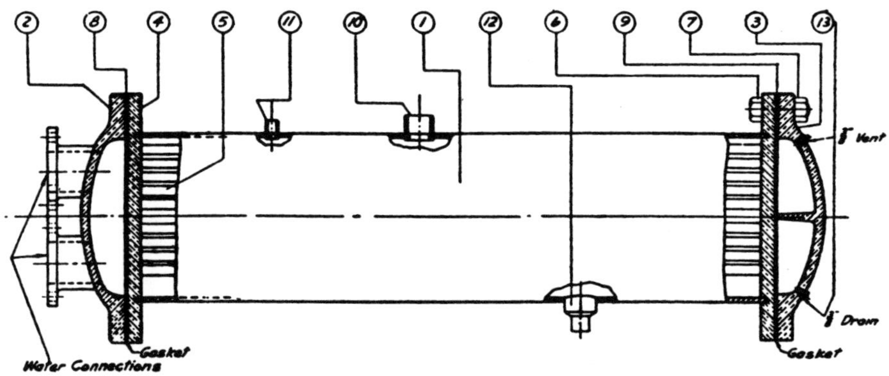

1 Condenser shell
2 Front head
3 Back head
4 Tube head
5 Tubes
6 Bolts
7 Hex nuts
8 Gasket
9 Gasket
10 Coupling, gas inlet
11 Coupling, purge
12 Coupling, liquid outlet
13 3/8-inch pipe plug

Figure 18-22. Condenser

Evaporators

Marine evaporators are usually either of the natural convection type or the forced convection type. Natural convection evaporators are bare or finned coils mounted on the walls or overhead of the refrigerated storage room. They are suitable for applications where low air velocities and minimum dehydration of the product are desired. Natural convection evaporators rely on the difference in density between cool and warm air to circulate air across the coils.

Forced convection evaporators are referred to as "unit coolers," "fan coil units," or "cold diffusers." They are in common use in ship's stores refrigeration, cargo refrigeration, and air conditioning systems. The units consist of a cooling coil with finned tubes, a motor-driven fan, and drain pan, all enclosed in a sheet metal casing. Units designed for air-conditioning applications may also contain a heating coil so that the same unit may be used for winter heating as well as air-conditioning. Units designed for freeze room applications are usually fitted with an electric resistance defrost system.

Liquid-Suction Heat Exchangers

Liquid-suction heat exchangers (heat interchangers) are commonly installed in the suction line to prevent carryover of liquid refrigerant to the compressor. It also subcools the liquid refrigerant flowing to the evaporators thus increasing the capacity and efficiency of the system. The heat exchanger is typically a shell-and-tube or shell-and-coil assembly with suction gas flowing through the tubes and the liquid flowing through the shell in a counterflow arrangement.

Thermostatic Expansion Valve

While there are a number of devices available to control the flow of refrigerant to the evaporator, such as the capillary tube, the float valve, and the constant-pressure expansion valve, the thermostatic expansion valve is the device most commonly found in marine systems.

The thermostatic expansion valve responds to the evaporator temperature and pressure and maintains a constant superheat at the outlet of the evaporator. As refrigerant is fed to the evaporator, the liquid boils off into a vapor. By the time the refrigerant gas reaches the end of the evaporator, it is superheated. Feeding more refrigerant to the evaporator will lower the superheat temperature, while feeding less refrigerant will raise the superheat temperature.

Figure 18-23 shows a typical thermostatic expansion valve. A feeler bulb partially filled with a volatile liquid, typically the system refrigerant, is clamped to the outlet of the evaporator so that the bulb and the power fluid inside closely assume the suction gas temperature. The pressure of power fluid is applied to the top of the valve diaphragm, and the evaporator pressure pushes on the bottom. A spring applies a force and balances the valve assembly as shown in Figure 18-24. For the valve to open and feed refrigerant to the evaporator, the bulb pressure must be higher than the evaporator pressure. Since the bulb pressure responds to the evaporator outlet temperature, this can only occur when the suction gas is superheated. The spring is normally adjusted by the factory to obtain about 10°F (6°C) of superheat in service. Valves set for less than this will not operate properly and may flood back liquid refrigerant to the compressor. Valves set for less than 10°F (6°C) will reduce the effectiveness of the evaporator and will reduce the system capacity.

Thermostatic expansion valves can be purchased as internally equalized or externally equalized. An internally equalized valve has a small port which admits the valve outlet pressure (i.e., the evaporator inlet pressure) to the bottom of the valve diaphragm. Some refrigeration systems have an appreciable pressure drop in the evaporator. Use of an internally equalized valve in such applications will result in unacceptably high superheat at the evaporator outlet. The solution is to use an externally equalized valve which has a fitting for connecting the evaporator outlet pressure rather than the inlet pressure to the bottom of the valve diaphragm. In general, an external equalizer should be used when the evaporator pressure drop is sufficient to change the saturation tempera-

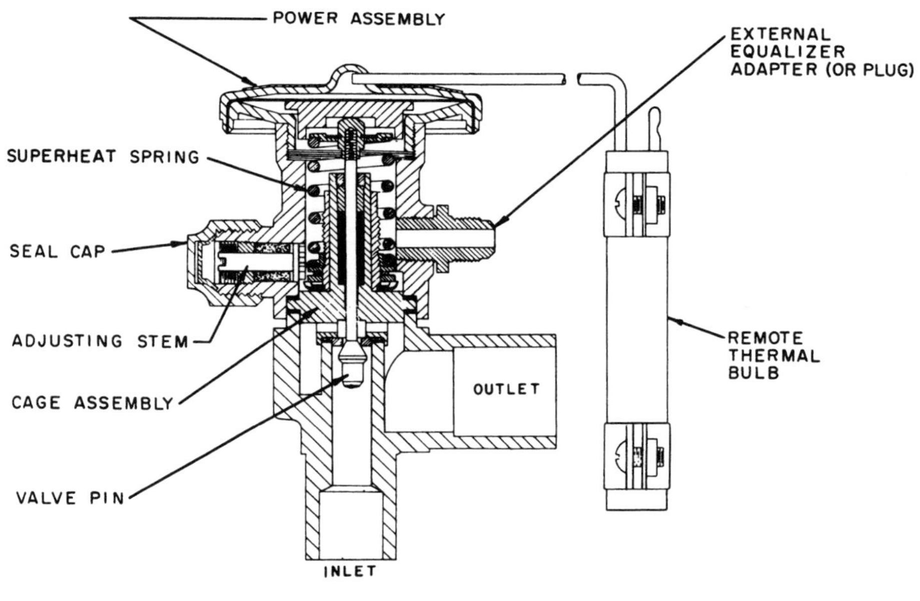

Figure 18-23. Thermostatic expansion valve. Courtesy Alco Controls

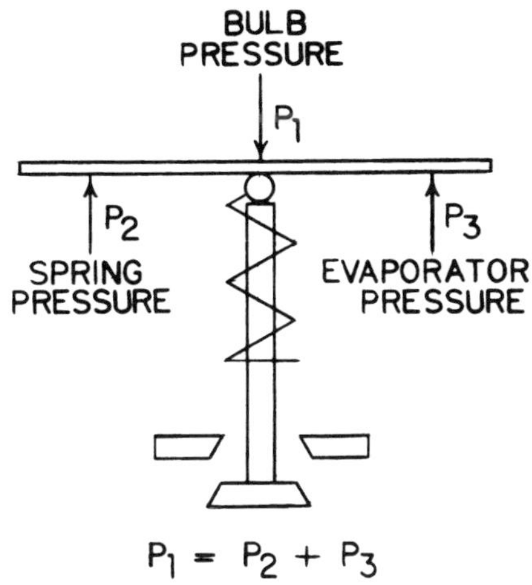

Figure 18-24. Thermostatic expansion valve equilibrium

ture 3°F (1.5°C) in air- conditioning applications, 2°F (1°C) in refrigeration systems, or 1°F (0.5°C) in low temperature (cryogenic) systems.

There are some specialized types of thermostatic expansion valves available. Pressure-limiting or gas-charged valves can be installed in systems to limit the compressor load during pulldown conditions. Cross-charged valves which have a bulb fluid different than the system refrigerant are used in low temperature applications. Because of the wide variety of valves available, it is important to replace a thermostatic expansion valve with one of the same type and size. Check the manufacturer's manual.

The superheat setting of most thermostatic expansion valves is adjustable. Remove the seal cap on the side of the valve and turn the adjusting stem with a screwdriver. Turning the stem to the right increases the spring pressure, reducing refrigerant flow, and increasing superheat. Turning the stem to the left decreases the spring tension, increasing refrigerant flow, and reducing superheat. Two turns of the stem will typically change the superheat about 1°F (0.5°C). One technique of measuring superheat is to use a contact thermometer to accurately measure the temperatures at the inlet and outlet of the evaporator. This method will yield good results only when the evaporator pressure drop is low. A second technique is to compare the evaporator outlet temperature with the saturation temperature corresponding to the compressor suction pressure. This will yield good results only when the pressure drop between the evaporator and the compressor is low. The best technique will involve taking pressure and temperature readings at the evaporator outlet. For example, the following data are taken for an R-12 system:

1. Evaporator outlet pressure (from gauge) = 37 psig (255 kPa)
2. Evaporator outlet temperature (contact thermometer) = 50°F (10°C)
3. 37 psig = 40°F (4.4°C) (Table 18-2)
4. Superheat = 50°F – 40°F = 10°F (5.6°C)

If a check of superheat indicates a problem, valve adjustment is indicated. It should be noted, however, that improper installation of the feeler bulb is a common cause of expansion valve problems. Figure 18-25 shows some recommended bulb installations. It is most important that the bulb make good thermal contact with the suction line. Clean the line thoroughly and clamp the bulb securely. If the bulb is installed in a location outside the cooled space, it will probably be necessary to insulate the bulb as shown in Figure 18-26.

High and Low Pressure Switches

These devices are very similar in construction and operation, but perform very different functions in the refrigeration system. The high pressure switch is a safety device. It is actuated by the compressor discharge pressure and stops the compressor in the event of high pressure. The low pressure switch is actuated by the compressor suction pressure. It is the primary control for stopping and starting the compressor during normal operation. When the suction pressure has been pumped down to the

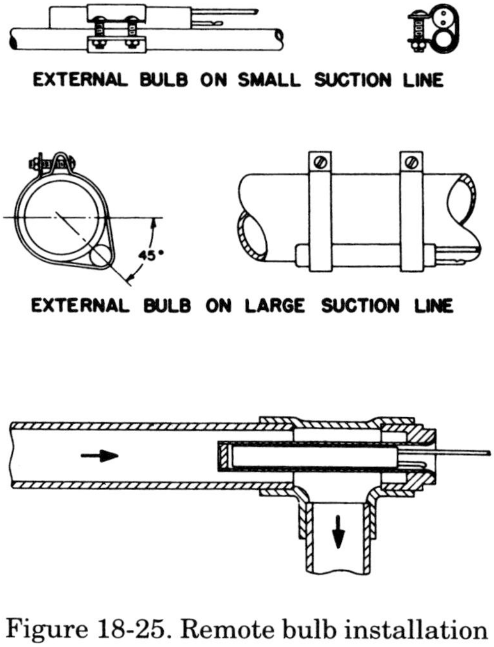

Figure 18-25. Remote bulb installation

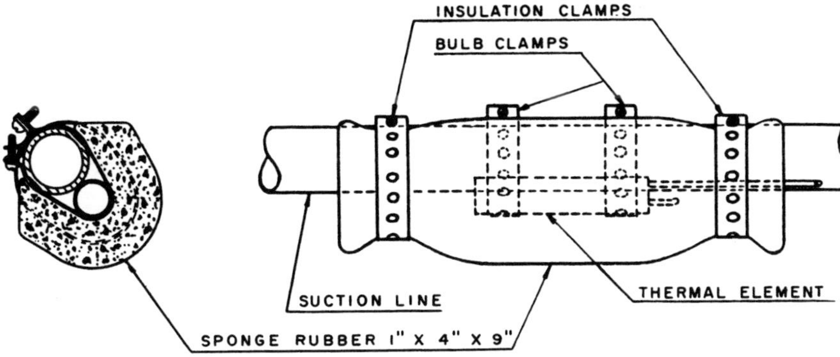

Figure 18-26. Expansion valve bulb insulation

desired level (the cut-out setting), the low pressure switch opens and stops the compressor. When the pressure rises to the desired level (the cut-in setting), the switch closes and the compressor starts.

The high and low pressure switches can be supplied as separate units or as a single unit usually called a dual pressurestat. Each switch consists of a sensing element (either a diaphragm or a bellows) and a snap action switch with a set point (range) adjustment and a differential adjustment. Figure 18-27 shows a typical pressure control switch.

Proceed as follows to set the switches. To set the high pressure switch:

1. Turn differential screw to minimum and range screw to maximum.
2. Start compressor and control discharge pressure by throttling condenser water flow.
3. Bring discharge pressure to cut-out point. Turn range screw until contacts open, stopping compressor.
4. Turn differential adjustment until contacts close, starting compressor when discharge pressure drops to cut-in point.

To set the low pressure switch:

1. Turn differential screw to maximum and range screw to minimum.
2. Start compressor and control suction pressure by throttling the compressor suction stop valve.
3. Lower suction pressure to about 10 psi (69 kPa) below desired cut-in point. Turn range screw until contacts open, stopping compressor. Allow suction pressure to rise to desired cut-in point and close suction valve to hold it there. Turn range setting slowly until contacts close, starting compressor. Open suction valve. The cut-in point is now set.
4. Lower suction pressure to desired cut-out point and decrease differential setting until contacts open, stopping compressor. The cut-out point is now set.

Thermostatic Switch

A thermostatic switch (or thermostat) is a temperature control switch used to open or close an electric contact based on sensing a temperature. The remote sensing bulb of the thermostat is filled with a volatile liquid charge. Changes in the sensed temperature cause changes in the pressure exerted by the remote bulb on a bellows which then operates the switch. On temperature rise, the pressure increases and the bellows closes the switch contact to complete the electric circuit. The circuit is interrupted on a reduction in temperature. A permanent magnet is typically part of the mechanism to impart a snap action to the switch and prevent excessive arcing at the contacts.

A typical thermostat will have a set point (range) adjustment and a differential adjustment. Changing the set point adjustment will affect the cut-in and cut-out points an equal amount. The differential adjustment affects the cut-out point only. To adjust a thermostatic switch proceed as follows:

1. Turn differential adjustment to maximum.
2. Turn range adjustment to minimum.
3. Bring the compartment down to the cut-in temperature and turn the range adjustment until the contacts close, then turn backwards until the contacts just open. This fixes the cut-in point.

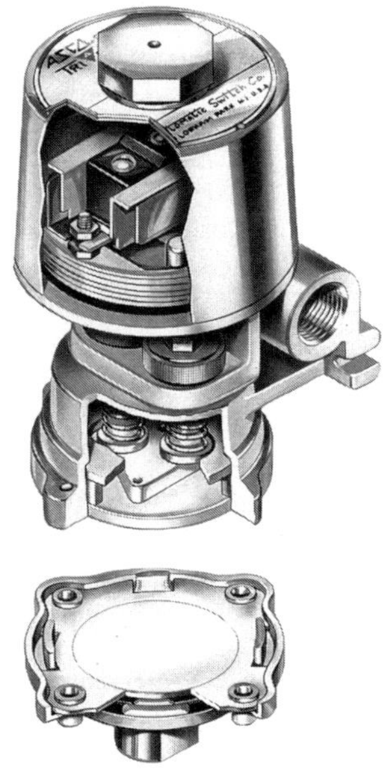

Figure 18-27. Pressure control switch. Courtesy Automatic Switch Co.

4. Lower the compartment to the cut-out temperature, then turn the differential adjustment until the contacts open. This fixes the cut-out point.

Solenoid Valve

A solenoid is used where there is a requirement to start and stop remotely the flow of refrigerant. Solenoid valves can be operated by a thermostatic switch, a float switch, a high pressure switch, a low pressure switch, or some similar device. The most common use of solenoid valves in reciprocating compressor systems is to control the flow of refrigerant to the evaporator using a thermostat. Figure 18-28 shows a cross section of a typical direct acting solenoid valve. When the coil is energized, the magnetic field draws a steel plunger towards the center of the coil, lifting the valve off its seat. When the coil is deenergized, the weight of the plunger, and in some designs a spring, causes it to fall and close the valve. The gravity-closing type must be installed with the plunger upright.

In larger capacities, a pilot-operated type is used to avoid large uneconomical coils. In this type of valve, the solenoid operates a small pilot valve which then uses refrigerant pressure to open the main valve. A small pressure drop across the valve is required for operation.

Oil Pressure Failure Switch

This device is a differential pressure switch designed to prevent operation of the compressor in the event of low oil pressure. It senses the pressure

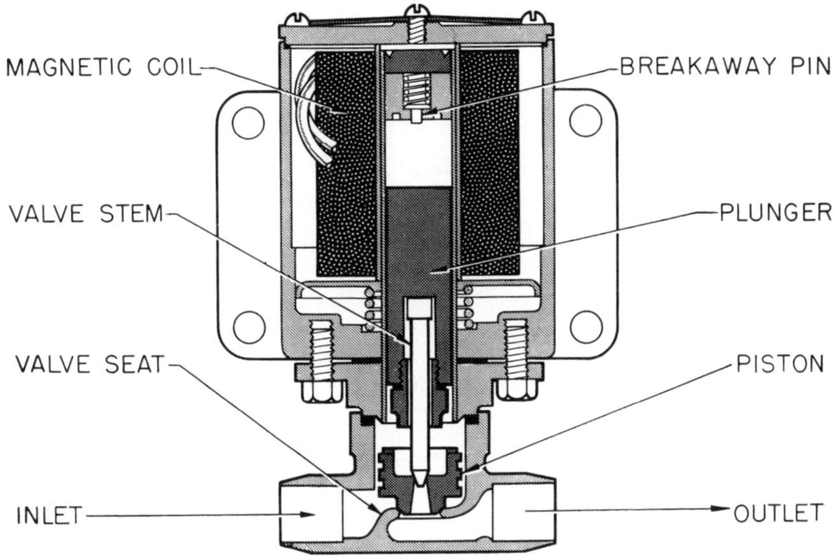

Figure 18-28. Solenoid valve

difference between the oil pump discharge and the crankcase (suction pressure). The switch stops the compressor motor in the event of low oil pressure. A time delay is incorporated into the control circuit to permit compressor start-up. A time delay is sometimes also incorporated in the cut-out mode. A manual reset is commonly installed. The switch is typically factory adjusted and if it does not operate properly, it must be replaced. Never operate the compressor if the oil pressure switch is inoperative. Correct the malfunction since improper lubrication will cause serious damage to the compressor.

Water Failure Switch

This switch is similar in construction to the low pressure switch. It stops the compressor in the event of loss of condenser cooling water supply. If the switch fails to function, the refrigerant pressure in the condenser increases to the point where the high pressure switch goes into operation.

Evaporator Pressure Regulator

Evaporator pressure regulators are commonly installed at the outlet of the higher temperature evaporators in a multibox system. Figure 18-29 shows a typical regulator. Its function is to prevent the evaporator pressure, and therefore its temperature, from falling below a predetermined minimum regardless of the compressor suction pressure. Note that the evaporator pressure regulator limits the minimum pressure, but does not maintain the evaporator pressure constant. The limiting of evaporator pressure and temperature allows the control of box humidity and prevents the dehydration of the stored product while permitting the compressor to satisfy the requirements of the coldest evaporator.

Suction Pressure Regulator

The suction pressure regulator or "holdback valve" is sometimes installed to limit the suction pressure at the compressor inlet to a predetermined maximum. This prevents overload of the compressor driver during pulldown or other periods of high evaporator pressure. The valve is similar in construction to the evaporator pressure regulator but senses the downstream rather than the upstream pressure.

Water-Regulating Valves

Most marine refrigeration systems use seawater-cooled condensers. Water-regulating valves are commonly installed to control the quantity of cooling water circulating through the condenser. Figure 18-30 is a typical water-regulating valve. The valve is actuated by the refrigerant pressure in the condenser. If the condenser pressure should increase, the valve will open, admitting more cooling water, and returning the pressure back to the set point. A decrease in condenser pressure will cause the valve to close

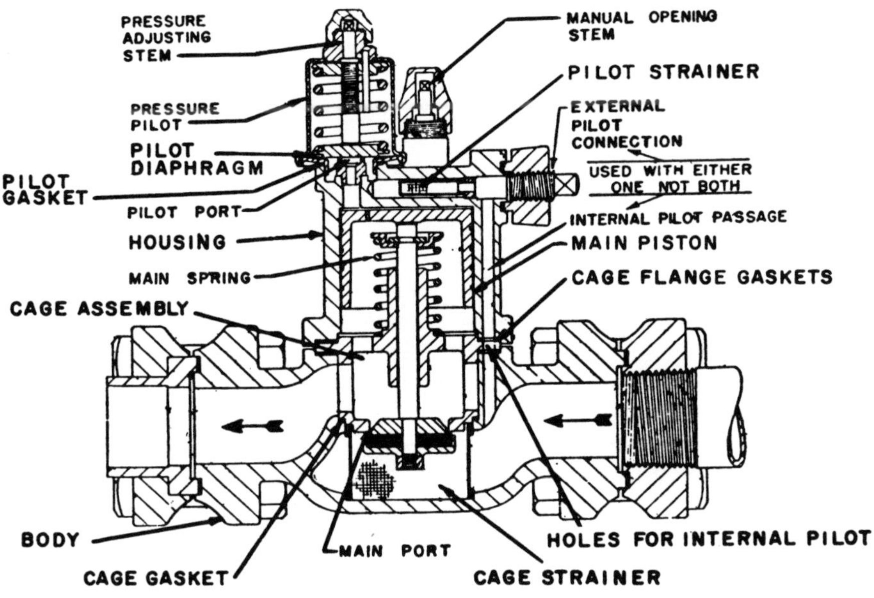

Figure 18-29. Evaporator pressure regulator. Courtesy Alco Controls

and reduce the water flow. Valves on R-12 systems are typically adjusted for about 125 psig (682 kPa). Consult the manufacturer's manual for the proper set points. When the compressor stops, the condenser pressure gradually decreases to the saturated vapor pressure corresponding to the ambient temperature. This decrease in pressure is sufficient to close the regulating valve and stop the flow of cooling water.

Relief Valve

Relief valves are installed in refrigeration systems to relieve unsafe system pressures. Marine systems will have at least one relief valve installed on the condenser with discharge overboard through a rupture disc. If there is an isolation valve between the condenser and the receiver, there will be a relief valve installed on the receiver with the outlet connected to the condenser. Small systems sometimes have a fusible plug installed in place of a relief valve. A fusible plug is simply a pipe plug which has been drilled and filled with a metal alloy designed to melt at a predetermined temperature. The design melting point depends on the pressure-temperature relationship of the system refrigerant.

Manual Valves

Manual valves are installed at various locations in the refrigeration system to facilitate system operation, to permit cutting units in and out, to isolate

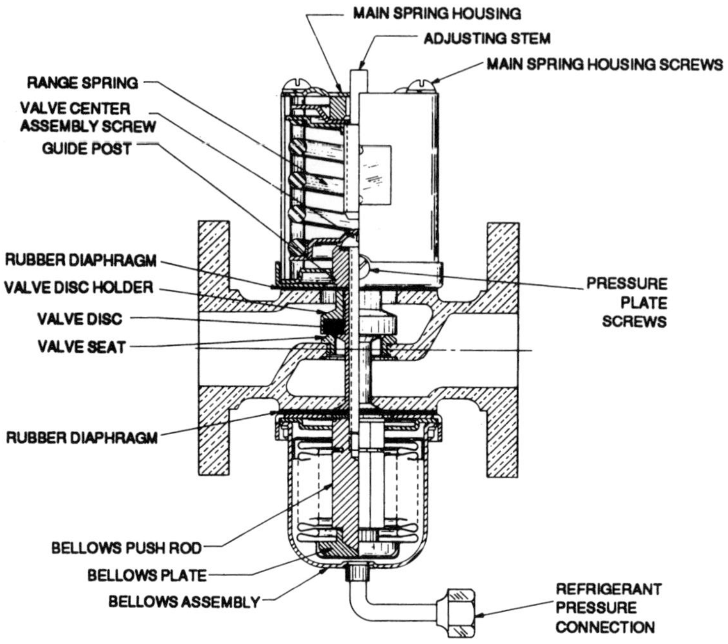

Figure 18-30. Water regulating valve. Courtesy Penn Valve

components for maintenance, and for other purposes. Two basic types of valves are used, packed and packless. Most of the valves used in refrigeration systems today are of the packless type as shown in Figure 18-31. Packless valves employ a diaphragm to isolate the handwheel assembly from the refrigerant to minimize the possibility of leakage. Packed valves are usually of the backseating type as shown in Figure 18-32. When in the open position, the valve is backseated to minimize the possibility of leakage. The compressor service valves are usually of the packed type and commonly include a connection for attaching a service gauge. Backseating the valve permits attaching a gauge without loss of refrigerant. To measure the system pressure, the valve must be in an intermediate position between the front and back seats.

Liquid Indicators (Sight Glasses)

Liquid indicators or sight glasses are commonly installed in the liquid line to indicate a proper refrigerant charge. Bubbles appearing in the liquid stream are an indication of a shortage of refrigerant. Some indicators also include a moisture indicator. A portion of the indicator will change color based on the relative moisture content of the refrigerant.

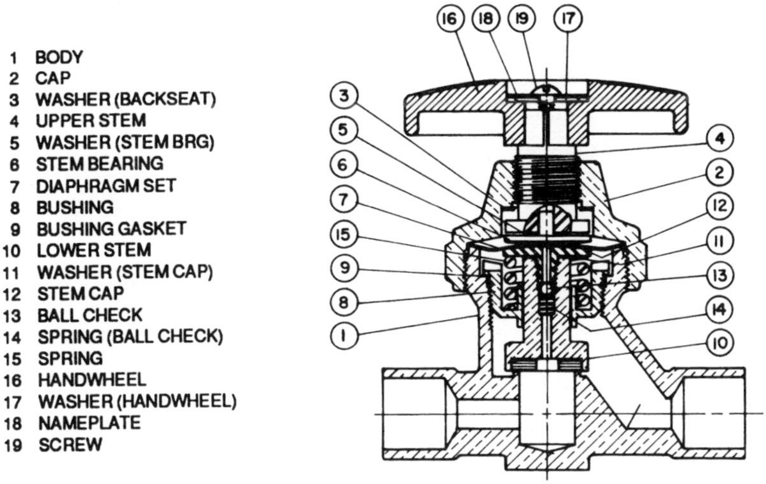

Figure 18-31. Packless valve. Courtesy Henry Valve Co.

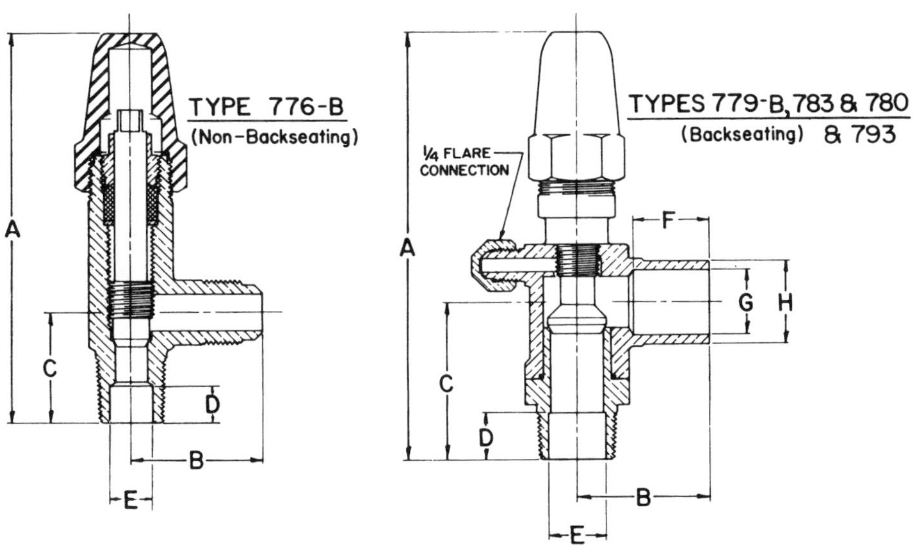

Figure 18-32. Packed valve. Courtesy Henry Valve Co.

Dehydrator

A dehydrator or drier is usually installed in a bypass line downstream of the receiver. This permits replacement of the cartridge without shutdown of the system. A typical dehydrator is shown in Figure 18-33. The shell is

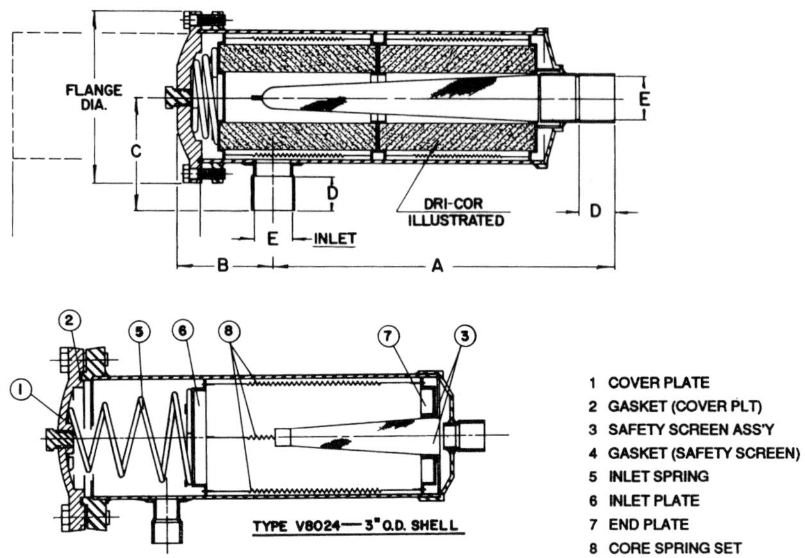

Figure 18-33. Dehydrator. Courtesy Henry Valve Co.

filled with a desiccant such as activated alumina or silica gel which adsorbs moisture and also acts as a filter. Even small amounts of moisture can cause problems such as frozen thermostatic expansion valves, so it is important to remove sufficient moisture to prevent the release of water in the low pressure portions of the system. Lower temperatures require lower moisture content. To obtain maximum service life, the dehydrator should only be in service during charging or when high moisture content is suspected.

Strainer

Strainers are commonly installed in the liquid line before the solenoid valve and expansion valve to remove foreign particles such as scale, metal chips, and dirt. Since the refrigeration system is a closed circuit and most refrigerants are excellent solvents, any solid impurities will be circulated and cause clogging and sticking of the automatic valves. Fine mesh strainers such as are shown in Figure 18-34 will remove these solid impurities. In addition to mesh strainers, filters with porous pad elements for the removal of very fine particles are available. The strainer element can be removed for cleaning without securing the system by isolating the strainer and then using the hand expansion valve.

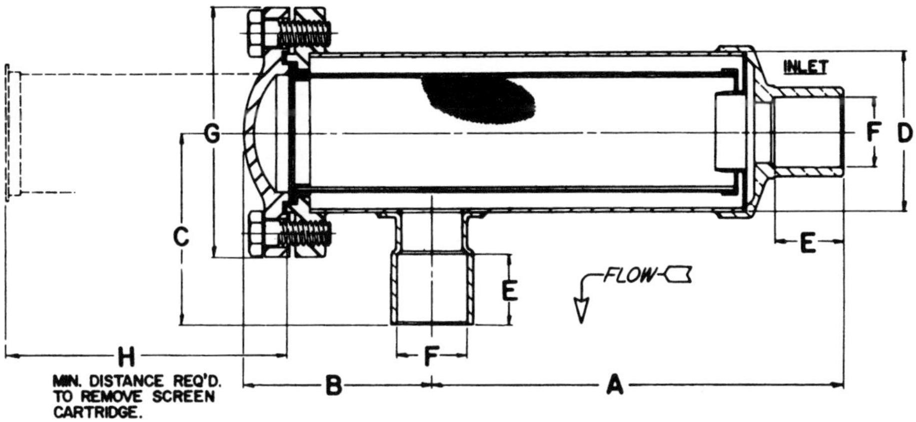

Figure 18-34. Strainer. Courtesy Henry Valve Co.

OPERATION AND MAINTENANCE

Safety Precautions

The following precautions are general in nature. They apply to typical reciprocating compressor systems and to CFC-12, HCFC-22, HFC-134a, and other fluorocarbon refrigerants used in such systems.

1. Inspect compressor oil level and check oil pressure periodically. Typical oil level is from one-half to three-quarters up on the sight glass. The typical oil pressure is 45 to 55 psi above the suction pressure.
2. Do not start a compressor without ensuring that shutoff valves between the compressor and the condenser are open.
3. Do not jack or turn the compressor by hand when the power is on.
4. Monitor compressor operation carefully during initial start-up. Check for proper lubrication, liquid floodback, severe vibration, or unusual noise.
5. Do not attempt to add oil to compressor crankcase while compressor is in operation.
6. Do not bypass or jump any protective device because it is operating improperly. Find the trouble and make the necessary corrections.
7. Do not wipe down near moving parts.
8. Open circuit switch in case of electrical fire, and extinguish with CO_2 (never use water).
9. Be sure power is turned off before working on electrical equipment and circuits. Tag circuit breaker to prevent accidental energizing of circuit.

10. Bleed piping section of liquid refrigerant prior to opening for repairs. Close the inlet valve, wait until the piping warms indicating liquid refrigerant removal, then close the outlet valve.
11. Do not open any part of the system which is under a vacuum, or air and moisture will be drawn in.
12. Drain the cooling water system to prevent a freeze-up during system shutdown in freezing weather.
13. Expel all air from the system section by admitting a small amount of refrigerant into the section prior to final tightening after inspection or repair.
14. Do not use a torch on a line that has not been bled of refrigerant since some fluorocarbons decompose into phosgene, a highly toxic gas, when exposed to high temperatures (above 1000°F [538°C]). The area should be well ventilated, and all isolation valves closed securely.
15. Always wear goggles when handling refrigerants. If liquid refrigerant accidentally comes in contact with the eyes, obtain medical help immediately. Do not rub or irritate the eyes, and give the following first aid treatment immediately: (a) Introduce drops of sterile mineral oil into the eyes to irrigate; (b) Wash the eyes with a weak boric acid solution, or a sterile salt solution not to exceed 2 percent sodium chloride if irritation continues at all.
16. Treat as if the skin had been frostbitten if liquid refrigerant comes in contact with the skin.
17. Do not work in a closed area where refrigerant may be leaking unless adequate ventilation is provided. Should a person be overcome due to lack of oxygen and high concentration of refrigerant, treat with artificial respiration.
18. Use care in handling and storing refrigerant cylinders. Don't subject cylinders to high temperatures. Take precautions to prevent mechanical damage.

Starting

To start up a system that has been secured, proceed as follows:

1. Check oil level in compressor crankcase.
2. Line up condenser seawater circulating system. Open valves in suction and discharge lines and close condenser water vents and drains.
3. Line up refrigerant system valves for normal system operation. Leave compressor suction stop valve closed.
4. Start seawater circulating through condenser. Vent air from condenser waterheads.
5. Start fans in refrigerated compartments (or pumps in brine or chilled water systems).
6. Check electrical power supply to compressor and solenoid valves.
7. Open compressor suction stop valve one full turn.
8. Start compressor in Auto mode.
9. Open suction valve slowly to prevent rapid pumping down of low pressure side to avoid oil foaming. If there is evidence of liquid floodback, throttle suction valve. If the compressor develops a knock, secure for five minutes, and restart with suction valve throttled.
10. Observe system operation carefully for five to ten minutes.

Operation

1. Follow carefully the "Preventive Maintenance Schedule" below, including a regular check of all pressures, temperatures, and crankcase oil level.
2. Adjust seawater flow to maintain proper discharge pressure.
3. Check for an open hand expansion valve and thermostatic valves for proper operation if frosting occurs. Frost on compressor cylinders and crankcase is caused by liquid refrigerant floodback.
4. Bypass the heat interchangers if the compressor discharge gas temperature exceeds 240°F. Otherwise, the unit should be kept in operation.
5. Crack valves and open gradually to avoid liquid floodback if hand expansion valves must be used to permit inspection or repair.
6. Change over to standby compressor at least weekly during normal plant operation.

Securing

To secure a condensing unit, proceed as outlined below:

1. Close main liquid line stop valve.
2. Let compressor run until it cuts out on low pressure switch.
3. Depress compressor motor controller Stop button.
4. Close compressor suction and discharge stop valves.
5. Secure condenser seawater circulating system.
6. Close appropriate refrigerant valves.

Defrosting

Hot gas defrosting is performed by pumping hot discharge gas from the compressor directly into the cooling coil circuit to be defrosted. The hot gas is cooled and condensed, and the frost melts off the coils. The resulting liquid refrigerant is expanded and evaporated in a second coil through a hand expansion valve. During defrosting, the plant should be under normal operation. To defrost a coil by hot gas, proceed as follows:

1. Close cut-out valve on outlet side of thermal expansion valve for coil to be defrosted.
2. Allow sufficient time for all liquid refrigerant in coil being defrosted to be pumped out by the compressor. Close coil outlet valve. Open hot gas supply valve for coil to be defrosted.
3. Stop compressor. Throttle condenser inlet valve, and open hot gas supply valve connecting compressor discharge to hot gas defrosting line. Throttling of condenser inlet valve should be regulated so as not to unduly prolong defrosting and also not to starve coils not involved in the defrosting.
4. Start compressor and slowly open hand expansion valve between coil being defrosted and coil used for expansion. Adjust hand expansion valve to prevent liquid refrigerant return to compressor. Close cut-out valve on outlet of thermal expansion for coil used for expansion.

5. Stop compressor when coil is defrosted. Close hot gas supply valve, and open condenser inlet valve fully.
6. Restart compressor.
7. Close hand expansion valve when all liquid refrigerant from defrosted coil has been removed and expanded.
8. Open coil outlet valve and both thermal expansion valve cut-out valves.

Refrigerant Charge

The initial refrigerant charge for the system will be given in the system instruction manual. This quantity, however, is an estimate, and the actual quantity must be determined by trial and error. A refrigerant overcharge is indicated by high head pressure. A refrigerant undercharge is indicated by one or more of the following: low head pressure, low receiver level, vapor bubbles in the liquid line sight glass, compressor running continuously, short cycling, and hissing at control valves. Test for leaks if a loss of refrigerant occurs during normal operation.

Charging Refrigerant

To add refrigerant to an operating system, proceed as follows:

1. Close drier bypass and inlet valves.
2. Check to see that solenoid valves are open.
3. Weigh refrigerant drum.
4. Connect refrigerant drum to charging valve with flexible charging line. Crack drum vapor valve before tightening line to blow out air. Observe regulations on venting CFCs and HCFCs.
5. Close main liquid line valve and pump down system.
6. Open charging valve and carefully open liquid valve on refrigerant drum. Liquid refrigerant will flow into the system.
7. Start compressor.
8. Continue charging until required amount of refrigerant has been charged. Check scale reading and observe liquid level in receiver. Close charging valve, close drier outlet valve, and open drier bypass valve. Open main liquid valve and observe liquid flow through sight glass. Bubbles indicate the need for further charging. If charge is complete, close drum valve, and disconnect charging line and drum. Store empty refrigerant drums for reuse.

Removing Refrigerant

Refrigerant must be removed from a system that has been overcharged or prior to performing many repairs. In order to recover sufficient refrigerant to perform most repairs, a dedicated recovery unit must be used. These units are discussed later in this chapter. If a small amount of refrigerant must be removed due to an overcharge, refrigerant may be transferred from the system to a reusable storage cylinder as follows:

1. Secure the liquid supply to the evaporators. Close the drier bypass, open drier inlet valve, and check that drier outlet valve is closed.
2. Weigh the refrigerant cylinder and verify that it can accommodate the refrigerant to be removed from the system.
3. Connect a hose between the refrigerant cylinder and the system high-side liquid connection. Purge the hose of air.
4. Open the cylinder liquid valve and the system liquid valve with the system compressor operating automatically. Liquid refrigerant will flow from the system to the cylinder. Continue to weigh the cylinder while transferring refrigerant. Do not overfill the refrigerant cylinder.
5. Close the system valve when the required amount of refrigerant has been transferred. Allow any liquid in the hose to drain into the cylinder and then close the cylinder valve. Disconnect the hose and store the cylinder. Restore the liquid supply to the evaporators.

Testing for Refrigerant Leaks

If a system being serviced is found to require a charge, leak detection procedures must be followed prior to the charging of additional refrigerant. A system that regularly requires a charge obviously has leaks. Adding refrigerant to such as system is a violation of the regulations. The following techniques can used to detect leaks in refrigeration systems:

1. Visual inspection and fluorescent dyes. A simple examination of the system looking for signs of leakage such as oil at joints or seals will often reveal leaks or locations for further investigation. The charging of a fluorescent dye into the system will enhance the sensitivity of this approach, as even small leaks can be identified by brightly glowing spots.
2. Soap bubble test. This procedure is very effective in finding larger leaks. By applying a solution of soap and water to the joints, seals, and fittings of a system under pressure, gas escaping from the system will cause bubbles to form.
3. Halide torch. This device consists of a burner, needle valve, suction tube, and a chimney with a copper reaction plate. To operate, adjust the flame so that the top of the blue flame cone is level with or slightly above the reaction plate. Move the suction tube slowly along the suspected area. If refrigerant is present, it will react with the copper plate, changing color to a vivid blue with large leaks and taking on a greenish tint with small ones. The halide torch will not detect many of the new alternative refrigerants.
4. Electronic halide detector. A variety of leak detectors are available that utilize an ionization cell or other specialized sensor. Most emit a clicking sound that increases in frequency as the sample probe draws in air containing refrigerant. Different models vary in sensitivity or their ability to detect different refrigerants. Many of the older units relied on detecting chlorine and will not work with HFCs. It is important to use a unit designed to detect the refrigerant used in the system.
5. Ultrasonic leak detector. This sensor detects the sound created by refrigerant leaking from a system under pressure or air leaking into a system under

vacuum. Background noise can interfere with the use of this device. One advantage is that it will work in areas where air currents would upset other methods like the electronic halide detector.

6. Standing pressure test. This technique can be used on a new system or one that has been evacuated of refrigerant. Inert gas is used to pressurize the system. If the system holds pressure, leaks are not present. If the pressure drops, an ultrasonic detector or soap bubbles can be used to find the leaks. One advantage of this method is that if leaks are present, refrigerant will not be vented to the atmosphere.
7. Standing vacuum test. This is a good test to use after servicing and before charging the system. The system is evacuated to a high vacuum with a vacuum pump. The ability of the system to hold vacuum is a good indication of a tight system. The procedure for evacuation and dehydration of a refrigeration system is described in the following section.

Evacuation and Dehydration

After initial installation or following extensive repairs, evacuation and dehydration of the system is required to prevent later troubles. Moisture in the system causes oil sludge and corrosion, and it is likely to freeze up the expansion valves of a low temperature system. Tests and field experience have shown that most troubles with seals and internal valves are caused by moisture in the system. Proper dehydration requires a vacuum pump capable of producing a vacuum of .02 inches Hg absolute (69 Pa) and an electronic vacuum gauge. Ambient temperature must be above 60°F (15.5°C) for proper dehydration. To evacuate and dehydrate a system proceed as follows:

1. Pressure-test the system to be sure it is free of leaks. Drain oil from the compressor crankcase. Replace with new oil after dehydration.
2. Release all refrigerant pressure from the system. Connect a tee to system-charging valve. Connect vacuum pump to one side of tee and vacuum gauge to the other. Provide shutoff valves in branch lines.
3. Open compressor stop valves and all line stop valves in system. Close all valves and connections to atmosphere. Be sure to open hand expansion valves, coil return valves, and any other line valves which will connect the high and low sides of the system and allow the pump to draw a vacuum on the entire system. If necessary, install a "jumper" line between the high and low sides of the system.
4. Open shutoff valve in vacuum pump suction line, start vacuum pump, and slowly open charging valve.
5. Open shutoff valve in gauge line occasionally and take a reading. Continue dehydrating until vacuum gauge indicates 1,500 microns (207 Pa). The operation will probably take 18 to 72 hours depending on system size and amount of moisture in the system. Failure to achieve a reading of 1,500 microns (207 Pa) may be due to one or more of the following problems: (a) Leak in system or connecting tubing; (b) Closed line valves; (c) Defective vacuum pump; (d) Defective vacuum gauge; and (e) Ambient temperature below 60°F (15.5°C).

6. Close refrigerant charging valve and vacuum gauge and pump valves, and stop the vacuum pump. Disconnect vacuum line from charging valve, and connect a drum of system refrigerant. Raise system pressure to 10 psig (69 Pa).
7. Disconnect refrigerant drum, and release system pressure to atmosphere. Reconnect vacuum line, gauge, and pump to charging valve. Repeat steps 4 and 5. Dehydrate system to 500 microns (69 Pa) instead of 1,500 microns (207 Pa). Close vacuum pump shutoff valve and stop pump.
8. Monitor vacuum gauge for 15 minutes to ensure system maintains vacuum. If system holds a vacuum, repeat step 6. The system is now ready for charging.

Testing for Noncondensable Gases

Air and noncondensable gases, if present in the system, are pumped through the system and discharged by the compressor into the condenser. These gases are trapped in the condenser and cause excessive condensing pressures. In order to check the condenser for the presence of air or noncondensable gases, it is essential that gauges and thermometers be accurate and that the system has sufficient charge so that the liquid refrigerant present in the receiver will seal the liquid line connection. To check for noncondensable gases, proceed as follows:

1. Close liquid line valve and allow system to pump down.
2. Shut off compressor and close suction line valves.
3. Determine the actual condensing temperature. A service gauge should be installed in the compressor discharge connection if a discharge pressure gauge is not already provided. An approximation of the actual condensing temperature will be reached when no further decrease is noted in the discharge pressure. The reduction in pressure can be accelerated on water-cooled condensers by permitting circulation of cooling water until discharge pressure is reduced. The thermometer in the liquid line at the receiver provided on most vessels will then indicate the actual condensing temperature. If a thermometer is not installed in an air-cooled condenser application, a thermometer should be placed near the condenser to record the ambient temperature at that location. When the pressure and temperature of the air-cooled condenser have lowered to the ambient temperature, a reading of the thermometer will approximate the actual condensing temperature.
4. Record the condensing pressure. Refer to Table 18-2 or a similar table of properties for the system refrigerant, and look up the saturation temperature that corresponds to the condensing pressure.
5. Subtract the temperature recorded in Step 3 from the temperature recorded in Step 4. If the difference between these two temperatures is more than 5°F (2.8°C), it is necessary to purge.

Purging Noncondensable Gases

If the above test indicates the presence of noncondensable gases, proceed as follows:

1. Stop the compressor for 10 to 15 minutes. Leave all valves in their normal position.
2. Open purge valve on top of condenser, and slowly release gases.
3. Since it is difficult to tell if excessive refrigerant is being purged with the noncondensables, purge slowly and check the condenser continually for the presence of noncondensable gases as explained above to minimize refrigerant loss while purging. Observe regulations on venting CFCs and HCFCs.

Compressor Oil Level

A certain amount of compressor oil will always circulate through the system because oil is miscible in refrigerant. To allow for oil circulation, systems requiring a large refrigerant charge will need the addition of oil in excess of the normal compressor crankcase oil charge. When the system is first placed in operation, closely observe the oil level in the crankcase. Add oil whenever the oil level drops below normal (halfway up on the bull's-eye sight glass). Allow sufficient time for the system to balance after adding oil, since some of the circulating oil may return to the compressor. Then, again check oil level. Generally, the addition of one quart of oil for every fifty pounds of refrigerant charge will be an adequate allowance for oil circulation. After adding oil, if the oil level in the crankcase still falls below normal, oil is not returning to the compressor. The oil is probably being trapped in the cooling coils by an improperly adjusted thermal expansion valve. After the compressor has been stopped for several minutes, the oil level in the compressor crankcase should be about halfway up on the bull's-eye sight glass. During operation, the oil level will be slightly lower but will appear higher when oil is foaming. Check the oil level hourly. Add or remove oil to bring the level in the crankcase to the middle of the sight glass during steady operating conditions.

Adding Oil

The method of adding oil given below, if properly followed, will prevent air and moisture from entering the system. Since refrigerant gas is heavier than air, and the crankcase is loaded with this gas, the position of the oil-charging hole is located to prevent the admission of air. Only clean oil from sealed containers should be used. Check the manufacturer's manual for the proper type of oil for the system.

1. Close the liquid line valve and pump down the system.
2. Remove oil filler plug slowly.
3. Add oil to center of bull's-eye sight glass using a clean, well-dried funnel or an oil pump and suitable connector.
4. Replace oil filler plug tightly.
5. Restart system.

Removing Oil

Proceed as follows to remove oil from a compressor crankcase:

1. Close liquid line valve and pump down the system.
2. Loosen crankcase drain plug. Since crankcase is under light pressure, do not fully remove drain plug. Allow required amount of oil to be drained to seep slowly around the threads of loosened plug.
3. Retighten drain plug.
4. Restart system.

Oil Pressure

Correct oil pressure will insure adequate compressor lubrication and satisfactory operation of the compressor capacity control system. Compressors with a forced lubrication system are typically designed to operate with a normal oil pressure of 45 to 55 psi (310 to 379 kPa) above suction pressure. For example, if the compressor suction gauge reading is 40 psig (276 kPa), the oil pressure gauge reading should be 85 to 95 psig (586 to 655 kPa). During start-up, observe the oil pressure gauge to be sure that oil pressure develops during the first few minutes. Always check oil pressure when starting. The oil pressure should be normal after steady operating conditions have been reached, and oil has stopped foaming. Oil foaming may last fifteen minutes or longer. Do not allow compressor to run longer than one minute if oil pressure of at least 15 psig (103 kPa) over suction pressure does not develop. Reasons for low oil pressure include the following:

1. Insufficient oil in crankcase.
2. Oil pressure regulator not seating properly.
3. Oil filter screen in bottom of crankcase clogged with dirt.
4. Oil pump worn or defective or rotating in wrong direction.
5. Faulty oil piping.
6. Rapid pulldown of suction pressure on start-up causing excessive oil foaming.

Compressor Overhaul and Repair

Consult the manufacturer's manual for detailed disassembly and reassembly procedures. The following are general procedures to be followed in any compressor overhaul:

1. Be sure that faulty operation of the plant is not caused by trouble in some other part of the system before dismantling a compressor.
2. Dismantle only the part of the compressor necessary to correct the fault.
3. Sweep clean the deck in the vicinity of the compressor prior to any dismantling. Remove from the area any spare parts, fittings, or tools not to be used.

Obtain clean buckets or boxes in which to place disassembled components. Have on hand a supply of clean rags, as lint free as possible.

4. Spread clean canvas or heavy paper on the deck to lay out the larger parts (cylinder heads, crankshaft, etc.).
5. Maintain cleanliness during the overhaul. Clean all parts with an approved solvent after disassembly. Use a stiff brush to remove dirt from grooves and crevices. Do not use carbon tetrachloride for cleaning polished steel parts because it removes the oil film, and the steel may rust. Coat all moving parts with compressor oil before reassembly.
6. Dip dismantled parts to be left overnight in clean compressor oil, and wrap them in oil-soaked rags to prevent rusting.
7. Use special tools furnished by the compressor manufacturer for the particular operation involved to dismantle or reassemble a compressor.
8. Avoid damage to gaskets or gasket-seating surfaces when disassembling the compressor. These gaskets or new gaskets of identical thickness and material must be employed when reassembling. The use of discharge valve plate and compressor cylinder gaskets of proper thickness is particularly important since the thickness of these gaskets determines the clearance between the top of the pistons and discharge valve plate.
9. Mark carefully when disassembling compressor parts, so that each part removed will be replaced in its original position when reassembling.
10. Avoid filing, scraping, and grinding wherever possible when making compressor repairs or adjustments because of the danger of introducing emery or metal particles into the compressor.
11. Place pistons upon reassembly on the proper rods facing the same direction as originally.
12. Make certain that the oil dipper on lower connecting rod bearing of splash lubricated compressors is in correct position for dipping up oil when machine is in operation.
13. Stagger the position of the ends of the piston rings so that all joints do not come on one side of the piston.
14. Clean compressor crankcase and provide fresh charge of proper oil.

Analysis of Faulty Compressor Valves

Before opening a compressor for valve inspection or replacement, it should be definitely determined that the faulty operation of the system is due to the improper functioning of the valves and not to some other problem. Before assuming that compressor valves must be serviced, carefully check the system for all other possible causes of faulty operation.

Faulty compressor valves may be indicated by either a gradual or a sudden decrease in the normal compressor capacity. Either the compressor will fail to pump or the suction pressure cannot be pumped down to the designed value. This will cause the compressor to run for abnormally long intervals or even continuously. Short shutdown periods may indicate leaky compressor valves provided the faulty operation is not due to some other fault in the system.

Testing Compressor Discharge Valves

The compressor discharge valve may be checked for leakage as follows:

1. Close liquid line valve and pump down system.
2. Stop compressor, and quickly close suction and discharge line valves.
3. If discharge pressure drops at a rate in excess of 3 psi per minute and crankcase (suction) pressure rises, there is evidence of discharge valve leakage. It may be necessary to pump down several times to remove refrigerant mixed with crankcase oil in order to obtain a true test.

If the discharge valves are found to be defective in any way, it is advisable to replace the entire valve assembly with a spare. If valve operation is faulty, chances are the discharge plate requires relapping. This process generally requires highly specialized machinery to produce a satisfactory surface and should not be attempted aboard ship except in an emergency.

Testing Compressor Suction Valves

The compressor suction valves may be checked for leakage as follows:

1. Start the compressor under manual control.
2. Close the suction line stop valve gradually, exercising care to prevent violent oil foaming.
3. With the suction line stop valve finally closed, if a vacuum of approximately 20 inches Hg (34 kPa abs.) can be readily pumped, the suction valves may be considered satisfactory. Do not expect the vacuum to be maintained after the compressor stops due to the release of refrigerant from the oil. New valve assemblies may require a break-in period of several days before being checked.

If the test indicates a possible problem, the compressor should be pumped down, opened, and the valves inspected. Defective valves should be replaced with spare assemblies. Be sure all small pieces of a broken valve are accounted for. If any pieces are not removed, the compressor may be damaged when put back in operation. Before installing a new suction valve assembly, the piston should be checked for damage. If marred, the piston must be replaced as well as the suction valve.

Alignment of Compressor Coupling

Couplings on direct drive units should be checked for proper alignment after repair or replacement. Both parallel and angular misalignment should be checked. Parallel misalignment in direct drive units is shown in Figure 18-35 and angular misalignment is shown in Figure 18-36. Angular misalignment can be checked with a feeler gauge or by clamping a dial

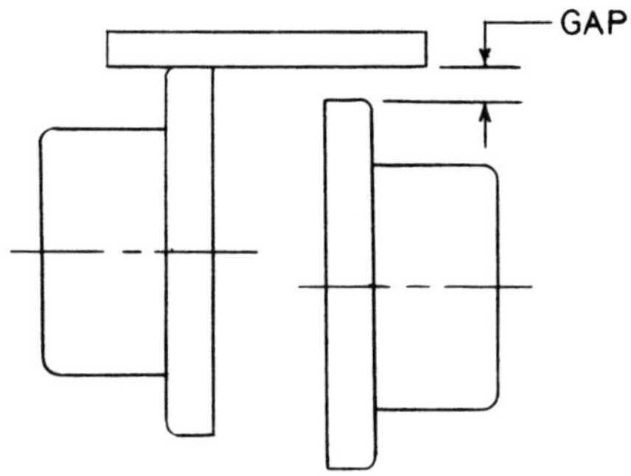

Figure 18-35. Parallel misalignment in a direct drive unit

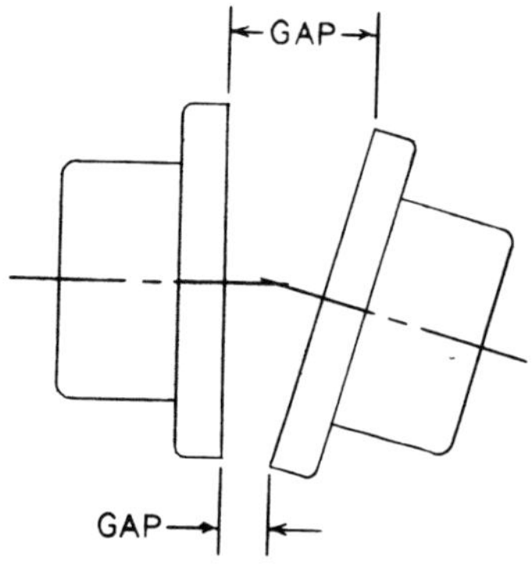

Figure 18-36. Angular misalignment in a direct drive unit

indicator to the motor flange and adjusting it so the stem will contact the inside face of the compressor flange. Rotate the motor flange through 360° and record readings at 90° intervals. To check parallel misalignment, use a steel rule held against the coupling periphery, or a dial indicator clamped to the motor coupling with the stem adjusted to run on the compressor coupling periphery. Shim motor feet and shift motor position as necessary to correct the misalignments. Tighten all hold-down bolts and recheck.

Belt Drive Adjustment

The belt drive must be aligned so that there is no angular or parallel misalignment. Both alignments can be checked with a straightedge or string. Parallel misalignment in belt drive units is shown in Figure 18-37, and angular misalignment is shown in Figure 18-38. Correct parallel misalignment by sliding the motor pulley on its shaft. Correct angular

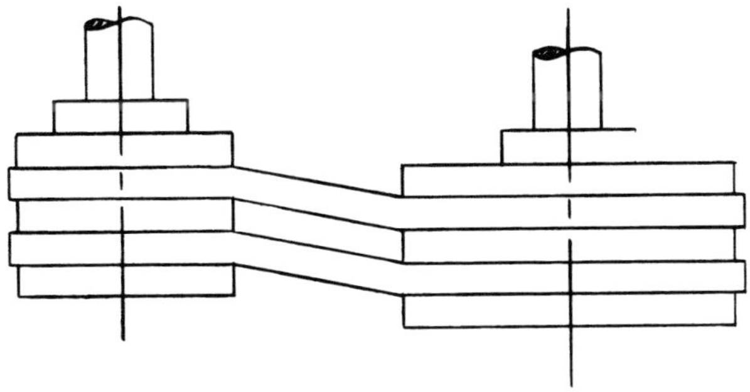

Figure 18-37. Parallel misalignment in a belt drive unit

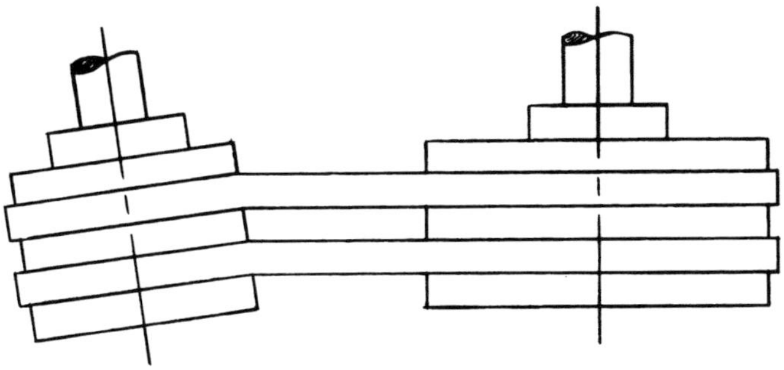

Figure 18-38. Angular misalignment in a belt drive unit

misalignment by loosening motor hold-down bolts and turning the motor frame. Check belt tension by depressing a single belt at the center of the span with one finger. A heavier belt with a 24-inch (61 cm) span should deflect ½ to ¾ inch (¼ to 2 cm). Lighter belts or longer spans should deflect proportionally more. Belts should always be replaced in sets, not singly.

Preventive Maintenance Schedule

Mechanical equipment can best be kept in good repair and operating at top efficiency by strict adherence to a planned maintenance schedule. A systematic check of operation and condition of refrigeration systems should be maintained. A recommended preventive maintenance schedule is outlined below. This schedule is intended only as a guide. It is not intended to replace the recommendation of the equipment manufacturer, and may be altered to suit operating and maintenance conditions peculiar to the individual installation.

Hourly

1. Enter all operating temperatures and pressures in a log. The log should contain columns for recording:
   (a) the time the check is made; (b) machinery room ambient temperature; (c) refrigerant suction pressure and temperature; (d) refrigerant discharge pressure and temperature; (e) oil pressure; (f) bull's-eye oil level (i.e., low, normal, high); (g) crankcase temperature (i.e., warm, cold, normal); (h) compressor noise (i.e., normal, knock, or other); (i) condenser seawater supply pressure; (j) condenser seawater entering and leaving temperatures; (k) liquid refrigerant temperature; (l) liquid refrigerant condition at sight glass (i.e., normal or vapor bubbles); and (m) temperature of refrigerated compartments.
2. Check motors for overheating. Get accustomed to the motor temperature by the way it feels to the hands.

Daily

Review hourly operating log. Note any significant changes in plant performance and take corrective action, if necessary.

Weekly

1. Test refrigerant system for leaks.
2. Check for noncondensable gases in condenser. Purge, if necessary.
3. Check operation of all solenoid valves.
4. Check glands on circulating pumps.

Monthly

1. Lubricate motor bearings, if necessary.
2. Blow dust out of motors and check lint screen.
3. Check contact points in motor controllers and control switches. Clean as required.
4. Check operation and settings of operating and safety control switches. Adjust as required.
5. Clean seawater strainers.
6. Check condenser zincs. Clean or replace as required.
7. If a brine is used, check brine density.

Quarterly

1. Clean condenser water side.
2. Check all motors and starters.

3. Clean refrigerant liquid line strainers.
4. Check alignment of compressor and motor.
5. Check tightness of all bolts on equipment.

Annually

1. Check calibration of all instrumentation.
2. Check onboard repair parts inventory and order parts as needed.

TROUBLESHOOTING THE SYSTEMS

The troubleshooting charts which follow are intended as a guide for locating malfunctions in a typical marine refrigeration or air-conditioning system with a reciprocating compressor. It is not intended to replace the detailed information provided by the manufacturer of a particular installation.

When a system is not operating properly, first determine the symptoms, then consult the chart for the possible causes and the recommended actions to determine the actual reason for the problem. Once a problem is uncovered, it should be corrected using the procedures outlined by the manufacturer or by using the general procedures in this chapter. It is usually necessary to use a process of elimination in locating the trouble. Try the easier possibilities first, then move on to the more difficult ones. The order in which the symptoms and possible causes are presented does not necessarily indicate a suggested troubleshooting sequence.

High Compartment Temperature

Possible Cause	*Action*
Controls not functioning	Check thermostat, solenoid valve, electrical circuits, and fuses. Check expansion valve and evaporator pressure adjustment and operation.
Coils frosted	Check and defrost, if necessary.
Insufficient airflow through fan coil unit	Check fan operation. Check filters. Check for obstructions.
Infiltration of warm air	Check for door ajar, poor door gaskets, or excessive traffic.
Recent loading of warm product	Normal pulldown. Start second condensing unit, if required.

High Compressor Discharge Pressure

Possible Cause	*Action*
Air in system	Check for noncondensables. Purge, if present.

Insufficient cooling water flow	Check water supply pressure. Check water regulating valve setting and operation. Check for obstruction or plugged strainer.
Dirty condenser or corroded shell baffles	Clean or replace as necessary.
Refrigerant overcharge (condenser tube partially covered)	Check refrigerant level. Remove any excess.
Compressor discharge stop valve partially closed	Check. Open fully.

Low Compressor Discharge Pressure

Possible Cause	*Action*
Excessive cooling water flow	Check water regulating valve setting and operation.
Liquid refrigerant flooding back	Check expansion valve superheat setting and operation. Check for open or leaking hand expansion valve.
Compressor suction stop valve partially closed	Check. Open fully.
Leaking compressor valves	Pump down, remove cylinder heads, and check.
Worn piston rings	Pump down, disassemble, and check.

High Compressor Suction Pressure

Possible Cause	*Action*
Overfeeding refrigerant	Check expansion valve superheat setting and operation.
Leaking compressor suction valves	Pump down, remove cylinder heads, and check.
Compressor unloading at too high pressure	Check setting and operation of capacity control system.

Low Compressor Suction Pressure

Possible Cause	*Action*
Low refrigerant charge	Check refrigerant charge. Add as required.
Restricted flow of refrigerant	Check expansion valve superheat setting and operation. Check solenoid valve. Check strainers for clogging.
Compressor not unloading	Check setting and operation of capacity control system.
Low pressure cut-out not stopping compressor	Check setting and operation of low pressure cut-out.

Cold (Sweating or Frosting) Compressor Crankcase

Possible Cause	*Action*
Liquid refrigerant flooding back	Check for open hand expansion valve. Check expansion valve superheat setting and operation.
Too much oil in circulation	Check for proper oil level.

High Crankcase Temperature

Possible Cause	*Action*
Clogged liquid line strainer	Pump down and clean.
Excessive suction temperature	Check expansion valve superheat setting and operation. Bypass heat interchanger.
Leaking compressor valves	Pump down, remove cylinder heads, and check.

Compressor Will Not Start

Possible Cause	*Action*
No power to compressor	Check breaker, main switch, fuses, and wiring.
Low voltage overload tripped	Reset. Check and eliminate cause of low voltage.
Safety control switch open	Check high pressure switch, water failure switch, and oil failure switch. Find and eliminate cause of switch operation. Check switch setting and operation.
Thermostat set too high	Reset.
Solenoid valve closed	Turn on power to solenoid. Check valve for proper operation.
Lack of refrigerant	Check for proper charge. If low, repair any leaks and recharge.

Compressor Runs Continuously

Possible Cause	*Action*
Low refrigerant charge	Check for proper charge. If low, repair any leaks and recharge.
Solenoid valve leaking	Check solenoid for proper operation.
Compressor valves leaking	Pump down, remove cylinder heads, and check.
Worn piston rings and/or cylinder liner	Pump down, disassemble, and then check.

Compressor Short-Cycles on High Pressure Switch

Possible Cause	*Action*
High pressure switch setting too low	Check and reset.
See "High Compressor Discharge Pressure" above for possible causes and action.	

Compressor Short-Cycles on Low Pressure Switch

Possible Cause	*Action*
See "Low Compressor Suction Pressure" above for possible causes and action.	
Reduced evaporator capacity	Check for frosted coils. Check for proper operation of cooling coil fan. Check for obstructions to airflow.
Relief valve leaking slightly	Test valve.
Internal leak in heat interchanger	Test for leaks.

Loss of Oil from Compressor Crankcase

Possible Cause	*Action*
Refrigerant flooding back to compressor	Check superheat setting and operation of expansion valve.
Worn piston rings or cylinder liners	Pump down, disassemble, and check.

Oil Not Returning to Crankcase

Possible Cause	*Action*
Expansion valve not flooding evaporator (oil trapped in coils)	Adjust expansion valve.
Oil return check valve stuck	Pump down. Check valve and oil return passage for plugging and valve for proper operation.

Low Oil Pressure

Possible Cause	*Action*
Insufficient oil in crankcase	Check oil level.
Faulty oil gauge	Check gauge.
Oil filter clogged	Pump down and clean.
Defective oil pump or improper rotation	Check direction of pump rotation. Disassemble pump and check for wear.
Defective oil relief valve	Test valve.
Worn bearings	Pump down, disassemble, and check.

Compressor Noisy

Possible Cause	*Action*
Loose mounting bolts	Check bolts for proper tightness.
Compressor drive loose, worn, or improperly aligned	Check flexible coupling and coupling flanges on direct drive units for looseness, wear, or misalignment. Check belts, pulleys, and flywheel on belt-drive units for looseness, wear, or misalignment.
Excess oil in circulation	Check for high oil level. Check for refrigerant floodback.
Liquid refrigerant flooding back	Check for open hand expansion valve. Check expansion valve superheat setting and operation.
Discharge stop valve rattling	Backseat valve fully.
Wear of compressor bearings, pistons, connecting rods, etc.	Pump down, disassemble, and check compressor for wear.

Loud Hissing Noise in Piping

Possible Cause	*Action*
Insufficient refrigerant charge	Check refrigerant charge.
Obstruction in liquid line	Check for plugged strainer. Check for partially closed valve. Check for other liquid line obstruction.

Thermo Expansion Valve—High Superheat

Possible Cause	*Action*
Superheat adjustment too high	Check superheat setting.
Power assembly failure	Check operation of power assembly.
Valve orifice plugged	Check for plugging. Wax or oil indicates improper oil type. Ice or dirt indicates need to cut in filter-drier.
Low inlet pressure	Check for proper head pressure toallow valve to pass rated flow.
External equalizing line plugged	Check line for plugging.

Thermo Expansion Valve—Low Superheat

Possible Cause	*Action*
Superheat adjustment too low	Check valve superheat setting.
Remote bulb not sensing evaporator outlet temperature correctly	Check for proper clamping of remote bulb to suction line. Insulate if exposed to high ambient temperature.

Loss of charge from remote bulb/power assembly	Check for loss of charge by checking valve response to different bulb temperatures.
External equalizing line plugged	Check line for plugging.
Valve frozen open	Apply hot rags to valve to melt ice. If valve operation is restored, cut in filter-drier to remove moisture.
Valve pin and seat worn	Check for tight shutoff. Check for worn parts.

Solenoid Valve Will Not Open

Possible Cause	*Action*
Electrical power	Check voltage and current at valve. If none or improper, check wiring, fuses, and breaker. Check thermostat setting and contacts.
Coil burnout	Check coil continuity and resistance. See "Solenoid Valve Coil Burnout" below for possible cause.
Valve improperly assembled	Check for proper assembly.
Improper pressure differential	Check system differential against rated maximum (valve closed) and minimum (valve open).
Plunger movement restricted	Check for corroded parts, foreign matter in valve, distorted enclosing tube or valve body, or oil trapped above piston.

Solenoid Valve Will Not Close

Possible Cause	*Action*
Plunger movement restricted	Check for corroded parts, foreign matter in valve, or distorted enclosing tube or valve body.
Electrical power	Check for voltage at coil. Voltage indicates contact not opening or power coming from another source. Check wiring.
Manual valve operator in open position	Check position of manual operator—should be backed out to allow valve to close.

Solenoid Valve Closes but Flow Continues

Possible Cause	*Action*
Foreign material lodged under seat	Check all internal parts.

Seat chipped, broken, or deformed	Check seat.
Valve improperly assembled	Check that valve assembly is set up according to manufacturer's instructions.

Solenoid Valve Coil Burnout

Possible Cause	*Action*
Electrical power	Check for low or high voltage. Check for improper wiring resulting in voltage drop preventing valve opening.
Ambient conditions	Check for high ambient temperature. Check for coil exposure to moisture or high humidity. Check for handling of fluids at temperature higher than valve rating.
Valve remains closed with electrical power supplied	Check for reason valve fails to open. Check for corroded parts, foreign matter, distorted parts, or excessive differential pressure.

REFRIGERANT CONSERVATION

The Montreal Protocol and the amendments to the Clean Air Act have mandated a number of changes to procedures for operating and maintaining refrigeration systems. Most of these changes are aimed at reducing the release of CFCs and HCFCs into the atmosphere. The regulations require that technicans involved in servicing refrigeration systems recover refrigerants rather than vent them to the atmosphere. Venting that takes place with the knowledge of the technician (other than good faith "de minimus" releases) are in violation of the regulations. The regulations also require that technicans performing maintenance, service, or repair procedures that could release CFCs or HCFCs into the atmosphere obtain certification through an EPA-approved testing organization.

Recover, Recycle, and Reclaim

The terms recover, recycle, and reclaim have become industry standard and their definitions and differences must be understood by those involved in the use of refrigerants. Recovery means the removal of refrigerant from a system and storage of it in an external container without necessarily testing or processing it in any way. Recycle refers to the processing of refrigerant for reuse by the removal of contaminants (moisture and particulates) through the use of devices such as filter-driers. This term usually applies to processes that can take place on site or at a local service shop. Reclaim describes the reprocessing of refrigerants back to new product specifications. This typically involves a variety of processes and procedures including distillation and is normally only available at a specialized reprocessing or manufacturing facility. Most technicians involved directly in the servicing of refrigeration systems will be using only recovery process. Details on procedures for recovery are discussed in the following section.

Refrigerant Recovery

The removal of refrigerant from the system can be accomplished by passive or active means. The traditional passive method using the pressure in the system to return refrigerant to the storage bottle was described earlier in this chapter. The limitation of this method is that it is not possible to remove sufficient refrigerant from the system to satisfy the current regulations. Prior to opening a system containing CFCs or HCFCs, the pressure must be reduced to between 0 inches mercury vacuum (for small HCFC-22 systems) and 15 inches mercury vacuum (for large CFC-12 systems). This can only be accomplished with a certified refrigerant recovery unit.

A refrigerant recovery unit is basically a portable condensing unit with a compressor and air-cooled condenser and related controls. These units are designed for removal of both liquids and vapors from the refrigeration systems and may also contain oil traps and filter-driers to accomplish limited purification of the refrigerant being recovered. They are designed to be used with special recovery tanks. These tanks are similar to the bottles used for the delivery of new refrigerant, but contain separate vapor and liquid connections with shutoff valves, and a high level cutout switch. The two separate connections permit simultaneous transfer of liquid and vapor between the unit and the system. The level switch automatically shuts the recovery unit off if the recovery tank becomes 80 percent full, preventing overfilling. Figure 18-39 shows the gas, liquid, and level switch connections on a recovery cylinder. Figure 18-40 shows a recovery cylinder connected to a portable recovery unit.

Figure 18-41 is a diagram of a typical recovery unit and recovery tank connected to a system for refrigerant recovery. Recovery normally consists of first recovering liquid from the unit, then proceeding with vapor recovery to lower the pressure to the required evacuation level. This unit has internal valves and controls to switch the unit from liquid recovery to vapor recovery without disconnecting the hoses. Some units do not have this capability; rather they require securing the unit after liquid recovery, reversing the hoses, and then proceeding with vapor recovery. When recovering liquid, the incoming liquid bypasses the recovery unit's compressor and is led directly to the recovery tank, and vapor exhausted from the tank is returned to the recovery unit for compression and condensation. When recovering vapor, the incoming vapor is led to the unit's compressor and condenser and the recovered refrigerant is delivered to the tank as a liquid. The change from liquid to vapor recovery is controlled by the float switch in the inlet tank. A high level in the inlet tank will switch the unit to liquid recovery by closing solenoid valve 2 and opening solenoid valves 3 and 4. A low level in the inlet tank will switch the unit to vapor recovery by opening solenoid valve 2 and closing solenoid valves 3 and 4.

Since each recovery unit is different, it is essential to read the instruction manual and carefully follow the manufacturer's operation procedures.

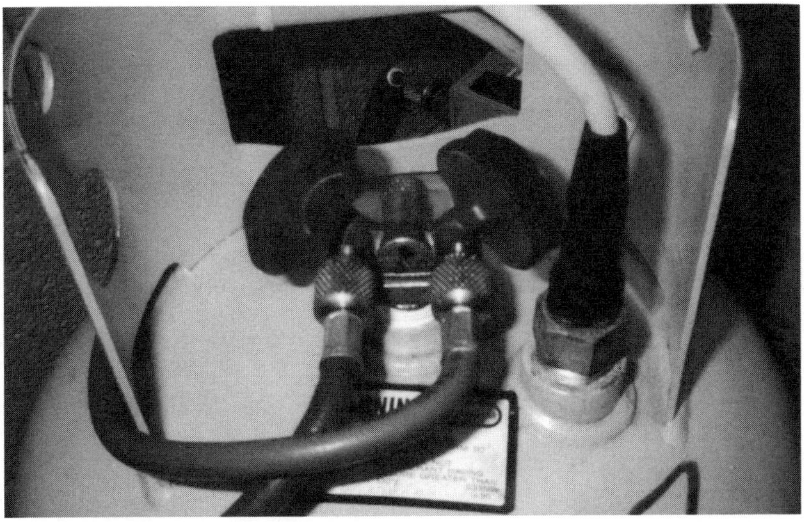

Figure 18-39. Connections to refrigerant recovery cylinder

Figure 18-40. Refrigerant recovery unit and recovery cylinder

Regardless of the particular recovery unit being used, the following general guidelines should be followed:

1. Only use a recovery unit designed for the refrigerant being recovered.
2. Recover liquid first, then vapor. Do not admit liquid into the inlet of the compressor of a recovery unit as damage can result.
3. Use only approved recovery cylinders. Never fill recovery cylinders beyond 80 percent of capacity. Never use disposable refrigerant cylinders to store recovered refrigerant.
4. Begin recovery with a clean, dry, evacuated recovery cylinder.
5. Always use proper safety equipment such as goggles and gloves when working with refrigerants.
6. Never mix different types of refrigerants. Mixed refrigerants cannot be separated and will have to be incinerated.

Certification of Technicians

All persons involved in the servicing of refrigeration systems that could result in the release of CFCs or HCFCs into the atmosphere must be certified through an EPA-approved certification program. Technicians must be certified for each type of equipment for which they work as follows:

Type I—small appliances.

Type II—high pressure systems, including marine systems using CFC-12 and HCFC-22.

Type III—low pressure systems, including chiller units using CFC-11.

Universal Certification—allows work on all the above system types.

Certification for each type of equipment requires passing a core multiple-choice test plus a separate multiple-choice test for each system type. Passing all tests results in Universal Certification. The EPA maintains a listing of approved certification programs.

Retrofitting Refrigeration Systems

The conversion of an existing refrigeration system to another refrigerant can be a complicated and expensive process. It depends on many factors such as the age and condition of the system, the type of refrigerant to be used, and the compatibility of the system components with the new refrigerant and lubricating oil. In some cases it may make economic sense to replace the entire system rather than undertake a retrofit. The system manufacturer must be consulted to obtain guidance on components that will need changing and procedures to be followed.

For existing systems using CFC-12, there are two main refrigerant choices for retrofits: HFC-134a and HCFC-22. HFC-134a has the advantage that its pressure-temperature relationships are very close to that of CFC-12. The disadvantages of an HFC-134a retrofit are that many systems components may be incompatible with the new refrigerant and lubricating oil and will need replacing. Careful flushing of the old oil charge

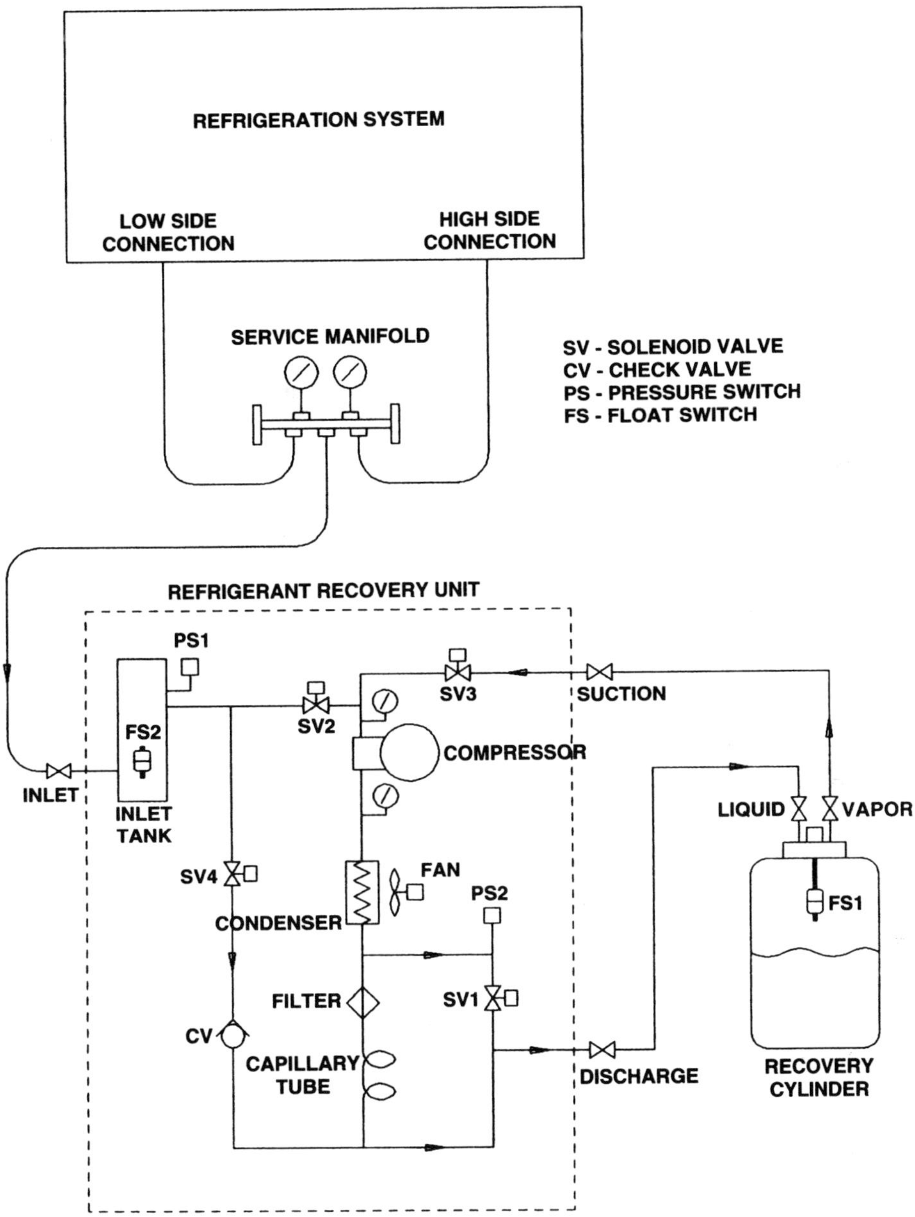

Figure 18-41. Typical refrigerant recovery unit

before replacement is essential to avoid problems. An HCFC-22 retrofit can be a good choice if the condensing unit was originally designed for both CFC-12 and HCFC-22. Because HCFC-22 will operate at higher pressures, the pressure rating of the condenser and receiver must be checked, and relief valves, burst disks, and expansion valves will need replacement. The compressor power requirements will increase due to the greater refrigerant density, and it may be necessary to slow the compressor speed by changing the belt pulleys or the motor to avoid motor overloading. The ability of the oil pump to properly operate at this lower speed must be verified. Permanently unloading cylinders is another possibility for reducing the compressor capacity and avoiding motor overloading.

The following procedures are typical of those required for retrofit of a CFC-12 reciprocating compressor system to HFC-134a:

1. Leak test the system and repair any leaks noted. Beginning the retrofit with a tight, properly functioning system will avoid problems during the conversion.
2. If schedule permits, begin changing the oil several weeks in advance of the conversion. At least three changes of the new POE oil will be necessary to reduce the mineral oil residual to 5 percent or less. POE oil has a substantial cleaning effect, and large volumes of debris and sludge will be dislodged from the system. The initial oil charge should remain in operation for no more than 48 hours before being changed. This initial oil charge will typically be gray-black in color with a significant amount of fine particles dispersed in it. Repeated changes are necessary until the oil remains clear and clean.
3. Recover the old refrigerant charge with a recovery unit into recovery cylinders. The weight of the refrigerant recovered should be noted to aid in recharging with the new refrigerant. Evacuate the system with a vacuum pump, and charge with clean dry nitrogen.
4. Make the hardware changes recommended by the system manufacturer. This may include the expansion valve, filter-drier, evaporator pressure, suction pressure regulator, and certain seals and gaskets.
5. Evacuate the system to a hard vacuum of at least 500 microns to insure dehydration.
6. Charge the system with the new refrigerant. Start with 90 percent of the charge removed and add additional as necessary.
7. Start system and make any necessary control and system adjustments. Carefully observe the system operation for the next couple of days. It is recommended that the oil and filter-drier be changed after 60–90 days of operation.

REVIEW

1. What is the latent heat of fusion for water?
2. How is one ton of cooling defined?
3. What is a refrigerant? What are the factors that should be considered in the selection of a refrigerant?

4. What is a CFC? What is an HCFC? What is an HFC?
5. Explain the industry standard refrigerant numbering system.
6. What is an azeotrope? What is a zeotrope?
7. What is the Montreal Protocol? What do the 1990 changes to the Clean Air Act require?
8. What new refrigerants have been proposed as alternatives to CFC-12 and CFC-11?
9. Describe the vapor-compression cycle.
10. What is the "pump-down cycle"?
11. What substances are commonly used as the secondary refrigerants in an indirect system?
12. What types of compressors are used in refrigeration systems?
13. Define the terms hermetic and semi-hermetic as applied to reciprocating compressors.
14. What is a compressor safety head?
15. What are the differences between a splash lubrication and a forced feed lubrication system?
16. Describe how a compressor cylinder unloading system works.
17. Describe the operation of a thermostatic expansion valve. Why are some valves externally equalized?
18. What pressure operates the low pressure switch? Describe the procedure to set the switch.
19. What is a common use of a solenoid valve in a refrigeration system?
20. What is the major difference between a suction pressure regulator and an evaporator pressure regulator?
21. Outline the procedure for starting a refrigeration system.
22. How is an evaporator defrosted using hot gas?
23. How is refrigerant added to a system?
24. How can refrigerant leaks be detected in a refrigeration system?
25. How can one determine if air is present in a refrigeration system?
26. Describe the steps to be taken if a refrigerated compartment temperature is high.
27. Describe the steps to be taken if the compressor is short-cycling.
28. Define the terms recover, recycle, and reclaim.
29. Describe a typical refrigerant recovery unit and how it is used to recover liquid and vapor from a refrigeration system.
30. How does a refrigerant recovery cylinder differ from a regular refrigerant storage cylinder?
31. Who must be certified through an EPA-approved certification program?

REFERENCES

Burghardt, M. D., and Harbach, J. A. 1999. *Engineering Thermodynamics*. 4th ed. Centreville, Md.: Cornell Maritime Press.

Dossat, R. J. 1996. *Principles of Refrigeration,* 4th ed. New York: Prentice Hall.

Handbook and Product Directory. 4 vols. American Society of Heating, Refrigerating, and Air Conditioning Engineers. Atlanta, Ga.

Naval Ships Technical Manual. September 1967. Refrigerating systems, chap. 9590. In: Navships 0901-590-0002. Washington, D.C.: Naval Sea Systems Command.

Stoecker, W. F., and Jones, J. W. 1982. *Refrigeration and Air Conditioning.* 2d ed. New York: McGraw-Hill.

ACKNOWLEDGMENTS

The author is indebted to the following companies engaged in the manufacture of refrigeration systems and equipment who so kindly contributed photographs, drawings, and other material used in the preparation of this chapter:

Carrier Corporation, Syracuse, New York
The Trane Company, LaCrosse, Wisconsin
Thermo King Corporation, Minneapolis, Minnesota
Alco Controls Division, Emerson Electric Co., St. Louis, Missouri
Henry Valve Company, Melrose Park, Illinois
Detroit Switch, Inc., Pittsburgh, Pennsylvania

CHAPTER 19

Heating, Ventilation, and Air-Conditioning

JAMES A. HARBACH

DEFINITIONS AND PRINCIPLES

HVAC (heating, ventilation, and air-conditioning) functions on modern vessels are performed by complex systems which satisfy the requirements of controlling the temperature, humidity, and purity of the air in the various spaces on the vessel. In the past, the heating, ventilation, and air-conditioning functions were treated separately, while today integrated HVAC systems handle all requirements. In this chapter, the various systems and components in common use today will be described. Emphasis will be on operation and maintenance of existing systems rather than the design of new systems. Readers interested in system design should consult the list of references at the end of the chapter.

Air-Conditioning

Air-conditioning is defined by ASHRAE (American Society of Heating, Refrigerating and Air Conditioning Engineers) as the process of treating air to control simultaneously its temperature, humidity, cleanliness, and distribution to meet the comfort requirements of the occupants of the conditioned space. Note that this definition includes not only cooling and dehumidification during warm weather, but the heating operation during cold weather, as well as the regulation of air velocity and removal of foreign particles. The system commonly referred to as the air-conditioning system is actually only part of the overall system and would more properly be called the cooling system.

Ventilation

Ventilation may be defined as the process of supplying outside air to and exhausting air from an enclosed space. This is done to prevent the accumulation of dangerous or annoying gases, to reduce the humidity of spaces containing sources of moisture, or to reduce the temperature of spaces not mechanically cooled such as the engine room. Ventilation requirements for cargo spaces and living spaces are normally based on a given number of air changes per hour. In public areas such as dining saloons and messrooms, ventilation is normally based on a given CFM (cubic feet per minute) per occupant at maximum capacity. Engine room ventilation is normally sized to maintain the temperature at 10°F to 15°F (5°C to 8°C) above ambient.

Heating

Heating may be defined as the process of adding heat to an enclosed space to maintain its temperature above ambient. The heat added makes up for losses due to heat transfer to the outside, and for losses due to ventilation or infiltration of outside air.

Heat Transfer

See Chapter 1 in Volume I for further information on the principles of heat transfer including conduction, convection, and radiation.

Psychrometry and the Psychrometric Chart

Psychrometry is the study of the properties of mixtures of air and water vapor. This subject is important to air-conditioning because the systems handle air-water vapor mixtures, not dry air. Some air-conditioning processes involve the removal of water from the air-water vapor mixture (dehumidification) while some involve the addition of water (humidification). A convenient way to represent the properties of air-water vapor mixtures is the psychrometric chart, Figure 19-1. On the chart, such properties as dry bulb temperature, wet bulb temperature, dew point, relative humidity, humidity ratio, specific volume, and enthalpy are presented in graphical form.

DB (Dry Bulb Temperature)

The temperature of the air as sensed by a standard thermometer is called the dry bulb temperature.

WB (Wet Bulb Temperature)

The temperature sensed by a thermometer whose bulb is wrapped in a water-soaked wick and moved rapidly in air is called the wet bulb temperature. A sling psychrometer is an inexpensive device containing a standard and a wet bulb thermometer mounted on a pivoting handle. Swinging the device rapidly produces the required air movement and allows determination of the wet and dry bulb temperatures.

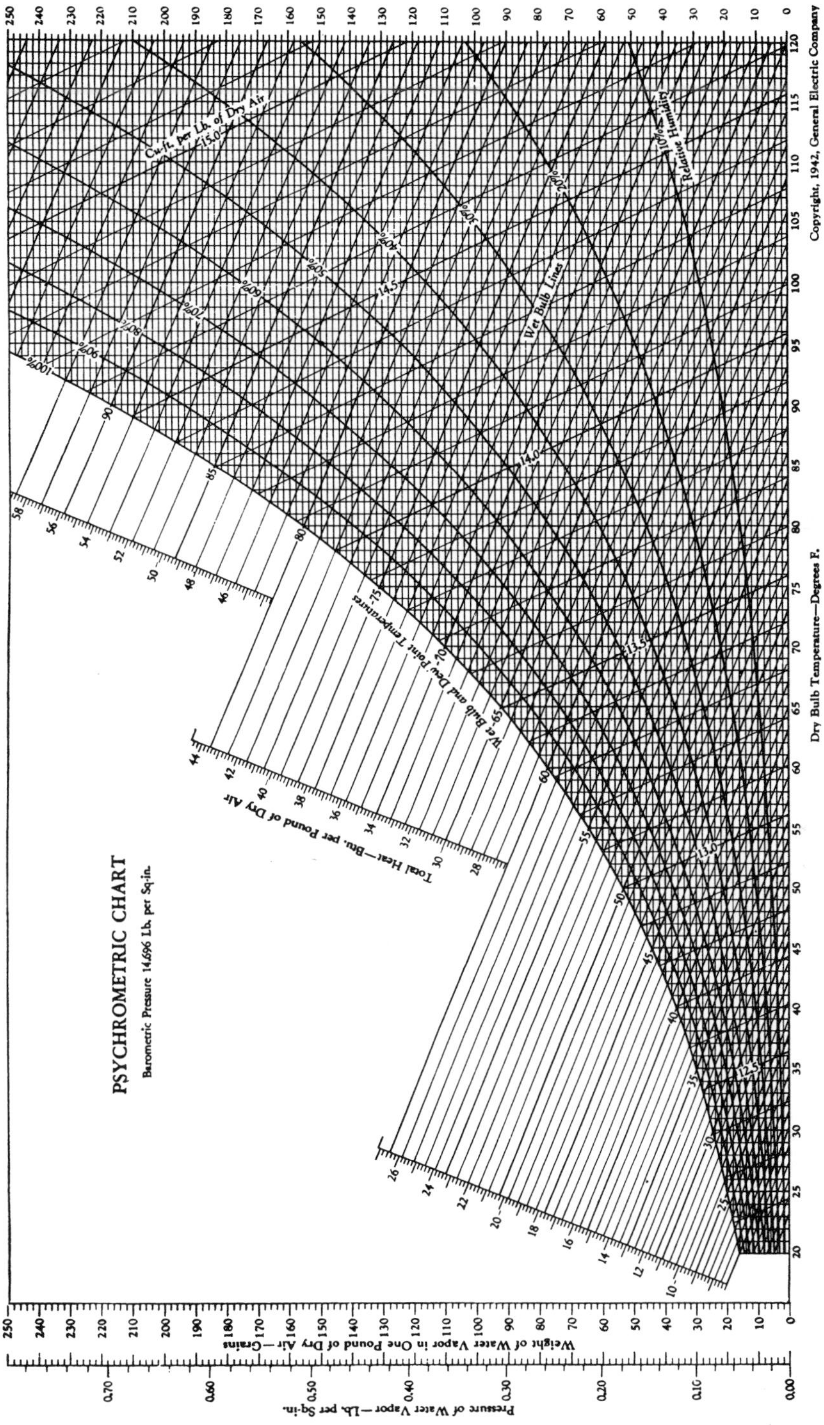

Figure 19-1. Psychrometric chart

DP (Dew Point Temperature)

Dew point temperature is the temperature at which water vapor will begin to condense if the air were cooled at constant pressure. At the dew point, the air is saturated.

RH (Relative Humidity)

Relative humidity is the ratio of the actual water vapor pressure to the water vapor pressure that would exist if the air were saturated at the same dry bulb temperature. It is expressed in percent.

W (Humidity Ratio)

Humidity ratio (also called absolute humidity) is the ratio of the weight of water vapor in the air to the weight of dry air. It is expressed in pound/pound dry air or grains/pound of dry air (7,000 grains = 1 pound), or kilogram/kilogram of dry air.

v (Specific Volume)

Specific volume is the volume occupied by one pound of air. The units are cubic feet/pound or cubic meters/kilogram.

h (Enthalpy)

Enthalpy is the thermal energy content of the air expressed in BTU/pound or kilojoule/kilogram.

Reading the Psychrometric Chart

In order to locate any condition of air on the chart, two independent properties must be determined. The air condition can then be plotted on the chart, and all other properties can be read from the chart. Dry bulb temperatures are read along the horizontal scale at the bottom of the chart, humidity ratio is read along the right-hand vertical scale, and the wet bulb temperature, dew point temperature, and enthalpy are read along the diagonal scale at the upper left. Lines of constant relative humidity and specific volume are labeled in the body of the chart. Refer to Figure 19-2 for assistance in plotting and reading air conditions on the psychrometric chart. For example, sling psychrometer measurements are taken as 78° DB and 65° WB. Determine the relative humidity and dew point. (Answer: RH = 50 percent, DP = 58°F).

HVAC SYSTEMS

Heating and Cooling Loads

One of the first steps in the design of an HVAC system is the calculation of the heating and cooling loads. While the detailed coverage of these

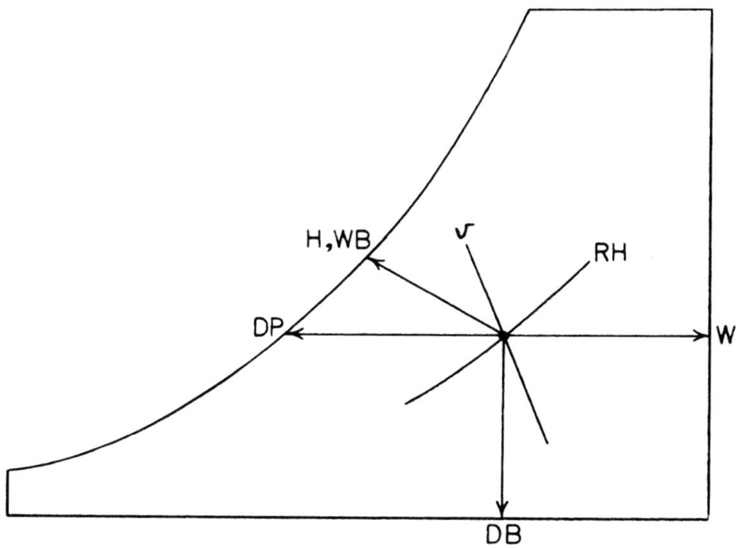

Figure 19-2. Reading the psychrometric chart

calculations is not within the scope of this book, a basic understanding of the various factors which affect the heating and cooling loads will be helpful to the operator of these systems. The references listed at the end of the chapter provide further details.

The purpose of the HVAC system is to maintain a safe and comfortable internal environment despite variable outside conditions. The system must have sufficient capacity in terms of heating and cooling to meet the extremes of outside conditions which are anticipated. Some of the factors which can affect the cooling and heating loads are (1) design outside air conditions; (2) design inside air conditions; (3) materials and insulation in bulkheads, overheads, and decks; (4) infiltration and ventilation air; (5) lights, equipment, and occupants inside space (cooling only); and (6) solar gain (cooling only).

The selection of design outside air conditions depends to some extent on the anticipated routes the ship will travel. Since it is difficult to predict the routes a ship will travel over its entire service life, it is common to base system design on 90° to 95° DB (32° to 35°C), 78° to 82° WB (26° to 28°C) cooling and 0° to 10° DB (−18° to −12°C) for heating. Inside design conditions are typically 68° to 75° DB (20° to 24°C) for winter and 75° to 80° DB (24° to 27°C) for summer, with relative humidity varying from 20 to 75 percent. Figure 19-3 shows a recommended comfort zone. It is somewhat larger

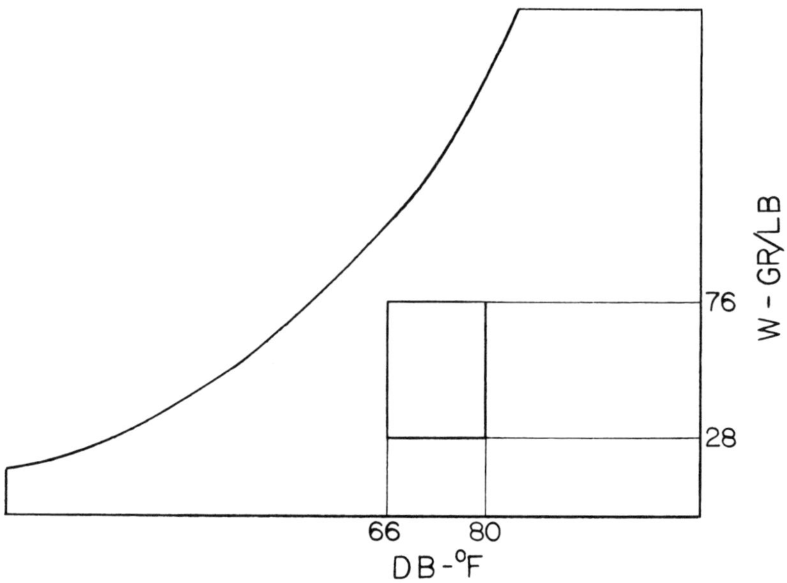

Figure 19-3. Design inside air conditions

than the ASHRAE comfort zone, permitting lower winter and higher summer temperatures, in consideration of energy conservation criteria.

The construction, and particularly the insulation used, of the bulkheads, decks, and overheads will affect heat transmission through them. Use of adequate insulation can significantly reduce the heating and cooling loads. The introduction of outside air to the conditioned space by ventilation and infiltration will increase loads. While ventilation is required in many applications, excessive ventilation should be avoided. Infiltration can be minimized by proper weather stripping. In cooling load calculations, the additional load imposed by lighting, cooking facilities, various operating equipment, the occupants, and solar heat gain must also be included.

Ventilation Criteria

The introduction of outside air is required in some instances to prevent the accumulation of certain gases and contaminants to unhealthy levels. In densely occupied spaces, the buildup of carbon dioxide (CO_2) is a problem. Table 19-1 lists some typical criteria for ventilation based on CFM per occupant. In staterooms and other sparsely occupied spaces, infiltration is normally sufficient to maintain air quality, and total air supply rates are normally based on a number of air changes per hour. Use of filtered recirculated air in these instances can significantly reduce heating and cooling loads. In those applications such as the engine room and certain

TABLE 19-1

Ventilation Requirements

Space	*CFM per person*
Staterooms	10-15
Dining rooms	15-20
Toilet rooms	20-25
Offices	20-40
Kitchens	20-40

other machinery spaces where ventilation is used for heat removal, the ventilation rate is based on a given temperature rise in the space, typically 10°F to 20°F (6°C to 12°C), or an air change time of 1 to 2 minutes, whichever is greater. Mechanical exhaust should be 110 to 120 percent of the supply to compensate for expansion of the air and to create an indraft to prevent the escape of heat into conditioned spaces. Cargo space ventilation is normally based on two to three air changes per hour. Consideration is given to control of humidity in the holds to avoid condensation damage and to remove CO_2 and other gases given off by some cargoes.

HVAC System Types

In order to maintain the desired temperature and humidity in the conditioned spaces, the HVAC system must add or remove thermal energy to or from the space. The heat transfer is accomplished typically by air or water. Some heating systems may use steam space heaters, and some cooling systems may use refrigerants directly. Air systems in common use are (1) terminal reheat, (2) dual duct, and (3) variable air volume. Water systems in common use are (1) one pipe, (2) two pipe, and (3) four pipe. These systems are designed to serve areas with more than one conditioned space or zone and where independent control of space air conditions is required. A single zone air system will be described first to illustrate the basic principles.

Single Zone System

Figure 19-4 shows a single zone air system with provisions for heating, cooling, humidification, dehumidification, and the introduction of outside ventilation air. The supply fan delivers conditioned air to the space. An exhaust fan is commonly installed to avoid excessive positive air pressure in the space. The dampers allow the control of ventilation and recirculation air quantities. Air control can be manual or automatic based on outside air temperature. Automatic control can be used to increase ventilation air when the outside air temperature is somewhat lower than the conditioned

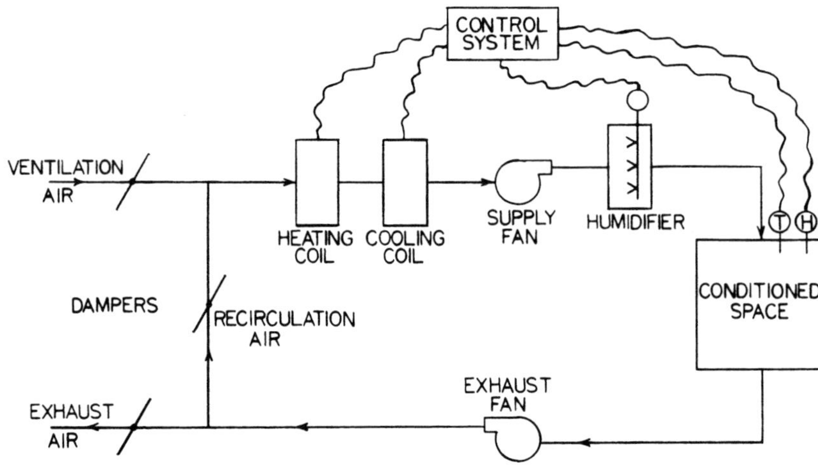

Figure 19-4. Single zone system

space, thus reducing the cooling coil load. At cold or hot outside air temperatures, the ventilation air will automatically be set at the minimum.

The cooling coil also performs the dehumidification function. If the coil temperature is below the air dew point, water vapor will condense and air humidity will be lowered. The humidifier admits steam directly into the air stream, increasing humidity with little increase in dry bulb temperature. The control system senses space temperature and humidity, then controls the cooling and heating coils and the humidifier. In cold weather, heating and humidification will be required. In hot weather, cooling and dehumidification are needed. The space humidity during cooling can be controlled in two ways: (1) varying the cooling coil temperature, or (2) maintaining a constant low cooling coil temperature and then using the heating coil to reheat the air. The second method wastes energy, and is normally only used on systems which require good humidity control or have high latent heat loads.

Multiple Zone Systems

It is usually not economically feasible to provide a separate system for each zone. A zone is a room or group of rooms controlled by a single thermostat. The basic principles of temperature and humidity control outlined for the single zone system are expanded in the systems below to meet the requirements of multiple zone operation.

Terminal Reheat System

Figure 19-5 is a diagram of a terminal reheat system. The fan delivers a fixed quantity of air, and the duct system distributes the air to the various conditioned spaces. During warm weather operation, the air leaving the cooling coil is maintained at a constant temperature, typically 55°F (13°C). The thermostat in each zone controls the reheat coil associated with that zone to maintain proper zone temperature. The reheat coil may be hot water, steam, or electric. The terminal reheat system provides good control of temperature and humidity over a wide range of loads, but is wasteful of energy since air is cooled, then heated. This can be reduced by raising the cool air temperature until the lightest loaded reheat coil turns off, or to use waste heat to operate the reheat coils. During cold weather, the reheat coils supply the heating requirements.

Dual Duct System

Figure 19-6 is a diagram of a dual duct system. The constant quantity of air from the supply fan is split into two streams. Part of the air flows across the cooling coil and part across the heating coil. Each zone thermostat controls a mixing box which proportions the warm and cool air delivered to the zone to maintain the desired zone temperature. The dual duct system is responsive to load changes and can accommodate heating in some areas and cooling in others. Since two supply ducts are required, the system is commonly designed as a high velocity air system. However, this increases fan power requirements. As with the terminal reheat system, there are periods of simultaneous heating and cooling. During periods of low (below 55°F, 13°C) outside air temperature, the cooling coil can be secured. During hot weather, the warm duct temperature can be lowered or perhaps the heating coil secured. Maintaining warm duct temperatures higher than that required to maintain comfort in all zones is a waste of energy. Some marine systems are arranged with the cooling coil and a reheater in series. While less efficient, this arrangement provides better dehumidification and reduces condensation problems.

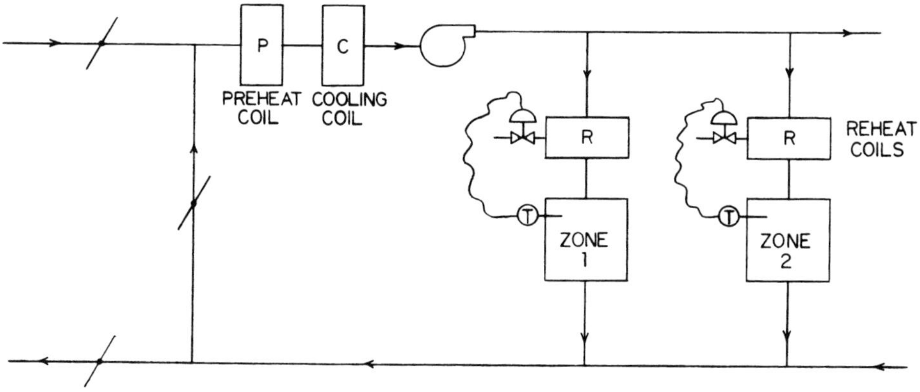

Figure 19-5. Terminal reheat system

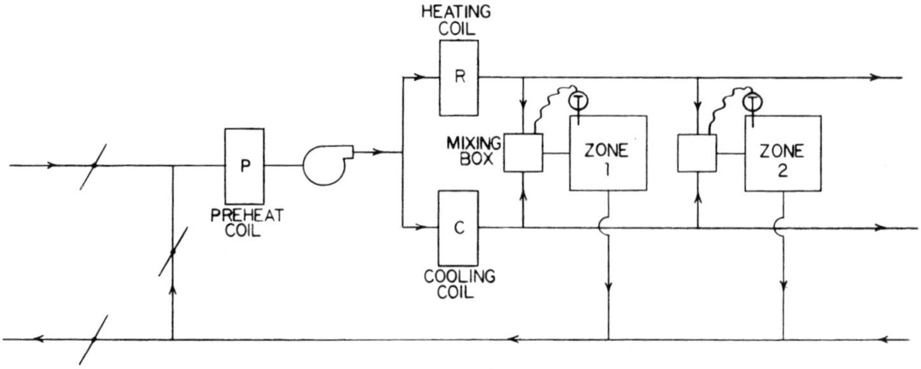

Figure 19-6. Dual duct system

Variable Air Volume Systems

The terminal reheat and dual duct systems described above waste energy, especially at light cooling or heating loads. VAV (variable air volume) systems reduce this waste. While there are a number of different VAV systems, all utilize dampers in the supply duct to each zone to vary the air delivered to the zone. Figure 19-7 shows a simple VAV cooling system. The problem with this simple system is that, at low loads, the air supply is reduced to the point where poor air distribution and ventilation result. The solution to this problem is to combine the VAV technique with a terminal reheat or dual duct system.

In a VAV reheat system, as zone load decreases, the damper will reduce air flow. Once a minimum flow (perhaps 25 to 30 percent of maximum) is reached, the reheat coil is energized, and air flow remains constant at the minimum. The VAV reheat system does incur some energy loss, but it is small compared to the constant volume terminal reheat system. It also overcomes the air distribution and ventilation problems of the simple VAV system.

In the VAV dual duct system, the mixing box characteristics are modified to reduce the cool or warm air flows before mixing the other stream. During cold weather, the warm air duct flow is modulating with no addition of cool air. At moderate outside temperatures, warm and cool air flows are reduced, and air mixing of the reduced flows will occur. Like the VAV reheat system, there is still some energy waste, but it occurs at low flows and is small compared to the constant volume system.

Water Systems

Water or hydronic systems accomplish heating and cooling through the distribution of hot and chilled water to the conditioned spaces. Water systems take up less space than air systems due to the higher density and specific heat of water. The required piping is much smaller than the

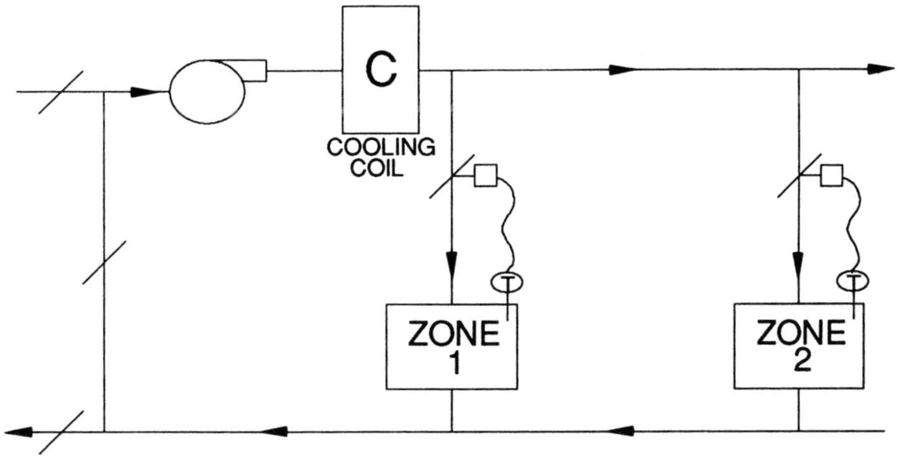

Figure 19-7. Variable air volume system

ductwork required for the same installation. While the initial cost of a water system may be less, the system usually lacks humidity control, ventilation is uncertain, and maintenance is higher.

A simple one-pipe hydronic system is shown in Figure 19-8. Figure 19-9 shows a two-pipe reversed return system. In each system, hot water or chilled water is circulated to the conditioned spaces, which have fan coil units, convectors, or similar devices for exchange of heat to or from the air in the space. The two-pipe system has the advantage that the water temperature at each terminal unit is the same. The term reversed return as applied to Figure 19-9 refers to the fact that the water path to and from each terminal unit is about equal. This avoids flow imbalance which can occur in a direct return system.

The disadvantage of the one- and two-pipe systems is that heating and cooling cannot occur simultaneously in different zones. The four-pipe system shown in Figure 19-10 overcomes that limitation. The terminal units are fitted with two coils, one for heating and one for cooling. The thermostat for each zone regulates the flow of hot and chilled water to the coils. Control is sequenced so that flow is not admitted to both at the same time.

Unitary Systems

Factory assembled and packaged air-conditioning systems are available from the major air-conditioning manufacturers. The units are available as a single package containing condenser, compressor, evaporator, fan, controls, and filters, or as a split unit with the condenser and compressor located separately. The units are low in cost due to mass production, but do not have the flexibility of a custom-designed system. On ships, they are

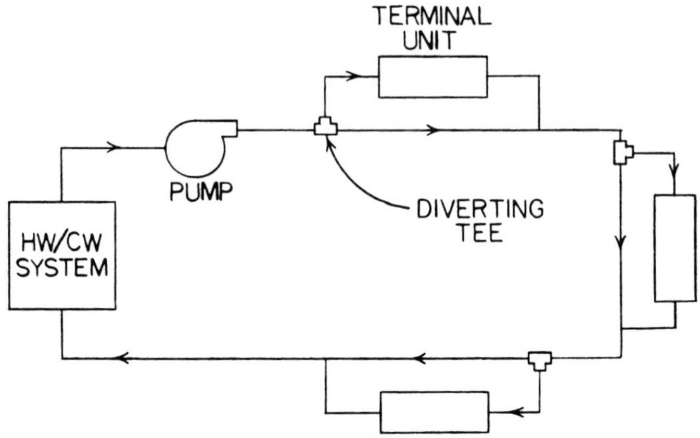

Figure 19-8. One-pipe water system

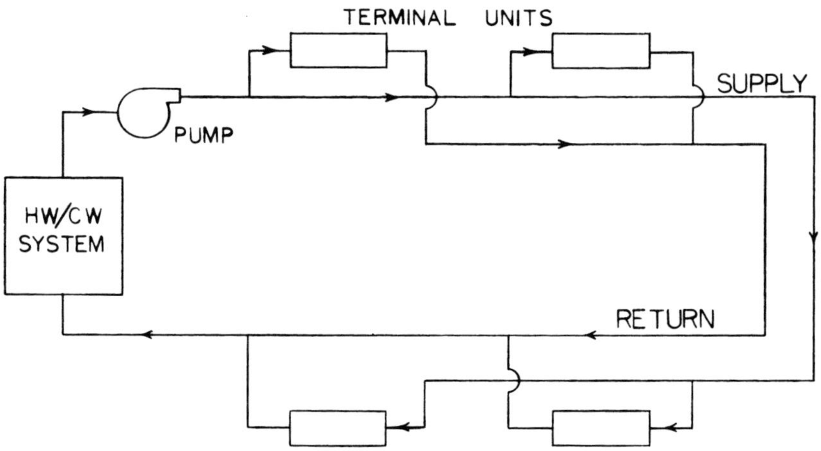

Figure 19-9. Two-pipe water system

sometimes found serving a single zone located remotely from the main conditioned spaces, such as the forward deckhouse on a tanker. Figure 19-11 shows a typical unitary system.

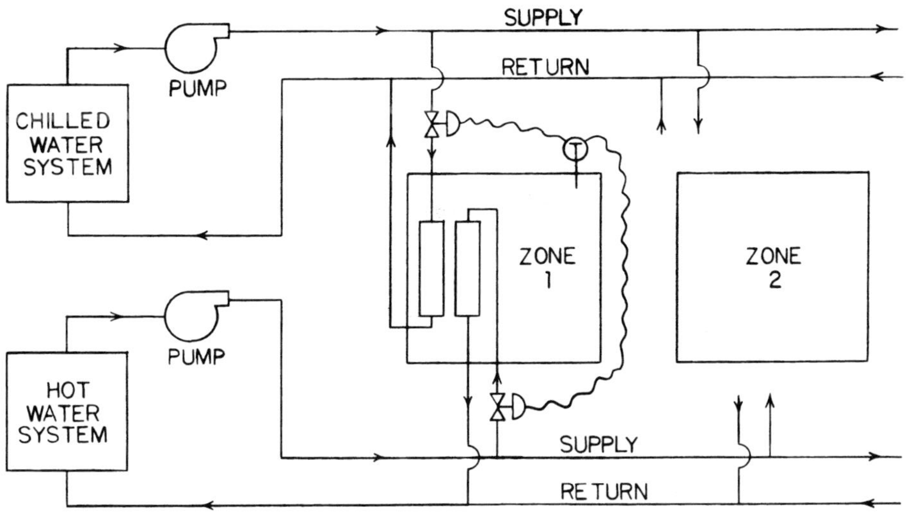

Figure 19-10. Four-pipe water system

Cargo Hold Dehumidification Systems

Cargo holds on dry cargo vessels are generally ventilated with mechanical supply and exhaust, using one air change in 20 to 30 minutes. This is done to reduce hold temperatures and condensation of moisture on hold surfaces and cargo. Condensation is reduced by introducing dry air into the holds and exhausting an equal quantity of humid air. One or more automatic dehumidification systems may be provided for reducing the moisture level in the holds.

Dehumidification systems utilize liquid or solid desiccants such as silica gel or lithium chloride. Figure 19-12 is a diagram of a liquid desiccant unit. Figure 19-13 shows a solid desiccant unit. A common feature of these systems is continuous regeneration of the water-laden desiccant by passing heated air through it. Moisture removed from the desiccant by the regeneration air is exhausted to the atmosphere. Since the moisture removal process involves the conversion of water vapor in the air to liquid in the desiccant, heat is released. A cooling coil is commonly installed to reduce the temperature of the dehumidified air.

The dew point temperature in any cargo hold is normally maintained 10°F (5.6°C) below the surface temperature of the cargo or ship structure. The cargo hold dew point temperature is usually measured in the exhaust trunk and is thus an average of the different hold levels. The ventilation and dehumidification system is designed to operate in one of four modes: ventilation without dry air, ventilation with dry air, recirculation without dry air, and recirculation with dry air. The mode of operation can be

Figure 19-11. Unitary system. Courtesy The Trane Company

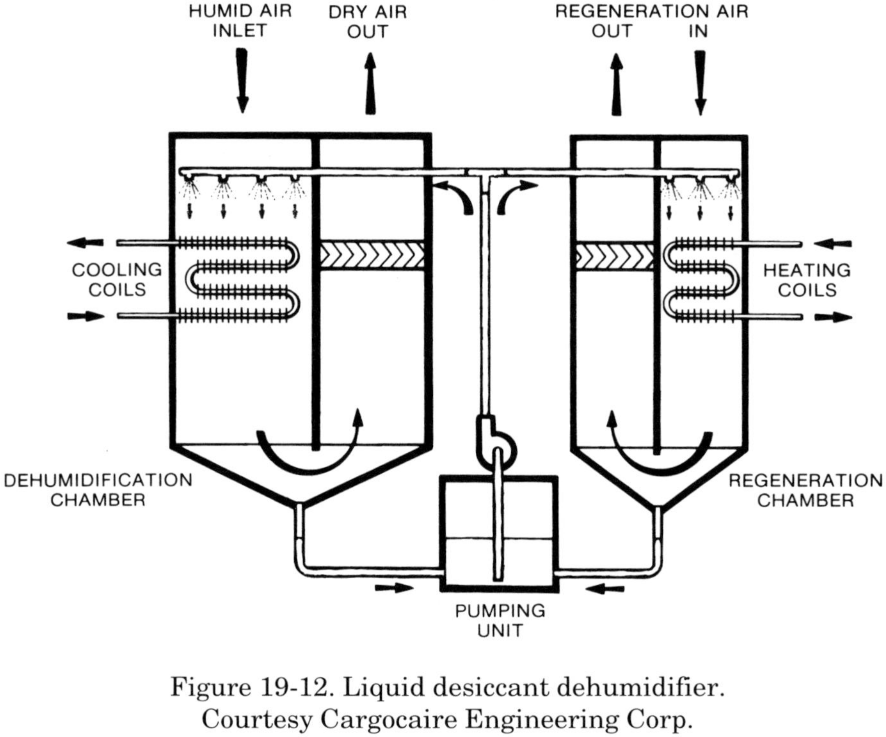

Figure 19-12. Liquid desiccant dehumidifier.
Courtesy Cargocaire Engineering Corp.

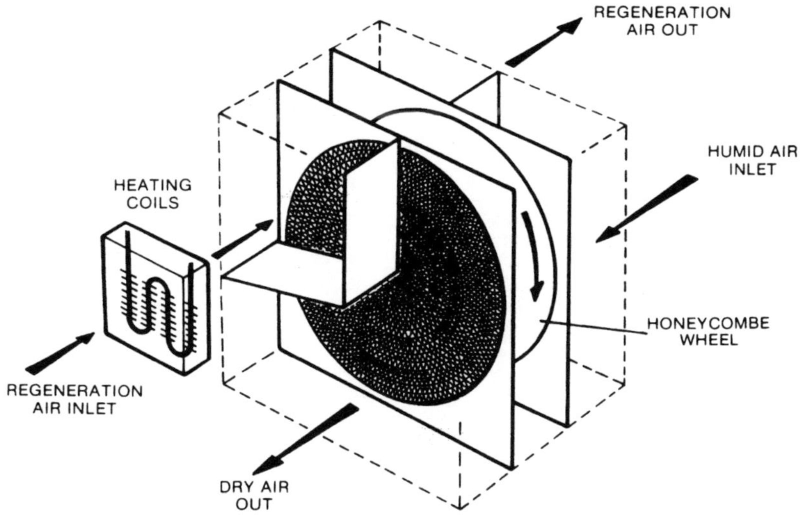

Figure 19-13. Rotary Honeycombe dehumidifier.
Courtesy Cargocaire Engineering Corp.

selected from the wheelhouse by remote operation of pneumatically or electrically actuated control dampers.

HVAC SYSTEM COMPONENTS

Fans

Fans are used to distribute air through the HVAC equipment and ducts to the air-conditioned spaces. Fans used in HVAC systems may be classified as either axial flow or centrifugal fans. In an axial flow fan, the air flow is in the same direction as the fan shaft. In centrifugal fans, the air discharge is radial relative to the fan shaft. Additional information on fans can be found in Chapter 12 of Volume I.

Axial flow fans are termed propeller, tube-axial, or vane-axial depending on their construction. A propeller fan (Figure 19-14) is merely a propeller-type wheel mounted on a plate. A tube-axial fan consists of a vaned rotor mounted in a cylinder. The vane-axial fan (Figure 19-15) is similar, but fixed guide vanes improve air flow in the fan. Axial flow fans are limited in their ability to develop pressure, and thus cannot be used in large duct systems. Vane-axial fans are used in ducted systems with modest pressure drops where compact size is important.

Centrifugal fans are the commonly used type in ducted air-conditioning systems. Centrifugal fans can be classified according to the shape of the impeller blades: forward curved, radial, or backward curved. In addition, high performance backward curved blade fans are called airfoil blade fans

Figure 19-14. Propeller fan. Courtesy Buffalo Forge Co.

Figure 19-15. Vane-axial fan. Courtesy Buffalo Forge Co.

due to the shape of the fan blades. Forward curved blade fans are characterized by low initial cost, lower efficiency, and an increasing power requirement with increasing air flow. Backward curved blade fans (especially airfoil types) are higher in initial cost, but have higher efficiency and thus lower operating costs, and have a nonoverloading head-capacity characteristic. Figure 19-16 shows a typical centrifugal fan.

Fan performance characteristics. The fan is installed in the HVAC system in order to maintain a certain flow of air through the conditioned spaces.

Figure 19-16. Centrifugal fan. Courtesy Buffalo Forge Co.

Because of fluid friction in the ducts and other components, the fan must develop pressure in order to move the air. The important fan performance characteristics are capacity (CFM), static pressure (inches of water), efficiency (percent), and power (BHP). Knowledge of fan performance is useful in fan selection, fan operation, and troubleshooting. Figures 19-17 and 19-18 are typical performance characteristics of a forward curved and backward curved blade centrifugal fan. Static pressure, efficiency, and brake horsepower are plotted as a function of flow rate. An inspection of the curves will reveal the following:

1. The static pressure curves for both types tend to peak in the middle range, and then tend to drop off with increasing flow.
2. The BHP for the forward curved blade fan increases with flow. The backward curved blade fan BHP increases gradually, peaks, then falls off.
3. Efficiency is highest in the middle ranges of flow. As previously mentioned, backward curved blade fans will usually have higher peak efficiency than forward curved blade fans. This, combined with the decreasing power requirements at high flow, rather than increasing as with forward curved types, explains the wide application of backward curved blade centrifugal fans in larger air-conditioning systems.

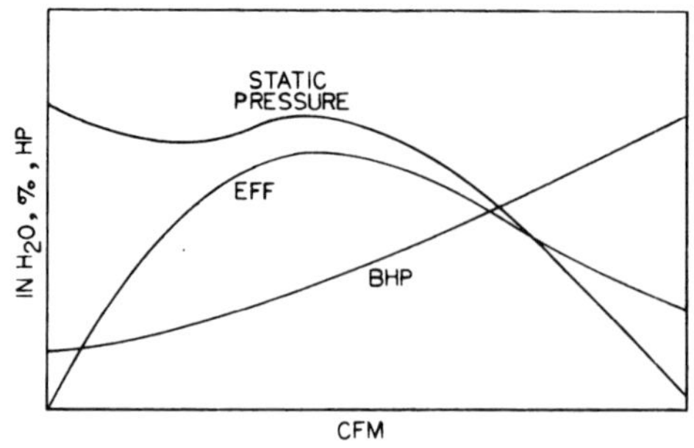

Figure 19-17. Forward curved blade centrifugal fan performance characteristics

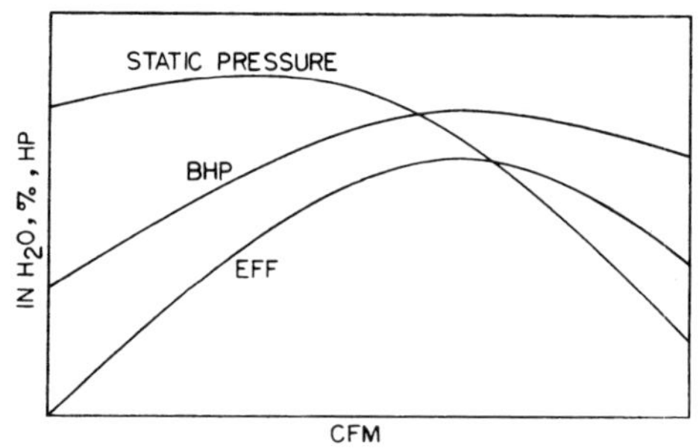

Figure 19-18. Backward curved blade centrifugal fan performance characteristics

Fan laws. There are several relationships which can be used to predict the behavior of fans at conditions other than those for which data are available. These relationships are:

$$CFM_2 = CFM_1 \times \frac{RPM_2}{RPM_1}$$

$$\text{Static Pressure}_2 = \text{Static Pressure}_1 \times \left(\frac{RPM_2}{RPM_1}\right)^2$$

$$BHP_2 = BHP_1 \times \left(\frac{RPM_2}{RPM_1}\right)^3$$

Example: A fan is rated to deliver 8,000 CFM at 900 RPM at 6.5 BHP. What will the RPM and BHP be at a flow of 9,000 CFM?

$$RPM_2 = RPM_1 \times \frac{CFM_2}{CFM_1} = 900 \times \frac{9{,}000}{8{,}000} = 1{,}010\ RPM$$

$$BHP_2 = BHP_1 \times \left(\frac{RPM_2}{RPM_1}\right)^3 = 6.5\left(\frac{1{,}010}{900}\right)^3 = 9.2\ HP$$

This large increase in horsepower may overload the fan motor. The suitability of the motor, fan, and wiring should be checked.

Fan capacity control. The capacity of a fan can be varied in several ways: (1) varying the fan speed, (2) using outlet dampers, and (3) using inlet vanes. The simplest means of changing fan speed is to change the belt pulley diameters. This is obviously unsuitable for a system that requires frequent capacity changes but is suitable for a permanent capacity change. Many types of variable speed drives are available for variable capacity applications. A recent development is solid-state variable frequency controls which permit the use of standard induction motors. The fan laws can be used to predict fan operation at different operating points. Variable speed operation is the most efficient means of controlling fan capacity, but can be expensive. Inlet vane control is preferable to outlet damper control from an energy efficiency point of view.

Filters

Air filters are used in HVAC systems to protect human health and comfort, to maintain room cleanliness, and to protect system equipment. Filters used in marine HVAC systems are of three basic types: (1) roll, (2) panel, and (3) viscous.

Roll-type air filters (Figure 19-19) with disposable dry media are popular for marine service. Their performance is better than the washable viscous type. Cleaning consists of rolling to a new clean section or replacing the roll when required. All filters on a particular ship are commonly selected with the same width. Heights of six feet can be accommodated.

Panel-type air filters utilize dry media similar to that used in roll-type filters. They are common in small systems where use of the roll-type filters is not practical or economical.

Viscous-type air filters utilize a washable element. The filter element is typically fabricated of bronze or copper mesh mounted in a hot-dipped galvanized frame. A common application of this type of filter is as a grease filter in galley ventilation systems and other locations where frequent replacement of disposable elements would be unacceptable.

Figure 19-19. Roll-type air filter. Courtesy The Trane Company.

Air filters should be installed so that they are protected from the weather and are accessible for servicing by operating personnel. The length of time between filter servicing depends on many factors such as filter type, port conditions, and types of cargo handled. Filters should always be checked after dusty cargoes are handled. Pressure drop across the filter is the best means of determining the need for replacement or cleaning. A typical filter will have a pressure drop of 0.1 inches of water and should be serviced when the resistance reaches 0.5 inches of water. Some newer roll-type filters will automatically roll to a clean section when the pressure drop increases to the set point on a pressure switch.

Heating Coils

Heating coils in marine HVAC systems may be steam, hot water, or electric resistance. Steam heating is generally cost effective; however, it poses certain problems such as difficulty in controlling small heating capacities, condensate return, and water hammer. Hot water and electric resistance heating eliminate many of these problems. Medium temperature hot water (250°-350°F, 121°-177°C) is replacing auxiliary steam in many applications. While not energy efficient, electric heating may be used wherever adequate generating capacity is available. Electric heating is particularly suited for electronics and other spaces where leakage from wet systems cannot be tolerated. Solid-state controls are available to provide excellent modulation of heat output.

Steam coils are usually of the header type (a header at each end of straight tubes). This avoids return bends which may trap condensate, causing water hammer and possible freezing. If possible, the tubes should be installed vertically. Each coil should be installed with a stop valve followed by a strainer and control valve on the steam supply and a dirt pocket or strainer, trap, and a cut-out valve on the condensate outlet. At least two feet of bare pipe for cooling should be provided between the coil and the trap. Condensate lines should be level or pitched in the direction of flow and typically are connected to an atmospheric or contaminated drain tank.

Hot water coils are generally of the serpentine type. Header type coils have lower tube velocities. This reduces the pressure drop, but adversely affects the transfer of heat. Coils may be either bare or finned tube. Air chambers and vents must be installed to eliminate air and noncondensable gases. They reduce heat transfer in the coil and cause cavitation and noise in the piping system.

The heating elements in electric resistance heating coils are usually of the tubular type and are Monel sheathed. On finned elements, the fins are Monel or aluminum alloy cast integral with the sheathing.

Figure 19-20. Hot water heating coil. Courtesy The Trane Company.

Cooling Coils

Either chilled water or evaporating refrigerant may be used in cooling coils. DX (dry expansion) is the name given to coils using evaporating refrigerant. Cooling coils are typically made of copper tubing with aluminum or copper fins arranged in a serpentine arrangement. Figure 19-21 shows a DX coil while Figure 19-22 shows a chilled water cooling coil. An even number of rows, usually 6 to 10, are used so that piping connections can be made to the same end of the coil. Bare tube coils are not common in cooling coils due to the small temperature difference. Fins increase the effective heat transfer area. A counterflow arrangement (Figure 19-23) is commonly used for chilled water coils. In this way, the coldest water is cooling the coldest air, resulting in increased heat transfer.

Radiators and Convectors

Radiators and convectors are heat exchangers used to transfer the heat from the circulating hot water or condensing steam to the room air. Radiators are not common on U.S. flag vessels. However, hot water radiators of embossed steel plate construction are used on many foreign flag ships. Convectors have a finned pipe heating element enclosed in a sheet metal cabinet. The elements are made of either copper or hot-dip galvanized steel. Room air enters through an opening in the bottom and

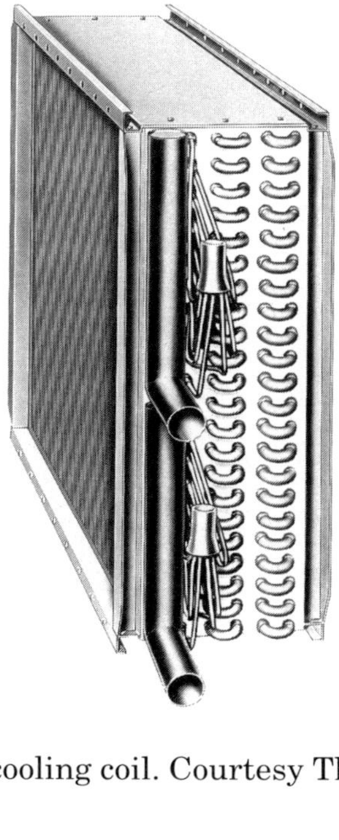

Figure 19-21. DX cooling coil. Courtesy The Trane Company

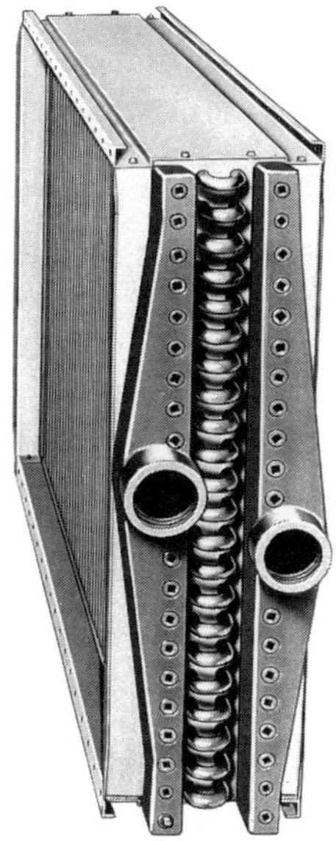

Figure 19-22. Chilled water cooling coil. Courtesy The Trane Company

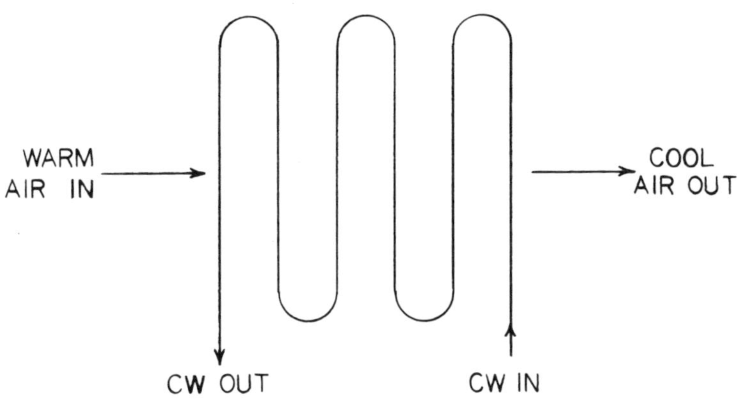

Figure 19-23. Counterflow cooling coil arrangement

leaves through an outlet grille at the top. Figure 19-24 shows some typical convectors. Convectors should be installed along weather-exposed bulkheads, under windows, and near the deck. This puts the heat where the heat loss is the greatest, and cold downdrafts are prevented. Elevating convectors off the deck decreases their capacity, and stratification due to insufficient air circulation can occur.

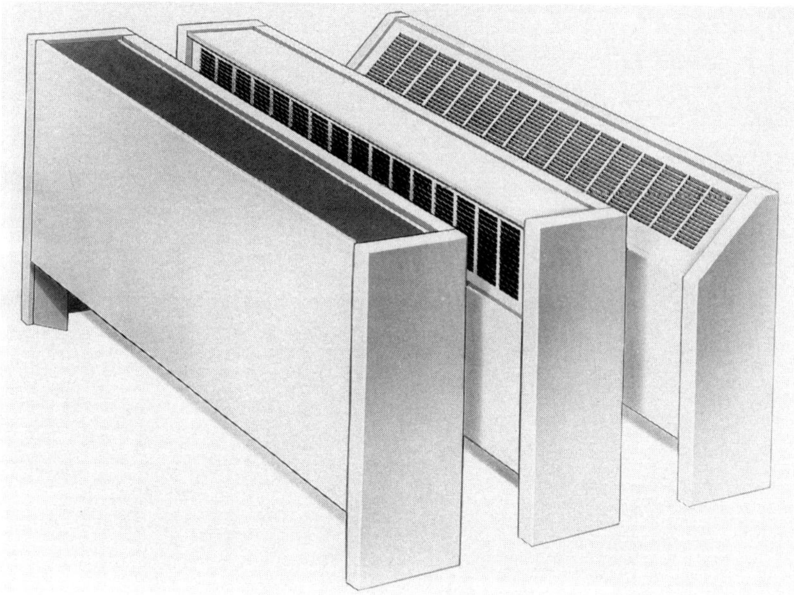

Figure 19-24. Convectors. Courtesy The Trane Company

Unit Heaters

A unit heater consists of a finned heating coil and a fan mounted in a compact assembly. See Figure 19-25. Forced circulation of air results in an increased rate of heat transfer. Unit heaters are most commonly used in steering gear rooms, shop areas, and machinery spaces of diesel ships (to prevent freezing when the main engine is secured). In large spaces where more than one is installed, unit heaters should be directed to circulate air in the space.

Fan-Coil Units

A fan-coil unit consists of a finned tube coil, a small centrifugal fan, and a filter mounted in a cabinet (Figure 19-26). Floor-mounted units are similar in appearance to convectors. Fan-coil units can be used for both cooling and heating by supplying either chilled or hot water to the coil, or by providing both a cooling and a heating coil. Fan-coil units may also be mounted horizontally at ceiling levels.

Ducts and Trunks

Conditioned air main ducts on ships are typically located in passageways with branches run to each space. The use of factory fabricated sheet metal ducting and fittings can reduce the cost of the installation. The ducting system is hung from existing structure. Small horizontal ductwork is

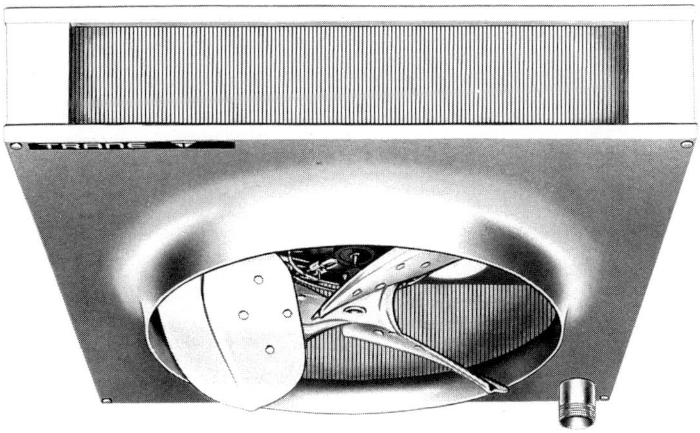

Figure 19-25. Unit heater. Courtesy The Trane Company

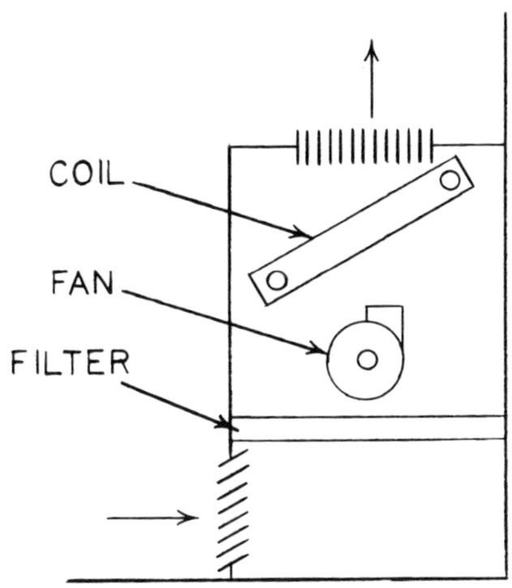

Figure 19-26. Fan-coil unit

commonly supported by sheet metal straps, while heavier ducts require angle iron cradles suspended from rods. Occasionally "built-in" trunks using ship's structure to form all or part of the air passage are employed in ventilation systems. The size of trunks must be large enough to permit access to the interior for inspection and maintenance.

Ducts in cargo spaces are usually made of steel plate to minimize the possibility of damage. Watertight ducts are constructed of the same material as the associated structure and must meet applicable test standards. Ducts carrying conditioned air are covered with thermal insulation such as glass fiber or mineral wool, and covered with a vapor barrier to eliminate condensation. Insulation is available in rigid board or blanket form. Ducts can be lined with acoustic insulation to absorb sound. The acoustic insulation can also serve as the thermal insulation.

Terminals

Conditioned air is distributed to the conditioned spaces by various types of supply terminals. Types in common use include (1) double-deflection registers, (2) fixed-pattern diffusers, and (3) perforated-panel distributors. Double-deflection registers consist of a frame with vertical and horizontal bars. The bars permit control of air direction. A volume control damper is also commonly fitted. Diffusers are available in round, square, or rectangular shapes. They consist of a series of concentric louvers to direct the air. A perforated-panel diffuser is merely a flat plate with a large number of holes in it. This type of diffuser is popular because it is inconspicuous when installed in hung ceilings.

Exhaust terminals are similar in construction to supply terminals. Simple grilles are common. The supply and exhaust terminals should be located to maintain proper circulation in the space to prevent hot or cold air collections and avoid short circuiting. Specifications typically call for a maximum ambient temperature differential of 2°F (1°C) and velocities of 35 to 50 feet per minute (11 to 15 meters per minute).

HVAC Control Systems

The HVAC control system has the task of varying the amount of heating, cooling, dehumidification, and humidification to automatically maintain the conditioned space in a comfortable condition. In addition, the system must control the HVAC system in an efficient manner to minimize energy consumption.

Figure 19-27 shows a simple closed loop control system. The temperature sensor senses the air temperature downstream of the coil. The control compares the sensed temperature with a set point and sends a signal to the control valve to vary the amount of hot water admitted to the coil. The change in air temperature caused by the change in valve position will be sensed and additional adjustments will automatically be made as necessary.

While a control system may appear complex, it can be reduced to a series of simple system elements. The system elements are assembled from standard control devices: sensors, controllers, relays, and actuators. Pneumatic control devices are the most common in HVAC systems because of their simplicity and low cost. Pneumatic controls are powered by compressed air at 15 to 20 psig (103 to 138 kPa), although occasionally higher pressures are used to operate large valves and dampers. Self-actuated control valves are used in a few applications, primarily for controlling

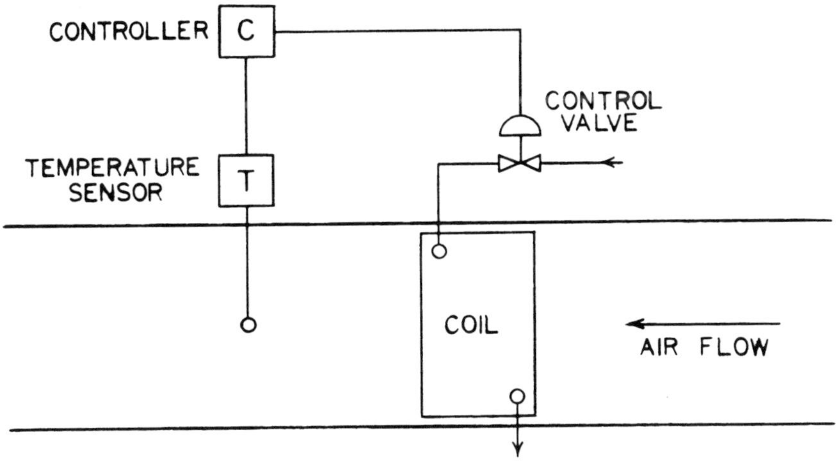

Figure 19-27. Simple closed loop control system

steam coils. Electric controls are occasionally used for simple applications where compressed air and/or suitable self-actuated devices are not available.

Electronic controllers have become increasingly popular for HVAC control systems as reliability and capability have increased and cost has decreased. Figure 19-28 shows a simple control system using a digital electronic controller to vary air flow to a space and maintain temperature. A thermocouple or resistance temperature detector (RTD) is used to sense the room temperature. The controller output is a 4-20 ma, 24-volt DC signal. Since the actuators used are still most commonly pneumatic, a current-to-pressure transducer is necessary to convert the current output from the controller into a standard 3-15 psig pneumatic signal that the actuator is designed for. More complicated systems with multiple inputs and outputs can be controlled by a special microprocessor-based controller or a computer running process control software. Both inputs and outputs to such systems are commonly 4-20 ma current signals.

Temperature sensors. The major types of temperature sensors include bimetallic, bellows, and bulb-and-capillary.

The bimetallic sensor is most commonly used in room thermostats. It consists of two strips of dissimilar metals bonded together. Since the two metals expand at different rates, a temperature change causes the element to bend. The bimetallic element may be also shaped in a spiral to provide a rotary motion output.

The bellows sensor is usually made of brass and filled with a vapor or gas. Temperature changes cause the bellows to expand and contract. A spring is used to adjust the set point.

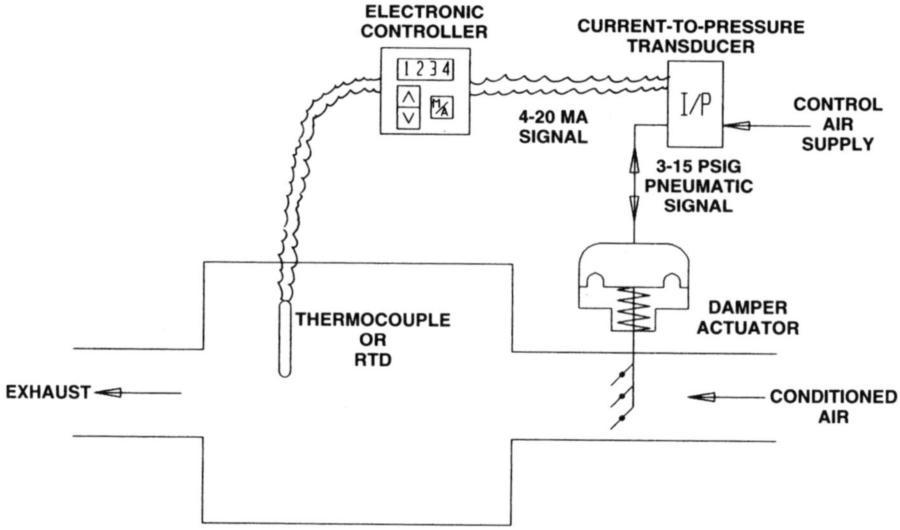

Figure 19-28. HVAC control system with electronic controller

Bulb-and-capillary elements are commonly used to sense temperatures in remote locations. The element consists of a bulb, a capillary tube, and a spring-loaded diaphragm head. The element may be filled with a liquid, a gas, or a refrigerant depending on the application. As the bulb temperature changes, the pressure exerted by the liquid or gas changes. The bulb pressure is transmitted to the diaphragm by the capillary, where the pressure change is translated into movement.

Thermocouples, RTDs, or thermistors are used for sensing temperature in control systems with electronic controllers.

Pressure sensors. The major types of pressure sensors include diaphragms, bellows, and bourdon tubes.

A diaphragm sensor is a flexible plate attached to a container. A spring balances the force caused by the pressure applied to the other side.

A bellows sensor is a diaphragm connected to a container by a series of convolutions to permit an increased degree of movement. The bellows acts like a spring, or an external spring may be added to change the sensitivity.

A bourdon tube sensor consists of a flattened tube bent in a circular shape. One end is connected to the pressure source, and the other end is free to move. An increase in pressure causes the tube to straighten out.

Strain gauge pressure transducers are commonly used in systems with electronic controllers. The strain gauges change resistance as a diaphragm deflects due to the applied pressure.

Humidity sensors. Humidity sensors utilize hygroscopic materials which change their size in response to humidity changes. One element is made of two strips of different woods—it operates similarly to a bimetallic temperature sensor. Elements using such materials as hair, special cloth, or animal membrane which change length in response to humidity changes are common. Electric sensors which use compounds that change in resistance with a change in humidity are also available.

Controllers. The purpose of a controller is to convert the output of the temperature, pressure, or humidity sensor into a pneumatic control signal, typically 3 to 15 psig (21 to 103 kPa). Controllers can be either direct acting or reverse acting. A controller is direct acting when an increase in the input parameter (temperature, pressure, humidity) results in an increase in the output air pressure. An increase in the input to a reverse-acting controller results in a decrease in the output air pressure.

A bleed-type controller is shown in Figure 19-29. The flapper is moved by a sensor and varies the amount of air exiting the nozzle. If the flapper is moved towards the nozzle, the flow is decreased, and the output pressure rises. The linkage between the sensor and flapper determines whether the controller is direct or reverse acting. When a bleed-type controller must deliver large air flows, it is commonly combined with a relay. The bleed-type controller acts as a pilot to vary the output of the relay.

Figure 19-30 is a simplified diagram of a nonbleed controller. It demands supply air only when the outlet pressure is increasing. A downward sensor motion will cause an inward movement of the diaphragm, raising the left end of the lever, and opening the supply valve. The incoming air will increase the chamber pressure, pushing up on the diaphragm to offset

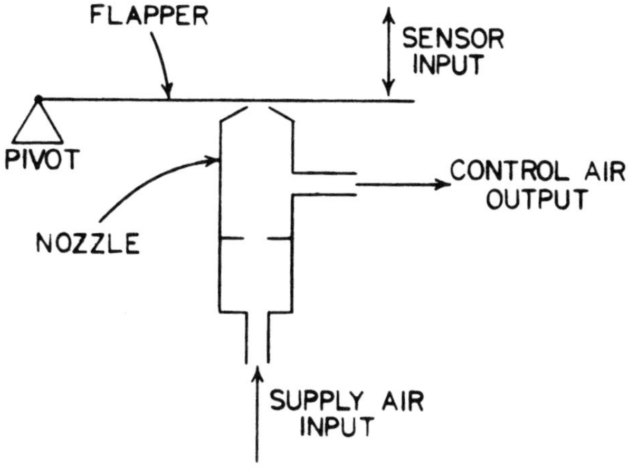

Figure 19-29. Bleed-type controller

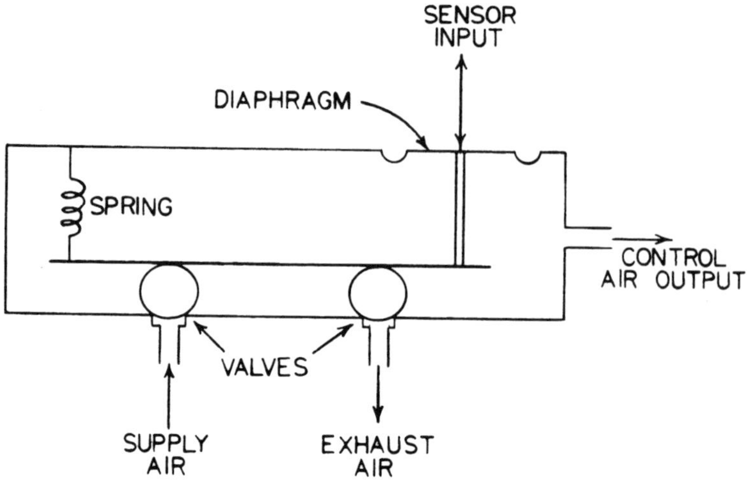

Figure 19-30. Nonbleed controller

the input from the sensor (negative feedback). When equilibrium is attained, the supply valve closes, and the new chamber pressure becomes the controller output pressure. A further downward sensor input will cause a rebalancing at some higher output pressure. An upward sensor movement will cause the bleed valve to open, and the controller will balance at a lower pressure. The type of linkage between the sensor and the controller determines whether the controller is direct or reverse acting.

Electronic controllers used in HVAC control systems are usually microprocessor-based digital units. The built-in programming has a variety of control modes included from simple on-off control to three-mode PID (proportional-integral-derivative) control. Autotune capability for the three-mode mode control is common. Electronic controllers can be designed for direct connection of various sensors or to accept standard 4-20 ma current signals as inputs. Outputs are typically 4-20 current signals.

Relays. Relays are used in a control circuit to perform a function that is beyond the ability or capacity of the controller. Most are variations of the nonbleed controller shown in Figure 19-30. Some of the functions that the various relays can perform include (1) amplification (or reduction) of input signal, (2) reversing of input signal, (3) difference (or sum) of input signals, (4) selection of higher (or lower) of two input signals, (5) average of two input signals, (6) conversion to on-off operation, (7) addition of reset (integral) action, and (8) sequencing of two or more actuators. Figure 19-31 is a diagram of a reversing relay. It may be used to change the output from a direct-acting controller to a reverse-acting output. Figure 19-32 is a diagram of a difference relay. Its output is the difference between the two input

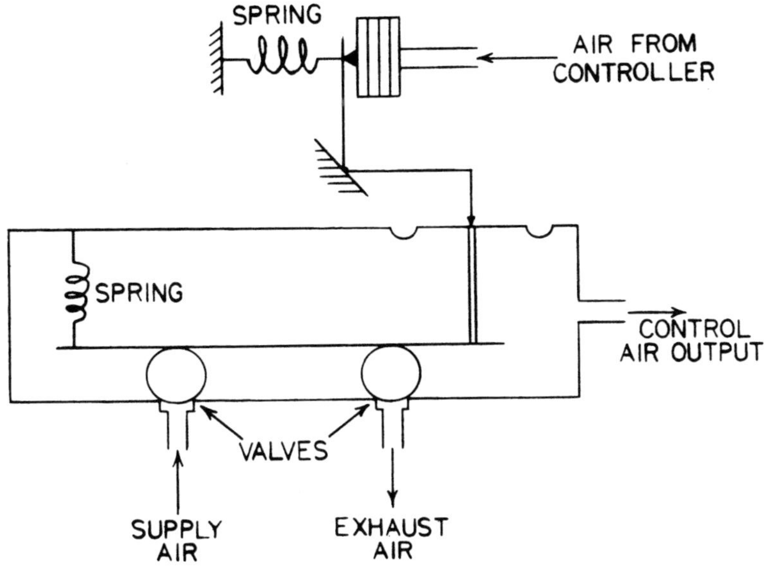

Figure 19-31. Reversing relay

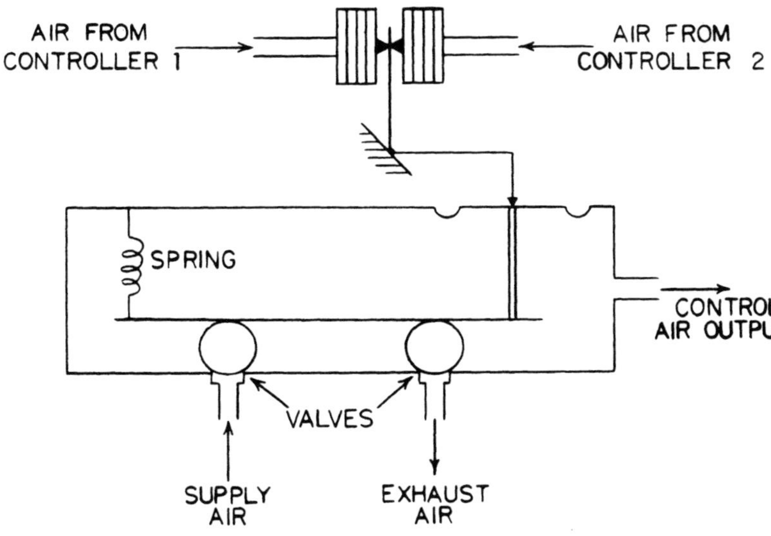

Figure 19-32. Difference relay

signals. These two examples illustrate the modifications that can be made to a basic nonbleed controller so that it will perform various functions.

Current-to-pressure (I/P) transducers. These devices convert a standard 4-20 ma current signal into a standard 3-15 psig pneumatic signal. Such devices are most commonly used to interface an electronic controller to a pneumatic actuator. Figure 19-33 is a simplified schematic of an I/P transducer. It can be described as a reducing valve with the outlet pressure setpoint varied by a coil and magnet assembly driven by the 4-20 ma current signal. In most transducers, the coil and magnet assembly actually moves a small pilot valve which then uses the supply air to position the main air valve. This reduces the force required from the coil and magnet assembly and improves performance. Zero and span adjustments are included for calibration of the transducer.

Actuators. The function of the actuator is to convert the pneumatic signal from the control system into a movement to position a fluid valve or an air damper. A pneumatic actuator is basically a piston or diaphragm on which the control air pressure acts on one side, and spring pressure acts on the other. See Figure 19-34. Increasing air pressure compresses the spring until the spring load and the load from the valve or damper balance. The spring range is typically set to provide full movement over the 3 to 15 psig (21 to 103 kPa) range, but other spring ranges are available.

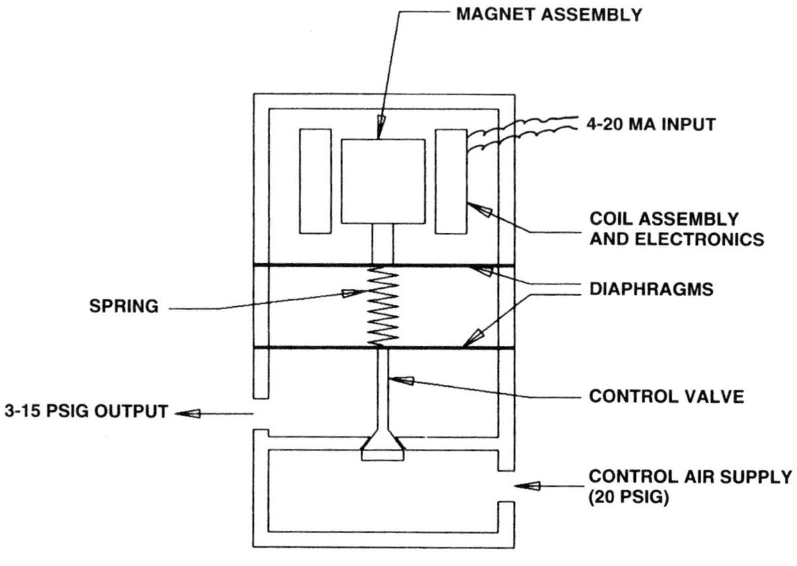

Figure 19-33. Simplified current-to-pressure transducer

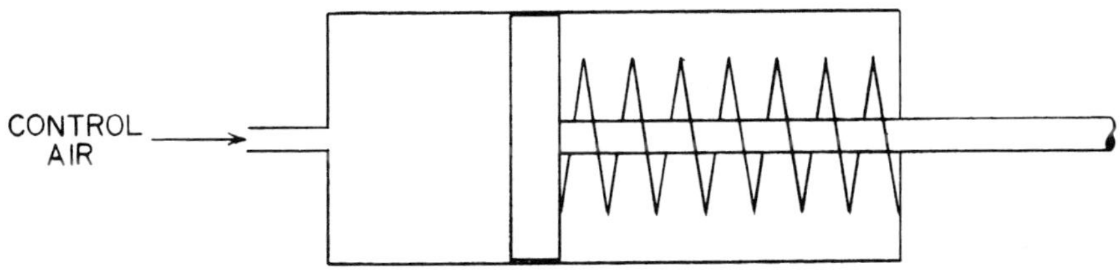

Figure 19-34. Piston actuator

Positioners. Under certain conditions such as friction in the actuator or load, or wind load on a damper, accurate positioning of a simple actuator may not be possible. The solution is to use a positioner. A positioner (or positive positioning relay) will accurately position the actuator in response to the input pressure rather than a balance of forces within the actuator. A positioner is essentially a modification of a nonbleed controller with feedback for actuator position. Use of a positioner assures full supply air pressure is available to move the actuator even though the input control pressure may change only slightly.

Dampers. Dampers are used to control the flow of air in HVAC systems. Most dampers are of the multileaf type. They can be described as a parallel- blade damper or an opposed-blade damper, and as normally open or normally closed. Figure 19-35 shows the operation of parallel-blade and opposed-blade dampers. Opposed-blade dampers are used in modulating applications while parallel-blade dampers are used in on-off applications. A normally open damper assumes an open position when no air pressure is applied to the actuator. A normally closed damper assumes a closed position when no air pressure is applied. The difference depends on the way the actuator is mounted, and on the way the linkage is connected to the damper.

In marine applications, outside air dampers should be made of stainless steel. Other dampers may be stainless or hot-dip galvanized steel. Damper construction should be integrally airtight without the use of nonmetallic seals.

Control valves. Control valves are used to control the flow of steam and water to heating and cooling coils. The size of a coil is normally based on some maximum design load, and the control valve must pass the required flow to meet that maximum load condition. However, in most cases, the system is operating at part load, and the valve must satisfactorily control the fluid flow over the entire range of load conditions. Control valves are available with a number of flow-versus-lift characteristics to suit different applications. Figure 19-36 shows three common characteristics: quick opening, linear, and equal percentage. Standard globe valves typically

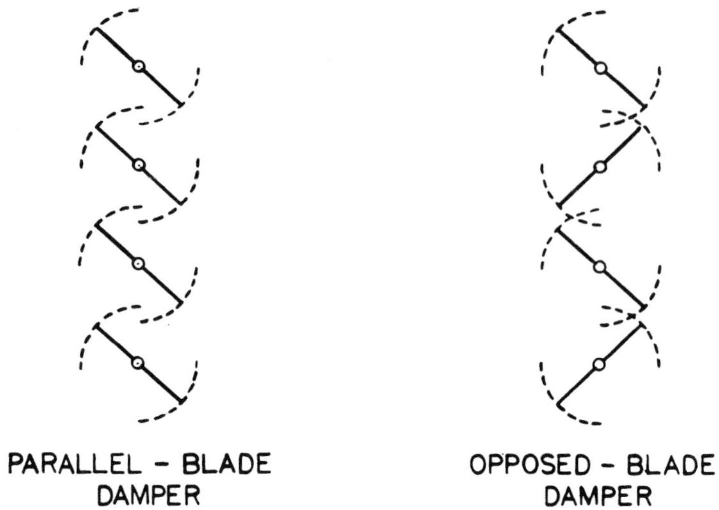

Figure 19-35. Damper types

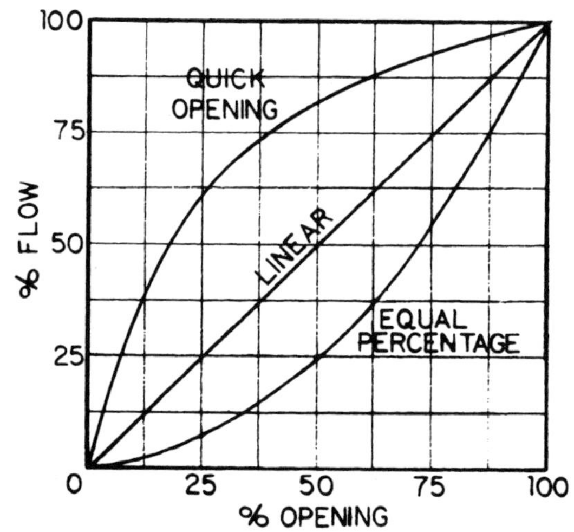

Figure 19-36. Control valve characteristics

have a quick-opening characteristic. Quick-opening valves do not modulate at low loads satisfactorily, and are primarily used in on-off applications. Valves with linear or equal-percentage characteristics are especially suitable for controlling heating and cooling coils. Valves of this type are usually of the ported or cage-guided construction. See Figure 19-37. By changing the shape of the ports in the plug, the flow-versus-lift characteristic of the valve may be changed.

Three-way valves are also available to control a heat exchanger. Three-way valves are available in two types: mixing and diverting. A mixing valve has two inlets and a single outlet. A diverting valve has one inlet and two outlets. Using either design in the wrong service will cause valve chatter and other problems.

Valves, like dampers, can be classified as normally open or normally closed. This characteristic depends on the way the actuator is connected to the valve body. A normally open valve will assume an open position with no air supply pressure to the actuator. A normally closed valve will assume a closed position with no air input.

Terminal reheat control system. In order to illustrate how the various control components may be connected into a complete system, a typical control system for a terminal reheat air-conditioning installation will be described. Refer to Figure 19-38. The steam preheater coil and chilled water cooling coil are controlled by thermostats in the conditioned supply duct. The hot water reheater at the inlet to each conditioned space is controlled by the individual room thermostat. The preheater thermostat is set several degrees below the design chilled air (off-coil) temperature to prevent simultaneous operation of the preheater and cooling coil. The supply duct temperature thus remains relatively constant during both summer and winter operation, and the room temperature is maintained in all seasons by the reheat coil. The quantity of outside air is controlled by duct thermostats in the outside air duct which positions the dampers in the exhaust, outside air, and recirculation ducts. By the use of normally closed dampers in the exhaust and outside air ducts and a normally open damper in the recirculation duct, the three dampers operate in sequence in response to a single control air input pressure. The duct thermostats are adjusted to reduce the quantity of outside air during hot or cold outside air

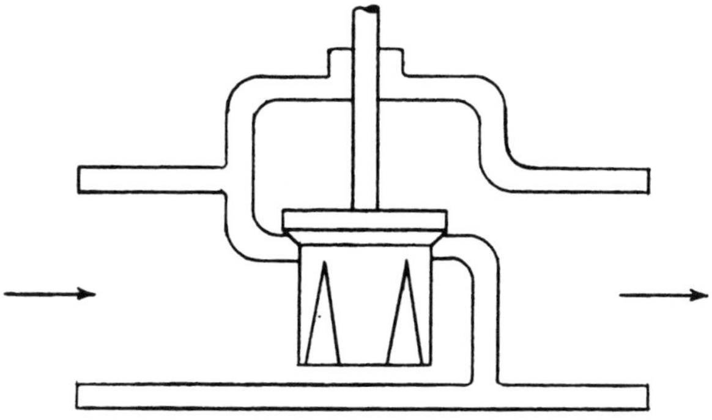

Figure 19-37. Ported valve control

conditions, and to increase the outside air during moderate conditions. This is done to reduce heating and cooling energy requirements.

The control systems for other types of air conditioning systems will utilize many of the techniques described above, but, obviously, must be adapted to suit the particular application. By breaking the system down into a series of simple elements, even the most complex control system can be understood. An understanding of the operation of the control system is essential to any attempt to adjust or troubleshoot the HVAC system.

SYSTEM TESTING AND BALANCING

The complexity of modern HVAC systems requires careful adjustment to achieve design performance. Failure to properly adjust and balance the system will result in uncomfortable conditions in the conditioned spaces and a waste of expensive energy. The balancing process requires an understanding of the HVAC principles described in this chapter and in the references at the end of the chapter. Organized procedures must be followed to insure a properly balanced system.

Instrumentation

The success of system testing and balancing depends on accurate, calibrated instrumentation. These instruments are used to measure such system parameters as temperature, pressure, flow, RPM, humidity, and electrical power.

Instruments are available with various degrees of accuracy. In most cases, an accuracy of 1 to 5 percent is adequate. The calibration of each in-

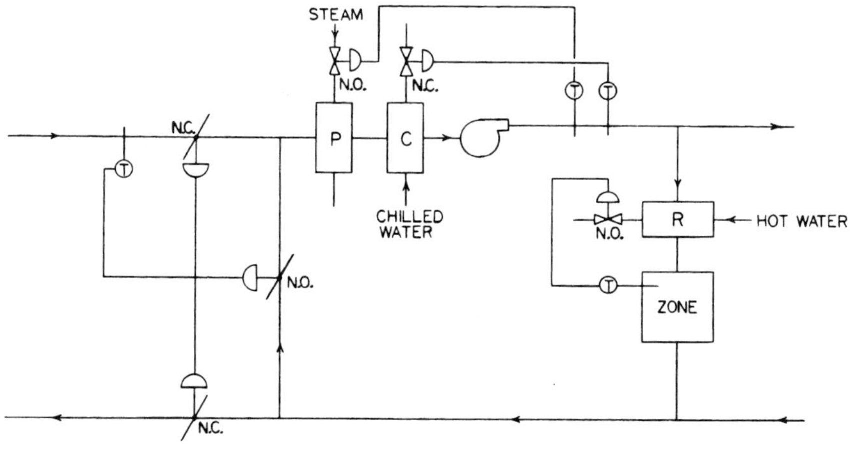

Figure 19-38. Terminal reheat control system

strument should be checked before and after use. This can usually be done by comparison with an instrument of known accuracy. The instrument can then be adjusted, or a calibration curve prepared, to insure accurate readings.

Temperature

Temperatures in HVAC systems are commonly measured by liquid-in-glass, bimetallic, or vapor bulb thermometers. For thermometers installed permanently in pipes, a well is used to allow replacement without securing the system. An inexpensive pocket dial (bimetallic) thermometer is convenient for checking temperatures where permanent instrumentation is not installed. Contact thermometers are available to conveniently determine the approximate fluid temperature in a pipe by reading the outside temperature of the pipe.

A recent development is portable digital instruments using thermocouple, thermistor, or RTD (resistance temperature detector) probes. Interchangeable probes are available for surface measurements, fluid measurements, and other applications. A permanent installation can utilize a number of probes, a selector switch, and a single readout, thus permitting the remote reading of many points from a central location. An RTD operates on the principle that the resistance of a coil of wire will change with temperature. The coil resistance is converted directly into a temperature by the readout circuitry. A thermocouple is a joint of two wires of different materials. The joint generates a very small voltage, and this voltage varies with temperature. The readout converts the voltage directly into temperature.

Pressure

Pressures in HVAC systems are commonly measured by manometers, bourdon tube pressure gauges, and diaphragm pressure gauges. Bourdon tube pressure gauges are used to measure higher pressures such as exist in piping systems, while manometers and diaphragm gauges are used to measure lower pressures such as in air ducting systems.

The simplest manometer is a U-shaped glass tube partially filled with water or mercury. A pressure difference between the two tube connections shows as a difference in height between the liquid in the two legs. If one end of the manometer is open to atmosphere, gauge pressure (or vacuum) is measured. By connecting the manometer to two different locations in the system, pressure drops can be measured.

Diaphragm pressure gauges can be used in place of manometers to measure gauge and differential pressures in duct systems. The pressure difference moves a diaphragm, and a mechanical linkage or magnetic field is used to move the readout pointer. Gauges which can read pressures as low as 0.01 inches water are available.

Bourdon tube pressure gauges use a small curved metal tube which changes shape to measure pressures. A linkage from the end of the tube moves the gauge pointer. It is most suitable for measuring pressures in hot or chilled water, steam, and compressed air systems. It is not suitable for measuring the small pressure differences that exist in air ducting systems.

Flow

An important part of the system balancing process is measuring the flow rate of air in ducts and water in pipes. The flow rate of air in ducts is usually done by measuring the air velocity using a Pitot tube or an anemometer. Flow rate can then be calculated using the velocity and duct cross-sectional area. A Pitot tube is used in conjunction with a manometer or diaphragm gauge. It consists of two concentric tubes. The opening of the inner tube is pointed in the direction of flow, and the outer tube has holes perpendicular to the flow. The manometer or diaphragm gauge thus measures the difference between the total and static pressure, i.e., the velocity head (pressure). This can be converted to velocity using the following equation:

$$V = 4{,}000\sqrt{H_v}$$

where V = air velocity, feet/minute
H_v = velocity head, inches of water

Rotating vane and deflecting vane anemometers can also be used to measure air velocity. The rotating vane anemometer consists of a small propeller and a dial which reads in feet. It is used in conjunction with a stopwatch to read feet per minute. In the deflecting vane anemometer, the air flow deflects a small vane, and the deflection is read on a scale calibrated in feet per minute.

The velocity measured by the Pitot tube or anemometer is converted to a flow rate using the continuity equation:

$$CFM = A \times V$$

where A = cross-sectional area, square feet
V = fluid velocity, feet/minute

The average fluid velocity, determined by a number of measurements at a number of locations across the duct should be used.

There are many types of devices available for measuring the flow of liquid in pipes. These include rotameters, nutating disc meters, turbine-type flowmeters, orifices, and venturis. If a flowmeter is not installed, the flow delivered by a pump can be estimated by measuring the head across the pump and referring to the manufacturer's head-capacity curve. The flow to a coil can be estimated in a similar manner if its pressure-flow characteristic is known. A recent development is portable ultrasonic doppler flowmeters. An ultrasonic transducer mounted on a pipe measures the fluid velocity inside by measuring the change in frequency of the signal reflected from particles or air bubbles in the moving fluid. Knowing the cross section of the pipe, the flow can be determined.

Humidity

The humidity in conditioned spaces is most commonly measured with a sling psychrometer. It consists of two thermometers, one of which has a cloth wick soaked in water wrapped around the bulb. The instrument is spun rapidly, and the wet bulb and dry bulb temperatures are read. The humidity can then be read off the psychrometric chart.

Hygrometers use a sensing element that changes shape with a change in humidity. A linkage converts the changes in the sensing element shape to a movement of a pointer. The dial is calibrated directly in percent relative humidity.

Electronic instruments are also available to measure humidity in conditioned spaces, ducts, and other locations. The sensing elements use a substance such as lithium chloride which changes resistance with changes in relative humidity. The readout instrument converts the resistance directly into a digital or analog display of the relative humidity.

RPM

It is often necessary to measure the operating RPM of rotating equipment such as fans, pumps, and compressors. If the end of the shaft is accessible, a handheld mechanical or electronic tachometer can be used. The tip of the instrument is pressed into the countersink, and the RPM is read directly. The instruments are also supplied with a wheel which can replace the tip. By bringing the wheel in contact with rotating shaft or coupling of the rotating machine and multiplying by the ratio of the diameters, the RPM can be determined.

A stroboscope can be used where contact with the rotating shaft is difficult. The instrument consists of a flashing light with an adjustable frequency. The light is directed at the rotating parts, and the frequency of the instrument adjusted until the equipment appears to stand still. This will occur when the frequency is equal to the speed of rotation, or a multiple or fraction thereof. Some care must be taken to read the proper frequency.

Electrical Power

The measurement of voltage, current, and electric power supplied to motors is commonly required. A standard portable multimeter can be used for checking voltages, currents, and resistances in HVAC circuits. A clamp-on ammeter is convenient for measuring the power delivered to motors. A

U-shaped jaw is opened and clipped around a single wire of the circuit. The instrument operates by measuring the magnetic field created by the current flow. Clamp-on wattmeters which measure current, voltage, and power factor, and display kilowatts directly are also available.

Balancing an Air System

Listed below are the recommended steps for balancing an air system:

1. Obtain a duct system drawing and list all design velocities and flow rates. If a system drawing is not available, prepare a one-line diagram. Show the location of all balancing dampers. If design velocities or flow rates are not available, Table 19-2, which lists typical design velocities in HVAC systems, can be used for estimating purposes.
2. Obtain design data for the system and all major components, i.e., fans, filters, coils, dampers, mixing boxes, etc.
3. Prepare data sheets.
4. Decide on required instrumentation and on measurement locations.
5. Check the operation of all system components and controls. Check all damper positions.
6. Start fans. Measure fan speeds and adjust RPM to design values.
7. Measure fan capacities (CFM) using a Pitot traverse of the main duct or an anemometer reading across the coils. If flow is not within 10 percent of design, check for possible problems such as closed dampers, equipment malfunction, etc. If system is operating properly, adjust the fan speed to obtain the proper CFM.
8. Measure flow in major supply and return ducts and adjust to design value (±10 percent) with dampers.
9. Measure and adjust the flow of each outlet terminal. Work either from the farthest outlet towards the fan or from the fan outwards. Use effective outlet areas when calculating flow rates. Repeat the cycle at least once, since later adjustments will affect earlier measured flows.
10. Record all data including flows, temperatures (including WB and DB temperatures at coil inlet and outlet), and pressures. Record all performance data for all operating equipment.

TABLE 19-2

Typical Velocities in HVAC Systems

Location	*Velocity*
Main ducts, living spaces	800-1500 FPM
Main ducts, other spaces	1500-2500 FPM
Branch ducts	800-1500 FPM
Coils	500-700 FPM
Filters	250-350 FPM

Balancing a Water System

Listed below are the recommended steps for balancing a water system:

1. Obtain (or prepare) a piping system drawing and record design flows, temperatures, and pressures.
2. Obtain design data for the system and all major components, i.e., pumps, coils, chillers, hot water heaters, and terminal units.
3. Prepare data sheets.
4. Decide on required instrumentation and on measurement locations.
5. Check operation of all pumps, valves, and controls. Check the position of all valves.
6. Start the circulating pump. Check the flow with a flowmeter or by recording the suction and discharge pressures and referring to the design head-capacity curve. Adjust flow with discharge valve to about 110 percent of design.
7. Do not balance the branches if automatic control valves are installed to vary the flow through the system coils and terminal units. If automatic control valves are not fitted, the flows in the various branches must be measured and adjusted to ±10 percent of design. Measurement of the water temperatures in and out of the coils and terminal units can be used as an additional check.
8. Measure and record all operating data including performance of all major equipment.

REVIEW

1. What are the properties of air that must be controlled by an HVAC system?
2. Define psychrometry. What is the psychrometric chart?
3. Explain the differences between dry bulb temperature, wet bulb temperature, and dew point.
4. Explain the difference between relative humidity and humidity ratio.
5. Use the psychrometric chart to find the relative humidity and dew point given a dry bulb temperature of 70°F and a wet bulb temperature of 60°F.
6. What are the factors that must be considered in determining ventilation supply and exhaust flow rates?
7. How does an air-type HVAC system control space temperature and humidity in hot weather? In cold weather?
8. Describe how terminal reheat and a dual duct HVAC systems can be operated to improve efficiency and reduce energy costs.
9. Describe how variable air volume (VAV) reduces energy waste when added to a terminal reheat system and a dual duct system.
10. Describe the differences between a one-pipe and two-pipe hydronic system. What advantage does a four-pipe system have over a one-pipe or two-pipe system?

11. Describe a cargo hold dehumidification system. How is the desiccant regenerated?
12. Why are axial flow fans not commonly used in larger ducted HVAC systems?
13. What are the differences in the performance characteristics between forward-curved blade and backward-curved blade centrifugal fans?
14. The speed of a centrifugal fan is reduced 25 percent. How much will its capacity, static pressure, and power be reduced?
15. What are the methods for controlling the flow of centrifugal fans?
16. What is a fan-coil unit?
17. Describe the common sensors used for measuring temperature, pressure, and humidity in HVAC systems.
18. Define the terms direct-acting and reverse-acting as applied to controllers.
19. What are some of the functions that can be performed by relays?
20. Define the terms normally-open and normally-closed as applied to actuators, dampers, and valves.
21. Describe the difference between a quick opening and an equal percentage control valve characteristic. Which would most likely be used to control chilled water flow to a cooling coil? Why?
22. Outline the procedure for balancing an air-type HVAC system.

REFERENCES

Dossat, R. J.1996. *Principles of Refrigeration.* 4th ed. NewYork: Prentice Hall.

Haines, R. W. 1971. *Control Systems for Heating, Ventilation, and Air Conditioning.* New York: Van Nostrand Reinhold Company.

Handbook and Product Directory. 4 vols. Atlanta, Ga.: American Society of Heating, Refrigerating, and Air Conditioning Engineers.

Harrington, R. L., ed. 1992. *Marine Engineering,* 3rd ed. Jersey City, N.J.: Society of Naval Architects and Marine Engineers.

Pita, E. G. 1997. *Air Conditioning Principles and Systems,* 3rd ed. New York: Prentice Hall.

Stoecker, W. F., and Jones, J. W. 1982. *Refrigeration and Air Conditioning.* New York: McGraw-Hill.

ACKNOWLEDGMENTS

The author is indebted to the following companies engaged in the manufacture of HVAC systems and equipment who so kindly contributed photographs, drawings, and other material used in the preparation of this chapter:

Carrier Corporation, Syracuse, New York
The Trane Company, LaCrosse, Wisconsin
Buffalo Forge Co., Rockville, Maryland
Cargocaire Engineering Corp., Amesbury, Massachusetts

CHAPTER 20

Cryogenic Cargo Systems

JAMES A. HARBACH

CRYOGENIC PRINCIPLES

CRYOGENICS is the study of matter at very low temperatures. The primary interest of the marine industry in cryogenics is for the transportation of liquefied gases. If you liquefy gases you reduce their volume several hundred times, thus permitting economical transportation in specially designed tankers. While the transportation of liquefied gases at normal temperatures has been considered, the cost of the required high pressure tanks has prevented their use in large scale commercial operations. The liquefied gas tankers in service today utilize insulated tanks which carry the liquefied gases at atmospheric pressure.

The properties of some liquefied gases at atmospheric pressure are shown in the table below:

Properties of Liquefied Gases at Atmospheric Pressure

	Formula	*Boiling Point*	*Liquid Density*
Nitrogen	N_2	–320°F (–196°C)	.81
Methane	CH_4	–259°F (–162°C)	.42
Ethylene	C_2H_4	–155°F (–104°C)	.57
Ethane	C_2H_6	–128°F (–89°C)	.56
Propane	C_3H_8	–44°F (–42°C)	.59
Butane	C_4H_{10}	+31°F (–0.5°C)	.60

Two common liquefied gas cargoes are LPG (liquefied petroleum gas) and LNG (liquefied natural gas). LPG is primarily propane (boiling point –44°F, –42°C) while LNG is primarily methane (boiling point –259°F, –162°C). The actual makeup of LPG and LNG depends on the source of the gas field, and varies widely from field to field. This chapter will concentrate on LNG cargo systems, but most of the information will apply to other, higher temperature, and more easily handled cargoes.

LNG tankers are usually built for dedicated service between specific ports such as Algeria (Arzew) to France (Le Havre) and the United Kingdom (Canvey), and Indonesia to Japan. The basic transportation cycle consists of liquefaction of the gas near the gas fields, marine transport of the liquefied gas in insulated tanks at atmospheric pressure, and vaporization of the liquefied cargo at the receiving terminal for distribution. Liquefaction facilities are normally not installed on the vessels. The gas produced by heat transfer through the tank insulation, termed boil-off, is used as fuel for the main engines.

Liquefaction Processes

The first step in the transportation cycle of LNG is liquefaction. Natural gas from the supply field is delivered via high pressure pipeline to the loading terminal. Prior to liquefaction, impurities such as water, carbon dioxide, and heavier hydrocarbons which would solidify during the liquefaction process are removed. After liquefaction, the LNG is stored for transfer to the LNG tankers.

There are a number of processes available for liquefying gases. These include the Linde-Hampson process, the cascade process, the auto-refrigerated cascade (ARC) cycle, the expander cycle, the open Claude cycle, and the closed Brayton cycle. Two of the processes in common use in large natural gas liquefaction plants, the cascade process and the ARC cycle, will be described below.

The cascade cycle. The cascade cycle consists of two or more vapor-compression refrigeration systems connected in series. Refer to Chapter 18 for a discussion of the vapor-compression cycle. The evaporator for the high temperature cycle serves as the condenser for the second cycle. The evaporator for the second cycle serves as the condenser for the third cycle, and so on. Figure 20-1 shows a simplified cascade system using propane in the high temperature cycle, ethylene in the second cycle, and liquefaction of methane in the third cycle. The use of several refrigeration cycles improves the overall system efficiency. Pressure differences between low and high sides are reduced and refrigerants are selected to suit the operating temperatures. The main disadvantage of this system is its complexity. Balancing and adjusting the system can be difficult, and maintenance costs are high.

The auto-refrigerated cascade (ARC) system. The ARC system is a simplification of the cascade cycle described above. A mixture of as many as six gases, including methane, ethane, propane, butane, pentane, and nitrogen, is used as the refrigerant. The composition of the refrigerant

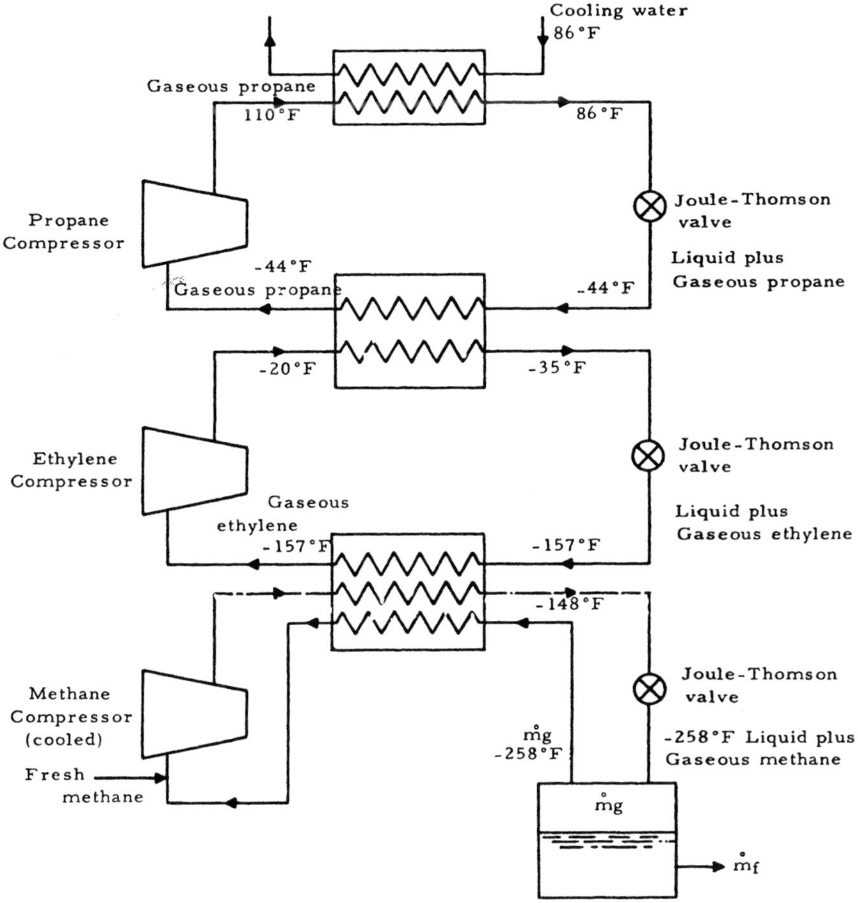

Figure 20-1. Cascade liquefaction cycle

mixture is chosen so that the partial condensation in any one stage corresponds exactly to the refrigeration requirements of the next stage. Partial condensation of the refrigerant takes place at the high side pressure, while evaporation and the resultant cooling occurs at the low side pressure. Figure 20-2 is a diagram of a simplified ARC cycle. The refrigerant is condensed in a series of condensers. The liquid is sprayed into the column at several elevations, lowering the temperature and liquefying the natural gas flowing countercurrent to it. The ARC cycle eliminates many of the limitations of the cascade cycle. The simplification of the compression process results in lower capital costs. By installing compressors in parallel, system reliability and flexibility of operation are increased. Failure of one compressor in the ARC cycle results in reduced capacity, while a compressor failure in the cascade system results in

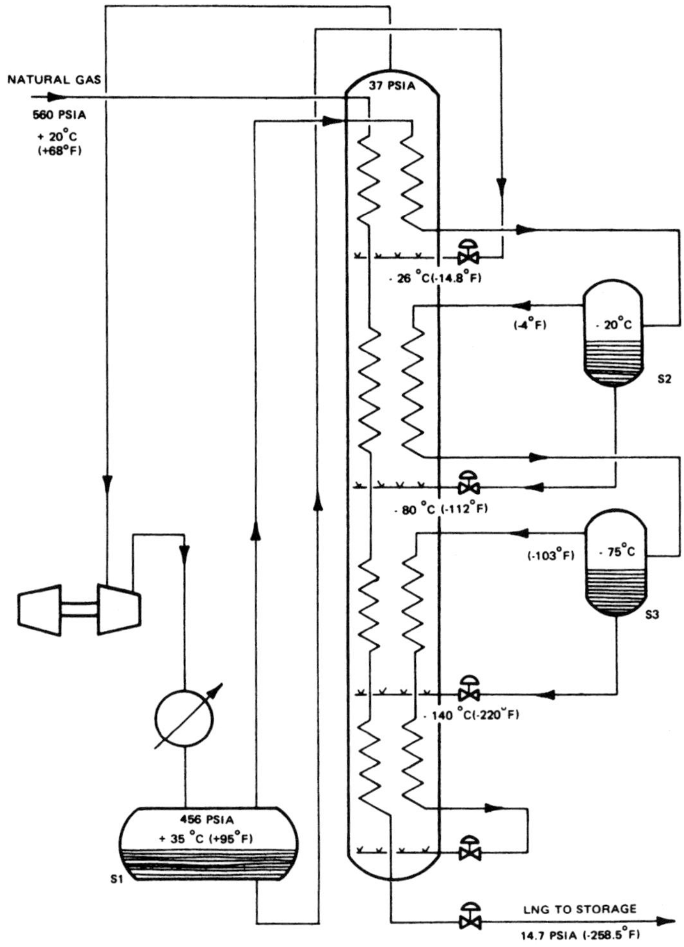

Figure 20-2. Auto-refrigerated cascade (ARC) liquefaction system

total shutdown. The main disadvantage of the ARC cycle is its higher power requirements compared to the cascade cycle.

Materials

All materials that come in contact with LNG must possess appropriate mechanical and thermal properties (strength, ductility, fatigue life, thermal coefficient of expansion) at the operating temperature of –260°F (–162°C). The most important characteristic is no transition from ductile to brittle fracture modes at the lowest operating temperature. At temperatures below the transition temperature, i. e., the brittle fracture region, the material loses most of its ability to absorb energy, and catastrophic failure is more likely. Nickel is the most effective alloy in increasing the toughness of steel. Materials suitable for LNG service include stainless steel, certain aluminum alloys, 5 percent and 9 percent nickel steel, and Invar (36 percent nickel steel). Final selection of materials must consider the proposed application (tank structure, tank membrane, piping, etc.), ease of welding, amount of required aftertreatment, and, of course, cost. Stainless steel is typically used for LNG piping because it is readily available and has superior fire resistance compared to aluminum. Free-standing tanks are typically made of 5083 aluminum alloy. Membrane tank linings are typically stainless steel or Invar.

LNG CARGO TANKS

Many different designs have been proposed for the carriage of LNG cargo. They can be broken down into two categories: the freestanding or self-supporting type, and the membrane or integrated type. The freestanding tank containment system incorporates tanks which are independent of the vessel's hull. The tanks are capable of supporting their own weight as well as withstanding the static and dynamic loads imposed by the liquid cargo. These loads are transferred to the ship structure through the tanks themselves and their key supporting structure. Membrane tanks, on the other hand, are made of thin metallic materials and are not self-supporting. The tanks are attached indirectly to the ship's structure through a load-bearing insulation.

There is a definite tendency among shipyards and ship operators to select a proven tank design. Considering the extremely high costs of LNG vessels, the use of a proven, reliable design rather than a new, untried design is understandable. Unless a new design has significant advantages over the existing designs, it is unlikely to find customers.

Safety is crucial when considering the advantages and disadvantages of the various tank designs. Classification societies have established regulations governing tank system design and construction. However, such considerations as survivability of collision or grounding are not covered. An important safety feature of most LNG tanks is the secondary barrier. It contains the LNG in the event of leakage of the primary containment, and prevents damage to the ship structure. Tanks designed as pressure vessels are not required to have a secondary barrier. Such features as ease of inspection can also affect the overall safety and reliability of a design.

Most LNG vessels of recent construction are of the 125,000 to 135,000 m^3 size. This produces a vessel of about 935 ft (285 m) length overall with a loaded displacement of 95,000 tons. Service speed is typically about 19 to 20 knots, requiring a propulsion plant of approximately 40,000 SHP. The cargo is carried in 4 to 6 tanks. The 4 most popular LNG cargo tank designs will be discussed in the following sections.

Conch

The Conch tank design as shown in Figure 20-3 is a development of the pioneering work of Conch International Methane Company in the marine carriage of LNG. The tank is a prismatic, free-standing design constructed of 5083 aluminum alloy. A longitudinal liquid-tight bulkhead and a transverse swash bulkhead provide structural strength and reduce free surface and dynamic loading effects. The tank thickness is approximately ½ inch (13 mm) but increases to 1½ inch (38 mm) at the chamfer panels. The tank is fitted to the hull shape and is located by keys and keyways at the bottom and top, allowing for the necessary expansion and contraction. The keys are made of permali, a reinforced hardwood. The tank is supported by balsa-plywood insulation panels which are uniformly distributed under approximately 30 percent of the tank bottom. The spaces between the panels are filled with polyurethane foam and fiberglass, each 6 inches (15.25 cm) thick.

The insulation system, which includes secondary barrier capabilities, consists of balsa-plywood frame panels along the corners of the cargo hold

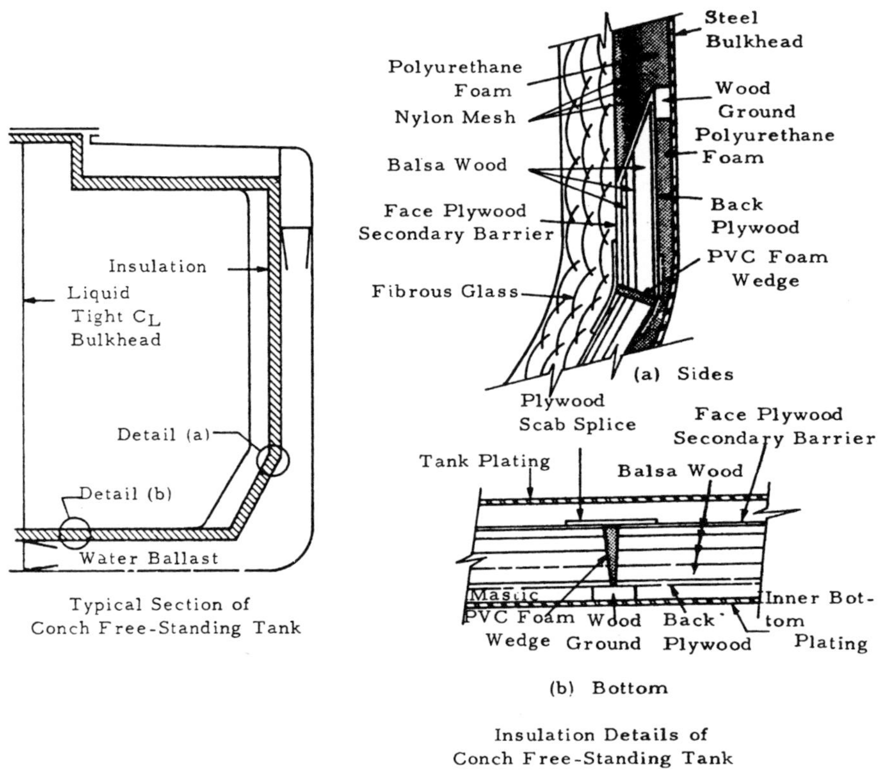

Figure 20-3. Conch containment system

with a sprayed-in-place high density polyurethane covering the inner hull between the frames. The polyurethane foam is approximately 6 inches (15.25 cm) thick. Inboard of the foam is an attached layer of fiberglass of 6 inch (15.25 cm) thickness. The space between the tank wall and the insulation is maintained inert with nitrogen gas.

Technigaz Membrane

The Technigaz Mark I membrane containment system as shown in Figure 20-4 consists of wooden support beams, wooden insulation panels containing a wood secondary barrier, and a stainless steel primary barrier. The wooden support beams called grounds are made of layers of Douglas fir and lauan plywood. They are mounted on the tank structure by steel studs and mastic. The space between the grounds is filled with glass wool to provide additional insulation and prevent formation of convection currents. Prefabricated laminated balsa and plywood insulation panels are glued to the grounds. Joints between insulation panels are filled with PVC, and ply-

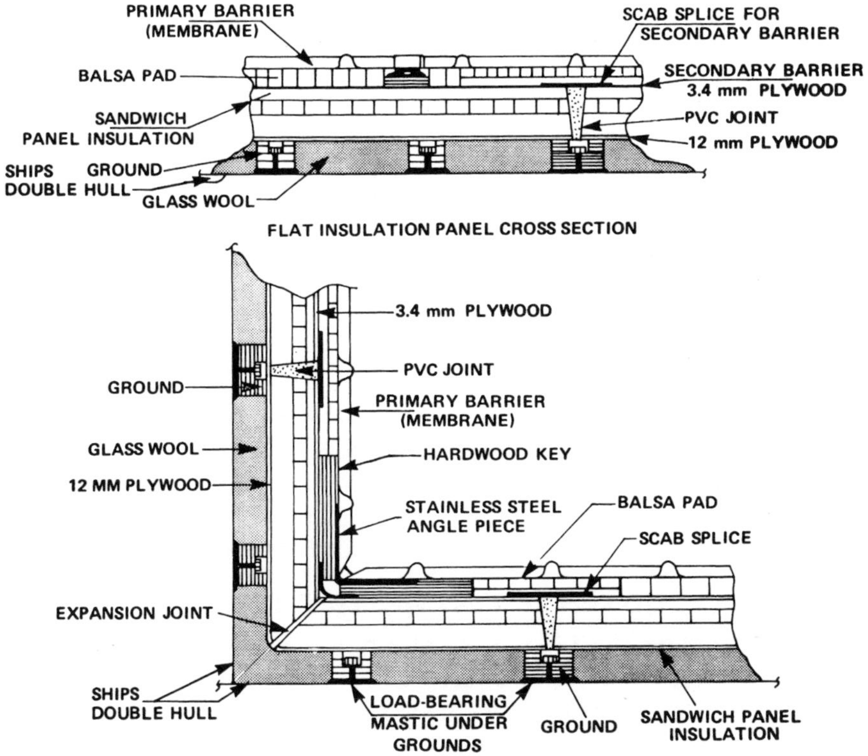

Figure 20-4. Cross section of Technigaz Mark I containment system. Courtesy Newport News Shipbuilding and Drydock Co.

wood pieces are glued over the PVC joints to complete the secondary barrier. Balsa pads and plywood corner pieces complete the insulation. Stainless steel anchor points are bonded to wood insulation to provide a suitable surface for mounting the stainless steel primary barrier.

The membrane is made of 1.2 mm (0.047 in.) low carbon stainless steel sheets. The sheets are corrugated to permit expansion and contraction. The intersection of the corrugations have a special geometry referred to as knots. Figure 20-5 taken during construction shows the details of the membrane. The individual membrane sheets are lapped and welded to complete the primary barrier. Figure 20-6 shows a completed tank.

The current Technigaz Mark III system incorporates the basic design features of the earlier Mark I system described above. The primary barrier is a waffled 1.2 mm 304L stainless steel membrane supported by load-bearing insulation. Due to the scarcity of balsa wood, the secondary barrier now consists of fiber-reinforced polyurethane insulation along with a membrane of a glass fabric aluminum composite material called Triplex. Boil-off rates as low as 0.09 percent per day are claimed for the Mark III system.

Gaz Transport Membrane

The Gaz Transport containment system consists of two 0.5 mm (0.02 in.) thick Invar membranes and an insulation system made of a series of subdivided plywood boxes, factory-made on a production line and filled with perlite (see Figure 20-7). Invar was selected for the membranes because it has a very low coefficient of expansion, and thus eliminates the need for expansion joints or corrugations. The membranes are fabricated from strips of Invar approximately 0.4 m (13.3 in.) wide. The edges are flanged, then welded to a tongue support attached to the insulation boxes. The system is unique in that the primary and secondary barriers and the primary and secondary insulation systems are identical.

The current Gaz Transport system (GT No. 96) is an evolutionary development of the original system described above. In the GT No. 96 system, the primary and secondary barriers are identical, consisting of flat 0.7 mm Invar membranes supported by a load-bearing insulation system consisting of plywood boxes filled with silicon-treated perlite.

Kvaerner-Moss Spherical Tank

The Kvaerner-Moss tank is a 5083 aluminum alloy, or 9 percent nickel steel, sphere supported by a cylindrical steel skirt attached to the equatorial ring of the tank. The skirt is welded to the ship structure at its base and bonded to the tank equatorial ring at its upper edge. No tank internal stiffeners or bulkheads are fitted (see Figure 20-8). The tanks and support skirts are insulated with a system of polyurethane foam panels and glass fiber. The panels and glass fiber are fixed to the tank by phenolic studs screwed into aluminum sockets welded to the tank shell. A rigid disk is threaded onto the outer end of the stud to secure each panel at six points. The panels and studs are oversprayed with a Hypalon rubber coating to minimize the permeation of moisture and other contaminants. Figure 20-9 illustrates the insulation system.

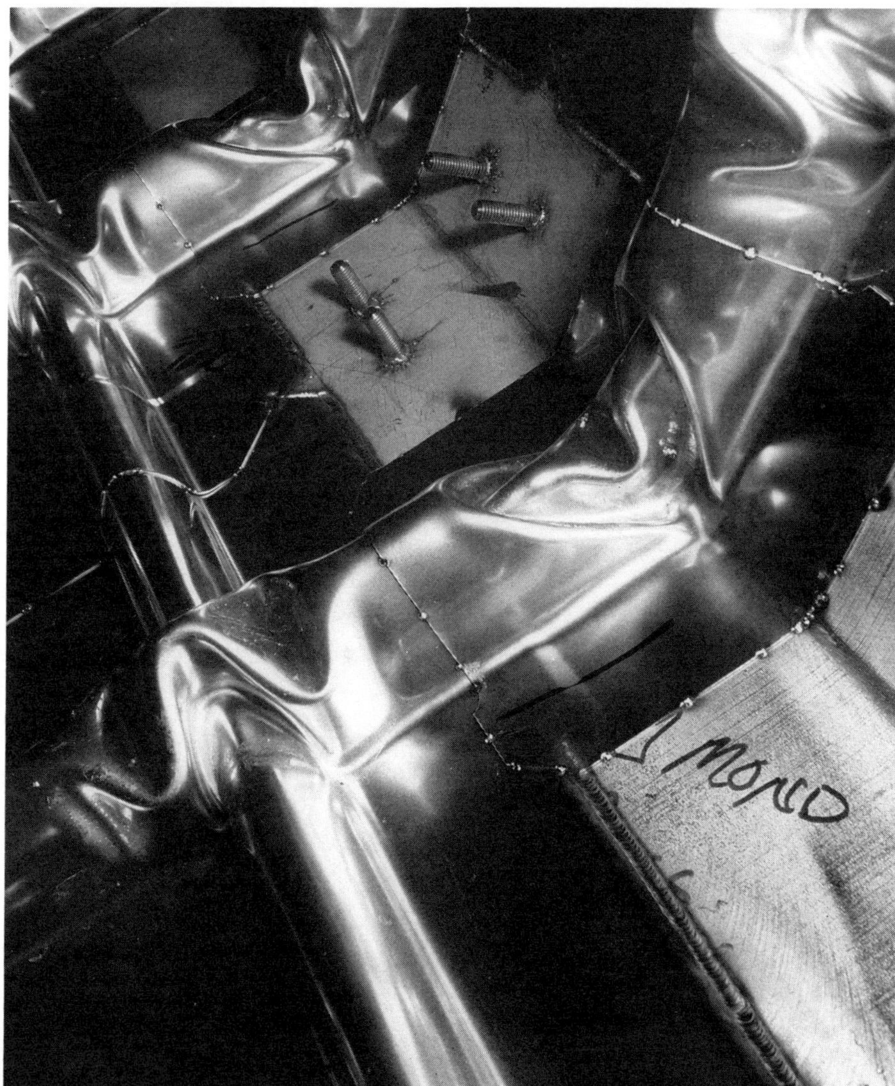

Figure 20-5. Technigaz membrane during construction. Courtesy Newport News Shipbuilding and Drydock Co.

Figure 20-6. Completed Technigaz tank. Courtesy Newport News Shipbuilding and Drydock Co.

A major advantage of this design is that it can be designed and analyzed according to pressure vessel standards. This permits the system to be built without a secondary barrier. However, a stainless steel drip pan, insulated from the hull structure, is installed under the tank as a precaution. The ability to pressurize the tank provides an emergency means of discharging cargo. Some disadvantages given for this design are the relatively poor visibility from the bridge due to the protruding tanks, a relatively high center of gravity, and poor hull volume utilization. The design is claimed to be less susceptible to collision damage than membrane designs.

Improvements to the Moss system have focused on reducing boil-off rates and increasing tank sizes, permitting the number of tanks on a typical vessel to be reduced to four versus the five or six on earlier designs. New

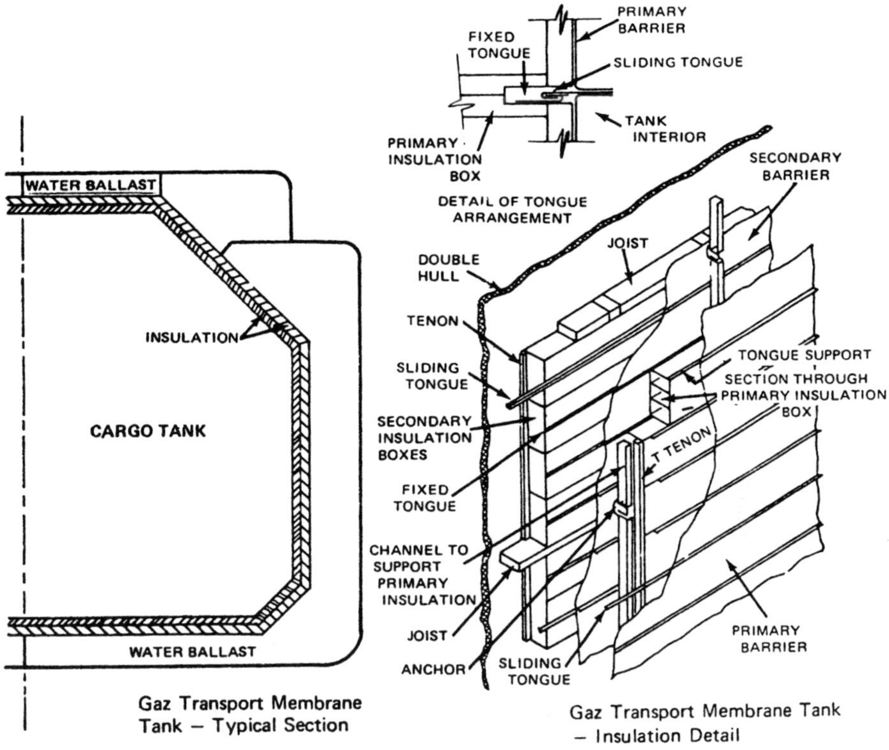

Figure 20-7. Gaz Transport containment system

tank insulation systems are more effective, reducing boil-off rate and requiring less labor than the original system. Panels are now kept in position by welded aluminum bolts, springs, bushings, and pressure distribution washers. The pressing ring system is no longer necessary. Improvements to the covering plate include the installation elastic joints to absorb the contraction of cargo tanks during loading of LNG. Heat conduction through the tank support has been reduced by the inclusion of a stainless steel ring strake. An aluminum–stainless steel bi-metallic insert permits the welding of the aluminum tank to the stainless steel "thermal brake." Boil-off rates of 0.1 percent per day are claimed for the latest designs.

LNG CARGO SYSTEMS

LNG and other cryogenic cargoes are loaded and discharged in a closed cycle as shown in Figure 20-10. Liquid and vapor lines with articulated loading arms are run between the ship and the shore terminal. Figure 20-11 shows a swivel joint used in the arms. During loading, shore terminal pumps deliver the LNG to the ship's tanks. Vapor displaced by the loaded

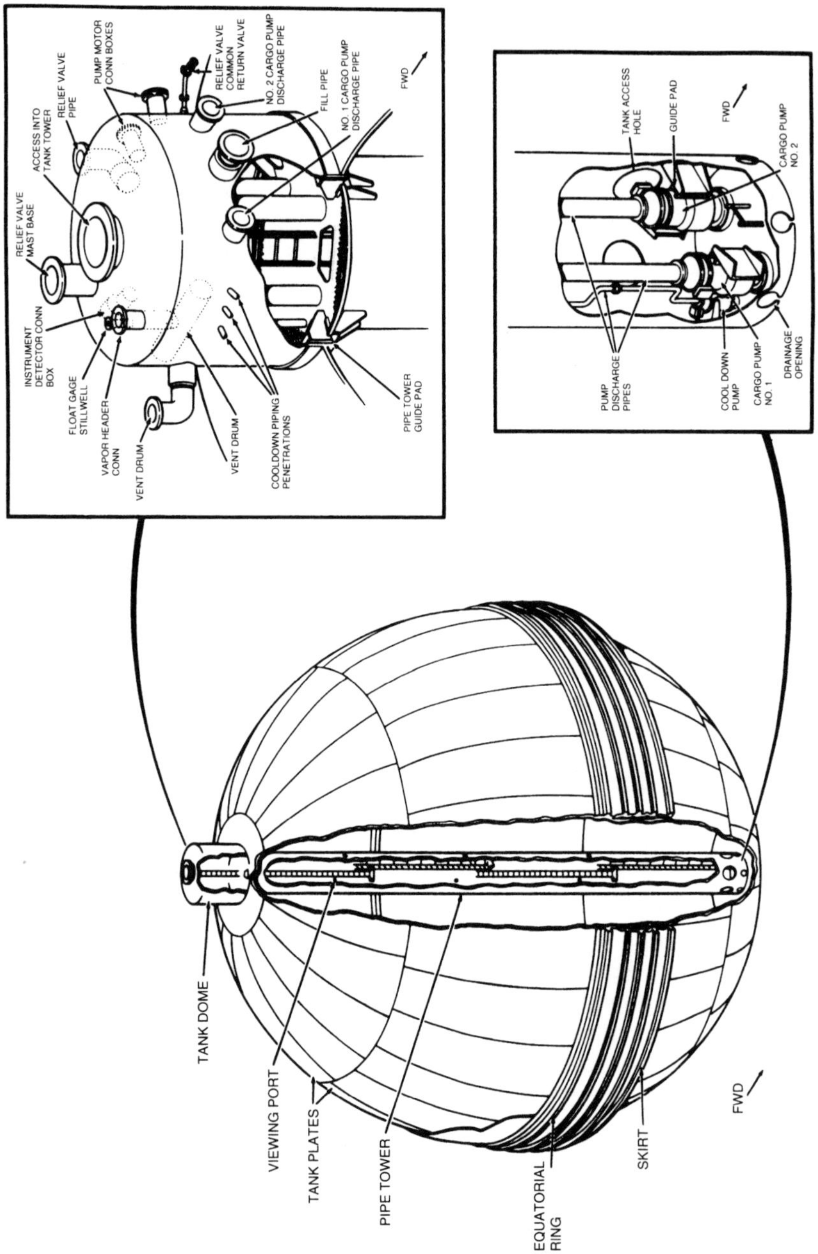

Figure 20-8. Kvaerner-Moss spherical tank arrangement

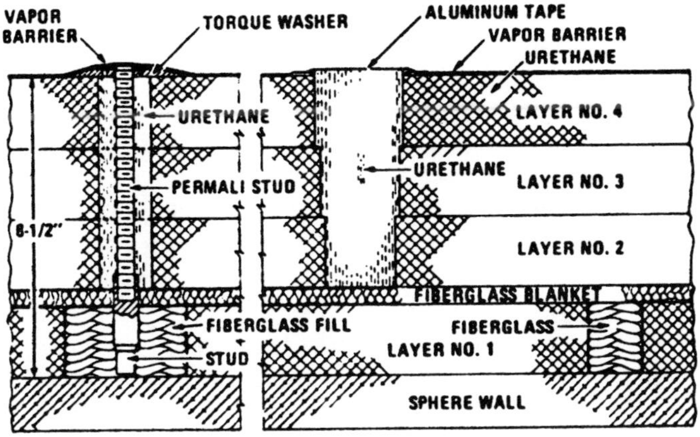

Figure 20-9. Kvaerner-Moss insulation system

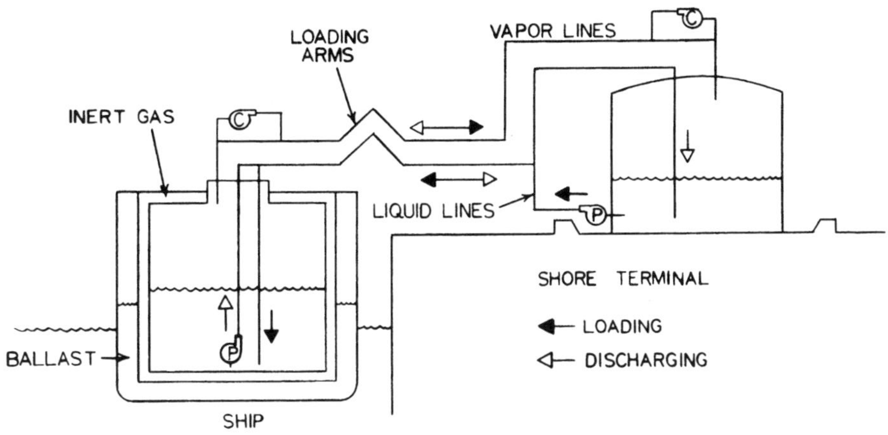

Figure 20-10. Closed cycle cargo loading and discharge

LNG is returned to the shore facility with the assistance of onboard compressors. Pressure in the tanks is maintained at a slight positive pressure at all times by control of the compressor capacity. Excess vapor produced by the loading operation is reliquefied or flared. Discharge operations are similar except shipboard cargo pumps and shore terminal compressors are used. During loading operations, ballast is discharged to maintain a constant draft. The ship is ballasted during discharging operations.

Additional features of a typical shipboard cargo system include:

1. Inerting of the space surrounding the cargo tank.
2. A means for controlled cool-down and warm-up of the tanks and piping.

3. Extensive instrumentation to monitor temperatures, pressures, tank levels, and other key parameters.
4. Provisions for expansion and contraction of cargo piping.
5. Relief valves to prevent overpressure of the tanks and vapor main. Liquid piping which can be isolated is also fitted with relief valves which relieve to the vapor main.
6. Provisions for emergency discharge of the cargo.
7. Provisions for use of boil-off as fuel in the boilers.

Liquid and Vapor Systems

Figure 20-12 is a simplified diagram of the liquid and vapor systems on a typical LNG vessel. The systems are designed to perform the following functions:

1. Loading and discharging of LNG cargo. These portions of the systems include the liquid and vapor headers, port and starboard manifold stations, loading lines to each tank, discharge lines with submerged or deep well pumps for each tank, vapor lines to each tank, and the gas compressors and heaters.
2. Cool-down of the tanks. This is accomplished by the spray nozzles in each tank served by submerged or deep well spray pumps in each tank and a cool-down header.

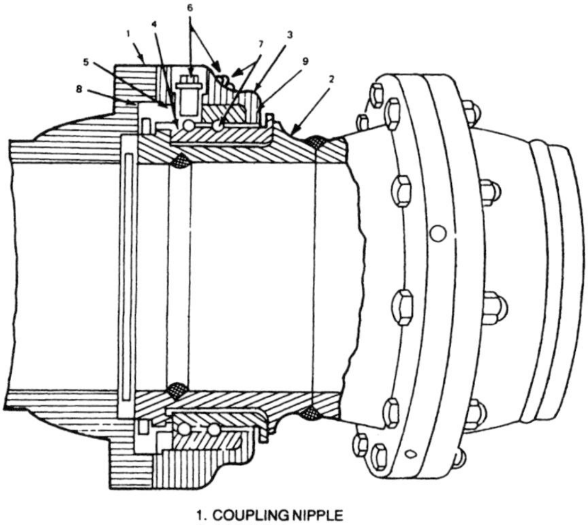

1. COUPLING NIPPLE
2. COUPLING
3. SWING NIPPLE
4. SPACER DISK
5. O-RING
6. BALL RETAINER PLUG
7. BALLS IN RACES
8. PACKING AND RINGS
9. ENERGIZER AND RETAINER

Figure 20-11. Loading arm swivel joint

3. Cool-down of the cargo lines. This is accomplished by the spray pumps and crosssovers between the liquid and cool-down headers.
4. Warm-up of the tanks. This is accomplished by using the compressors and heaters to deliver heated vapor to the tanks.
5. Vaporization of LNG or LN_2 (liquid nitrogen). This is accomplished by the vaporizer which is connected between the cooldown and vapor headers.

LNG Cargo and Spray Pumps

The pumps used for the transfer of LNG can be either the deep well or submerged type. The submerged type is the most common on LNG tankers.

The deep well pump consists of an electric motor mounted on top of the tank. It operates a multistage centrifugal pump mounted on a long shaft extended deep into the tank. The drive shaft is supported by bearings inside the discharge pipe. Problems with binding of the drive shaft due to differential expansion and bearing contamination led to the development of suitable submerged pumps.

Submerged electrically driven pumps can be used because the tanks are maintained at a slight positive pressure to exclude air. LNG has high electrical resistance, and the low temperature reduces the electrical resistance of the motor windings, thus increasing motor efficiency. The pump bearings are lubricated by the LNG. Submerged pumps are less expensive than the deep well type and are capable of pumping the tank down to less

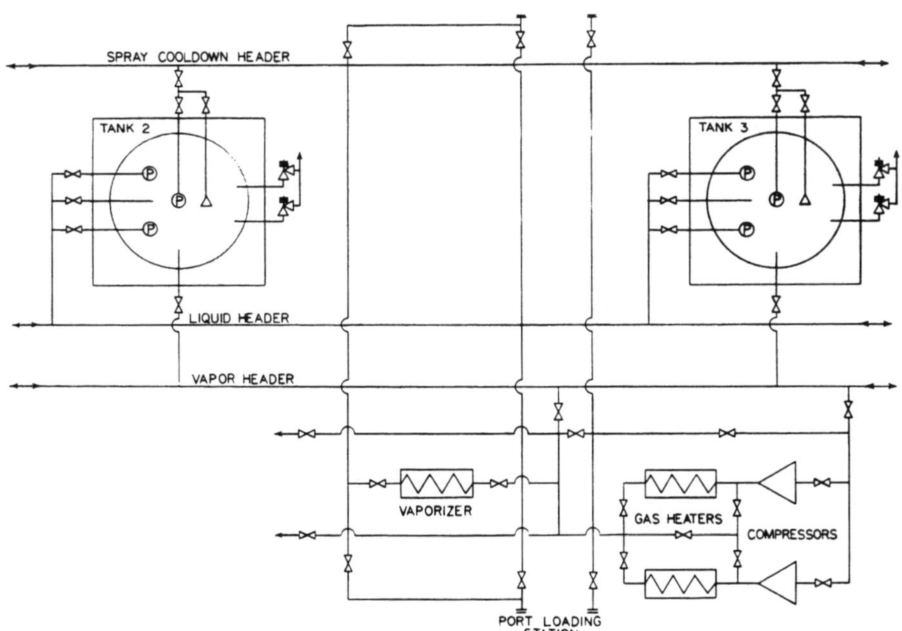

Figure 20-12. Liquid and vapor cargo systems

than 2 inches of LNG. No mechanical seal or packing gland is required since the pump is completely exposed to LNG. A typical single stage submerged pump is shown in Figure 20-13. A unique feature of this pump is the inlet inducer which is responsible for its low NPSH (net positive suction head).

Gas Compressors

Two or three compressors are usually installed on a typical LNG tanker. The low capacity unit is commonly termed a boil-off compressor while the high capacity unit is termed a high duty or warm-up compressor. The compressor can be centrifugal or positive displacement driven by an electric motor or steam turbine. The compressor shafts are sealed with nitrogen to prevent gas leakage. The nitrogen seal pressure is typically adjusted to 20 psi (138 kPa) above the compressor casing pressure. Electric motor-driven compressors are normally installed with a bulkhead between the motor and compressor with a seal fitted where the drive shaft passes through the bulkhead. Compressor discharge pressures are typically in the range of 5 to 15 psig (35 to 105 kPa).

Gas Heaters and Vaporizers

Gas heaters are installed for heating boil-off gas for use in the boilers, and for heating gas during cargo tank warm-up. Gas enters the heaters from

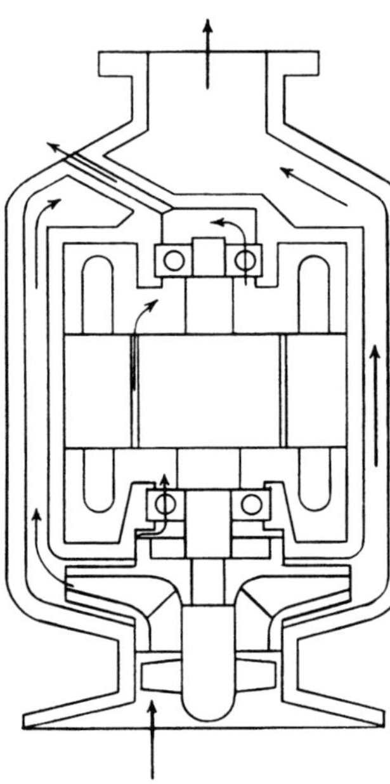

Figure 20-13. Submerged-type cargo pump

discharge of the gas compressors. The gas is heated either directly by steam in the heater or by a hot glycol-water mixture circulated through a steam-to-water heat exchanger and the gas heater. Figure 20-14 is a diagram of a typical direct steam-heated gas heater. Steam is maintained at constant pressure in the heater shell, and gas is circulated through the tubes. Gas temperature is controlled by bypassing cold gas around the heater. A low temperature alarm and control on the condensate prevent freeze-up.

Vaporizers are installed to perform the following functions:

1. Displace inert gas from the cargo tanks with vaporized LNG prior to cool-down and cargo loading.
2. Maintain cargo tank pressure during discharge when insufficient vapor is being returned from shore.
3. Supplement natural boil-off.
4. Vaporize liquid nitrogen.

Vaporizers are similar in construction to gas heaters. Most are of the steam-heated shell-and-tube type. Separators are commonly installed at the vaporizer outlet to prevent liquid carryover into the vapor header.

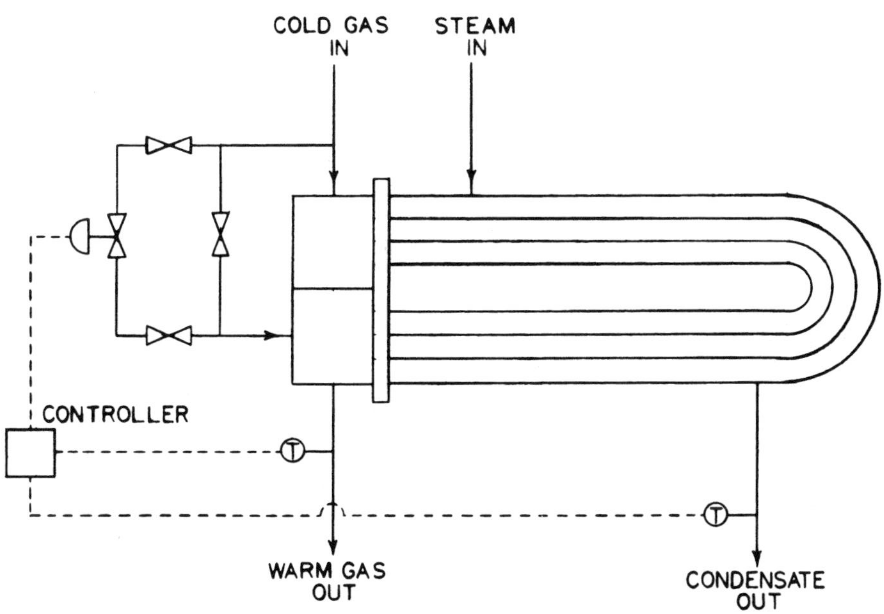

Figure 20-14. Gas heater

Control Valves

Control valves used in cryogenic systems must be designed for reliable operation at the very low temperatures encountered. The valve must be designed so that liquid cannot be trapped, or rupture may occur on warm-up. Most control valves in cryogenic service are pneumatically operated with a long warming extension. The warming extension allows the actuator to operate at close to ambient, rather than cryogenic, temperatures. Figure 20-15 shows a typical cryogenic control valve while Figure 20-16 shows a typical pneumatic controller used with the valve.

Nitrogen and Inert Gas Systems

The purpose of the nitrogen and inert gas systems on LNG tankers is to supply gases for the inerting of various spaces aboard the vessel. These systems include the hold spaces around the LNG storage tanks, cargo piping prior to loading or discharge, and the cargo tanks prior to gas freeing or preparation for cargo loading.

The nitrogen system consists of insulated liquid nitrogen storage tanks, fill connections at the port and starboard loading stations, the vaporizer, and distribution piping to the various locations requiring nitrogen. Nitrogen gas is supplied automatically by a pressure control valve. The annular space is protected from possible overpressurization by relief valves. This is especially important on vessels with membrane tanks since overpressurization can cause membrane damage.

The inert gas system employs an inert gas generator similar to that used on many crude carriers for cargo tank inerting. Low sulphur (less than 1 percent) light marine diesel oil is burned with air at close to ideal (stoichiometric) conditions to produce an inert mixture of primarily nitrogen and carbon dioxide. Oxygen levels are typically 0.5 percent. The combustion gases are passed through a scrubbing tower to cool them and to lower the SO_x content. The gases then pass through a refrigerant and/or desiccant dryer to remove water vapor and lower the dew point to about –50°F (–45°C). The system is fitted with extensive automatic controls to permit unattended continuous operation. The output from the inert gas plant is used primarily for inerting of the main cargo tanks.

Fuel Boil-Off Systems

Because of the normal heat leakage through the insulation of the cargo tanks, a certain amount of LNG cargo will vaporize. The gas produced is called boil-off and amounts to about 0.25 percent of the tank capacity per day for older tank designs and down to about 0.1 percent for newer tank designs. This boil-off is used as fuel in the ship's boilers. The boiler fuel system is designed to use both boil-off and Bunker C fuel. Most systems are designed to maintain a minimum oil flow called pilot oil. At normal sea speed,

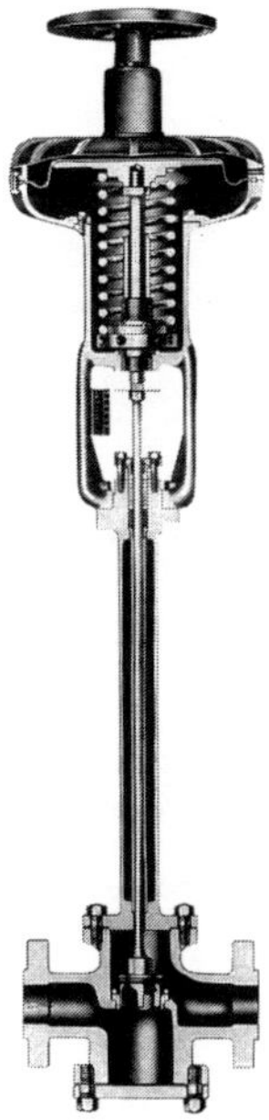

Figure 20-15. LNG control valve. Courtesy Leslie Co.

the pilot oil plus boil-off is not sufficient to maintain boiler pressure, and additional Bunker C must be burned. In port, the boil-off plus pilot oil is more fuel than is required, and excess steam is dumped into the main condenser. A spray-type desuperheater is used to lower the steam temperature.

Boil-off gas is removed from the cargo tanks by one of the compressors via the vapor header. Discharge pressure is 10 to 15 psig (69 to 103 kPa). The vapor is discharged to one of the gas heaters where the gas is heated to a controlled temperature of 70°F (21°C). The compressor maintains the cargo tank pressure constant by monitoring the vapor header pressure and varying compressor capacity accordingly. The gas supply piping in the machinery space is double-walled ventilated pipe. Gas is delivered through the inner pipe, and exhaust fans continuously vent the annulus between the inner and outer pipes. If the vent fans fail or gas is detected in the vent, the master shutoff valve will close automatically.

Boilers designed for dual-fuel operation require some minor modifications. The main change is in the superheater temperature control. Since methane burns with less luminosity than oil, less energy is transferred by radiation in the furnace. This results in higher superheater outlet temperatures with gas firing than with oil. Although some boilers have been designed with double superheaters, most employ single superheaters with the control desuperheater, control valve, and piping sized to handle the 60°

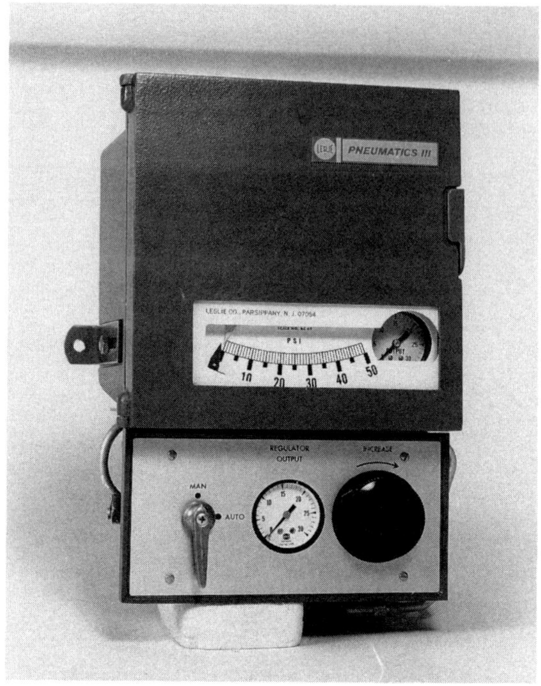

Figure 20-16. Pneumatic controller. Courtesy Leslie Co.

to 70°F (33° to 39°C) of additional temperature control required with gas firing, or with the superheater sized for gas firing and thus operating at less than design outlet temperature with oil firing.

The registers must be modified to burn gas as well as oil. A series of gas nozzles are installed around the burner cone. A large number of nozzles are necessary to accommodate the large volume of gaseous fuel as compared to liquid fuel. The combustion control system for a dual-fuel system is obviously more complex than for a standard oil-fired boiler.

Instrumentation

Cargo operations are generally controlled from a central cargo control room. Remote instrumentation is available for measurement of many parameters including cargo and ballast tank levels, cargo density, and cargo tank pressures and temperatures. Control and monitoring of cargo and ballast valves are also provided. Some of the parameters monitored in the cargo control room on modern LNG carriers include:

1. Levels of the cargo tank, both remote and local. Remote instruments include nitrogen bubbler and capacitance gauges. Local instruments include float

gauges and viewing ports with graduated scales. Reliability and redundancy are paramount in monitoring this key parameter because of the disastrous consequences of an LNG spill.

2. Pressures of liquid, vapor, and cool-down headers; cargo tanks; LNG pump discharge; insulated and void spaces; inert gas system; compressor discharge; liquid nitrogen tank; compressed air system; and heating steam system.
3. Temperatures of cargo tanks and liquid cargo at various levels to monitor cool-down; secondary barrier for detecting primary barrier and insulation failure; nitrogen system; and vaporizer and gas heater outlets.
4. Monitoring of gas concentrations in insulation barriers, voids, and machinery spaces. Portable instruments are available for testing combustible gas, oxygen, carbon dioxide, dew point, and halogen levels.

LNG CARGO OPERATIONS

The procedures described in this section are general and do not apply to any particular class of vessel. They are not intended to be a substitute for the detailed procedures contained in the ship's cargo manual. Reference to Figure 20-12 will aid in understanding the procedures outlined.

Normal Transport Cycle

An LNG vessel will travel with a small amount of LNG in the cargo tanks during the ballast leg. This maintains the cargo tanks at low temperatures and ready for loading without inerting operations. Tank temperatures are monitored and if a significant difference is noted between the top and bottom of the tank, the spray pump is started and residual LNG is sprayed into the top of the tank to equalize temperatures.

Upon arrival at the loading terminal, the liquid and vapor loading arms are connected, the lines are inerted and cooled, and loading begins. Ship compressors are used for return of vapor to the shore terminal. After loading, the lines are drained, typically using a nitrogen gas purge, and the arms disconnected. During the loading operation, deballasting will maintain a constant draft.

During the loaded voyage leg, tank pressure is maintained by the fuel boil-off system. Upon arrival at the discharge terminal, the liquid and vapor loading arms are connected, the lines are inerted and cooled, and discharge begins. The ship cargo pumps are started and vapor is returned by shore compressors. After loading, the lines are drained, and the arms disconnected. During the discharge operation, ballasting will maintain a constant draft.

Preparing a Cargo Tank for Loading

Prior to loading LNG in a warm noninerted tank, the tank must be prepared. The preparation consists of three basic steps: (1) purging with inert gas, (2) purging with natural gas, and (3) cool-down with LNG. The initial purging is normally done using the inert gas generator, though liquid nitrogen from shore can be vaporized for the task. After inerting, the natural gas purging is accomplished by vaporizing LNG from shore and returning the vapor to shore. After purging with vaporized LNG, LNG is sprayed into the tank until the cool-down is complete.

Preparing a Cargo Tank for Entry

Prior to entry for maintenance or inspection, the cargo tank must be warmed, inerted, and aerated. After stripping the tank of LNG, the tank is warmed by pulling vapor from the tank with the compressor, heating it with the gas heaters, and discharging the heated gas back to the tanks. The tanks are then inerted using the inert gas generator or vaporized liquid nitrogen. After inerting, the tank is purged with air.

DETAILED OPERATING PROCEDURES

Listed below is more detailed information on the operating procedures (which have been outlined above) for the normal transport cycle, preparation for loading, and preparation for tank entry.

Spraying Tanks during Ballast Voyage

Some LNG is left in each tank for spraying inside the cargo tanks to keep them cold during the ballast voyage. Spray pumps are started to discharge LNG to the cool-down header to supply the spray nozzles in each tank. Vapor produced during the spraying operation is led through the vapor header to the boil-off compressor. The gas is then heated, and delivered to the engine room for use as fuel in the boilers.

Cargo Line Inerting and Cool-Down

During the transit voyage, the cargo lines are isolated from the cargo tanks. Prior to loading or discharge, the lines are inerted with vaporized liquid nitrogen from the LN_2 storage tank. After inerting, the spray pumps are used to deliver LNG to the warm liquid lines for cool-down. The pumps discharge to the cool-down header, and the crossover between the liquid and cool-down header and the fill valve on each tank is opened. Liquid for cooling flows through the lines back to the tank. Cooling will continue until the line temperatures are close to that of liquid methane. If not enough liquid is available in the ship's tanks, LNG will be supplied from shore.

Loading LNG

After liquid line cool-down, the loading of LNG can begin. The ship and shore terminal isolation valves are opened, and liquid is delivered to the ship's tanks. The vapor generated during the fill operation and that displaced by the loaded liquid is returned to shore via the vapor header and the ship's compressors. Fill valves on each tank are closed automatically as the preselected level is reached. When all tanks are loaded, the shore isolation valves are closed. An emergency quick-closing valve system will stop filling automatically if the high tank levels are exceeded.

Draining LNG Lines

After completion of a loading or discharging operation, the liquid in the lines will be returned to the tanks. The fill valves on each tank are closed, and the crossover between the liquid and cool-down headers is opened. A nitrogen purge is applied to the high point of the liquid fill lines, and the LNG is forced back through the spray nozzles to the tanks. The pressure at the spray nozzle indicates when liquid is no longer flowing. As an alternative, the nitrogen purge can be eliminated. Heat leakage will vaporize some of the LNG, pressurizing the line and forcing the remaining liquid back to the tank.

Discharging LNG

After inerting and cool-down of the cargo lines, the cargo pumps are started. When pump operation is verified, the pump discharge valves are opened. LNG vapor is taken from shore during the discharging operation to maintain cargo tank pressure. The vapor valve to each tank can be used for throttling the gas supply to control tank pressure. If shore pressure is low, a gas compressor can be started to speed up discharging. If LNG vapor is not available from shore, vapor may be generated by vaporizing a small amount of LNG. At the end of the discharging operation, the cargo pumps are stopped and the discharge valves closed.

Inerting Cargo Tanks

This procedure applies to tanks that are being prepared for loading LNG. It does not apply to tanks that have been maintained cool and oxygen-free by LNG spraying operations.

The inert gas plant is started and discharges to the tank via the vapor header and the gas heater. The gas is heated to about 150°F (66°C) to minimize the time and gas required. The tanks are vented to the atmosphere via the liquid header. The compressors may be used to aid in the venting process. Purging continues until the exhausting gas has an oxygen content of less than 3 percent. In lieu of using the inert gas generators, liquid nitrogen from shore may be vaporized and heated in the LNG vaporizer. Tank pressure must be carefully monitored during this process.

Purging Cargo Tanks with Vaporized LNG

After inerting, the tanks are purged by vaporizing LNG from shore in the ship's LNG vaporizer. The vapors are delivered to the tanks via the vapor header. Purged vapors are exhausted through the liquid fill line and liquid header and returned to shore. Tank pressure is controlled by the LNG vaporizer inlet control valve. The gas being returned to shore is monitored for CO_2 content. When the CO_2 content reaches 1 percent, the purging operation is stopped. The ship's compressors may be used to aid in gas return.

Cargo Tank Cool-Down

After inerting and purging with vaporized LNG, liquid LNG is routed to the spray header. The spray line valves on each tank are opened and LNG is discharged from the spray nozzles to cool the tanks. The spray rate will depend on the vapor-receiving capacity of the shore facility and the allowable cool-down rate of the tank, typically about 18°F (10°C) per hour. The spray rate is controlled by adjusting the spray control valves. Initially the tank pressure will decrease due to the cooling effect of the LNG. Spraying may have to be stopped for a period if the tank pressure goes too low. When the tank pressure increases, a compressor is started to assist in returning vapor to shore.

Cargo Tank Warm-Up

The first step in preparing a tank for entry is warm-up. The compressors pull vapor from the tank via the vapor header. The vapor is delivered to the gas heaters, heated to about 140°F (60°C), and returned to the tank via the liquid header. Excess vapor is burned in the main boilers as usual. Warming continues until the tank temperatures are approximately ambient, thus insuring that freezing will not occur during the inerting operation.

Purging with Inert Gas

After warm-up, the inert gas plant is started. Inert gas is delivered to the tank via the gas heater and liquid header. Gas is removed via the vapor header. Gas may be delivered to the boilers until the methane concentration drops below 50 percent, then it is vented to the atmosphere. Purging continues until the methane concentration drops to 5 percent. Liquid nitrogen from shore can be used instead of inert gas.

Cargo Tank Aerating

After warm-up and inerting, the inert gas plants are used to deliver dry air to the tanks. Dry air is delivered via the vapor header, and the gas is exhausted to the atmosphere via the liquid header. Compressors may be used to aid in the venting process. Purging continues until the oxygen content in the exhausted gas reaches 18 percent.

LNG TANKER PROPULSION SYSTEMS

By far the most common propulsion system for large LNG carriers is the geared steam turbine power plant. While the slow and medium speed diesel has become the plant of choice for most other commercial vessels, they have not found use on LNG carriers. The reason for this is simple—boil-off. The LNG carrier must somehow deal with the cargo that is vaporized by heat transfer though the tank insulation, and the boilers of the steam propulsion system can easily utilize this as fuel. Dual-fuel boilers are a well proven technology and the ability to dump steam from the boilers to the condenser provides an easy way to process boil-off at low loads. The boil-off greatly reduces the quantity of fuel oil that must be burned and takes away much of the fuel cost advantage of the diesel.

Several developments are causing the designers of LNG carriers to take a second look at other propulsion systems. One is that the newer tank designs have better insulation systems, reducing the typical boil-off from 0.25 percent per day in older designs to 0.1 percent per day in newer designs. This reduction in boil-off means more fuel oil must be burned and increases the fuel costs of the steam plant in comparison to more efficient plants like the diesel. A second development is that the diesel manufacturers have more experience with dual-fuel engines in offshore and shoreside power generation applications, reducing the perceived risk of using this technology on LNG carriers. A third development is that the reduced quantity of boil-off from the better insulated cargo tanks makes onboard reliquefaction of the boil-off more attractive. These plants can now be smaller and therefore less costly to install and operate.

This does not mean that the steam power plant is no longer a viable option, but the simple inefficient cycles used on older vessels no longer make economic sense. While there is an endless variety of plant combinations that could be considered, the main choices are the following: A reheat steam power plant, dual-fuel slow speed or medium speed diesels, slow speed or medium speed diesels burning fuel oil with onboard liquefaction, and gas turbines (dual-fuel or combined cycle). A discussion of each alternative follows.

Reheat Steam Plant

The reduced availability of boil-off makes a more efficient steam plant attractive. A reheat steam cycle will be about 10 percent more efficient than a five-heater non-reheat cycle and about 20 percent more efficient than a two-heater non-reheat cycle. Information on these steam power cycles can be found in Chapter 3 (Steam Power Plants) of Volume I of this text. This alternative is a logical extension of the proven technology of the current LNG steam turbine power plants. It retains the ability to easily utilize boil-off for the generation of electricity in turbo-generators and process excess boil-off at low loads by dumping steam to the condenser at low loads.

Dual-Fuel Diesel Engine

Slow speed and medium speed diesel engines can be easily modified to burn natural gas as well as fuel oil. The main change is the addition of a gas injection valve to the head. The fuel oil injector is retained. Methane has poor ignition quality, requiring the injection of a small quantity of fuel oil (called pilot oil) whenever natural gas is being used. The combustion of the pilot oil initiates the combustion of the natural gas.

There are two type of dual-fuel diesel injection systems—low-pressure and high-pressure. In a high-pressure system, gas at approximately 250–350 bar is injected along with the pilot oil near TDC at the end of the compression stroke. In a low-pressure system, gas at about 3 bar is injected into the cylinder just before the start of compression and the air and gas are compressed. Because of the low ignition quality of the natural gas, the gas does not ignite until the pilot oil is injected. High-pressure injection permits better control of the combustion process but the high-pressure gas compressors consume significant power, approximately 6–7 percent of the power produced by the engine, and the high-pressure gas system must be carefully engineered and maintained to avoid gas leakage into the engine room. The compressor power requirements are much lower with low-pressure injection and because of lower cylinder temperatures and the efficient combustion of the high quality gas, engine maintenance is reduced and service intervals can be extended.

There are a number of alternatives for generating electricity with dual-fuel diesel power plants. Power takeoff (PTO) generators can be used at sea. Diesel generators using either fuel oil or boil-off as fuel are also viable choices. Dual-fuel diesel generators have the advantage of utilizing boil-off when the main propulsion engines are shut down. Even so, an additional means of handling boil-off is necessary as natural gas cannot be vented in port. The common solution is an incinerator. Figure 20-17 shows a natural draft unit which operates without the need for fans, only requiring power for ignition and controls.

Gas Turbine and Combined Cycle Plants

Gas turbines have been operated on natural gas in shoreside applications for many years. Dual-fuel gas turbines can be operated on all gaseous fuel, all liquid fuel, or a mixture of the two. Chapter 7 of Volume I of this text describes the changes that must be made to the engine fuel system and to the ship fuel supply system to permit dual-fuel operation. Typical gas pressures are 24 to 28 bar for a high-pressure ratio aircraft-derivative engine. While lower than that required for high-pressure injection in a diesel, compressor power requirements are still 3 to 5 percent of the engine output.

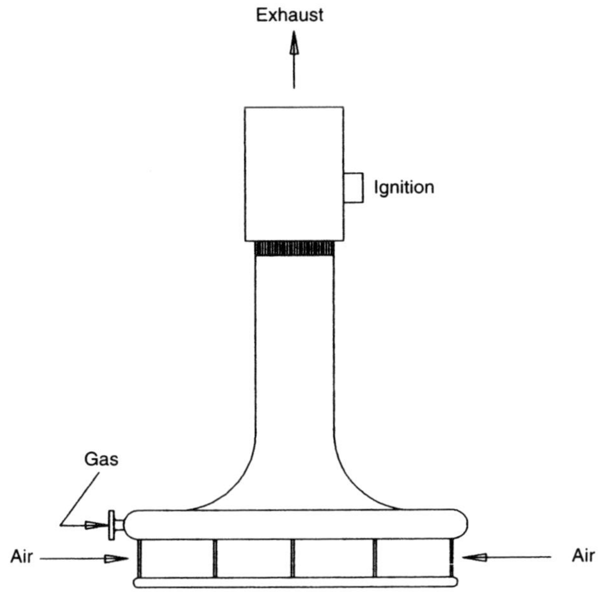

Figure 20-17. Natural draft incinerator

With the lower boil-off available from the newer design storage tanks, additional fuel is necessary to maintain design power. This can be made up be either liquid fuel (in a dual-fuel engine plant) or forcing boil-off. Since an aircraft-derivative gas turbine cannot burn residual fuel, either distillate fuel oil or forced boil-off will be a higher cost alternative. A combined cycle power (COGAS—combined gas turbine and steam turbine) plant makes sense in this situation because it improves the cycle efficiency and maximizes the power produced by the available boil-off. Figure 20-18 is a simplified diagram of such a system. A waste heat boiler is installed in the exhaust of the gas turbine and produces steam to power a steam turbine. A dual-fuel auxiliary boiler is also available to provide steam to supplement the steam produced by the waste heat boiler. One advantage of this plant configuration is that low cost residual fuel can be burned in the boiler to make up any power deficit with the steam turbine. Another is that the excess boil-off can be burned in the auxiliary boiler and the steam dumped to the condenser, just as is currently done in conventional steam turbine power plants. A third is that a conventional astern steam turbine can be used for astern operation, eliminating the need for a CRP propeller or a reverse reduction gear.

Reliquefaction Systems

Liquefaction systems were discussed earlier. A reliquefaction system takes the vapor from the cargo tanks, cools and recondenses it, and then

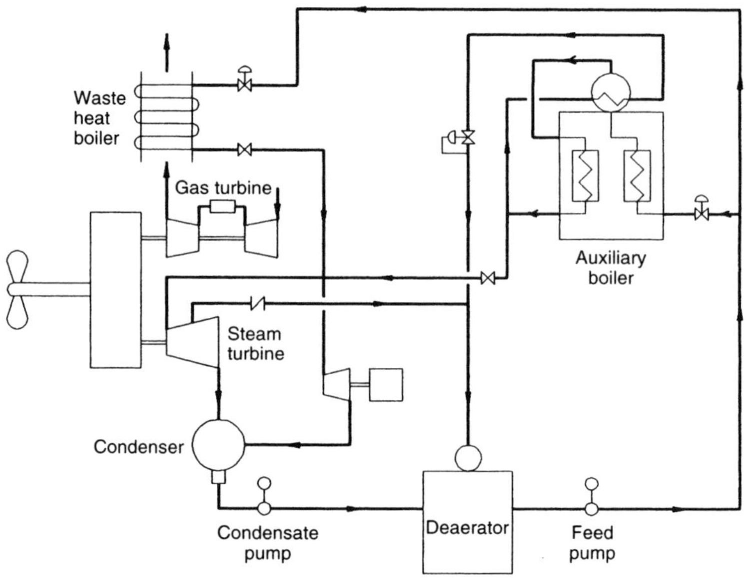

Figure 20-18. COGAS combined cycle power plant

returns the liquid to the tank. In the size range of an onboard system for a typical LNG carrier, a nitrogen expander system is commonly recommended. Figure 20-19 is a simplified diagram for a nitrogen expander reliquefaction system. Nitrogen is compressed and cooled several times—first in a diesel or steam turbine powered compressor and then in a turbo-compressor. The high-pressure nitrogen is then expanded in the turbine of the turbo-compressor, and the cold nitrogen exhaust is then used to condense the natural gas.

These systems are expensive to install, operate, and maintain. Installed costs of a typical reliquefaction system can be $10,000,000 or more. Power requirements for the main compressor can range from 3,000 to 7,500 horsepower for a 125,000 m^3 carrier, with power requirements decreasing with lower boil-off rates achieved with more efficient cargo tank insulation. The fuel necessary to operate this system will obviously be a significant cost factor. One advantage of a reliquefaction system is that it will increase the amount to cargo delivered to the discharge terminal, thus increasing revenue. Another is that it permits the use of standard diesel engines burning HFO for main propulsion, electric power generation, and powering the reliquefaction system's main compressor. The use of efficient diesel engines, the ability to use lower cost fuel, and the increased cargo revenue can offset the costs of the reliquefaction plant. The lack of shipboard experience with reliquefaction plants remains a concern by shipowners considering this alternative.

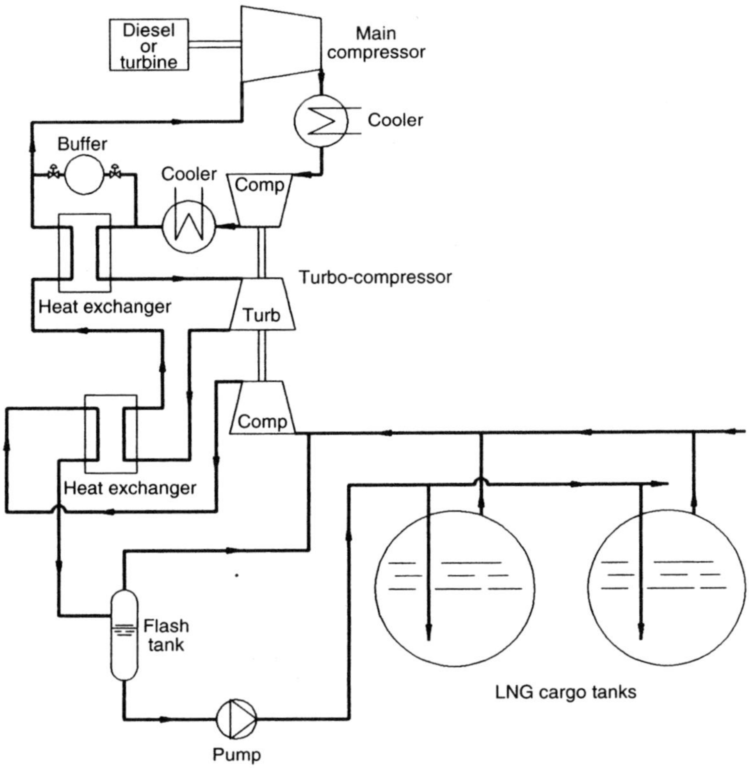

Figure 20-19. Nitrogen expander reliquefaction system simplified diagram

REVIEW

1. Define cryogenics.
2. Why are gases liquefied for transport? Why are liquefied gases stored at low temperatures rather than normal temperatures?
3. What are the two most common liquefied gas cargoes? What are their temperatures at normal atmospheric pressures?
4. Describe the cascade cycle used for liquefaction of gases.
5. What are the two basic categories of LNG cargo tanks? What is an example of each type? What are the advantages and disadvantages of each?
6. Describe the liquid and vapor systems for handling LNG cargoes. What functions must these systems perform?
7. What are the two types of pumps commonly used for the transfer of LNG?
8. What is boil-off? What is it commonly used for?
9. What parameters are commonly monitored in an LNG cargo system?

10. Describe the normal cycle for the transport of LNG including loading, discharging, and operation during the loaded and ballast legs.
11. Describe the procedure for preparing a warm aerated cargo tank for receiving LNG.
12. Describe the procedure for preparing a tank containing LNG for entry to permit maintenance.
13. How are cargo tanks cooled down prior to loading LNG?
14. Describe the nitrogen expander system for onboard reliquefaction.
15. What are the advantages and disadvantages of the various choices for LNG vessel propulsion plants?
16. How is excess boil-off handled with each of the different propulsion plant alternatives?

REFERENCES

Curt, R. P., and Delaney, T. D. 1972. *Marine Transportation of Liquefied Natural Gas.* Kings Point, N. Y.: National Maritime Research Center (NTIS PB249014/LK).

Ffooks, R. 1979. *Natural Gas By Sea.* London: Gentry Books.

Howard, J. L., and Kvamsdal, R. S. 1982. *Energy Efficient LNG Carriers.* Jersey City, N.J.: Society of Naval Architects and Marine Engineers.

Maritime LNG Manual. 1974. Kings Point, N. Y.: National Maritime Research Center (NTIS COM-75-10136/LK).

Van Langen, J. R. 1973. *LNG Liquefaction Plants and Associated Shoreside Operations.* Kings Point, N. Y.: National Maritime Research Center (NTIS COM-10469/LK).

ACKNOWLEDGMENTS

The author is indebted to Newport News Shipbuilding and Drydock Company, Leslie Company, the National Maritime Research Center, and Energy Transportation Corporation for their assistance in the preparation of this chapter.

CHAPTER 21

Hull Machinery

THOMAS P. MACKEY AND ALAN L. ROWEN

INTRODUCTION

HULL machinery normally includes all power-driven equipment located outside of the machinery spaces that is not associated with the main propulsion plant. Steering gear, anchor windlasses, and winches and capstans are discussed in this chapter. Cargo-handling gear, stabilizers, thrusters, elevators, and other, often specialized, hull machinery are not included, although they may share many of the features and operating principles with the equipment described.

The historic development of hull machinery began with the man-powered mechanical advantage devices on sailing ships. The evolution to steam-powered vessels led to the common use of steam-powered engines for all hull machinery applications. Steam-driven deck machinery can still be found on older tank vessels, but electrohydraulic and direct electric drive are now used for new construction and retrofit.

GENERAL REQUIREMENTS FOR HULL MACHINERY

Hull machinery must have high reliability; that is, the machinery must be ready to operate when needed and must continue to operate without excessive out-of-service time for maintenance. The design and construction must be simple and rugged. Since the machinery location is frequently exposed to the sea environment, the machinery must be weatherproof. Deck machinery is often designed with the mechanical power device located below deck to improve weather resistance. Simple design facilitates maintenance and repair, which are important considerations. To ensure the longest possible trouble-free life, owners and operators should consider the following items when selecting hull machinery.

Hull machinery must be designed (mechanically, electrically, and hydraulically) to withstand the pitch, roll, trim, and list of the ship. The following conditions are normally used for design: a pitch of 10° (bow up to bow down), a permanent trim of 5° by either the bow or stern, a roll of 30° to each side, and a permanent list of 15° to either side. No two of these conditions are to be considered to occur simultaneously.

Corrosion Protection

Structural parts of deck machinery exposed to the weather must be designed to be free of pockets that collect rain and seawater. Where appropriate, corrosion resistant materials should be selected. Fasteners should be galvanized or manufactured from corrosion resistant material. Access covers must have reliable watertight seals. Metal parts must be protected with a durable coating system, including a white metal blast preparation, an appropriate primer, and a final epoxy coating. Watertight enclosures should be provided with automatic drain fittings. Exposed hydraulic piping should be of stainless steel or have high wall thickness to provide a corrosion allowance.

Lubrication

Adequate lubrication of moving parts must be ensured by design, with oil and grease fittings that are visible and accessible. Lubricating oil sumps should have clearly labeled and screened fill openings, and drains for oil replacement. Selection and use of the proper lubricant is another requirement for low maintenance.

Mechanical Assemblies

Bearings should be designed for easy shipboard replacement, and spares must be carried. Fittings must have locking devices or setscrews and keys to ensure proper assembly and to keep parts from coming adrift. The watertight integrity of enclosures must be carefully designed and maintained. Access for inspection and adjustment is essential. Modular construction simplifies construction and subassembly replacement.

Electrical Assemblies

Electric motors for hull machinery should have conservative nameplate ratings. Electric circuits and control systems must be designed and operated to protect the motor. Motors must be located below deck or have watertight enclosures for deck installation. Watertight motors should be equipped with automatic drainage fittings. Cooling must be adequate. Motor bearings should be permanently lubricated and sealed. Terminal wiring should always be clearly marked.

Hydraulic Assemblies

Cleanliness of the hydraulic fluid is essential. Fill screens, pump strainers, and filters are necessary, all with adequate provision for cleaning and replacement. The hydraulic pumps and motors must have conservative design ratings. Cooling of the hydraulic fluid must be considered. Hydraulic sump heaters may be required to allow start-up in very cold temperatures. The capacities of hydraulic oil sumps should be large enough to accommodate all of the oil in a system to enable it to be drained to the sump for maintenance.

STEERING GEAR

The steering gear provides the force necessary to position and hold the rudder against the fluid forces of the sea. In general, the forces acting on a rudder are proportional to the rudder angle, the speed of the vessel squared, and the density of the water. The greater the rudder angle, the higher will be the force needed to turn and hold the rudder in position. The rudder forces are transferred to the rudderstock as radial loads that are, in turn, transferred to the rudderstock bearings, and as a torque that is transmitted to the rudder actuator. The forces are large, and rules generally require that the major parts of the steering gear, including the tiller and the rams, have strength equivalent to that of the rudderstock, and that they be of steel.

A rudder carrier bearing supports the weight of the rudder and rudderstock. The upper rudderstock radial bearing and the rudder carrier bearing are often built into the steering gear.

In modern practice there are three types of steering gear in common use, all electrohydraulic: the ram type; the rotary-vane type; and the clevis-mounted, hydraulic-cylinder type. There are generally two types of pumping systems used to provide the hydraulic oil pressure: variable-displacement pumps and fixed-delivery pumps with directional control valves. A third element comprised by the steering gear, in addition to the actuator and the pumping system, is the control system. These systems are discussed in the following sections.

Electrohydraulic Ram Type Steering Gear

A typical arrangement for a four-cylinder Rapson-slide steering gear is shown in Figures 21-1 and 21-2. Two-cylinder Rapson-slide arrangements are also used, particularly on naval ships. Link-type steering gears, which drive two rudders with a single set of hydraulic cylinders, are often used on twin-rudder naval auxiliary vessels.

Normally, dual electric-motor-driven pumps are fitted to deliver the hydraulic fluid, which is a high-grade mineral oil. The pumps are further described below. In most cases only one pump is used at a time to operate the steering gear, with the second pump in reserve.

The oil, typically under a pressure of 1,500 to 2,500 psi (100 to 175 bars), passes through control valves to either of the opposed cylinders or sets of cylinders to turn the rudder. The pump in service is driven continuously in one direction by an electric motor, which runs at light load except when called upon to move the rudder. With variable-delivery pumps an electronic or mechanical feedback controller operates the stroke controls of the pump to begin delivery of the oil to the cylinders, to move the rudder in direct relationship to the steering wheel or other steering control, and to stop the oil delivery when the rudder is in the ordered position.

On four-cylinder Rapson-slide steering gears the four single-acting hydraulic rams act in pairs to provide the force acting on the tiller. Each pair of opposing cylinders is in line, to operate a pair of rams connected at their juncture, which carries trunnions for the swivel block. In Figure 21-2, the square swivel blocks and circular upper trunnions are visible in the slots in the tiller crosshead. The action of the swivel block is shown schematically in Figure 21-3, where it is identified as the "Rapson slide." The movement of the rams is transmitted to the tiller through the trunnions to the swivel blocks, which move longitudinally with the rams, but turn on the trunnions and slide in the slots of the tiller crosshead as the rudder is turned. The Rapson slide has the desirable characteristic of increasing the torque

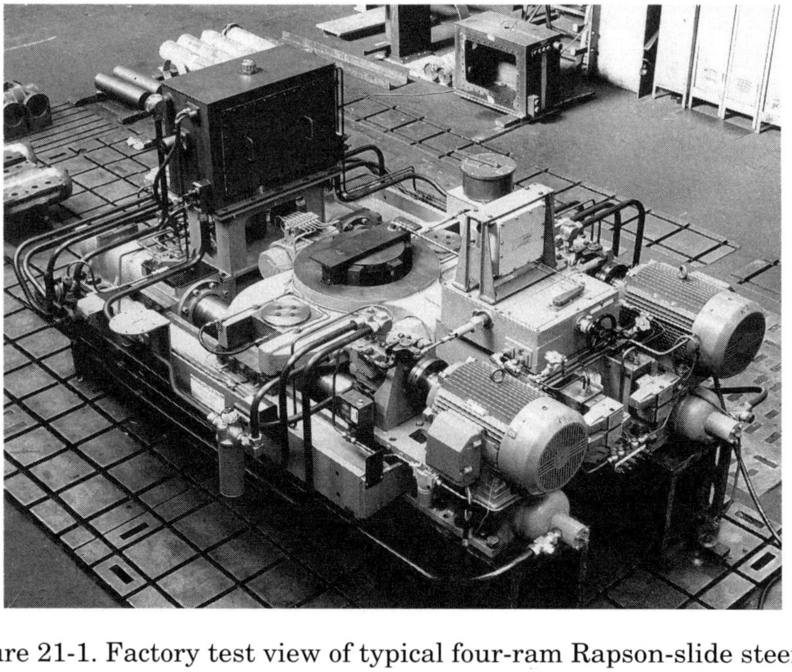

Figure 21-1. Factory test view of typical four-ram Rapson-slide steering gear

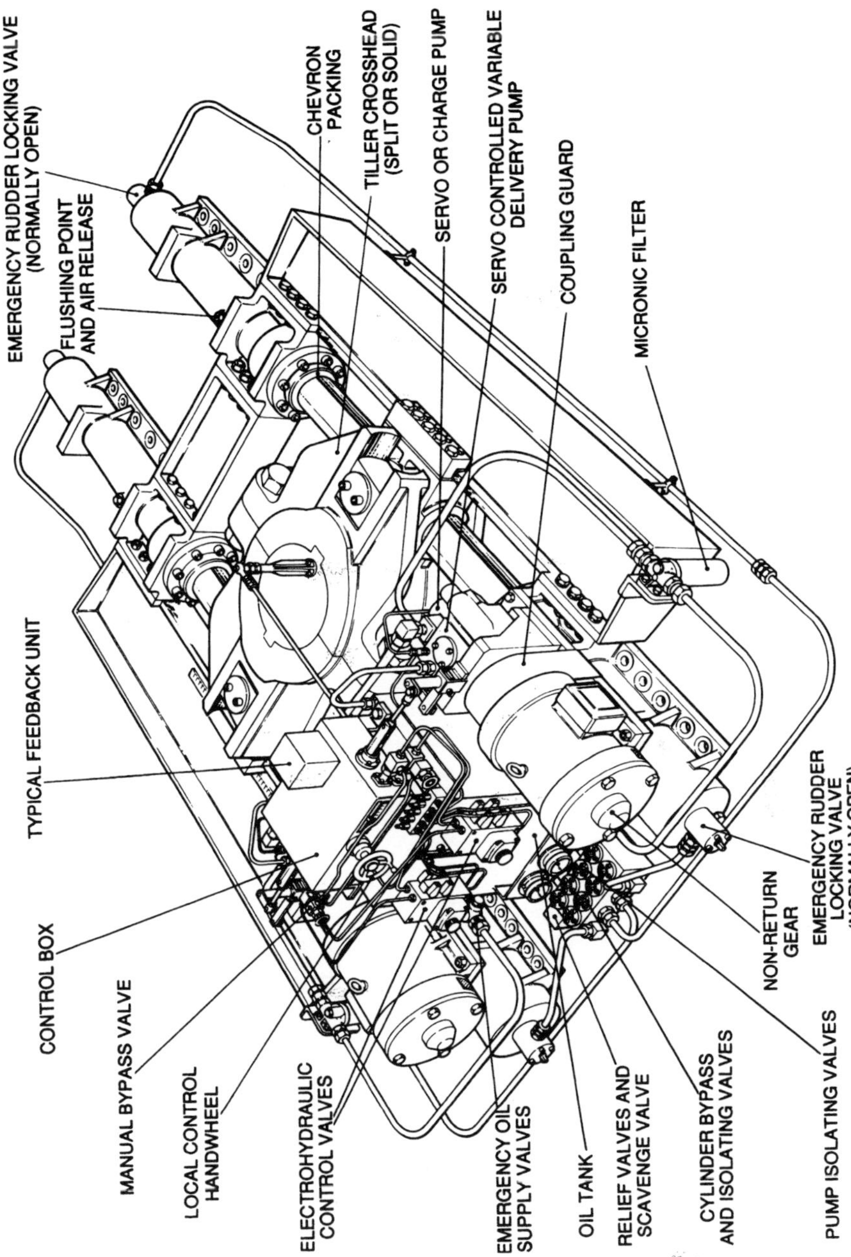

Figure 21-2. Four-ram Rapson-slide steering gear

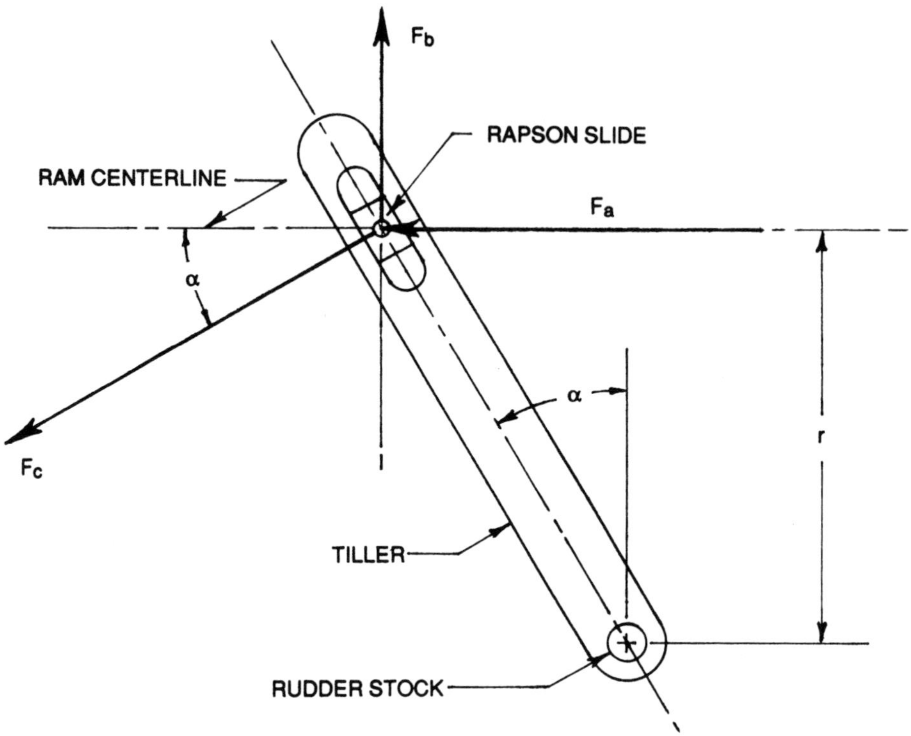

Figure 21-3. Rapson-slide mechanism

available to move the rudder as the rudder angle increases. Two-cylinder Rapson-slide and link-type steering gears have only one pair of rams.

The tiller crosshead may be cast, forged, or fabricated steel. The tiller is bored and keyed to fit the rudderstock. Single-piece tillers may be shrunk onto the rudderstock. Two-piece tillers are bolted together around the rudderstock as shown in Figure 21-2. The rams are usually made of forged steel, and the cylinders of cast steel with bronze bushings. The side loading of the Rapson slide, identified as Fb in Figure 21-3, is transmitted through the bushings and cylinders to the foundation. In Figure 21-2, the substantial side beams used to resist this force are visible.

Rotary-Vane Steering Gear

Rotary-vane steering gear is shown schematically in Figure 21-4. The gear consists of a housing or stator, which is fixed to the foundation and contains two or three vane cavities, and a rotor with vanes attached, which acts as a tiller. The pressure of the oil, which is delivered to the vane cavity on one side or the other of each of the rotary vanes, acts on the side area of the vane to turn the rudder. Figure 21-4 shows this principle: the paired

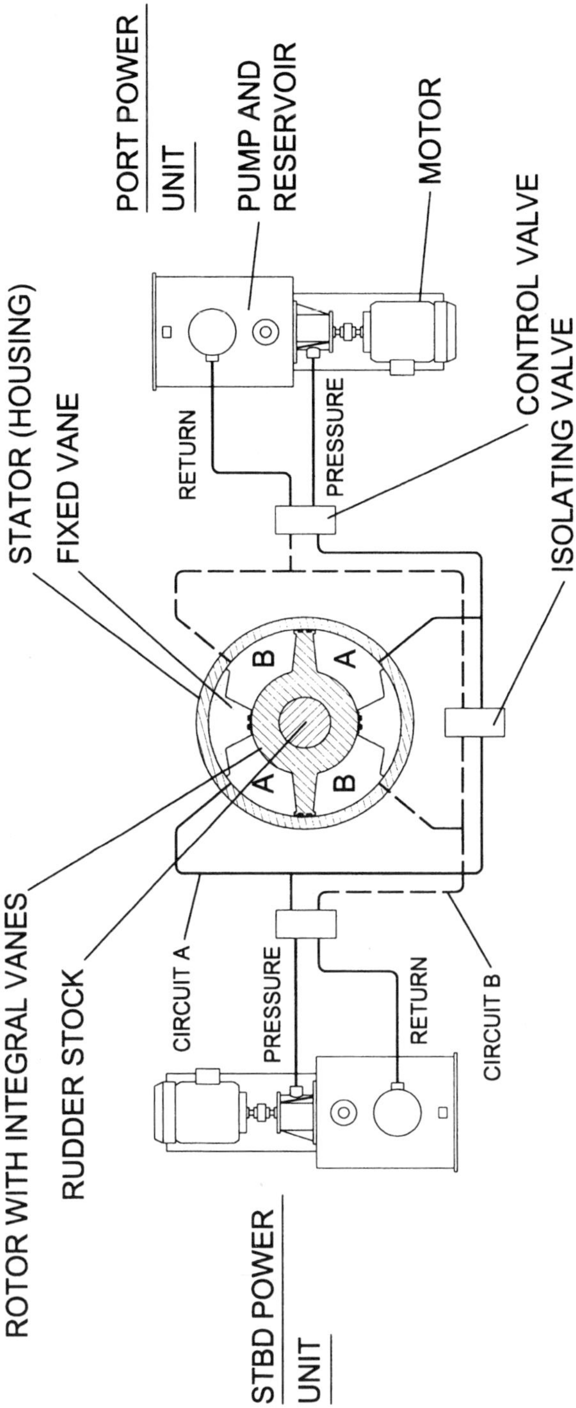

Figure 21-4. Schematic plan view of rotary-vane steering gear

chambers A-B are the port and starboard vane cavities; oil under pressure from either pump, supplied via circuit A to the A-ends of the cavities, rotates the rudder counterclockwise, while the B-ends drain through the return lines to the reservoirs. When the control valves are repositioned to connect the pressure sides of the pumps to circuit B and the return lines to circuit A, the rudder will be rotated clockwise.

In Figure 21-4, the rotating vanes are integral with the rotor, but in the unit of Figures 21-5 and 21-6 the fixed vanes are secured to the stator by high-tensile steel dowel pins and cap screws to form the vane cavities, while the moving vanes are secured to the rotor by cap screws and keys. In either arrangement, specially designed sealing strips are fitted in grooves along the sealing faces of the fixed and moving vanes. When the rudder is hard over, the vanes act as rudder stops. The actuator is provided with a relief valve normally set at the required 125 percent of the maximum working pressure.

Rotary-vane actuators develop a constant torque throughout their full travel and are suitable for any torque rating. The torque is equal to the product of the applied pressure, the vane area, the effective moment arm

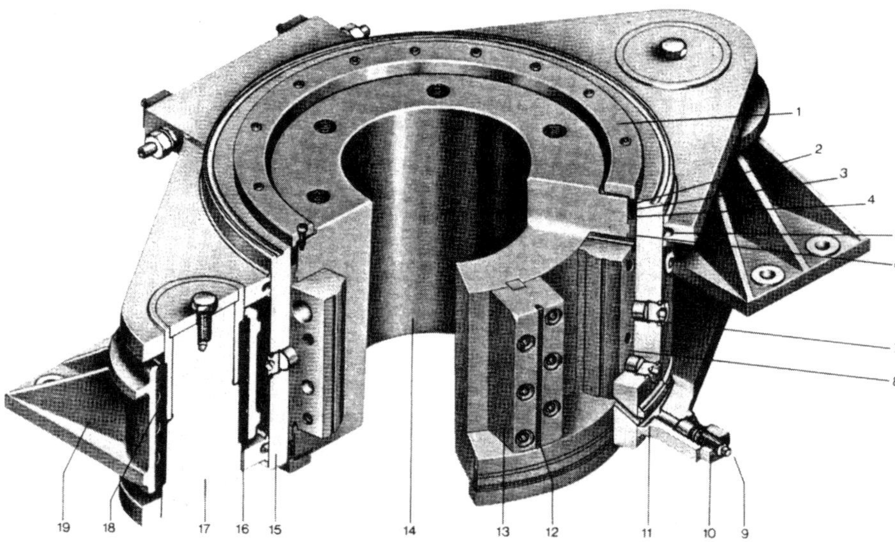

1 GLAND
2 AUTOMATIC PRESSURE RING
3 AUTOMATIC LIP RING
4 RUBBER SUPPORT
5 UPPER MANIFOLD
6 LINER
7 CYLINDER DOWEL
8 FIXED VANES (STOPPERS)
9 SHUT-OFF VALVE SPINDLE
10 SHUT-OFF VALVE FLANGE
11 LOWER MANIFOLD
12 STEEL SEALING STRIP
13 ROTARY VANE
14 ROTOR
15 STATOR
16 SHOCK ABSORBER
17 ANCHOR BOLT
18 PRELOADING BUSH
19 ANCHOR BRACKET

Figure 21-5. Cutaway view of rotary-vane steering gear

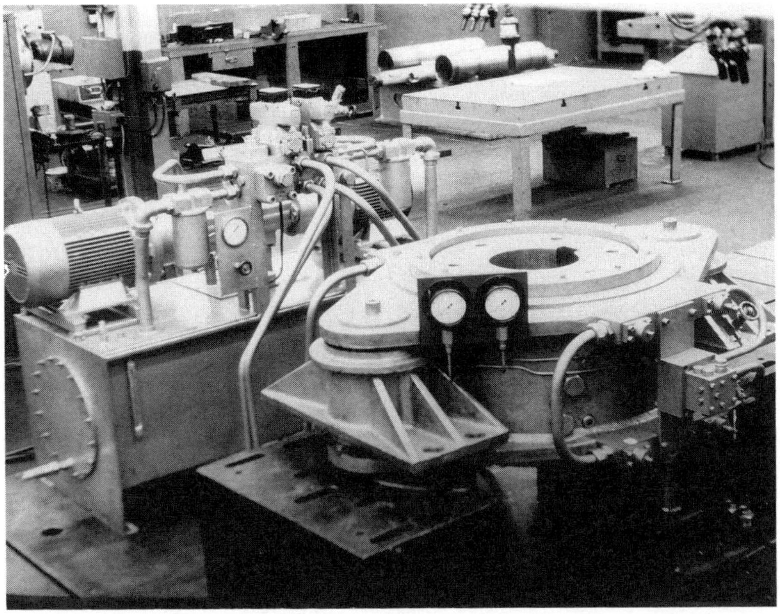

Figure 21-6. Factory test view of typical rotary-vane steering gear. The actuator is mounted in the foreground, with the power units, control valve rack, and reservoir behind.

(the radius from the rudderstock centerline to the center of pressure of the vane), and the number of vanes. Since the vanes impart only torque to the rotor and stator, there are no radial loads, and wear is minimal.

Maximum attainable rudder angle, either side of amidships, is normally 40° for three vane actuators and 65° for two vane actuators, well in excess of the 35° normally required. Rotary-vane actuators are often used for very large rudder torque requirements and when high-maneuverability rudders, requiring larger rudder angles, are specified.

Rotary-vane actuators are normally supplied by fixed-delivery pumps. Pressures are lower than in ram-type steering gears, generally 850 to 1,450 psi (60 to 100 bars). A directional control valve controls the fluid supply and direction of flow. Modulating flow control valves may be used with the fixed delivery pumps to provide smoother operation of the actuator. An electronic or mechanical feedback controller positions the valves to begin delivery of the oil to one end of the vane cavities, drawing from the other end, to move the rudder in direct relationship to the steering control, and stops the oil delivery when the rudder is in the ordered position. Fixed delivery pumps can be of the gear, vane, or screw type. The flow is controlled by either direct-acting or pilot-operated solenoid valves.

Most rotary actuator designs incorporate the upper rudderstock radial bearing and the rudder carrier bearing, which support the horizontal and vertical loads of the rudder and rudderstock, within the actuator housing. The actuator is connected directly to the rudderstock by a keyway, hydraulic nut, clamp rings, or a hydraulic coupling.

Clevis-Mounted Cylinder Steering Gear

Clevis-mounted electrohydraulic steering gear is used extensively on smaller vessels, but has also been installed on larger cargo ships. Figure 21-7 shows the arrangement fitted to a Great Lakes bulk carrier. Clevis-mounted steering gear consists of pumping units connected to operate one or more single- or double-acting cylinders. The cylinder end is pinned to ship's structure and the piston rod end is attached to the tiller, also by a pinned connection. To allow for the movements of the cylinder the oil connections to the cylinders must be flexible hoses. These hoses require routine inspection and replacement to ensure steering system reliability.

In the system shown in Figure 21-7 fixed displacement pumps are used. The fixed piping terminates at the stop valves. Additional stop valves are fitted at the ends of the hoses to facilitate isolation and replacement. The many isolating valves and cross connections enable either or both power units to serve either or both cylinders.

Pumping Systems for Electrohydraulic Steering Gear

Electrohydraulic steering gear requires motor-driven, constant-speed, unidirectional, continuously operating pumps to supply oil under pressure to the actuators. Variable/reversible-delivery pumps are commonly used with ram-type actuators, and sometimes with rotary-vane actuators and clevis-mounted cylinders, supplying the fluid at pressures that reach 1,500 to 2,500 psi (100 to 175 bars). Newer rotary-vane actuators and most clevis-mounted cylinders are normally supplied by fixed-delivery pumps operating at pressures of 850 to 1,450 psi (60 to 100 bars), used in conjunction with solenoid valves to control and direct the flow.

Variable/reversible-delivery pumps are of the rotary or axial-piston type, discussed and illustrated in Chapter 12. In these pumps the rotor comprises a set of six or more cylinders, each of which houses a single-acting piston. Each piston or piston rod has a slipper pinned to it, which rides on the inside diameter of a stationary circular floating ring in a radial-piston pump, or on the surface of a stationary tilt plate (also called a swash plate or cam plate) in axial-piston pumps. As long as the floating ring is concentric with the rotor or the plate is perpendicular to the shaft, the pistons do not move in the cylinder as the rotor turns; however, when the ring is moved to one side, or the plate is tilted, the pistons reciprocate through two strokes (one in, one out) in each revolution of the rotor. Moving the ring or tilting the plate to the maximum produces the longest strokes of each

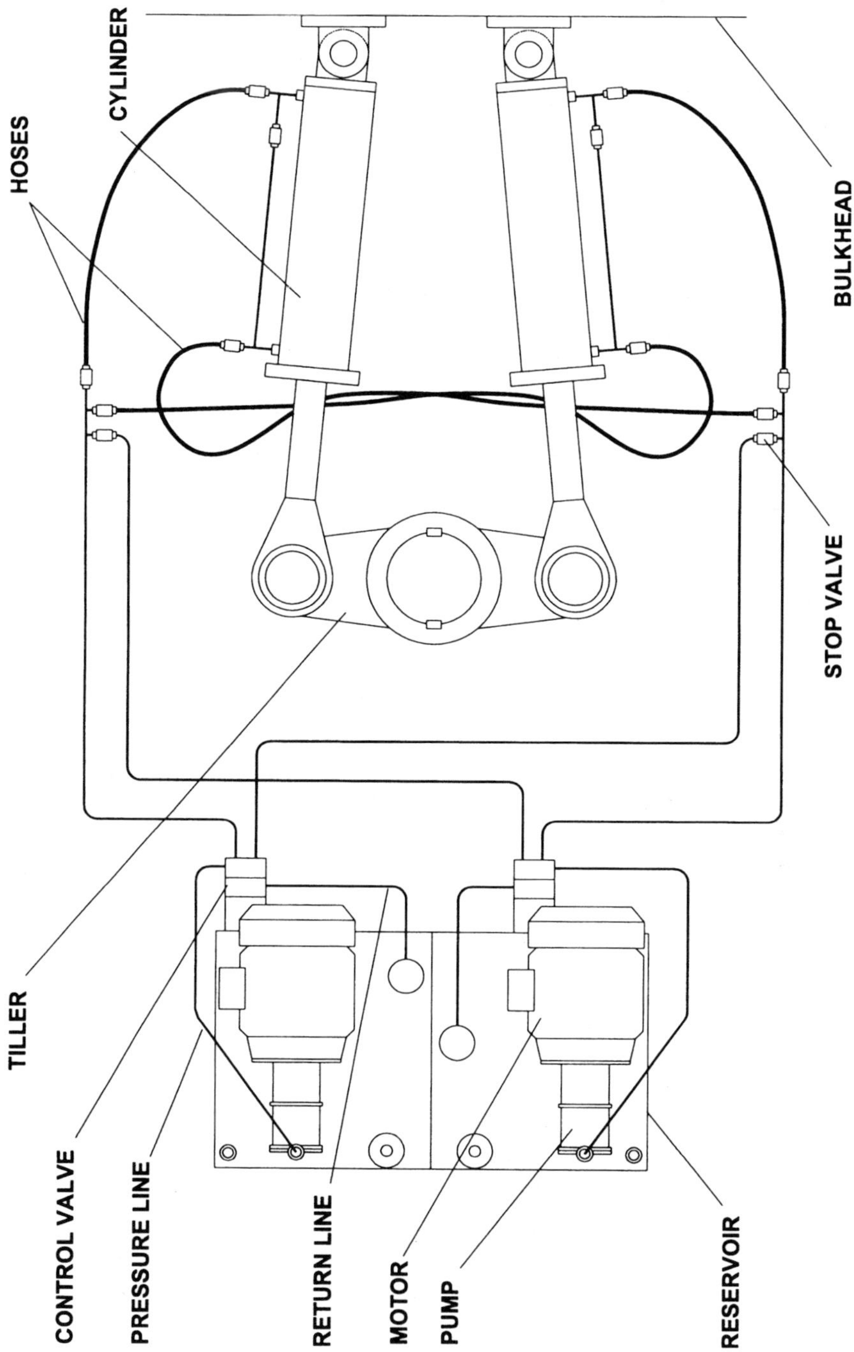

Figure 21-7. Schematic plan view of clevis-mounted steering gear

piston and the maximum flow rate. Moving the ring to the other side, or tilting the plate in the opposite direction, will reverse the direction of fluid flow. These pumps thus have the ability to provide continuously variable flow, from maximum in one direction, through to neutral, to a maximum in the opposite direction, while the motor is driving the pump in one direction at constant speed. The pump runs in neutral, except when called upon to move the rudder. The pump is controlled by an actuator operated by the steering gear control system.

Fixed-delivery pumps may be of the screw, gear, or vane type, also discussed and illustrated in Chapter 12. The pump runs continuously, and until called upon to move the rudder, the oil discharged by the pump at high pressure passes through an unloading valve to return to the reservoir and the pump suction. When a rudder movement is required, the directional control valve opens to connect the pressure line from the pump and the return line to the reservoir to the appropriate connections of the rudder actuator, while the unloading valve simultaneously closes.

A servo pump driven by the motor of the pump may be used to provide hydraulic pressure for the actuators of variable-delivery pumps, or for the solenoid valves used with fixed-delivery pumps.

CONTROL SYSTEMS FOR ELECTROHYDRAULIC STEERING GEAR

The control system is the means by which rudder orders are transmitted from the steering stations to the rudder. Most control systems comprise a rudder order transmission system from the bridge to the steering gear and a local control unit at the steering gear that controls the flow of fluid to the rudder actuator. The classic transmission system, still widely used with manually operated steering wheels, is the hydraulic telemotor, but electric, electronic, and mechanical transmission systems are also used. Autopilots require electric or electronic transmission systems. Mechanical transmission, usually using cables or chains, is generally practical only where the steering station is in close proximity to the steering gear. With few exceptions, the rudder order received at the steering gear then serves as input to a local follow-up control unit, which may be a mechanical differential or a floating lever, or an analogous electrical or electronic unit. The follow-up control passes the rudder order to operate the pumps or the directional control valves to move the rudder, but it also receives the actual rudder angle, which it subtracts from the rudder order until the rudder order is matched by the rudder angle. Most ships are fitted with two control systems, independent of each other and physically separated from the bridge to the steering gear compartment.

In its simplest configuration, the hydraulic telemotor is a pair of fluid-filled, double-acting piston/cylinders connected by tubing. One cylinder, the transmitting unit, is built into the steering stand or console and connected to the steering wheel or lever through a rack and pinion or a linkage. The other cylinder is the receiver, which is located at the steering gear. When the transmitter piston is repositioned by movement of the steering wheel or lever, fluid is displaced from one end of the cylinder and drawn in at the other, through the tubing, to and from the corresponding ends of the receiving cylinder, the piston of which therefore assumes a position duplicating that of the transmitter piston. The receiver piston rod is connected to the follow-up control to provide the rudder-order input to the pumping system. The system relies on the incompressibility of the hydraulic fluid, which must therefore fill the cylinders and tubing, and be free of air. The telemotor circuit is therefore connected to a reservoir by relief valves, and by equalizing and replenishing valves which operate automatically each time the transmitter is at the amidships position. The telemotor hydraulic circuit is completely independent of the steering gear circuit.

An electric transmission system may use synchros to transmit and receive the rudder order, actuating the pumping system by a mechanical input to a follow-up differential or a floating lever. In another form of electric control the steering wheel or lever moves a potentiometer to produce a proportional analog signal which is transmitted to a balanced-bridge circuit in the local control unit at the steering gear, which also receives the rudder angle as a proportional signal from a potentiometer operated by the tiller. The balanced bridge acts as a follow-up unit and operates the steering gear power unit through an electric or hydraulic servo. Electronic steering control can transmit an analog or digitized rudder order to a microprocessor, which also receives an analog or digitized signal for the rudder angle, and produces a digitized signal to operate the steering gear through a signal converter, powered servos, or solenoid valves.

International Agreement on Steering Gear Requirements

Requirements for steering gear capabilities are set by governmental regulatory authorities and by classification societies, but generally in accord with the International Convention for the Safety of Life at Sea of 1974 (SOLAS 74), which has been revised and amended, but remains in effect. The requirements are summarized below by paraphrasing and extracting from the American Bureau of Shipping Rules for Building and Classing Steel Vessels. This summary does not include all of the requirements or alternatives, nor the most current amendments, nor does it address vessels with azimuthing or similar propulsion systems. These rules apply to electrohydraulic steering gear, but steering gear of other types must meet equivalent standards.

- Vessels 90 meters in length or over are to be provided with power-operated means of steering. Such means is to be supported by duplication of power units, and by redundancy in piping, electrical power supply, and control circuitry. Steering is to be capable of being readily regained in the event of a failure.

- Vessels carrying large quantities of dangerous cargo, such as oil carriers and other vessel types over 70,000 tons deadweight, are additionally to be provided with duplication of rudder actuators. For such vessels, in the event of a failure, steering capability is to be regained automatically within 45 seconds.
- The steering gear compartment is to be protected from the weather. It is to be readily accessible and separated from the machinery spaces. The compartment is to be provided with visual compass readings.
- The steering gear is to be capable of putting the rudder from 35° on one side to 35° on the other with the vessel running ahead at the maximum continuous rated shaft output and at the summer load waterline, and from 35° on either side to 30° on the other side in not more than 28 seconds. With one power unit inoperative, the steering gear is to be capable of putting the rudder from 15° on one side to 15° on the other side in no more than 60 seconds with the vessel running ahead at the summer load waterline at one half of maximum ahead service speed or seven knots, whichever is greater.
- For passenger vessels the steering gear is to be capable of the first criterion cited above with any one of the power units inoperative.
- For tankers and gas carriers of 10,000 gross tons or higher, the steering gear is to comprise either two independent and separate power actuating systems, each capable of the first criterion cited above, or two or more identical power actuating systems which, acting simultaneously in normal operation, and which may be interconnected hydraulically, are capable of meeting the criterion. (The power actuating system includes the pumping system, the rudder actuator, and the associated piping and fittings; it does not include the tiller or rudderstock.)
- Steering gear may be composed of a single rudder actuator for all vessels except tankers and gas carriers of 10,000 gross tons and above but less than 100,000 tons deadweight, which may have a single actuator, subject to additional design requirements for mechanical strength, and for tankers and gas carriers of 100,000 tons deadweight and above, which must have two or more identical rudder actuators.
- The hydraulic system is to be designed so that after a single failure in the piping system or one of the power units, the defect can be isolated so that the integrity of the remaining part of the system will not be impaired and the steering capability can be maintained or speedily regulated. For tankers and gas carriers of 10,000 gross tons

or higher, the steering gear is to be so arranged that in the event of loss of steering capability due to a single failure in any part of one of the power actuating systems, steering capability is regained in not more than 45 seconds. (These rules are cited as "single failure criteria.")

- Steering gear is to be served by at least two exclusive electrical supply feeder circuits, fed directly from the main switchboard; however, one circuit may be supplied through the emergency switchboard. The circuits are to have adequate rating for supplying all motors, controls, and instrumentation normally connected to them and operated simultaneously. The circuits are to be separated throughout their length as widely as practically possible. Each circuit is to be provided with short-circuit protection at the switchboard. Provision of motor overload circuit breakers is restricted, but alarms are to be provided.
- Where the required rudderstock diameter is over 230 millimeters (9 inches), an emergency power supply, sufficient to supply at least one power unit and its controls and rudder angle indicator, is to be provided automatically within 45 seconds. Under this power supply the power unit is to be capable of putting the rudder from 15° on one side to 15° on the other side in no more than 60 seconds with the vessel at the summer draft while running at one half of maximum speed ahead or seven knots, whichever is greater. This power supply is to have a capacity of at least 10 minutes of continuous operation, and in every vessel of 10,000 gross tons and upwards, of 30 minutes of continuous operation.
- There are to be two independent steering control systems, operated from the bridge. These control systems are to provide, on the bridge, for the starting and stopping of steering gear motors and the rapid transfer of steering and control between units. Cables and piping of these systems are to be separated throughout their length. If these control systems are electric, each system is to be served by its own separate circuit supplied from a steering gear power circuit in the steering gear compartment, or directly from the bus bars of the switchboard supplying that power circuit. Circuits supplying power to steering gear controls are to be provided with short-circuit protection only.
- Where the control system consists of a hydraulic telemotor, a second independent control system need not be fitted, except in tankers and gas carriers of 10,000 gross tons and above.
- Local steering gear control is to be provided in the steering gear compartment. Means are to be provided in the steering gear compartment to disconnect the steering gear controls from other control systems when local control is to be used.

- Rudder angle indicators, running indicators for power unit motors, and alarms for power unit power supply, power unit motor overload or phase failure, control power failure, and hydraulic oil reservoir low level are to be provided on the bridge and in the steering gear compartment. Rudder angle indication is to be independent of the control system and visible from the control position. An alarm is to be provided at the bridge, activated when the position of the power unit control does not correspond to the rudder order.
- Communication is to be provided between the bridge and the steering gear compartment, and between these spaces and the main propulsion control station.
- Steering gear components transmitting forces to the rudder are generally to be of steel as are pressure retaining components; gray cast iron is generally not to be used. Bearings are to be permanently lubricated or provided with lubrication fittings. Oil seals are to be of the metal upon metal type or an equivalent type.
- Hydraulic systems are to include isolation valves, filters, and relief valves. Piping design pressure and relief valve setting pressure is to be 1.25 times maximum working pressure. A fixed, permanently connected hydraulic oil storage tank, independent of the power unit reservoir, is to be provided, with a capacity and means to recharge at least one of the power-actuating systems from a point in the steering gear compartment.

Operation of Steering Gear

Requirements for the operation of steering gear are also set out in the International Convention for the Safety of Life at Sea (SOLAS), generally as follows:

- Where navigation requires special care, more than one power unit shall be in use.
- Within 12 hours prior to departure, the steering gear and linkage shall be visually inspected and tested over the full movement of the rudder of which the gear is capable by the ship's crew. The tests shall include the main and auxiliary steering gear, remote steering gear control systems, steering positions on the bridge, the emergency power supply, rudder angle indicators (relative to actual rudder position), power failure alarms, automatic equipment, including automatic isolating arrangements, and communication means between the bridge and the steering gear compartment. For ships regularly engaged in short voyages, these inspections and tests shall be conducted weekly.
- Operating instructions with a block diagram showing changeover procedures shall be permanently displayed on the bridge and in the

steering gear compartment. Ship's officers shall be familiar with the operation of the steering systems, including procedures for changing from one system to the other. Drills shall be conducted at least every three months and shall include direct control from the steering gear location and the use of alternative power supplies.
- Checks, tests, and drills shall be recorded in the logbook.

Maintenance of Steering Gear

The following daily inspections are recommended:

- Check piping and glands for leaks.
- Ensure that pressures are normal.
- Check oil levels in all tanks and other reservoirs.
- Check motor and pump temperatures.

With the steering gear in operation, the following weekly routine is recommended:

- Check communications systems.
- Check filters, air release valves, grease cups, and reservoirs of automatic lubricators.
- Change over to alternative steering control and test its operation.
- Change over to alternative systems and test their operation.
- Change over power units of main steering gear.
- Confirm rudder angle indicators (relative to actual rudder position).
- Test power failure alarms and emergency power supply.

WINDLASSES, WINCHES, AND CAPSTANS

Windlasses are used to raise and lower ships' anchors. Winches and capstans are used to handle wire or rope lines. These machines have in common a prime mover, which may be a reciprocating steam engine on older ships, or a hydraulic or electric motor, a speed-reducing gear, and a wildcat, drum, or head to haul on the chain or line. Brakes and clutches are included to suit the application. A control system is also necessary.

Prime Movers

Almost all modern deck machinery is driven directly by electric motors or by hydraulic motors supplied by electrohydraulic pumps, but many older ships, especially tankers, remain in service with steam-driven deck machinery. Prime mover selection is largely determined by owner's preference, although the fire hazard associated with electrical equipment generally precludes direct electric drive for exposed deck machinery on tankers.

Figure 21-8 shows the configuration of steam engine commonly used for windlasses and winches. The cylinder is on the right. The connecting rod drives a pinion via the eccentric, driving the bull gear that dominates the figure. Directly below the cylinder is a spool or piston valve, normally operated by a vertical lever to set the winch or windlass direction and speed. The availability in recent years of inexpensive and reliable hydraulic machinery and of solid-state power electronics for AC motors (enabling simple speed control) has made steam-driven deck machinery obsolete.

Hydraulic systems for deck machinery may utilize a central hydraulic system or a unit hydraulic system. With unit hydraulic systems, a hydraulic motor for a winch or windlass is driven by its own nearby pumping unit. In a central hydraulic system two or more pumps, each driven by an electric motor, are mounted on a common foundation with reservoirs, control, and relief valves, and supply hydraulic oil at constant pressure to a number of hydraulic motors via supply and return mains. Additional pumping units start automatically to maintain pressure during periods of high demand. Central systems are commonly used for tankers and gas carriers because of the fire hazard, but the advantage of locating the equipment in a single, accessible location has led to more general application, despite the long runs of hydraulic piping that are necessary. On many tankers the central hydraulic system is used to drive hydraulic cargo pumps as well as windlasses and winches.

Pumps for deck machinery hydraulic systems are generally of the fixed-delivery type fitted with unloading valves, but for some unit systems, variable-displacement, axial or radial piston pumps of the type described above under steering gear may be used.

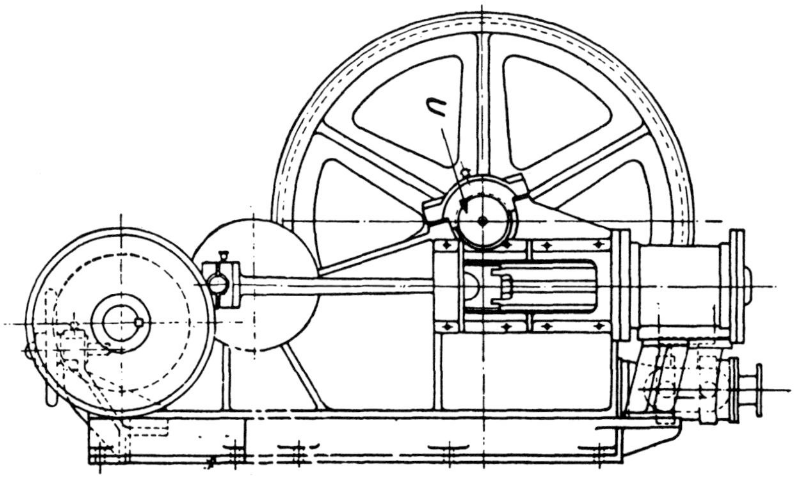

Figure 21-8. Steam-driven windlass, showing steam engine

Hydraulic motors for deck machinery are usually of the fixed displacement type, with directional control and speed control obtained by a spool valve. With unit pumps of the variable-delivery type, control is achieved at the pump. The spool valve or variable-delivery pump can be locally and manually operated, or it may be fitted with an electric or hydraulic actuator for remote control. The motors are generally of the radial-piston, axial-piston, or vane type. Figure 21-9 shows a vane motor. The vanes follow the shape of the casing because of centrifugal force, supplemented by springs, or by pressure from oil led to the inside of the radial slots. Figure 21-10 is an assembly drawing of a fixed-displacement, radial-piston motor. A close look at Figure 21-10 will show that the crankshaft block (GA) and its enclosing drum (GM) are eccentric to the shaft axis, so that the force of each piston (E) and connecting rod (F) acting in turn will rotate the shaft. In these motors, reversing the oil flow will reverse the motor, while throttling the flow will reduce the motor speed.

Windlasses

Windlasses are used for hoisting (or "heaving") anchor chain, and the drum takes the form of a sprocket, called the "wildcat," which is shaped to grasp the links of the chain, as in Figure 21-11, in alternating whelps and pockets cast into the sprocket. This example is a simplex windlass, with only one wildcat, and therefore capable of serving only one anchor. A second, identical, windlass would be provided in a typical ship for the other anchor. The chain is not stored on the windlass but passed through a chain pipe in the main deck to a chain locker below. In Figure 21-11, from left to right, are visible the reduction gear housing, the wildcat, and a duplex band brake. This windlass is driven by a hydraulic motor and is capable of lifting 190 fathoms (570 feet) of anchor chain at a hauling rate of 30 feet per minute. It

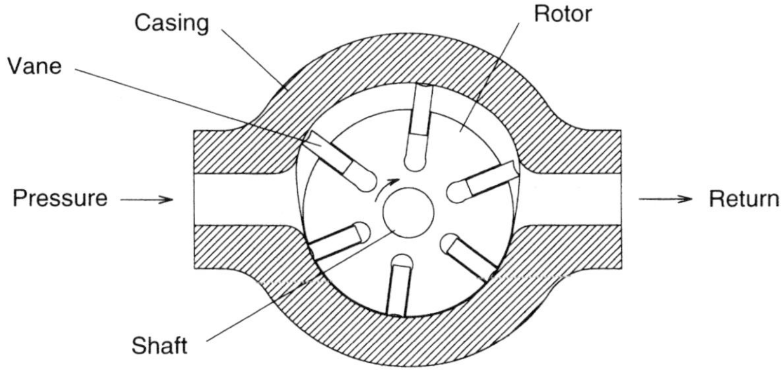

Figure 21-9. Schematic section of vane-type hydraulic motor

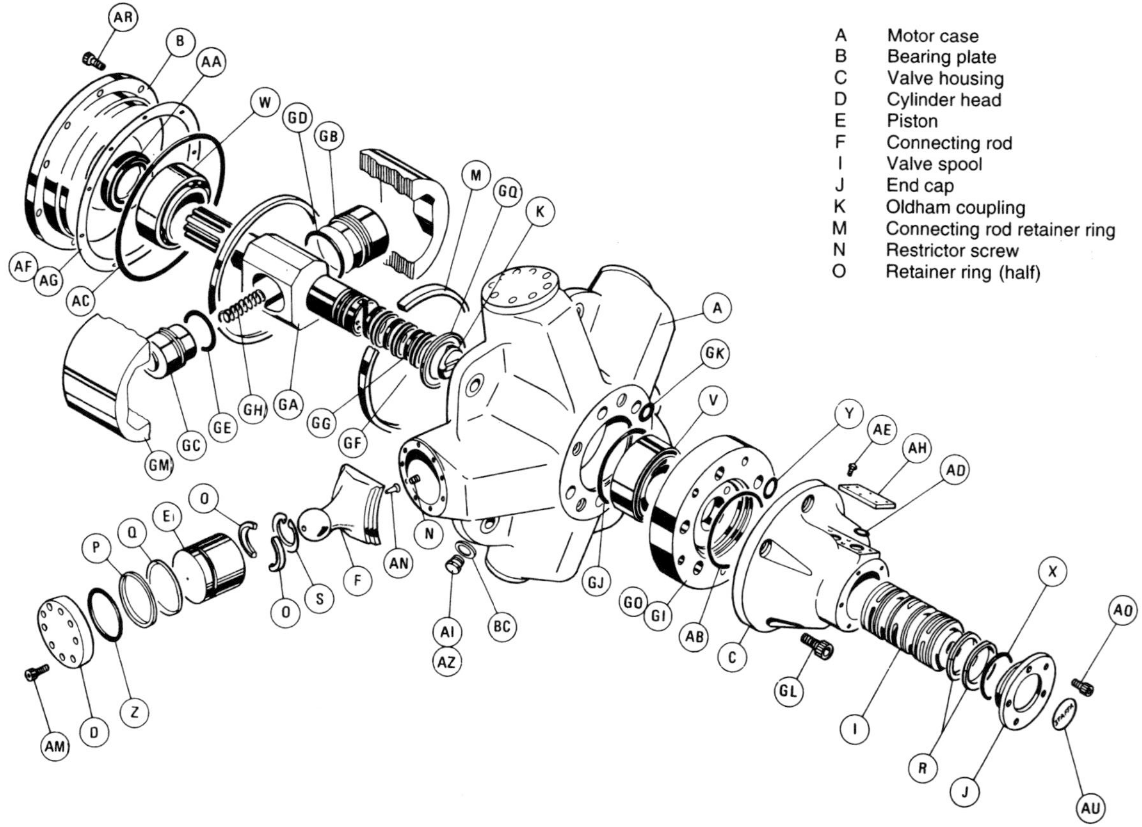

A	Motor case	P	Piston backing ring
B	Bearing plate	Q	Piston ring
C	Valve housing	R	Valve ring
D	Cylinder head	S	Circlip
E	Piston	V	Rear bearing
F	Connecting rod	W	Front bearing
I	Valve spool	X	'O' ring
J	End cap	Y	'O' ring
K	Oldham coupling	Z	'O' ring
M	Connecting rod retainer ring	AA	Shaft seal
N	Restrictor screw	AB	'O' ring
O	Retainer ring (half)	AC	'O' ring
		AD	'O' ring
		AF	Shim, plastic
		AG	Shim, steel
		AI	Drain plug
		AM	Socket cap screw
		AN	Filter
		AO	Socket cap screw
		BC	Bonded seal
		GA	Crank block and shaft
		GB	Shift piston (large)
		GC	Shift piston (small)
		GD	Large shift piston ring
		GE	Small shift piston ring
		GF	Shaft ring
		GG	Shaft backing ring
		GH	Spring
		GI	Valve spacer plate
		GJ	'O' ring
		GK	'O' ring
		GL	Socket cap screw
		GM	Drum
		GQ	Rear spacer plate

Figure 21-10. Assembly drawing of a five-cylinder, fixed-displacement, rotary-piston hydraulic motor. Pistons (E) drive crankshaft via connecting rods (F) and eccentrically mounted drum (GM) and crank block (GA).

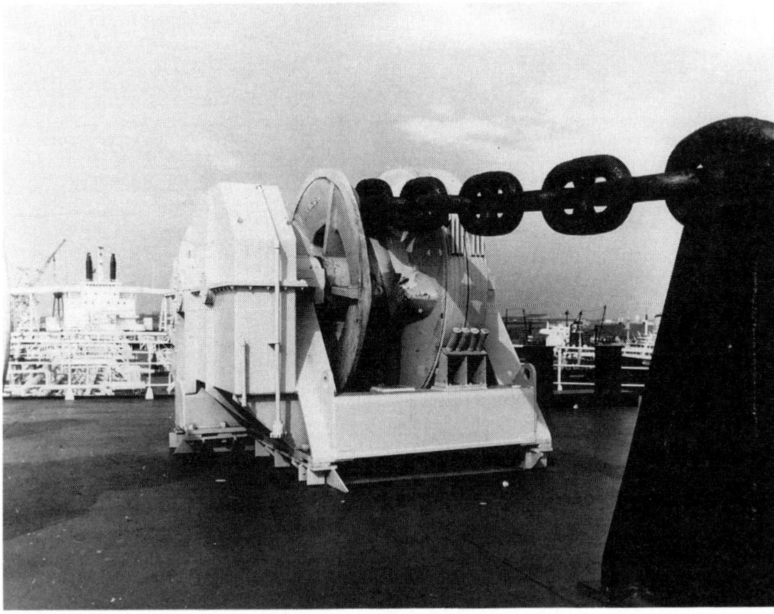

Figure 21-11. Anchor windlass for 4½-inch chain

should be noted that the reference to 4½-inch chain refers to the thickness of the links, which are therefore about 18 inches wide.

Figure 21-12 shows this windlass in plan view and side elevation. In the plan view the reduction gear is at the top, and below it, on the left, is a bearing, then the wildcat, the brake, and another bearing; at the right end of the reduction gear housing is the hydraulic motor. The elevation shows the band brake, engaged by the handwheel at lower right. The brake must be capable of controlling the fall of the anchor when it is "dropped" or "let go." Figure 21-13 is a horizontal section of the unit. The four-step speed reduction is necessary because of the relatively high RPM of the hydraulic motor, which connects to the small diameter input shaft at the top. A disk brake is fitted to this same shaft (the caliper is visible at the top, bolted to the gear case). Notice in Figure 21-13 that the wildcat shaft passes through the bronze bushings of the bearings, and also through a bronze bushing in the wildcat, to which it is connected only by the clutch plate, which slides axially on keys that are visible inboard of the bearing on the right side. Only when the clutch plate is moved to the left to engage the dogs (visible on the wildcat inside the brake drum) will the motor turn the wildcat. The clutch is engaged manually by a lever or by the small hydraulic cylinder shown in both views of Figure 21-12. The brake drum is integral with the wildcat. This construction, with wildcat and brake drum integral and free-wheeling on the shaft unless a clutch is engaged, is typical of most

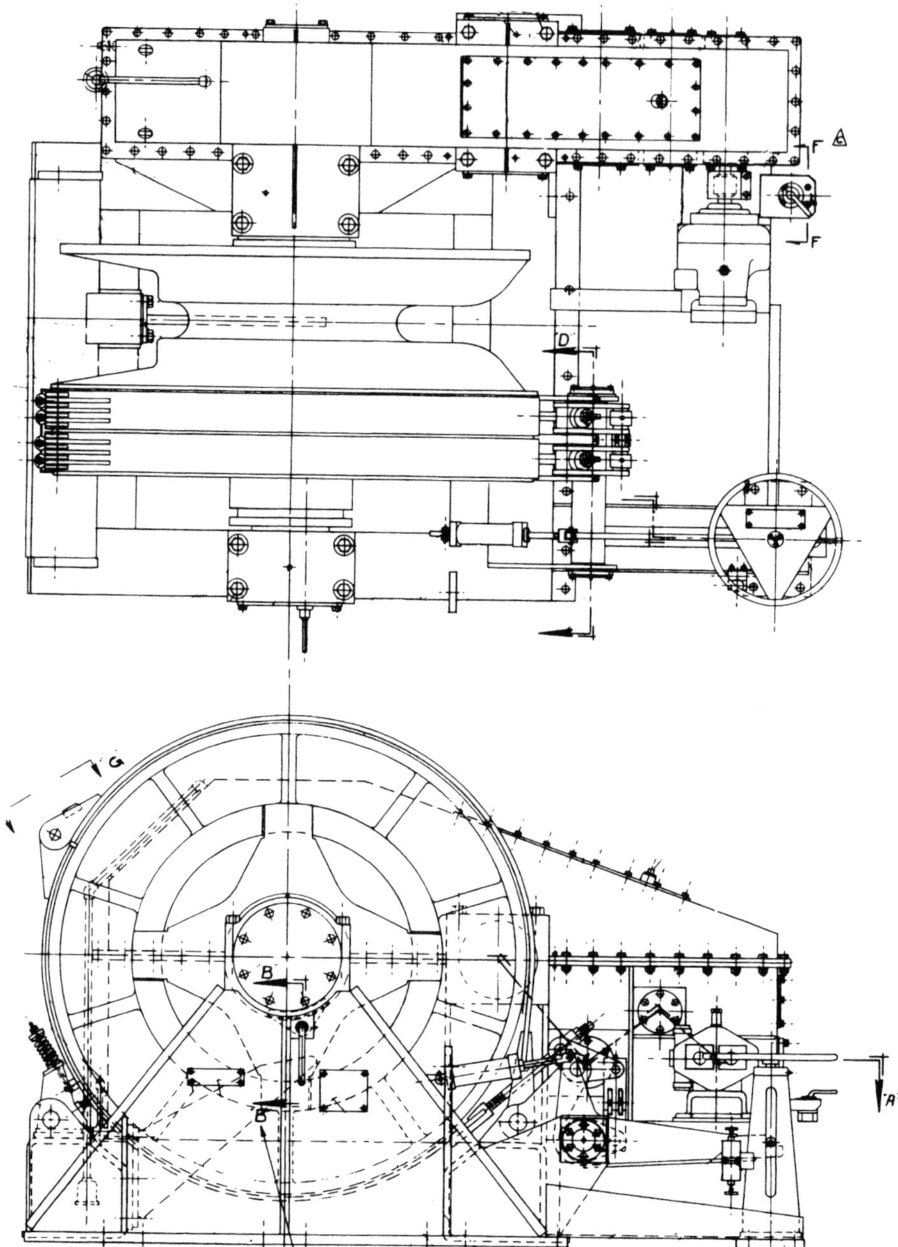

Figure 21-12. Assembly drawing of anchor windlass for 4½-inch chain

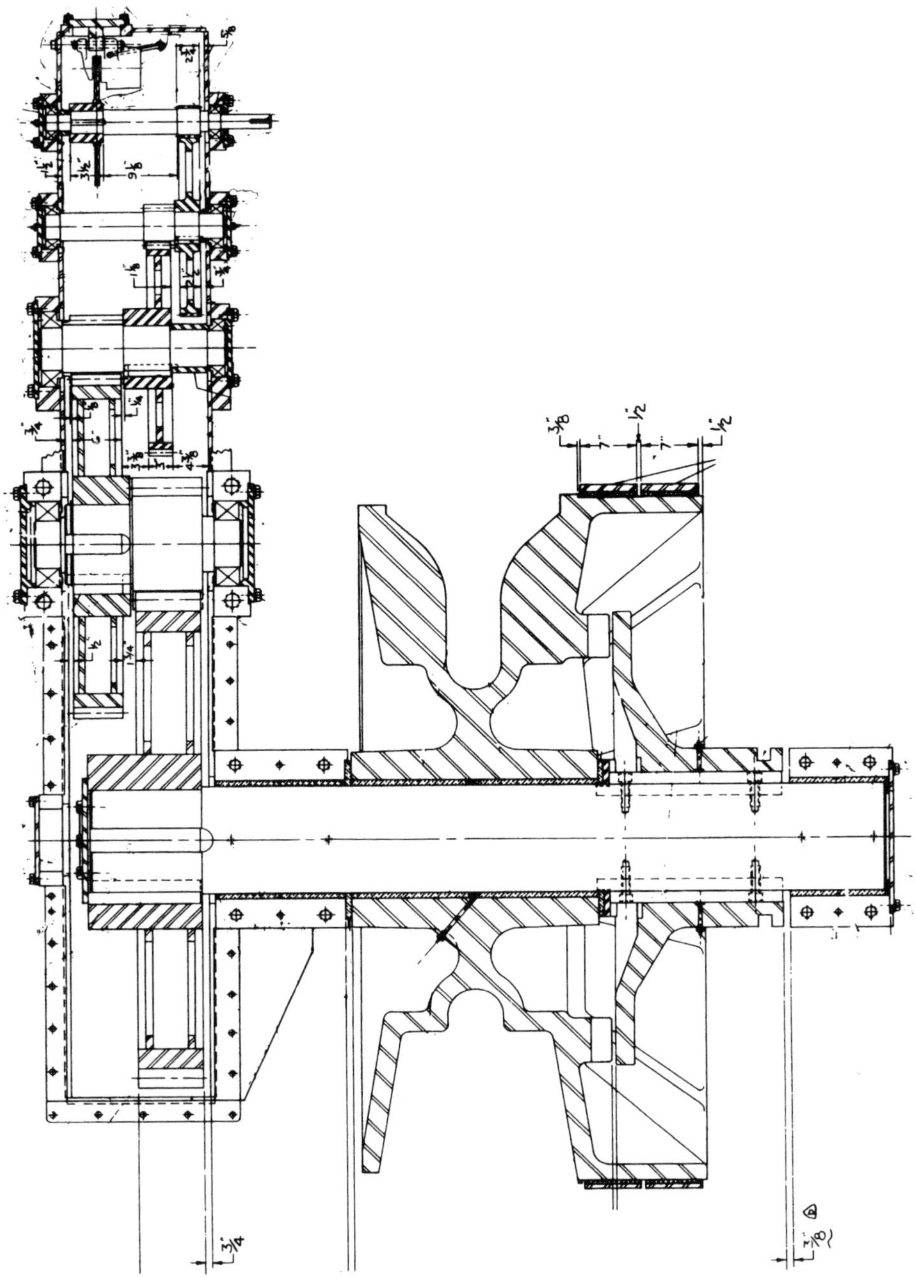

Figure 21-13. Horizontal section of anchor windlass for 4½-inch chain

windlasses. In operation, with the clutch disengaged, the anchor will fall under its own weight as the chain turns the wildcat on the stationary shaft, until the band brake is engaged to control the fall. To hoist the anchor, or to lower the anchor under power, the clutch is engaged. The bronze bushings of the wildcat and its shaft bearings, and the keys and dogs for the clutch plate, are grease lubricated. The enclosed reduction gear is oil lubricated.

The windlass is not used to secure the anchor at sea: once the anchor is heaved up and firmly seated at the foot of the hawsepipe, a chain stopper is used to lock the chain in place.

Figure 21-14 shows a duplex anchor windlass used to hoist both anchors. The hydraulic motor is left of center, at the bottom in the figure, driving through the centrally located gear case. The small handwheel operates the valve for the motor. The two long levers operate the two clutches to engage the two wildcats. The large handwheels operate the brakes, separately for each anchor. The windlass can hoist both anchors simultaneously at low speed, or each anchor separately at higher speed. This windlass is also fitted with warping heads (see below) driven from the intermediate shaft in the reduction gear train. These heads are not clutched, and turn whenever the motor is driving the windlass, but are used only when both of the wildcats are disengaged and locked by the brakes.

Figure 21-14. Plan view of duplex anchor windlass

Although both of the windlasses illustrated are hydraulically driven, identical windlasses are driven by electric motors. Further, although both examples are horizontal, vertical windlasses are used when deck space is limited, as on naval vessels. In the vertical arrangement only the wildcats are above the deck, and the gearbox, brakes, and motor are sheltered below.

Winches

Winches are used to handle wire or fiber rope lines. Figure 21-15 shows two typical modern winches. These examples are driven directly by electric motors, via the reduction gears in the gear cases on which the motors are mounted. Similar units are driven by hydraulic motors.

Each winch in Figure 21-15 has a single split drum, with a clutch and hand-operated band brake visible on the near end, and, on the opposite side of the gear case, a warping head (see below). Notice on each the lever for the clutch and the handwheel for the brake. Split drums are advantageous in that they prolong the lives of the rope lines by allowing the length of line not in use to remain loosely spooled on the drum on one side of the dividing flange, while the active part of the line is worked by the other part of the drum. The tensioned part of the line is thereby kept from working in among the coils of stored line, which could then become frayed and damaged.

Winches are normally operated under manual control, which may be local or remote, to bring the mooring line to sufficient tension to hold the ship in place at the berth, at which point the brake is set tightly enough to hold the ship, but loosely enough to slip should line tension rise as the ship is loaded or discharged or as the tide changes. Therefore, while a ship is moored to a berth, a mate will periodically adjust the lines. Automatically tensioned mooring winches (also called self-tensioned or constant-tension winches) are used to keep a ship tightly alongside a berth as the tide changes or cargo is loaded or discharged. After the mooring operation is complete the automatic feature may be selected. A load cell or other sensor fitted to the winch is then used to control the winch automatically, paying line out as tension on the line is sensed to rise, and taking line in when slack. Although the automatic feature reduces the frequency with which lines must be adjusted, it does not eliminate the need completely, and the oversight of a mate is still necessary. Because of the continual working of the line, automatically controlled mooring winches often have split drums.

The winches in Figure 21-15 are mounted directly to the deck, but winches often have their own independent bedplates.

Capstans and Warping Heads

Modern capstans and warping heads (also called gypsy heads) are similar in construction, purpose, and use, the difference being in the orientation: capstans are vertical while warping heads are horizontal. While warping heads are often fitted to shafts of windlasses and winches as noted above and as illustrated in Figures 21-14 and 21-15, capstans are most often driven by independent motors. Figure 21-16 shows a capstan driven by an electric motor via reduction gearing, in an arrangement fitted to a cut-out in the upper deck, to which it is bolted and sealed through the flange visible at A, with the motor therefore sheltered below deck. Similar capstans are driven by hydraulic motors. In another common configuration the motor and reduction gear are mounted to the deck below, and drive the capstan head via an extended vertical shaft fitted with flexible couplings, which penetrates the upper deck. A similar arrangement is used to enable a capstan to be driven by an intermediate shaft from a vertical windlass. In yet

Figure 21-15. Two electric-motor-driven winches with split drums and warping heads

another arrangement, the motor is housed within the capstan head, allowing the entire unit to be mounted above deck.

Capstans and warping heads are used to assist in the manual handling of fiber rope lines: the person handling the line wraps it once or twice around the rotating head, and then, standing behind the head, applies sufficient tension to enable the head to grasp the line by friction. The operator pulls the line hand-over-hand to maintain the tension, coiling the line on deck or passing it to a bin or storage spool below through an open hatch. Because the line must be handled manually, the rotation of the head cannot be too fast, and a stop switch or other stop control is located nearby. A two-speed motor or another means of achieving a high speed for hauling slack lines, and a reversing capability, are useful features and are often provided.

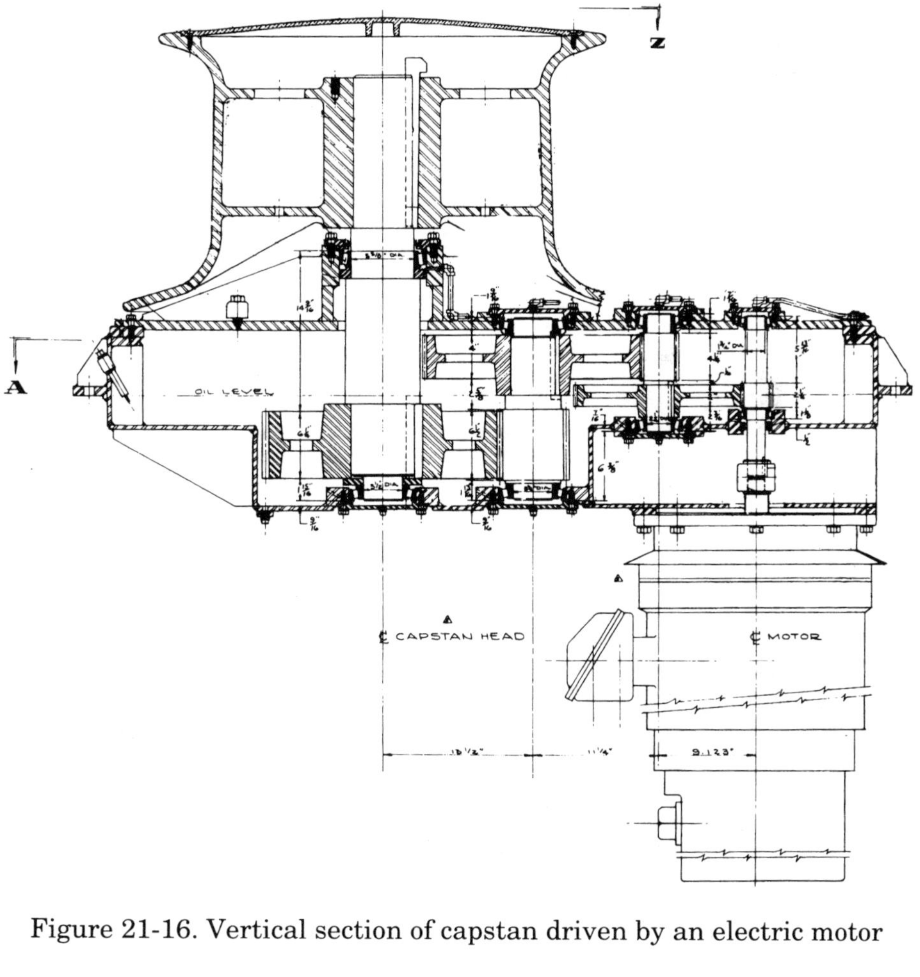

Figure 21-16. Vertical section of capstan driven by an electric motor

Arrangement of Deck Machinery

Figure 21-17 shows the foredeck of a typical modern tanker. There are two simplex windlasses immediately aft of the mast, angled to match the hawse pipes, with the anchor chains led to the hawse pipes over the chain stoppers. Aft of these are three duplex, split-drum mooring winches, each with a warping head at the aft end. All of these windlasses and winches are

Figure 21-17. Foredeck of a ship, with two anchor windlasses and three mooring winches

hydraulically driven. More winches may be located along the main deck of the ship, with another two or three grouped together at the stern.

Maintenance of Windlasses, Winches, and Capstans

A windlass, winch, or capstan can be expected to operate with little maintenance if the exposed parts of the unit receive regular lubrication. Oil sumps and grease cups should be checked regularly. Oil sumps should be sampled annually for water, which should be drained off, and for oil condition. Periodic inspections should be made for worn parts, including bearings, bushings, gears, and brake linings, and for leaking covers or casing joints that might allow ingress of sea water. Mounting bolts and stoppers should be checked for tightness. Listening at gearboxes and bearings for abnormal noise when a unit is in operation, and feeling exposed bearings for overheating, can provide early warnings of problems. With steam-driven machinery, water must be drained from the cylinders and valve chests prior to starting. For all units, clutches and brakes should be tested sufficiently in advance of use to allow time for repair. Electric and hydraulic control systems should be checked according to manufacturer's instructions.

REVIEW

1. Explain how rudder torque is developed by a simple ram-type steering gear. Use a sketch with correctly named components in answering the question. Do the same for a rotary-vane gear.
2. Use a sketch to show how, for equal torque, equal tiller length, and an equal rudder angle of 30°, a Rapson slide will require less force application than that needed from a clevis-mounted cylinder.
3. Explain why the footprint of a rotary-vane actuator can remain compact even for very high torque requirements.
4. Use sketched vector diagrams to show why the cylinders of ram-type steering gear must be rigidly mounted while the cylinders of clevis-mounted gear can be free to pivot.
5. Sketch a hydraulic telemotor, and show how it can be used to transfer a motion from the bridge to the steering gear.
6. What is the purpose of a follow-up mechanism? Describe how a follow-up mechanism might work.
7. Explain how a windlass differs from a winch. Cite the common and different components by name.
8. List the advantages and disadvantages of a central hydraulic pumping system for deck machinery, when compared to unit pumping systems.
9. Cite the principles and features that enable a windlass with a warping head to drive the warping head even while the winch drum is locked by the brake. How do these principles and features differ from

or resemble those that enable a duplex windlass to be used to drop one anchor without dropping the other? Explain.

10. Explain why the hand brake drums of winches and windlasses are so large in diameter.
11. Sketch a simple section across the casing and rotor of a vane-type hydraulic motor with an even number of vanes.
12. Explain how a capstan differs from a winch. Cite the common and different components by name.

REFERENCES

Additional information on deck machinery, including information on cargo gear, stabilizers, and other equipment not covered in this chapter, may be found in the references listed below.

American Bureau of Shipping. 2000. *Rules for Building and Classing Steel Vessels.* Houston, Texas: American Bureau of Shipping.

Beattie, D. H., and Somerville, W. M. 1978. *Deck Machinery.* MEP Series, Vol 2, Pt 16. London, England: Institute of Marine Engineers.

Harrington, R. L. 1992. *Marine Engineering*, Chapter XVIII. Jersey City, New Jersey: Society of Naval Architects and Marine Engineers.

Paulin, W. S., and Fowler, D. J. 1975. *Steering Gear*. MEP Series, Vol 2, Pt 9. London, England: Institute of Marine Engineers.

U.S. Coast Guard, *46 Code of Federal Regulations,* Subchapters F and J. Issued annually. Washington, D. C.: U. S Government Printing Office.

ACKNOWLEDGMENTS

Material and illustrations used in this chapter were provided by the following:

Brown Brothers & Co., Edinburgh, Scotland
Hyde Marine, Cleveland, Ohio
Kawasaki Heavy Industries, Japan
Lake Shore Inc., Iron Mountain, Michigan
Lindenau GmbH, Kiel, Germany
Rolls-Royce Oy Ab, Rauma, Finland
Transmission Technology Co., Fairfield, New Jersey

CHAPTER 22

Marine Electrical Systems

CONRAD C. YOUNGREN

SHIPBOARD ELECTRICAL DISTRIBUTION SYSTEMS

Introduction

REGULATIONS for electrical installations on shipboard are promulgated by governmental regulatory agencies (e.g., U.S. Coast Guard), the Intergovernmental Maritime Organization (IMO), and classification societies (e.g., American Bureau of Shipping). In making any modifications to the existing electrical installation, the ship's engineer should always make reference to those regulations that may be applicable, such as Title 46 (Shipping) of the *Code of Federal Regulations*, Subchapter J, which in turn references sections of NFPA 70 (National Electrical Code) and IEEE Standard No. 45 (Recommended Practice for Electrical Installations on Shipboard).

With respect to applicable regulations, the U.S. Coast Guard has divided vessels into the five following groups: (1) oceangoing vessels that navigate on any ocean or the Gulf of Mexico more than 20 miles offshore; (2) ocean-going vessels that navigate on any ocean or the Gulf of Mexico but 20 miles or less offshore; (3) vessels navigating on the Great Lakes only; (4) vessels navigating on bays, sounds, and lakes other than the Great Lakes; and (5) vessels navigating on rivers only.

System Voltages

The following voltages are recognized as standard:

	Alternating Current (volts)	*Direct Current (volts)*
Generation	120-208-230-240 450-480-2400-4160	120-240
Power	115-200-220-230 440-460-2300-4000	115-230
Lighting	115	115

The following systems of distribution are recognized as standard with a frequency of 60 Hz standard for all AC lighting and power systems: (1) two-wire (single-phase AC or DC); (2) three-wire (single-phase AC or DC); (3) three-phase, three-wire AC; and (4) three-phase, four-wire AC.

On small vessels with little power apparatus, 120V three-phase alternators may be utilized. Power of 115V and lighting distribution may consist of three-phase feeders. Lighting distribution may consist of 115V single- phase distribution balanced at the main switchboard to provide approx-imately equal load on each phase. On DC vessels of comparable size, 120V generators with 115V power and lighting distribution systems would be utilized.

On intermediate size vessels with considerable power apparatus, 240V (or 230V) three-phase generators supplying three-phase power loads at the 230V (or 220V) utilization level and 115V three- or single-phase lighting distribution may be utilized. Four-wire 208Y/120 volt (or 200Y/115) systems are another possibility for this application.

Typical of today's large merchant ships is the installation of 450V (or 480V) three-phase generators to supply 440V (or 460V) power consumers and 115V, three-phase lighting distribution via a bank of step-down transformers. Four-wire, 208Y/120 lighting distribution, though rare, is another possibility. Although any of the above AC systems may be grounded or ungrounded, most 450V, three-phase shipboard systems are ungrounded. Consequently, while four-wire 480Y/277 volt distribution (277V lighting) is a common shoreside industrial distribution scheme, it is not a common shipboard system.

Certain vessels with very large electrical systems may require generation at higher voltages (4160V or 2400V) with some power utilization at 4000V or 2300V, three-phase, and the remainder of the system at 440V (power) and 115V (lighting). The lower voltages are provided through three-phase transformer banks. Occasionally 440V or 460V floodlights may be used.

System Arrangements

Ship's service distribution. Electrical power is generally routed through the ship's electrical service system in an arrangement known as a *radial* system. That is, from any consumer there is only one continuous path back to the source of electrical power. Other arrangements utilized in shoreside commercial and industrial applications, such as primary loop, primary selective, secondary selective, and secondary network, are generally not found on commercial vessels. This is not to say that certain specified equipment cannot be energized from an alternative source in an emergency, but that typically only one protective device or disconnect is necessary to isolate a piece of equipment or section of the system.

Figure 22-1 is a schematic overview of a typical cargo ship's power and lighting circuit distribution system. Under normal operating conditions all power emanates from the generator section of the main switchboard. In this example it is assumed that the ship's service generators are independently driven by diesel or steam turbine prime movers. Main engine-

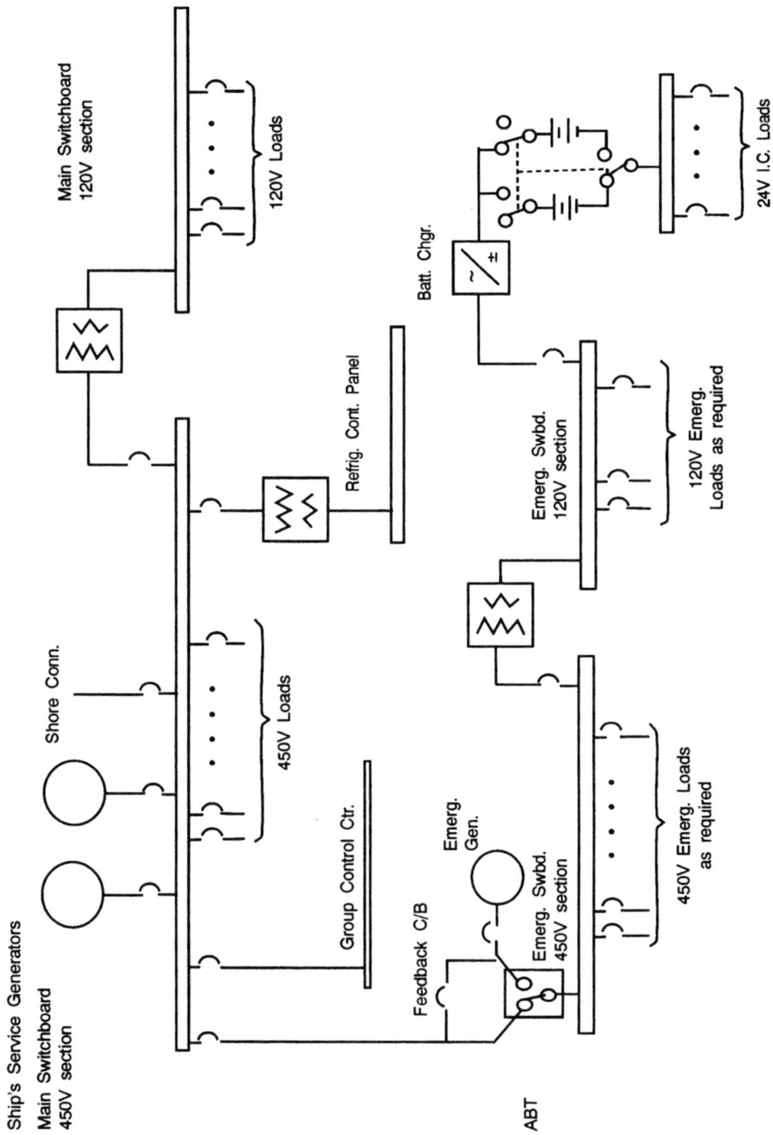

Figure 22-1. Electrical general arrangement

driven generators will be addressed in a later section of this chapter. Individual consumers may be connected directly to the main switchboard, or to distribution panels that are in turn connected to the main switchboard. Only two voltage levels are shown, although additional transformer banks could provide 240V service if it were required. Isolation transformers are shown segregating a refrigeration container power supply panel from the rest of the system. This is often done so that unintentional grounds in the container equipment do not affect the rest of the system. On ships where there is an exceptional lighting load, additional lighting transformers and lighting load centers may be included and located remote from the machinery spaces.

A detailed map of the electrical power and lighting system is known as the "electrical one-line diagram" or sometimes the "electrical general arrangement." This drawing is a schematic layout of the electrical plant circuitry, with a one-line route signifying a possible path of electrical power flow regardless of the number of conductors actually carrying the current. Other information found on the one-line diagram includes the generator rating, voltage frequency, and number of phases; circuit number or name including cable designator and connected load (amperes) or other basis for sizing cable; circuit protective devices with frame size, continuous rating, trip settings, and any special features such as interlocks, undervoltage trips, etc.; motor ratings and starter data such as NEMA (National Electrical Manufacturers Association) size, full voltage or reduced voltage, LVR or LVP, etc.; transformer primary and secondary voltages, continuous KVA ratings, and winding connection; panelboard identification with circuits supplied and rating of consumers served; and ground protection/detection equipment. Figures 22-2 and 22-3 show the level of detail available on a typical electrical one-line diagram.

Emergency power distribution. To ensure a dependable independent power source with the capacity to supply all those services necessary for the safety of passengers, crew, and other persons in an emergency, an emergency power distribution system is incorporated in all vessels equipped with an electric plant. For all cargo vessels over 1,600 tons and certain passenger vessels with restricted operations, an automatically connected storage battery bank or automatically started generator is required as a final emergency power source. For passenger vessels operating on oceans or the Great Lakes, on international or coastwise voyages, a temporary emergency source of limited capacity, that can carry for a short time selected emergency loads, is required in addition to the final source described above.

Figure 22-1 shows an emergency bus tie connecting the 440V section of the main switchboard to the 440V section of the emergency switchboard through an automatic bus transfer (ABT) switch. Voltage failure at the

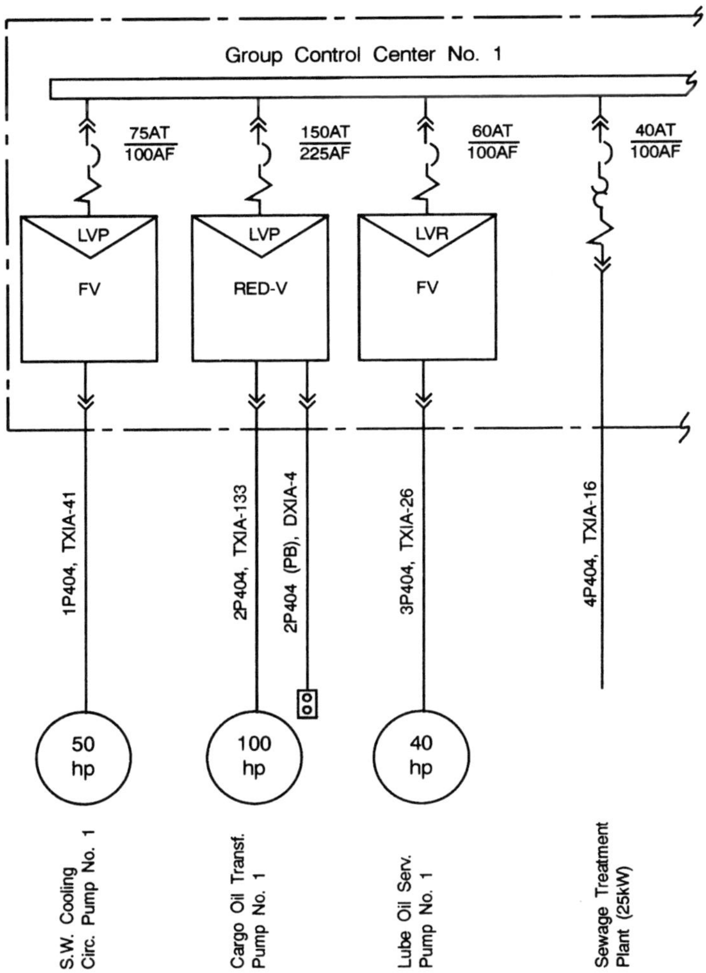

Figure 22-2. One-line diagram, power circuits

main switchboard would automatically start the emergency generator and simultaneously connect the emergency bus to the final emergency power source. Notice that the arrangement of the ABT prohibits the paralleling of emergency and ship's service sources. A feedback breaker is fitted to temporarily bypass the ABT for operations such as connecting to the main switchboard for dead ship start-up. The feedback breaker is designed to open automatically upon overload of the emergency power source before the emergency generator is tripped off the line from overload.

Where a temporary emergency source is required, it must be arranged to supply the following:

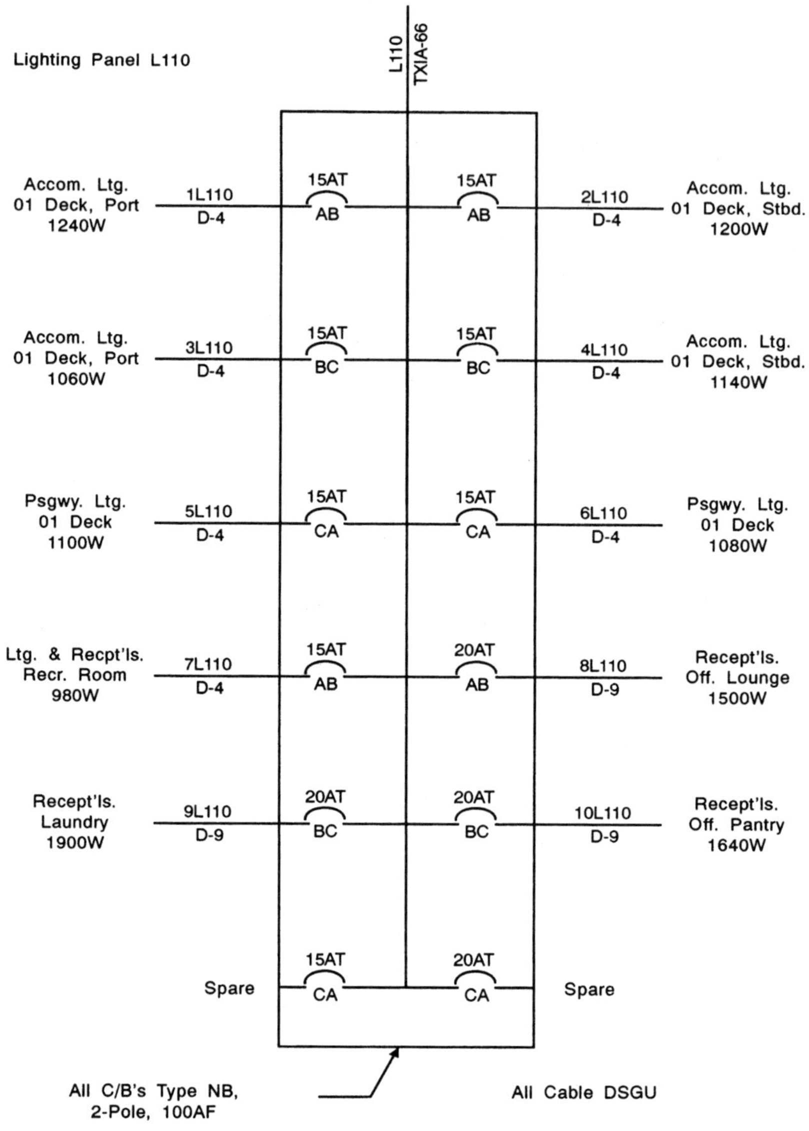

Figure 22-3. One-line diagram, lighting circuits

1. Emergency lighting, including navigation lights; sufficient illumination in machinery spaces for operations essential to restoring normal power; illumination of passageways, stairwells, lifeboat and life raft assembly, embarkation, and launching points, and watertight doors; at least one lamp in each public and working space.

2. Electric communications systems that do not have a dedicated storage battery source; emergency loudspeaker systems.
3. Power-operated watertight doors and fire screen door release systems.
4. Smoke detector and fire detection systems.
5. Each daylight signaling light.
6. Each electrically controlled or powered ship's whistle.

The final emergency source, in turn, must be arranged to service the following:

1. The entire temporary load.
2. Each elevator on a passenger vessel.
3. Charging panels for temporary emergency batteries, emergency generator starting batteries, and general alarm batteries.
4. Where required by regulation, one bilge pump, one fire pump, a sprinkling system pump, lube oil pump for each propulsion turbine and ship's service generator requiring external lubrication, one steering gear feeder.
5. Each rudder angle indicator and steering gear failure alarm.
6. Navigation equipment such as radio direction finders, radars, Loran, gyrocompass, and depth sounders.

Figure 22-4(a) illustrates a typical arrangement of final/temporary source transfer. Figure 22-4(b) details suggested arrangements for AC/DC source transfer. Where equipment requiring emergency supply operates on power that has characteristics other than those available from the emergency sources, the motor-generators, converters, rectifiers, or other apparatus supplying this equipment must automatically start and assume the load upon establishment of emergency supply.

A navigating and signal light panel installed in the wheelhouse combines an automatic or semiautomatic telltale navigating light section for audible and visible alarm and control of the masthead, range, side, and stern lights, and a signal light section for control of anchor, not-under-command, and towing lights. The navigating light section (Figure 22-5) is arranged to indicate failure of each primary lamp or filament and is provided with a switch for transfer to the secondary lamp or filament.

Main engine-driven auxiliaries. The concept of generating electricity by utilizing the main engine as a prime mover is not new. When DC installations were prevalent aboard ships, the shaft generator was quite common, as voltage control was easily obtained. With the transition to AC-driven auxiliaries, however, because of the problem of maintaining system frequency with varying main engine RPM, and the lack of economic justification due to low fuel costs, the shaft generator disappeared from ships in favor of independently driven alternators. Today, economic pressures favor installations that are fuel efficient. Since the production of

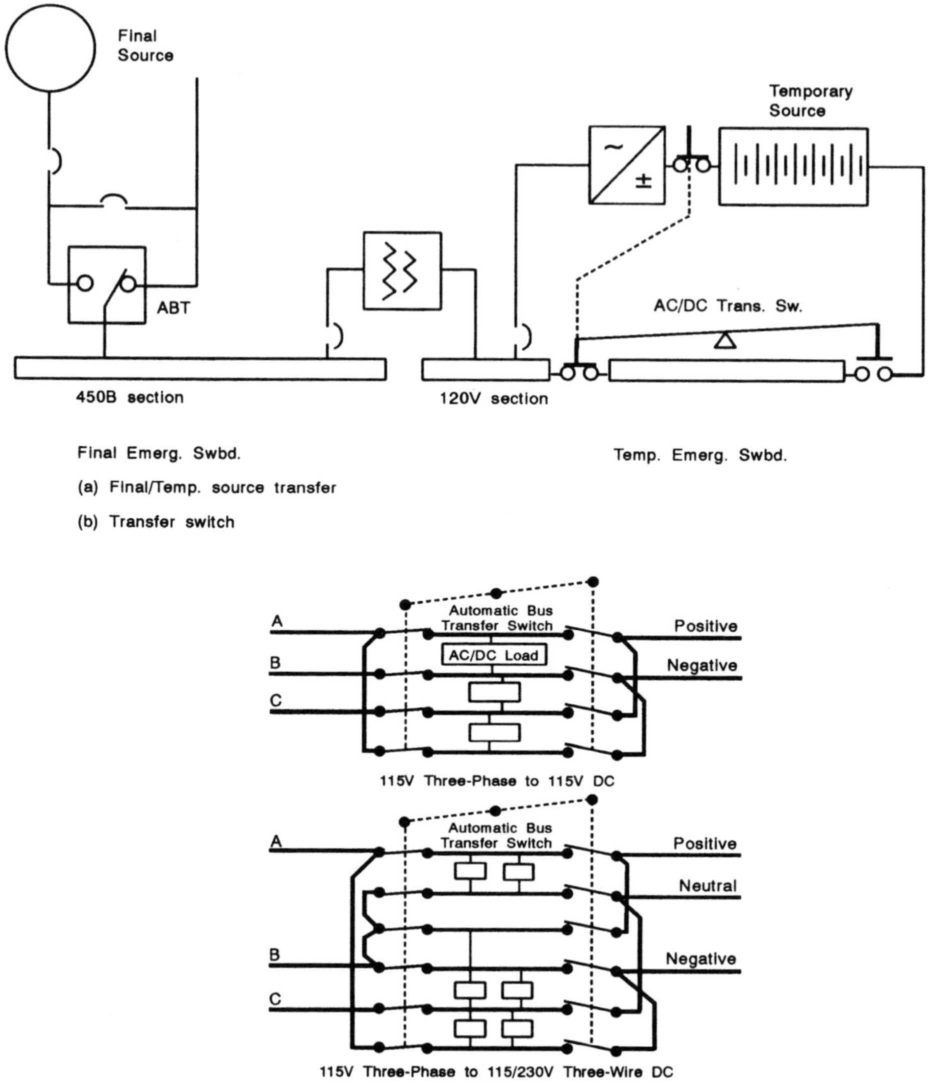

Figure 22-4. Final and temporary emergency power distribution

electricity is, next to propulsion, the largest consumer of energy sources, attention again has focused on main engine-driven generators. This enables the production of electricity to be based on the consumption of heavy fuel used in the main engines. Four arrangements involving the main engine in the production of electrical power are discussed in the following paragraphs.

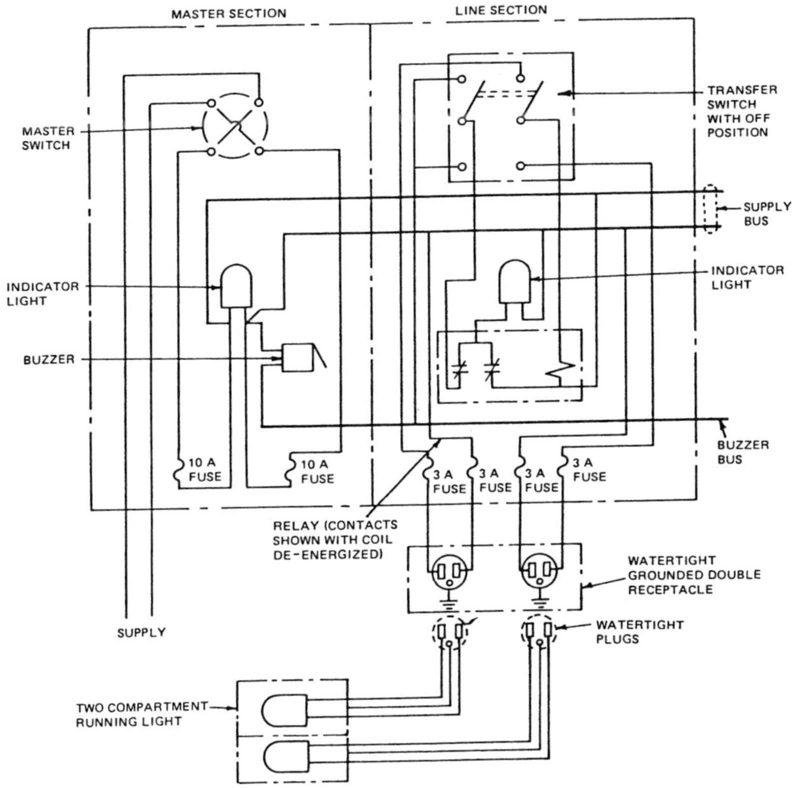

Figure 22-5. Semiautomatic navigation light panel circuit

Constant speed engines. The use of a controllable pitch (CP) propeller theoretically allows a power take-off (PTO) generator to operate at constant speed and therefore constant frequency. While this arrangement is simple, there are losses in propulsion efficiency and the inevitable frequency variations (especially in heavy weather) that prohibit parallel operation with diesel- or turbine-driven alternators. Figure 22-6 is a schematic of the electrical generation plant aboard a 125,000 DWT North Sea shuttle tanker. Two PTO generators are associated with each propulsion engine. One of the main engines drives two 2200 kW, 1200 RPM shaft generators which service nonfrequency-sensitive loads.

During cargo loading, the propeller is uncoupled and the main engine is used to control cargo pump output. The other main engine drives a similar 2200 kW generator and a 960 kW generator with a frequency converter at speeds ranging from 850 to 1200 RPM. A diesel generator is incorporated in the plant for in-port operation only.

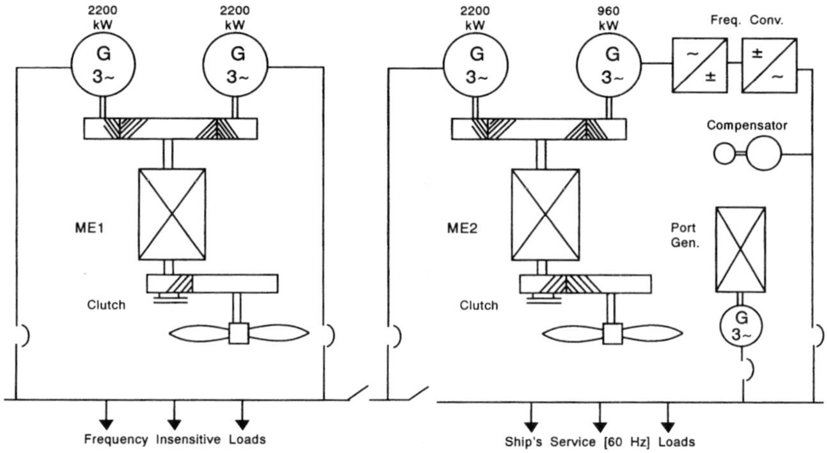

Figure 22-6. Constant speed PTO

Variable speed drives with frequency converters. With the main engine driving a fixed pitch (FP) propeller, a frequency converting device is required to supply ship's service mains with constant frequency at varying engine speeds. These frequency converters may be of the static type that employs a solid-state rectifier and inverter, or of the rotating type that utilizes a DC PTO generator and DC-to-AC motor-generator set.

The electrical plant of a 65,000 DWT product carrier is illustrated in Figure 22-7. Here a 500 kW exhaust gas-generated steam turbine supplies the base load, augmented by an integral shaft-driven alternator. The shaft generator (S/G) can supply 800 kW between 70 and 123 RPM and reduced output in proportion to speed down to 35 RPM. The DC-link converter comprises a three-phase bridge rectifier utilizing uncontrolled silicon diodes, a thyristor switch in the intermediate DC circuit to protect the rectifier and inverter, and the solid-state, AC mains-commutated inverter consisting of a thyristor bridge. The inverter can only supply active power to the ship's mains, and thus the reactive load must be supplied by a synchronous compensator. The compensator, a standard brushless synchronous machine, is accelerated to near-synchronous speed via a squirrel-cage starting motor flanged to the generator. At synchronous speed it operates at a power factor near zero, supplying the reactive power to the ship's mains and also the power required by the inverter for proper commutation. In addition, the compensator provides the almost sinusoidal form of the system output voltage and the short circuit current required for selective tripping of the system's protective devices. Maintenance in this type of system is minimal, as there are no moving parts in the power

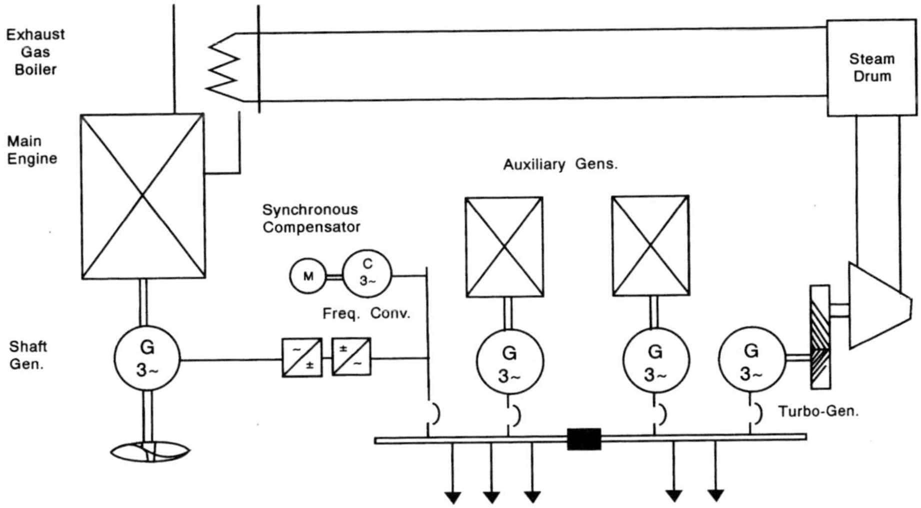

Figure 22-7. Variable speed PTO with solid-state frequency converter

circuitry, and the converter equipment may be installed anywhere in the engine spaces.

The schematic of a rotating converter or generator-motor-alternator (G-M-A) system is shown in Figure 22-8. The DC generator runs at engine speed and is typically designed to provide constant output with speed variations of ±15 percent. Speeds outside this range would initiate the automatic start-up and takeover by selected diesel-driven auxiliary generating sets. A conventional shunt-wound DC motor drives a synchronous alternator to provide ship's service power. Although the maintenance costs are slightly higher than for static converters, the efficiency of rotary converters is higher and the technology is more familiar to ship's personnel. No compensator is required since the alternator is capable of supplying both real and reactive power.

Variable ratio mechanical coupling. The use of a variable ratio epicyclic gear train to provide constant speed generator drive (CSGD) from a variable speed main engine PTO is another arrangement allowing the main engine energy source to be utilized in generating electricity. In the British Shipbuilders' version, shown diagramatically in Figure 22-9, the CSGD consists of an input clutch coupling attached to the PTO shaft, a single-stage epicyclic train with fixed displacement hydraulic units, the generator, and the variable displacement hydraulic unit in coaxial configuration. The variable displacement hydraulic unit is driven at constant speed (1800 RPM) by the output (sun gear) side of the train. If the input speed is nominally 1050 RPM, the planet carrier is held stationary by the

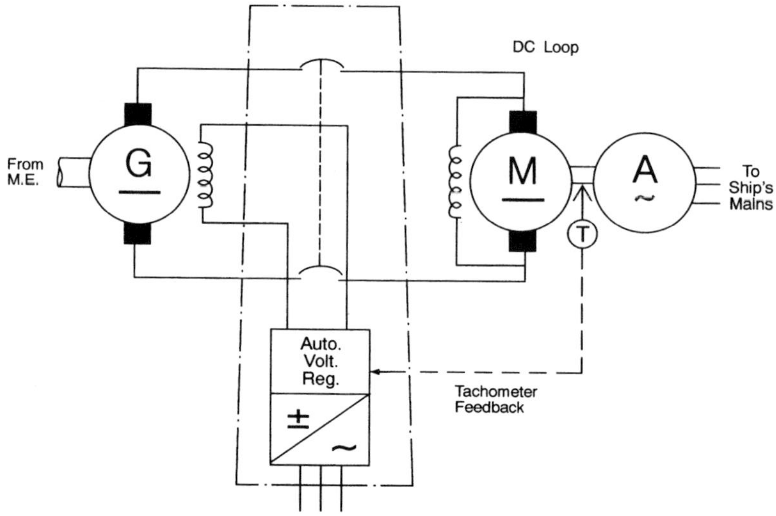

Figure 22-8. Variable speed PTO with rotary converter

hydraulic units and the system operates as a fixed ratio drive. If the input falls below 1050 RPM, power is developed by the variable displacement unit which, acting as pump, transmits power to the fixed displacement

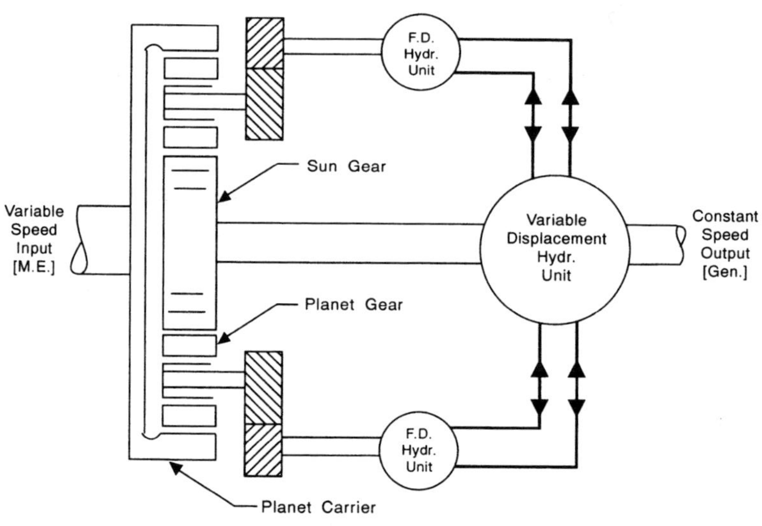

Figure 22-9. Variable ratio geared PTO

units, which in turn act as motors to turn the planetary carrier against its developed torque. Conversely, if the input rises above 1050 RPM, the fixed displacement units extract power from the planet carrier, which is allowed to rotate under its developed torque. The variable displacement unit then absorbs the power as a motor driven by the fixed displacement pumps, and delivers it to the generator.

Augmented generator drive. A PTO attached to a variable speed main engine may be used to augment the mechanical power required to drive a ship's service alternator. In the arrangement shown in Figure 22-10, a geared turbogenerator, supplied with steam from an exhaust gas boiler, is mounted in tandem with a DC motor. The DC motor in turn is fed through a three-phase silicon diode bridge from a variable frequency PTO alternator. The supplementary supply has to provide the difference between the required electrical load and the available turbine power after priority steam consumers with fluctuating loads have been supplied from the exhaust gas boiler.

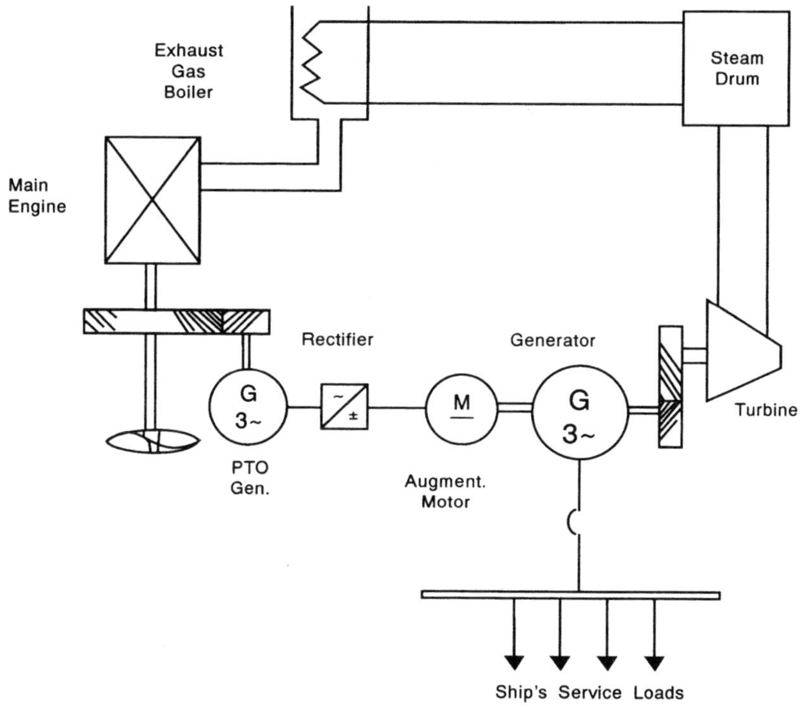

Figure 22-10. Augmented generator drive

Hazardous Areas

Commercial practice classifies hazardous areas into three zones: *Zone 0,* in which an explosive gas-air mixture is continuously present or present for long periods. Interior spaces of cargo oil tanks would be an example of a Zone 0 area. *Zone 1* areas would include enclosed or semi-enclosed spaces on the deck of a tanker, battery rooms, and in general any space where an explosive gas-air mixture is likely to occur during normal operation. *Zone 2* areas, such as open spaces on the deck of a tanker, are those where an explosive gas-air mixture is not likely to occur during normal operation or, if it occurs, will be of short duration. Areas not classified as Zone 0, 1, or 2 are considered to be nonhazardous or "safe" areas. Current maritime practice is to simply designate areas as "dangerous spaces" or "normally safe" as determined by the nature of the flammable cargo of the tanker; however, electrical equipment is manufactured on the basis of such zones. Tankers are classified as Type A, B, C, or D and for each classification specific areas are designated as dangerous spaces. *Type A* tankers carry nonboiling oil cargoes having a flash point of 60° or less. *Type B* tankers transport nonboiling oil cargoes with a flash point in excess of 60°C. *Types C and D* tankers are LPG or LNG carriers and chemical carriers respectively.

Flammable atmospheres in which electrical equipment may have to operate are grouped according to the spark energy required to ignite the gas. Gasses associated with the mining industry are classified as *Group I;* all other industrial gasses are described as *Group II* with *IIA, IIB,* and *IIC* as subcategories. Equipment certified for use in a Group IIC (hydrogen) environment may be used in Group IIB and IIA atmospheres as well, but Group IA certified equipment may not be used in any other group.

Although the terms "flameproof" and "explosion-proof" have been used for some time, and the construction techniques associated with flameproof enclosures, for example, are well established, more recent developments in electronics and materials have led to an internationally agreed code for identifying explosion protective equipment. The following is a list of the types of protection and brief explanations. The types are designated by the symbol "Ex" followed by a code letter.

Exd: Flameproof Enclosure. A flameproof enclosure will withstand, without damage, the ignition of the flammable gas within itself, and prevent transmission of any flame that could ignite the gas in the surrounding atmosphere.

Exi: Intrinsically Safe. These are circuits from which no spark or any thermal effect produced under a prescribed test is capable of causing ignition of a given explosive atmosphere. Generally this means limiting the circuit conditions to less than 30V and 50 mA. Cable for intrinsically safe circuits aboard ships should be separated from power cables to prevent magnetic induction from compromising the intrinsically safe circuit, and metallic sheaths surrounding intrinsically safe cabling should be grounded at the supply end only.

Exe: Increased Safety. This was originally a German construction specification calling for tight tolerances on the control of surface temperatures and the elimination of "open sparking" at relay and switch contacts, for example.

Exn: Non-Sparking. This is the British version of Exe, but less rigorous and therefore more restricted in its application.

Exo: Oil Immersed. Equipment is placed in oil-filled tanks to isolate any arc from the surrounding atmosphere.

Exp: Pressurized Enclosure. Air or an inert gas is supplied to the equipment at slightly above atmospheric pressure to prevent entry of a hazardous atmosphere.

Exq: Sand Filled. This is similar to the oil immersion concept, except it uses quartz crystals (sand), as in some fuses.

Exs: Special Protection. Precautions other than those described above must be taken to prevent an explosion in a given environment.

Only Exi and Exs (when specifically certified) may be used in Zone 0 areas. Exd, Exi, Exe, Exp, and Exs may be used in Zone 1 and Zone 2 areas. Exn, Exo, and Exq may only be used in Zone 2 areas. For example, a fixture marked "Exd IIC T4" (flameproof, certified for a hydrogen atmosphere, maximum surface temperature 135° C) is acceptable in a battery room.

The United States has no governmental certifying organization. However, Underwriters Laboratory (UL) and Factory Mutual Research Corporation (FM) test equipment and issue listings of approvals based on their published standards. In the United Kingdom, BASEEFA, in Canada, C.S.A., and in Germany, P.T.B. are the national testing and certification authorities.

SYSTEM COMPONENTS

Switchboards and Distribution Panels

Switchboard construction/requirements. Ship's service switchboards are located so as to be accessible front and rear, with the space in front of the switchboard convenient for operation, maintenance, and removal of equipment, and the space in the rear enclosed if it would be accessible to unauthorized personnel. An insulating mat or grating is provided on the deck in the front and rear, a nonconducting handrail is attached to the front of the switchboard, and guardrail protection is provided at the rear. Switchboards are installed in a dry location, away from steam, water, and oil pipes, and fitted with drip-proof covers for the top and protective covers for the ends of the structure.

Electrically, buses and primary connections are arranged so that, for three-phase AC, the phase sequence is A, B, C (and for DC switchboards, positive, neutral, negative), from front to back, left to right, and top to bottom, as viewed from the front of the switchboard.

The following equipment is required for AC generator switchboards:

1. A draw-out or plug-in generator circuit breaker for each generator. Low-voltage, trip-free circuit breakers must include both short-time delay and long-time overcurrent trip units. Medium-voltage circuit breakers need not include a short-time feature if the long-time pickup characteristic provides sufficient protection. When *three* or more generators are capable of being operated in parallel, generator breakers should also include instantaneous trips set in excess of the maximum short circuit contribution of the individual generator. When *two* or more generators are to operate in parallel, each machine must be protected against reverse power flow.
2. An ammeter for each generator with selector switch to read the current in each phase.
3. A voltmeter for each generator with a selector switch to read each phase voltage on the generator and one phase on the bus. One of the voltmeter switches should be capable of providing a reading for each phase of voltage at the shore connection.
4. An indicating wattmeter for each generator.
5. A temperature detection instrument with a selector switch for each generator rated 400 kW or above.
6. A pilot lamp permanently connected to the source side of each generator circuit breaker.
7. Control for each prime mover speed.
8. A synchroscope and synchronizing lamps with selector switch to provide for paralleling in any combination. These must be located so as to be readily visible from the position from which the operator controls the incoming generator during paralleling.
9. A voltage regulator (unless internal with the exciter), complete with all accessories for each generator: a switch for cutting out the regulator and transferring from automatic to manual control mode for each generator, an automatic control rheostat for each generator, and a manual control rheostat for each generator. Voltage adjusting rheostats must be in the vicinity of their respective voltmeters.
10. Ground detection means for each bus of the switchboard: for ungrounded systems, three ground lamps and a closed spring return-to-normal switch or push button in the ground connection; for grounded systems, a ground current ammeter, ammeter switch, and current transformer if required, in the ground connection circuit.

Group control panels. Grouped motor control, or motor control centers, provide a means for centralizing motor control and related control equipment. Made up of standardized vertical sections that house vertical and main horizontal bus, vertical wiring troughs, and compartmentized control units (Figure 22-11), the sections are bolted together to form a single line-up. Each section may contain all controller units, all distribution units, or a combination of both. Controller units contain a circuit breaker or fused load switch, main and auxiliary contactors for across-the-line, reduced voltage, reversing, and/or multispeed starter types, overload relays, indicating lamps, ammeter, push buttons, etc. Distribution units contain a circuit breaker and possibly interlocking relays, and metering and other miscellaneous devices.

The distribution/control groups may form an integral part of the main switchboard or a separate group starter board, and sectionalizing may be obtained with circuit breakers, load switches, or disconnect links. Circuit breakers are normally used when operation with a separate busbar system, such as a diesel generator and a shaft generator busbar system, is desired;

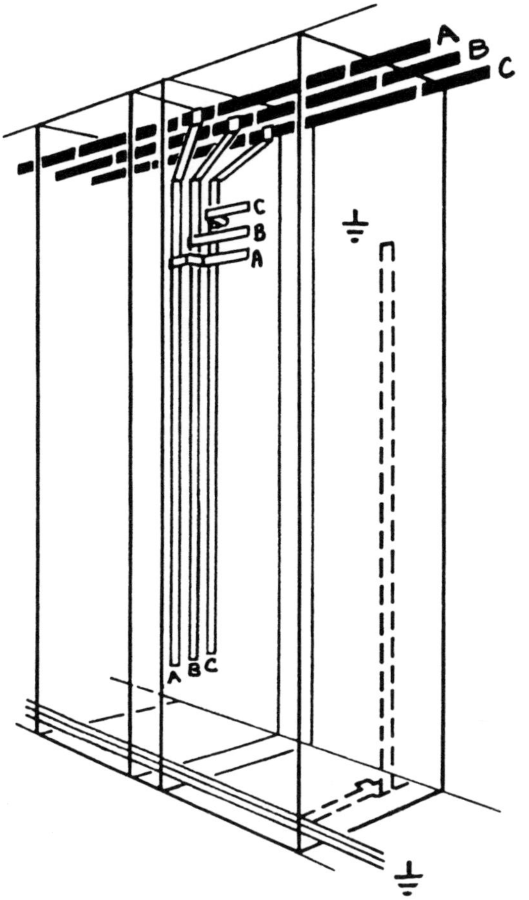

Figure 22-11. Group-control section

or when large motors are to be run solely on one generator; or for the purpose of reducing the available short circuit current. Load switches, often motor operated, are used in automated preferential tripping schemes, where a number of nonessential loads may be dumped simultaneously in the event of generator overload. Disconnectors are typically used in the horizontal busbar system for isolating one part of the switchboard when it is desired to service it under no-voltage conditions.

The National Electrical Manufacturers Association (NEMA) classifies control centers by the degree of control system engineering required. NEMA Class I motor control centers are essentially a mechanical grouping of combination motor control, feeder tap, and/or other units arranged in a convenient assembly. They include connections from the common horizontal bus to the units but not interwiring or interlocking between units or to remotely mounted devices. Only diagrams of the individual units are required to be supplied by the manufacturer. Class II motor control centers consist of a grouping of units to form a complete control system. They include the necessary interwiring and interlocking between units and interlocking provisions to remotely mounted devices. The control manufacturer is required to supply a suitable diagram to illustrate operation of the control associated with the motor control center.

Panelboards for distribution to motor, appliance, lighting (including emergency), receptacle, or other branch circuits are fitted with switches that have a rated capacity of not less than 30 amps, with one pole that includes a fuse or other overload device per conductor. The number of branch circuits controlled by a single panelboard should not exceed 18 for three-phase branch circuits or 26 for single-phase or two-wire DC service. Figure 22-12 shows a typical arrangement of single- and three-phase branch circuits in a three-wire, three-phase panel. Figures 22-13 and 22-14 show typical arrangements in four-wire, three-phase and three-wire, single-phase panelboards respectively. As with switchboards, buses and primary connections should be arranged so that for three-phase assembled distribution panels and panelboards the phase sequence is A, B, C, from front to back, top to bottom, and left to right as viewed from the front of the panel. Panelboards not exposed to moisture, and particularly flush-mounted panelboards in way of joiner work, may be of UL standard construction with cables entering the bottom and sides through bushings or clamps and through terminal tubes on the top. Elsewhere, panelboards must be of drip-proof or watertight construction as required by location.

Marine Cabling

Nomenclature. A variety of cable constructions, designed to meet the recognized special environmental, installation, and reliability

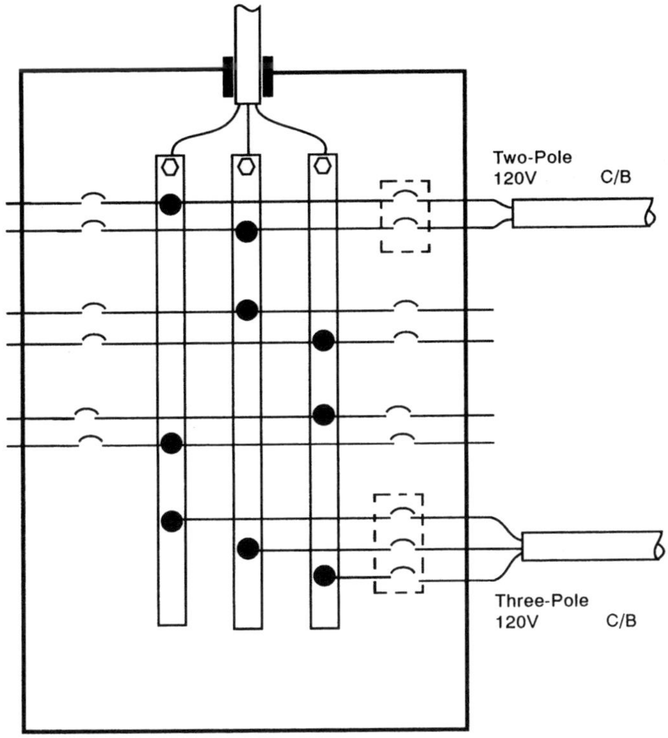

Figure 22-12. Three-wire, three-phase 120V panelboard

requirements of marine service, are found on shipboard. The ship's engineer should be familiar with the various conductor insulations, jacket types, ampacities, and designation schemes when making repairs and installing new equipment.

In general, marine cabling is identified by a series of letters and numbers signifying the type of service, number of conductors and conductor size, and types of insulation, moisture-resistant jacket, and armor as follows:

Cable type (service)	S	Single conductor—distribution
	D	Double conductor—distribution
	T	Three conductor—distribution
	F	Four conductor—distribution
	C	Control cable
	TP	Signal (twisted pair)
	TPS	Signal (twisted pair, shielded)

Insulation	E	Ethylene propylene rubber
	X	Cross-linked polyethylene
	T	Polyvinyl chloride (PVC)
	GTV	Impregnated glass varnished cloth
	M	Mineral
	S	Silicone
Jacket	I	Moisture-resistant jacket (type T: thermoplastic PVC) (type N: thermosetting polychloroprene) (type CP: thermosetting chlorosulfonated polyethylene)
Armor	A	Aluminum
	B	Bronze
		Unarmored

Distribution cables of 600V are recommended for the supply power equipment and lighting circuits up to their voltage rating. Occasionally these are also used in communication, control, and electronic circuits as well. In

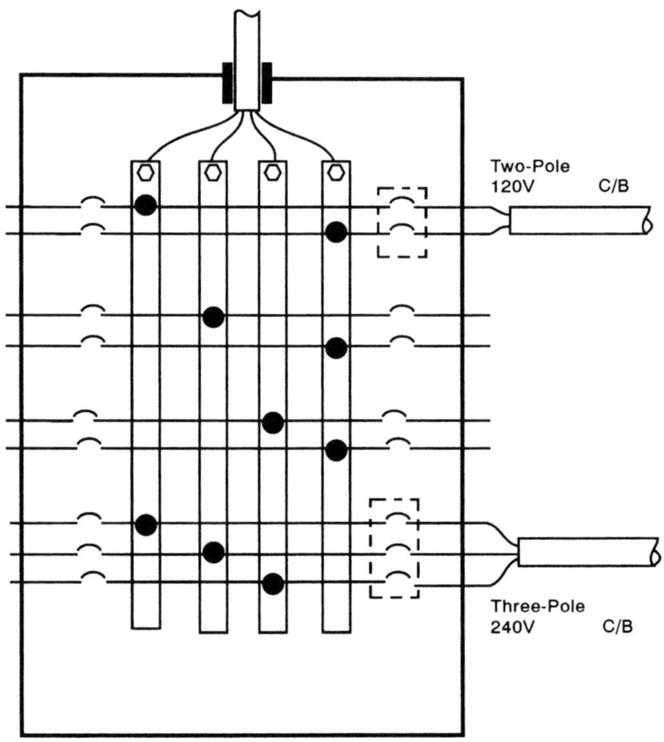

Figure 22-13. Four-wire, three-phase 208/120V panelboard

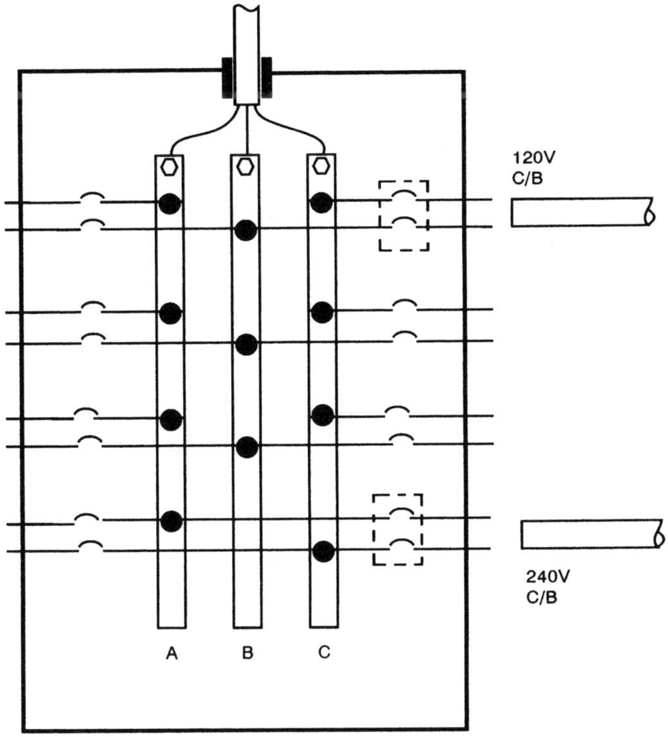

Figure 22-14. Three-wire, single-phase 240/120V panelboard

general, multiple-conductor cable is installed for all AC lighting and power circuits (packaging all current-carrying conductors of any circuit in a single cable tends to neutralize inductive effects). Where the rating of a circuit exceeds the rated capacity of any one multiple-conductor cable (see Table 22-1), two or more multiple-conductor cables may be connected in parallel (banked). In this arrangement one conductor of each phase should be contained in each cable. For relatively short runs of very high capacity circuits, single-conductor cables may be installed utilizing one or more cables per phase. This is often the case in the run from generator to switchboard.

The use of armored cable is typical, although unarmored cable may be used provided special precautions are taken during installation. Excessive pulling force, sharp bends, abrasion against metal surfaces or other cables in wireways may cause damage, particularly to the watertight jacket. Armor (or metal sheath in the case of mineral-insulated, metal-sheathed cable) should be continuous from outlet to outlet and grounded at each end. Final subcircuits, which may be grounded at the supply end only, are an exception. Although splicing is permitted under certain conditions per

TABLE 22-1*
Distribution Cable Data

Distribution Cables
Single Banked
Maximum Current-Carrying Capacity
(Types T, E, X, S, and GTV — 45 °C Ambient)

			Single-Conductor Cable			Two-Conductor Cable			Three-Conductor Cable		
AWG	mm²	Circular Mils	T 75 °C	E, X 90 °C	S, GTV 100 °C	T 75 °C	E, X 90 °C	S, GTV 100 °C	T 75 °C	E, X 90 °C	S, GTV 100 °C
14	2.1	4410	28	34	37	24	29	31	20	24	25
12	3.3	6530	35	43	45	31	36	40	24	29	31
10	5.3	10 400	45	54	58	38	46	49	32	38	41
8	8.4	16 500	56	68	72	49	60	64	41	48	52
7	10.6	20 800	65	77	84	59	72	78	48	59	63
6	13.3	26 300	73	88	96	66	79	85	54	65	70
5	16.8	33 100	84	100	109	78	92	101	64	75	82
4	21.1	41 700	97	118	128	84	101	110	70	83	92
3	26.7	52 600	112	134	146	102	121	132	83	99	108
2	33.6	66 400	129	156	169	115	137	149	93	111	122
1	42.4	83 700	150	180	194	134	161	174	110	131	143
1/0	53.5	106 000	174	207	227	153	183	199	126	150	164
2/0	67.4	133 000	202	240	262	187	233	242	145	173	188
3/0	85.0	168 000	231	278	300	205	245	265	168	201	218
4/0	107.2	212 000	271	324	351	237	284	307	194	232	252
	127	250 000	300	359	389	264	316	344	217	259	282
	152	300 000	345	412	449	296	354	385	242	290	316
	177	350 000	372	446	485	324	387	421	265	317	344
	203	400 000	410	489	533	351	419	455	286	342	371
	253	500 000	469	560	609	401	479	520	329	393	428
	271	535 000	485	579	630	415	496	538	340	407	443
	304	600 000	521	623	678	450	539	585	368	440	478
	327	646 000			715						
	380	750 000	605	723	786	503	602	656	413	494	537
	394	777 000			804						
	507	1 000 000	723	867	939						
	562	1 110 000			1003						
	633	1 250 000	824	990	1072						
	706	1 500 000	917	1100	1195						
	1013	2 000 000	1076	1292	1400						

			DC Ratings		
			75	90	100
		750 000	617	738	802
		1 000 000	747	896	964
		1 250 000	865	1038	1126
		1 500 000	980	1177	1276
		2 000 000	1195	1435	1557

NOTES: (1) Current ratings are for ac or dc except for sizes one-conductor 750 000 circular mils and larger.

(2) For service voltage 601 V to 5000 V, Type T should not be used.

(3) Current-carrying capacity of four-conductor cables where one conductor is neutral, is the same as three-conductor cables listed in Table A6.

(4) The above values are based on ambient temperatures of 45 °C and maximum conductor temperatures not exceeding 75 °C for type T insulated cables, 90 °C for types X and E insulated cables, and 100 °C for types GTV and S insulated cables.

(5) If ambient temperatures differ from 45 °C the values shown above should be multipled by the following factors:

Ambient Temperature	40 °C	50 °C	60 °C	70 °C
Type T insulated cables	1.08	0.91	—	—
Type X and E insulated cables	1.05	0.94	0.82	—
Type GTV and S insulated cables	1.04	0.95	0.85	0.74

(6) The above current-carrying capacities are for marine installations with cables arranged in a single bank per hanger and are 85% of the ICEA calculated values [see Note (7)]. Double banking of distribution-type cables should be avoided. For those instances where cable must be double banked, the current-carrying capacities in the above table should be multiplied by 0.8.

(7) The ICEA calculated current capacities of these cables are based on cables installed in free air, that is, at least one cable diameter spacing between adjacent cables. See IEEE Std-135-1962 [31].

regulatory body restrictions, all cables should be continuous between terminations.

The designation of distribution cables includes, in addition to cable symbol, the conductor size in thousandths of circular mils (MCM). Some examples:

TXIA-66: AWG #2 (66,400 CM), three-conductor, X-linked polyethylene insulated, moisture-resistant jacket, with aluminum armor.
(6)SEIB-300: six (two per phase), 300,000 CM, single-conductor, ethylene propylene rubber-insulated, moisture-resistant jacket, with bronze armor.

For "medium voltage" cables (2000V or 5000V) a "-2kV" or "-5kV" suffix is added to the designation (e.g., TXIB-5kV-83).

One further note: On many "one-line diagrams" found among the available ship's blueprints, a note may be found stating ". . . all cables type XIB unless otherwise indicated." In this case cable designations may be abbreviated T-16 or D-4, indicating number of conductors and cable size only. It is advisable to avoid using terms like "number four cable" as it may lead to confusion between 4110 CM (AWG #14) and AWG #4 (41,700 CM) cables.

Control and signal cable. Rated at 600V (see Table 22-2), *control cables* are used for control, indicating, communication, electronic, and similar circuits where multiple parallel conductor cables are required. Particular attention should be paid to voltage drop considerations when selecting control cabling.

Signal cable, rated at 300V, is used for signal transmission where twisted pairs of conductors are desired. Twisted pair signal cable comes shielded or unshielded (see Table 22-3) and the shielded variety includes a drain (ground) wire for each twisted pair.

In designating control or signal cable, the conductor size (in AWG#, not MCM) is embedded in the cable symbol following the service designator. The suffix is the number of conductors (for control cable) or number of conductor pairs (for signal cable). Hence:

C14TIB-19: AWG #14, 19-conductor control cable, thermoplastic insulated, with moisture-resistant jacket and bronze armor.
TPS18TIA-22: AWG #18 signal cable, 22-twisted pairs (shielded), thermoplastic insulated, with moisture-resistant jacket and aluminum armor.

Designation as to application of cabling in signal and communications systems is accomplished by utilizing the system designators (per Table 22-4) as well as cable type designations. In both control and signal cable, the individual conductors are color coded (see Table 22-5).

TABLE 22-2*

600V, Multi-Conductor Control Cable Data

600V Multiconductor Control Cable

	Conductor Size AWG 14				Conductor Size AWG 16				Conductor Size AWG 18			
Number of Conductors	Max Diam (in)	Weight (lbs per 1000 ft)	Ampacity Type T Conductors 75 °C	Ampacity Type X or E Conductors 90 °C	Max Diam (in)	Weight (lbs per 1000 ft)	Ampacity Type T Conductors 75 °C	Ampacity Type X or E Conductors 90 °C	Max Diam (in)	Weight (lbs per 1000 ft)	Ampacity Type T Conductors 75 °C	Ampacity Type X or E Conductors 90 °C
2	0.50	155	13	17	0.46	140	11	14	0.44	115	8	11
3	0.52	175	11	14	0.48	145	8	11	0.46	125	7	9
4	0.56	200	9	11	0.52	170	7	9	0.49	145	6	7
7	0.68	295	8	10	0.60	225	6	8	0.56	190	5	7
10	0.82	410	8	10	0.75	330	6	8	0.71	280	5	7
14	0.89	500	8	10	0.81	395	6	8	0.76	325	5	7
19	1.02	670	8	10	0.89	485	6	8	0.83	395	5	7
24	1.16	815	8	10	1.06	640	6	8	0.94	485	5	7
30	1.22	945	7	8	1.11	735	5	7	1.04	600	4	6
37	1.31	1120	7	8	1.19	860	5	7	1.11	690	4	6
44	1.46	1300	6	7	1.31	1000	4	6	1.22	805	4	5
61	1.61	1680	6	7	1.47	1270	4	6	1.34	1010	4	5

Notes: (1) Weights given are for cables with Type T insulated conductors. Those with Type X or E insulated conductors will be from three to five percent lighter depending on the number of conductors.

(2) Ampacities are average current capacities for all conductors in the cables. No individual conductor should be permitted to carry more than 1.5 times these values.

(3) All ampacities are for double-banked cables in trays, 45 °C ambient temperature.

(4) For ambient temperatures other than 45 °C, the ampacity values shown should be multiplied by the factors in Table A6, Note (5).

*Reprinted from ANSI/IEEE Std 45-1983, © 1983 by The Institute of Electrical and Electronics Engineers, Inc., with permission of the IEEE Standards Department.

TABLE 22-3*

One-Line Diagram, Lighting Circuits

		Pairs Shielded				Pairs Unshielded			
No of Pairs	No of Con-ductors	Maximum Diameter (in)			Approximate Weight (lbs/1000 ft)	Maximum Diameter (in)			Approximate Weight (lbs/1000 ft)
AWG#		16	18	20	16	16	18	20	16
1	2	0.44	0.42	0.40	125	0.44	0.42	0.39	125
2	4	0.72	0.67	0.60	260	0.59	0.56	0.52	195
3	6	0.76	0.71	0.66	295	0.68	0.60	0.56	245
4	8	0.82	0.76	0.71	345	0.77	0.72	0.67	295
5	10	0.89	0.83	0.77	405	0.79	0.74	0.68	345
6	12	1.01	0.90	0.83	505	0.82	0.76	0.71	370
7	14	1.01	0.90	0.83	500	0.82	0.77	0.71	385
8	16	1.09	1.01	0.89	560	0.91	0.84	0.78	440
9	18	1.16	1.08	1.00	630	1.03	0.91	0.85	540
10	20	1.25	1.15	1.06	690	1.06	0.93	0.86	560
11	22	1.25	1.15	1.06	720	1.08	1.00	0.88	590
12	24	1.28	1.19	1.10	755	1.10	1.02	0.90	625
13	26	1.32	1.21	1.12	800	1.14	1.05	0.93	660
14	28	1.35	1.24	1.15	835	1.18	1.09	1.01	710
15	30	1.40	1.28	1.18	890	1.22	1.12	1.04	740
16	32	1.42	1.31	1.21	920	1.22	1.13	1.04	770
17	34	1.48	1.36	1.25	985	1.23	1.14	1.05	800
18	36	1.50	1.38	1.27	1030	1.27	1.17	1.08	830
19	38	1.50	1.38	1.27	1020	1.29	1.19	1.10	865
20	40	1.55	1.43	1.31	1070	1.32	1.22	1.12	895
21	42	1.57	1.45	1.33	1110	1.35	1.24	1.14	930
22	44	1.63	1.50	1.38	1160	1.37	1.26	1.16	960

NOTE: Ampacities for signal cable conductors should not exceed 3 A for cables installed double-banked in a 45 °C ambient temperature.

"Navy" Cable. Where the above cable designations, dimensions, and ampacities are often referred to as "IEEE" cable, occasionally cable manufactured and tested in accordance with MIL C-915-19 is installed. Use of such "Navy" cable leads to designators such as:

TSGA-60: 60,090 CM, three-conductor, shipboard, general use, (aluminum) armored cable.

Navy cable is silicon insulated and contains an impervious sheath. In addition, Navy cable sizes do not match IEEE sizes exactly (e.g., T-60 versus T-66) and are generally somewhat smaller and lighter per unit length for the same ampacity. Consequently, it is fairly common to find D-9 Navy cable utilized in lieu of D-10 IEEE cable in certain lighting branch circuits.

Adding to the alphabet soup is the occasional use of special-purpose Navy cable; for example:

TTRSA-12: (Twelve) twisted-pair, telephone, radio, shielded, armored cable.

*Reprinted from ANSI/IEEE Std 45-1983, © 1983 by The Institute of Electrical and Electronics Engineers, Inc., with permission of the IEEE Standards Department.

MHOF-24: Multiple-conductor, heat- and oil-resistant, flexible cable (24 conductors).

SHFS-2.5: Single-conductor, heat- and flame-resistant, switchboard cable (2613 CM).

TABLE 22-4

Signal and Communication Systems Designations

Code	*System*
A	Call bells
AC	Autocombustion control
AN	Nurse's call light indicator
BL	Boiler remote water level
CE	Electric clocks
CO	CO2 release alarm
D	Docking telegraph
DA	Docking announcing
DE	Electric door control (other than watertight doors)
E	Sound-powered telephone and voice tube call
EC	Lubricating oil low pressure alarm
EL	Lubricating oil low level alarm
ES	Echo depth sound
EW	Diesel circulating water high temperature alarm
F	Fire alarm
FA	Fuel oil high level alarm
FB	Fog alarm
FD	Fire door release
FS	Fuel oil filling signal system
FT	Resistance thermometers
FW	Feedwater low level alarm
G	General alarm
GB	Burglar alarm
HB	Anchor telegraph
HD	Wind direction indicator
HE	Wind intensity indicator
J	Telephone, manual or automatic
1JV	Telephone, sound powered, ship control
2JV	Telephone, sound powered, engineer's
3JV	Telephone, sound powered, miscellaneous
K	Shaft revolution
L	Steering telegraph

TABLE 22-4 (continued)

Code	*System*
LA	Steering gear alarm
LC	Gyrocompass
LP	Automatic pilot
MB	Engine order telegraph
MC	Emergency announcing
ME	Boiler telegraph
MP	Motion pictures
MT	Turbogenerator telegraph
N	Rudder angle indicator
PB	Pyrometer
R	Radio
RA	Refrigerator alarm (stores)
RB	Radio broadcast receiving antenna distribution
RE	Radio broadcast receiving and entertainment
R-ER	Radar navigation
RH	Refrigerator alarm (cargo)
SB	Salinity indicator
SM	Smoke indicator
ST	Sewage tank alarm
T	Wet and dry bulb indicator
V	Voice tubes
W	Whistle operator
WD	Watertight door control
X	Tank level indicator
Y	Underwater log

System Protection

Introduction. Reliable circuit protection is automatic protection that minimizes the danger of smoke and fire, not only in the equipment, but in the cabling leading to and from the equipment. Secondly, besides protecting the cabling and equipment, the protector must isolate the fault from the power source so that nonfaulted circuits can continue functioning in a normal manner. These objectives may not always be achieved by a single protective device, and indeed coordination between protective devices, correct sizing of wire, and safe routing of circuits all contribute to a safe circuit protection philosophy.

The entire electrical system, including power source, distribution cabling, switching devices, and equipment, must be adequately protected. Generally a circuit-protecting device is located at any point in the system where the conductor size is reduced, unless the immediate upstream

TABLE 22-5*

Color Code: Twisted-Pair Control and Signal Cables

Conductor number	Base color	Tracer color	Tracer color
1	Black		
2	White		
3	Red		
4	Green		
5	Orange		
6	Blue		
7	White	Black	
8	Red	Black	
9	Green	Black	
10	Orange	Black	
11	Blue	Black	
12	Black	White	
13	Red	White	
14	Green	White	
15	Blue	White	
16	Black	Red	
17	White	Red	
18	Orange	Red	
19	Blue	Red	
20	Red	Green	
21	Orange	Green	
22	Black	White	Red
23	White	Black	Red
24	Red	Black	White
25	Green	Black	White
26	Orange	Black	White
27	Blue	Black	White
28	Black	Red	Green
29	White	Red	Green
30	Red	Black	Green
31	Green	Black	Orange
32	Orange	Black	Green
33	Blue	White	Orange
34	Black	White	Orange
35	White	Red	Orange
36	Orange	White	Blue
37	White	Red	Blue
38	Brown		
39	Brown	Black	
40	Brown	White	
41	Brown	Red	
42	Brown	Green	
43	Brown	Orange	
44	Brown	Blue	

*Reprinted from ANSI/IEEE Std 45-1983, © 1983 by The Institute of Electrical and Electronics Engineers, Inc., with permission of the IEEE Standards Department.

protector provides adequate protection for the smaller wire. Consideration must also be given to the fact that circuit components such as motors, transformers, rectifiers, and electronic circuits have significantly different overload-withstanding characteristics than those of wire and cable. For example, many electronic circuit components require extremely fast clearing devices to provide adequate protection from thermal damage.

Fuses. Cartridge fuses of the nonrenewable type are permitted for shipboard use. Although typically utilized in control and electronic circuitry, they may be used as motor overload protection up to 200A (see Table 22-6). Where fuses may often offer an adequate and low-cost method of protection, there are some troublesome characteristics that mitigate their use in power circuits on U.S. merchant ships. For one thing, the replaceable nature of fuses and the ease of overfusing may result in inadequate protection. Within certain minimum and maximum ratings, fuse dimensions are usually the same; and it is possible to substitute a 20-ampere fuse where a 5- or 10-ampere fuse should be. Single-phasing may be another problem although today some fuse designs incorporate a pin or "target" that is expelled when the fuse blows, actuating a switch trip mechanism to open all phases.

A less well known but still troublesome characteristic is fuse element deterioration, caused by physical and chemical stresses produced in the fuse element during repeated short-duration overloads. During acceleration, motor inrush current, for example, often exceeds the trip rating of the fuse. The inrush current is not present long enough to blow a correctly sized fuse; however, deterioration of the element, resulting from repeated motor starting, often causes mysterious fuse failures.

When semiconductor devices, such as SCRs and power diodes, are subjected to high current overloads, thermal damage occurs proportional to $I^2 T$. As a result, fast-blow, silver-link fuses have been developed with performance characteristics that are similarly $I^2 T$ dependent. Power matching of fuse to semiconductor results in very effective protection. The high speed action of "current limiting" semiconductor fuses comes from a silver link with a small link section joining a substantial size sheet of silver. The silver provides a maximum of thermal conductivity, and short circuit protection is provided when the rate of rise of heat in the small link exceeds the rate of thermal conduction away from the link. As the link melts, the voltage across it rises and arcing begins. Arc quenching is aided by silica sand crystals, which effectively lengthen the arc path. The "clearing time," which includes the melting and arcing periods, and the "peak let-through current" are the most significant parameters of current-limiting fuse action (see Figure 22-15). The voltage rating of the fuse is also important. Sometimes an operator is tempted to apply 250V fuses in 120V circuits in the mistaken belief that this will achieve greater safety. Voltage ratings should not be interchanged, however, since a high voltage rating may provide a less desirable I^2T rating.

TABLE 22-6*

Overcurrent Protection for Motors

	For Running Protection of Motors		Maximum Allowable Rating of Branch Circuit Fuses		
Full-Load Current Rating of Motor	Maximum Rating of Fuses	Maximum-setting of inverse time protective device	Single-Phase and Squirrel-Cage and Synchronous (full-voltage reactor and resistor starting)	Squirrel-Cage and Synchronous (auto-transformer starting) High reactance Squirrel-cage	Direct Current and Wound-rotor alternating Current
Amperes	Amperes	Amperes	Amperes	Amperes	Amperes
1	2	1.15	15	15	15
2	3	2.3	15	15	15
3	4	3.45	15	15	15
4	6	4.6	15	15	15
5	6	5.75	15	15	15
6	8	6.9	20	15	15
7	10	8.05	25	20	15
8	10	9.2	25	20	15
9	12	10.35	30	25	15
10	12	11.5	30	25	15
11	15	12.65	35	30	20
12	15	13.8	40	30	20
13	15	14.95	40	35	20
14	20	16.1	45	35	25
15	20	17.25	45	40	25
16	20	18.4	50	40	25
17	20	19.55	60	45	30
18	25	20.7	60	45	30
19	25	21.8	60	50	30
20	25	23.0	60	50	30
22	30	25.3	70	60	35
24	30	27.6	80	60	40
26	30	29.9	80	70	40
28	35	32.2	90	70	45
30	35	34.5	90	70	45
32	40	36.8	100	70	50
34	40	39.1	110	70	60
36	45	41.4	110	80	60
38	45	43.7	125	80	60
40	50	46.0	125	80	60
42	50	48.3	125	90	70
44	60	50.6	125	90	70
46	60	52.9	150	100	70
48	60	55.2	150	100	80
50	60	57.5	150	100	80
52	60	59.8	175	110	80
54	70	62.1	175	110	90
56	70	64.4	175	110	90
58	70	66.7	175	125	90
60	70	69.0	200	125	90
62	80	71.3	200	125	100
64	80	73.6	200	150	100
66	80	75.9	200	150	100
68	80	78.2		150	110
70	90	80.5		150	110
72	90	82.8		150	110
74	90	85.1		150	125
76	90	87.4		150	125

*Reprinted from ANSI/IEEE Std 45-1983, © 1983 by The Institute of Electrical and Electronics Engineers, Inc., with permission of the IEEE Standards Department.

TABLE 22-6 (continued)

	For Running Protection of Motors		Maximum Allowable Rating of Branch Circuit Fuses		
Full-Load Current Rating of Motor	Maximum Rating of Fuses	Maximum-setting of inverse time protective device	Single-Phase and Squirrel-Cage and Synchronous (full-voltage reactor and resistor starting)	Squirrel-Cage and Synchronous (auto-transformer starting) High reactance Squirrel-cage	Direct Current and Wound-rotor alternating Current
Amperes	Amperes	Amperes	Amperes	Amperes	Amperes
78	90	89.7		175	125
80	100	92.0		175	125
82	100	94.3		175	125
84	100	96.6		175	150
86	100	98.9		175	150
88	110	101.2		175	150
90	110	103.5		200	150
92	110	105.8		200	150
94	110	108.1		200	150
96	125	110.4		200	150
98	125	112.7		200	150
100	125	115		200	150
105	125	120.75			175
110	150	126.5			175
115	150	132.25			175
120	150	138			200
125	150	143.75			200
130	150	149.5			200
135	175	155.25			
140	175	161			
145	175	166.75			
150	175	172.5			
155	200	178.25			
160	200	184			
165	200	189.75			
170	200	195.5			
175		201.25			
180		207			
185		212.75			
190		218.5			
195		224.25			
200		230			
210		241.5			
220		253			
230		264.5			
240		276			
250		287.5			
260		299			
270		310.5			
280		322			
290		333.5			
300		345			
320		368			
340		391			
360		414			
380		437			
400		460			
420		483			
440		506			

NOTE: Fuse ratings in excess of 200 A are not recommended for motor branch circuit protection, in accordance with Section 21, Motor branch circuit overcurrent protection by circuit breakers is covered by 21.33.

Molded-case circuit breakers. Assembled as an integral unit in a supporting and enclosing housing of insulating material, molded-case circuit breakers may be of the adjustable or nonadjustable type. Adjustable breakers are generally factory set, however, in accordance with an overall protection and coordination scheme, and should not be adjusted by the plant operator without careful consideration of the effects on the entire system. The overcurrent and tripping means of molded-case circuit breakers may be of the thermal type, the magnetic type, or a combination of both. In the following example (see Figure 22-16), the overcurrent and tripping mechanisms of a thermal-magnetic, molded-case circuit breaker are described.

Thermal circuit breaker elements (in this case a bimetallic strip) are dependent for actuation upon temperature rise in the sensing element. Deflection of the thermal sensing element causes the circuit to open by rotating the trip latch counterclockwise, activating the common tripper bar. In turn this releases the mechanism latch that separates the contacts. Sufficient deflection to operate the breaker occurs when a predetermined calibration temperature is reached. Temperature rise in the sensing element is caused principally by load current I^2 R heating. The thermal element also integrates heating or cooling effects from external sources, and tends to derate or uprate with ambient temperature. Some models include an ambient temperature compensating element (usually electrically isolated from and independent of the current-carrying thermal trip

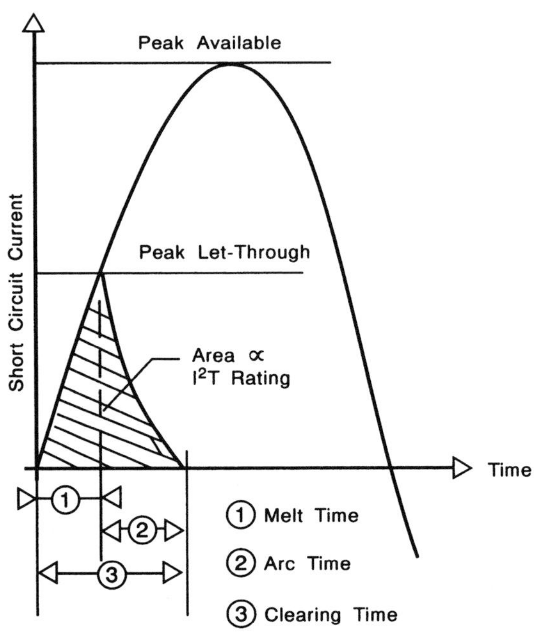

Figure 22-15. Current-limiting fuse action

element), which acts only when a change in ambient temperature occurs. To protect components from high fault level currents, an electromagnet mechanism is added to cause faster tripping. During a short circuit fault,

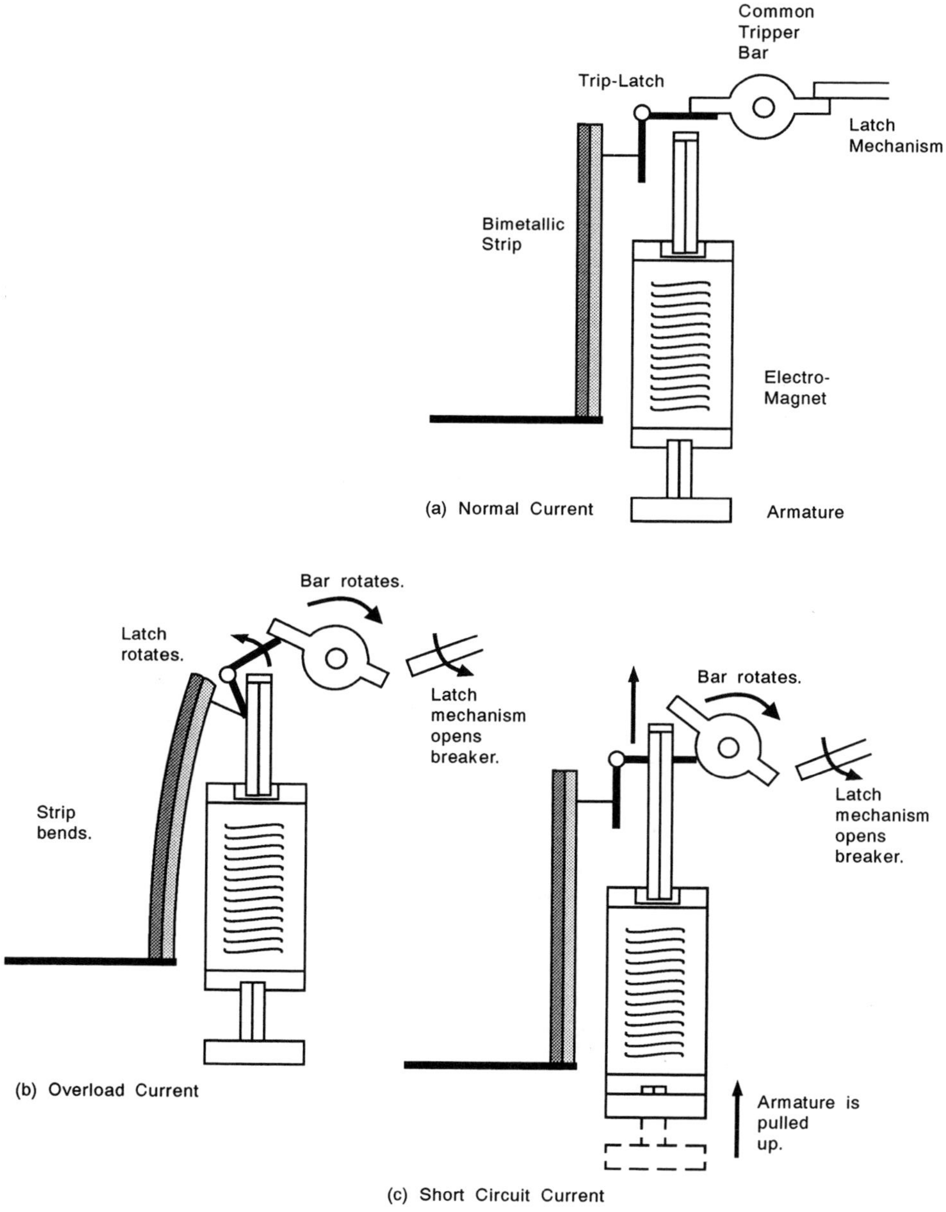

Figure 22-16. Thermal-magnetic circuit-breaker operation

current passing through the circuit causes the electromagnet to attract its armature, the common tripper bar being rotated in this instance by the upward movement of the armature plunger. The "instantaneous" (magnetic) trip levels are typically 10 times (1,000 percent) the nominal breaker rating and often 2,000 percent for breakers rated at 20 A or less.

A typical tripping curve for a molded-case circuit breaker with thermal and magnetic trip elements is shown as Figure 22-17. It shows how the time to trip the breaker changes as the overload current increases. Modest overloads of 15 to 20 percent can take many minutes to cause a breaker trip. Overloads of 1,000 percent (10 times rated) will cause a trip in fractions of a second to a few seconds. Higher overloads will cause a trip in as few as 2 cycles (0.033 seconds). In larger molded-case circuit breakers, the thermal element setting can be adjusted to calibrate the long-time (overload) trip setting and delay time, and the magnetic element setting can be adjusted to calibrate the short-time (short circuit) trip setting. Some also include adjustments for the short delay time and, if the breaker is fitted with one, for the setting of the instantaneous current limiter. Adjustment ranges vary from manufacturer to manufacturer and are a function of rating of the breaker. The long-time current trip setting can typically be adjusted from 0.5 to 2 times rated current while the delay time can be adjusted from 3 to 40 seconds. The short-time current trip setting can be adjusted from 2 to 12 times rated current while the delay time can be adjusted from 0.08 to 0.5 seconds. The setting of the instantaneous current limiter can typically be adjusted from 2 to 15 times rated. Smaller molded-case circuit breakers are not adjustable.

Some molded-case breakers include a magnetic time-delay mechanism in lieu of the thermal apparatus. In these units a movable solenoid core is held with a spring in a tube, and damped with a fluid. As the magnetic field of a series coil moves the core toward a pole piece, the reluctance of the magnetic circuit containing the armature is reduced. The armature is then attracted, causing the mechanisms to trip and open the contacts on an overload condition. Typically the minimum current that will provide a reliable trip of the breaker is 115 percent of the rating, independent of ambient temperature. The instantaneous trip current is that value of current required to trip the circuit breaker without causing the core to move inside the tube. This is possible because excessive leakage flux in the magnetic circuit, caused by extreme overloads or faults, will attract the armature and trip the breaker immediately. The instantaneous trip point is also independent of ambient temperature.

Instantaneous trip circuit breakers have no intentional time delay, and may be thought of as the thermal-magnetic type described above with the thermal sensing element removed. These are utilized where overload protection is available in another form (such as overload relays incorporated in motor control circuits) and so is not required in the circuit breaker.

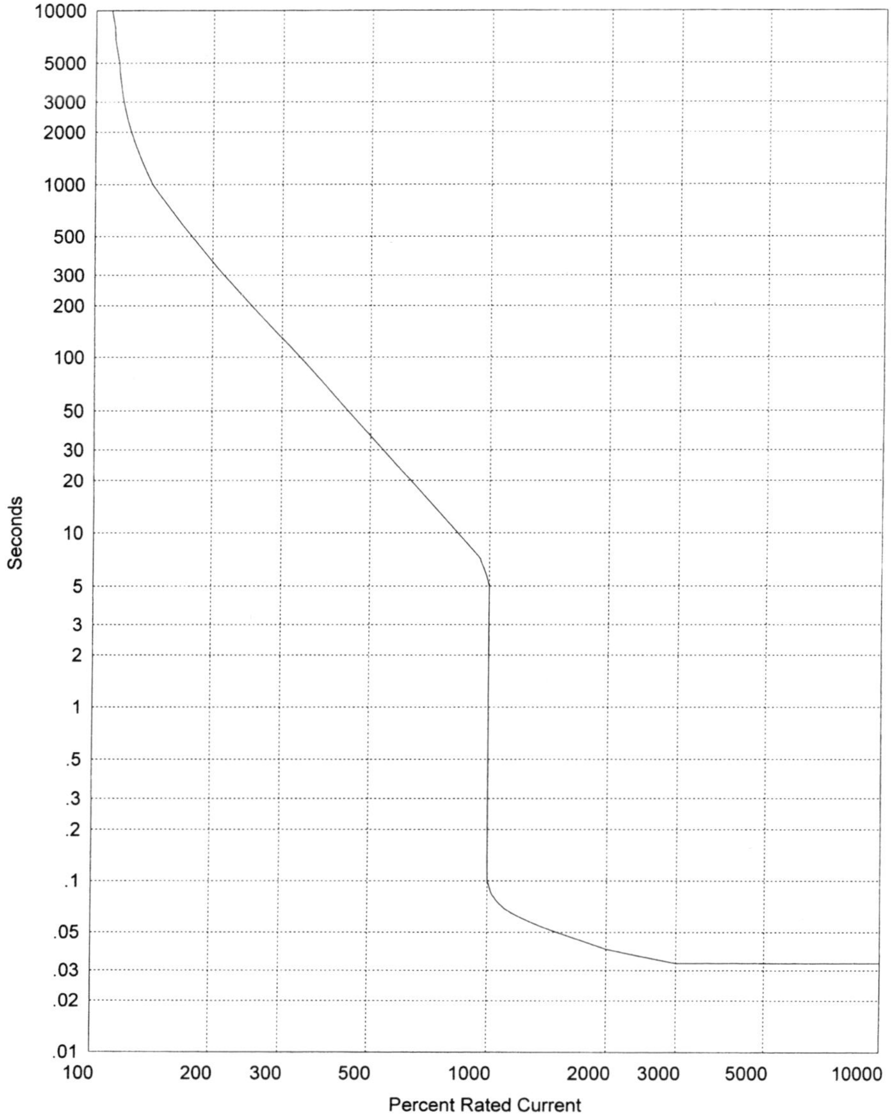

Figure 22-17. Typical circuit breaker tripping curve

Steering gear motor supply feeders, for example, are required by U.S.C.G. regulations to be protected by instantaneous trip elements (set at 200 percent of the motor locked rotor current) only. Note that in steering gear applications, overcurrent relays trigger an alarm, but do not trip the breaker.

Solid-state molded-case circuit breakers are available today, and do not contain a bimetallic strip or electromagnetic apparatus. Instead they employ a current transformer and electronic circuitry that compares sensed current load with a preprogrammed time-inverse characteristic.

Metal-clad switchgear. Circuit breaker protection involves four major components: (1) A sensing device, such as a current transformer, which monitors the quantity to be controlled; (2) a relay that interprets the sensing device signal and initiates power circuit breaker response; (3) an actuating mechanism to open and close contacts; and (4) a main interrupter, including arc controlling. A metal-enclosed low voltage air circuit breaker switchgear, as the name indicates, is a metal-enclosed assembly containing air circuit breakers, together with bare buses, connections, control wiring, and accessory devices.

In shipboard use, metal-clad air circuit breakers are reserved for larger size (amperage) applications and are likely to have separate relays and sensing devices, adding to their flexibility. They may have independently adjustable long-time delay (LTD), short-time delay (STD), and instantaneous trip settings to customize the time-current characteristics of a particular breaker for coordination purposes. Breaker accessories may also include under-voltage relays, phase-balance relays, reverse-power relays, mechanical-electrical interlocks, and load-shedding relays. Air circuit breakers may be operated manually or electrically (remote operation) and are mounted in individual metal compartments. They may be of the stationary type (rigidly mounted on bases within the enclosure) or the draw-out type. Draw-out type switchgear contains breakers so arranged that they may be easily withdrawn from or inserted into their housing. As breakers are removed or inserted, they are automatically disconnected from or connected to their power and control circuits by means of self-aligning contacts. Interlocking is provided so that breakers cannot be removed or inserted when they are closed.

U.S.C.G. regulations require that all shipboard circuit breakers be of the "trip-free" variety, that is, breakers whose poles cannot be maintained closed when carrying overload current that would normally trip the breaker to the open position.

Transformers

Construction and operation. A simple means for providing systems with multiple voltage levels is one of the great advantages of AC distribution. The device that facilitates this arrangement is the transformer. Containing no moving parts, transformers require little maintenance and the larger capacity units operate at efficiencies in the 98 percent range at full load. Two-winding transformers consist of two coils, electrically isolated, wound around a common laminated iron core. The winding connected to an alternating voltage source is called the *primary*. The winding connected to the load circuit is called the *secondary.* When the primary winding is energized, an alternating magnetic flux is created that circulates in the core and cuts the conductors of both the primary and the secondary windings. Since the turns of both windings are cut by the same flux, the emf produced in each turn of the primary and secondary windings is the same. The ratio of primary to secondary voltages, therefore, is equal to the ratio of primary to secondary coil turns. That is:

$$\frac{V_1}{V_2} = \frac{N_1}{N_2}$$

It follows that when the winding with the greater number of turns is connected as the primary, a lower voltage is produced at the secondary terminals. In this mode of operation the transformer is called a *step-down* transformer. With the fewer-turns winding connected to the supply voltage source, the voltage produced at the secondary terminals is greater than the primary voltage. Such a connection is termed a *step-up* transformer. The standard transformer terminal markings are H1, H2 . . . for the high voltage windings and X1, X2 . . . for the low voltage windings. Any transformer may be connected as either a step-up or a step-down transformer, depending on the application, as long as the voltage ratings are consistent with the application. In shipboard systems, the generated voltage is typically the highest voltage level, and thus step-down applications predominate.

Under no-load (secondary circuit open circuited) conditions, only a small amount of ("magnetizing") current flows in the primary winding. This is because the self-induced voltage, or "counter-emf" in the primary winding is practically equal to the applied voltage. Transformer heating under no load is primarily due to hysteresis and eddy-current phenomena in the core; thus the small amount of power consumed under no-load conditions is called *core loss.* When a load is connected to the secondary windings, a secondary current is produced that is inversely proportional to the load impedance. The secondary coil current, in accordance with *Lenz's law,* opposes the action that caused the (secondary) emf to be produced, and tends to reduce the flux circulating in the transformer core. However, the resultant decrease in counter-emf allows more primary current to flow, restoring the core flux to the original value. Since in the magnetic circuit the primary ampere turns (I_1N_1) are (approximately) equal to the secondary ampere turns (I_2N_2), it follows that the ratio of primary to secondary currents is inversely proportional to the turns ratio. So, since

$$\frac{I_1}{I_2} = \frac{N_2}{N_1} = \frac{V_2}{V_1}$$

it would seem that the volt-ampere output of the transformer is equal to the volt-ampere input. Indeed, this is a very close approximation, and useful in determining primary/secondary voltage and current relationships in practice. For example, the output rating or capacity of a transformer is given in kilovolt-amperes (kVA). This value, along with the voltage ratings of the transformer, may be used to determine the maximum primary and secondary currents the transformer can carry without exceeding the allowable temperature rise (at a given frequency). A 480V/120V, 30 kVA transformer, when operating at full load, will carry 62.5 A (30,000/480) in the high voltage winding and 250 A (30,000/120) in the low voltage winding. Note that the power factor of the load determines the *kilowatt* output of the transformer at any percent of (kVA) load.

In actual transformers, the relationships differ slightly from the above ideal characteristics. As already mentioned, the presence of magnetizing current leads to a reduction of efficiency to below 100 percent by introducing core losses. Also, the resistance of the conductors making up the primary and secondary windings gives rise to copper losses. Transformer efficiency is determined by dividing output kilowatts by input kilowatts. The input is the sum of the output plus the total losses (core plus copper). When operated at or near rated voltage, the core losses are constant (independent of load); the copper losses, however, vary with the square of the load. Hence transformer efficiency varies with percent load as well as load power factor according to the following formula:

$$\text{Percent efficiency} = \frac{k \times \text{kvar} \times \text{pf}}{k \times \text{kvar} \times \text{pf} + P_{core} + k^2 \times P_{cu}} \times 100\%$$

where kvar = rated kVA

k = percent of full load: kVA_{act} / kVA_{rated}

pf = load power factor

P_{core} = core losses (at rated voltage)

P_{cu} = core losses (at rated current)

"Percent impedance," which incorporates the effects of flux leakage (leakage reactance) as well as winding resistance, is found among the nameplate data on transformers. It is a value determined by the manufacturer by operating the transformer with the secondary shorted and the primary energized with just enough voltage to drive rated current through the transformer windings. The ratio of this "short circuit voltage" to the rated current is the winding impedance (in ohms). It is usually expressed, however, as a percent of some standard, or "base," impedance. This base value is:

$Z_{base} = V_{rated} / I_{rated}$
or $Z_{base} = V_{rated}^2 / (kVA \times 1{,}000)$
hence % impedance = $(Z_{act} / Z_{base}) \times 100\%$

To reduce the number of transformer designs (differing only in voltage rating), manufacturers may divide the high and/or low voltage windings into two identical sections as shown in Figures 22-18(a), (b), and (c). The possibility of connecting the windings in either series or parallel enables the user to obtain either of two optional voltages from the split secondary winding, and thus the design could be suitable for three-wire, single-phase service. If the transformer has a center-tapped secondary as shown in Figure 22-18(a), the secondary is specified as "whole-half" voltage (240-120V, for example). If the secondary coils are physically separated they will work satisfactorily in series or parallel, but not when used in three-wire service,

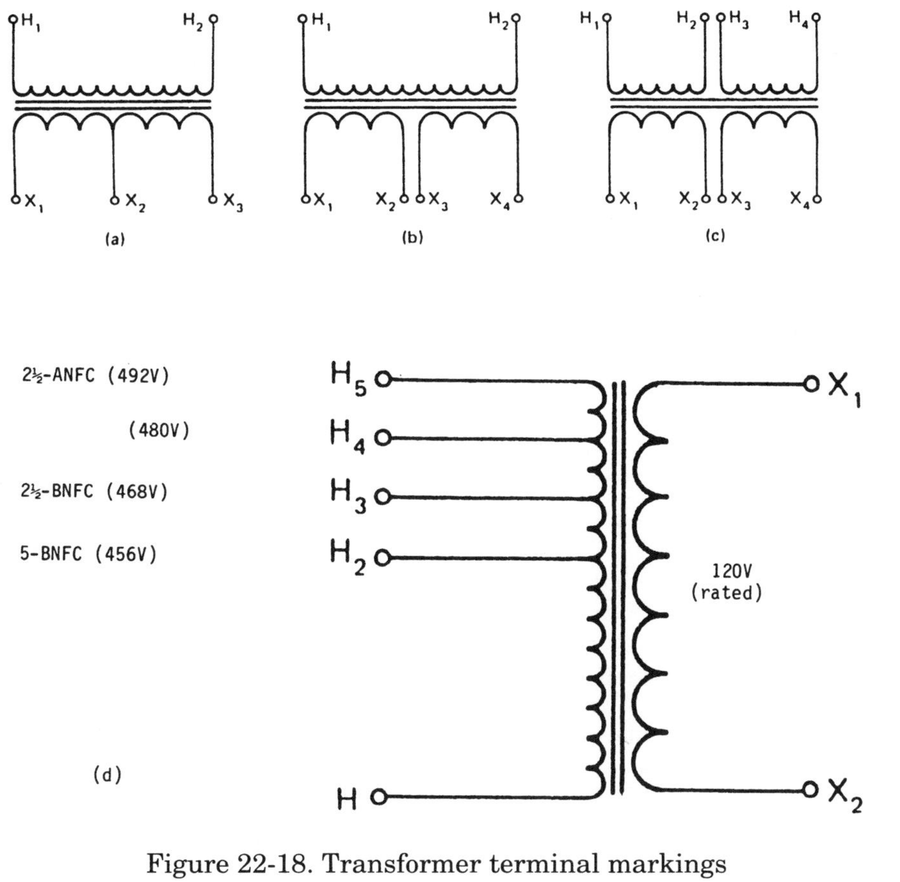

Figure 22-18. Transformer terminal markings

as unbalanced loading will lead to unbalanced voltages at the terminals. For three-wire service the coils must be interweaved with one another, and thus are distributed over approximately the same portion of the magnetic circuit. Transformers specifically designed for three-wire service are slightly more expensive than those intended for only series or parallel connection and are designated in the form "parallel-series" volts (e.g., 120-240V). Transformers not intended for three-wire service are designated using the "parallel × series" format (e.g., 240 × 480V). Two optional supply voltages are made possible by splitting the primary winding. The transformer illustrated in Figure 22-18(c) may be suitable for up to six single-phase arrangements. Standard numbering of dual voltage transformer terminals requires that the lower numbers be simultaneously positive in all coils. This is important when making the series or parallel connections.

Since the available supply voltage does not always match the nameplate rating of the primary, many transformers have taps arranged on the high voltage winding so that rated secondary voltage may be obtained when the primary voltage is less than the nominal rating. Standard distribution transformers have taps arranged in 2.5 percent steps so that rated secondary voltage can be obtained when the primary supply is 0, 2.5, 5, or 7.5 percent below the nominal primary rating. A primary winding with three BNFC (below normal full capacity) taps is shown in Figure 22-18(d). If the primary were rated at 480V, then connecting terminals H_1 and H_2 (7.5 percent BNFC) to a 444V supply would yield a secondary voltage of 120V at full load.

Three-phase connections. Three two-winding, single-phase transformers may be connected in four possible arrangements to make a three-phase bank. These are the delta-delta, delta-wye, wye-wye, and wye-delta connections. In shipboard use, with ungrounded systems, the delta-delta is the most common arrangement. If connected as shown in Figure 22-19(a), the primary and secondary voltages will be in phase with each other. If a center tap is present on one transformer secondary, the bank may be used to supply a four-wire delta system (e.g., a 240-120V three-wire single-phase, two two-wire 240V single-phase, and a 240V three-phase load). Most importantly with a delta-delta connection, if one transformer fails in service, the remaining two can be operated as an open delta (Figure 22-19(b)) at 57.7 percent of the original capacity of the bank. If a spare transformer is available, the defective phase can be replaced and the capacity restored. In a complete three-transformer delta-delta bank, transformers tend to share the load in inverse proportion to their internal impedances, thus identical transformers should be used.

With medium voltage generation, the 4160V primary line voltage can be stepped down to 480V using standard 2400/480V single-phase transformers connected in a wye-delta (Figure 22-19(c)). However, the second-

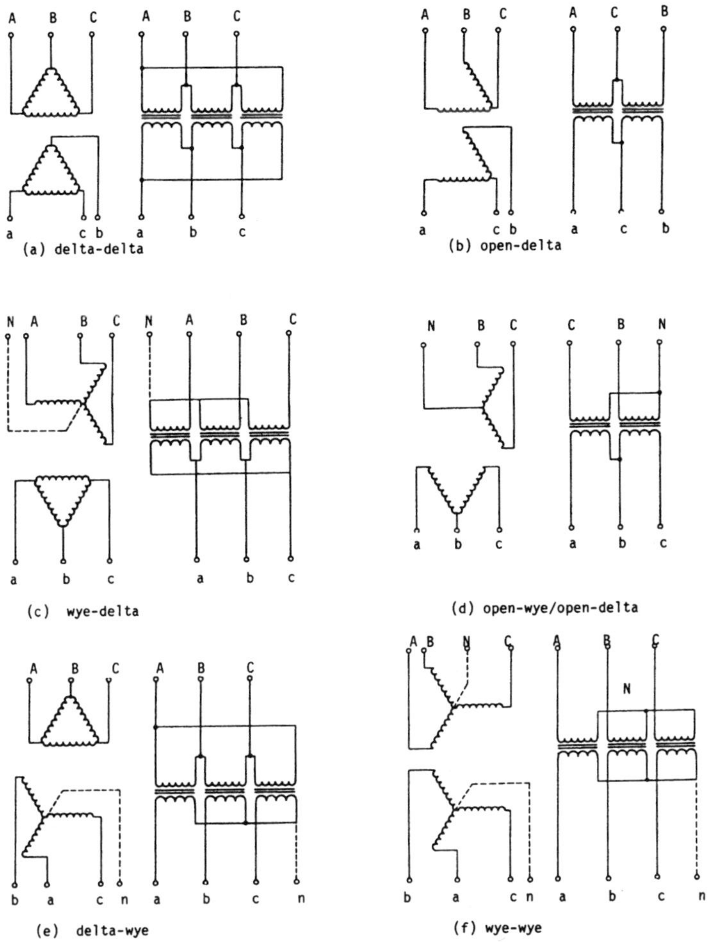

Figure 22-19. Three-phase transformer connections

ary line voltages are always shifted 30 degrees from the nearest primary, which is of importance if banks of transformers are to be paralleled. Also, the presence or absence of a primary neutral makes for significantly different operating characteristics. Without a primary neutral, this transformer arrangement does not require identical transformers, as differences in magnetizing currents are offset by a small circulating current around the delta. No appreciable voltage imbalance appears and the transformers uniformly share a balanced load even if the units do not have equal internal impedance. However, if one primary line becomes open (e.g., because of a blown line fuse), three-phase motors operating on the secondary side will continue to run, drawing normal current in two lines, but about twice the normal current in the third. Three overload devices are required to adequately protect motors operating on this type of system. With a primary neutral, the transformer bank can still deliver three-phase power with one transformer entirely disconnected from the other two. This open-wye/open-delta arrangement, shown in Figure 22-19(d), however, will be capable of supplying only 57.7 percent of the original bank. Furthermore, wye-delta connected units tend to share load inversely with their internal impedance, requiring the use of identical transformers.

A delta-wye transformer bank, illustrated in Figure 22-19(e), can supply a four-wire, three-phase (208-120V) secondary system like those used on some ferries and passenger liners. These systems typically incorporate a grounded neutral, thus a ground fault occurring in the secondary will immediately trip the faulty circuit's protective device. This interruption of supply leads to rapid identification of the faulty branch circuit. The delta-wye connection also introduces a 30-degree phase shift (either leading or lagging) between primary and secondary line-to-line voltages.

Wye-wye connections, illustrated in Figure 22-19(f), are common in high voltage transmission and distribution networks, but rarely found aboard ship.

Three-phase transformer assemblies, with a common magnetic core to all three winding pairs, offer economy of size, weight, and cost per kVA. The first two of these advantages are of particular importance in aircraft. However, the inability to take one phase out of service and maintain a three-phase secondary supply (open delta), prevents their use on shipboard.

Autotransformers, often described as "single-winding" transformers, actually only require that primary and secondary circuits share a common terminal. This may be accomplished by use of a single tapped winding (Figure 22-20(a)), with the intermediate tap connected to the primary for step-up (boost) operation or to the secondary for step-down operation. But conventional two-winding transformers may be connected as shown in Figure 22-20(b) to obtain the desired secondary voltage. In the illustration, a 480-120V two-winding transformer is connected to provide 600V service from typical 480V mains. Since the load may still receive rated secondary winding current at this higher voltage, the kVA which may be delivered to the load is $[1 + (N_1/N_2)]$ times the original rating. (In this case a 30 kVA transformer, used as an autotransformer, can deliver 150 kVA to a 600V load without overheating.) Autotransformers cost less, have better voltage regulation, are smaller and lighter in weight, and are more efficient than two-winding transformers of the same kVA capacity. However, they have two major disadvantages. The autotransformer does not isolate the secondary from the primary circuit, which may be unsuitable for grounding purposes. Also, in the event the common winding becomes open, the primary voltage can still feed through the extension winding to the load. With

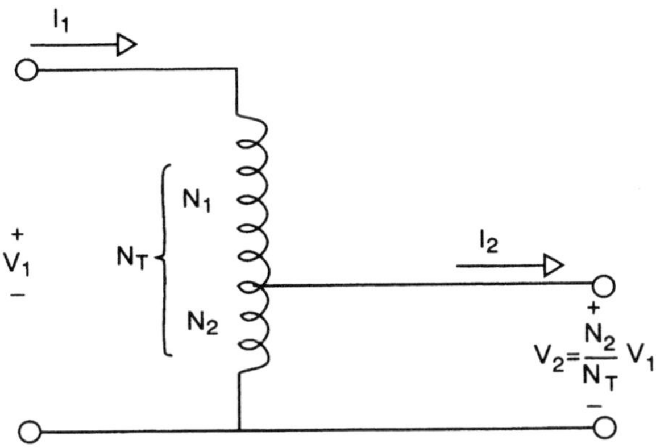

(a) Single-Winding Autotransformer (Step-Down Configuration)

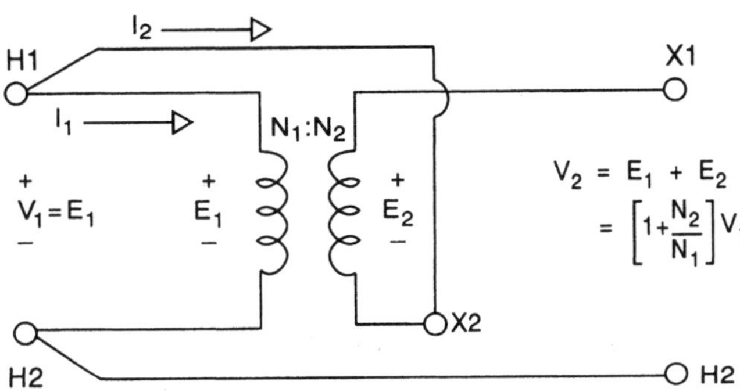

(b) Two-Winding Transformer Configured as Step-Up (Boost) Autotransformer

Figure 22-20. Autotransformer configurations

a step-down transformer, this could mean a serious over-voltage condition for the secondary equipment and a safety hazard, particularly if the step-down ratio is high. Occasionally autotransformers are used on ships as series line boosters for exceptionally long runs (large bow thruster feeder, for example), but primarily autotransformer use is reserved for fluorescent lamp ballasts and reduced-voltage motor starters (discussed in Chapter 23).

Instrument transformers. Instruments, meters, and control apparatus are rarely connected directly to high voltage or high current circuits. Instrument transformers are utilized (1) to act as *ratio devices,* making possible the use of standardized low voltage and low current instrumentation;

and (2) to act as *isolating devices,* to protect apparatus and personnel from direct contact with high voltage or high current mains.

Potential transformers supply a voltage—directly proportional to the line voltage—to the terminals of potential monitoring instruments. Secondaries of potential transformers are typically rated at 120V and thus may be utilized in conjunction with 120V (or 150V) rated instruments. The burden on these transformers is very low (a 50,000 ohm 120V voltmeter will draw less than 3 mA, for example) and hence ratings of 200, 600, and 1000 VA are common. Operating on exactly the same principles as the power transformer, a 5000/120V potential transformer produces 100V at the secondary with a primary line voltage of 4160V. Thus a 120V full-scale voltmeter, with a 0-5000V scale behind the movement, will indicate the actual line voltage to the operator.

The other type of instrument transformer is the *current transformer.* As the name implies, a "CT" delivers a current-proportional line current to a current-monitoring instrument (usually rated at 5 A). Stepping down current, of course, means stepping up voltage. Thus a 1000/5 A current transformer will step up voltage 200:1! However the primary winding consists of few windings of very low impedance and the primary current is a function of the load in the primary circuit rather than the load on the CT secondary. Thus a primary voltage of perhaps a fraction of a millivolt is stepped up to a few dozen millivolts in the secondary which is compatible with, say, a 50mV meter movement.

Under normal operating conditions, the magnetomotive force (mmf) established by the current in the secondary circuit opposes the primary mmf and limits the flux density in the transformer core. This demagnetizing mmf would disappear should the secondary circuit be opened while the primary circuit is still energized, resulting in a very high core flux density and dangerously high voltages induced in the secondary winding. Most current transformers are provided with a device for short-circuiting the secondary winding while an instrument is being removed or inserted in the secondary circuit. Thus anyone working with the secondary of a current transformer should always make certain that the short-circuiting device is closed before the secondary winding is disturbed.

ELECTRIC PROPULSION

Overview

Electric drive propulsion systems offer (1) adaptability to multiple use of prime mover generator sets such as cargo pumps or dredge operations as well as propulsion; (2) the use of multiple prime movers to power the propeller to meet varying torque requirements such as in ice-breaking; (3) flexibility of arrangement, as the prime mover-generator set is mechanically independent of the propulsion motor-propeller combination; and (4) ease of control, which may be effected from a number of different locations, and which may easily incorporate time lags or other limits to inhibit abusive operation of propulsion machinery. Electric drive is therefore most often applied on vessels requiring a high degree of maneuverability, large amounts of special-purpose power, the utilization of non-reversing, high speed, multiple prime movers, or combinations thereof.

Considering the use of turbines (either steam or gas) or diesel engines as prime movers, and either AC or DC machines as propulsion motors, four basic categories of electric drive propulsion plants are possible. Examples of each will be discussed in the following paragraphs. Traditionally, however, the use of AC drives has been associated with turbine prime movers, as conventional DC machines are incompatible with these high speed machines and reduction gears must be employed. Likewise, the frequency changing methods of motor speed control appropriate to AC propulsion motors do not match the torque characteristics of diesel engines at reduced speeds; thus there is the risk of stalling the engines during critical maneuvering if sophisticated control restrictions are not employed.

AC Turboelectric Drive

The classic T-2 tanker of World War II utilized this type of drive, but at 6000 hp those plants were well below the 10,000 to 60,000 hp per shaft deemed appropriate for this arrangement today. The typical system consists of a single turbine driving a directly connected high speed alternator, which supplies power to a single low speed synchronous motor directly coupled to the propeller shaft. With a two-pole synchronous alternator turning at 3600 RPM supplying 60 Hz, three-phase AC to a 60-pole synchronous motor, a propeller speed of 120 RPM is obtained. The pole ratio of motor to generator is in effect a fixed speed ratio of the prime mover to propeller similar to a geared turbine drive. Varying the prime mover speed from 900 to 4500 RPM will therefore produce propeller speeds over the 30 to 150 RPM range.

Figure 22-21 shows the arrangement of a typical AC turboelectric drive with a two-lever control station. To illustrate the operation of this plant consider the following operational sequence and plant response:

1. Reversing lever at *Stop,* throttle in *Maneuver* position. (Turbine is idling at minimum speed, excitation is at zero volts with M-G set running, contacts F, R, and G2 open, D and G1 closed.)
2. Reversing lever moved to *Ahead* position. (F contacts close, propulsion regulator increases generator excitation to maintain rated volts/Hz or pre-determined maximum level, propulsion motor accelerates as an induction motor, contacts E close and contact D opens via an automatic synchronizing system designed to apply motor field at proper slip and phase angle to obtain maximum pull-in torque.)

3. Throttle moved to desired speed setting. (Turbine governor set point resets to level consistent with throttle lever position, prime mover power increases until speed reaches governor set point.)

To reverse from full ahead to full astern:

1. Throttle moved to *Maneuver* position. (Turbine governor set point resets to minimum speed, synchronous motor decelerates.)
2. Reversing lever moved from *Ahead* to *Astern.* (Propulsion regulator drops excitation to zero, E contacts open, D contact closes, F contacts open, R contacts close, propulsion regulator increases generator excitation to maintain rated volts/Hz or to predetermined maximum level, propulsion motor stops, reverses, and accelerates as an induction motor, contacts E close and contact D opens via automatic synchronizing system.)
3. Throttle is moved to desired propeller speed. (Turbine governor set point resets to level consistent with throttle lever position, prime mover power increases until speed reaches governor set point.)

DC Turboelectric Drive

As mentioned earlier, the use of a reduction gear is required to match the characteristics of a high speed, nonreversing steam turbine to speeds compatible with DC machinery. Often twin or double armatures are utilized to take advantage of the higher speed at which the smaller units may be run,

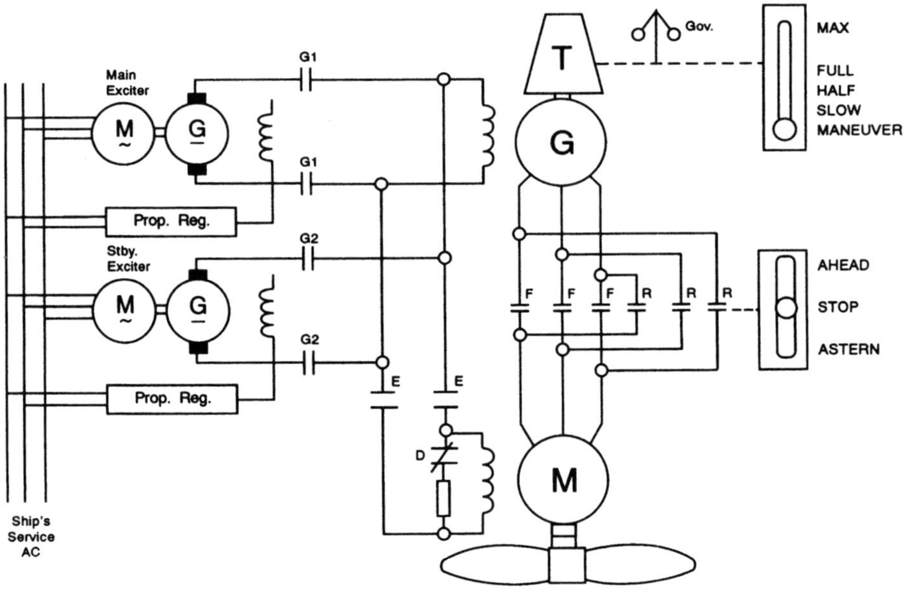

Figure 22-21. AC turboelectric drive

and, running at constant speed, the propulsion turbine may also be used to drive a ship's service generator. An example of a twin-screw turboelectric DC drive employing exactly this arrangement is the cable-laying ship *Long Lines*. Varying the magnitude and polarity of the double armature 900 RPM propulsion generator through excitation control to regulate the speed and direction of a single armature direct drive DC propulsion motor provides the excellent maneuvering capabilities required during cable-laying operations. The propulsion motors, capable of developing 4,250 hp at 135 RPM, can be operated from the bow, the stern, three locations on the bridge, and the engine room.

Employing high capacity silicon controlled rectifier circuits is another possibility for matching a high speed turboelectric generator with DC propulsion. Known as AC/DC drive, or rectified DC drive, this system is illustrated in Figure 22-22. Losses between prime mover and propeller in an AC rectified system are less than in a DC drive system because of the higher efficiency of the high speed AC generator. In this arrangement, the speed of the propulsion motor is controlled by varying the AC generator excitation; the direction of rotation is determined by the polarity of the motor field. Some alternatives include connecting 6-phase or 12-phase rectifiers directly to the generator and/or inserting a transformer between AC

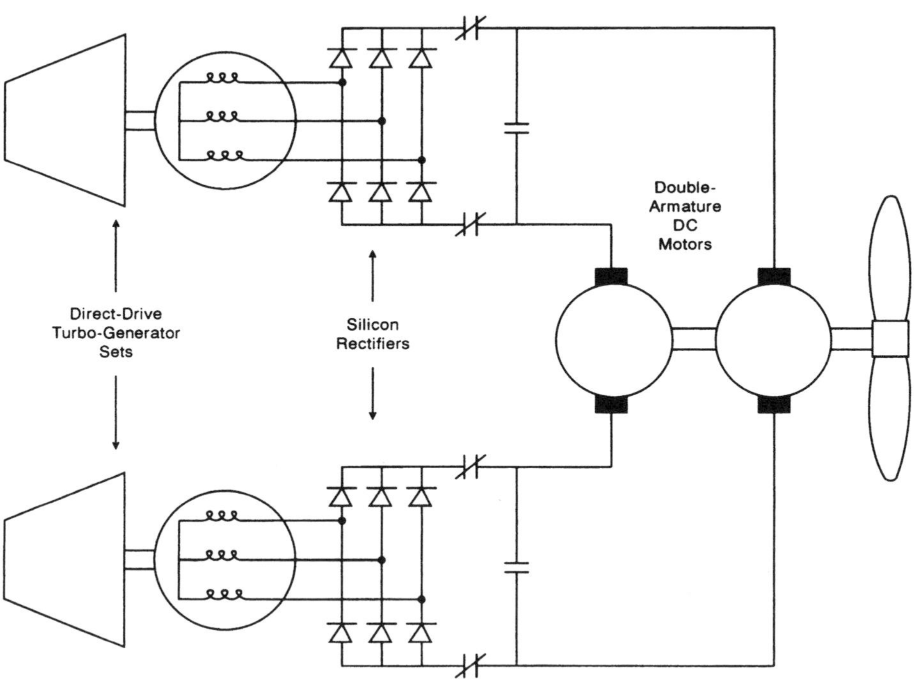

Figure 22-22. Rectified DC turboelectric drive

generator and rectifier. Unlike conventional DC drives, however, the rectifier blocks the transfer of power back to the generator during engine reversal; therefore the armature-rectifier circuit must be capable of absorbing the pump-back power. In some cases resistors are switched into the armature circuit to absorb this energy and limit armature currents.

DC Diesel-Electric Drive

A schematic diagram of a two-generator, two-motor, DC diesel-electric drive system is shown in Figure 22-23. Control of the system is accomplished by varying the voltage and polarity of the main generators. Considerable redundancy is built into this arrangement, which is currently employed aboard a 3,000 hp oceanographic survey ship whose special- purpose mission demands extraordinary maneuverability. Switching to operate with only one generator in the DC loop is easily accomplished via the setup contacts (G1 and S1 for main engine 1); the gas turbine can be switched into the DC loop via the A1 contacts if both diesels needed to be shut down; and either motor armature can be removed from service in an emergency via movable links (not shown). The gas turbine is normally used to power a 350 hp bow thruster, but an auxiliary diesel can also perform this service. The vessel also contains a 250 hp AC propulsive rudder electrically connected to a 300 kW AC generator on either the gas turbine or the auxiliary diesel shaft.

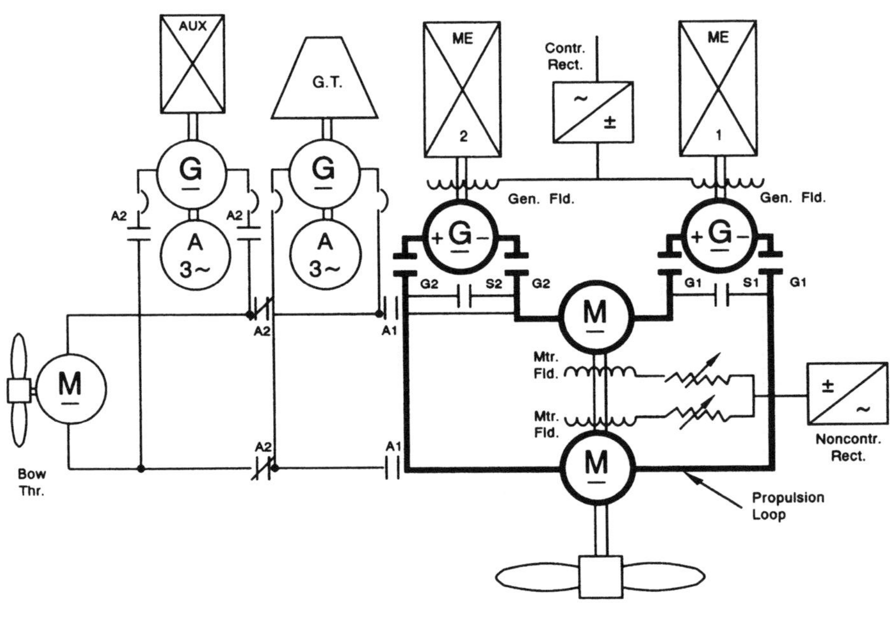

Figure 22-23. DC diesel-electric drive

AC Diesel-Electric Drive

Though the fundamental principles of speed control and reversal are the same as for synchronous turboelectric drive, the inherent problems of producing sufficient motor starting torque (which is maximized at low frequency and hence low prime mover speed) have rendered this combination of prime mover and electric drive rare. However, when there is a large demand for auxiliary power concurrent with little or no demand for propulsive power, the arrangement may be applicable. One example of this application is the plant aboard some U.S. Navy submarine tenders (Figure 22-24). This propulsion system consists of six 2000 kVA, 70.8 Hz, 3300V, diesel generator sets connected to a 15,000 hp, 157 RPM, 3300V synchronous propulsion motor. From three to six engines may be lined up to the propulsion motor to obtain desired ship speed and power. A single exciter supplies field power to both the propulsion motor and the propulsion generators. Up to three of the generator sets can be devoted to ship's service power or to an alongside submarine. In this mode they are operated at 720 RPM, producing a rated 1875 kVA at 2800V and 60 Hz. Three 2000 kVA transformer banks step down the 2800V to 450V for ship's service.

Above 35 percent rated speed (55 RPM) the generator and propulsion motor are synchronized and there is a fixed ratio between engine and propeller speed. During starting and reversing, however, the motor, fitted with a heavy amortisseur winding, runs as an induction machine. With three generators, the system torque is sufficient to accelerate the propeller to synchronous speed at about 40 percent of rated speed and then to increase to about 70 percent of rated speed. Reversing from full ahead, however, requires considerably more torque. In fact, the system torque with six generators running is insufficient to reverse the propeller until ship headway is reduced to about 65 percent. By the use of dynamic braking (the braking energy absorbed in a resistor, not the motor windings), the propeller speed can quickly be reduced to a few RPM ahead. With the brake applied, the ship will lose headway until the available system torque is capable of pulling the motor into synchronism astern.

With the increased availability of reliable and economically viable solid-state frequency converters, the operation of AC propulsion motors off a fixed frequency bus (and therefore constant RPM prime movers) becomes more feasible. An example of a large plant utilizing this technique is illustrated in Figure 22-25. Nine 10.5 MW diesel generator sets in various combinations provide propulsion and ship's service power at 10 kV and 60 Hz. On this twin-screw vessel, four generator sets and one 44 MW synchronous propulsion motor are located in one engine room, the remaining five generator sets and propulsion motor in the other. The 10 kV buses in the separate engine rooms are electrically connected through a bus tie which may be broken in either machinery space. At speeds above 18 knots, the

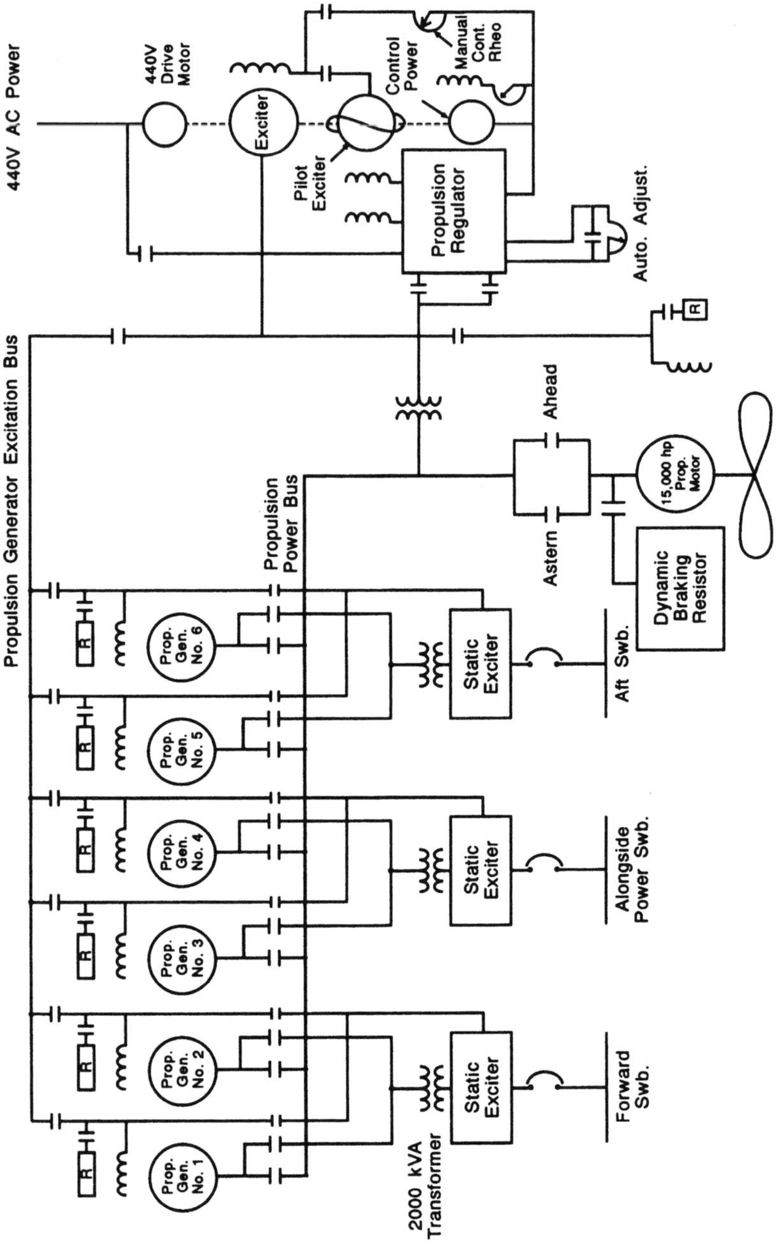

Figure 22-24. AC diesel-electric drive

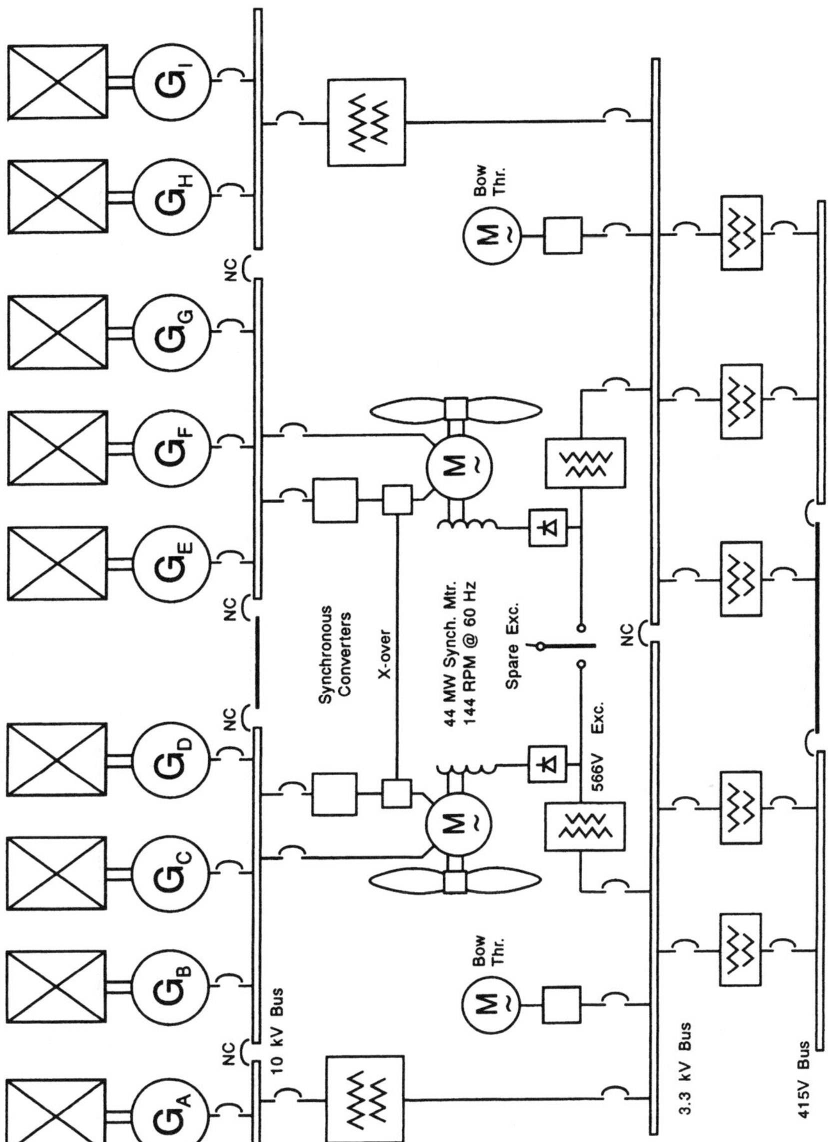

Figure 22-25. Passenger liner AC diesel-electric arrangement

propulsion motors are powered through 60 Hz mains and ship speed is controlled via propeller pitch. Under 18 knots, propeller pitch is fixed and speed is controlled by supplying the motor with variable frequency AC through a synchronous converter. The 14-pole synchronous propulsion motors therefore rotate at 144 RPM above 18 knots and down to 72 RPM under converter supply. Cross-connection is provided so that either motor may be supplied through either converter. Excitation is provided through silicon controlled rectifiers fed from the 3.3 kV bus. To segregate ship's service supply from propulsion supply, the outer generator sets may be electrically disconnected from the propulsion bus. This may be advantageous in rough weather where severe torque variations on the propeller may result in frequency disturbances on the propulsion bus.

CIRCUIT CALCULATIONS

Circuit Elements

Active elements. Circuit elements may be classified as *active* or *passive.* Active elements have the capacity to produce electrical energy and may be idealized as sources of either voltage or current. Some examples of voltage sources (not necessarily ideal) are batteries and generators. Some electronic circuits can behave essentially as current sources as can exciters under the control of an automatic voltage regulator (AVR).

Passive elements. Passive elements absorb electrical energy and are classified according to the manner in which the energy is absorbed. *Resistors* dissipate energy in the form of heat, *inductors* store energy in the form of a magnetic field, and *capacitors* store energy in the form of an electric field.

The voltage across a resistor is proportional to the current through it. This is true for DC, AC, and any other form of voltage and current. Expressed mathematically:

$$v = i \times R$$

where this proportionality constant, R, is called *resistance.* The value of this constant depends on the type of material, specifically its *resistivity,* its geometric shape (length and cross-sectional area), and the temperature of the material. This resistance value, expressed in *ohms,* is often referred to as the "size" of the resistor, and is not to be confused with its physical size, which depends on the anticipated amount of heat to be dissipated. To illustrate the relationship between material, size, length, and resistance, consider 150 feet of AWG #10 copper conductor. The resistivity of copper at 50°

C is approximately 12 ohm-CM/ft, thus the resistance of a 150-foot run of copper with cross-sectional area of 10,400 CM (circular mils) is:

$$\frac{12\ \text{ohm} \backslash \text{CM}}{\text{ft}} \times \frac{150\ \text{ft}}{10{,}400\ \text{CM}} = 0.1731\ \text{ohms}$$

If the length were increased to 300 feet, the resistance would double to 0.3461 ohms; if AWG #7 (20,800 CM) were used instead, then the resistance of the 150 feet would be halved; and if an aluminum conductor of 20 ohm-CM/ft were used, a wire size of 17,330 CM would be necessary to maintain the same 0.1731 ohm resistance over 150 feet. Occasionally, a reciprocal relationship is used to describe the voltage-current characteristic of a resistive circuit:

$$i = G \times v$$

where G, *conductance,* is 1/R and is measured in siemens (i.e., 25 ohms is equivalent .04 siemens or 40 mS).

The voltage across an inductor is proportional not to the current, but to the *rate of change of current,* expressed mathematically as:

$$v = L \times (di/dt)$$

where di/dt is the rate of change of current in amps per second, and the proportionality constant, L, is called *inductance.* An abrupt change in the current through an inductive circuit can produce a very high voltage. This opposition to a change in current is a characteristic of inductive circuits and is the reason (for example, when current is suddenly interrupted to an inductive circuit such as a motor) an arc is drawn at the interrupting contacts. Note that in a steady state DC circuit, the current does not change (i.e., di/dt = 0) and thus there is no voltage across an inductor in a steady DC circuit regardless of the magnitude of the current. In an AC circuit the rate of change of current is proportional to the frequency.

$$di/dt = 2\pi f \times I$$

The $2\pi f$ term is combined with the inductance, L, to form XL, *inductive reactance.* Inductance in measured in henries and inductive reactance in ohms, since it is a ratio of AC volts to AC amps.

The current through a capacitor is proportional not to the voltage, but to the *rate of change of voltage,* expressed mathematically as:

$$i = C \times (dv/\ dt)$$

where dv/dt is the rate of change of voltage in volts per second, and the proportionality constant, C, is called *capacitance.* An abrupt change in the voltage across a capacitor can produce a very high current. This opposition to a change in voltage is a characteristic of capacitive circuits and is the reason a capacitor should be discharged before removing it from a circuit board, for example, when making repairs. Note that in a steady state DC circuit, the voltage does not change (i.e., dv/dt = 0) and thus there is no current through a capacitor in a steady state DC circuit regardless of the magnitude of the voltage. In an AC circuit the rate of change of voltage and current is proportional to the frequency:

$$dv/dt = 2\pi f \times V$$

The $2\pi f$ term is combined with the capacitance, C, to form $X_C = \frac{1}{2\pi fC}$ *capacitive reactance.* Capacitance is measured in farads (or in practical circuits, microfarads) and capacitive reactance in ohms, since it is a ratio of AC volts to AC amps.

The following table summarizes the electrical characteristics of passive elements.

Circuit Element	*v-i Relationship*	*DC Circuit* $di/dt = 0$ $dv/dt = 0$	*AC Circuit* $di/dt = 2\pi f \times I$ $dv/dt = 2\pi f \times V$
Resistance	$v = R \times i$	$V = R \times I$	$V = R \times I$
Conductance	$i = G \times v$	$I = G \times V$	$I = G \times V$
Inductance	$v = L \times (di/dt)$	$V = 0$	$V = X_L \times I$ $(X_L = 2\pi fL)$
Capacitance	$i = C \times (dv/dt)$	$I = 0$	$V = X_C \times I$ $(X_C = \frac{1}{2\pi fC})$

Combinations of Circuit Elements

When resistors are connected in series their values are added to determine the equivalent resistance of the series circuit.

$$R_{eq} = R_1 + R_2 + R_3 + \ldots$$

The same is true of inductances connected in series:

$$L_{eq} = L_1 + L_2 + L_3 + \ldots$$

When capacitors are connected in series, however, their values are added inversely to obtain an equivalent capacitance value.

$$1/C_{eq} = 1/C_1 + 1/C_2 + 1/C_3 + \ldots$$

A handy relationship for calculating the equivalent capacitance of *two* series capacitors is:

$$C_{eq} = \frac{C_1 \times C_2}{C_1 + C_2}$$

Note that the smaller value dominates and the equivalent capacitance is always smaller than either of the original values (Figure 22-26(a)).

Resistors and inductors connected in parallel combine in the manner of capacitors in series. That is:

$$1/R_{eq} = 1/R_1 + 1/R_2 + 1/R_3 + \ldots$$

$$1/L_{eq} = 1/L_1 + 1/L_2 + 1/L_3 + \ldots$$

And similarly, for only two of these elements in parallel, the relationships determine the equivalent values.

$$R_{eq} = \frac{R_1 \times R_2}{R_1 + R_2} \text{ and } L_{eq} = \frac{L_1 \times L_2}{L_1 + L_2}$$

Capacitors in parallel are added to determine the equivalent capacitance of the parallel combination:

$$C_{eq} = C_1 + C_2 + C_3 + \ldots$$

Using these parallel and series combination techniques sequentially can result in significant simplification of the circuit under inspection (Figure 22-26(b)). In series or parallel combinations only, resistors may be combined with resistors, inductors with inductors, etc., although the *order* in which they appear in the original circuit may be rearranged (see Figure 22-26(c)).

In AC circuits, resistances (R) and reactances (X_L or X_C) are combined geometrically or as *vectors*. Resistances are represented as horizontal segments whose lengths are proportional to their ohmic value. Reactances are represented as vertical segments oriented either upward (X_L) or downward (X_C). Mathematically, the symbol j is used to designate "up" and $-j$ to designate *down;* thus an impedance described as (40 +j30) ohms consists of 40 ohms resistance and 30 ohms (inductive) reactance. These components are combined into a single impedance value using the geometric relationships of the *impedance triangle* (Figure 22-27). The length of the impedance side of the triangle (hypotenuse) is found using the Pythagorean theorem (Figure 22-27(a)). Impedances are also described by documenting the total impedance value, Z, and the "phase angle," θ. Thus (40 +j 30) and (50 ∠36.9°) describe the same impedance. Note that since θ indicates the phase difference between voltage and current, positive phase angle (j X) implies that the current *lags* the voltage, which is of course true in an inductive circuit. When θ is negative (Figure 22-27(b)), the reactive component is capacitive (–jX) and the current leads the voltage. Impedances are

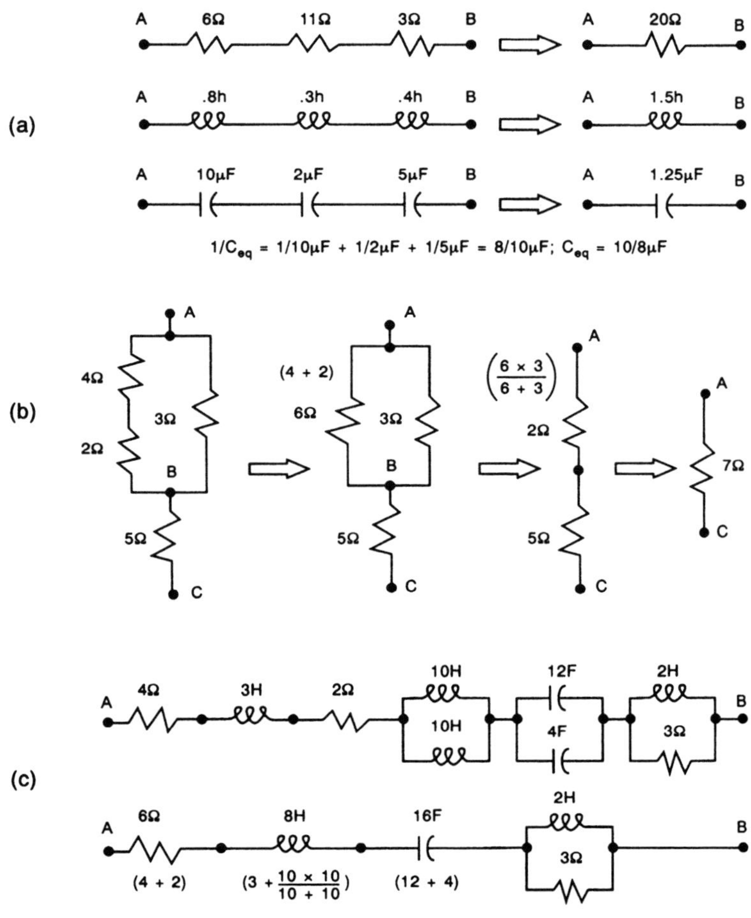

Figure 22-26. Circuit element combinations

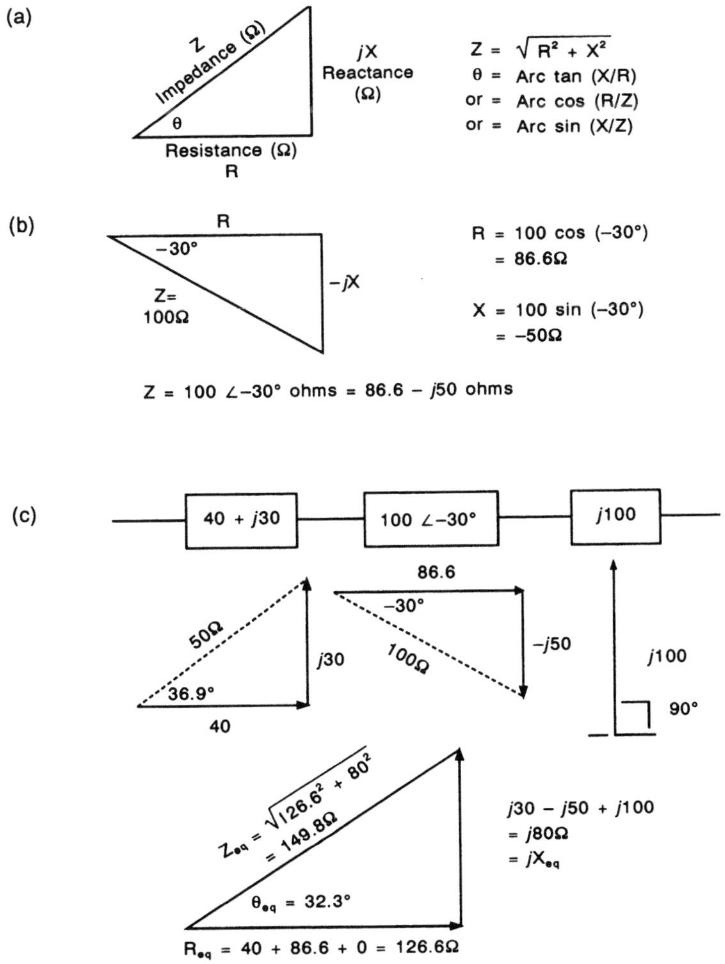

Figure 22-27. AC circuit impedance calculations

combined in series by adding their resistive and reactive components separately (Figure 22-27(c)), and *not* the overall impedance values.

Power Calculations

In DC circuits energy is dissipated in resistors only and may be determined by knowing the current through or voltage across the resistor as follows:

$$P_{dc} = VI = I^2R = V^2/R = V^2G$$

In AC circuits the phase angle plays a role in determining the amount of power actually dissipated in a given impedance. *Power triangle* calculations are similar to impedance triangle calculations with the horizontal component representing the *true power* in watts and the *j* component representing the *reactive power* in vars (volt-ampere reactive). Reactive power is the rate at which energy is being moved into or out of storage in capacitors and/or inductances, but not consumed as heat or actual work. The vector combination of true and reactive power is the *apparent power* in volt-amps, since it is the direct product of effective voltage and current in the circuit. As in the impedance triangle, a negative phase angle indicates capacitive elements in the circuit, and results in a negative reactive power. The relationships between impedance and AC power are illustrated in Figure 22-28(a). The phase angle is often denoted indirectly as cos θ or *power factor.* Power factor (pf) may be thought of as that fraction of the apparent power actually doing work, since it is the ratio of true power to apparent power (e.g., kW/kVA). Since the cosines of 30° and –30° are the same value, power factors are identified as *lagging* ($\theta > 0°$) or *leading* ($\theta < 0°$) to differentiate between the presence of positive or negative vars. In three-phase networks, true, reactive, and apparent powers are calculated using *line* voltages and currents and a factor of $\sqrt{3}$ or 1.73. Again, note that *apparent power* times pf equals *true power* (Figure 22-28(b)).

AC loads therefore may be described by identifying (1) any two sides of the power triangle; (2) one side and the phase angle; or (3) one side and the power factor. To illustrate this, consider Figure 22-29. Shown is a 440V, three-phase distribution panel with four loads described by various schemes. The power triangle for each load is shown, and the total power that must be supplied to the panel is determined. Note that the total kVA supplied is not equal to the sum of the individual load kVA (902 kVA versus 1341 kVA) but considerably less. This is because the leading power factor of Load 2 has canceled out much of the lagging reactive power of the other loads. Leading pf loads such as capacitor banks, synchronous condensers, or overexcited synchronous motors are often used in this manner to improve (increase) the power factor of the overall plant. The distribution panel power factor (880/902 = .976) in this example is considerably higher than any of the individual loads. The feeder current is then calculated based on the panel kVA, not the sum of the individual load currents.

Voltage Drop Calculations

Although sized initially by ampacity ratings, classification society and regulatory body requirements specify the maximum voltage drop that is permitted in a given service (3 percent in lighting circuits, for example, 5 percent in motor circuits, and 1 percent in bus ties). These requirements may dictate use of cabling larger than that which simply meets ampacity requirements at full load. The voltage drop in percent of nominal circuit voltage is calculated as follows: (1) Two-conductor circuit (DC or single-phase AC).

$$\%VD = \frac{200 \times I \times R \times L}{CM \times V}$$

(2) Three-conductor, three-phase circuit

$$\%VD = \frac{173 \times I \times R \times L \times CF}{CM \times V}$$

where I = full load current (amps)
R = conductor resistivity (ohm–CM/ft) for copper conductors, nominal value of R = 12
L = single length of circuit (ft)
CF = correction factor for cable sizes AWG #2 or larger (see Table 22-7)
CM = conductor cross-sectional area (circular mils)
V = circuit voltage (volts)

TABLE 22-7

Correction Factors for Voltage Drop Calculations

Cable size		*Load Power Factor*							
AWG No.	*MCM*	*1.00*	*0.95*	*0.90*	*0.85*	*0.80*	*0.75*	*0.70*	*0.65*
2	66	1.00	1.01	0.99	0.96	0.92	0.84	0.76	0.68
1	83	1.00	1.03	1.01	0.98	0.95	0.88	0.80	0.71
0	106	1.00	1.05	1.04	1.02	0.99	0.93	0.85	0.77
00	133	1.00	1.07	1.07	1.05	1.03	0.98	0.91	0.84
000	168	1.00	1.10	1.11	1.10	1.09	1.04	0.98	0.92
0000	212	1.00	1.13	1.16	1.06	1.15	1.12	1.07	1.01
	250	1.00	1.17	1.21	1.22	1.22	1.20	1.16	1.10
	300	1.00	1.21	1.26	1.28	1.29	1.28	1.25	1.21
	350	1.00	1.24	1.31	1.34	1.36	1.37	1.35	1.31
	400	1.00	1.32	1.39	1.43	1.45	1.46	1.44	1.40

As an example, consider the 115V lighting panel of Figure 22-29, to be supplied by a 250-foot, three-phase feeder. The total load supplied is 15560 W and may be considered to be a balanced three-phase load (AB: 5180 W, BC: 5170 W, CA: 5180 W). Assuming all-fluorescent lighting with a .8 pf, the feeder line current is 97.6 amps. A type TXIB-52 (AWG #3), rated at 99 amps, would seem to be adequate for this service. However:

$$\%VD = \frac{173 \times 12 \times 97.6 \times 250}{52{,}600 \times 115} = 8.3\%$$

which is too large a voltage drop for a lighting circuit. Allowing for a 1.5 percent drop in the longest branch circuit (which must be verified) a CM value of greater than 294,000 CM is required to limit the feed drop to less than 1.5 percent. Checking the results with a 300 MCM feeder (CF = 1.29 at .8 pf):

$$\%VD = \frac{173 \times 12 \times 97.6 \times 250 \times 1.29}{300{,}000 \times 115} = 8.3\%$$

This might suffice if none of the branch circuits had a voltage drop of greater than 1.1 percent.

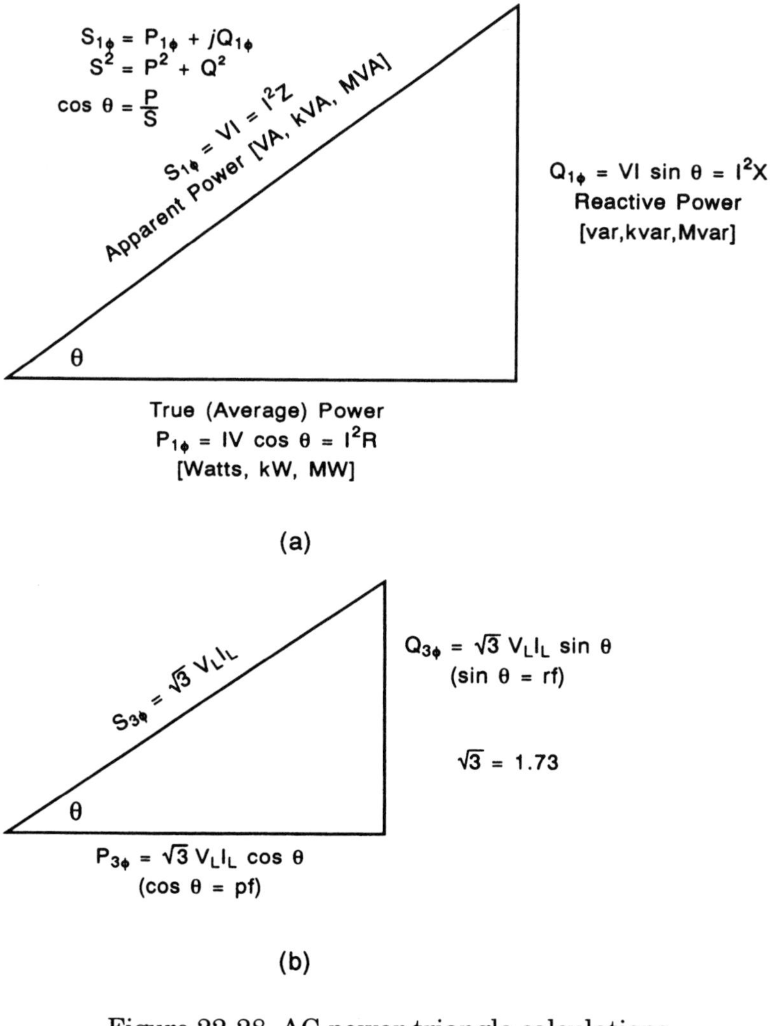

Figure 22-28. AC power triangle calculations

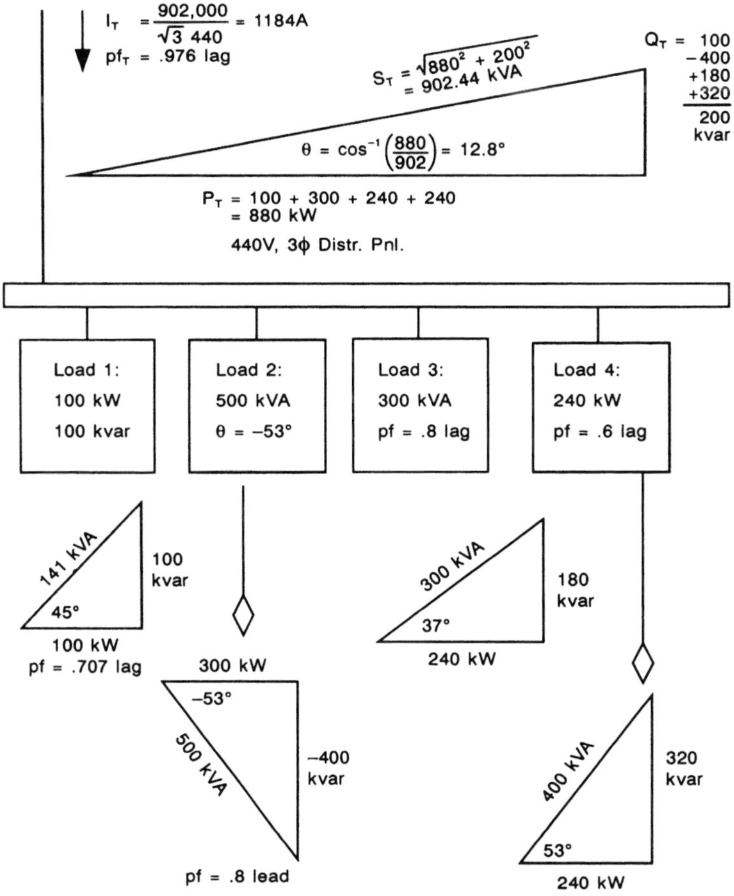

Figure 22-29. Power factor correction

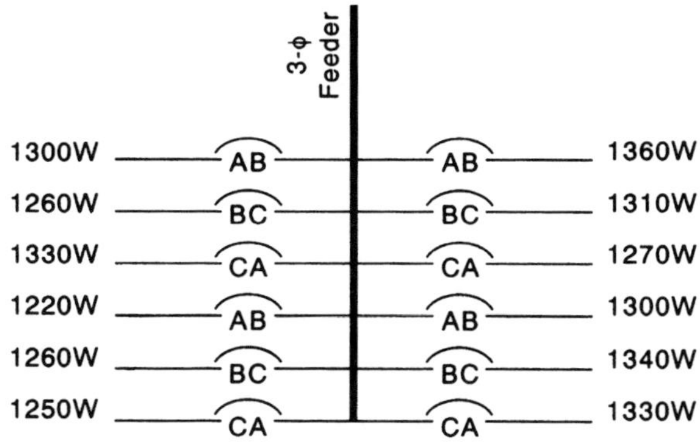

Figure 22-30. Data for feeder sizing calculations

GLOSSARY

All the definitions in the glossary are taken from IEEE Standard 45-1983.

General Terms

Machinery spaces. Those spaces primarily used for machinery of any type, or for equipment for the control of such machinery, as boiler, engine, ship's service generator, and evaporator rooms.

Direct current (DC). A unidirectional current in which the changes in value are zero or so small that they may be ignored.

Alternating current (AC). A periodic current, the average value of which over the period is zero.

Cycle. The complete series of values of a periodic quantity that occurs during the period.

Frequency. Of a periodic quantity, the number of cycles completed in a unit of time.

Hertz. The unit of frequency; one cycle per second equals one hertz (Hz).

Single-phase circuit. A circuit energized by a single alternating electromotive force.

Three-phase circuit. A combination of circuits energized by alternating electromagnetic forces that differ in phase by one-third of a cycle (120 degrees).

Slip. In an induction machine, the difference between its synchronous speed and its operating speed. It may be expressed in the following forms: (1) as a percent of synchronous speed, (2) as a decimal fraction of synchronous speed, or (3) directly in RPM.

Embedded temperature detector. A resistance thermometer or thermocouple built into a machine for the purpose of measuring the temperature.

Capacitor. A device, the primary purpose of which is to introduce capacitance into an electric circuit. Capacitors are usually classified, according to their dielectrics, as air capacitors, mica capacitors, paper capacitors, etc.

Capacitance (capacity). That property of a system of conductors and dielectrics that permits the storage of electricity when potential differences exist between the conductors. Its value is expressed as the ratio of a quantity of electricity to a potential difference. A capacitance value is always positive.

Reactor. A device, the primary purpose of which is to introduce reactance into a circuit. A reactor introduces reactance into a circuit for purposes such as starting motors, paralleling transformers, and controlling current.

Reactance. The reactance of a portion of a circuit for a sinusoidal current and potential difference of the same frequency is the product of the sine of the angular phase difference between the current and potential differences times the ratio of the effective potential difference to the effective current, there being no source of power in the portion of the circuit under consideration.

Direct current commutating machine. A machine that comprises a magnetic field excited from a DC source or formed of permanent magnets, an armature, and a commutator connected thereto. Specific types of direct current commutating machines are DC generators, motors, synchronous converters, boosters, balancers, and dynamotors.

Synchronous machine. A machine in which the average speed of normal operation is exactly proportional to the frequency of the system to which it is connected.

Asynchronous machine. A machine in which the speed of operation is not proportional to the frequency of the system to which it is connected.

Induction machine. An asynchronous AC machine that comprises a magnetic circuit interlinked with two electric circuits, or sets of circuits, rotating with respect to each other, and in which power is transferred from one circuit to another by electromagnetic induction. Examples of induction machines are induction generators, induction motors, and certain types of frequency converters and phase converters.

Corrosion-resisting materials. Silver, corrosion-resisting steel, copper, brass, bronze, copper-nickel, certain nickel-copper alloys, and certain aluminum alloys are considered satisfactory corrosion-resisting materials within the intent of the foregoing.

Corrosion-resistant treatments. Treatments that, when properly done and of a sufficiently heavy coating, are considered satisfactory corrosion-resistant treatments, within the intent of the foregoing, and that include electroplating of cadmium, chromium, copper, nickel, silver, and zinc; sherardizing; galvanizing; and dipping and painting (phosphate or suitable cleaning, followed by the application of zinc chromate primer or equivalent).

Flame-retarding. Flame-retarding materials and structures should have such fire-resisting properties that they will neither convey flame nor continue to burn for longer times than specified in the appropriate flame test.

Multicable penetrator. A device consisting of multiple nonmetallic cable seals assembled in a surrounding metal frame for insertion in openings in decks, bulkheads, or equipment enclosures and through which cables may be passed to penetrate decks or bulkheads or to enter equipment without impairing their original fire or watertight integrity.

Lug. A wire connector device to which the electrical conductor is attached by mechanical pressure or solder.

Generator Classifications

Electric generator. A machine that transforms mechanical power into electric power.

Shunt-wound generator. A DC generator in which ordinarily the entire field excitation is derived from one winding consisting of many turns with a relatively high resistance. This one winding is connected in parallel with the armature circuit for a self-excited generator, and to the load side of another generator or other source of direct current for a separately excited generator.

Stabilized shunt-wound generator. A stabilized shunt-wound generator is the same as the shunt-wound type, except that a series field winding is added, of such proportion as not to require equalizers for satisfactory parallel operation. The voltage regulation of this type of generator must comply with that given for shunt-wound generators.

Compound-wound generator. A DC generator that has two separate field windings: one of them, supplying the predominating excitation, is connected in parallel with the armature circuit, and the other, supplying only partial excitation, is connected in series with the armature circuit and is of such proportion as to require an equalizer connection for satisfactory parallel operation.

Magnetoelectric generator. An electric generator in which the magnetic flux is provided by one or more permanent magnets.

Exciter. The source of all or part of the field current used to excite an electric machine.

Pilot exciter. The source of all or part of the field current used to excite another exciter.

Synchronous generator. A synchronous AC machine that transforms mechanical power into electrical power. (A synchronous machine is one in which the average speed of normal operation is exactly proportional to the frequency of the system to which it is connected.)

Induction generator. An induction machine driven above synchronous speed by an external source of mechanical power.

Brushless exciter. An AC (rotating armature type) exciter whose output is rectified by a semiconductor device to provide excitation to an electric machine. The semiconductor device would be mounted on and rotate with the AC exciter armature.

Brushless synchronous machine. A synchronous machine having a brushless exciter with its rotating armature and semiconductor devices on a common shaft with the field of the main machine. This type of machine with its exciter has no collector, commutator, or brushes.

Motor Classification

Electric motor. A machine that transforms electric power into mechanical power.

Universal motor. A series-wound or compensated series-wound motor designed to operate at approximately the same speed and output on either direct or single-phase alternating current of a frequency not greater than 60 Hz and of approximately the same root-mean-square voltage.

Shunt-wound motor. A DC motor in which the field circuit is connected either in parallel with the armature circuit or to a separate source of excitation voltage.

Stabilized shunt-wound motor. A shunt-wound motor that has a light series winding added to prevent a rise in speed (or to obtain a slight reduction in speed) with increase of load.

Series-wound motor. A commutator motor in which the field circuit and the armature circuit are connected in series. (It operates at a much higher speed at light load than at full load.)

Compound-wound motor. A DC motor that has two separate field windings: the first, usually the predominating field, is connected in parallel with the armature circuit, and the second is connected in series with the armature circuit. (The characteristics regarding speed and torque are intermediate between those of shunt and series motors.)

Induction motor. An AC motor in which a primary winding on one member (usually the stator) is connected to the power source, and a polyphase or squirrel-cage secondary winding on the other member (usually the rotor) carries induced current.

Squirrel-cage induction motor. A motor in which the secondary circuit consists of a squirrel-cage winding suitably disposed in slots in the secondary core.

Wound-rotor induction motor. An induction motor in which the secondary circuit consists of polyphase winding or coils whose terminals are either short-circuited or closed through suitable circuits. (When provided with collector or slip rings, it is also known as a slip ring induction motor.)

Synchronous motor. A synchronous machine that transforms electric power into mechanical power. Note: Unless otherwise stated, it is generally understood that a synchronous generator (or motor) has field magnets excited with direct current.

Amortisseur. A permanently short-circuited winding consisting of conductors embedded in the pole shoes of a synchronous machine and connected together at the ends of the poles, but not necessarily connected between poles. This winding, when used in salient-pole machines, sometimes includes bars that do not pass through the pole shoes, but are supported in the interpole spaces between the pole tips.

Squirrel-cage winding. A permanently short-circuited winding, usually uninsulated, that is chiefly used in induction machines, and has its conductors uniformly distributed around the periphery of the machine and joined by continuous end rings.

Converters

Dynamotor. A form of converter that combines both motor and generator action, with one magnetic field and with two armatures or with one armature having separate windings.

Motor-generator set. A machine consisting of one or more motors mechanically coupled to one or more generators.

Synchronous converter. A converter that combines both motor and generator action in one armature winding and is excited by one magnetic field. It is normally used to change AC power to DC power.

Rectifier. A converter unit in which the direction of average power flow is from the DC circuit to the AC circuit.

Static converter. A unit that employs static rectifier devices such as semiconductor rectifiers or controlled rectifiers (thyristors), transistors, electron tubes, or magnetic amplifiers to change AC power to DC power or vice versa.

Motor Speed Classification

Constant-speed motor. A motor whose speed of normal operation is constant or practically constant—for example, a synchronous motor, an induction motor with small slip, or an ordinary DC shunt-wound motor.

Multispeed motor. A motor that can be operated at any one of two or more definite speeds, each being practically independent of the load—for example, a DC motor with two armature windings, or an induction motor with windings capable of various pole groupings.

Adjustable-speed motor. A motor whose speed can be varied gradually over a considerable range, but when once adjusted remains practically unaffected by the load, such as a DC shunt-wound motor with field resistance control designed for considerable range of speed adjustment.

Base speed of an adjustable-speed motor. The lowest speed obtained at rated load and rated voltage at the temperature rise specified in the rating.

Varying-speed motor. A motor whose speed varies with the load, ordinarily decreasing when the load increases, such as a series-wound or repulsion motor.

Adjustable varying-speed motor. A motor whose speed can be adjusted gradually, but when once adjusted for a given load will vary in considerable degree with change in load; an example is a DC compound-wound motor adjusted by field control or a wound-rotor induction motor with rheostatic speed control.

Ventilation of Machines

Open machine. A machine having ventilating openings that permit passage of external cooling air over and around the windings.

Self-ventilated machine. A machine having its ventilating air circulated by means integral with the machine.

Separately ventilated machine. A machine that has its ventilating air supplied by an independent fan or blower external to the machine.

Enclosed self-ventilated machine. A machine having openings for the admission and discharge of the ventilating air, which is circulated by means integral with the machine, the machine being otherwise totally enclosed. These openings are so arranged that inlet and outlet ducts or pipes may be connected to them.

Enclosed separately ventilated machine. A machine having openings for the admission and discharge of the ventilating air, which is circulated by means external to and not a part of the machine, the machine being otherwise totally enclosed. These openings are so arranged that inlet and outlet duct pipes may be connected to them.

Totally enclosed machine. A machine so enclosed as to prevent the free exchange of air between the inside and outside of the case, but not sufficiently enclosed to be termed airtight.

Totally enclosed fan-cooled machine. A totally enclosed machine equipped for exterior cooling by means of a fan or fans integral with the machine but external to the enclosing parts.

Enclosure of Equipment

General-purpose enclosure. An enclosure that primarily protects against accidental contact and slight indirect splashing but is neither drip-proof nor splash-proof.

Semiguarded enclosure. An enclosure in which all of the openings, usually in the top half, are protected (as in the case of a guarded enclosure) but the others are left open.

Guarded enclosure. An enclosure in which all openings giving direct access to live or rotating parts (except smooth rotating surfaces) are limited in size by the structural parts or by screens, baffles, grilles, expanded metal, or other means to prevent accidental contact with hazardous parts. The openings in the enclosure shall be such that they will not permit the passage of a rod larger than a half-inch in diameter, except where the distance of exposed live parts from the guard is more than four inches, in which case the openings may be of such shape as not to permit the passage of a rod larger than three-quarters of an inch in diameter.

Dustproof enclosure. An enclosure so constructed or protected that any accumulation of dust that may occur within the enclosure will not prevent the successful operation of, or cause damage to, the enclosed equipment.

Drip-proof enclosure. An enclosure in which the openings are so constructed that drops of liquid or solid particles falling on the enclosure at any angle not greater than 15 degrees from the vertical either cannot enter the enclosure, or if they do enter the enclosure, will not prevent the successful operation of, or cause damage to, the enclosed equipment.

Splash-proof enclosure. An enclosure in which the openings are so constructed that drops of liquid or solid particles falling on the enclosure or coming toward it in a straight line at any angle not greater than 100 degrees from the vertical cannot enter the enclosure either directly or by striking and running along a surface.

Waterproof enclosure. An enclosure constructed so that any moisture or water leakage that may occur into the enclosure will not interfere with its successful operation. In the case of motor or generator enclosures, leakage that occurs around the shaft may be considered permissible provided it is prevented from entering the oil reservoir and provision is made for automatically draining the motor or generator enclosure.

Oil-proof enclosure. An enclosure constructed so that oil vapors, or free oil not under pressure, that may accumulate within the enclosure will not prevent successful operation of, or cause damage to, the enclosed equipment.

Explosion-proof enclosure. An enclosure designed and constructed to withstand an explosion of a specified gas or vapor which may occur within it, and to prevent the ignition of the specified gas or vapor surrounding the enclosure by sparks, flashes, or explosions of the specified gas or vapor that may occur within the enclosure. Note: An explosion-proof apparatus should bear Underwriters Laboratory approval ratings of the proper class and group consonant with the spaces in which flammable volatile liquids or highly flammable gases, mixtures, or substances may be present.

Dust-tight enclosure. An enclosure constructed so that dust cannot enter the enclosing case.

Watertight enclosure. An enclosure constructed so that a stream of water from a hose not less than 1 inch in diameter under a head of 35 feet from a distance of 10 feet can be played on the enclosure from any direction for a period of 15 minutes without leakage. The hose nozzle shall have a uniform inside diameter of one inch.

Oil-tight enclosure. An enclosure constructed so that oil vapors or free oil not under pressure that may be present in the surrounding atmosphere cannot enter the enclosure.

Submersible enclosure. An enclosure constructed so that the equipment within it will operate successfully when submerged in water under specified conditions of pressure and time.

Torque Classification

Locked rotor torque. The minimum torque that a motor will develop at rest for all angular positions of the rotor, with rated voltage applied at rated frequency.

Pull-out torque. The maximum sustained torque a synchronous motor will develop at synchronous speed with rated voltage applied at rated frequency and with normal excitation.

Breakdown torque. The maximum torque a motor will develop with rated voltage applied at rated frequency, without an abrupt drop in speed.

Torque margin. The increase in torque above rated torque to which a propulsion system may be subjected without the motor pulling out of step with the generator.

Duty Classification

Continuous duty. A requirement of service that demands operation at a substantially constant load for an indefinitely long time.

Intermittent duty. A requirement of service that demands operation for alternate periods of (1) load and no load, (2) load and rest, or (3) load, no load, and rest, such alternate intervals being definitely specified.

Control Apparatus and Switchgear

Electric controller. A device, or group of devices, that serves to govern, in some predetermined manner, the electric power delivered to the apparatus to which it is connected.

Full magnetic controller. An electric controller having all of its basic functions performed by devices that are operated by electromagnets.

Semimagnetic controller. An electric controller having only part of its basic functions performed by devices that are operated by electromagnets.

Manual controller. An electric controller having all of its basic functions performed by devices that are operated by hand.

Drum controller. An electric controller that utilizes a drum switch as the main switching element.

Starter. An electric controller for accelerating a motor from rest to normal speed, and for stopping the motor. (A device designed for starting a motor in either direction of rotation includes the additional function of reversing and should be designated a controller.)

Automatic starter. A starter in which the influence directing its performance is automatic.

Across-the-line starter. A device that connects the motor to the supply without the use of a resistance or autotransformer to reduce the voltage. It may consist of a manually operated switch or a master switch, which energizes an electromagnetically operated contactor.

Master switch. A switch that dominates the operation of contactors, relays, or other remotely operated devices.

Resistor. A device the primary purpose of which is to introduce resistance into an electric circuit. A resistor, as used in electric circuits for purposes of operation, protection, or control, commonly consists of an aggregation of units. (Resistors, as commonly supplied, consist of wire, metal, ribbon, cast metal, or carbon compounds supported by or embedded in an insulating medium. The insulating medium may enclose and support the resistance material as in the case of the porcelain tube type, or the insulation may be provided only at the points of support as in the case of heavy-duty ribbon or cast iron grids mounted in metal frames.)

Overcurrent protection (overload protection). The effect of a device operative on excessive current (but not necessarily on short circuit) to cause and maintain the interruption of power in the main circuit.

Undervoltage or low voltage release. The effect of a device, operative on the reduction or failure of voltage, to cause the interruption of power to the main circuit, but not to prevent the reestablishment of the main circuit on return of voltage.

Constant torque resistor. A resistor for use in the armature or rotor circuit of a motor in which the current remains practically constant throughout the entire speed range.

Fan duty resistor. A resistor for use in the armature or rotor circuit of a motor in which the current is approximately proportional to the speed of the motor.

Overload relay. An overcurrent relay that functions at a predetermined value of overcurrent to cause the disconnection of the power supply. Note: An overload relay is intended to protect the motor or controller and does not necessarily protect itself.

Step-back relay. A relay which operates to limit the current peaks of a motor when the armature or line current increases. A step-back relay may, in addition, operate to remove such limitations when the cause of the high current has been removed.

Temperature-compensated overload relay. A device that functions at any current in excess of a predetermined value essentially independent of the ambient temperature.

Molded-case circuit breaker. A circuit breaker assembled as an integral unit in a supporting and enclosing housing of insulating material, the overcurrent and tripping means being of the thermal type, the magnetic type, or a combination of both.

Rated continuous current. The designed limit in root-mean-square (rms) amperes or DC amperes, which a switch or circuit breaker will carry continuously without exceeding the limit of observable temperature rise.

Circuit breaker interrupting rating (rated interrupting current). The highest rms current at a specified operating voltage that a circuit breaker is required to interrupt under the operating duty specified and with a normal frequency recovery voltage equal to the specified operating voltage.

Control circuit. The circuit that carries the electric signals of a control apparatus or system directing the performance of the controller but does not carry the main power circuit.

Indicating circuit. That portion of the control circuit of a control apparatus or system that carries intelligence to visible or audible devices which indicate the state of the apparatus controlled.

Types of Circuits and Terms

Lighting branch circuit. A circuit supplying energy to lighting outlets only. (Lighting branch circuits also may supply portable desk or bracket fans, small heating appliances, motors of one-quarter horsepower (186½ W) and less, and other portable apparatus of not over 600 W each.)

Appliance branch circuit. A circuit supplying energy to one or more outlets to which appliances are to be connected; such circuits are to have no permanent connected lighting fixtures not a part of an appliance.

Motor branch circuit. A branch circuit supplying energy only to one or more motors and associated motor controllers.

Communication circuits. Circuits used for audible and visible signals and communication of information from one place to another aboard the vessel.

Feeder. A set of conductors originating at a main distribution center and supplying one or more secondary distribution centers, one or more branch circuit distribution centers, or any combination of these two types of equipment. (But the circuits between generator and distribution switchboards, including those between main and emergency switchboards, are not considered feeders.)

Branch circuit. That portion of a wiring system extending beyond the final overcurrent device that protects the circuit.

Outlet. A point on the wiring system at which current is taken to supply utilization equipment.

Lighting outlet. An outlet intended for the direct connection of a lamp-holder or a lighting fixture.

Receptacle. A device installed in a receptacle outlet to accommodate an attachment plug.

Receptacle outlet. An outlet where one or more receptacles are installed.

Attachment plug. A device which, by insertion in a receptacle, establishes connection between the conductors of the attached cord and the conductors connected permanently to the receptacle.

Distribution center. A point at which is located equipment generally consisting of automatic overload protective devices connected to buses, the principal functions of which are subdivision of supply and the control and protection of feeders, subfeeders, or branch circuits, or any combination of feeders, subfeeders, or branch circuits.

Switchboards. A device that receives energy from the generating plant and distributes it directly or indirectly to all equipment supplied by the generating plant. A subdistribution switchboard is essentially a section of the generator and distribution switchboard (connected thereto by a bus feeder and remotely located for reasons of convenience or economy) which distributes energy for lighting, heating, and power circuits in a certain section of the vessel.

Distribution panel. A piece of equipment that receives energy from a distribution or subdistribution switchboard and distributes that energy to energy-consuming devices or other distribution panels or panelboards.

Panelboard. A distribution panel enclosed in a metal cabinet.

Demand factor. The ratio of the maximum demand of a system or part of a system to the total connected load of the system or of the part of the system under consideration.

Vital services. Those services required for the safety of the ship and its passengers and crew. These may include propulsion, steering, navigation, firefighting, emergency lighting, and communications functions. Since the specific identification of vital services is influenced by the type of vessel and its intended service, this matter should be specified by the design agent for the particular vessel under consideration.

REVIEW

1. What are the standard system voltages for generation, power, and lighting on ships? What is most common on large merchant ships?
2. What is shown on an electrical one-line diagram?
3. What services must be supplied by the emergency generator on large merchant vessels?
4. What are the advantages of generating electric power using a power takeoff (PTO) generator from the main engine? Describe the four basic arrangements for producing constant frequency power from the PTO generator.
5. How are hazardous areas classified?
6. What equipment is required for an AC generator switchboard?
7. What are the troublesome characteristics of fuses when utilized to protect power circuits on ships?
8. What should the maximum rating of the fuses be for a motor with a full-load current of 40 amps?
9. Describe the operation of a thermal-magnetic molded-case circuit breaker.
10. Where are metal-clad air circuit breakers typically used in shipboard electrical systems?
11. A transformer has a primary-to-secondary coil turn ratio of 2-to-1. If the supply voltage current is 240 volts and 20 amps, what will the secondary voltage and amperage be?
12. What are the four possible arrangements for three-phase transformers? What is the most common in shipboard use?

13. What is an autotransformer?
14. What are the four basic categories of electric propulsion?
15. Describe the procedure for reversing an AC turbo-electric drive system.
16. Describe how solid-state frequency converters can be used with a constant frequency bus in an AC propulsion system.
17. Define resistance and conductance. What are the common units of each?
18. Define capacitance and capacitive reactance. What are the common units of each?
19. Define inductance and inductive reactance. What are the common units of each?
20. Define impedance? Explain how the impedance triangle is used to calculate impedance?
21. Compare true power, reactive power, and apparent power. What are the units of each?
22. Define power factor. What is power factor correction?
23. How are cables sized for shipboard lighting and power services? What are the maximum voltage drops permitted for each?
24. List the different types of enclosures for electrical equipment per IEEE Standard 45.

REFERENCES

National Fire Protection Association. *National Electrical Code,* NFPA 70. Quincy, Mass.

Institute of Electrical and Electronics Engineers. *Recommended Practice for Electrical Installations on Shipboard,* IEEE Standard No. 45.

Harrington, R. L. (ed.). 1992. *Marine Engineering,* 3rd. ed. Jersey City, N.J.: Society of Naval Architects and Marine Engineers.

Young, W. A. 1999. *Marine Electrical Basics Workbook,* 3rd ed. Rockville, Md.: ABS Group Inc., Government Institutes Division.

CHAPTER 23

Electrical Machinery

CONRAD C. YOUNGREN

SYNCHRONOUS ALTERNATORS

Introduction

THE bulk of electric power for everyday use is produced by polyphase synchronous alternators, which are the largest single-unit electric machines in production. As ashore, three-phase electrical power on shipboard is generated with synchronous machines. The name *synchronous machine* comes from the fact that these machines are generally operated at a constant (synchronous) speed, independent of load and directly proportional to line frequency:

$$N_s = \frac{120f}{P}$$

In this formula N_s is the synchronous speed in RPM, f is the line frequency, and P is the number of poles. (Note: As in all rotating machinery, stator and rotor poles are identical in number.)

Like most rotating machines, synchronous machines are capable of being operated to produce electrical power as generators (at frequency f, when rotated by prime mover at N_s) or as motors (rotating at N_s when supplied from mains at frequency f). Synchronous motors will be discussed in another section, but for both motors and generators operating at *60 Hz* the following table illustrates the relationship between machine poles and RPM.

# poles	2	4	6	8	10	12	14	...	36	...	80
N_s (RPM)	3600	1800	1200	900	720	600	514		200		90

In synchronous alternators, as in all electromagnetic devices, voltage is determined by relative motion between conductors and lines of magnetic flux. In small units the armature (load carrying conductors) may be rotated in way of a magnetic field. In the majority of synchronous machines, and in all of the larger units, magnetic flux is produced in the rotor poles and swept across stationary armature windings. These armature windings consist of coils embedded in slots on the periphery of the stator. The number of slots spanned by an armature coil is called the coil pitch. The pole pitch is the total number of stator slots divided by the number of machine poles. The coil pitch is usually slightly less than the pole pitch and is in units of *electrical degrees,* since by definition poles are separated by 180 degrees:

$$\text{Coil pitch} = 180^\circ \times \frac{\text{(coil pitch in slots)}}{\text{(pole pitch in slots)}}$$

Note that for a two-pole machine, electrical degrees and mechanical degrees are identical. In a four-pole machine one quarter of the distance around the periphery (90 *mechanical* degrees) constitutes 180 *electrical* degrees, and in general:

$$\text{electrical degrees} = \text{mechanical degrees} \times \frac{\text{\# poles}}{2}$$

Construction

Stator windings. Figure 23-1 shows a developed view of a two-layer, three-phase stator winding. In this view the cross section of the cylindrical stator is laid flat; in the actual stator, slot 18 is adjacent to slot 1. In a two-layer winding each slot holds two coil sides (the "left" side of one coil and the "right" side of another) and, since each coil has two sides, the number of stator slots is equal to the number of stator coils. Each phase has the same number of coils (18/3 = 6), and each phase has groups of coils corresponding to the number of poles (e.g., the "A-group" and the "a-group" for phase A). Therefore the number of coils per group equals the number of "slots per pole per phase." In the example winding, each coil spans 8 slots (A-group coil #1: slots 1 and 9; coil #2: 2 and 10; coil #3: 3 and 11, for example) and the center of each group of a particular phase is separated by 180 degrees (A-group is centered at slot 6, a-group is centered at slot 15—9 slots away). Thus when peak positive voltage is induced in the A-group, peak negative voltage is induced in the a-group. The senses of the A-group and a-group windings are opposite, so that with the groups connected in series the A-phase sinusoidal voltage is twice the group

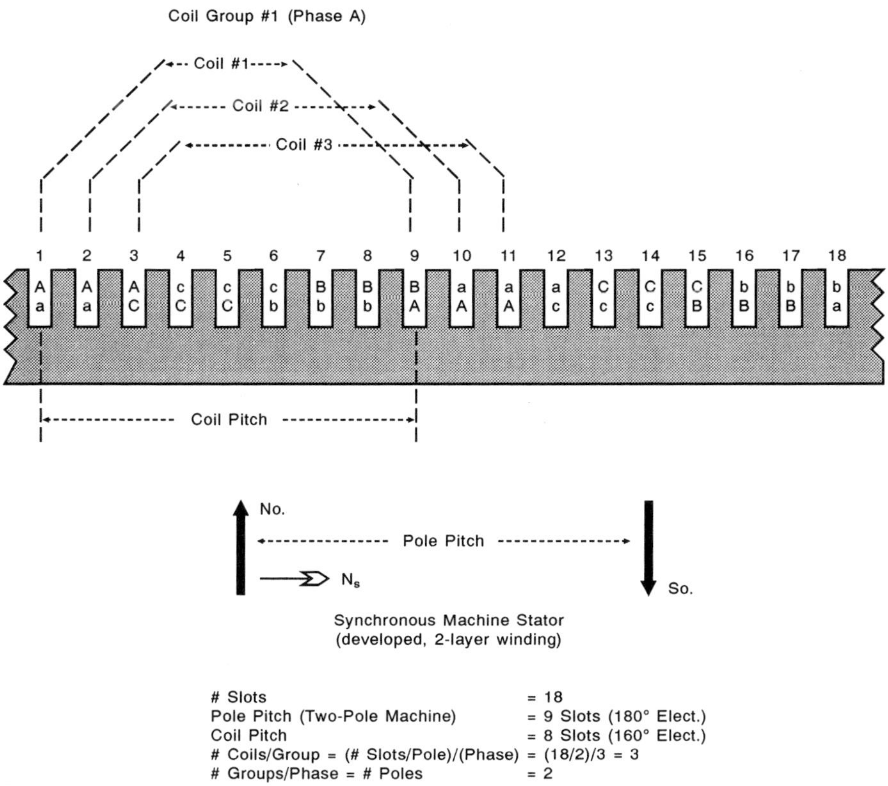

Figure 23-1. Synchronous machine stator construction

voltage. The B-phase windings are centered at slots 12 and 3, *120 degrees away* from the centers of the A-phase groups; hence peak values will be reached in these groups one-third of a cycle after the A-phase groups have peaked. Similarly, the C-phase groups are centered at slots 18 and 9, 120 degrees after the B-phase and 120 degrees before the A-phase groups. This physical distribution of the phase windings results in the production of three sinusoidal voltages (of equal amplitude and frequency) out of phase with one another by exactly one-third cycle. The phase winding leads may be connected in either a delta or a wye, yielding line voltages equal to phase voltages (delta) or 1.73 times the phase voltage (wye). Wye connected windings must be sized to carry full line current, and delta connected windings carry 86.6 percent (1/1.73) of line current under balanced conditions. Wye connected machines allow for four-wire, three-phase service as the neutral terminal is accessible. Embedded temperature detectors are generally provided in the stator windings of machines rated over 500 kVA.

Rotor windings. Rotor windings on synchronous machines are "concentrated" windings (as opposed to the "distributed" windings that cover the periphery of the stator), which carry DC current in an appropriate direction so as to determine the polarity of each individual pole. On high speed machines, typically 3600 RPM or 1800 RPM turbine-driven alternators (turbogenerators or TG), the cylindrical construction pictured in Figure 23-2(a) is utilized. This compact construction is necessary because the high centrifugal stresses make the bulkier salient-pole construction (Figure 23-2(b)) difficult to design. On lower speed machines, such as the 10-pole (720 RPM) rotor shown in Figure 23-2(c), rotor designs have projecting laminated pole pieces mounted on the rim of a steel spider. The spider is in turn keyed to the rotor shaft. The field coils are fitted over the pole pieces and connected so that adjacent poles have opposite polarity when energized by the DC excitation current. Amortisseur or dampening bars are embedded in the pole faces and connected to one another by conducting straps that serve to dampen rotor oscillations during abrupt load changes.

Excitation systems. Regardless of rotor type, cylindrical or salient pole, a source of DC current is needed to excite the field of the alternator. For many machines, field power is supplied by a DC generator called an *exciter,* which itself may be separately excited by a *pilot exciter.* These exciters may be driven by separate prime movers or their rotors may be on the same shaft as the field poles of the alternator, as shown in Figure 23-3. In either case, DC power must be extracted from the armature of the exciter through a commutator and brushes and injected into the alternator field circuit through a set of slip rings and brushes. Note that Figure 23-3(a) shows three sets of sliding contacts: (1) amplidyne (pilot exciter) armature to exciter field connection; (2) exciter armature to stationary (exciter) terminal connection; and (3) stationary terminal to alternator field connection. Alternator output voltage and current are monitored by potential and current transformers respectively. The voltage regulator serves to maintain the alternator line voltage at the desired level by compensating for the effects of generator loading. By adjusting the pilot exciter's field, the exciter voltage is altered, and control of DC excitation to the alternator rotor is thereby accomplished.

As brush/commutator/slip ring arrangements tend to be a significant maintenance burden, many modern alternators are fitted with a brushless excitation system as illustrated in Figure 23-3(b). Here the pilot exciter is a small permanent magnet alternator whose output voltage, together with a feedback voltage from the synchronous alternator armature terminals,

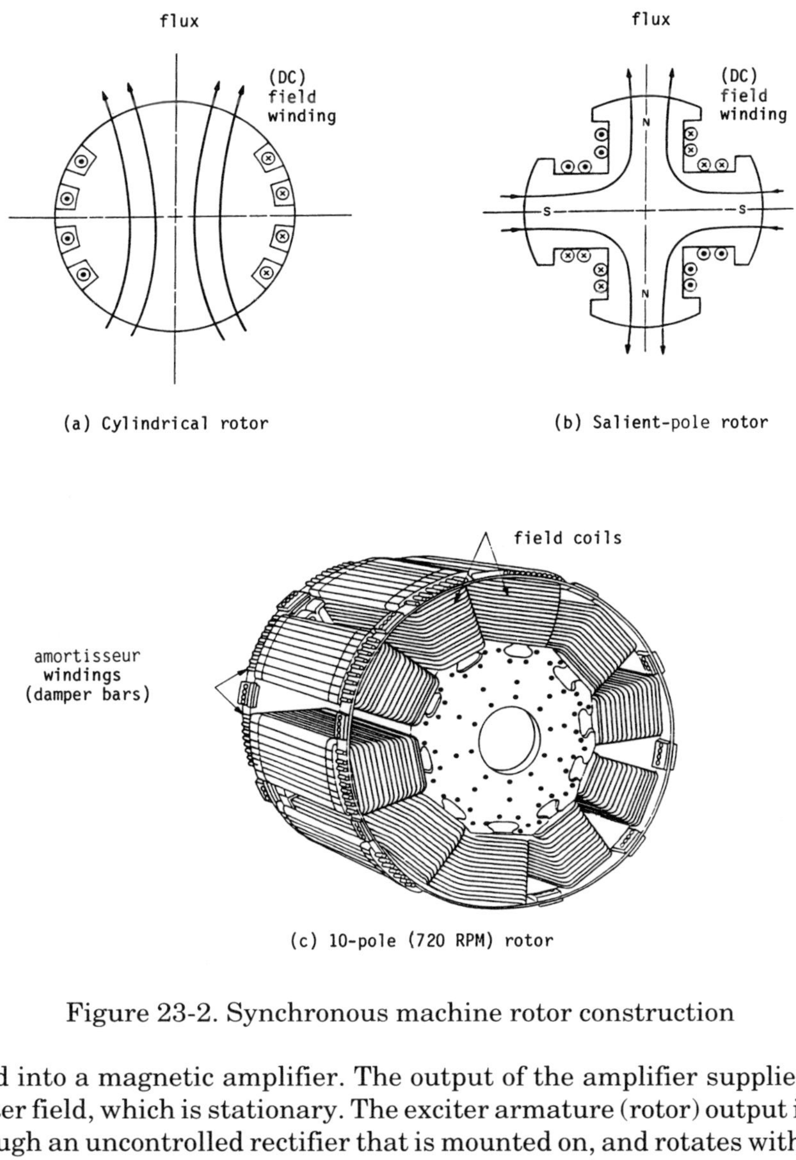

Figure 23-2. Synchronous machine rotor construction

is fed into a magnetic amplifier. The output of the amplifier supplies the exciter field, which is stationary. The exciter armature (rotor) output is fed through an uncontrolled rectifier that is mounted on, and rotates with, the shaft common to pilot exciter permanent magnet, exciter armature, and alternator field poles. The rectifier supplies the alternator field with its DC excitation. Control is accomplished through regulating the magnetic amplifier output. Another version of this arrangement includes controlled rectification of the alternator's output directly to the (stationary) exciter field, eliminating the pilot exciter.

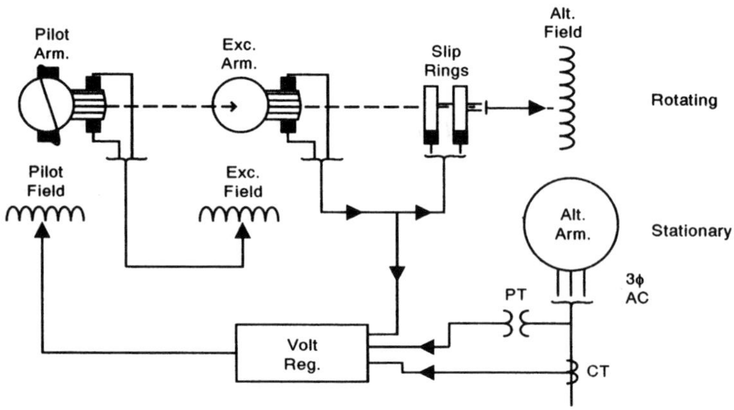

(a) DC Exciter

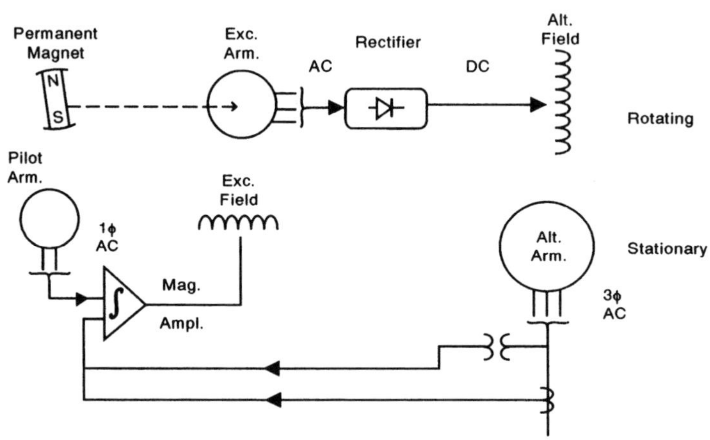

(a) Brushless Exciter

Figure 23-3. Exciter arrangements

Operation

Alternator control. Control of synchronous alternator output is accomplished through the manipulation of (1) prime mover speed (or more precisely, torque) and (2) exciter output (and thereby rotor field strength). The first of these is normally performed by adjusting the prime mover governor set point. While this action determines the no-load speed of the prime mover, the speed (and hence frequency) of the alternator under load is also a function of plant status (i.e., total load, number of machines paralleled, load share, etc.). Exciter output, for a given speed, determines no-load voltage. Again the system voltage is a function of plant status as well. In the *automatic* mode, adjustment of the automatic voltage regulator (AVR) set point accomplishes this control. In the *manual* mode, direct control of exciter field current is available. Manipulation of both governor and AVR set points may have different purposes depending on whether the affected alternator is operating alone or in parallel. These cases will be treated separately in the following paragraphs.

Single alternator operation. When an alternator is operating alone, its load is determined by the aggregate demand of all plant equipment on line. Power factor is a function of the reactive/resistive characteristics of the load items and cannot be adjusted by manipulating the alternator control devices. In this mode the governor set point control has a direct effect on generator speed and therefore bus frequency. Thus any deviation from the nominal system frequency may be corrected by raising or lowering the governor set point. With the exciter output under the control of the AVR, any slight speed adjustment will have a negligible effect on system voltage. System voltage may be adjusted by raising or lowering the AVR set point. System voltage does have a slight effect on load characteristics, as does frequency. However, since the goal is to maintain rated voltage and frequency (e.g., 450 volts, 60 Hz), it is not the normal function of the governor or AVR to manipulate load characteristics.

The following procedure may be used as a guide for single alternator operation:

1. Make a visual inspection to insure that all repairs have been completed, and that unit is free of tools or other debris and in general is in a condition satisfactory for operation.
2. Insure that disconnect links are in place. (Note: Disconnect links should be opened or closed with an insulated wrench and always with the alternator circuit breaker open.)
3. Start prime mover.
4. Switch voltage regulator to *Automatic*.
5. Adjust frequency to rated value with governor set point control.
6. Use AVR set point control to obtain rated voltage. Check A-B, B-C, and C-A voltages (they should be equal), and line-to-ground voltages where possible.
7. Close circuit breaker. (Note: In some installations where parallel operation is possible, the synchroscope switch is interlocked with the circuit breaker switch so that even under single generator operation the synchroscope switch must be on *Gen. No. 1* in order to close the circuit breaker on generator number 1. The synchroscope should be turned off once the alternator is on line.)

Parallel operation

Synchronizing. Off line, the governor and AVR set point controls have the identical functions as described under single alternator operation. Thus the procedure for parallel operation is the same as above *up through step 6.* For connection in parallel the goal is to have the incoming generator output be in phase with and identical to (or slightly greater than) the bus voltage in amplitude and frequency. Here is a guide to complete the paralleling procedure:

7. Adjust incoming generator voltage to slightly above bus voltage. It is preferred to utilize the same voltmeter (i.e., a calibrating voltmeter) to compare voltages rather than two different meters on the on-line and incoming machines.
8. Switch the synchroscope to the incoming machine and adjust frequency until the synchroscope pointer revolves slowly in the *fast* direction. Pointer position indicates relative phase of incoming machine and bus voltages: 12 o'clock identifies a zero phase difference and 6 o'clock identifies a 180-degree phase difference. The *rate* of rotation indicates the frequency differential, with 1 RPM (the speed of a clock's second hand) identifying a difference of 1⁄60 Hz.
9. Check synchroscope indication against synchronizing lamps to insure proper operation of synchroscope. Indicating lamps should be alternating bright and dark in unison and at a rate equal to synchroscope rotation, with 12 o'clock corresponding to the middle of the dark period, and 6 o'clock corresponding to the brightest condition of the lamps. (Note: If lamps are not in unison, phase rotation of oncoming machine is opposite that of bus—*Do not parallel machines.*)
10. Close circuit breaker at a few degrees before 0 degrees. (Because of a slight time lag—15 to 30 cycles—for the closing mechanism to complete its operation, it is good practice to anticipate closing by a few degrees so that actual closing occurs at 0 degrees.) Closing of the breaker with voltage out of phase by more than 15 degrees will cause dangerous crosscurrents between the machines paralleled. Out-of-phase closing at 180 degrees is identical to short-circuiting both machines!

A few additional comments on step 9 are required. Any discrepancy between synchronizing lamps and synchroscope usually indicates a defective synchroscope. This assumes proper lamp connection and use of identical wattage lamps with the specified voltage rating.

Load sharing. The total plant load is the sum of the true power (kW) and reactive power (kvar) load on each machine operating in parallel. The load is balanced when the kW and kvar loads are distributed in proportion to the ratings of the machines paralleled. For purposes of this illustration, it is assumed that two machines of equal capacity (rating) are supplying a common load.

The prime mover governor position has a direct effect on the steam flow to a turbine-driven machine (fuel flow to a diesel generator). Considering a turbogenerator example, the aggregate steam flow to both machines, operating at nominal bus voltage and frequency, is determined by the plant kW load. The *division* of kW load between generators is in approximate proportion to the fraction of total steam flow to each unit. Increasing the governor set point of, say, generator #1 while *simultaneously* decreasing the governor set point on generator #2 will shift some of the kW load to generator #1 from #2. Since the total steam flow has not changed, the bus frequency remains unchanged. It should be noted that generator #1 speed is not increased nor is #2 speed decreased, but rather at a constant speed (frequency) prime mover #1 is producing more torque (torque × RPM = horsepower) and #2 less. If the governor set point of #1 were increased without a corresponding decrease of the #2 setting, generator #1 would take on an increased share of the kW load; but since the total steam flow had been increased, *both* machines would run faster and the bus frequency would rise. Increasing the governor set points of both machines simultaneously would similarly raise the bus frequency without changing the kW load share percentage of either machine. Corresponding arguments could be made when decreasing the governor set points of either (or both) machines.

The aggregate excitation (which may be thought of in terms of either total flux or total rotor field current) determines the system voltage for a given load demand. Reactive (kvar) loads are shared between machines in approximate proportion to the percentage of total excitation supplied to each machine by its AVR (assuming kW loads are equal). Therefore, increasing the AVR set point of generator #1, while *simultaneously* decreasing the AVR set point of #2, shifts kvar load to #1 from #2. And, since the total excitation has not changed, the bus voltage remains constant. Increasing the excitation to #1 only will increase its share of reactive load; but, since the total excitation has been increased, the system voltage will rise. When balancing kvar load, remember that increasing the kvar load *decreases* the power factor of that machine. Since most ship's switchboards do not include varmeters, equal power factor meter readings indicate reactive power balance (assuming kW load has already been balanced). Another indication of equalized reactive power division is amperage of each machine. With the kW load balanced (say, equally), equal reactive load will be realized when the amperage of both machines is the same.

In the following numerical example, two machines of equal capacity are operating in parallel, with generator #1 having just been put on the line. The status in Figure 23-4(a) is that unit #1 is supplying 40 kW and 40 kvar, #2, 760kW and 560 kvar (total kW load is 800, total kvar load is 600). Note that there is not a great disparity in power factor, yet the amperages differ widely. Also the total amperage is 73 plus 1211—1284 amps.

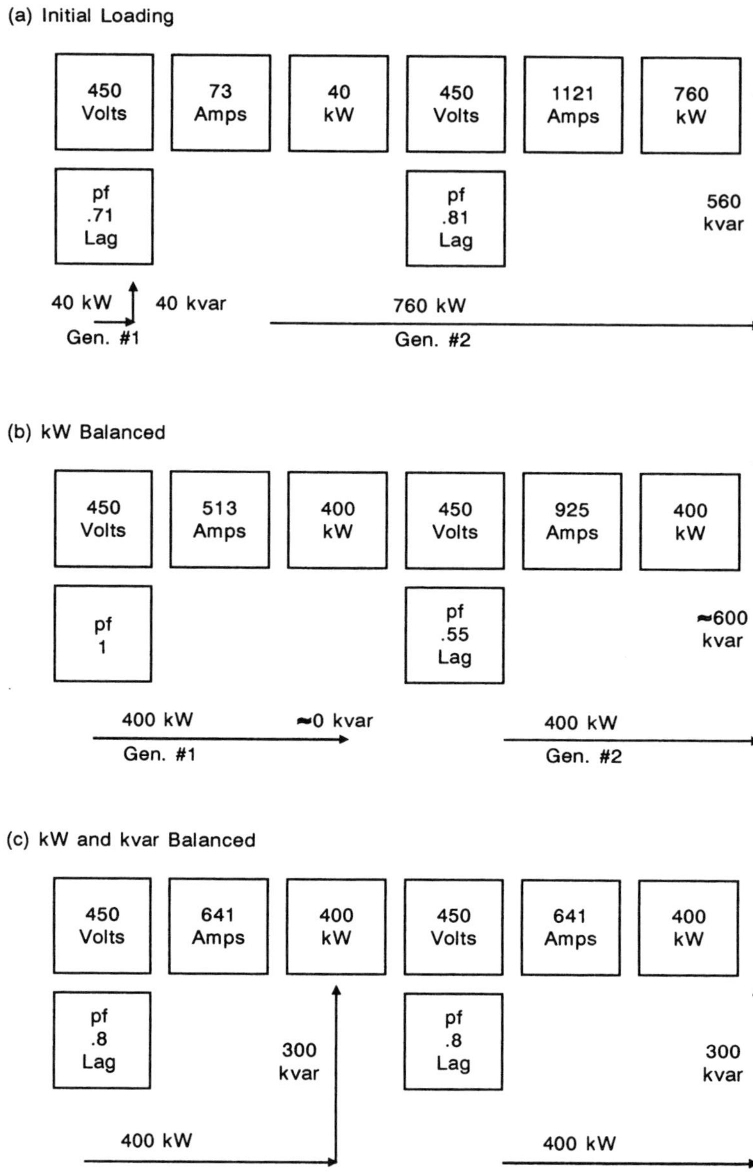

Figure 23-4. Alternator load sharing

The governor set point of generator #1 is increased as the set point on #2 is decreased a corresponding amount to keep the bus frequency at 60 Hz and until equal kW loading is achieved (400 kW). At this increased kW load, generator #1 is now grossly underexcited (higher pf, lower amperage) while #2 is overexcited (low, lagging pf, higher amperage). The total amperage is now 513 plus 925—1438 amps (Figure 23-4(b)).

Finally, the AVR set point of generator #1 is increased while simultaneously the set point is lowered on unit #2, maintaining the system at 450 volts while shifting 260 kvars to #1 generator. The result is equal kW, equal kvar, equal amps, and equal pf (Figure 23-4(c)). Total current in the balanced condition is 641 plus 641—1282 amps, which is the minimum value for any distribution of an 800 kw, 600 kvar system load.

The above is an oversimplification of a complicated set of phenomena. In particular, increasing kW load on a synchronous alternator without increasing field strength causes some shedding of kvar to the other machine and may result in a leading power factor for the machine gaining kW. An even greater over-simplification, though occasionally useful, is the statement: "The governor controls watts, the exciter controls amps."

Maintenance

Introduction. Electrical maladies in AC machinery generally consist of either opens (discontinuity where there should be continuity), shorts (low resistance paths bypassing circuits/devices intended to be energized), and grounds (low resistance paths to ground or component frame or housing). Controller malfunctions, of course, can be a maintenance problem, as can mechanical failures of bearings, couplings, ventilation, etc. The following sections, however, will deal with locating opened, shorted, and grounded windings in the stator and rotor of synchronous machines. A breakdown of the winding insulation is a root cause of any of these, and excessive heat, moisture, or mechanical abrasion will lead to insulation deterioration. Keeping machinery clean and dry is an obvious preventive maintenance priority, and periodic monitoring of bearing and coupling performance will lessen the probability of mechanical damage due to reduced clearances. A synchronous alternator troubleshooting chart is presented as Table 23-1.

TABLE 23-1

Synchronous Alternator Troubleshooting Chart

Problem	*Possible Cause*	*Remedy*
No output voltage	Defective voltmeter	Check with properly operating voltmeter and replace if defective.
	Open/shorted PT	Check potential transformer (PT) primary and secondary voltages; replace PT if defective.
	Open field coil(s)	Test and replace defective coil(s).
	Exciter failure	Test and repair (see Table 23-5).
	Shorted slip rings	Disconnect field coils and check slip ring insulation resistance with megger.
	Blown diodes	Test and replace if defective.
	Defective regulator	Check fuses, connections, instrument transducers.

Problem	*Possible Cause*	*Remedy*
Low output voltage	Low speed	Check frequency; adjust governor set point.
	Insufficient excitation	Check regulator output; adjust/repair as necessary; troubleshoot exciter (see Table 23-5).
	Shorted/grounded field coil(s)	Test and repair.
	Shorted/grounded armature coil(s)	Test and repair.
High output voltage	High speed	Check frequency; adjust governor set point.
	Overexcitation	Check regulator output; adjust/repair as necessary; troubleshoot exciter (see Table 23-5).
	Open armature winding (one leg of delta connection)	Repair/replace defective armature coil(s).
Fluctuating voltage	Poor brush/slip ring contact	Clean slip rings and brushes; replace brushes.
	Faulty regulator	Check regulator output; regulator sensing transducers, loose connections.
	Irregular speed	Check frequency; adjust governor if required; check prime mover operation.
Fluctuating frequency	Faulty governor	Adjust governor; check lube oil pressure for fluctuation.
	Irregular excitation	Check regulator performance; check exciter connections.
	Prime mover performance	Check fuel system (DG); steam supply, back pressure (TG).
Overheating	Overload	Reduce load.
	Phase imbalance (extreme)	Check for open phase connection; open/shorted/grounded armature winding.
	Insufficient ventilation	Remove restrictions; clean windings; check C.W. supply.
	Bearings	See "Noisy operation" below.
	Uneven air gap	See "Noisy operation" below.
	Shorted armature winding	Test and repair/replace.
	Shorted field coil	Test and repair/replace.
Noisy operation	Bearings	Check for worn, loose, dry, or overlubricated bearings; replace worn or loose bearings; lubricate dry bearings; relieve overlubricated bearings.
	Uneven air gap	Check for bent shaft, bearing condition, gear alignment.
	Coupling	Check alignment; re-align and/or tighten coupling.
	Loose laminations	Tighten bolts; dip in varnish and bake.

Stator faults

Shorted stator coils. A higher than normal current, particularly in one phase, is an indication of a shorted stator coil. Visual inspection of the stator windings for evidence of burning or overheating may be all that is necessary to identify the offending coil. When discoloration is not evident, a "growler" may be used to locate the shorted coil. A growler consists of a coil wound on an iron core. It is connected to an AC source and placed against the armature core. Either a flexible feeler built into some portable growlers, or a separate feeler fashioned from a hacksaw blade, will vibrate (causing a buzzing sound) when located over a slot containing a shorted coil. The alternating flux produced by the growler induces a voltage in each armature coil as the feeler is moved from slot to slot around the periphery of the stator. If a coil is shorted, the completed circuit allows current to flow in that coil and the resulting magnetic field encircling the shorted coil attracts and repels the feeler, causing the telltale buzzing (growling).

Grounded stator coils. "Megging" between stator terminals and the machine frame will indicate whether or not there is a grounded stator coil, but will not identify the particular coil. Often the path to ground is through minute cracks in the insulation that have become filled with conducting contaminants; and moisture absorbed in the windings or condensed on the insulation surface will result in low insulation resistance readings. Thus cleaning and drying of the stator windings will often remedy the problem. For trending purposes it is advisable to take periodic insulation resistance readings at regular intervals and under similar conditions, usually immediately after shutdown when the windings are still warm and, presumably, dry. On large machines, however, good insulation has considerable capacitive effect between conductors and frame, and a large static charge can be built up during normal operation of the machine. Thus the windings of large machines should be grounded for about 15 minutes prior to taking "megger" readings. The response of the megohmmeter to good insulation is an initial downward dip, followed by a gradual climb toward the true resistance value. Slight downscale dips during this climbing period are indicative of dirty insulation, thus careful observation of the pointer movement during the test is important.

Visual inspection may identify the problem coil. As an aid in spotting the grounded coil, the somewhat sarcastically named "smoke test" is employed. After first insulating the stator from ground with wood blocks or other heavy insulating material, a reduced voltage is applied between stator winding and machine frame as shown in Figure 23-5. The current limiting rheostat should be adjusted to allow sufficient current to circulate through the defective coil to frame to cause overheating and perhaps smoking at the grounded location.

Open stator coils. An open in a single-circuit stator (one conducting path per phase) will prohibit a three-phase motor from starting and result in severe phase imbalance in a running motor or generator. If the open occurred under load conditions the resulting arcing may have caused visible damage to the insulation in the vicinity of the open. If not, the following sequence of tests is a guide to determine the exact location of the defective coil:

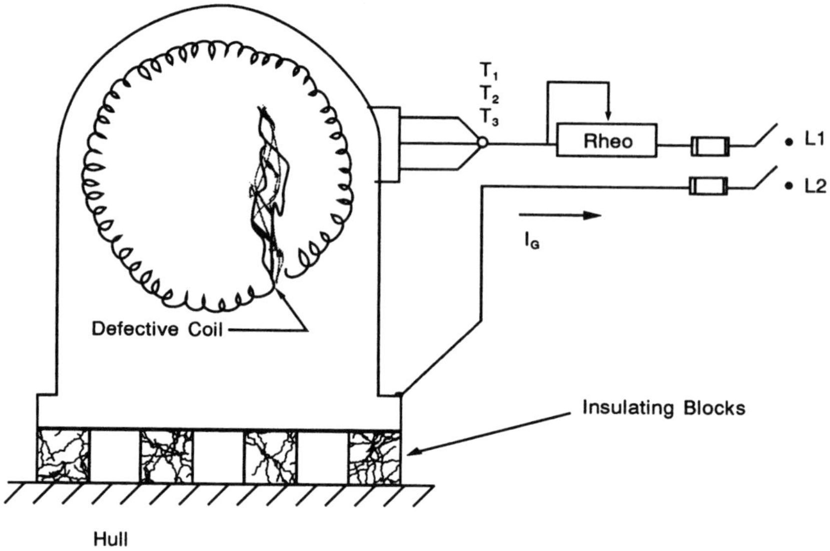

Figure 23-5. Smoke test

1. Determine if winding is wye or delta. Meg between each pair of terminal leads; three zero readings indicate a delta, one zero and two high readings indicate a wye with the offending phase identifiable by the lead common to the two high readings (Figure 23-6(a)).
2. If delta, isolate phase. Disconnect and carefully tag for identification the nine leads (two per phase plus three line leads). By megging each pair, the two leads to the phase containing the defective coil can be identified. In Figure 23-6(b), an infinity reading will result when megging between a1 and any other terminal, and a similar result will occur with a2. With the other four phase leads, however, there will be one other lead to which there is continuity (b1-b2, c1-c2).
3. Pinpoint fault. With the offending phase leads isolated, one lead of the megohmmeter is clamped to one of the phase leads and a needlepoint probe on the other megger lead is moved from coil junction to coil junction until a high reading is indicated. The faulty coil is located between junctions that yield zero and high readings, as illustrated in Figure 23-7(a).

Large machines may have multiple parallel paths in each phase, in which case each path must be isolated, tagged, and tested as in step 2 above before the fault can be pinpointed as in step 3.

Emergency repairs. Operating a machine with one coil out of service may have serious effects on a machine with only a few coils per phase.

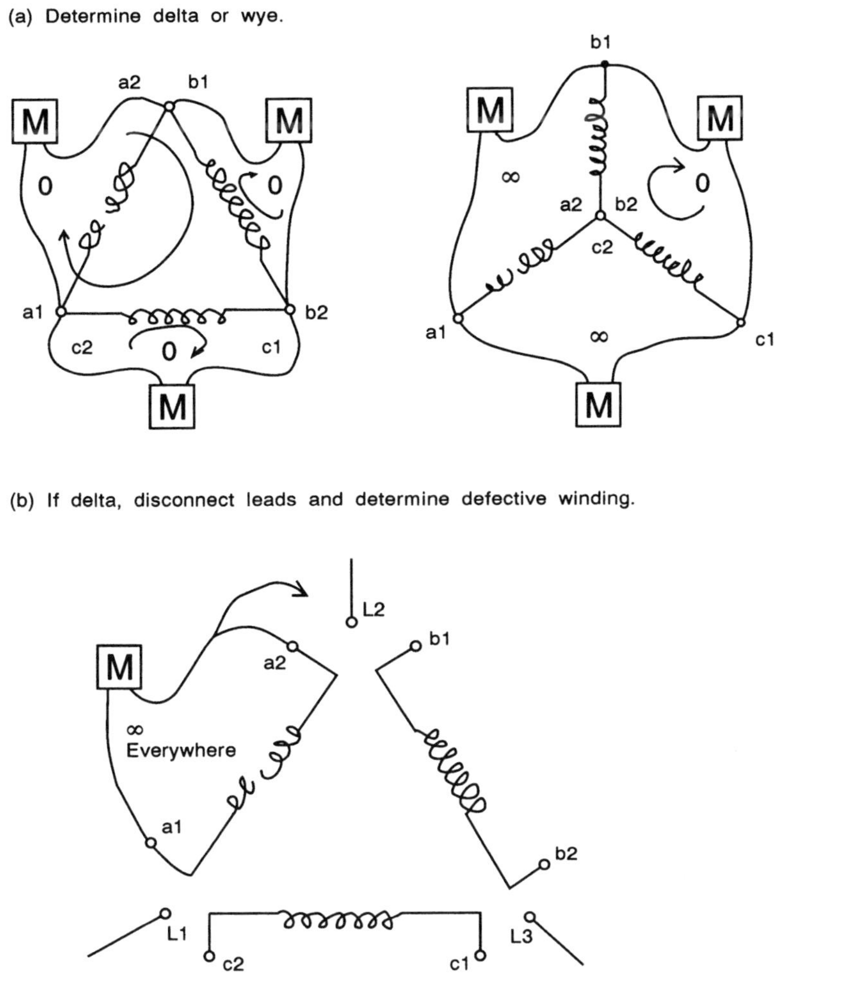

Figure 23-6. Identifying open stator phase

Large machines, however, may operate satisfactorily at reduced loads with a number of coils cut out. An emergency repair—cutting a defective coil in half and jumping adjacent coils to complete the stator circuit (Figure 23-7(b))—may be appropriate depending on the number of stator coils in the machine.

Rotor faults

Shorted field coils. By applying 120 or 240V AC (60 Hz) to the collector rings of a salient-pole synchronous alternator (or motor), and measuring the voltage drop across each individual field coil, it is possible to determine

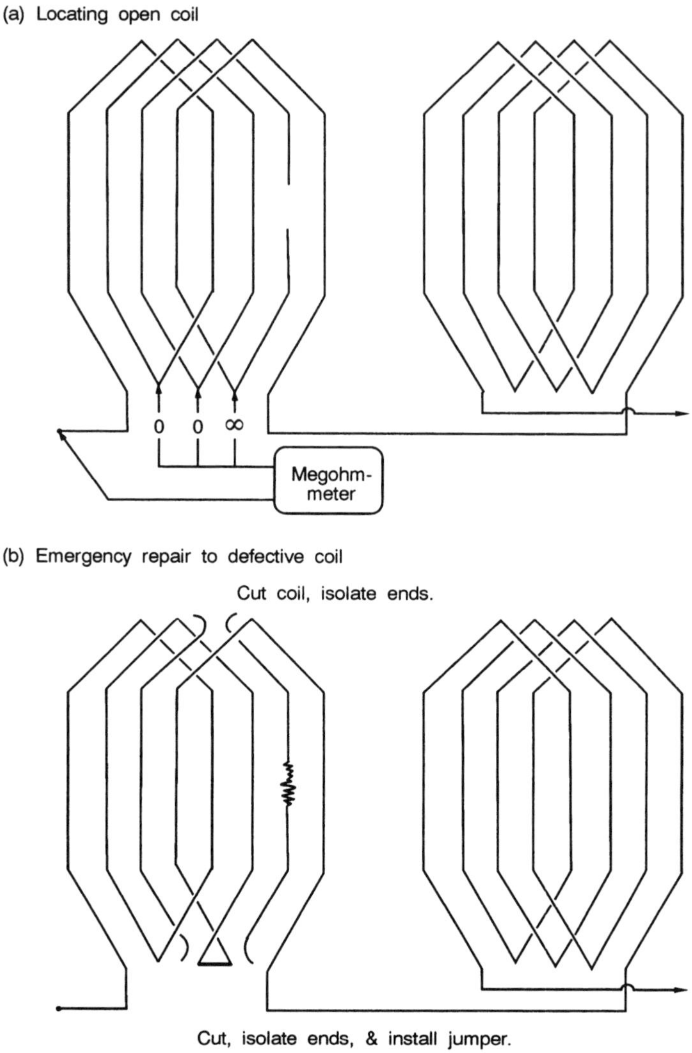

Figure 23-7. Identifying open stator coil

the defective coil (Figure 23-8). If all the coils were electrically identical (same impedance), the voltage drop across each should be 240 divided by the number of poles. On the presumption that a shorted coil will exhibit a lower impedance, simple voltage division dictates that the shorted coil will yield a smaller voltage drop than the rest. On brushless machines the diode circuitry must be first isolated from the field circuit and the AC voltage applied to the terminals of the field circuit.

Grounded field coils. A grounded coil may be identified with a megohmmeter after disconnecting each coil from the others (Figure 23-9). Each coil is then tested coil lead-to-shaft, a low (zero) reading indicating a grounded coil. On brushless machines care should be taken to isolate the shaft-mounted rectifier before using a megger on the rotor windings, as the high voltage generated by the megger could damage the diodes.

Open field coils. Open field coils may be identified by applying rated (DC) voltage or less to the field circuit. A voltmeter is connected from one field lead to successive coil junctions (Figure 23-10). The voltmeter will read zero until the defective coil is bridged, at which time full voltage will be indicated. Alternatively a megohmmeter could be used to locate the discontinuity. Open or grounded field coils should be replaced; shorted field coils should eventually be replaced too, but in an emergency may be left and the machine operated, provided overheating and/or vibration is not present.

Reversed field coils. Since adjacent field poles should be of opposite polarity, a simple way of testing for correct polarity is with a compass. Alternatively two bolts may be put in contact with the faces of adjacent poles and an attempt made to bridge the gap between poles by bringing the bolt heads together. Strong attraction of the bolt heads indicates opposite polarity of the adjacent poles; repulsion between the bolts indicates identical polarity and, hence, a reversed pole.

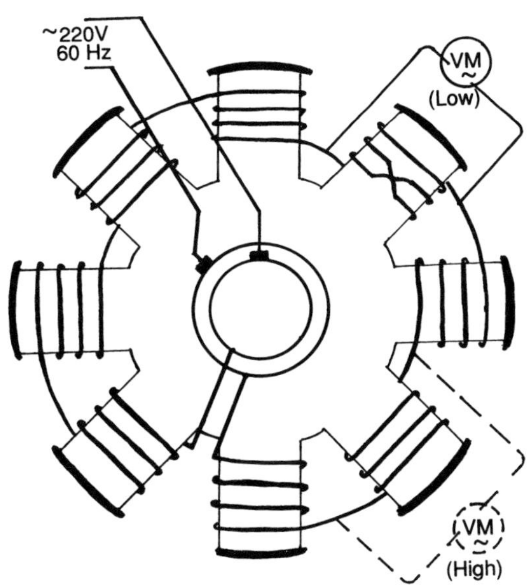

Figure 23-8. Identifying shorted field coil

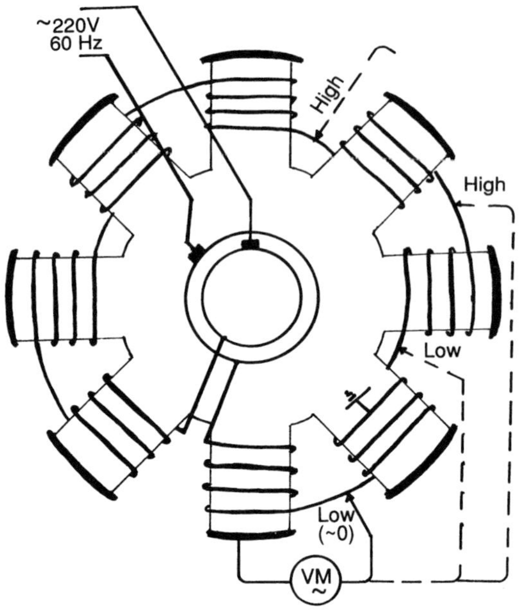

Figure 23-9. Identifying grounded field coil

SYNCHRONOUS MOTORS

Description

The description of the synchronous machine in the previous section is appropriate to both motor and generator operation. In fact, there are operations where the same machine is used for both applications: a synchronous machine as a motor driving a DC generator to charge a bank of batteries which, in turn, could supply the DC machine, acting as a prime mover for the synchronous machine in the role of an alternator; and a synchronous motor driving a pump in a large pump-storage facility during off-peak hours, acting as a generator driven by the pump/water turbine to supply peak electrical power. The fundamental difference between the two modes of operation is, of course, that as an alternator the machine is spun at synchronous speed by a source of mechanical power (prime mover), is supplied with DC excitation, and delivers electrical power in the form of three-phase AC. As a motor the stator is energized with three-phase AC, the rotor again is DC excited, and mechanical power (torque at synchronous speed) is produced.

Rotating Magnetic Field

The key to understanding the operation of polyphase motors (both synchronous and induction machines) is comprehension of the production of a magnetic field, rotating at synchronous speed (f × 120/P), by the stationary windings of the machines. The stator (armature) windings, as described previously, consist of three independent (phase) circuits distributed around the periphery of the stator such that each of the phase windings is separated from the other two by 120 degrees (electrical). When each

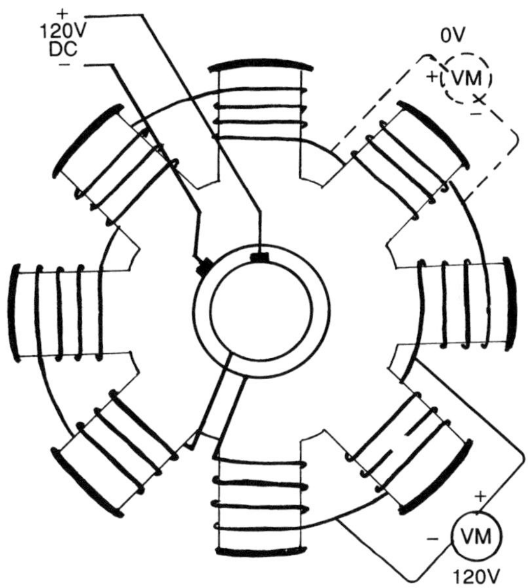

Figure 23-10. Identifying open field coil

winding is fully energized (at intervals separated by one-third of a cycle) the magnetic field produced by the armature windings will be oriented along the axis of that particular phase winding. Thus the direction of the stator-produced magnet field shifts around the stator, returning to its original orientation in one full cycle of armature current (see Figure 23-11). In a two-pole machine (two coil groups per phase) this constitutes one complete revolution in 1/60 second (60 rps = 3600 RPM). In a four-pole machine, although the initial magnetic orientation is repeated after one cycle, in following a particular pole around the stator, it takes *two* cycles for that pole to return to its original position (2⁄60 sec/rev yields 30 rps, which equals 1800 RPM). In summation: *a* north pole returns to its original position f × 60 times per minute no matter how many north poles are

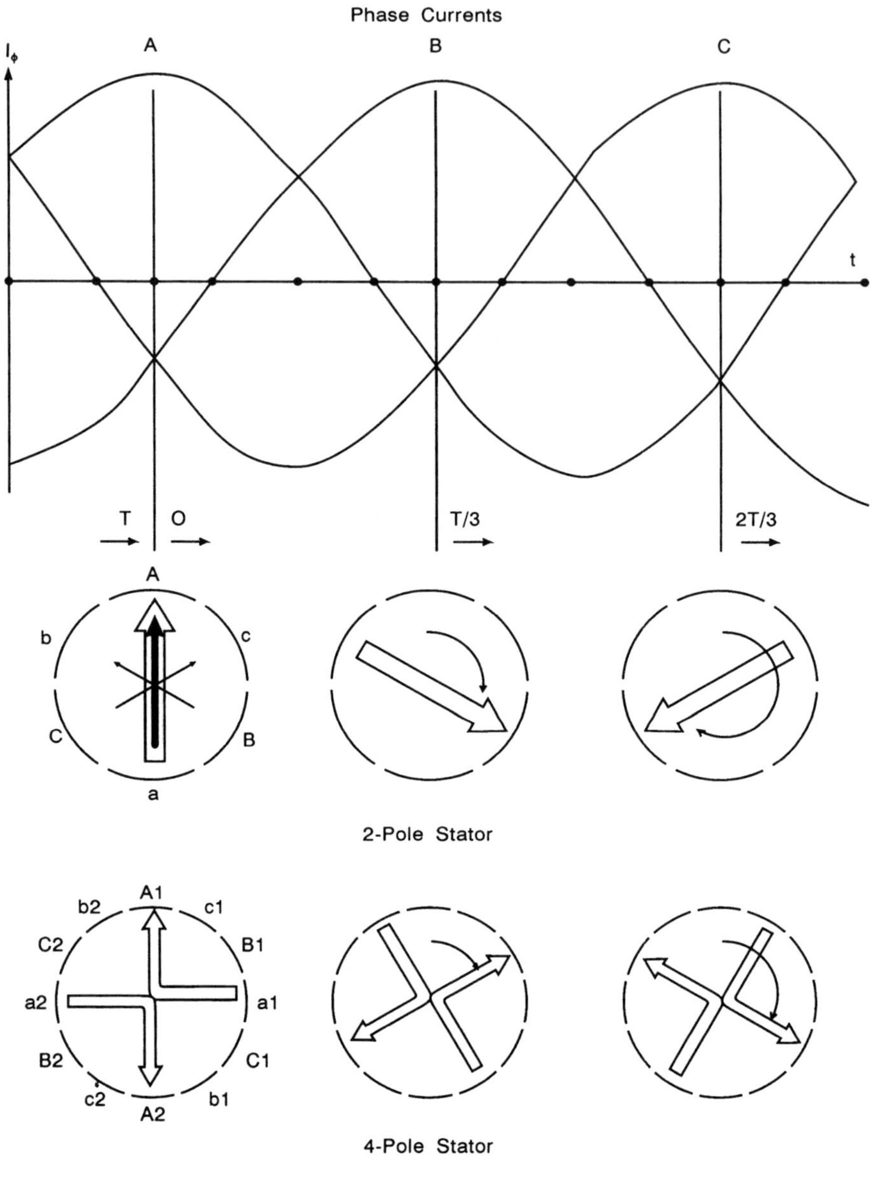

Figure 23-11. Rotating magnetic field

produced by the stator windings; but, since there are P/2 north poles, *the* original north pole is back at its original position in P/2 cycles. Thus the synchronous speed formula—Ns = f × 60 / (P/2) or f × 120/P—describes the rate of rotation of the stator magnetic field.

Operation

Synchronous speed. A synchronous motor rotates at synchronous speed because the rotor electromagnet is attracted by the rotating stator field. The rotor and stator magnetic fields are not perfectly aligned, however. The rotor lags the stator-produced rotating magnetic field by a few degrees. This is not to be confused with "slip"; rotor and stator fields are revolving at the same speed, but not at exactly the same position. The stronger the rotor electromagnet, the closer the two fields are in alignment. With constant excitation, the greater the mechanical load on the motor, the greater the degree of lag between rotor and stator fields. For a given mechanical load, this lag can be decreased by increasing the excitation to the rotor. There is a point at which, for a given level of excitation, an increase in mechanical load will cause the rotor to fall out of synchronism with the stator field. This phenomenon is called "slipping a pole" and results in severe vibration and oscillation in stator current. This is a situation to be avoided and may be remedied by decreasing the mechanical load or increasing the level of excitation. Exciter failure will, of course, cause this condition, requiring that the machine be shut down. This and other symptoms of maloperation are outlined in Table 23-2.

TABLE 23-2

Synchronous Motor Troubleshooting Chart

Problem	*Possible Cause*	*Remedy*
Fails to start	Open line(s)	Check fuses/circuit breaker. Note: Will not start with one line open.
	Open phase	Test armature windings for open.
	Field excited	Check controller relay coils and contacts. Note: Field should not be excited until rotor is spinning at near-synchronous speed.
	Overload	Relieve mechanical load on motor; check for bearing seizure or other mechanical restriction.
	Low voltage	Check supply.
Overheating	Overload	Check status against nameplate ratings; relieve mechanical load.
	Insufficient ventilation	Remove restrictions; clean windings; check C. W. supply.
	Insufficient excitation	Check pf; if too lagging, increase excitation.
	Excessive excitation	Check pf; if too leading, reduce excitation.
	High voltage	Check supply.
	Uneven air gap	Check gap clearances; check for evidence of rotor/stator contact; realign if necessary.
	Open/shorted/grounded stator coil(s)	Test and repair/replace as required.
	Open/shorted/grounded field coil(s)	Test and repair/replace as required.
Runs fast	Frequency high	Check supply.
Runs slow	Frequency low	Check supply.

Problem	*Possible Cause*	*Remedy*
Pulls out of synchronism	Overload	Check status against nameplate ratings; relieve mechanical load.
	Insufficient excitation	Check pf; increase excitation; check field coils; check exciter output, exciter field rheostat.
Will not synchronize	Controller failure	Check controller relay coils and contacts.
	Exciter failure	Check exciter output, slip ring/brush assembly; check for blown diodes (brushless types) (see Table 23-5).
	Overload	Check status against nameplate ratings; relieve mechanical load.
	Open/shorted/grounded field coil(s)	Test and repair/replace as required.
Noisy operation	Out of synchronism	See above.
	Bearings	Check for worn, loose, dry, or overlubricated bearings; replace worn or loose bearings; lubricate dry bearings; relieve overlubricated bearings.
	Uneven air gap	Check for bent shaft, bearing condition, gear alignment.
	Coupling	Check alignment; re-align and/or tighten coupling.
	Loose laminations	Tighten bolts; dip in varnish and bake.

Power factor correction. From an electrical point of view the relationship between load and excitation is indicated by the machine's power factor. An underexcited machine manifests a *lagging* power factor. Excessive lagging may cause the machine to slip a pole. For lightly loaded machines this level may be as low as a .4 or .5 lagging power factor; at rated loads the rotor may fall out of synchronism at power factors as high as .8 lagging. Increasing excitation will bring the power factor to unity, and overexciting the rotor yields a leading power factor. If optimizing the synchronous motor's operation were the only consideration, tuning the excitation to produce unity power factor would be desirable. At unity power factor the stator current draw is minimized for a given power output. Over- or underexciting the rotor results in more current to produce the same horsepower. At unity power factor the motor consumes kW proportional to the mechanical load, but no kvar; underexcited, the motor consumes lagging (positive) kvar; overexcited, the motor consumes leading (negative) kvar. Synchronous propulsion motors are typically the only load on the propulsion generator and thus operating at unity power factor is advantageous to both motor and generator. In a plant containing a synchronous motor and other inductive loads (induction motors, for example), the *overall* power factor of the plant may be improved by operating the synchronous machine in the overexcited state. The negative kvar consumed by the synchronous machine would cancel out some, or all, of the lagging kvar demanded by the inductive loads. A synchronous motor not loaded mechanically (and thus consuming only a small amount of kW) and deliberately overexcited to consume leading kvar (which may be thought of as *supplying* lagging kvar to those plant loads demanding them) is called a *synchronous condenser.* Banks of capacitors can accomplish the same purpose.

Starting. Synchronous motors do not develop any starting torque. This is because, although the stator-produced rotating magnetic field is instantly rotating at synchronous speed when the stator is energized, the rotor must *accelerate* to synchronous speed. At the instant of starting, the rotor poles are alternately attracted and repelled by the rotating stator poles, north poles followed by south poles 1/120 second later, resulting in zero net torque. It is necessary, therefore, to have the rotor spinning at near synchronous speed *before* energizing the rotor field. On some very large machines this is accomplished with an auxiliary starting motor. In most applications, however, the amortisseur windings serve as the starting device. These damper bars, imbedded in the rotor pole faces, form a squirrel-cage rotor, and the synchronous rotor is thereby accelerated as an induction machine. When the rotor is at near synchronous speed (approximately 1 percent slip or less) the field poles are excited with DC current and the rotor locks into synchronism. At synchronous speed no currents are induced in the starting (amortisseur) windings. The rotor windings may be energized manually by the operator upon witnessing the rotor speed approaching synchronous speed and the concurrent drop in stator current or it may be accomplished automatically. The DC source of excitation may be a separate source, from the AC mains via a rectifier or a separate DC generator, or an exciter mounted on the motor shaft.

To provide application of DC excitation at the proper moment, automatic synchronous motor starters typically incorporate a *polarized frequency relay* (PFR). The PFR illustrated in Figure 23-12 includes a static core supporting two windings, one DC and one AC. The direction of the flux produced by the DC coil is constant, but that of the AC coil alternates. If a voltage decaying in amplitude and frequency is applied to the AC coil, at some point the flux in the armature is no longer adequate to hold it closed. The threshold at which this occurs is on the negative portion of the cycle. Thus both magnitude and phase govern the timing of the relay closure. The PFR is appropriate for synchronous motor starting because, as the motor accelerates, voltages induced in the *rotor* windings decay in magnitude and frequency. Thus, applying the induced rotor voltage to the AC coil of the PFR makes the relay sensitive to both rotor speed and the relative position of the rotor field coils to the synchronous stator flux.

In the starting circuit schematic illustrated in Figure 23-13, the AC coil of the PFR is shunted by a reactor that acts as a short at very low frequencies. Upon depressing the start button, the M contacts close, energizing that stator and inducing stator-frequency AC in the rotor

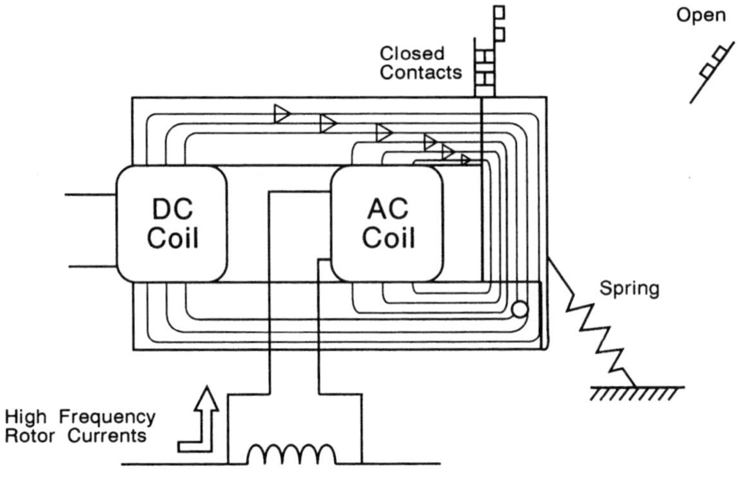

(a) High Slip Operation

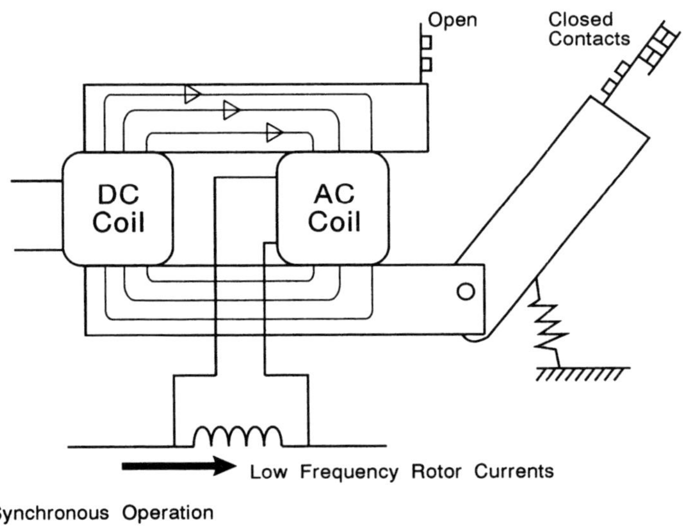

(b) Synchronous Operation

Figure 23-12. Polarized frequency relay

circuit. The PFR contacts are held open, preventing energization of the FC coil. As the rotor accelerates and the frequency drops, the reactor takes on a greater percentage of the induced rotor current. The PFR contact will fall back closed when the rotor nears synchronous speed, allowing the FC contacts to close and introducing the DC excitation current to the rotor field windings. The *out-of-step relay* (OSR) is included in the field circuit

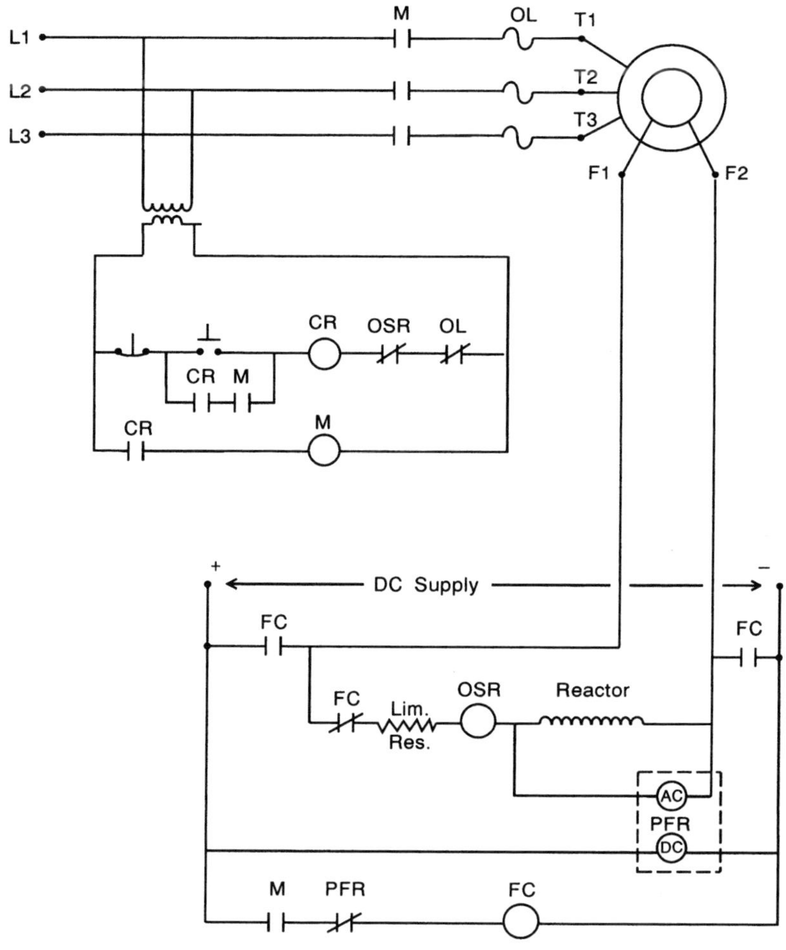

Figure 23-13. Synchronous motor starter

during acceleration so that if synchronization does not occur within a preset time (5 to 10 seconds), the stator will be de-energized and the motor shut down.

THREE-PHASE INDUCTION MOTORS

Principles of Operation

Synchronous motors and, for that matter, DC motors require external sources of electrical power to energize both their stator and rotor windings to produce torque. The stator of an induction machine is identical to the synchronous machine stator, producing a magnetic field that rotates at synchronous speed ($N_S = f \times 120/P$) when energized from a three-phase AC supply. The rotor of an induction machine, however, is not connected to any external power source. As the name implies, the principle of electromagnetic induction is responsible for rotor current. This is the same action that produces secondary voltage in a transformer; accordingly, stator and rotor circuits of induction machines are known as *primary* and *secondary* windings respectively. Two types or rotor construction differentiate classes of induction motor: the squirrel-cage rotor and the wound rotor.

Squirrel-Cage Motors

Motor construction. The squirrel-cage induction motor is the most widely used industrial motor in the world. A rugged, easily constructed, economical, and highly effective rotor, the *squirrel cage* is illustrated in Figure 23-14. Laminations, punched from magnetic steel sheets and including slot cutouts, are sandwiched on the rotor shaft. Conducting bars of copper or aluminum are inserted in the slots and cast end rings form a low-resistance short circuit. Rotor bar cross section will determine, to some extent, the operating characteristics of the motor as will be explained later.

Motor characteristics

Slip. The stator-produced rotating magnetic field induces voltages in the rotor bars as it sweeps across the cage. Currents circulate between bars under north and south poles with a frequency and amplitude proportional to the rate at which the stator-produced rotating field is moving relative to the bars of the cage. These circulating currents produce a rotor field that imparts torque to the rotor. As the rotor accelerates, the magnitude and frequency of the induced rotor currents decrease as the relative velocity of rotating magnetic field (N_S) and rotating squirrel cage (N_r) decreases. If the rotor were to spin at synchronous speed (N_S), no rotor currents would be induced and no rotor torque developed. Thus under normal (torque producing) conditions, induction motor speed (N_r) *is always less than* synchronous speed (N_S). The difference ($N_S - N_r$) is called the "slip speed" expressed in RPM. More often this difference is expressed as a decimal fraction (or percent) of synchronous speed called "slip" (or "percent slip").

$$\text{Slip (s)} = \frac{(N_S - N_r)}{N_s}$$

$$\text{Percent slip (\%s)} = \frac{(N_S - N_r) \times 100}{N_s}$$

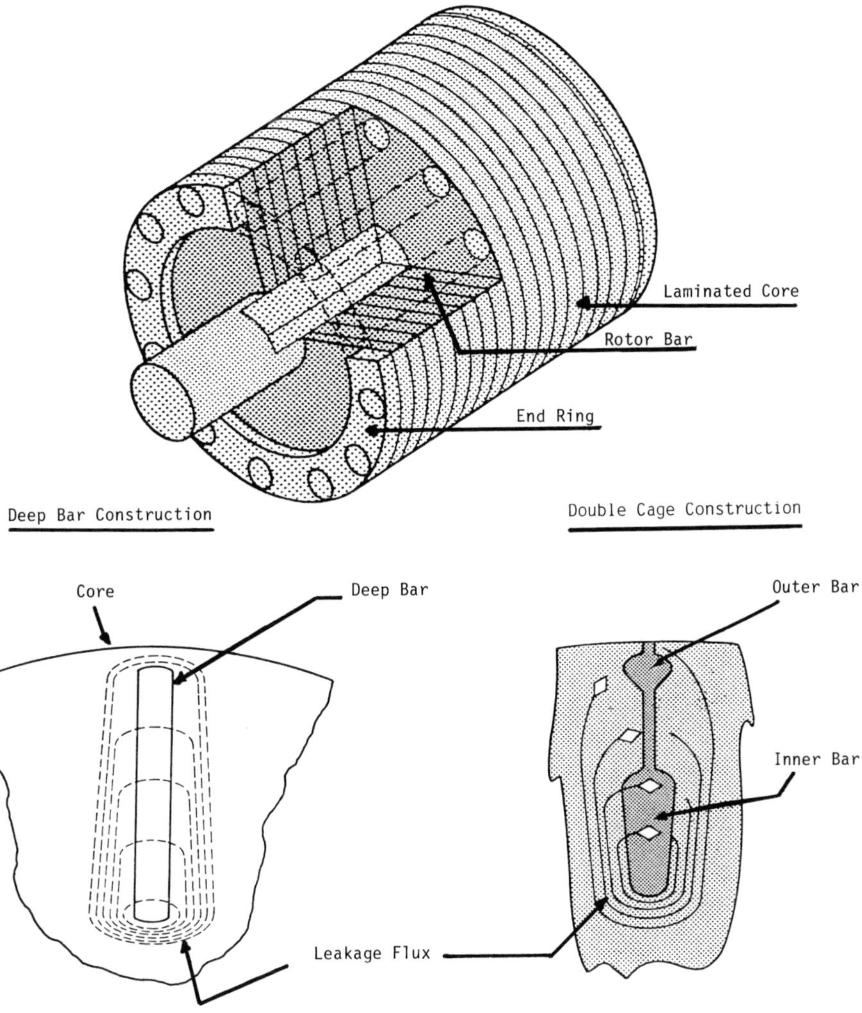

Figure 23-14. Squirrel-cage rotor

For example, a four-pole induction motor (N_S = 1800 RPM) spinning at 1746 RPM has a slip speed of 54 RPM and a slip of .03 (3 percent).

Torque. Induced rotor voltage and frequency are proportional to slip: $E_r = s \times E_s$ and $f_r = s \times f_s$. Thus at near synchronous speed, rotor current (and therefore developed torque) is small, although almost in phase with rotor voltage. At large slips (zero RPM is 100 percent slip) the induced rotor currents are greatest, but because of the greater rotor frequencies, the rotor reactance is higher and therefore power factor is lower. Since developed torque is also proportional to rotor power factor, *maximum* developed torque is neither at the lowest nor highest slips, but somewhere in between. Mathematically it can be shown that "somewhere in between" is the slip at which rotor resistance and reactance are equal. A generic squirrel-cage motor torque/speed characteristic is illustrated in Figure 23-15. The points of particular interest are (1) rated torque and speed (the operating range of the motor being between this point and synchronous speed), (2) maximum or "pull-out" torque (the motor will stall if the load demand exceeds the pull-out torque), (3) starting torque, and (4) the "pull-in" torque (minimum torque during acceleration). The operating speed of the motor is determined by the intersection of the motor characteristic and the load or system curve. The area between the two curves to the left of the operating point determines the acceleration during starting. The dotted line in Figure 23-15 illustrates what the characteristic would be if it were not for displacement currents. In deep bar rotors, leakage flux surrounding

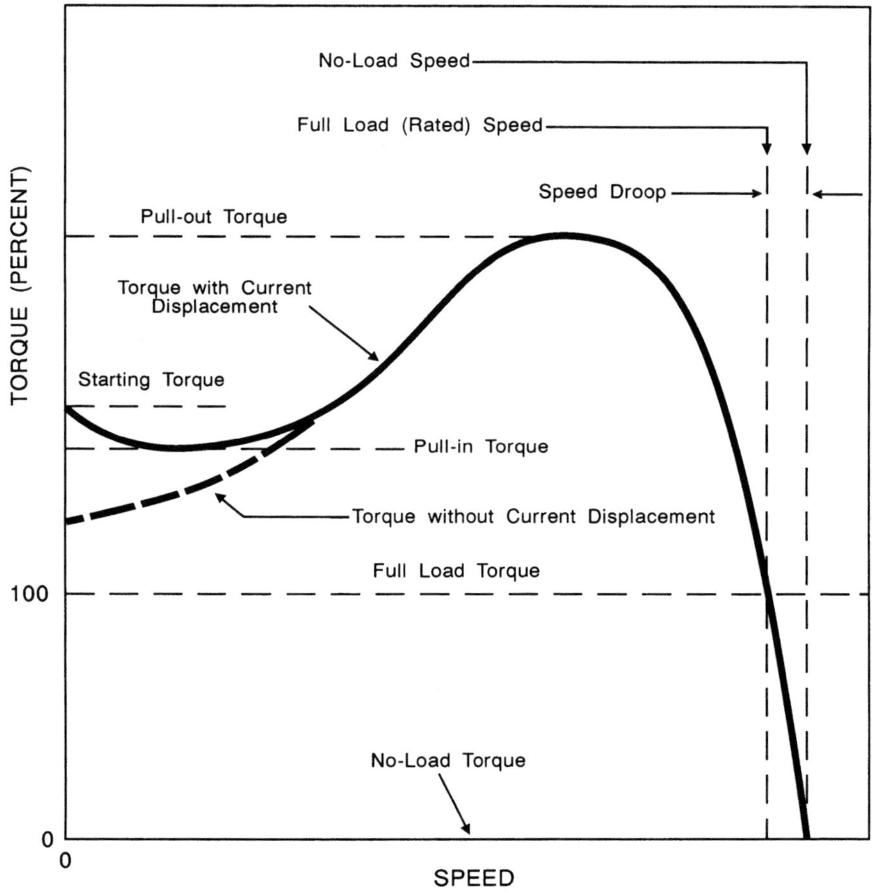

Figure 23-15. Induction motor torque/speed characteristics

the current-carrying bars produces a pattern illustrated in Figure 23-14. The inner end of the bar, looped by the most flux, will have the greater inductance. At low frequencies (i.e., low slip) this is not of much significance. However, at higher frequencies the inductive reactance of the inner portion of the rotor bar tends to choke off current at that end of the bar, displacing the current to the outer end of the bar. This action raises the effective value of rotor resistance which, simultaneously, raises starting torque while decreasing starting current.

The inductance of a squirrel-cage rotor is determined primarily by the depth at which it is buried in the rotor iron. The resistance is determined by the cross-sectional area and the resistivity of the material used for the bars and end rings. Thus a variety of rotor bar shapes, sizes, and materials can be used to massage the generic characteristic into one suitable for a particular application. Double squirrel-cage rotors, for example, consist of a high resistance, low inductance outer cage and a low resistance, high (by virtue of its deeper position) inductance inner cage. At high slip, most of the current flows in the outer cage, providing good starting torque; at low slip, most of the current will be in the low resistance inner bar, providing high efficiency and low slip at rated power.

For standard induction motors there are four lettered NEMA design classes: A, B, C, and D. Although characteristics within a design class vary with power and speed ratings, typical torque/speed characteristics based on NEMA MG-1 are shown in Figure 23-16. The classes are distinguished by their starting torque and current, and slip at rated load. Classes A, B, and C produce rated output at 5 percent slip or less and have pretty much the same characteristics within the operating range. Locked rotor currents for Class A motors, however, are approximately 8 × FLA, where 5 to 6 × FLA is more typical for Classes B and C. Class B, usually incorporating a deep bar rotor, is the general-purpose machine used for most centrifugal pumps and ventilation blowers. Class C is a double cage type utilized where its increased starting torque is applicable: compressors without unloaders, hydraulic machinery, conveyers started under loaded conditions, etc. Class D motors, employing high resistance rotor bars (sometimes of brass), provide a very high starting torque but sacrifice speed at full load (typically 12 percent to 15 percent slip at rated torque). Applications of Class D motors include hoists, punch presses, and machines with large flywheels.

Dual-voltage motors. If a motor is reconnected to get twice as many parallel paths in the stator winding, it will have precisely the same operating characteristics it had before the change, providing only half the line voltage is applied. The current per winding will remain the same but the line current will double in the low voltage mode. A dual-voltage motor has the necessary winding leads accessible at the motor terminal box. Manufacturers commonly build motors suitable for both 460 and 230 volt supplies since both these levels are commonly available. Terminal mark-

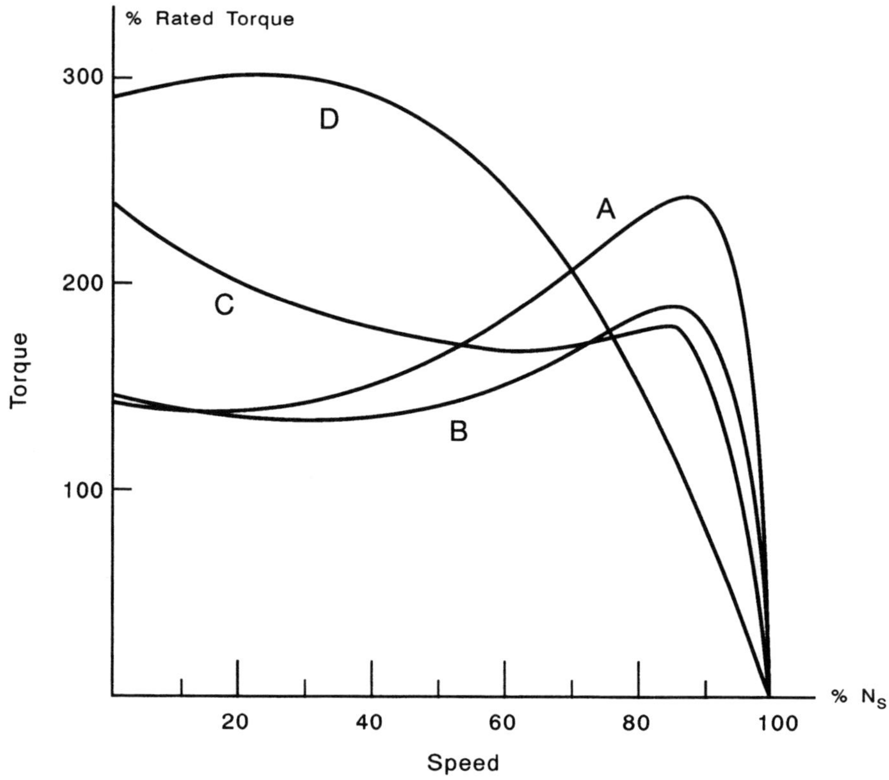

Figure 23-16. NEMA design class characteristics

ings T1 through T6 are utilized by NEMA to standardize the connection of wye and delta wound motors for dual-voltage service. The diagrams illustrated in Figure 23-17(a) or (b) are typically incorporated on the nameplate of dual-voltage motors. The connections are not to be confused with those for part-winding starters or consequent-pole, two-speed machines, which will be discussed in the following section.

Speed variation

Classification. Motors are often classified based on the variability of speed in the following categories:

Constant speed motors rotate at constant, or practically constant, speed over the normal load range. Classes A, B, and C squirrel-cage motors, DC shunt motors, and of course synchronous motors fall in this category.

Varying speed motors operate at a speed which decreases as load increases, as do DC series motors and some Class D squirrel-cage machines.

Adjustable speed motors' RPM may be adjusted over a continuous range for a given load but, once adjusted, may vary with subsequent load changes. Field rheostat control of a DC compound motor or secondary rheostat control of a wound-rotor induction motor fall in this category.

Multispeed motors operate at one of two or more speeds which may be selected, and run at essentially the selected speed regardless of load changes. Squirrel-cage motors with provisions for changing the number of stator poles fall in this category.

Speed regulation. Of the above categories, the first two describe a motor characteristic rather than a means of control. The term *speed regulation* is used to describe the difference between no-load and full-load speed (speed "droop") as a percent of rated speed. Since the full-load speed is by definition the rated speed, these two terms are related by the formula:

$$SR = \frac{\text{speed droop}}{\text{rated speed}}$$

$$\%SR = \frac{(N_{nl} - N_{fl}) \times 100}{N_{fl}}$$

Note that speed regulation as a *motor characteristic* is different from slip, which is a *parameter of operation.* Of course, by knowing the slip at no load and full load, one could calculate the speed regulation.

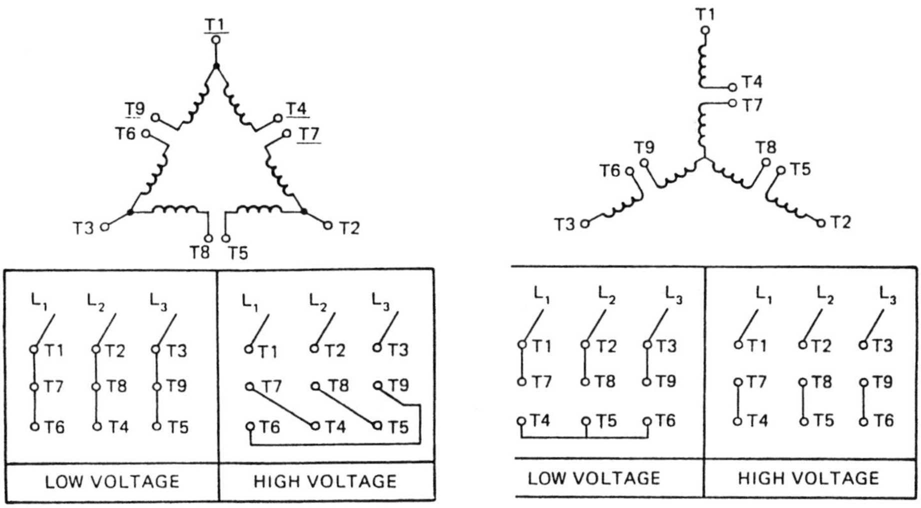

Figure 23-17. Dual-voltage motor terminal connections

Adjustable speed control. The latter of the two above categories actually describes means for controlling speed. Considering the relationship among rotor speed, synchronous speed, and slip ($N_r = [1 - s] \times N_s$), it is obvious that changing the synchronous speed or altering the shape of the torque/speed characteristic of the motor will effect a speed change of the motor. *Adjustable speed control* of induction motors may be accomplished by varying the stator frequency (hence synchronous speed) over some, possibly wide, range. Varying the stator voltage alters the torque characteristic by changing the torque produced at any speed in proportion to the voltage squared. This effectively changes the slip at which a given load will be met. This method is often utilized in electronic controllers (Figure 23-29).

Multispeed control. Multispeed (pole changing) controllers may be of two types: consequent-pole (single winding) or multiple winding stators. The ability to change the number of poles with a single winding is illustrated for a single-phase stator in Figure 23-18. In shifting from conventional to consequent-pole operation it is only possible to double the number of poles, hence halve the synchronous speed. Thus two-speed, single winding (2S1W) motors must have speed ratios of 2:1. There are three types of consequent-pole motors:

Constant horsepower motors run with the windings delta connected for high-speed operation and two parallel wye connections for low speed. With rated power available at both speeds, the torque at the lower speed is twice that at the higher speed. Figure 23-19 shows the NEMA standard winding terminal markings and connections.

Constant torque motors have the same full-load torque available at both speeds, therefore the horsepower ratings are proportional to the synchronous speeds produced by each winding arrangement. These run with the windings connected parallel wye for high speed and series delta for low speed (Figure 23-20).

Variable torque motors produce a full-load torque which is proportional to the synchronous speed of the two winding arrangements and thus the power is proportional to synchronous speed *squared.* This is accomplished by connecting the windings parallel wye for high speed and series wye for low speed (Figure 23-21).

If a motor is provided with two stator windings that are wound for a different number of poles (not necessarily in the ratio of 2:1), two-speed operation can be obtained by energizing either of the windings separately (2S2W). For three-speed operation, generally one consequent-pole and one single-spaced winding are utilized (3S2W); for four-speed machines, two consequent-pole windings are fitted (4S2W). More than two windings, and therefore four speeds, are generally not practical. By properly designing the second winding, two-winding machines may be constant horsepower, constant torque, or variable torque. Also, whenever two stator windings

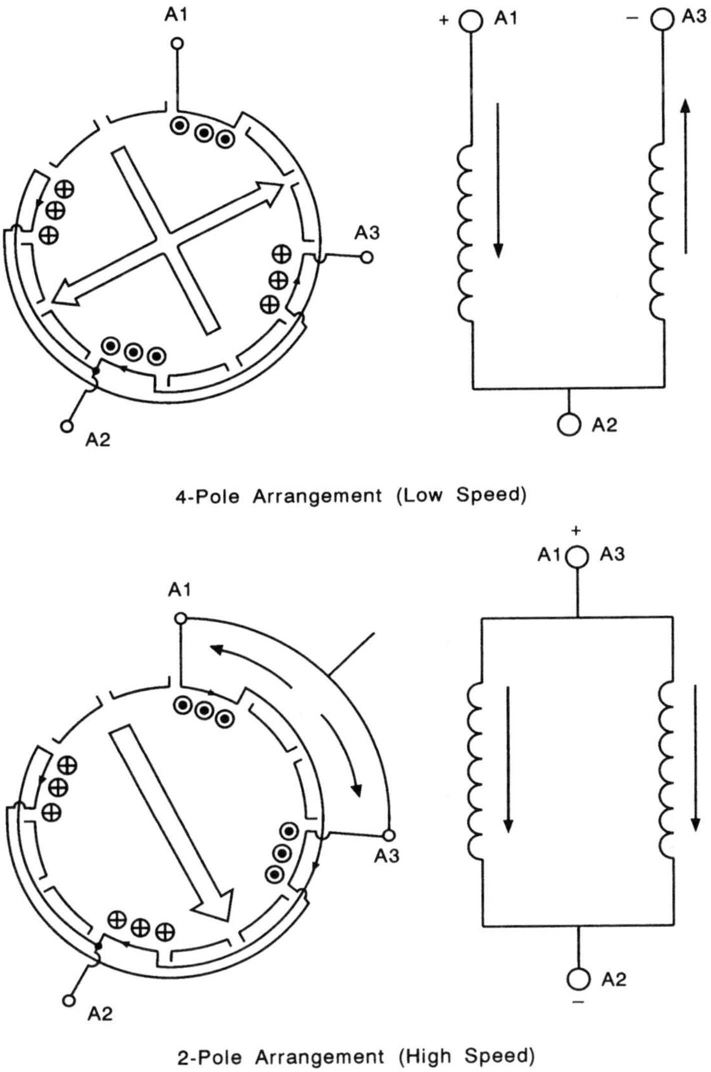

Figure 23-18. Consequent-pole windings

are installed, voltages will be induced by the rotating magnetic field in the unenergized winding. If the unused winding has closed loops (e.g., parallel paths or a closed delta connection) currents will circulate in the winding, causing loss of power and rapid overheating. For this reason windings must be series wye or broken delta. Two wye connected windings are preferred in 2S2W machines because extra contacts in the controller are necessary

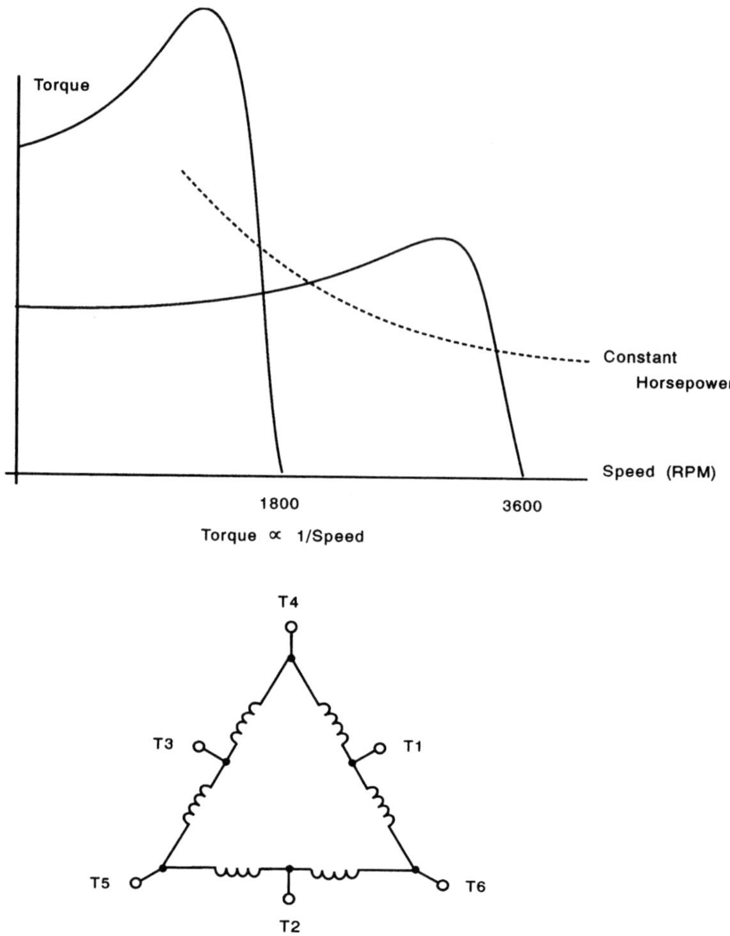

Speed	Line Leads L1-L2-L3	Tie
Low	T1-T2-T3	T4-T5-T6
High	T6-T5-T4	

2-Speed, Single-Winding (2S1W)
Constant Horsepower Drive

Figure 23-19. Constant-horsepower drive

to break the delta. The purpose of the T7 terminal in Figure 23-22 is to break the closed loop when the second winding is energized.

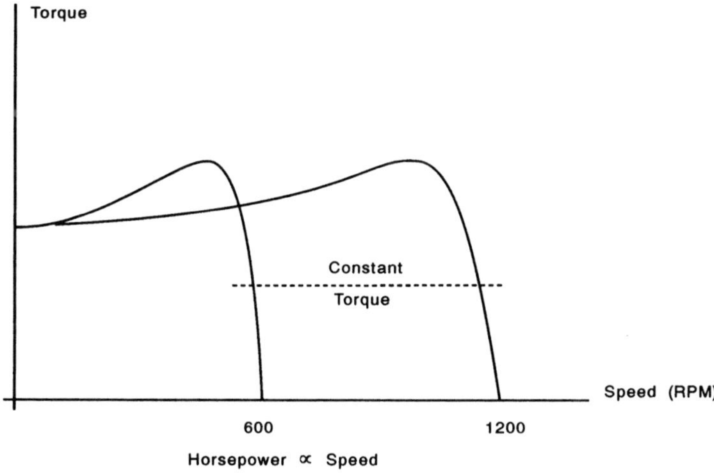

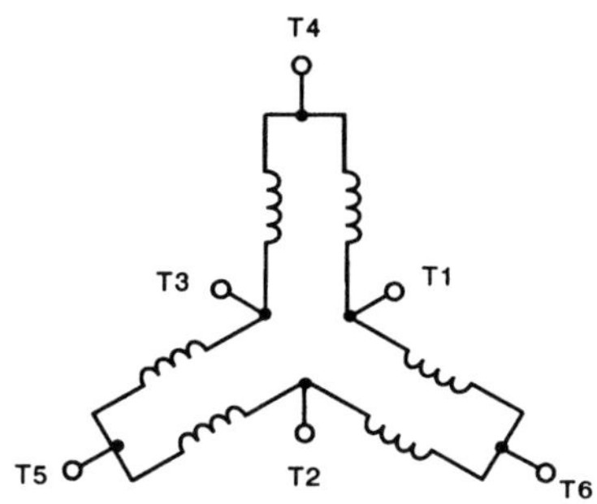

Speed	Line Leads L1-L2-L3	Tie
Low	T1-T2-T3	
High	T6-T5-T4	T1-T2-T3

(a) 2-Speed, Single-Winding (2S1W)
Constant Torque Drive

Figure 23-20. Constant-torque drive

Starting

Across-the-line starting. Unlike DC machines, where only the resistance of the motor windings is available to oppose current inrush during a full voltage start, induction motor impedance is sufficient to limit the starting current to 400 percent to 600 percent of rated stator current when started at full voltage. Thus, where system capacity permits, squirrel-cage

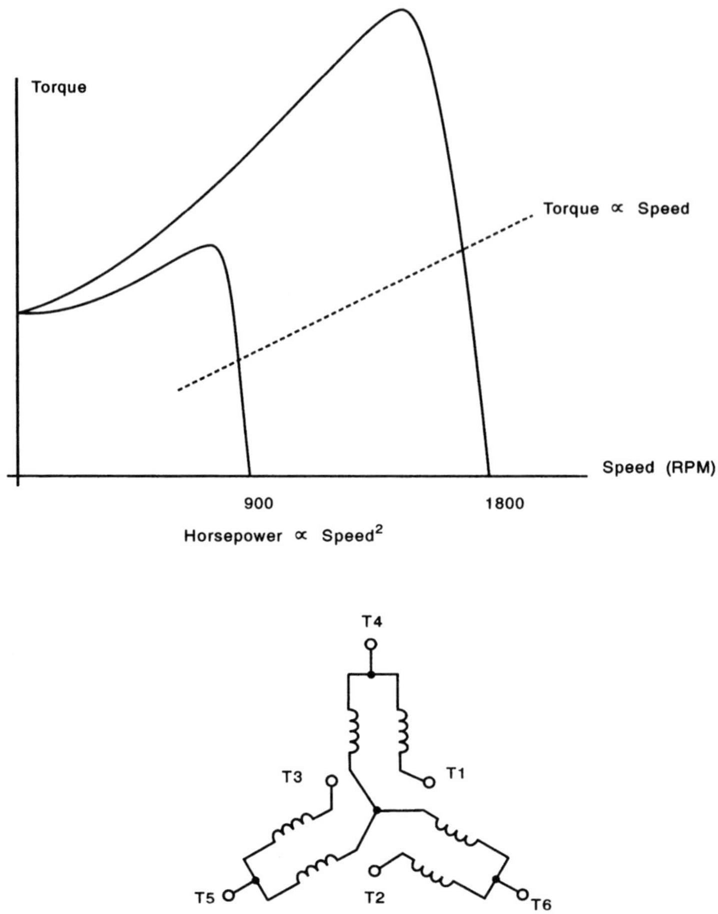

Speed	Line Leads L1-L2-L3	Tie
Low	T1-T2-T3	
High	T6-T5-T4	T1-T2-T3

2-Speed, Single-Winding (2S1W)
Variable Torque Drive

Figure 23-21. Variable-torque drive

motors may be started by simply energizing the main contactor coil (Figure 23-23). The control circuit is generally at a lower voltage than the power circuit, is supplied via a control transformer, and is often grounded. The energizing mechanism may be (1) a two-point starter (manual select switch or automatic pressure, temperature, or float switch) that allows restart upon resumption of voltage after voltage failure (low voltage release, LVR);

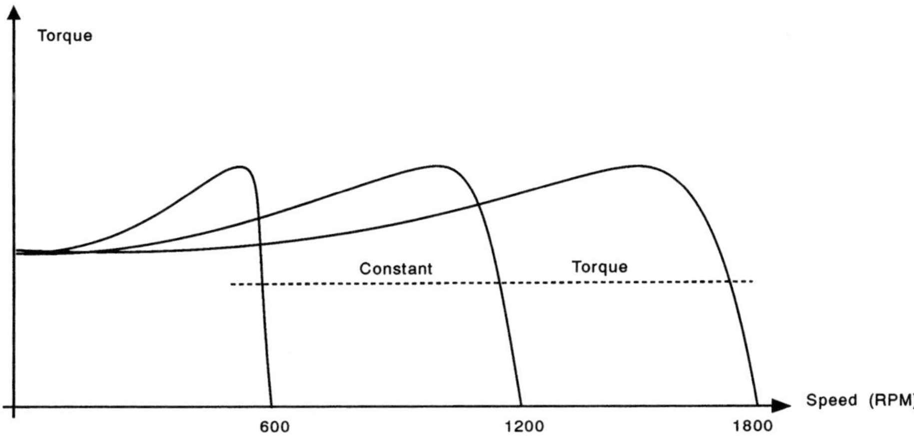

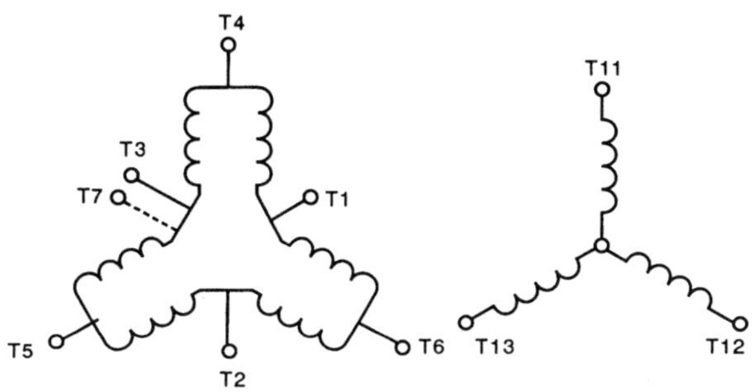

Speed	Line Leads L1-L2-L3	Tie	Open
Low	T1-T2-T3		
Medium	T6-T5-T4	T1-T2-T3	
High	T11-T2-T13		T3-T7

3-Speed, 2-Winding (3S2W)—Constant Torque

Figure 23-22. Three-speed, two-winding, constant-torque drive

or (2) a three-point starter that utilizes momentary contact push buttons and a sealing contact to provide low voltage protection (LVP). Full-voltage starters may be reversing or non-reversing, include plugging (energizing in one direction while rotating in the other) or antiplugging features, include a jogging circuit, and may be used in multispeed applications

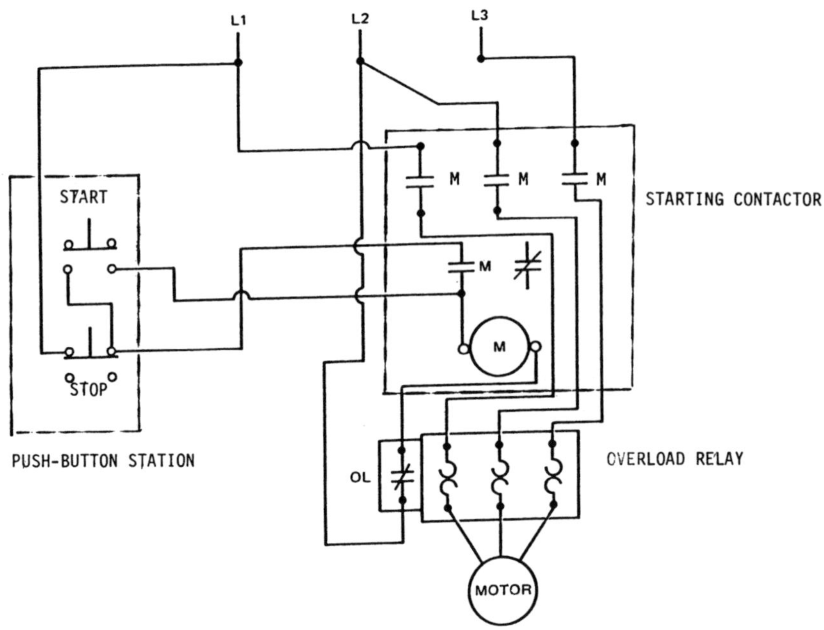

Figure 23-23. Full voltage starter

(Figure 23-24). On larger machines the main contactor may be energized indirectly via a control relay.

Reduced voltage starters. Although squirrel-cage motors of almost any size can themselves withstand starting inrush currents, starting large machines across the line may have a debilitating effect on the rest of the system. Taking into consideration the capacity of the system supply, the size of the motor to be started, the motor load already on line, and response time/capacity of the voltage regulator/exciter systems, the amount of voltage dip upon starting a large motor may be determined. If the dip is severe enough to possibly drop out voltage-sensitive equipment (e.g., contactors in LVP starters) then a means of reducing the inrush current to the motor in question must be provided. One of several common methods may be employed.

In *(primary) resistance starters,* starting current is reduced by inserting resistances (or reactances) in series with the stator windings, thus bolstering the starting impedance of the motor. In the starter illustrated in Figure 23-25, the normally closed starting contacts (1A) are energized prior to the main contacts (M), inserting the starting resistances. The starting contactor has a built-in time delay on de-energization (TDODE) so, although de-energized by the M contactor, the 1A contacts will fall back to their normally closed position after a predetermined accelerating interval. Re-

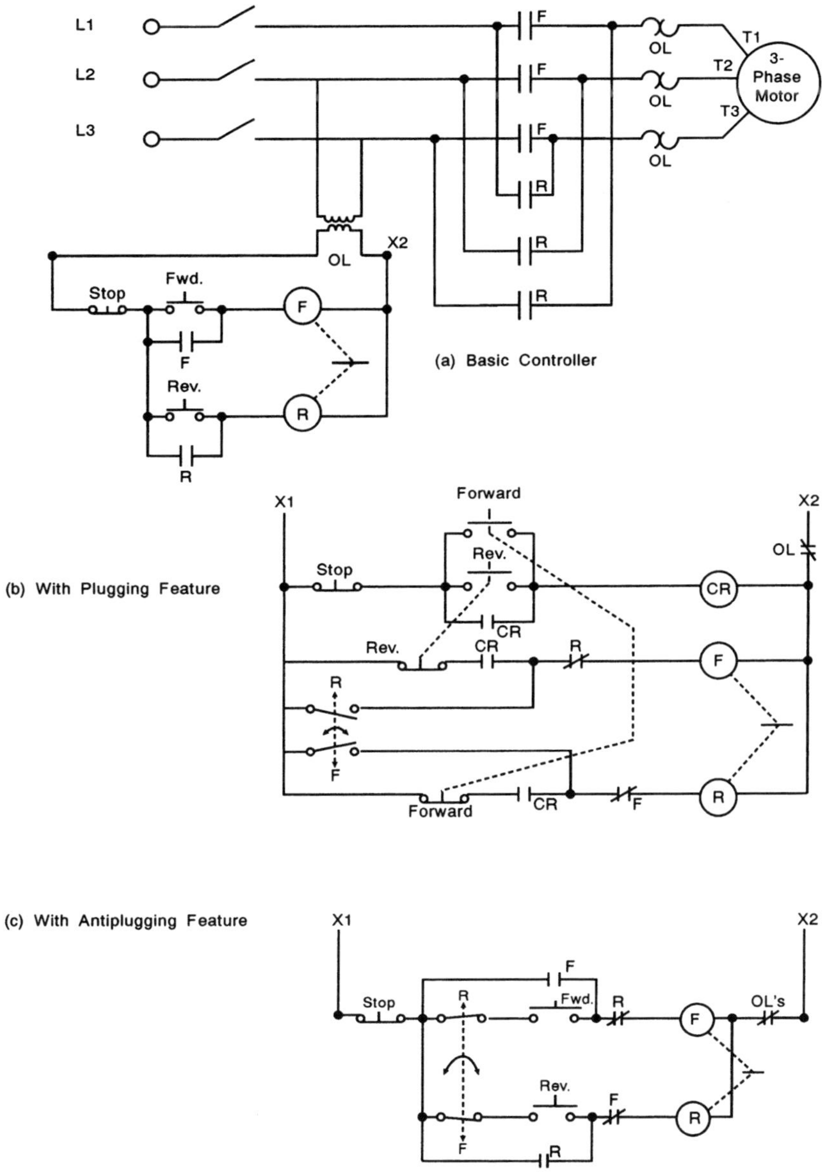

Figure 23-24. Reversing controllers

sistance starters are, in a sense, self-regulating since the reduced voltage at the motor terminals is a result of the IR drop across the resistors. As the motor accelerates and the inrush current abates, the motor terminal

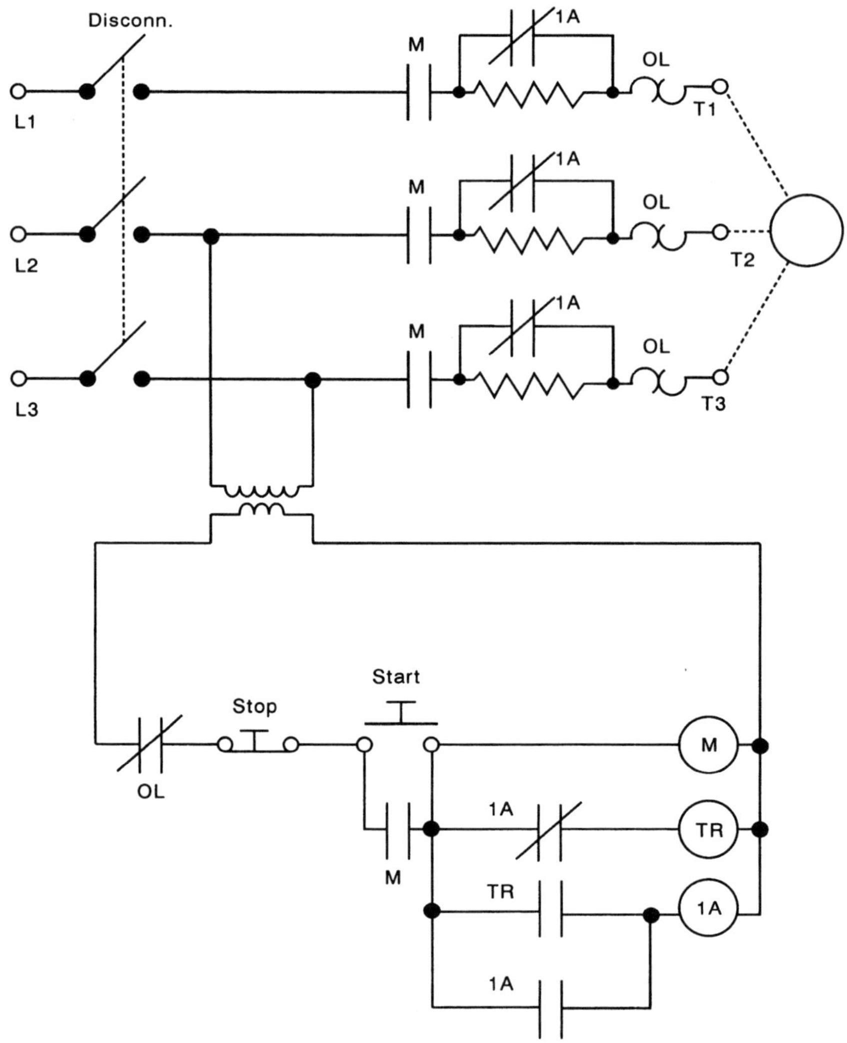

Figure 23-25. Primary resistance start

voltage rises, assuring smooth starting. Starting resistors are sized to produce a specific (reduced) voltage at the motor terminal at the instant of maximum inrush. At 80 percent of normal line voltage, for example, the motor will draw 80 percent of the starting current it would draw under full-voltage starting conditions. Other typical values are 65 percent and 50 percent of line voltage. Unfortunately, the starting torque, proportional to the applied voltage *squared,* is reduced to 64 percent (.8 × .8), 42 percent, and 25 percent of the full-voltage starting torque with 80 percent, 65 percent, and 50 percent reduced voltage starters respectively. The goal in choosing a reduced voltage starter is to limit the inrush current to a level that does not adversely effect the rest of the system, yet provides the motor being started with sufficient torque to accelerate to operating speed promptly.

Similar to primary resistance starters in sequence of operation and current and torque reductions, *primary reactor starters* are sometimes employed with large motors where heat dissipation from starting resistors is a problem. The reduction in voltage dip, however, is not as great as the reduction in starting current, since reactance starters tend to reduce motor power factor (which is already low) on starting.

The autotransformer starter is an even more common method of reduced voltage starting. This type of starter may utilize either three wye-connected or two open delta-connected autotransformers (compensators) to provide reduced voltages (typically 80 percent, 65 percent, or 50 percent) at the motor terminals. The voltage remains constant across the motor throughout the starting period (and so motor torque is reduced to 64 percent, 42 percent, or 25 percent of full voltage start) and full voltage is applied after a set delay determined by the accelerating time required of the particular application. While the motor starting current is always proportional to the voltage tap used, the turns ratio of the transformer reduces the starting *line* current even further. Thus on the load side of the autotransformer starter operating on the 80 percent voltage tap, the motor draws 80 percent of the starting current it would draw under across-the-line conditions; but the system need only supply 64 percent of the full voltage inrush current. Autotransformer starters and wye-delta starters are the ones that reduce current and starting torque in the same proportion. With all other types, the reduction in starting torque is greater than the reduction in inrush current. The starter illustrated in Figure 23-26(a), the starting contacts (S) are opened *before* the run contacts (R) are closed, briefly leaving the motor unenergized. This arrangement is called *open transition* and may cause a voltage flicker at changeover, particularly when the motor is driving a high friction, low inertia load. This effect may be overcome with a *closed transition* arrangement illustrated in Figure 23-26(b) known as a Korndorffer starter. In this scheme contactors S and 1A pick up, and, after a suitable accelerating interval, control relay CR releases 1A and R picks up in rapid succession. Autotransformers are usually sized for short-time duty only and may overheat if the motor is started too frequently.

The starters described above may be applied to any appropriately sized motor independent of the particular stator winding arrangement. A *part-winding starter,* however, requires a motor designed specifically for this purpose. The stator must include two parallel paths (either wye or delta) with separate leads accessible for each path. As illustrated by the starter

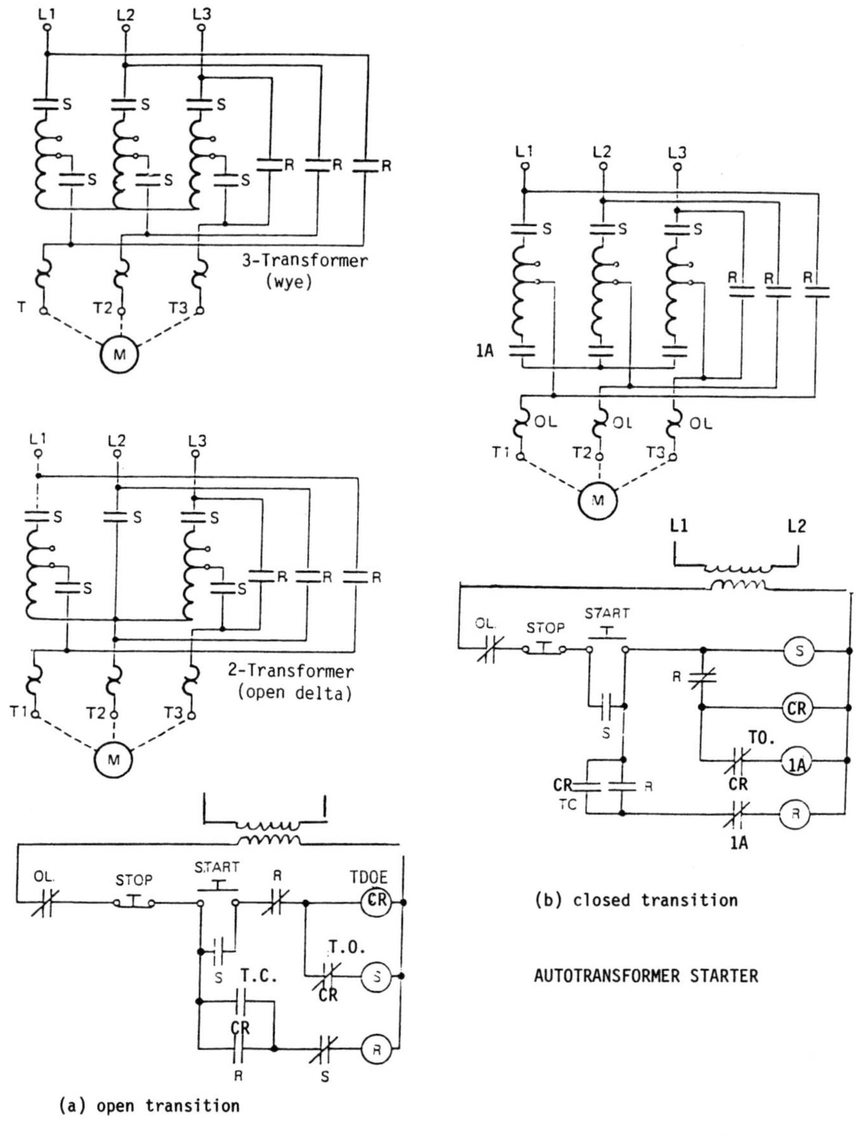

Figure 23-26. Autotransformer starters

in Figure 23-27, the S contactors pick up first, energizing half the stator winding. After a suitable interval contactor R picks up, including the second parallel path in the stator circuit. This type of starter is inherently closed transition.

The two parallel paths in the stator winding are generally identical (but not necessarily so) and it might be expected, therefore, that the inrush

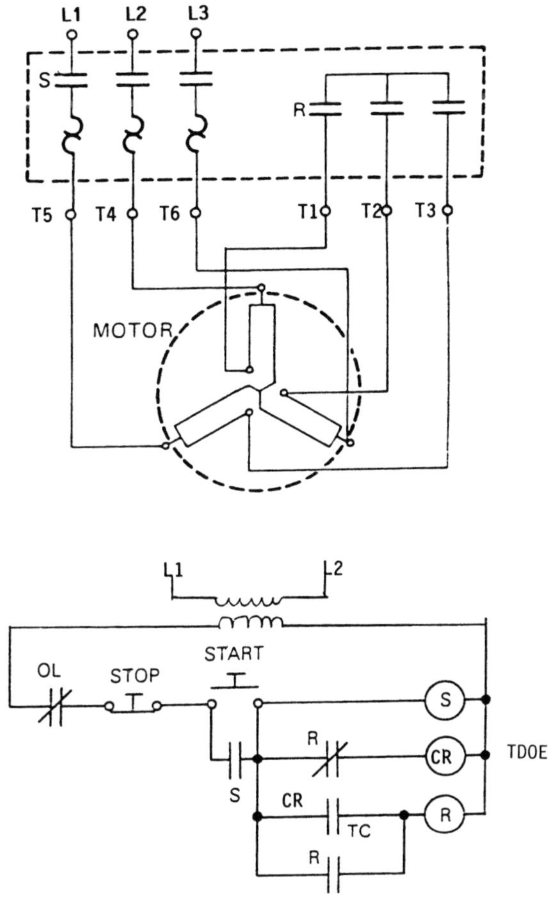

Figure 23-27. Part-winding starter

current be halved. However, the mutual inductance between the parallel windings yields an additional impedance not present when either winding is energized alone. Typically the starting current is reduced to 65 percent of the full-voltage value with the accompanying 42 percent torque reduction.

Most part-winding starters have a separate set of overload relays for each half of the motor winding (not shown in Figure 23-27), as a single set of relays sized for parallel operation would not protect the motor against burnout if contactor R failed to pick up. Also note that the terminal markings in the example are identical to the wye connected dual-voltage motor. Indeed, it is easy to connect an ordinary dual-voltage motor to a part-winding starter on a 230V system. However, if the motor is not designed specifically for part-winding start, it may have a severe torque dip at about half speed. This means that the motor will start to accelerate, then stall, then re-accelerate when the second winding is energized. This may be accompanied by excessive heating and noise upon starting.

Not applicable to just any three-phase motor, *wye-delta starters* require a motor designed to operate as a delta, with separate leads available at each end of each phase winding. The starter illustrated in Figure 23-28 is of the closed transition type, although open transition wye-delta starters are available also. In the example controller, contacts M and S1 pick up together, organizing the windings in a wye and starting the motor. After a suitable delay supervised by timing relay CR, S2 picks up, S1 releases,

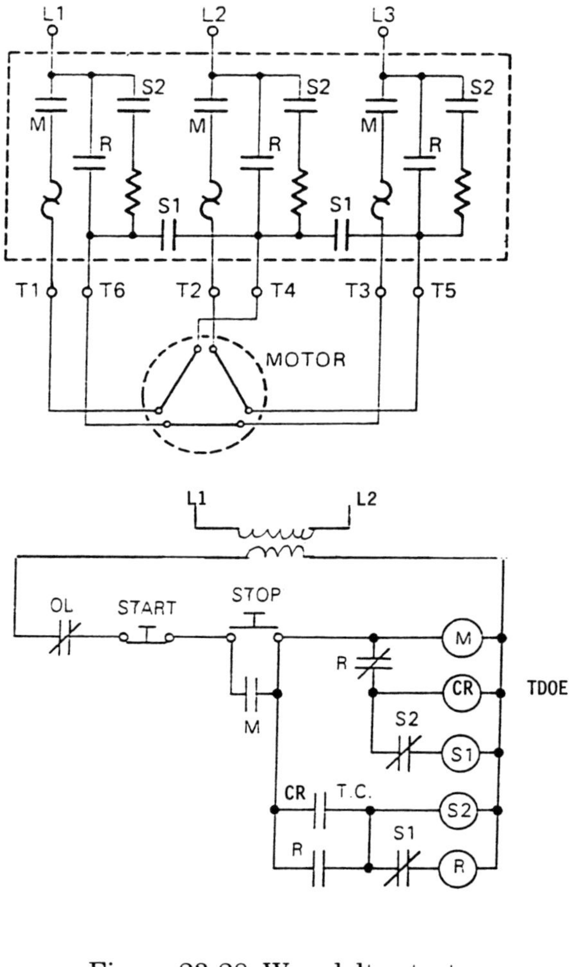

Figure 23-28. Wye-delta starter

and R picks up, in that order and in rapid succession. Since the equivalent impedance of a wye connected load is three times that of a delta arrangement, the inrush current (and torque) is 33 percent in the wye mode of what it would be in the running delta mode. This low torque makes the wye-delta starter inadequate for many applications. But for high inertia, low friction loads, such as variable pitch, motor-driven bow thrusters—started at zero pitch—they provide suitable accelerating and current limiting service.

Solid-state control. As previously mentioned, solid-state electronic devices can be used to alter the torque/speed characteristics of induction motors and thereby offer adjustable speed control. This may be accomplished by (1) stator voltage control, (2) frequency control, (3) stator current control, or (4) combinations of the first three.

Stator voltage control is effective since torque is proportional to voltage squared and proportional to slip in the operating range of most induction motors. Therefore, varying the voltage under a constant load torque will affect the slip. The range of control, however, depends on the slip at pull-out torque, and is narrow for most induction motors (Figure 23-29(a)). With this limited speed range requirement, AC voltage controllers (and, in rare instances, voltage-fed DC link inverters or pulse-width modulated [PWM] inverters) are used to provide the voltage control. The AC voltage controllers are simple, but the harmonic contents are high and the input power factor of the controller is low. These are used in low power applications such as small fans, blowers, and centrifugal pumps where the starting torque is low. They are occasionally used as reduced voltage starters for larger induction machines.

If the voltage is held fixed at its rated value while the frequency is reduced below its rated value, the flux will increase quickly to the saturation level. At low frequencies the motor reactances decrease and stator currents become unacceptably high. This type of *frequency control* is seldom used. Increasing the frequency above the rated value weakens the flux and speeds up the motor but reduces the torque proportionally (Figure 23-29(b)). In this type of control, the motor is said to be in the "field-weakening mode" applicable to constant horsepower type loads.

If the ratio of voltage to frequency is kept constant, the air gap flux remains constant as does the pull-out torque (Figure 23-29(c)). This type of *voltage and frequency control* is known as "volts/Hertz" control. Three possible circuit arrangements for obtaining variable voltage and frequency are shown in Figure 23-30. In the first, the DC voltage remains constant and PWM techniques are applied to vary with the voltage and frequency in the inverter. Due to the diode rectifier, regeneration (dynamic braking) is not possible and the inverter generates harmonics into the AC supply. In the second, a chopper varies the DC voltage to the inverter and the inverter controls the frequency. Due to the chopper, harmonic injection into

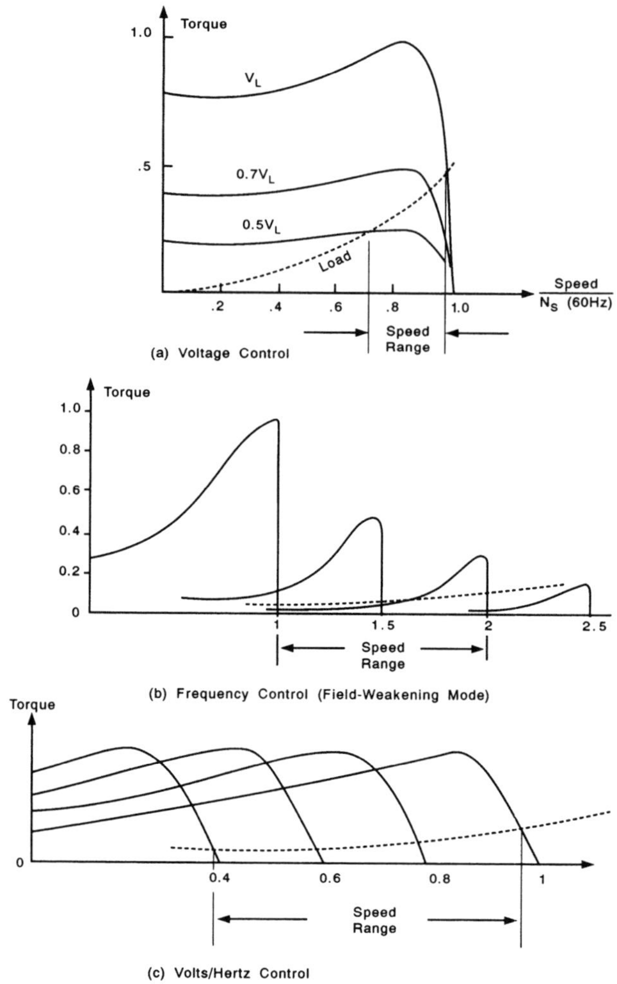

Figure 23-29. Solid-state control

the AC supply is reduced. In the third arrangement, the DC voltage is varied by a dual converter (permitting regeneration) and frequency is controlled within the inverter. Input power factor to the converter is low, however, especially at low speeds.

With stator current control, a constant current is supplied by three-phase current source inverters. The current-fed inverter has the advantages of (1) fault current control and (2) that the current is less sensitive to motor parameter variations. However, the variations generate harmonics and torque pulsation. Two possible arrangements of current-fed inverter drives are shown in Figure 23-31. In the first, an inductor acts as

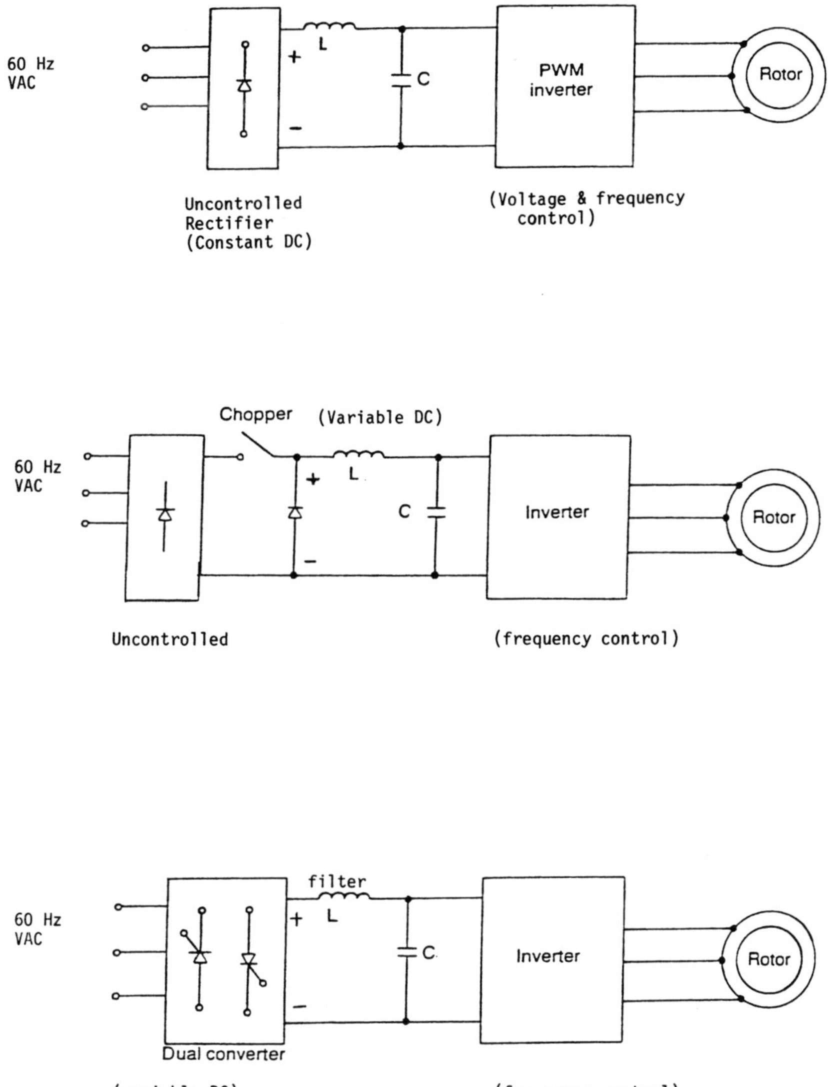

Figure 23-30. Volts/Hertz control methods

a current source and the controlled rectifier controls the current source. The input power factor of this configuration is very low. In the second arrangement, the chopper controls the current source and the power factor is higher. In either case to keep the air gap flux constant and avoid saturation due to high voltage, the motor is operated on the positive slope region of the torque/speed characteristic (Figure 23-31(c)) with voltage

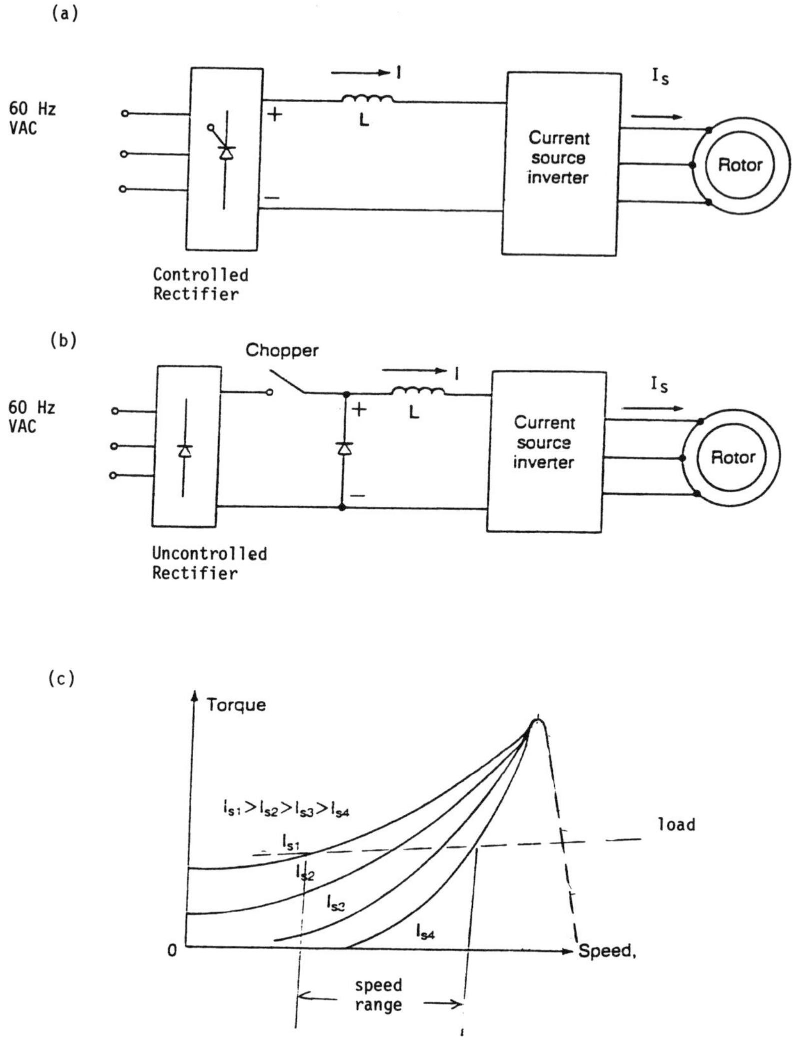

Figure 23-31. Current control

control. As speed increases, the stator voltage rises and the torque increases. This is the unstable region of the characteristic and this closed-loop control is appropriate.

For the widest range of speed control, it is necessary to vary the voltage, frequency, and current to meet the torque/speed requirements illustrated in Figure 23-32. In the first region (below base speed), the speed can be varied by voltage (or current) control, keeping the torque constant. In the second region the motor is operated at constant current and the slip is varied. In the third region the speed is controlled by frequency at a reduced

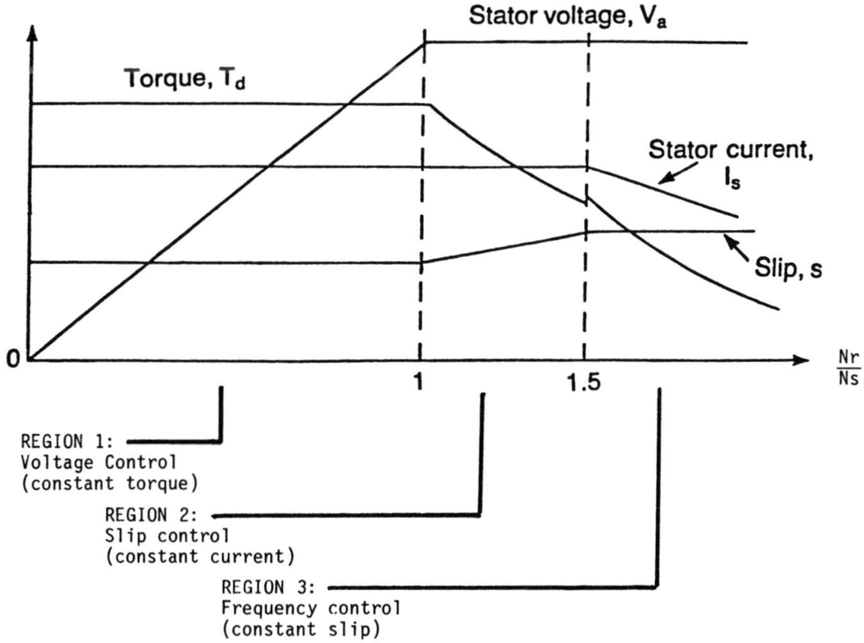

Figure 23-32. Voltage/current/frequency control

stator current (and torque). Closed loop control is normally required to satisfy the steady state and transient performance specifications of these AC drives.

Wound-Rotor (Slip Ring) Induction Motors

Description. The second type of rotor is the wound rotor, in which insulated conductor coils are inserted in rotor slots to form a polyphase winding, with the ends of each phase connected to a slip ring. The rotor must be wound for the same number of poles as the stator, but not necessarily the same number of phases (although it usually is). The number of rotor turns seldom matches the number of stator turns. Note that stator winding pole-changing methods are not appropriate for speed control of wound-rotor motors.

The wound-rotor motor operates under exactly the same principles as the squirrel-cage machine: a stator-produced rotating magnetic field induces currents in the rotor windings (as long as there is relative motion between rotating field and rotor conductors, i.e., slip). For similar motor parameters (rotor and stator resistances and reactances) the torque/speed characteristics would be the same. However, the slip rings give access to the rotor circuit, thereby allowing control of the running characteristics by altering the rotor circuit impedance externally. If the slip rings are shorted, the motor runs essentially like a low slip squirrel-cage type. Connecting resistance between the slip rings, however, has the effect of increasing rotor resistance and "stretching" the torque/speed curve toward increased slip. The amount of maximum torque is not changed, but the slip at which it occurs is increased—and may be increased all the way to 1.0 (standstill) so that the maximum possible starting torque is obtained.

The advantages of slip ring motors are (1) high starting and accelerating torque (and heat generation in the control resistors where it may be more readily dissipated); (2) with external resistance set for maximum starting torque, starting current is much less (1.5 × FLA) than squirrel-cage motors; and (3) varying the external rotor resistance under load offers adjustable speed control over a fairly wide range. As for disadvantages: (1) since efficiency is approximately proportional to operating speed, by operating at slips near 0.5, efficiency is virtually cut in half; (2) since the controller cannot change the no-load speed significantly, speed regulation is poor for low speed settings; and (3) wound-rotor motors are bulkier and more expensive, and are a greater maintenance burden than squirrel-cage types.

Wound-rotor motor control

Manual control. By connecting a three-phase rheostat to the brushes that ride on the slip rings (Figure 23-33), the operator can dial in the appropriate resistance for maximum starting torque, and adjust the speed

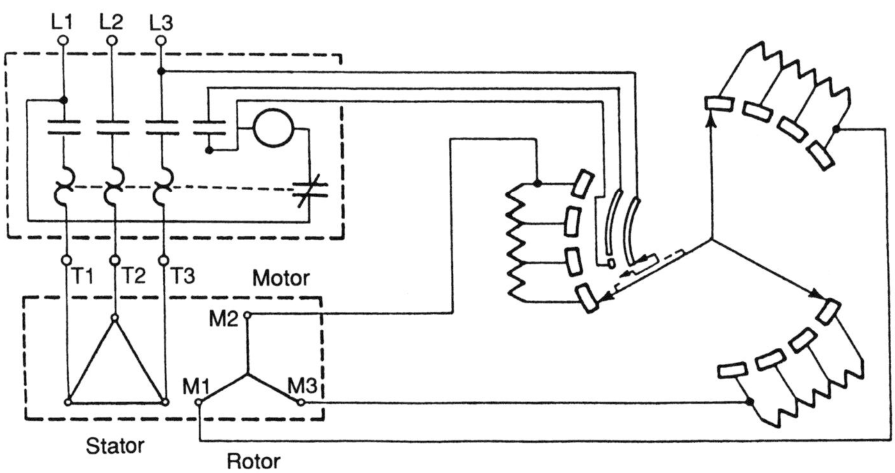

Figure 23-33. Wound-rotor motor manual control

under load. Although adding rotor resistance does reduce starting current, adding too much will not obtain maximum torque (Figure 23-33(b)).

Utilizing resistor banks and a drum controller to activate accelerating (and reversing) contacts, directly or through relays, is another method of manual control. Reversing is accomplished by interchanging any two stator leads; manipulation of rotor circuit resistance will yield the exact same torque/speed characteristic regardless of direction of rotation. This method usually results in multispeed, rather than adjustable speed, control and is applicable to winch drives.

Automatic control. The starting sequence may be automated through the use of time-delay relays. This is most common where increased rotor resistance is to be utilized as a starting device (secondary resistance start) and, once running, the motor is intended to operate at one speed setting. Definite (fixed) timers or compensated timers may be used to control acceleration. Definite timers, which may be pneumatic, dashpot, or electronic relays, are set for the highest load current and remain at the same setting regardless of load. The operation of a compensated timer is based on the applied load; the motor will be allowed to accelerate faster for a light load and slower for a heavy load.

The frequency relay is one such type of timer and its application is illustrated in Figure 23-34. Two accelerating coils (A1 and A2) are connected in parallel across one leg of the rotor circuit. A capacitor is connected in series with A1, which will block current in that branch if the frequency is low enough. At start the voltage and frequency in the rotor circuit are maximum and both A1 and A2 pick up, inserting full resistance. As the rotor accelerates, the induced rotor voltages decrease in magnitude and frequency in proportion to the slip. At some point, the frequency is low enough to cause sufficient voltage division between the capacitor and the A1 coil to drop out A1. Part of the rotor resistance is cut out and the motor continues to accelerate. When the slip and, hence, induced rotor voltages become low enough, A2 drops out, leaving minimum external rotor resistance. Because normally closed contacts are used in this example, the secondary cannot be totally shorted or else the accelerating contacts would never pick up upon starting. The sequence is totally dependent on the mechanical aspects of acceleration which, in turn, depend on the mechanical load on the motor when started.

Solid-state control. The three-phase external resistance can be replaced by a three-phase diode rectifier and chopper as illustrated in Figure 23-35. The gate turn-off thyristors (GTO) operate as the chopper, and the inductance as a current source. The effective external rotor resistance depends on chopper duty cycle, k, and is equal to $R \times (1 - k)$; thus the duty cycle may be used to control the speed of the motor. In large power applications where a limited range of speed control is required, other, more sophisticated, solid-state arrangements may be utilized. One such is called "static

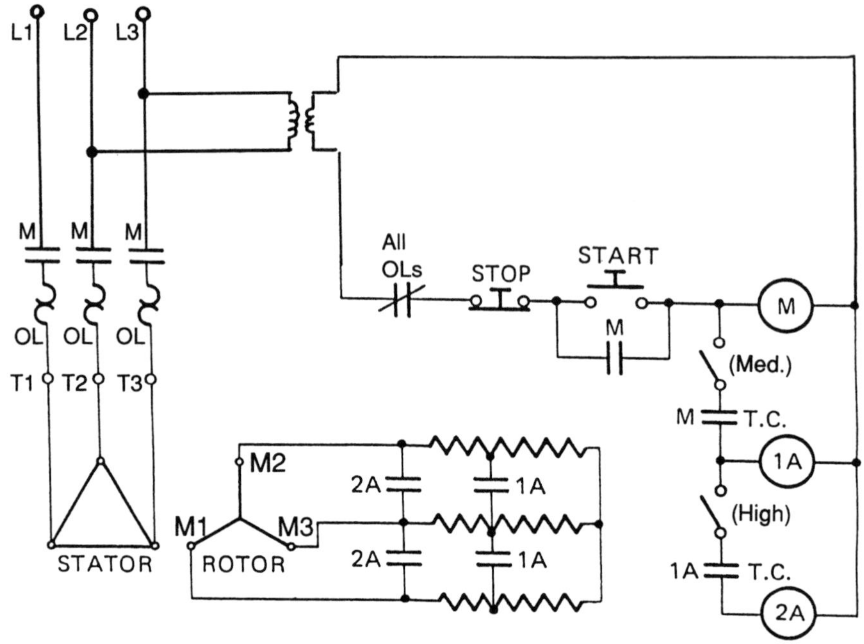

Figure 23-34. Wound-rotor motor automatic control

Kramer drive," in which the chopper described above is replaced by a three-phase full converter and transformer, delivering the slip power back to the AC mains. Another version, called "static Scherbius drive," connects the rotor circuit directly back to the stator three-phase supply through three, three-phase dual converters (or cycloconverters), permitting bidirectional flow of slip power. This arrangement is sometimes called a "doubly fed" machine, as a rotor current at frequency f_r is being fed to the rotor to produce a slip equal to f_r/f_s.

Self-Synchronizing (Selsyn) Motors

Another use of the three-phase wound rotor is in self-synchronizing devices. In these applications the rotor windings of two (or more) machines are connected together via their slip rings and brushes. Currents will circulate between rotor windings if unequal voltages (in either magnitude or phase) are induced in them. This occurs if the rotor windings differ in position (or rate of change in position, i.e., speed) given identical stator excitation.

In a single-phase selsyn motor system, the stators of all machines in the system are energized by the same single-phase source. On one unit (the transmitter), the rotor is turned and locked in place; rotors on the other units (indicators) will chase the transmitter rotor until their positions are once again synchronized. Transmitter and indicator rotors only differ in that the transmitter rotor must be capable of being locked in place (e.g., engine order telegraph handle), or connected to something massive enough not to be turned by an out-of-synch indicator rotor (e.g., rudder angle indicator). Indicator rotors are generally connected to pointers, mounted on an appropriate scale, and equipped with mechanical dampening to prevent them from oscillating or spinning.

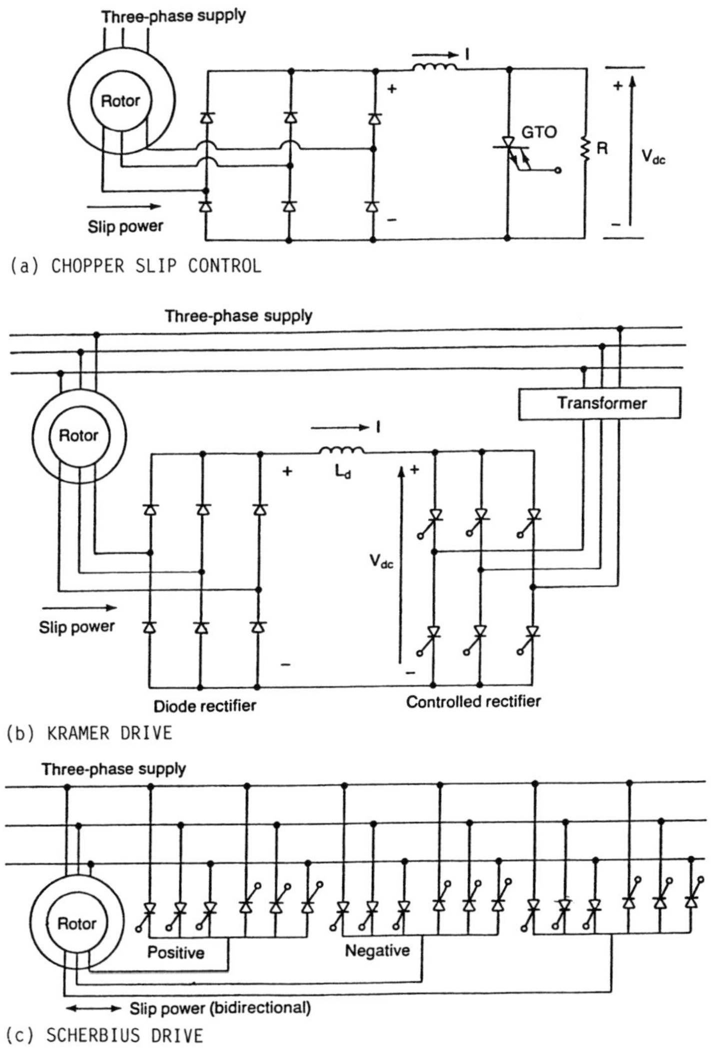

Figure 23-35. Wound-rotor motor solid-state control

Power selsyn motor stators are energized by three-phase AC producing the rotating magnetic field. For current not to circulate between rotors, the two rotors must be in step (same speed and same angular position). Thus two (wound-rotor) induction motors may be locked in step from start-up to full speed. Such an arrangement is applicable, for example, to large conveyers powered by more than one motor.

Figure 23-36 shows the rotor and stator circuits for singe-phase and three-phase selsyn motor systems.

Maintenance

Stator maintenance. Since induction motor stators are virtually identical to those on synchronous machines, refer to the section on synchronous machine maintenance for procedures to identify stator opens, shorts, and grounds.

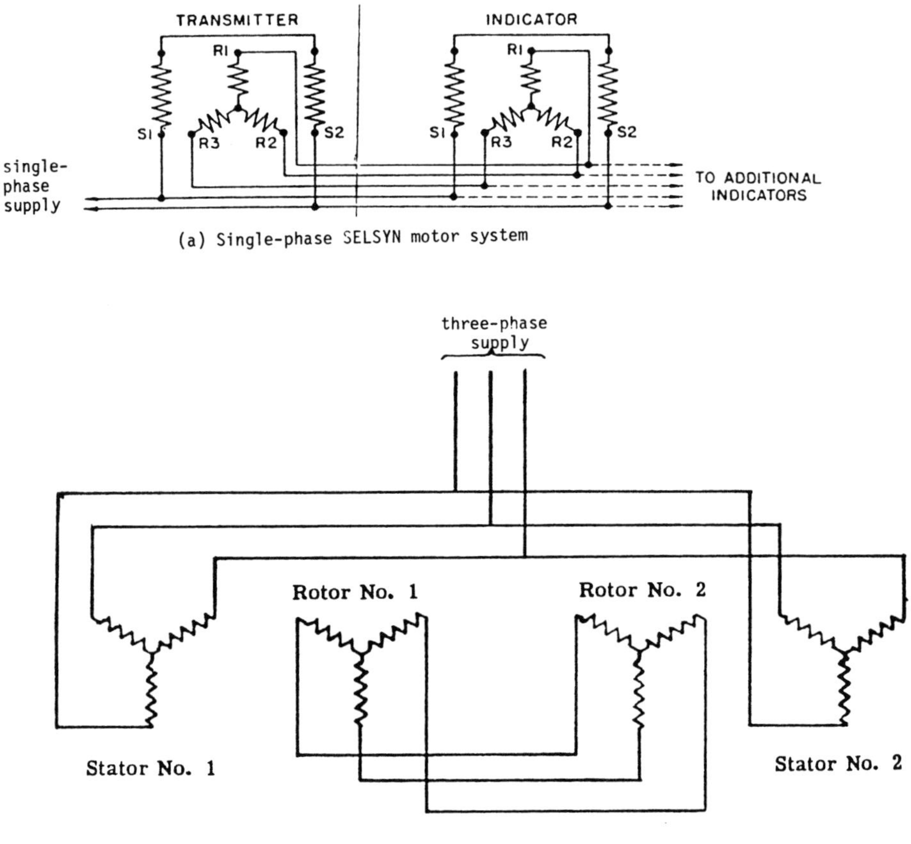

(a) Single-phase SELSYN motor system

(b) Three-phase (Power) SELSYN motor system

Figure 23-36. Selsyn rotor and stator circuits

Rotor maintenance. Broken squirrel-cage rotor bars will cause the machine to fail to come up to speed even at rated voltage and frequency and under normal load. Noisy operation is also a symptom. Rebrazing the defective rotor bar to the end ring is usually the effective emergency repair. In the event that the open bar is not readily visible, applying a (low) single-phase voltage to two motor terminals will allow determination of whether the rotor is defective. By turning the rotor slowly by hand and noting any variations in stator current, the number of open bars can be detected (number of variations per revolution = broken bars × number of poles per phase).

If a wound-rotor motor fails to start, there may be a problem with the external circuit (open in rheostat, loose terminal connections, poor brush-ring contact, etc.). If shorting the slip rings will start the motor, the problem is in the external circuit (note that for such a low resistance start, the motor should be unloaded). If not, there are internal rotor (or stator) problems.

Bearing maintenance. There are four types of bearings commonly used on motors: (1) sleeve bearings (made of steel, bronze, oil-impregnated bronze, or bronze with graphite-filled recesses); (2) ball bearings (either oil packed or grease packed); (3) needle bearings; and (4) thrust bearings (either of the needle type or, in smaller machines, steel, bronze, and even nylon thrust washers).

Sleeve bearings are the most common, the least expensive, and the most likely to give trouble. The most common problem is lack of lubrication. A dry sleeve bearing will make a scraping or sliding sound when turned slowly and may squeal at operating speed. Oil-impregnated bronze and graphitized bearings are sometimes referred to as "self lubricating." However, all types of sleeve bearings generally require lubrication. Motors having sleeve bearings usually have felt packings on one end of each bearing. The packing is kept saturated with oil, which seeps into the bearing surfaces. Lubricating oil is either applied directly to the felt wick or, more commonly, through a small hole in, or tube protruding from, the motor housing. It is advisable to use 10-weight motor oil (20- or 30-weight will also suffice) and only enough to saturate the wick. Too much oil (and oil in the wrong places) is as bad as no oil; it impedes the dissipation of heat and accelerates the deterioration of the winding insulation. Detergent oils should not be used because the additives can attack the winding insulation.

Ball bearings are more durable, less likely to seize, require less maintenance, and are more expensive than sleeve bearings. Grease-packed bearings are often sealed, do not require periodic maintenance, and are less noisy than oil-packed ball bearings. In cold weather, however, acceleration can be sluggish. Oil-packed bearings require periodic lubrication as do needle-type bearings. The latter are used where severe space restrictions exist and, although noisier, provide about the same performance as oil-packed ball bearings.

Worn balls or races in a ball bearing, or excessive clearances between shaft and inside of a sleeve bearing, require bearing replacement. Ball bearings often may be removed by simply tapping the motor end-bell until the bearing drops out. Sleeve bearings usually have to be pressed out or, using a cylindrical fitting, forced out by hammering the fitting. A new bearing can be pressed in the same way, with care taken not to deform or burr the new bearing. Occasionally, freezing the bearing overnight and preheating the bearing housing facilitates bearing insertion. Sometimes sleeve bearings must be reamed to size the bearing appropriately for free rotation of the motor shaft. In reassembling a motor with sleeve bearings, care should be taken in tightening the end-bells evenly to avoid binding and misalignment.

A sleeve bearing operated dry for an extended period can actually melt and fuse to the shaft. "Frozen bearings" may occasionally be broken loose from the shaft by hammering. If not, a propane torch may be necessary to remove the bearing from the shaft (using an oxyacetylene torch increases the chance of damaging the shaft). In any case, a scarred, pitted, or grooved shaft must be dressed before being reused. This can be accomplished by filing and polishing or, in the case of deep pits or grooves, building up shaft by welding or with steel epoxy, and turning down to size on a lathe.

Finally, gummy residues often build up in motor bearings from combinations of oil and dust, improper oil and heat, or just prolonged unattended use. Gummy bearings should be washed with an appropriate solvent (e.g., lacquer thinner), and care taken to keep solvent off the motor windings.

See troubleshooting chart (Table 23-3) for induction motor diagnostic procedures.

TABLE 23-3

Induction Motor Troubleshooting Chart

Problem	*Possible Cause*	*Remedy*
Fails to start	Open line(s)	Check fuses/circuit breaker. Note: Will not start with one line open.
	Open phase	Test stator windings for open.
	Controller failure	Check controller relay coils and contacts; repair/replace as required.
	Overload	Relieve mechanical load on motor; check for bearing seizure or other mechanical restriction.
	Low voltage	Check supply.
(wound-rotor motors)	Open in rotor circuit	Check slip ring/brush rigging; check external resistance/rheostat; test rotor windings and repair as required.

Problem	*Possible Cause*	*Remedy*
Overheating	Overload	Check status against nameplate ratings; relieve mechanical load.
	Insufficient ventilation	Remove restrictions; clean windings; check C. W. supply.
	High voltage	Check supply.
	Low voltage	Check supply.
	Low frequency	Check supply.
	Uneven air gap	Check gap clearances; check for evidence of rotor/stator rubbing.
Runs fast	High frequency	Check supply.
Runs slow	Low frequency	Check supply.
	Low voltage	Check supply.
	Overload	Check status against nameplate ratings; relieve mechanical load.
	Open/shorted/grounded stator coil(s)	Test and repair/replace as required.
(squirrel-cage motors)	Broken rotor bar	Rebraze defective bar(s) to end ring. Note: Cast aluminum rotors can seldom be repaired.
(wound-rotor motors)	Open in rotor circuit	Check slip ring/brush rigging; check external resistance/rheostat; test rotor windings and repair as required.
	Secondary resistance high	Check rheostat position/rating; check contacts on machines with fixed-resistance starters.
Noisy operation	Bearings	Check for worn, loose, dry, or overlubricated bearings; replace worn or loose bearings; lubricate dry bearings; relieve overlubricated bearings.
	Uneven air gap	Check for bent shaft, bearing condition, gear alignment.
	Coupling	Check alignment; re-align and/or tighten coupling.
	Loose laminations	Tighten bolts; dip in varnish and bake.

SINGLE-PHASE MOTORS

Classification

Sometimes referred to as "fractional horsepower" motors, these small (typically 10 W to 1 kW) machines are found in appliances, hand power tools, and a myriad of other applications. Single-phase motors may be loosely grouped in one of three categories: induction types, synchronous types, or commutator types. And, although many different varieties of these have been produced over the years, competition has effectively narrowed the field to squirrel-cage types, synchronous types, and the universal motor.

Induction Types

General description. Single-phase stator windings produce an *oscillating* magnetic field, as opposed to the *rotating* magnetic field produced by polyphase stators. This oscillating field will induce currents in a squirrel-cage rotor, and these currents will produce a torque once the rotor is spinning. However, a single-phase stator winding will not, by itself, generate any torque at rotor standstill. Thus, the means of providing starting torque is the factor used to classify this category of single-phase motor.

The rotors of single-phase squirrel-cage motors are essentially the same as the three-phase types described earlier. The stator core, with the exception of shaded pole motors, is also very similar. A single-phase stator can be wound for any even number of poles, with two, four, and six being the most common. However, where the slots of three-phase machines are all of the same size, single-phase stator cores may have two or three different size stator slots. The stator consists of two separate windings displaced from one another by 90 electrical degrees, with one usually having fewer turns of smaller wire than the other. In motors that operate with both windings energized, the heavier winding is known as the main winding and the other the auxiliary winding. If the motor operates with the auxiliary winding opened after starting, they are generally referred to as running and starting windings. The principle is to arrange for the current in the auxiliary winding to be out of phase with the main winding current—producing, in effect, a two-phase machine supplied from single-phase mains.

Split-phase motor. The starting winding of these windings is about one quarter of the size of the main winding and thus has a lower X/R ratio. This results in a starting winding current which leads the main winding current by about 30 degrees, producing a rotating magnetic field. These motors normally have a centrifugally operated starting switch (Figure 23-37) that cuts out the starting winding when the motor reaches about 75 percent of synchronous speed (and recloses at about 25 percent). The starting winding in these motors heats up rapidly during starting. If the load has too much inertia, or if the centrifugal switch malfunctions, the winding will quickly burn out.

The rotational direction of a split-phase motor may be reversed by interchanging the starting winding connections. The torque/speed characteristics are illustrated in Figure 23-38.

Capacitor start motor. In every sense capacitor start motors (and dual-capacitor motors and PSC motors) are split-phase machines. That actual term, however, has been reserved for the machine described above where the X/R ratio is lowered by boosting the resistance. Inserting a capacitor in series with the auxiliary winding achieves the same result by reducing the reactance (X). The greater phase differential achieved by using starting winding conductors half the size of the main winding conductors, plus an electrolytic capacitor, results in a higher starting torque and less starting current than split-phase motors.

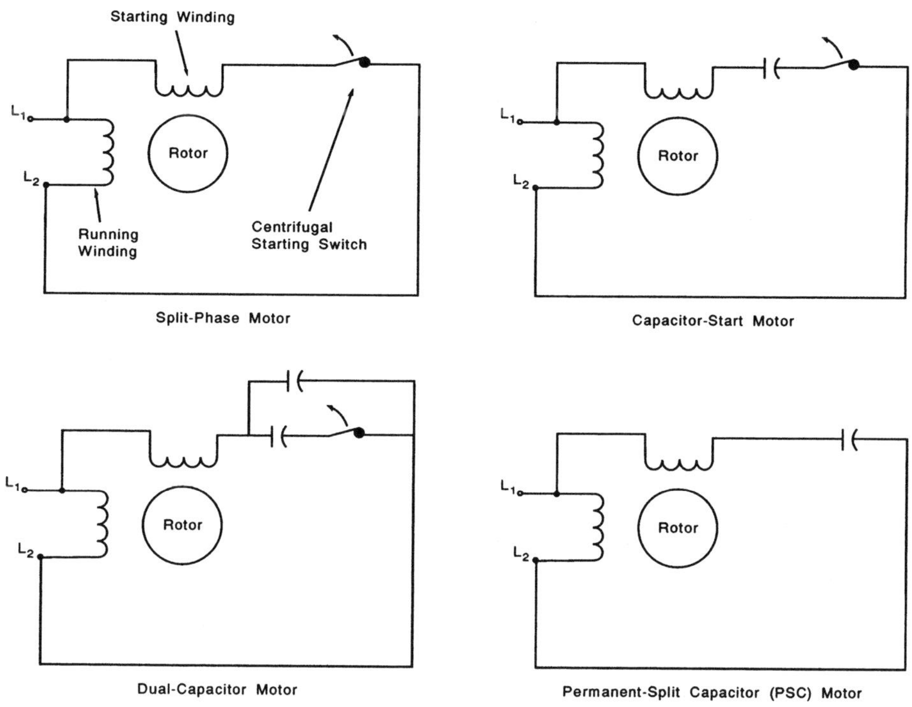

Figure 23-37. Single-phase squirrel-cage motors

The electrolytic capacitor is generally rated for short time duty only, and starting winding burnout can still occur. However, the higher starting torque will accelerate higher inertial loads in sufficient time.

Dual-capacitor motor. Also known as the "capacitor-start/capacitor-run" motor, this has an additional continuous-duty, oil-filled capacitor (of somewhat smaller capacitance) in the auxiliary winding circuit (Figure 23-37). The motor continues to run with both windings energized after the electrolytic capacitor is cut out by the centrifugal switch at about 75 percent of synchronous speed. The advantages of this arrangement are (1) higher breakdown torque as seen in Figure 23-38, (2) quieter operation (less vibration) due to maintaining a rotating magnetic field during operation, and (3) improved power factor by virtue of the leading current in the auxiliary winding.

Permanent-split capacitor (PSC) motor. The auxiliary winding of this machine usually has the same size conductors and as many turns as the main winding. A continuous-duty, oil-filled capacitor is connected in series with the auxiliary winding (Figure 23-37), sized to cause a 90-degree phase

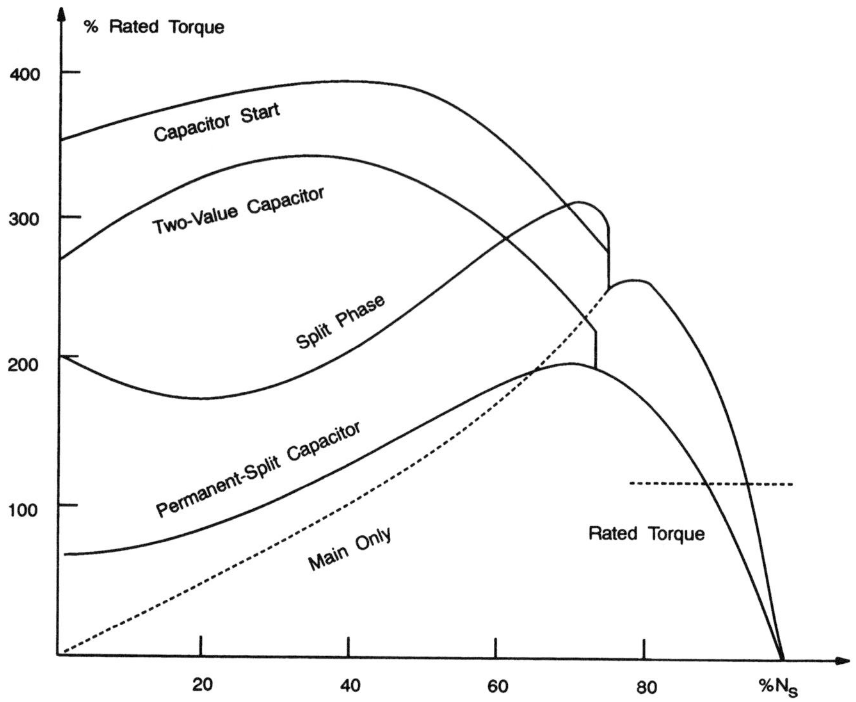

Figure 23-38. Single-phase motor torque/speed characteristics

differential in main and auxiliary currents and rated load. The actual phase difference varies with load, and, at start, the (small) capacitor size results in a low starting torque (Figure 23-38). A permanent-split capacitor motor operates quietly, with high power factor and low current, and, with no starting switch, is most suitable for wide-range speed control.

Shaded-pole motor. The construction of the stator pole pieces (Figure 23-39) includes a closed copper loop enclosing an axial strip of about one quarter of the pole width. Currents induced in the loop oppose the main winding flux, causing the flux to sweep across the pole face. This effect is amplified by designing the pole piece so that the air gap is wider at the pole sides opposite the shading coil. Shaded-pole motors are very economical for low power applications and are compatible with wide-range speed control. However, they are not reversible, and their efficiency and power factor are low.

Synchronous Types

General description. Obviously named for their constant (synchronous) speed operation, these usually very low power (15 W) motors are used wherever timing is essential, such as in clocks and tape drives. Most

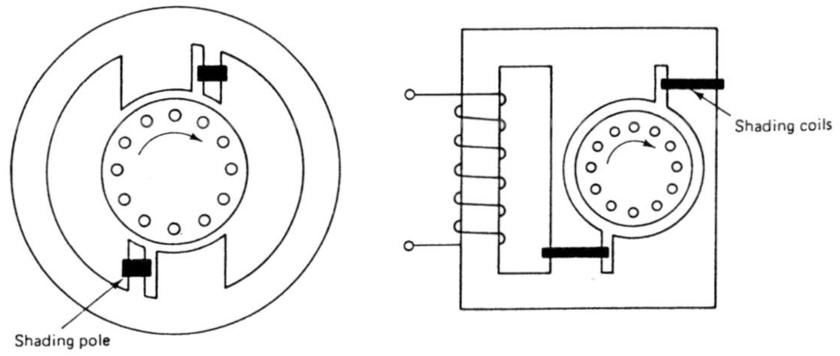

(a) Shaded-Pole Motors

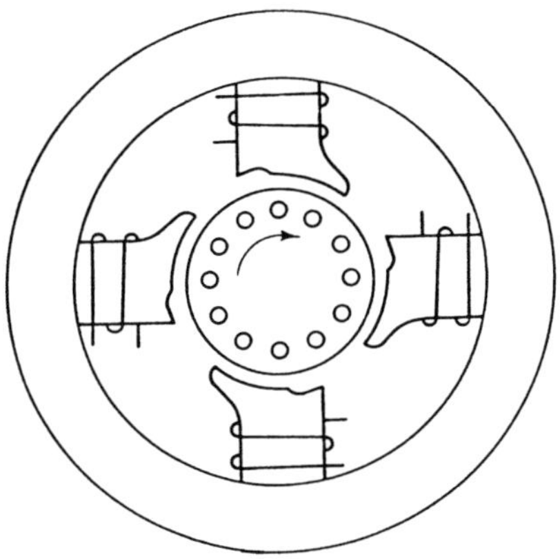

(b) Reluctance Type Synchronous Motor

Figure 23-39. Shaded-pole and reluctance type single-phase motors

single-phase synchronous motors use shaded-pole stators and rotors that can lock in step with the rotating magnetic field by becoming essentially a permanent magnet. There are two common types.

Hysteresis motor. The rotor on this type consists of a hollow cylinder of a hard ferromagnetic material with appreciable hysteresis. When operating below synchronous speed, the magnetizing of the rotor and the delay (due to hysteresis) in remagnetizing in the opposite direction develop torque in the rotor. Pulling into synchronism gradually, the induced rotor poles tend to remain permanently located (due to hysteresis) once synchronous speed is reached.

Reluctance motor. The reluctance motor rotor is similar to squirrel-cage rotors with portions cut away (Figure 23-39), leaving projecting iron poles.

The motor starts as a squirrel-cage motor, but when it gets very close to synchronous speed, the rotor poles snap into synchronism with the stator poles. The acceleration is not as smooth as that of the hysteresis type and below synchronous speed the speed is unstable, but they are inexpensive to manufacture.

Commutator Types

General description. A wound rotor, with rotor coil terminals connected to commutator segment (essentially a DC armature), distinguish this category of single-phase motors consisting of the universal (AC series) motor and the repulsion family of motors.

Universal motor. Electrically identical to a DC series motor, the universal motor is specifically designed to run on DC or AC of equivalent voltage and any frequency up to 60 Hz. The only design changes are a minimal number of field turns and the laminating of the entire magnetic circuit. Some universal motor arrangements have the field poles distributed in stator slots as opposed to the concentrated winding shown in Figure 23-40, and some have distributed compensating windings also. The principle of operation is that, since the direction of armature current and field pole polarity are reversed at exactly the same time during each half-cycle, the torque remains in the same direction. To reverse the direction of rotation, the armature leads must be reversed.

Although they are designed to run on either AC or DC, universal motors have characteristics that are not exactly the same on each type of voltage. In both modes the starting torque is high, as is the no-load speed. With universal motors, direct coupling or geared drives are used to prevent runaway at no load. When operated on AC, saturation of the magnetic circuit during peak current tends to raise the motor speed. On the other hand, inductive reactance tends to choke motor current and reduce torque when the motor is operated on AC.

The universal motor is generally designed to operate at very high speed (12,000 to 24,000 RPM), and speed control may be obtained by either varying the applied voltage or inserting resistance in series with the motor. Because they run at such high speeds, the motors are smaller and lighter for the same rated horsepower, which makes them particularly suitable for portable hand tools.

Repulsion-type motors. A repulsion-type motor consists of a split-phase-type stator with the rotor a typical lap-wound DC armature. There is no electrical connection, however, between rotor and stator. Instead, the brushes that contact the commutator are shorted together. The physical position of the brushes determines the axis of the rotor field and the direction of rotation, if any. There is no rotation if the brushes are aligned

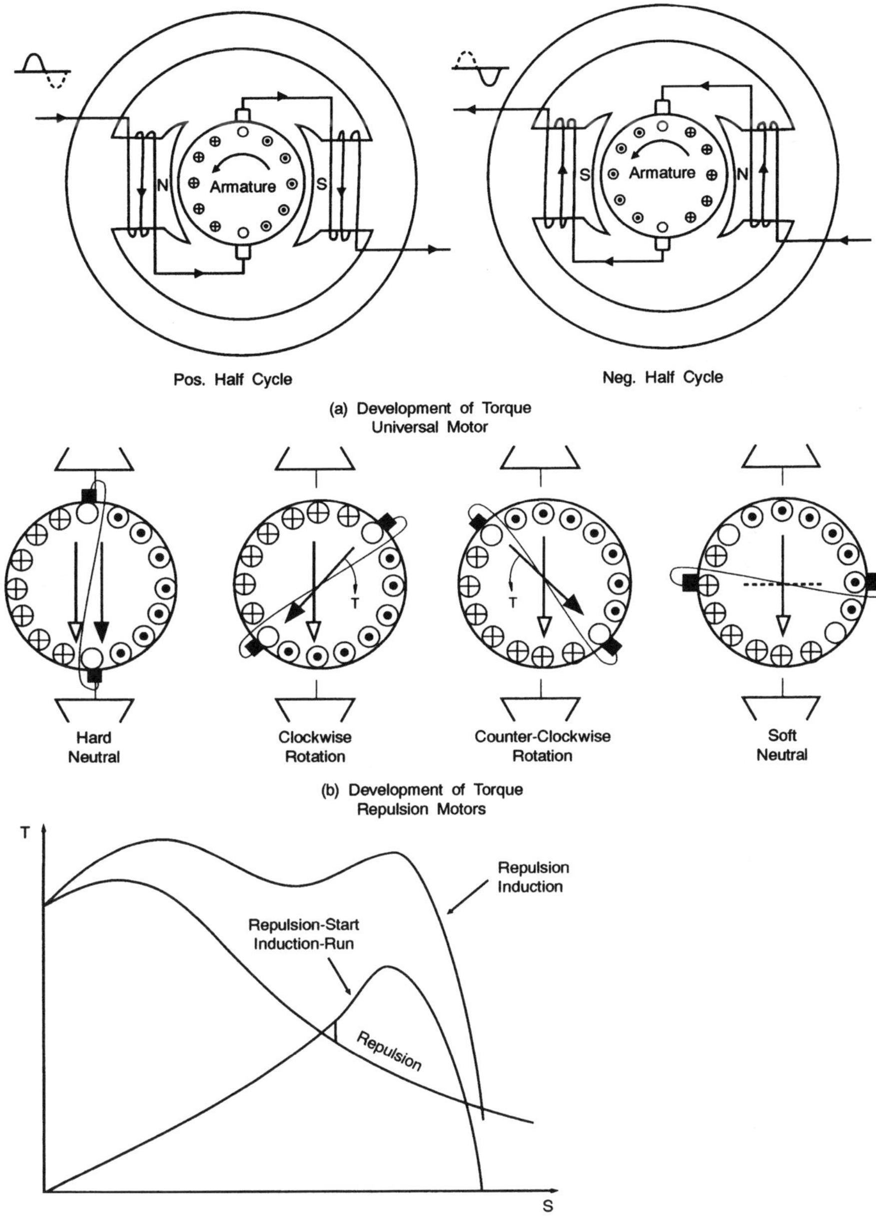

Figure 23-40. Commutator type single-phase motors

with the alternating magnetic field produced by the stator (called the *hard neutral*) and no torque is produced with the brush axis perpendicular to the stator field either (*soft neutral*). Maximum torque is produced at approximately 15 to 25 degrees on either side of hard neutral; the side determines the direction of rotation as shown in Figure 23-40. The members of the repulsion family include (1) *repulsion motors,* which operate on the repulsion principle described at start and under normal operation; (2) *repulsion start motors,* in which a centrifugally operated mechanism shorts all the commutator bars, leaving the motor to operate as an induction machine (in some cases the repulsion brushes are lifted by the same centrifugal device that shorts the commutator); and (3) *repulsion-induction motors,* in which the rotor has two independent sets of windings—repulsion and conventional squirrel-cage—and both are continuously operative, producing a characteristic curve (Figure 23-40) in which the torque is the sum of the repulsion and induction torques.

Repulsion-type motors, in particular repulsion-induction motors, have been made as large as 10 hp, but are expensive and very rare today.

Single-Phase Motor Control

Starting. Across-the-line starting is standard for single-phase motors. For those with starting windings, a mechanically activated centrifugal switch is typically responsible for opening the auxiliary circuit. Hall-effect sensors are also occasionally used to monitor RPM and open the starting winding at the appropriate time. Centrifugal switches have some drawbacks: Their maloperation is the most significant maintenance burden associated with single-phase machines; they inherently prohibit plug-stop operation (required for electric chain hoist operation); and, in hermetically sealed refrigeration compressors, the arc drawn in the switch can break down the refrigerant chemically. For those applications where centrifugal (or internal) relays are inappropriate, external starting relays may be employed. A thermistor in series with the starting winding, or a thermal starting relay whose heating element is designed to open the starting contact after about two seconds, is common in small air-conditioning units. They must be allowed to cool before they can reclose the starting circuit and are therefore not suitable for plugging or short-cycling types of operations. A current-actuated starting relay which picks up with main winding inrush current and releases when the current drops off at operating speed is another possible arrangement. These relays are suitable for plugging and repetitive starting. On capacitor start motors, a voltage-sensitive relay connected across the starting winding can detect the sharp rise in starting winding voltage as the motor approaches synchronous speed and thus can open the starting contacts. When the motor is running, the voltage generated in the unenergized starting windings keeps the relay open. This arrangement is not applicable to split-phase machines.

Speed control

Variable voltage speed control. In motors that do not employ a centrifugal switch (i.e., shaded-pole and PSC motors), a considerably wide range of speed control can be realized by varying the applied voltage. This may be accomplished by a tapped inductance in series with the motor winding, a tapped autotransformer, or electronic manipulation of the line voltage wave form. In the latter method a triac (triode AC switch) is often used to produce the varying voltage. In the circuit shown in Figure 23-41, a diac (diode AC switch) is used in the phase shifting (or "triggering") circuit, and the R-C time constant (determined by the variable-resistance/fixed-capacitor combination) provides the delay-time control. Note that some triac voltage controllers, designed to be used as dimmers for incandescent lamps, conduct on only one half of the cycle, producing, in effect, DC. These should not be used on inductive loads and can burn out single-phase stators. Universal motors may employ tapped series resistors or triac voltage controllers to vary speed.

Two-speed control. Changing stator lead connections can also be used to control speed in discrete steps, often at a ratio of 2:1. Tapped stator windings on shaded-pole motors offer the possibility of altering the volts/turn ratio (Figure 23-41) and hence torque. Since the actual speed depends on load, however, this method has no effect on no-load speed.

Two-speed PSC motors are rare, but two-speed starting switch types of single-phase motors are fairly common. If the starting switch is centrifugally operated, the speed range seldom exceeds 2:1, suggesting the use of consequent-pole main winding connections. The starting winding is typically not of the pole-changing type but wound for the *lower* number of poles. Thus the motor starts on its high speed winding and the starting switch changes the main winding to consequent-pole arrangement (if the low speed terminals have been energized) as well as cutting out the starting winding. If the two speeds are not in the ratio of 2:1, the motor will have two windings (usually four and six poles). Some of these motors have two starting windings (one for each speed), but often one starting winding is employed, working in conjunction with the high speed main winding. These machines are started on high and switched to the low speed winding either by the starting switch or manually.

Maintenance

Once it has been determined whether failure of a single-phase motor is due to problems within the starting winding or main winding, techniques to pinpoint the offending coil or bar are similar to those described for three-phase machines. The problem, of course may also lie in the centrifugal switch mechanism or starting relay, or capacitor, if one is employed.

By spinning the rotor rapidly by hand and closing the line switch, it is possible to isolate the problem to either winding. If the motor accelerates to normal speed, the trouble is with the starting circuit; if it cycles, accelerating and decelerating, the trouble is in the main winding. Capaci-

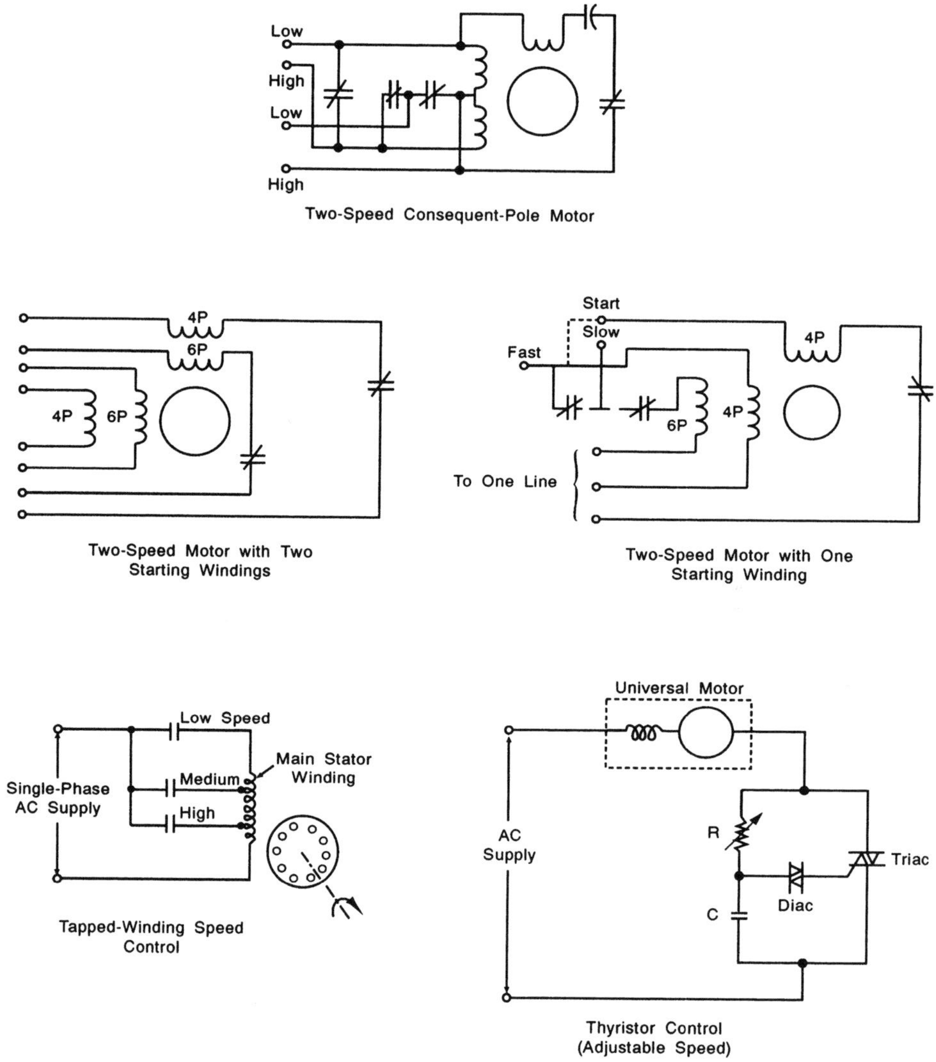

Figure 23-41. Single-phase motor speed control

tors, starting relays, and centrifugal mechanisms should be checked if the starting circuit is at fault.

If the speed control unit is at fault, the cause is often a defective SCR, triac, or diac. An SCR may be checked with an ohmmeter by connecting the positive probe to the anode and the negative to the cathode; an infinite resistance should be indicated. With these connections held fast, the gate should be momentarily shorted to the anode; the resistance reading should drop to a low value and remain there. With a triac (which is a back-to-back SCR) the same procedure is executed, but then the probes are reversed and the test is repeated. A diac can only be tested with an ohmmeter for shorts, as it requires 25 to 35 volts to trigger. Occasionally, a triac comes with an internal diac and, for the same reason, it can only be checked for shorts. To determine if you have an open triac or one with an internal diac, you must consult the information available from the manufacturer of replacement units or guides for selecting parts. For that matter, the same method must be used to determine whether a component is a transistor, an SCR, or a triac, as the three may have the same external appearance.

A single-phase motor troubleshooting chart is displayed as Table 23-4

TABLE 23-4

Single-Phase Motor Troubleshooting Chart

Problem	*Possible Cause*	*Remedy*
Fails to start	Blown fuse	Check and replace.
	Open/shorted/ grounded stator coil(s)	Check and repair/jump defective coil(s).
	Defective centrifugal mechanism	Disassemble, clean, inspect, adjust, repair, or replace parts.
(split-phase types)	Open in starting winding	Check connections; check and repair/jump defective coil(s).
	Defective capacitor	Replace.
(commutator types)	Brush contact	Inspect; adjust brushes for proper grade, pressure, seating.
	Brush position	Inspect; align brush rigging.
Overheating	Overload	Reduce load.
	Voltage high/low	Check supply and adjust voltage.
	Shorted/grounded stator coil(s)	Check and repair/jump defective coil(s).
(split-phase types)	Starting winding not cut out	Inspect centrifugal switch.
	Defective centrifugal mechanism	Disassemble, clean, inspect, adjust, repair, or replace parts.
	Capacitor shorted	Check and replace capacitor (capacitor-run type motor).
(repulsion types)	Improper brush position	Inspect and align brush rigging.
(series, reluctance, and shaded-pole types)	Shorted/grounded field poles	Check and repair/replace field poles.
Runs slower than normal	Voltage low	Check supply and adjust.
	Overload	Reduce load; check for mechanical binding.
	Starting winding not cut out	Check centrifugal mechanism.
	Shorted/grounded stator coil(s)	Check and repair/jump as required.
	Open/shorted/grounded rotor coil(s)	Check and repair as required.

Problem	*Possible Cause*	*Remedy*
Noisy operation	Bearings	Check for worn, loose, dry, or overlubricated bearings; replace worn or loose bearings; lubricate dry bearings; relieve overlubricated bearings.
	Uneven air gap	Check for bent shaft, bearing condition, gear alignment.
	Coupling	Check alignment; re-align and/or tighten coupling.
	Excessive end play	Adjust end-play takeup screw or add thrust washers to shaft.
	Loose mountings/ accessories	Tighten all loose components.

DC MACHINERY

Basic Principles

The principal phenomenon in all DC machine operation is the behavior of a moving charged particle in a magnetic field. From the submicroscopic point of view, this takes the mathematical character of the expression:

$$F = q\,(v \times B)$$

where F is the "induced" force, q is the amount of charge in motion, v is the velocity, and B is the magnetic field density. F, v, and B are *vector* quantities, having both a magnitude and a direction assigned to them (i.e., the strength and the direction of force F depend on the direction and speed of **v** and the strength and orientation of B). There are more than a handful of "hand rules"—left-hand rules, right-hand rules, etc.—with bewildering sets of instructions as to when and where to use them. One of the most useful is the "right-hand rule," which indicates the relative directions of F, v, and B. In Figure 23-42(a), particles of charge q are packed in a conductor. The conductor is being moved down in a magnetic field which is oriented into the page (i.e., a north pole is above the page, a south pole under the book). If the fingers of your right hand are first aligned with v, and then curled into B, your right thumb should point to the right-hand margin. (This mathematical operation is pronounced as "v *cross* B.") The *electromagnetic* force (emf) is the voltage produced in a generator, and your thumb is pointed toward the positive end of the conductor—the direction conventional current would flow if a closed external circuit permitted. If, instead of moving the *conductor* with velocity v, the charges were pumped down the conductor (to the right) with velocity v, then "v cross B" yields an induced force in the upward direction, tending to move the conductor up. This is, of course, the principle of motor action (*exactly* the same as the phenomenon that produces voltage in a generator—the only difference being the orientation of v). A load on the generator (moving down and producing an emf) will cause a current to flow to the right, inducing a force "up" which opposes the original motion of the conductors. This countertorque, which is proportional to load current, must be overcome by the prime mover to maintain the machine at its rated RPM. This is why the prime mover must produce a higher output (and consume more fuel or steam) at higher electrical loads. As a motor, the upward motion of the conductor, *caused* by pumping current to the right, *induces* an emf to the left. This counterelectromotive force (cemf) opposes the original applied voltage. Thus, on the submicroscopic scale, torque and countertorque, emf

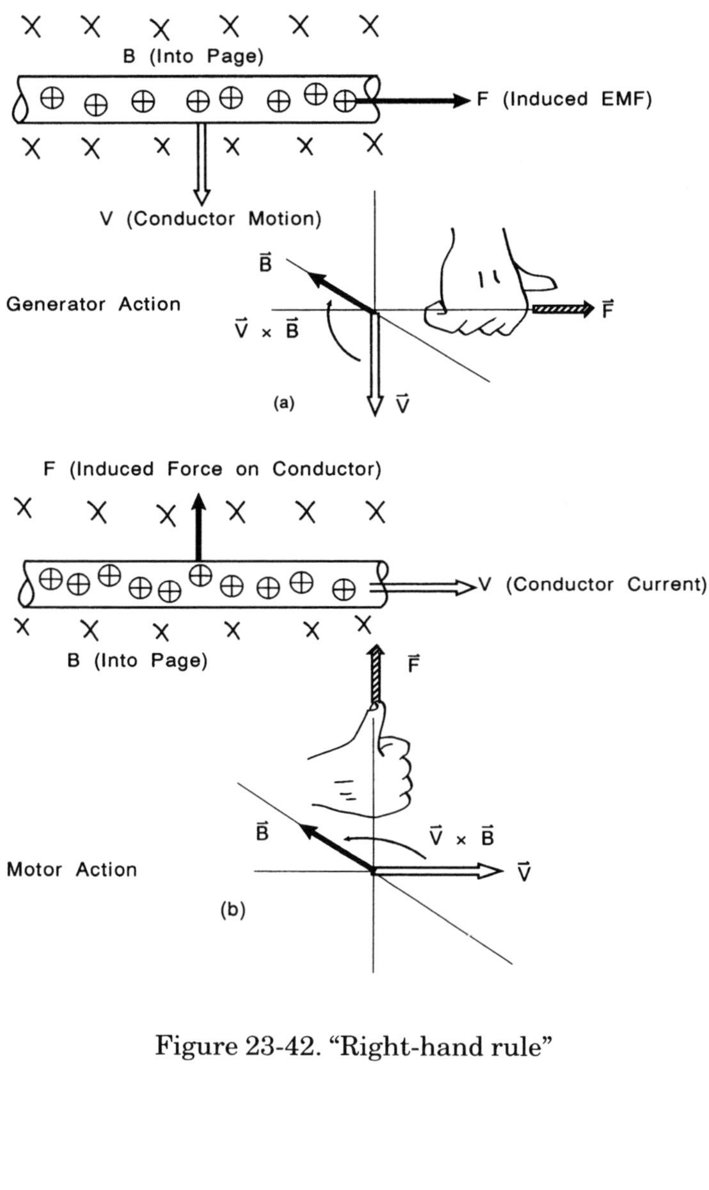

Figure 23-42. "Right-hand rule"

and cemf, motors and generators, all are explainable in terms of "v cross B."

On a slightly larger scale, instead of considering the induced forces on individual charges, the aggregate emf or force on a conductor of length l (oriented perpendicularly to the magnetic flux) is Blv or Bli respectively. Stepping back still further, and considering all the conductors present in a DC machine, the production of armature voltage (or counter-emf) and rotor torque is described by the two fundamental relationships:

$$E = K \times \phi \times RPM$$

$$T = K \times \phi \times Ia$$

Here K is the machine constant and depends on the number and arrangement of armature windings; ϕ is the flux per pole produced by the field circuit; and $K\phi$ is common to both expressions—indicating that $T \times RPM$ (mechanical power) should equal $E \times Ia$ (electrical power). Of course there are mechanical losses so this T is not the torque at the coupling, and there are electrical losses so this E is not the terminal voltage, but somewhere in the middle of the machine—in the "air gap" between stator and rotor—the conversion between electrical and mechanical energy is accomplished.

Construction

Overview. A DC machine may be considered as an "inside-out" synchronous machine and is physically identical when utilized as a motor or as a generator. Its field circuit consists of coils mounted on stationary pole pieces similar to salient-pole rotors. The coils are connected in series with one another with the sense of winding reversed in adjacent coils to produce opposite polarity. The field circuit terminals are labeled F1 and F2. In some machines a heavier winding of a few turns is wound around the outside of each field coil (in the same direction as the inside coil), connected in series with one another, and the leads are brought out as S1 and S2. Smaller pole pieces, with heavy conductor and few turn windings, are found in between the main field poles of large machines. These are known as *interpoles* or *commutating poles,* the purpose of which will be described later.

A DC rotor consists of a distributed winding of armature coils fitted in slots on the periphery of the rotor core. The ends of each coil are connected to commutator bars or segments mounted on the rotor shaft, and insulated from the shaft and each other by a sheath of mica. (With the commutator, the similarity between synchronous armature [stator] and DC armature [rotor] ends.) Each armature coil is connected to two bars, and, in most cases, each bar is connected to an end of two coils, so the number of commutator bars equals the number of armature coils. On two-layer armatures, with two coil sides per slot, the number of slots, coils, and bars is identical. There are several ways, however, in which armature coils may be connected to one another, via a common commutator bar.

Armature windings

Classification. The number of commutator bars between connections to a particular armature coil is called the commutator pitch and may be expressed in bars or degrees (mechanical or electrical). It is the commutator pitch that defines the armature as lap- or wave-connected and the degree of multiplicity (m) and the degree of reentrancy (r).

Lap-wound armatures. Lap windings have a rather small commutator pitch and, in their simplest form, result in the ends of a given armature coil being connected to adjacent commutator bars. This is known as a *simplex* winding and has a multiplicity of one. Lap windings with commutator pitches of two, three, or four bars are known as duplex, triplex, and quadruplex respectively (m = 2, 3, 4 respectively). Regardless of the number of poles or degree of multiplicity, a lap winding can be accomplished with any number of coils and commutator bars. The multiplicity is always equal to the commutator pitch, and the number of parallel paths is always m times P (the number of poles) as illustrated in Figure 23-43(a).

The principal advantage of lap windings is the relatively large number of parallel paths that may be created. Since there are practical limits on the amount of current that can be carried by a single coil, generators with large current ratings generally require a large number of parallel paths. The drawback is that, in each parallel path, the coils in that path "see" only one pair of poles. If the flux per pole is not identical for each pole (and air gap tolerance, pole piece reluctance, and field coil resistance inevitably contribute to inequalities in practical machines), circulating currents can develop among the paths, giving rise to additional heating and poor commutation. This effect can be remedied with *equalizer connections.* (Note: Equalizer connections in lap-wound armatures are not to be confused with equalizing buses required of compound generators operating in parallel, which will be discussed later.) An equalizer is a direct, low resistance path between points in the circuit that should be at the same potential. In the four-pole armature in Figure 23-43, coils 2 and 8 are identically situated from an electromagnetic point of view. An equalizer can therefore be installed between corresponding points of the two coils, either at the commutator end or the load end or at the rotor. If equalizers were connected between coils 1 and 7, 2 and 8, and so on, to 6 and 12, the armature would be said to be 100 percent (or fully) equalized. Due to space limitations, this cannot always be the case; if only half the possible connections were made (as symmetrically as possible) the armature would be 50 percent equalized.

Wave-wound armatures. The wave winding avoids the problem of uneven parallel path currents by insuring that all coils in any path are

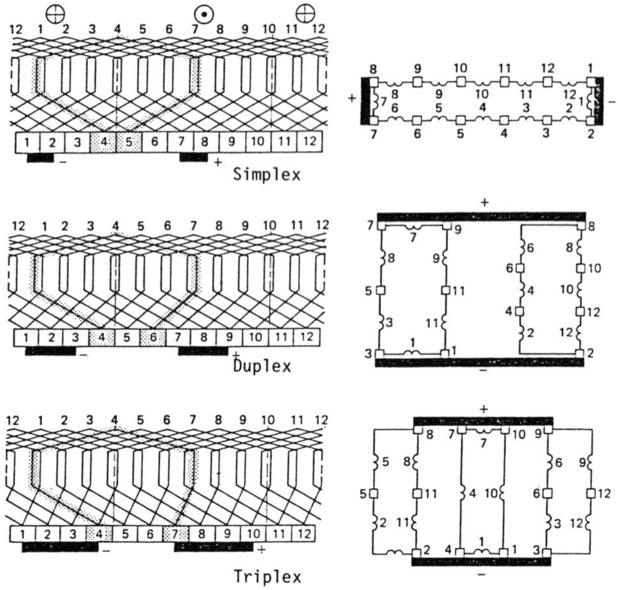

(a) Two-pole machine

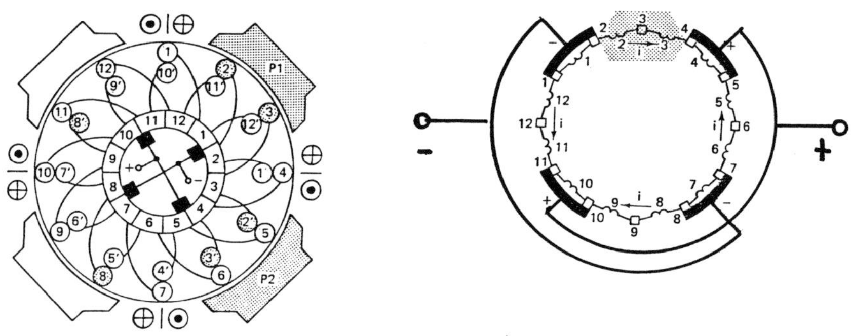

(b) Four-pole machine

Figure 23-43. Lap-wound armatures

distributed under all the poles. Therefore a simplex wave winding has only two parallel paths regardless of the number of poles. The commutator pitch in wave winding is close to 360 electrical degrees, which on a two-pole machine is not fundamentally different from a lap winding. Two-pole wave windings are virtually nonexistent. A single simplex (m = 1) wave winding is connected to the commutator at each north pole (360 degrees electrical) and returns to the commutator bar adjacent to where it "started." Ending two commutator bars away results in a duplex winding (m = 2), three commutator bars away, a triplex (m = 3). Wave-wound machines seldom have a degree of multiplicity greater than 3, and the number of parallel paths is always 2 × m, no matter how many poles. Figure 23-44(a) shows a duplex wave winding: Starting with commutator bar 1, the winding passes through coil 1-1′, ending at bar 8, and continues through coil 8-8′, ending at bar 3, two bars away from where it "started"; hence, m = 2. Note that the number of parallel paths is 4 (2 × m) and that the same voltage would be obtained between brushes B1 and B2 whether or not brushes B3 and B4 are present.

The number of brush sets is normally equal to the number of poles and *must* be so for nonequalized lap windings and frog-leg windings. Wave-wound armatures and armatures with fully equalized lap windings can be operated with only *two* brush sets, one positive and one negative, located 180 electrical degrees apart.

The term "reentrancy" refers to the number of closed loops formed by the completed armature circuit; it must always be at least one (singly reentrant: r = 1), but may be two (doubly reentrant: r = 2), or three (triply reentrant: r = 3), etc. Reentrancy is equal to the largest common factor between the number of elements and the degree of multiplicity, which usually equals m. This is significant during certain armature troubleshooting procedures.

Frog-leg windings. A frog-leg winding is a combination of a lap and a wave winding connected to the same commutator. It generally results in better commutation than either lap or highly multiplexed wave windings. With both lap and wave elements having the same number of turns and the same coil span, and each being equal in number to the number of commutator bars, the windings must have the same number of parallel paths to produce equal voltages. This means that if the lap winding is simplex (and it usually is), the wave winding must have a degree of multiplicity equal to P/2. Figure 23-44(b) shows a frog-leg-wound armature with 16 commutator bars, lap elements, and wave elements, a coil span of 3, and lap and wave commutator pitches of 1 and 7 respectively.

Armature reaction. Under no-load conditions (no armature current) the only air gap flux present is that produced by the field poles. Under load, however, the armature generates a magnetic field that is not aligned with the stator field. The resultant skewed, or distorted, field will cause a problem as the magnetic neutral (an imaginary line through the center of rotation, perpendicular to the net flux) is shifted. The brushes are intended to be located "on the magnetic neutral" (*brush neutral*) so that the coil(s) shorted by a given brush are exactly those in which no emf is generated. Having the brushes "off neutral" results in poor commutation (sparking)

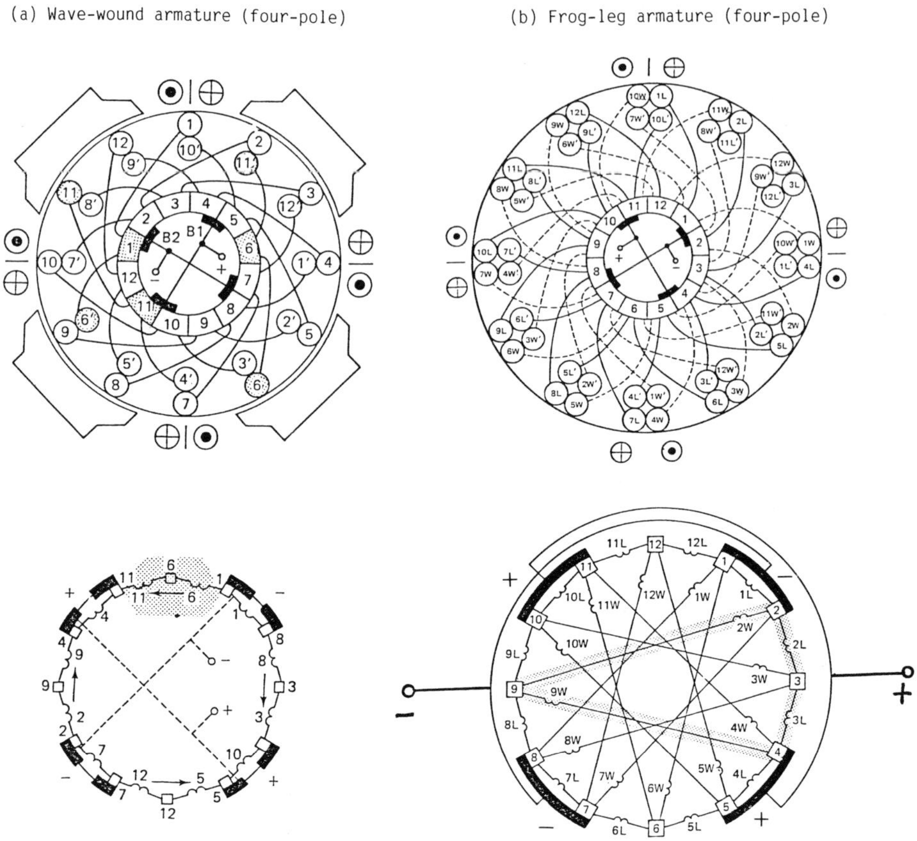

Figure 23-44. Wave-wound and frog-leg armatures

and less than full voltage produced by generators or less than full torque produced in motors. Shifting the brushes is a possible remedy, but the brush neutral changes whenever the load changes. Armature reaction cannot be eliminated, but its effects can be limited by canceling out the rotor-produced magnetic field with a third field. This is the responsibility of the *interpoles.* Interpoles are electrically connected in series with the armature so that their strength varies approximately with the distorting armature field. Armature/interpole connections are usually made internally and the series combination brought out as leads A1 and A2. Figure 23-45 shows the arrangement for either generator or motor operation. Note that for armature current to be in the same direction, generator and motor rotation must be opposite. Thus the polarity of the interpoles with respect to the direction of rotation in generators and motors is opposite. A reversed commutating pole, instead of limiting the effects of armature reaction, will

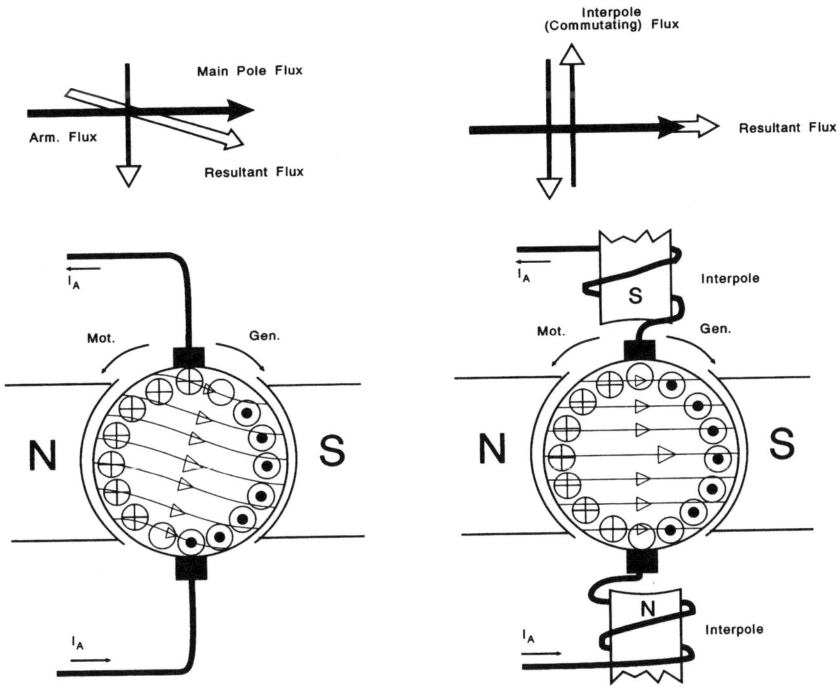

Figure 23-45. Interpoles and armature reaction

exacerbate the situation. In some small machines, only one interpole per main pole pair is installed. In some very large DC machines, *compensating windings,* buried in the main pole faces parallel to the armature conductors (and likewise connected in series with the armature) are utilized.

DC Generators

Voltage characteristics. The voltage produced in a rotating armature, as described above, is proportional to the speed of rotation (RPM) and the flux per pole (ϕ). The voltage available at the generator terminals, however, is less than this voltage due to (1) the internal IR drop in the armature circuit and (2) the effects of armature reaction. Thus:

$$V_L = E_g - I_a R_a - AR$$

where E_g is the product of $K \times \phi \times$ RPM and AR is the armature reaction voltage drop. At no load, the terminal voltage V_L *is* the internal voltage E_g. A plot of no-load voltage (E_g) versus field current I_f (Figure 23- 46) is known as the no-load characteristic, input characteristic, magnetization curve, or saturation curve. Note in this characteristic that (1) with zero field current

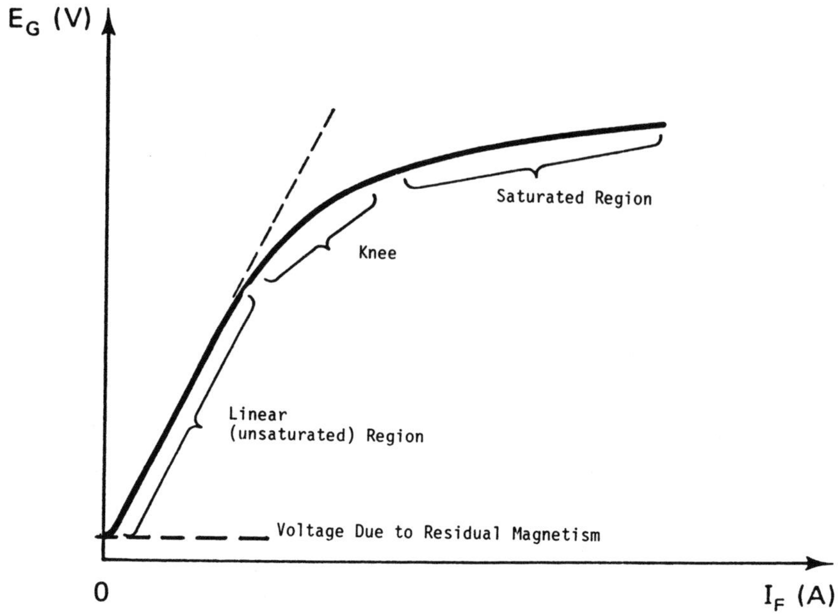

Figure 23-46. Magnetization curve

there is some voltage generated due to *residual magnetism;* (2) there is a *linear range* of voltages that are proportional to field current; and (3) there is a *saturation region,* where little increase in voltage results from increasing excitation. The characteristic is obtained at some constant base speed but the curve may be rescaled for any RPM without changing its shape.

Generator types

Separately excited generators. Generator types are designated by the arrangement of their flux-producing components (fields) relative to their voltage-producing component (armature). In a *separately excited generator,* an independent DC source is connected to the field circuit. This results in E_g being constant throughout the load range of the generator. The voltage droop has a linear component (IR) and a nonlinear component (armature reaction) as shown in Figure 23-47 and Figure 23-48(a). Normally operated at constant speed, voltage control is accomplished by regulating field circuit supply voltage or field circuit current via a rheostat. Separately excited DC generators are used as exciters in synchronous machines; the (separate) exciter field current is supplied from the AC mains via a rectifier, or from the output of an amplidyne pilot-exciter.

In fact the *amplidyne* is an example of a special-purpose separately excited machine. Also known as a rotary amplifier, the amplidyne can

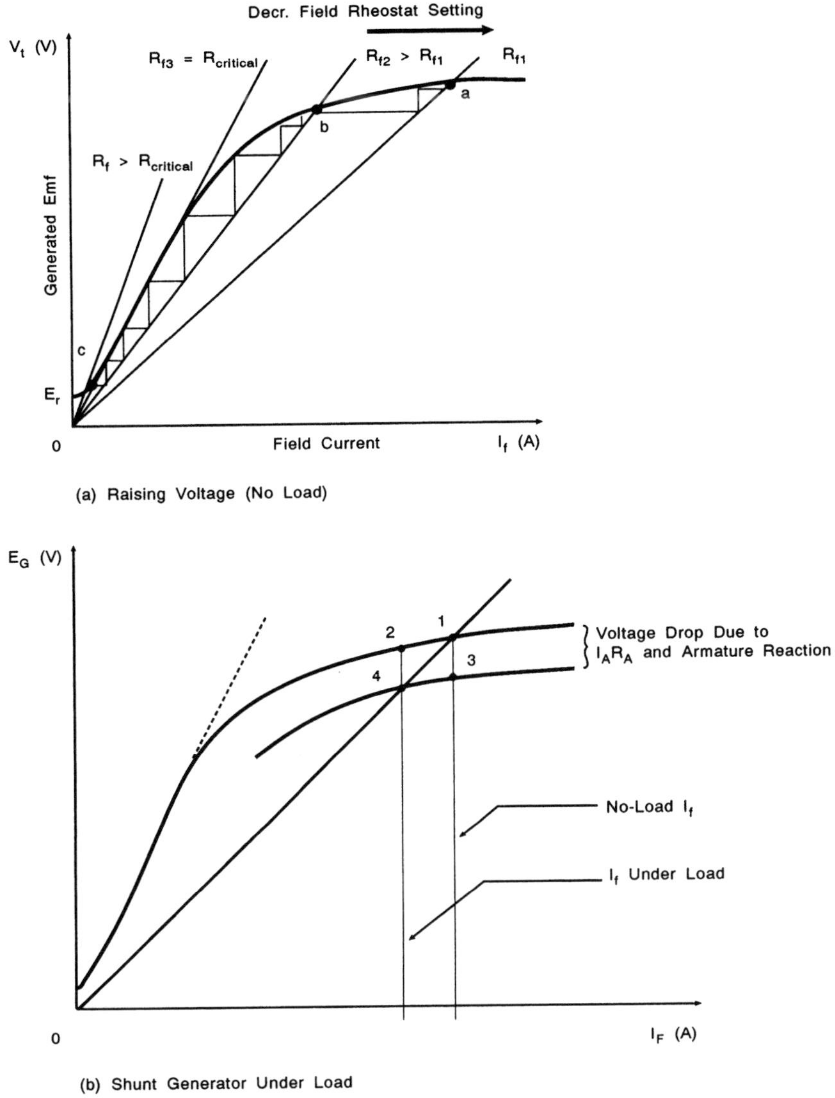

Figure 23-47. Shunt generator input characteristics

produce a ratio of output (armature) power to input (field) power as high as 20,000:1! This level is achieved by exploiting the effects of armature reaction. By *shorting out* the brushes, large armature currents can be made to circulate at low voltages (still in the linear range of the magnetic curve). These large currents produce an armature (reaction) field, and a second set of brushes on the direct axis taps the voltage induced due to this field.

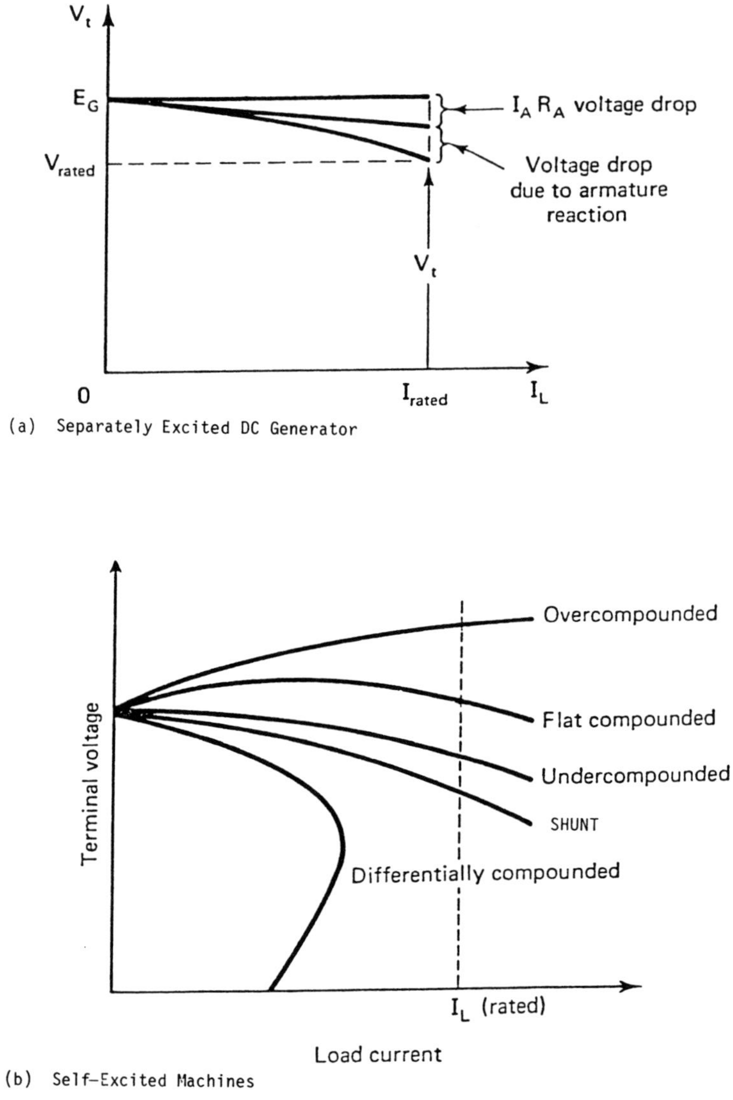

Figure 23-48. DC generator output characteristics

By connecting a load to this second set of brushes (and carefully designing a compensating winding to eliminate a second armature field) the amplidyne produces voltage very sensitive to changes in field current and nearly constant over the load range. In fact, in lieu of a single field winding, several independent windings may be placed on the field poles' cores. Each winding, in turn, can be excited by an electrical signal proportional to some physical quantity—torque, speed, acceleration, line voltage, load current—under the control of the amplidyne.

DC generators in which the electrical output is used to energize its own field windings are called *self-excited generators.* These include the shunt, series, and compound types.

Shunt generators. In a shunt generator the armature circuit and field circuit are connected in parallel, resulting in the armature terminal voltage being applied across the field circuit. Remember, the armature circuit consists of armature windings, commutator, bushes, and interpole windings. The field circuit consists of the main field windings and a field rheostat (or other control device). The no-load voltage produced by a shunt generator is easily obtained from the saturation curve if the field circuit resistance (coils plus rheostat) is known. Since under zero external load, I_a equals I_f and E_g equals V_f, the intersection of the saturation curve (E_g versus I_f) and the resistance line (V_f versus I_f) is the equilibrium point defining the current and voltage in both circuit branches. Decreasing the field resistance defines a new equilibrium point at a higher voltage. With the generator operating at rated speed and the field and armature circuits isolated, the armature terminal voltage is determined by the amount of residual magnetism. Connecting the field and armature circuits in parallel via a field switch suddenly impresses that (small) armature voltage across the field circuit. The resultant field current produces additional flux and a higher armature voltage which, in turn, causes an increase in field current. The process repeats itself, "climbing the stair" (Figure 23-47(a)) until, after several seconds, the equilibrium point is reached. For a fixed field rheostat setting, the external voltage characteristic has a greater droop than the separately excited type. The additional droop component is due to the fact the E_g is not constant, but falls with reduced I_f. This can be shown on the input characteristic (Figure 23-47(b)) where the dotted line indicates the difference between E_g and V_L due to IR and armature reaction, similar to the separately excited machine. The intersection of the dotted line and the resistance line is the new equilibrium point under load. Point 1 is the no-load voltage; point 2 corresponds to E_g under load; point 3 is the terminal voltage of a separately excited machine under this load; and point 4 is the terminal voltage of the shunt machine, which is below point 2 by an amount corresponding to the difference between points 3 and 1. Shunt machines are used on some ships as service generators and are capable of being paralleled.

Series generator. Field current and armature current are the same quantity in a machine that has a series connection between armature and field circuit. A series field winding would necessarily be wound with a few turns of a conductor sized to carry full load current; simply reconnecting the many turns of low ampacity shunt field windings would not suffice. Series generators have no practical application as DC power supplies, but their *rising characteristic* is important in terms of the more practical compound generators discussed later. Series generators are sometimes found as exciters on AC welding machines. The voltage characteristic of the series machine is virtually the same as the input characteristic, increasing with load (i.e., field) current.

Compound generators. DC generators with both shunt field windings and series field windings are called compound machines. The actual field-armature connections may be either *long shunt* or *short shunt* (Figures 23-49(d) and (e)) depending on whether the series field current is to be armature current or load current ($I_a - I_f$). It is, however, the relative strengths of the series and shunt fields (not the specific connection) that determine the voltage characteristic of the compound generator. The series field strength increases with load, tending to increase the generator output voltage. Thus compound generators inherently have a smaller voltage droop than shunt machines (Figure 23-48(b)). If the series field is strong enough to produce a voltage at rated load greater than the no-load voltage, the machine is said to be *over-compounded.* If the full-load voltage is lower than the no-load voltage, the machine is *under-compounded.* A *flat-compounded* machine has zero voltage regulation (droop divided by full load voltage), although the characteristic curve is not flat (Figure 23-48(b)). Compound generators with a weak series field and a characteristic droop only slightly less than a shunt machine are referred to as *stabilized shunt* generators, and the light series winding is referred to as the *stabilizing field.* All of the above compound generator descriptions assume that the series field coils have the same sense of winding as the shunt field coils. That is, both coils, on the same pole piece, are producing magnetic fields of the same polarity. This is known as *cumulative compounding* and is assumed to be the case unless it is specifically stated that the machine is *differentially compounded,* that is, has series field windings opposing shunt field windings in the production of pole flux. Differentially compounded generators have a huge voltage droop characteristic and are rare except for use in some DC welding machines.

To provide three-wire DC for dual voltage systems, the DC generator apparatus must include a *balance coil.* A balance coil is essentially an autotransformer with a central tap. The generator armature windings are tapped symmetrically (one tap per parallel path) and leads are brought out to a pair of slip rings. Each slip ring is connected to a group of armature taps at the same potential (there are two such groups). The resultant voltage between the two slip rings is alternating, with a peak value equal to the DC potential between positive and negative armature leads. Brushes riding on these slip rings are connected to the outer terminals of the balance coil; thus the center tap of the autotransformer is always at a potential halfway between slip rings (AC) and halfway between armature lead (DC) potential. The neutral is connected to the center tap of the

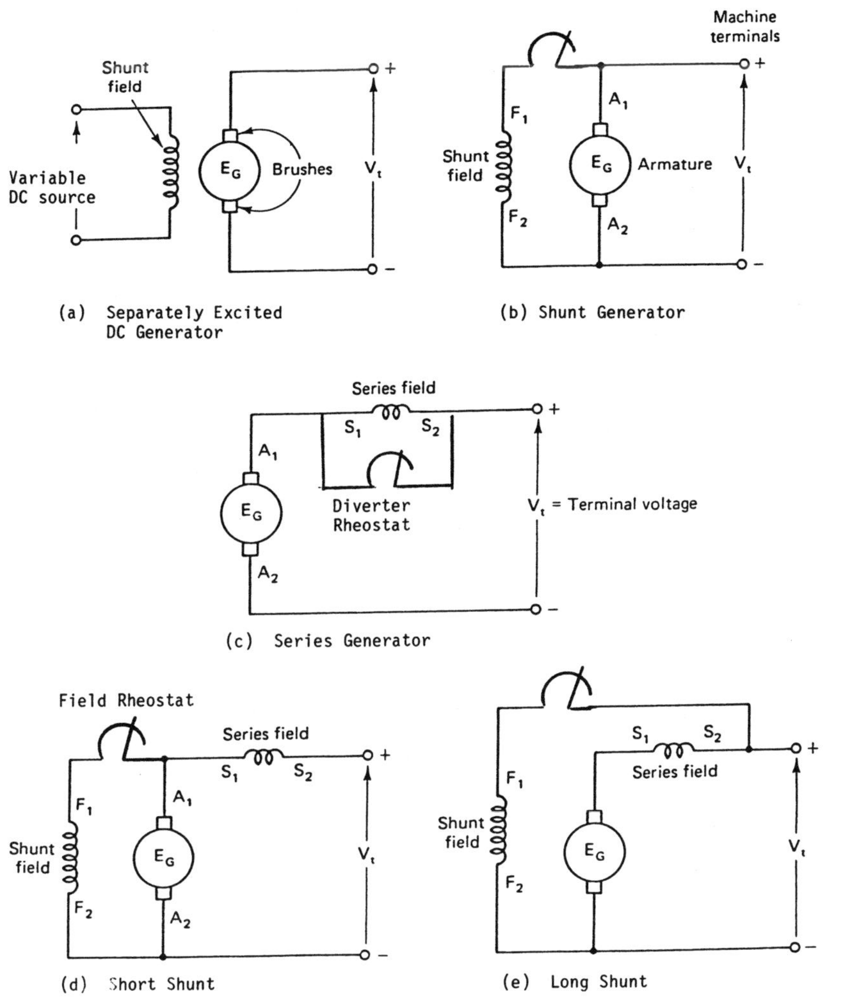

Figure 23-49. DC generator winding connections. *Note:* (d) and (e) are compound generators.

balance coil as shown in Figure 23-50. The neutral current is the difference between positive and negative amps, which would be zero if there were only 240-volt loads or if the 120-volt loads were exactly the same on either side of the neutral. Unfortunately, exact balancing of positive-to-neutral and neutral-to-negative loads is hardly ever accomplished in a practical system. The balance coil then returns to the generator any excess of

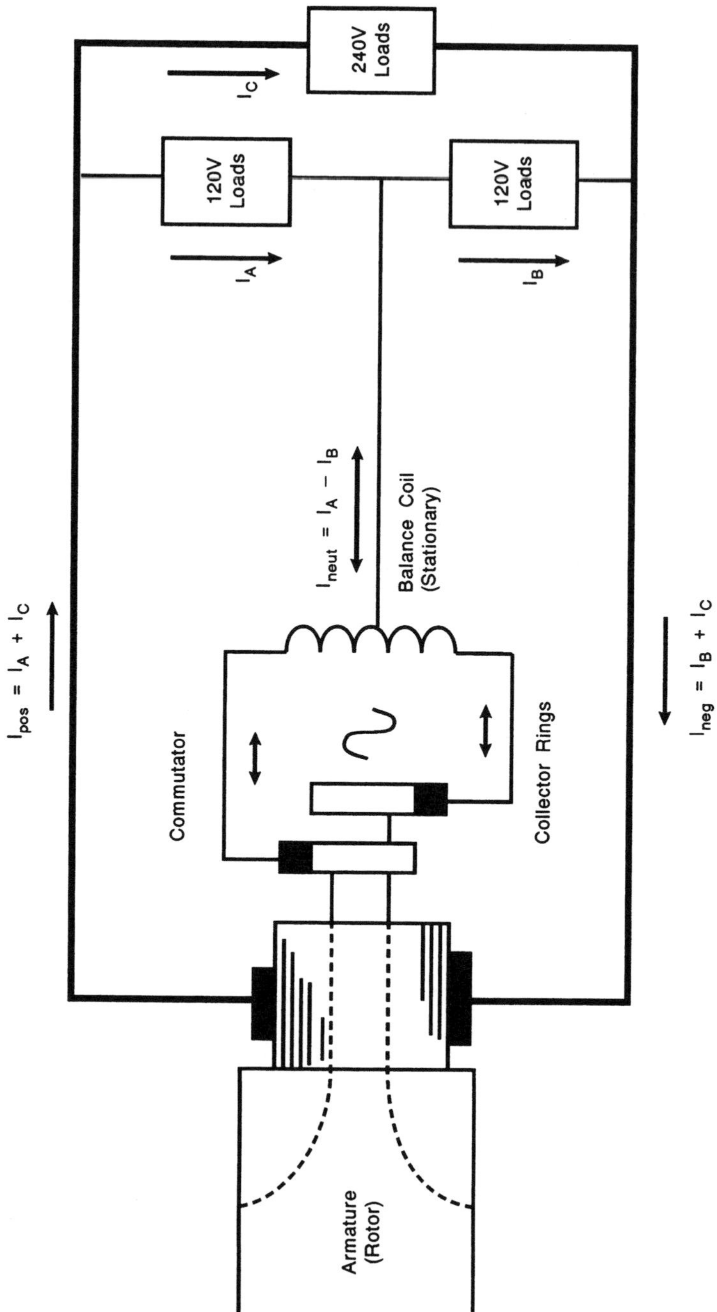

Figure 23-50. Balance coil operation

positive over negative amperes, or supplies from the generator any deficiency if the negative-to-neutral load is greater. What the balance coil does *not* do is balance load between generators in parallel or balance (make equal) positive and negative amps. What it does is "balance the books" with respect to the total current entering and leaving a particular generator in a three-wire DC system. Balancer sets are a mechanical means of accomplishing the same thing using motor-generator (M-G) sets. These were once used aboard ships, but are no longer allowed.

Parallel operation. Any number of DC generating sets of similar (and drooping) voltage characteristics may be paralleled. This drooping requirement limits parallel operation to shunt (and stabilized shunt), under-compounded, and separately excited machines. The machines may be of the two-wire or three-wire type. Considering only self-excited machines supplying a three-wire system, Figures 23-51(a) and (b) show standard line connections for stabilized shunt and compound generators respectively. The most significant difference is the requirement for *equalizing buses* when compound machines are to be operated in parallel. Equalizing buses connect together the terminals of the series fields of compound machines supplying a common load. In effect, one master series field is serving all of the paralleled compound generators. Without equalizing buses, the machines would not be stable in parallel. Rather, the machine with a slightly larger load would have a stronger series field and thus would take more of the load—further increasing its (series) field strength. Equalizing buses do *not* equalize (balance) the load distribution between generators, but they allow *stable* load division, controlled by the relative *shunt field* strengths of the paralleled machines. Equalizing buses are interconnections between generators only, and are not connected to any load. They are sized, however, no smaller than the main line leads. In the figures referenced above, the series fields are split and connected on either side of the armature. This arrangement requires *two* equalizing buses, and five-pole circuit breakers and disconnect switches (or links). Most three-wire systems are arranged this way. In two-wire systems, the entire series field is connected on one side of the armature (Figure 23-52(a)), requiring one equalizing bus and three-pole switchgear. An alternative three-wire arrangement is shown in Figure 23-52(b) that does not require algebraic (comparative) type reverse current and overload relays.

Figure 23-53(a) shows the idealized voltage characteristics of two DC generators operating in parallel, with the load equally divided. The characteristic of generator #1 is shown in mirror image to demonstrate better the degree of load sharing. Note that the no-load voltages of the two identical machines were set equal. In Figure 23-53(b) the no-load voltage of generator #1 has been raised by decreasing *shunt* field resistance and thereby increasing I_f. The load has shifted to that machine, and the system

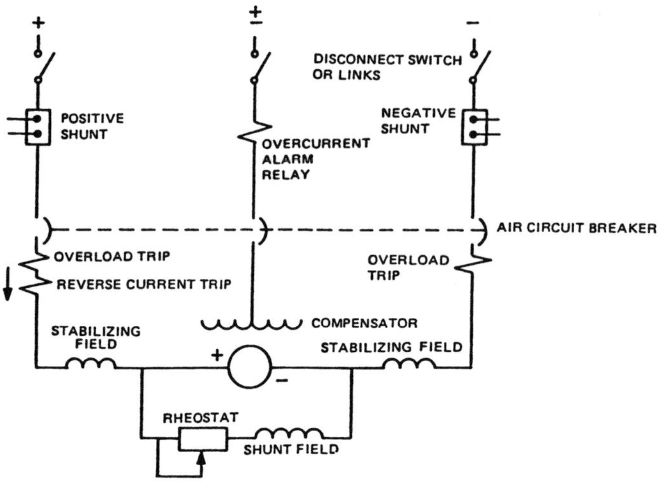

(a) Three-Wire Shunt-Wound Generator for Parallel Operation

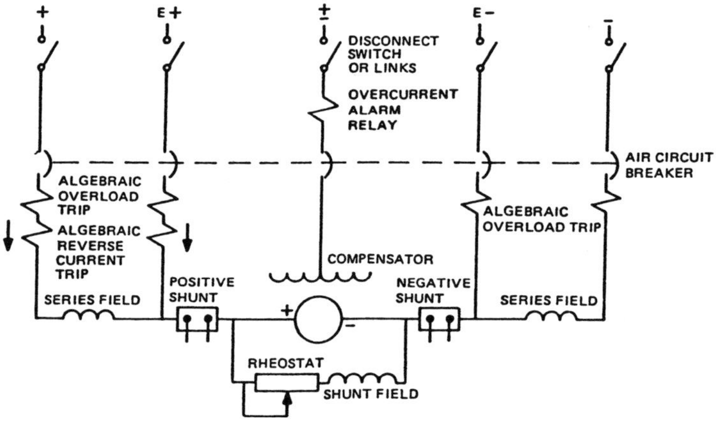

(b) Three-Wire Compound-Wound Generator for Parallel Operation

Figure 23-51. Three-wire DC generator line connections. Reprinted from ANSI/IEEE STD 45-1983, © 1983 by The Institute of Electrical and Electronics Engineers, Inc., with permission of the IEEE Standards Department.

voltage has increased. In Figure 23-53(c) the field current (hence no-load voltage) of generator #2 has been lowered, shifting the loads further to the left and, in this case, lowering the system voltage back to where it was originally. Thus the kW load on DC generators operating in parallel may be divided by manipulating the shunt field current. The series field strengths are equalized at any division of load by the equalizing buses. By simultaneously increasing excitation on one machine while decreasing excitation on the other(s), load may be shifted among generators while maintaining constant line voltage. Ideally, load division should be in proportion to the rating of the machines on line (e.g., two 600 and one 300 kW machines should share a common load 40 percent, 40 percent, and 20 percent respectively).

The following procedure may be used as a guide for DC generator parallel operation:

1. Make a visual inspection to insure that all repairs have been completed, that the unit is free of tools or other debris, and in general is in a condition satisfactory for operation.
2. Insure that disconnect links are in place. (Note: Disconnect links should be opened or closed using an insulated wrench and always with the generator circuit breaker open.)
3. Start prime mover.
4. Close field switch.
5. Raise voltage by decreasing field rheostat resistance or, in *Auto*, increasing the voltage regulator set point to a few volts above line voltage (verify with calibrating voltmeter).
6. Check line-to-ground voltages to detect any grounds on oncoming generator.
7. Close circuit breaker; generator should pick up some load.
8. Balance loads appropriately with field rheostats or, in *Auto*, voltage regulator adjusting pots (potentiometers).

DC Motors

Speed characteristics. In DC motors E_g is the counter-emf (cemf) and is less than the line voltage by I_aR_a. This fundamental relationship can be expressed in a number of different forms:

$$V_L = \text{cemf} + I_aR_a \quad (\text{where cemf} = K \times \phi \times \text{RPM})$$

$$I_a = \frac{(V_L - \text{cemf})}{R_a} \quad (\text{current equation})$$

$$\text{RPM} = \frac{(V_L - I_aR_a)}{(K \times \phi)} \quad (\text{speed equation})$$

Together with the torque equation ($T = K \times \phi \times I_a$), the above relationships can be used to describe the behavior of DC motors. For example, the cycle of events upon a sudden increase in load would be (1) deceleration due to demand/supply torque imbalance; (2) drop in cemf proportional to

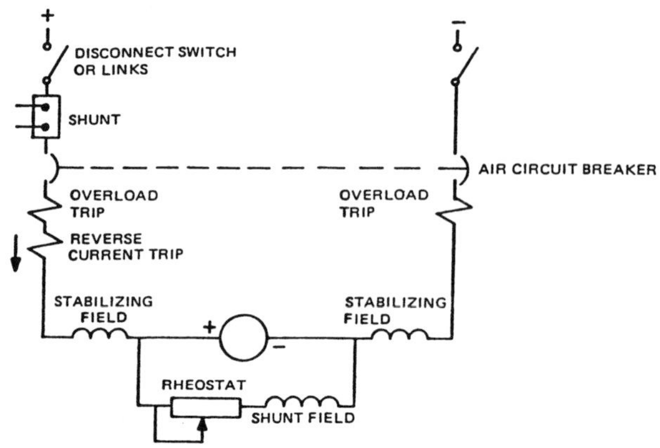

(a) **Two-Wire Stabilized Shunt-Wound Generatòr for Parallel Operation**

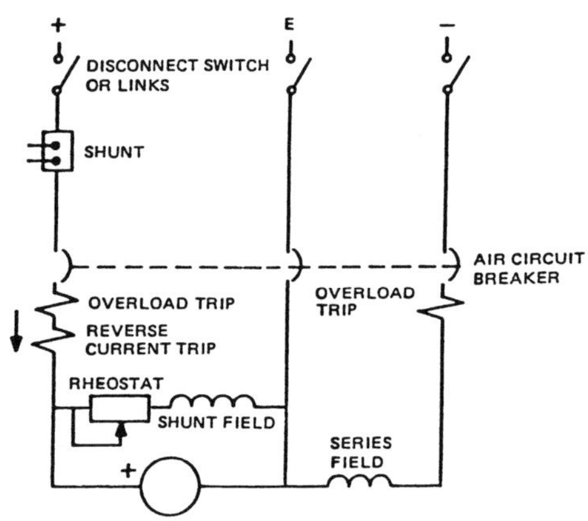

(b) **Two-Wire Compound-Wound Generator for Parallel Operation**

Figure 23-52. Two-wire DC generator line connections. Reprinted from ANSI/ IEEE STD 45-1983, © 1983 by The Institute of Electrical and Electronics Engineers, Inc., with permission of the IEEE Standards Department.

speed; (3) increase in current with drop in cemf (current equation); and (4) increase in torque with current to meet increased load demand—if not, deceleration continues until equilibrium is met.

Motor types

Shunt motor. DC motor and generator nomenclature is essentially the same, with motor classifications being derived from the manner of field

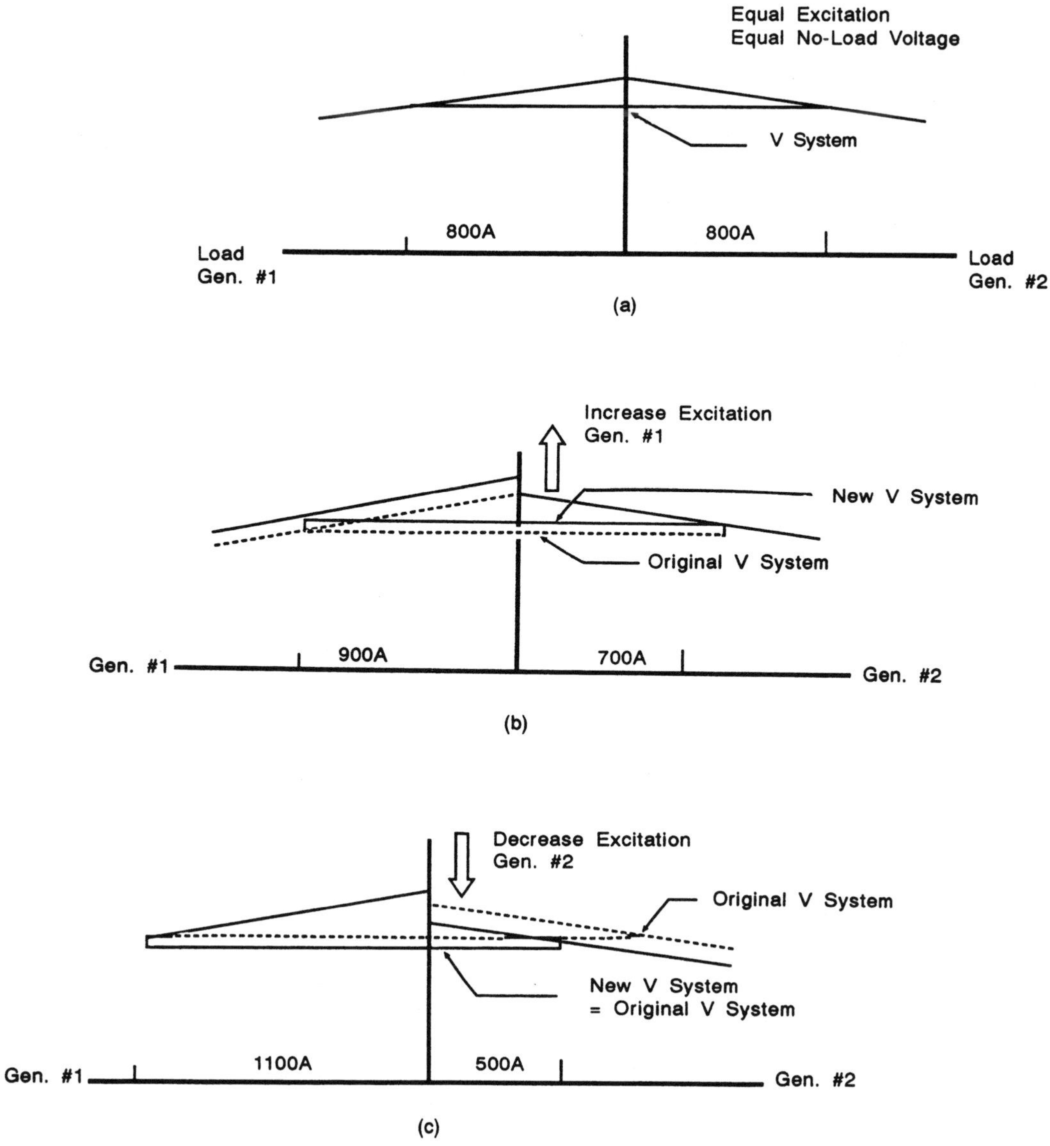

Figure 23-53. DC generator load sharing

armature connection. As with DC generators, the *shunt* arrangement is defined by connecting the main field circuit in parallel with the armature circuit. In fact the main field winding is usually referred to as the shunt field winding. Often the field circuit contains a rheostat. The result of this combination is a motor with constant field strength (ϕ) throughout the load range, very little speed droop, and a torque characteristic that is proportional to armature current. The starting torque is relatively low.

Series motors. At the other end of the spectrum, the series motor produces a torque in proportion to nearly the square of the armature current (since I_a is responsible for ϕ). Thus, tripling the armature current yields almost nine times the original torque; a similar demand on a shunt motor would require the current to increase ninefold. However, the denominator in the speed equation varies with $K \times \phi$, which is constant in a shunt machine but varies with I_a in a series machine. This yields a speed reduction, in the above example, in excess of 67 percent. At light loads, a series motor runs very fast and can run away if completely unloaded. Therefore series motors are always directly connected to their loads, either close coupled or via gear drives. For example, a series motor is never connected to a belt-driven load.

Compound motor. With careful design, cumulative compound motors can be built with characteristics that approach either the shunt machine or the series machine without the drawbacks of low starting torque or no-load runaway. The windings may be connected in the short-shunt or long-shunt fashion but, again, the motor characteristic depends more on the relative strengths of the series and shunt fields. Lightly compounded motors, with relatively weak series fields, behave in a manner similar to shunt machines. The series winding improves the starting torque and provides a definite droop, which is good for stability purposes. These general-purpose *stabilized shunt* machines are suitable for centrifugal pumps, ventilation blowers, and other applications where the motor will not be started under appreciable load. Moderately compounded DC motors have a stronger series field, greater starting torque, and a greater speed droop. In general use, the term *compound motor* applies to this group, although stabilized shunt machines are technically compound motors. These motors are suitable for positive displacement pumps, compressors without unloaders, valve actuators, conveyors, and the like. *Winch* or *crane motors* are heavily compounded, with the shunt field's role simply to guarantee an acceptable no-load speed. Their application is for those high-torque demand loads where a large speed droop is acceptable. DC motors may, theoretically, be *differentially compounded,* but this is usually not practical. A motor connected with the series field reversed can be very dangerous. A sudden increase in load torque can cause the series field to cancel out the shunt field flux, resulting in runaway or even reversal of direction. DC motor torque/speed characteristics are summarized in Figure 23-54.

Speed control

Field resistance control (FRC). The disadvantages of DC machines are significant. They are larger and more costly for the equivalent horsepower

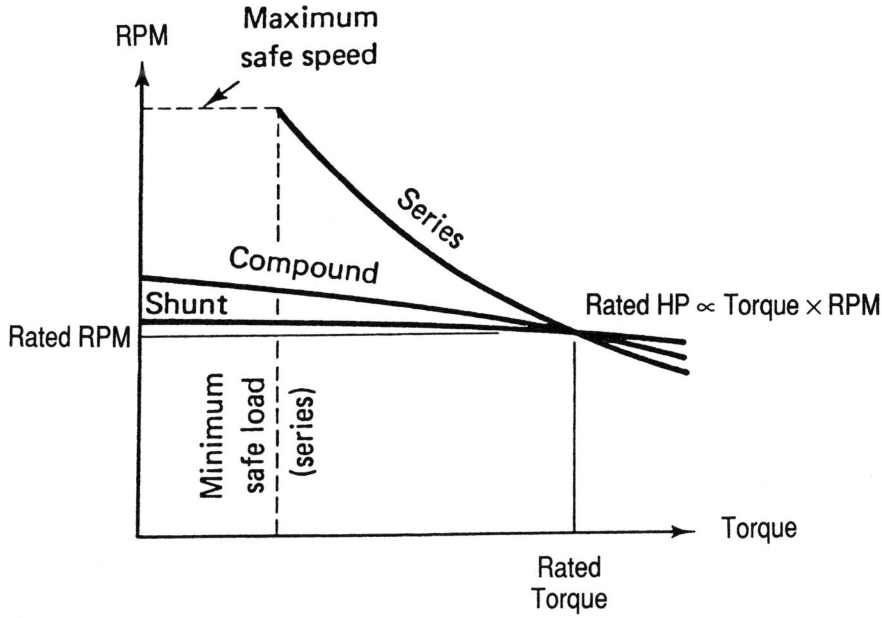

Figure 23-54. DC motor torque/speed characteristics

compared with induction motors; the brush and commutator arrangement adds considerably to the maintenance burden; and since DC power is not often readily available, special-purpose supplies are required to operate DC machinery. They have one major advantage, however, over most other types of electrical drives—the ease with which wide range adjustable speed control may be accomplished. From the speed equation, it is obvious that V_L, R_a, and ϕ are all candidate parameters to be manipulated in a speed control scheme.

With the voltage across the shunt field circuit fixed (usually at nominal line voltage value), the field current, and hence ϕ_f, may be reduced by increasing field circuit resistance. Since speed is *inversely* proportional to ϕ, increasing shunt field resistance increases speed. The lowest speed possible under this method of control is at maximum I_f, which occurs with the field circuit resistance reduced to that of the field coils themselves. This may be an appreciable amount, from a few dozen ohms in large machines to several hundred ohms in smaller machines. Anyway, minimum field resistance corresponds to base speed and FRC can only increase RPM above base speed. Theoretically there is no upper limit: increasing field resistance decreases I_f, increasing RPM, ad infinitum. However, the practical limits are dictated by acceptable armature current. For a given load torque ($K \times \phi \times I_a$), I_a must increase to offset the lower ϕ. These limits range from 2:1 for large machines to 8:1 for integral horsepower machines.

The reciprocal relationship of flux and speed means that the cemf ($K \times \phi \times RPM$) remains relatively constant. Since at rated current, constant power ($cemf \times I_a$) is developed over the entire speed range, this arrangement is known as constant hp drive. FRC is best suited for loads requiring the highest torque at lower speeds.

Buck / boost field control. If the series field terminals are not connected to the armature circuit, but instead brought out and connected to a separate variable DC supply, they may be used to make fine adjustments in the amount of field flux. This method is appropriate for DC/AC rotary inverters (M-G sets) where the buck/boost field is under the control of a governor.

Armature resistance control (ARC). Increasing resistance in the armature circuit under load will decrease the numerator of the speed equation and reduce speed. Maximum speed is achieved with no additional armature circuit resistance and ARC can only reduce speed below the base speed. With the flux remaining constant, cemf is proportional to speed and torque is constant for a given load current regardless of speed. This is known as constant torque drive. Problems with this method include (1) virtually no speed control at light loads and (2) large armature IR^2 losses, especially when running at low speeds for long periods of time. When combined with FRC, however, a very wide range of speed control can be obtained with (high) constant torque at low (below base) speeds and constant hp characteristics above base speed. At one time DC winches used such a control combination.

Armature voltage control (AVC). AVC requires two DC sources, one (fixed) field supply and one (variable) armature supply. Motor RPM is controlled by varying the V_L in the speed equation. With ϕ held constant by the fixed field supply, AVC, like ARC, is an example of constant torque drive.

The Ward-Leonard system is an example of this type of motor control. Used originally to accommodate DC motor-driven winches in AC ship's service supply systems, the Ward-Leonard arrangement consists of three rotating machines (Figure 23-55(a)). An M-G set (three-phase induction motor/DC generator) supplies a variable DC voltage to the armature of a DC winch motor. The low slip induction motor provides nearly constant generator RPM and the separately excited DC generator output is controlled by varying the generator field current. At rated output (i.e., maximum generator field current) the generator supplies the winch motor with rated armature voltage. With the DC motor excited (separately) with maximum (rated) field current, and the armature supplied with rated voltage, the motor will run at *base speed.* Between minimum and base speed the motor field is energized at the maximum level while the armature voltage is increased by increasing generator excitation (AVC). The motor speed may be further increased by decreasing the motor field strength. This arrangement yields an extremely wide range of speed control with the characteristic of constant torque drive below base speed and constant horsepower drive above base speed (Figure 23-55(b)).

Solid-state control. Replacing the M-G set described above with a solid-state converter and utilizing a second converter for the motor field supply will maintain the torque/speed characteristics of the Ward-Leonard drive. Figure 23-56 shows the circuit schematic for the four basic *single-phase* converter drives: (a) half-wave converter drive, (b) semiconverter drive, (c) full converter drive, and (d) dual converter drive. For larger machines, three-phase converters may be used. The type of converter is determined by the size of the motor to be driven and the mode (quadrant) of expected operation. For example, only *dual converter drive* arrangements are capable of four-quadrant operation: forward powering, forward braking (regeneration), reverse powering, and reverse braking. The field circuit in such an arrangement is a single- or three-phase full converter to allow field current reversal. Single-phase, dual converter drives may be used to power motors up to 15 kW, three-phase dual converter drives up to 1500 kW.

Two-quadrant operation is possible with *full converter drives.* Again the field circuit should be of the full converter type so that the field current may be reversed during regeneration. Power limitations of this type of drive are similar to that of the dual converter types.

Semiconverter drives and *single-phase, half-wave converter drives* are limited to one-quadrant operation only and applications up to 115 kW for three-phase semiconverter drives, or 15 kW for single-phase semiconverter drives. A freewheeling diode is required across the armature circuit and semiconverters are required as the field circuit supply. Single-phase, half-wave converter drives typically require a large inductance in series with the armature to reduce the ripple in the current and are limited to applications up to 0.5 kW. Three-phase, half-wave converter drives are capable of two-quadrant operation and could be used for applications up to 40 kW. However, since DC components can be passed into the AC supply during regeneration, this arrangement is not generally used in industrial applications.

With a fixed DC source available, *chopper drives* are often used to supply a variable voltage to the armature. In addition to AVC, chopper drives can provide regenerative braking by either returning power to the supply or to a resistor connected across the armature circuit. Actually the former method is technically regenerative and the latter rheostatic braking; often a combination of the two is utilized with chopper-driven DC motors.

DC motor starters

Overview. The current equation dictates the need for external apparatus when starting all but the smallest DC motors. With only the

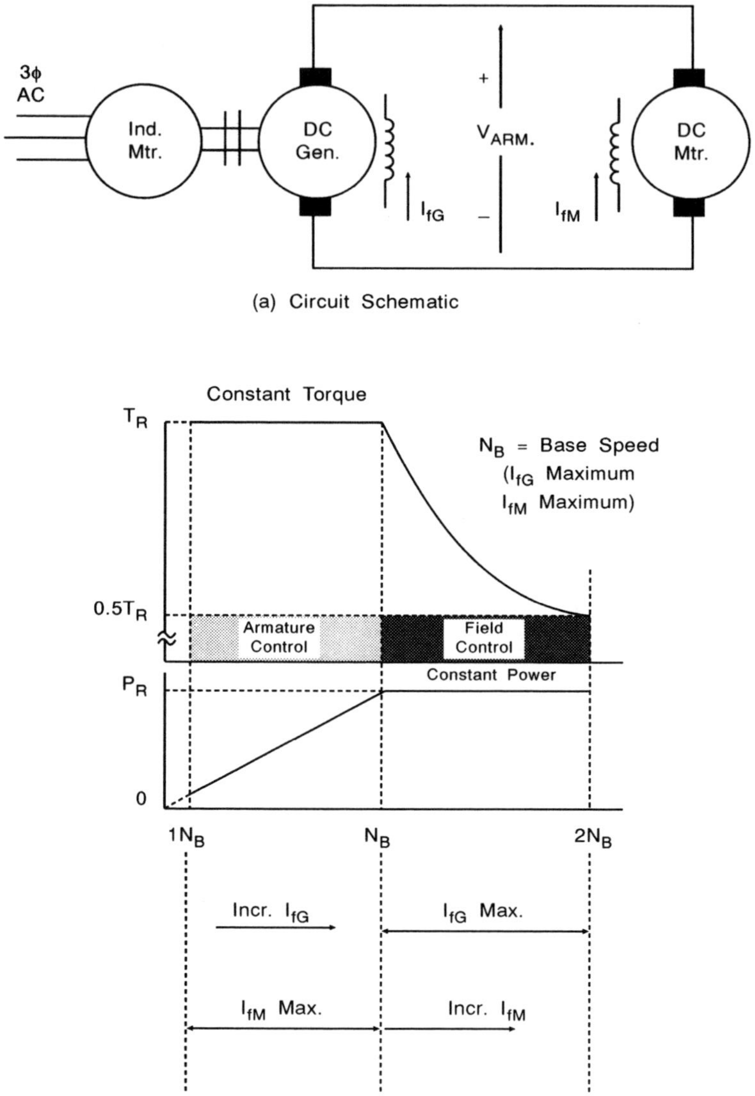

(b) Control Scheme

Figure 23-55. Ward-Leonard drive

armature winding resistance in the denominator (less than an ohm for even moderately sized machines, and milliohms in very large units), and no cemf upon starting, the current draw could be hundreds of times the rated current of the machine if started across the line. Ward-Leonard systems and electronic controllers solve the problem by reducing the applied armature circuit voltage at start. With a fixed DC voltage supply, however, the solution to limiting starting current to a reasonable amount

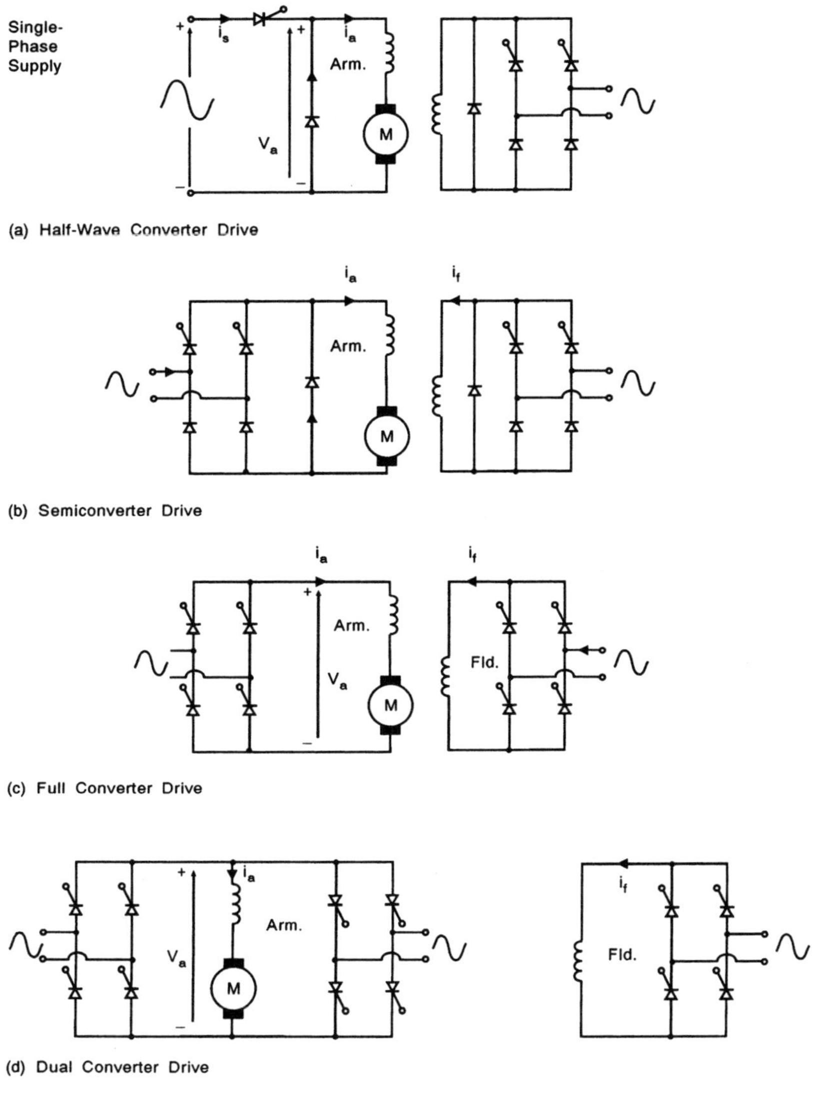

Figure 23-56. Single-phase converter drives

(two to four times the rated current) is by inserting *starting resistors* in series with the armature circuit. This starting resistance may be removed when the cemf has built up enough to protect the armature circuit. The starting resistance may be removed all at once or partially removed in distinct steps. The ohmic size of the resistors is determined by the maximum allowable starting current, and the number of steps is dependent on the *minimum* allowable current (and, hence, minimum starting torque) during acceleration. For large machines, the difference between maximum and minimum currents during acceleration is relatively small and a large number of steps is required. In the examples below, three-step starters are illustrated where: in step 1 all starting resistance is in and armature current rises quickly to the maximum value, then decreases, and RPM and accompanying cemf increase to the minimum allowable value; in step 2, part of the starting resistance is removed (bypassed by an accelerating contactor) and the current rises up again to maximum, then decreases again as the motor continues to accelerate; and in step 3, the remaining resistance is removed (leaving only the winding resistance itself) and the current rises for the third time. The final current level is a function of motor load (Figure 23-57). Various methods may be utilized in determining the

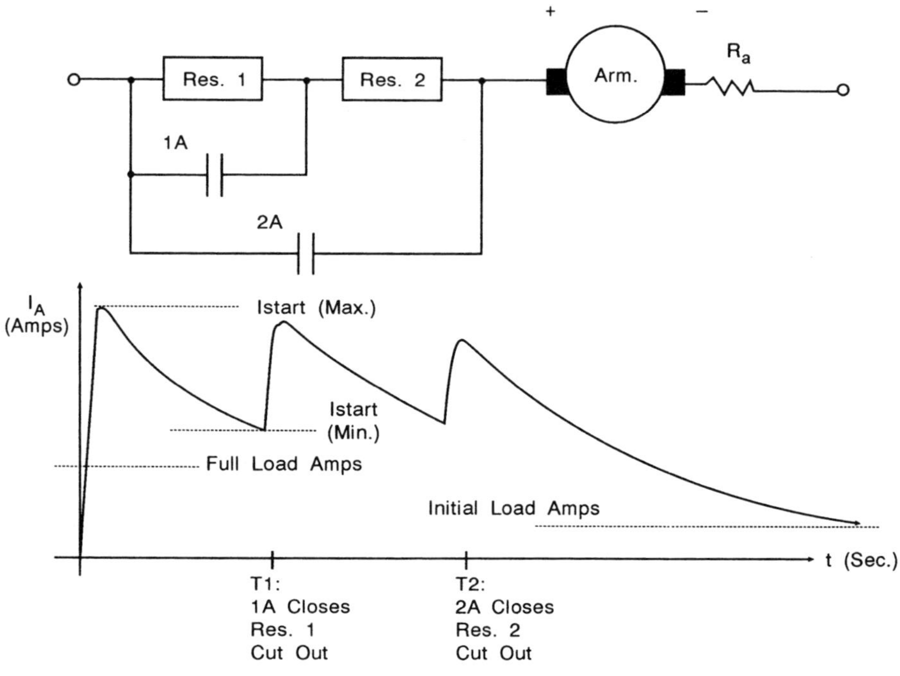

Figure 23-57. Three-step starter characteristic

rate at which the starting resistance is removed, including manual starters where the operator gradually removes the starting resistance as the motor accelerates. Three types of automatic DC starters are described below.

Definite-time starters. If the load characteristics are known and the load conditions will be practically identical whenever the motor is started (such as centrifugal pumps that are always started with no flow and shut-off head), then the time interval between steps may be precalculated. This enables the use of timing relays to close the accelerating contacts. The timing device may be pneumatic, hydraulic (dashpot), mechanical ratchet, or electronic. The control circuitry of Figure 23-58 is arranged so that the accelerating contacts, 1A and 2A, are normally closed and are energized upon starting prior to energizing the main, M, contacts. A normally closed auxiliary M contact opens, de-energizing the 1A coil. The accelerating contactors are TDODE (time delay on de-energization) and thus the contacts fall to their normal position after a set delay period. This action shorts part of the starting resistance (beginning the second step) and de-energizes coil 2A which will, after another delay period, bypass all the starting resistance. Another auxiliary M contact seals the start button. Replacing the momentary-contact push buttons with a maintain-contact selector switch will result in a controller with the LVR feature rather than LVP, but will not alter the starting sequence.

Cemf starter. The problem with a definite-time starter is that if the motor fails to accelerate sufficiently in the time allotted, there will not be enough counter-emf to protect the armature from excessive currents when the starting resistance is removed. Starting under excessive load or mechanical binding of motor or pump shaft will cause such a problem. With a cemf starter, successive steps do not take place until the cemf is sufficient to limit starting current to an acceptable level. This level is greater than the rating of the machine, of course; so eventually the machine will trip out on overload if sufficient acceleration does not take place. However, the timing of the steps is dependent on load and therefore cemf starters are suitable for applications where a variety of starting conditions may exist. In Figure 23-59 the normally open accelerating contacts 1A and 2A insure sufficient starting resistance in the armature circuit until accelerating *relays* 1AR and 2AR pick up, energizing 1A and 2A respectively. 1AR and 2AR are designed to pick up at successively higher voltages, insuring the proper sequencing. The disadvantage of cemf starters is that line voltage fluctuations may cause the accelerating relays to release, inserting starting resistance during otherwise normal operation.

Series lock-out starter. Sometimes called a current limiting starter, the series lock-out starter incorporates current-sensitive relays in the starting circuit that prevent (lock out) the accelerating contacts from picking up until the armature current falls below a preset value. In Figure 23-60, the initial inrush activates relay 1S which will only drop out when the motor

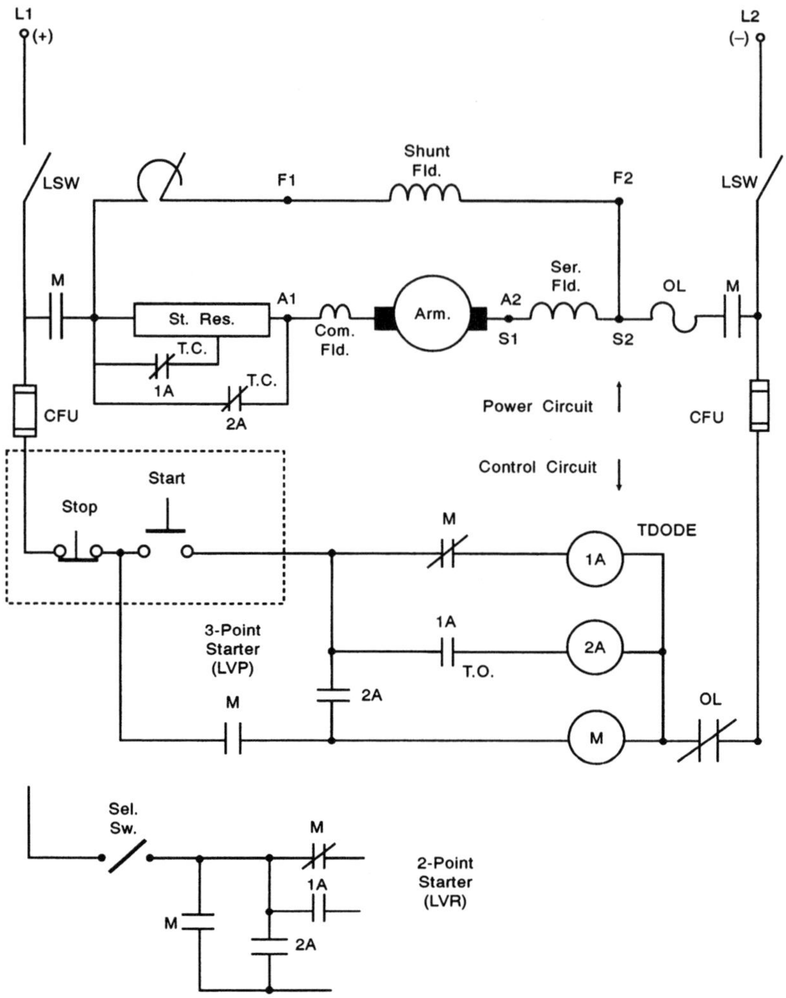

Figure 23-58. Definite time starter

speed (and hence cemf) is sufficient to limit the current during the next step. Should the motor not build up speed, the process is held at that step, drawing the maximum allowable starting current, until the overload relay aborts the start. Like the counter-emf starter, lock-out starters are utilized in applications where a variety of starting conditions may be encountered.

DC Machine Maintenance

Troubleshooting. The techniques for pinpointing defects in DC machinery are very similar to those appropriate for synchronous machines. In fact,

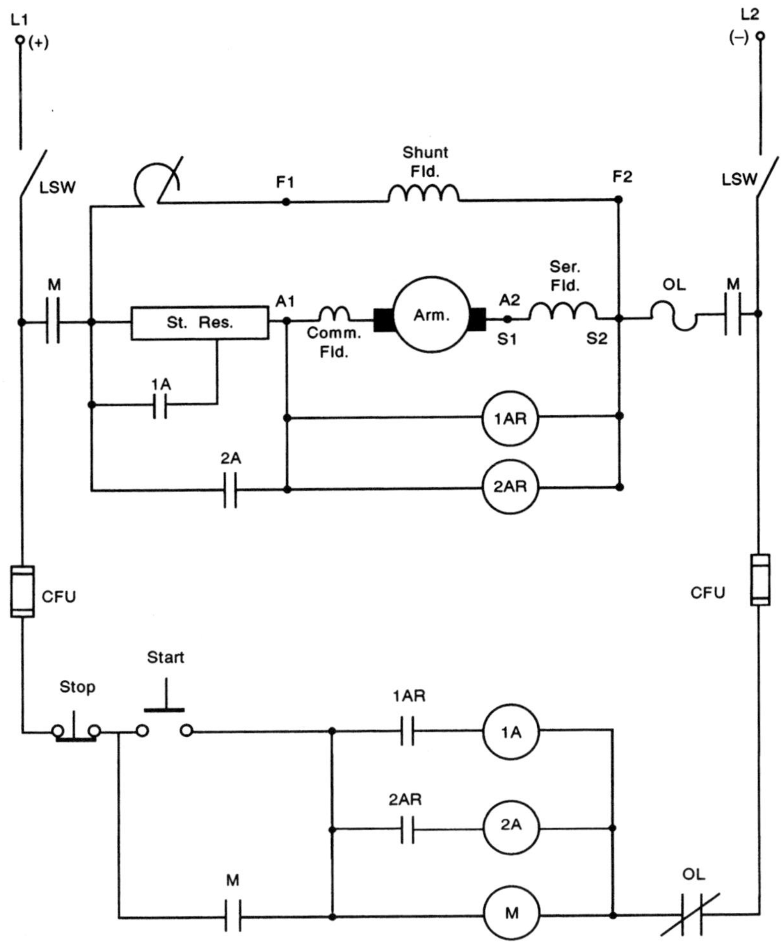

Figure 23-59. Cemf starter

the procedures for identifying open, shorted, or grounded field (stator) coils in a DC machine are identical to those described earlier for salient-pole synchronous rotor coils and will not be repeated here. Also the use of a growler to identify shorted armature coils is apropos to DC machines.

Identification of open or grounded armature coils is facilitated by the fact that, in DC machines, access to coil ends is available at the commutator.

A troubleshooting chart for DC generators is given in Table 23-5, for DC motors in Table 23-6.

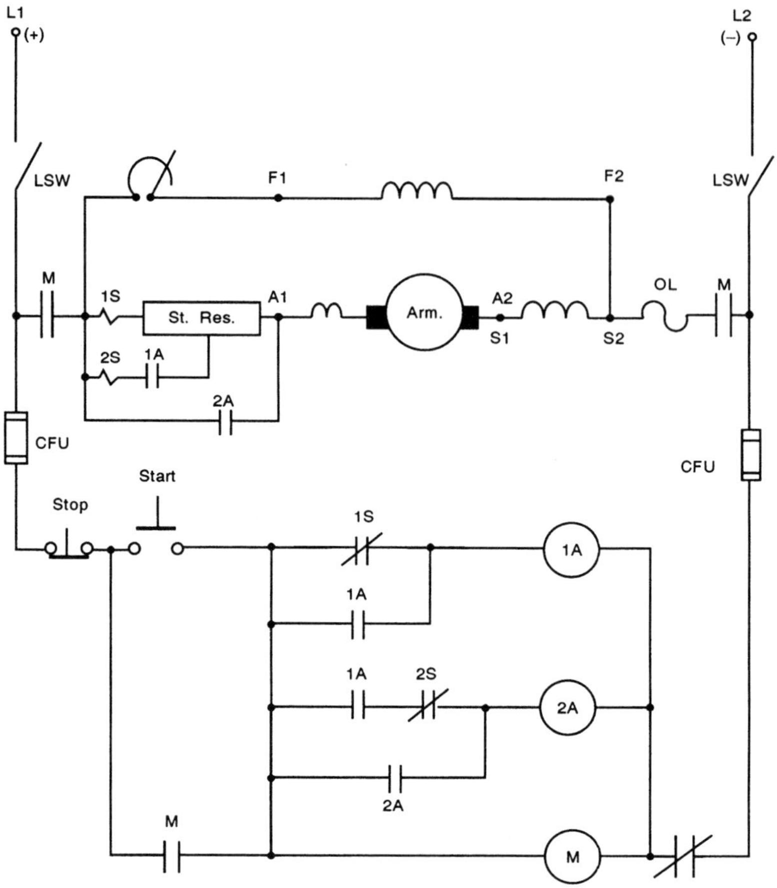

Figure 23-60. Series lock-out starter

TABLE 23-5

DC Generator Troubleshooting Chart

Problem	*Possible Cause*	*Remedy*
No output voltage	Open in field circuit	Check field circuit switch for closure; check field coils for opens or loose connections; check field rheostat for opens or loose connections.
	Open in armature circuit	Check armature connections (including brush rigging, series field, and interpole coils) for continuity; tighten loose connections; repair/replace defective coils; replace open ammeter shunt if required.
	Loss of residual magnetism	Lift brushes and flash field.
	Defective filter	Check for open choke or shorted capacitor and replace as necessary.

Problem	*Possible Cause*	*Remedy*
Low output voltage	Low prime mover speed	Check speed with tachometer; adjust governor setting as required.
	Poor commutation	Replace defective brushes; insure proper seating and brush pressure; clean commutator; undercut mica; adjust brush rigging to brush neutral if necessary.
	Shorted/grounded field pole(s)	Check and repair/replace as required.
	Shorted/grounded armature coil(s)	Check and repair/jump as required.
	Reversed field pole	Check and connect properly.
	Faulty voltage regulator	Adjust or repair/replace as required.
High output voltage	High prime mover speed	Check with tachometer and adjust governor setting.
	Faulty voltage regulator	Adjust or repair/replace as required.
Overheating	Overload	Reduce load.
	Insufficient ventilation	Remove restrictions; clean windings; check C. W. supply.
	Bearings	See "Noisy operation" below.
	Uneven air gap	See "Noisy operation" below.
	Shorted armature winding	Test and repair/replace.
	Shorted field coil	Test and repair/replace.
	Brushes off neutral	Adjust brush rigging as required.
	High ambient temperature	Check cooler operation; adjust as required.
Sparking at brushes	Dirty commutator	Clean, stone, and burnish commutator as required.
	Wrong brush grade	Check with manufacturer's guidelines; replace as required.
	Brush maladjustment	Check brush pressure; check rigging alignment.
	Commutator eccentric	Stone or cut on lathe as required.
	High mica	Undercut mica.
	Open armature coil	Repair or jump coil as required.
	Shorted or reversed commutating pole(s)	Replace/reconnect as required.
	Vibration	See "Noisy operation" below.
Noisy operation	Bearings	Check for worn, loose, dry, or overlubricated bearings; replace worn or loose bearings; lubricate dry bearings; relieve overlubricated bearings.
	Uneven air gap	Check for bent shaft, bearing condition, gear alignment.
	Coupling	Check alignment; re-align and/or tighten coupling.

Open armature coil. The arcing that occurs when a brush comes in contact with commutator bars connected to an open coil leaves a discoloration normally detectable by visual inspection. In fact, the cause of the open may be failure of the soldered connection of coil lead to commutator bar riser, which also may be detected visually. If not, a low voltage may be applied to the armature through two strips of copper conductor taped to commutator bars one pole apart. The open coil may then be detected by measuring voltage between commutator bars. Bridging the open coil would yield a reading equal to the applied voltage. An emergency repair may be made by bridging over the affected bars with a jumper of the same size as the armature windings. Lap-wound armatures will have only one burned spot on the commutator for each open, regardless of the number of poles.

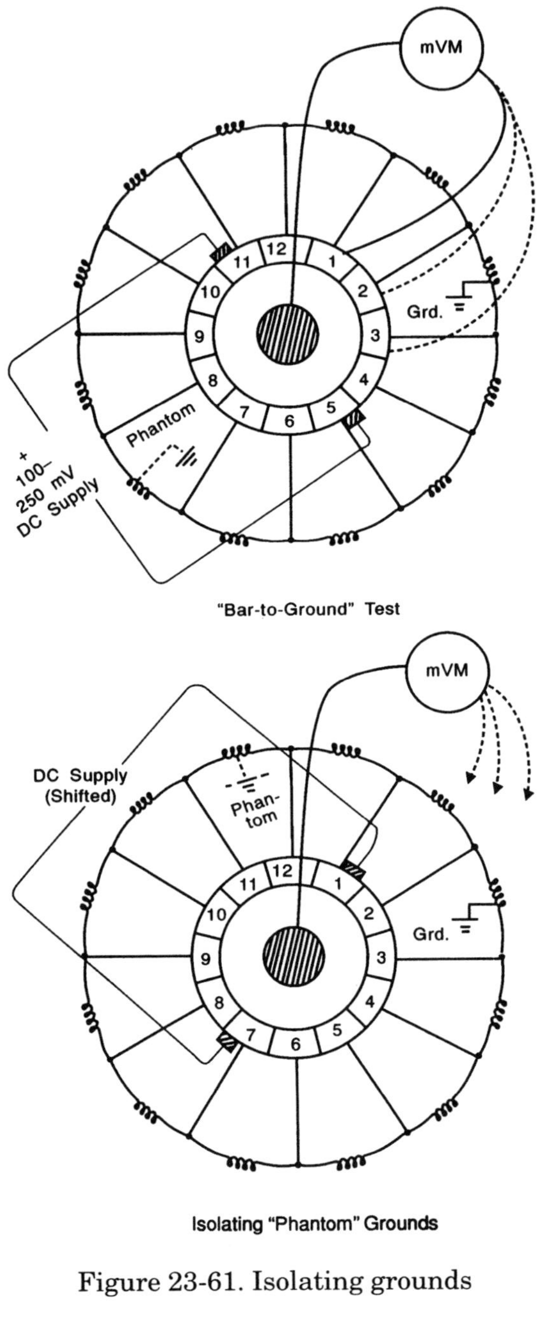

"Bar-to-Ground" Test

Isolating "Phantom" Grounds

Figure 23-61. Isolating grounds

In wave-wound machines of four poles or more, the number of discolored spots will be P/2 and any one (but only one) of the spots should be bridged.

Grounded armature coil. By connecting a low voltage source to the commutator as described above, grounded armature coils may be detected with a millivoltmeter in a bar-to-ground test. This assumes that megger readings have indicated that an armature ground does exist (all bar-to-ground voltages will be zero in an ungrounded armature). With one lead of the voltmeter connected to the rotor shaft, the other lead is moved around the commutator one bar at a time. Every bar that yields a zero or near zero reading should be marked with chalk. Because of the symmetry of the armature circuit, coils in equivalent positions to the affected coil in other parallel paths will also yield low (zero) readings. These are called phantom grounds. By shifting the voltage source connection points, the symmetry is changed and the phantom grounds will shift in location while the actual ground locations, of course, will remain the same (Figure 23-61).

TABLE 23-6

DC Motor Troubleshooting Chart

Problem	*Possible Cause*	*Remedy*
Fails to start	Open in armature circuit	Check fuses; check armature circuit for loose connections, brush contact; check starting resistances, series field, and interpole windings for opens; reconnect, repair, or replace as required.
	Open in field circuit	Check field circuit for continuity; tighten connections; repair/replace open field coil or field rheostat.
	Mechanical binding	Check for seized bearings, pump impeller, etc.
	Controller malfunction	Check contactor and relay coils; replace if necessary, If tripped on overload, allow thermal element to cool, and reset.
Overheating	Overload	Reduce load.
	Insufficient ventilation	Remove restrictions; clean windings; check C. W. supply.
	Bearings	See "Noisy operation" below.
	Uneven air gap	See "Noisy operation" below.
	Shorted armature winding	Test and repair/replace.
	Shorted field coil	Test and repair/replace.
	Brushes off neutral	Adjust brush rigging as required.
	Voltage too high or low	Check supply and adjust as required.
Sparking at brushes	Dirty commutator	Clean, stone, and burnish commutator as required.
	Wrong brush grade	Check with manufacturer's guidelines; replace as required.
	Brush maladjustment	Check brush pressure; check rigging alignment.
	Commutator eccentric	Stone or cut on lathe as required.
	High mica	Undercut mica.
	Open armature coil	Repair or jump coil as required.
	Shorted or reversed commutating pole(s)	Replace/reconnect as required.
	Vibration	See "Noisy operation" below.

Problem	*Possible Cause*	*Remedy*
Runs faster than normal	Voltage high	Check supply and adjust as required.
	Reversed series field	Check connections and reverse S1-S2 leads if necessary.
	Open/shorted/grounded field coil(s)	Check windings and repair/replace defective coils. Note: Motor will not start with open field coil, but a lightly loaded compound machine will run—fast—and probably hot.
	High field rheostat resistance	Check rheostat setting and rating.
Runs slower than normal	Low voltage	Check supply and adjust as required.
	Overload	Reduce load.
	Short/grounded armature coil(s)	Check windings and repair/jump as required.
	Brushes off neutral	Adjust brush rigging; check interpole integrity and polarity.
	Starting resistance inserted	Check accelerating relays for proper operation.
Noisy operation	Bearings	Check for worn, loose, dry, or overlubricated bearings; replace worn or loose bearings; lubricate dry bearings; relieve overlubricated bearings.
	Uneven air gap	Check for bent shaft, bearing condition, gear alignment.
	Coupling	Check alignment; re-align and/or tighten coupling.

REVIEW

1. Define the term "*synchronous machine*" as it pertains to a three-phase alternator.
2. Describe the difference between mechanical degrees and electrical degrees in a synchronous machine.
3. Compare a salient pole rotor to a cylindrical pole rotor. Where is each commonly used?
4. What is excitation in an alternator? What does the exciter vary?
5. What are the two things that must be controlled to insure proper output from an alternator under varying loads?
6. Describe the procedure for the startup of an alternator.
7. Describe the procedure for paralleling an alternator with a second one.
8. Describe the procedure for balancing the loads once the alternators are paralleled. What alternator control is used to vary the true power? What control is used to vary the reactive power?
9. What are the three common electrical problems in AC machinery?
10. What are the possible causes for no output voltage of an alternator?
11. What are the possible causes for high output voltage of an alternator?
12. What are the possible causes of noisy operation of an alternator?
13. What is a "growler" and what is it used for?
14. What is a "megger" and what is it used for?

15. How can shorted field coils be located?
16. How can open field coils be located?
17. What is the relationship between frequency, RPM, and the number of poles in an AC synchronous or induction motor?
18. What is "slipping a pole" in a synchronous motor and what is the remedy?
19. A synchronous motor has no starting torque. Why? How is a synchronous motor started?
20. What are the two common types of three-phase induction motors? What is the major difference between the rotor of a synchronous motor and an induction motor?
21. What is "slip" in an induction motor? Why can't an induction motor run without slip?
22. What are the four standard NEMA classifications of induction motors? What are the main characteristics of each? What types of applications is each most suitable for?
23. A multispeed induction motor can operate at two or more speeds. How is this accomplished?
24. Compare across-the-line starting with reduced voltage starting. When can each be used?
25. How can solid-state electronic devices be used to alter the torque/speed characteristics of induction motors?
26. Compare a wound-rotor induction motor to a squirrel-cage induction motor. What are the advantages and disadvantages of the wound-rotor motor?
27. What is a selsyn motor system? What are two common applications for selsyn systems on ships?
28. What types of bearings are found on electric motors?
29. What are the three categories of single-phase motors?
30. How does the magnetic field of a single-phase induction motor differ from that of a three-phase induction motor? How is starting torque generated?
31. What are synchronous single-phase motors commonly used for?
32. What is a universal motor?
33. What are the probable reasons for a single-phase motor failing to start?
34. Explain the "right-hand rule" used to describe electromagnetic force and current?
35. Compare the construction of a DC machine to an AC synchronous machine.
36. Compare shunt, series, and compound DC generators.
37. Define the terms under-compounded and over-compounded.
38. Outline the procedure for paralleling DC generators.
39. Compare the characteristics of a shunt DC motor and a series DC motor.

40. What are the disadvantages of DC motors compared with induction motors? What is their major advantage?
41. Explain why reducing the current through the field coils of a DC motor results in a higher motor speed.
42. What are the common causes for sparking of commutator brushes?
43. Describe the procedure for detecting an open armature coil.

REFERENCES

Del Toro, V. 1985. *Electric Machines and Power Systems,* Englewood Cliffs, New Jersey: Prentice-Hall.

Hubert, C. I. 1991. *Electric Machines: Theory, Operation, Application, Adjustment, and Control,* New York: Merrill (Macmillan).

Hubert, C. I. 1969. *Preventive Maintenance of Electrical Equipment,* 2nd ed., New York: McGraw-Hill.

CHAPTER 24

Shipboard Central Operating Systems

AARON R. KRAMER

A HISTORY

THE operation of a ship's machinery space from a central location requires instrumentation hardware to transmit information for the operation and control of engine room equipment. The central control concept began in the late 1940s and early fifties with the introduction of the Bailey board, shown in Figure 24-1. Instrumentation hardware was based on relatively large case instruments. Each instrument was directly connected to a process sensing point. The operator could only interface with one control loop at a time, e.g., the combustion control system or the refrigeration system. With the development of pneumatic transmission techniques, centralized control became possible, gradually permitting more control hardware to be placed on a central control room panel. However, the receiving instrumentation was still fairly cumbersome and dedicated to the measurement and control of only one process variable at a time.

By the late 1950s, miniaturization of the receiver instrumentation was another step towards central control. Although largely pneumatic, the individual central control room instrument size had decreased to 6 × 6 inches, then 3 × 6 inches, and finally the 2 × 6-inch standard was achieved. About this time, electronic instrumentation hardware based on transistor technology became available for the control applications.

The development of electronic transmission made possible the centralization of many more instruments and functions on one control panel.

During the early 1960s the use of digital computation was introduced to process control, adding peripheral hardware to the control room. New interface hardware, such as printers, typewriters, cathode ray tube (CRT) screens, and keyboards, made the control room somewhat complex, because this new hardware required backup by conventional analog equip-

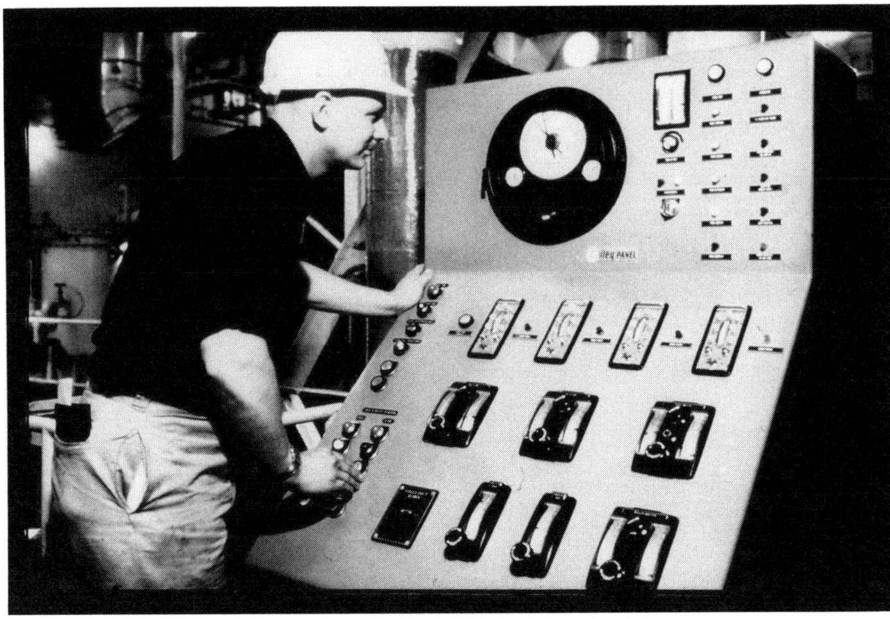

Figure 24-1. Typical Bailey board.
Courtesy Bailey Controls Company, Wickliffe, Ohio

ment. Thus, the operator had to learn new techniques, while using old familiar ones in emergencies. This was the state of the art of control panel design until the 1970s.

The seventies produced a revolution in man-machine interface design with the introduction of systems based on microprocessor hardware. This new hardware digitized the usual analog signals and made applicable new methods of control and monitoring. The ease of communications between component parts of the system enabled an increase in flexibility of control and monitoring. All information on these new control consoles was visible and within easy access of the operator.

Microelectronics: The Impact

Some understanding of the revolution in microelectronics that has occurred over the past 15 to 20 years is also necessary for a full appreciation of the development of distributed control systems.

A microcomputer on a single chip, comparable in performance to some of the early computers, is possible only because of the development of large-scale integrated circuit (LSI) metal oxide semiconductor (MOS) technology.

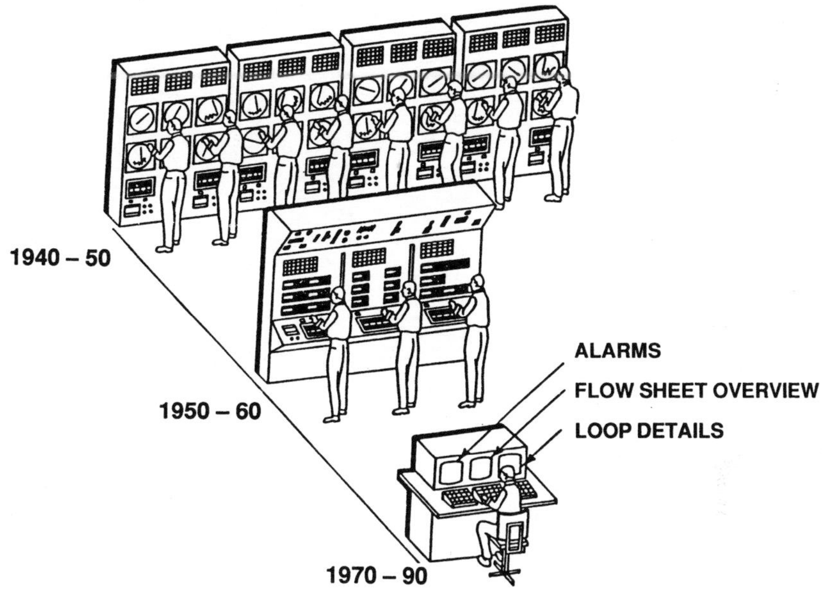

Figure 24-2. Evolution of control panel design

Microelectronic development has progressed exponentially in very recent history. It evolved from the earlier development of the germanium transistor, then the basic silicon integrated circuit. The metal oxide semiconductor technology of the medium-scale integrated circuit (MSI) progressed to the current LSI; most recently, the very large-scale integrated circuit (VLSI) is the current state of the art. The growth of this technology has provided the basis for the modern central operating systems aboard vessels today and in the foreseeable future. Figure 24-2 shows this evolution of central operating systems (COS) from 1950 through 1990.

CENTRAL ENGINE ROOM OPERATING SYSTEM COMPONENTS

Any central operating system consists of the following components.

Sensing Devices

Sensing devices that measure temperature, pressure, flow, salinity, position, torque, speed, and level must be selected, installed, and maintained with great care. These are the instruments that are required to indicate the necessity for corrective action at the central operating system. (Figure 24-3 shows typical pressure, temperature, and flow devices.)

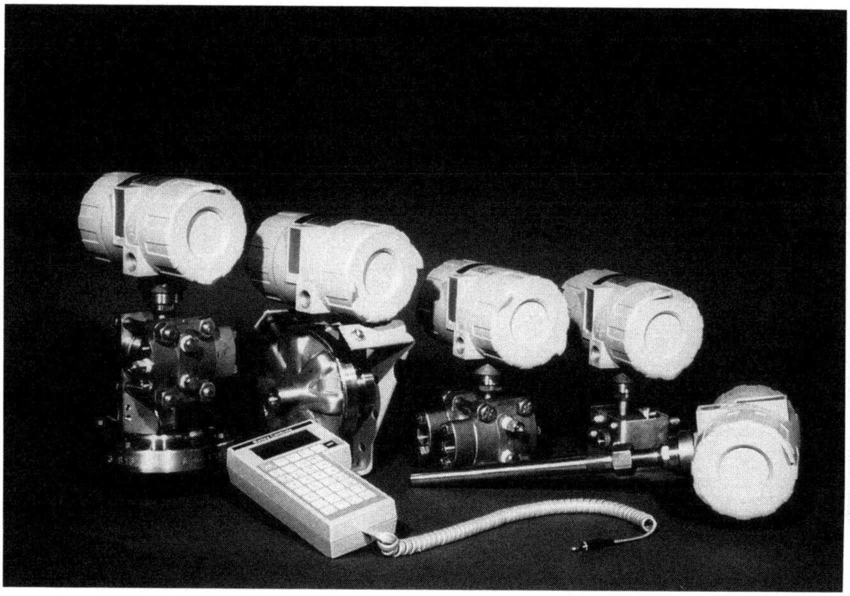

Figure 24-3. Typical pressure, temperature, and flow devices.
Courtesy Bailey Meter Co., Cleveland, Ohio, a McDermott company.

Transmitting Devices

Transmission devices convert the information sensed by the sensing devices installed in the line (i.e., steam line, shaft speed, etc.) into a physically realizable common low energy signal, so that it can be transmitted to receiving or control devices. Standard transmission signals and media are:

1. Pneumatic: 3 psi-15 psi; 3 psi-27 psi; 0-60 psi
2. Electric: 1 volt-5 volt, −10 volts-0 volts-+10 volts
   10 mA-50 mA, 4 mA-20 mA*
3. Digital transmission: Digital transmission varies as to type of transmission and receiving devices used. A rate can be selected from 110 to 19,200 bits/second. Bits/second digital transmission is defined as a baud rate.
4. Fiber optics: This is a method of direct transmission of light through cables. The optics can be incorporated within the sensing device and installed as a single unit. Some items that can be installed as a unit are differential pressure devices, temperature and pressure devices, flow devices, and level devices.

* Becoming the industry standard

Variables that may require separate transmission and sensing devices are torque, salinity, pH, smoke, and crankcase vapor.

Care must be taken to insure compatibility with the total COS system to insure proper and accurate transmission of information.

Receiving Devices

Receiving devices are those which convert the transmitted signal into useful information for observation by the operator. The devices that perform these functions are items such as bellows, pneumatic amplifiers, electric amplifiers, digital converters, and optical encoders (which convert light to a series of electrical impulses). These receiving devices are connected to observation mechanisms (pens, lights, pointers, etc.).

Indicators and Recorders

Indicators and recorders are devices that are manipulated by receivers so that operators may observe the information relating to the function of the engine spaces. These devices take the form of lights, strip indicators, chart recorders, gauges, CRTs, and teleprinters, and provide sound and sight information to the watch engineer and maintenance crew relating to plant operation.

Control Devices

Control devices regulate or maintain the proper operating condition of the systems in the engine spaces. Drum level, combustion control, engine RPM, refrigeration temperature, starting air systems, etc., all require control devices. Failure of this equipment generally requires a transfer from automatic to manual operation until the failure is corrected. Application of sensing, transmitting, receiving, indicating, and controlling devices requires a thorough understanding of the process in which these items are used. Proper engineering principles must be followed if the central operating system is to function as intended.

DESIGN CONSIDERATIONS

The design of a shipboard central operating system must consider the following requirements.

System Environment

1. Operator (crew) capability to perform operation, maintenance, and repair activities
2. Operator (crew) availability (number, location, duties)

3. Ambient conditions (temperature, humidity, cleanliness, corrosion, explosion hazards, etc.)
4. Relative location of control system components

System (Machine) Characteristics

1. Interrelationship of controlled events
   a. Dependent and independent operation (safety interlock, etc.)
   b. Requirement for logic sequencing
   c. Requirement for computed outputs
2. Characteristics of controlled variables
   a. Number and characteristics of inputs
   b. System response (time factors, stability, etc.)
   c. Permissible deviation from ideal conditions (pressure, temperature, force limitations, etc.)
3. Effect of failure

Elements of Component Selection

Instrument selection requires consideration of the following areas.*

Static performance characteristics

1. Ambient temperature effects
2. Steady state error
3. Hysteresis
4. Friction
5. Repeatability
6. Reproducibility
7. Resolution and threshold
8. Error band
9. Linearity

Dynamic performance characteristics

1. Frequency response
2. Response time
3. Damping
4. Temperature effects
5. Motion effects
6. Installation

Reliability characteristics

1. Life cycle operational

* An explanation of each of these terms may be found in *Process Instruments and Controls Handbook,* Douglas M. Considine, New York: McGraw-Hill Book Co., 1985.

2. Overrange protection
3. Life storage

Factors that depend on characteristics of input variables

1. Range
2. Overload protection
3. Dynamic response (transient and resonant)
4. Purpose of the measurement (observation and/or computation)

Factors affecting transducer input/output relation

1. Accuracy
2. Linearity
3. Sensitivity
4. Resolution
5. Repeatability
6. Reproducibility
7. Friction
8. Hysteresis
9. Threshold, noise, etc.
10. Zero drift
11. Motion

Factors relating to the system of which the transducer is part

1. Output characteristics
2. Size and weight
3. Power requirements
4. Needed accessories
5. Mounting requirements
6. Environment of transducer
7. Crosstalk
8. Effect of presence of transducer on measurand
9. Need for corrections or modifications dependent on other transducers
10. Need for corrections or modifications dependent on external information

Factors relating to measurement reliability

1. Ease and speed of calibration and testing
2. Time available for calibration prior to and/or during use
3. Mission time
4. Stability against drift
5. Vulnerability to sudden failure (mean time between failures, or MTBF)

6. Failure safety (Will transducer failure represent or cause a system failure?)
7. Failure recognition (Will transducer failure be immediately apparent to system or observer?)
8. Level of technical competence of all personnel expected to handle, install, use, and/or service the device

Factors relating to procurement

1. "Off-the-shelf" availability (Is development work necessary for an operational transducer?)
2. Cost
3. Availability and delivery
4. Previous experience with vendor
5. Availability of spare parts
6. Availability of calibration and test data from vendor
7. Operational familiarity

CENTRAL OPERATING SYSTEM TYPES

Central operating systems are classified as direct connected, pneumatic transmission, electric/electronic, or digital.

Direct Connected Systems

A direct connected system brings sensing equipment to a common point, usually the operating floor. The sensing equipment, which is also the indicating device, is connected directly to the lines through which the operating medium is flowing (e.g., steam lines, flue gas, cylinder exhaust temperature, etc.). The direct connected system exposes the operator to engine room hazards if a failure occurs in any of the connecting lines to the indicators. Figure 24-4 shows a direct connected panel.

Pneumatic Systems

Pneumatic systems utilize low pressure air to transmit a signal proportional to the measured variable. The receiver or indicator also functions with the same low pressure air. The sensing device is connected to the operating high pressure lines at a location remote from the operating personnel. Personnel are therefore not directly exposed to failure of operating lines.

Pneumatic systems eliminate the hazards of electrical equipment when the operator is monitoring the variables associated with liquefied natural gas, liquefied petroleum gas, and other volatile chemicals. Figure 24-5 shows a pneumatic system and its associated central control panel.

Figure 24-4. Direct connected panel

Electromechanical Systems

Electromechanical systems combine the essentials of mechanical sensing and electrical transmission devices with electric receiving and control devices. Electromechanical systems offer similar advantages to that of the pneumatic system in that multiple indicators and control functions may be utilized with the same transmission device. The advantages of electric transmission are the ease of installation, convenience of electric cable, and the speed of transmission of the measured signal. The significant disadvantage is the inherent danger of its application in explosive hazardous areas, such as aboard LNG, LPG, or volatile chemical transport vessels. Figure 24-6 shows a typical electromechanical system.

Digital Systems

All of the previously discussed systems were analog in nature. An analog system consists of a separate signal conditioner (transmitter) and a separate recording and indicating channel for each measured variable. The ease of conversion from the analog to the digital world makes possible the application of computers to the ship systems.

Digital systems convert continuous analog signals from the sensing and transmission devices into discrete bits of information that can be understood only by a digital computer. These bits of information are then

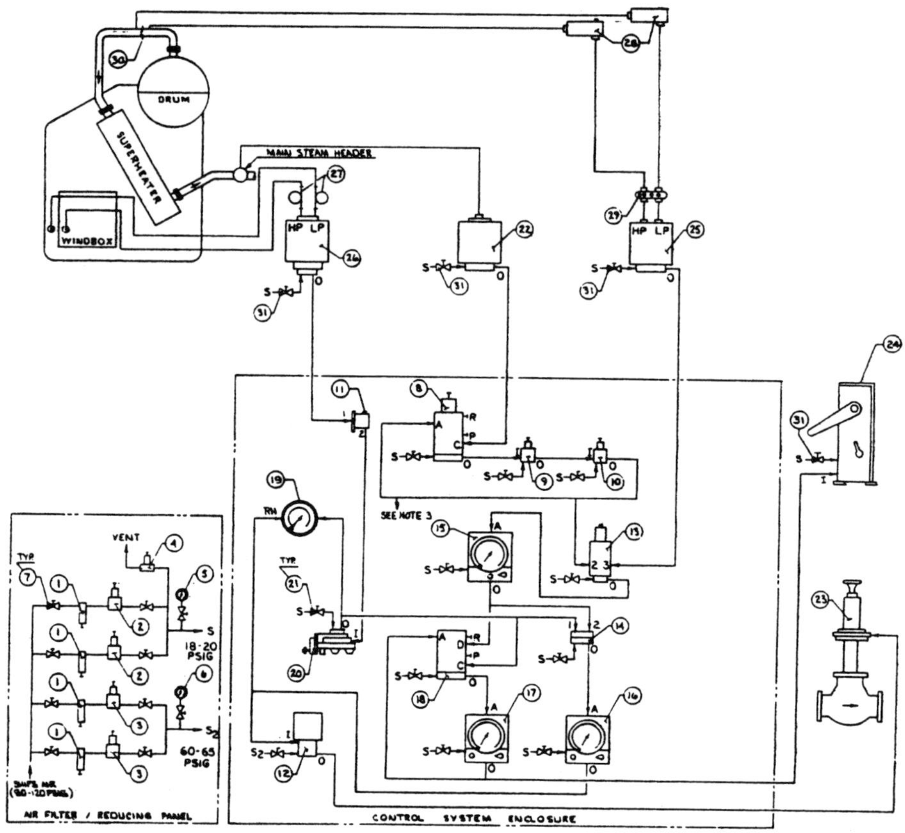

Figure 24-5. Pneumatic system schematic. Courtesy General Regulator Division of Forney Engineering, Carrollton, Texas.

translated into useful functions by various computer languages for further processing by the system. Clarification of analog and digital concepts may be provided by envisioning two common musical instruments. A piano is a digital device, because there are only 88 keys related to 88 distinct, discrete notes. A violin has only 4 strings but the fingers can be positioned on these strings in an infinite number of continuous (analog) positions. A Simpson voltmeter is an analog device.

DIGITAL SYSTEM COMPONENTS

System Communications

A digital centralized control and monitoring system requires communication with the analog information (flow, temperature, pressure, etc.) and

Figure 24-6. Typical electromechanical system panel. Courtesy General Regulator Division of Forney Engineering, Carrollton, Texas.

digital information (switches, pulses, contacts, etc.). The conversion of this information to a common form, understandable to the central operating digital computer, requires components called the input/output section or "front end." A general configuration for each component connection is shown in Figure 24-7. The components are:

1. Multiplexer
2. Filters
3. Analog to digital converter
4. The computer
5. Digital to analog converter
6. Man-machine interface
7. Memory (RAM, ROM, PROM, EPROM, E^2PROM)

Multiplexer (multiplexing). A multiplexer is a high speed solid-state switching device that has the ability to scan a large number of sensors in a short period of time (20,000-30,000 points/second). It is controlled by the computer in a programmed manner so that when the sensing point is addressed, all other necessary connections operate on that specific point. Examples of these operations may be switching a light, reading the variable on the CRT, or storing the information for future processing. The multiplexer may be installed in the same location as the computer or in

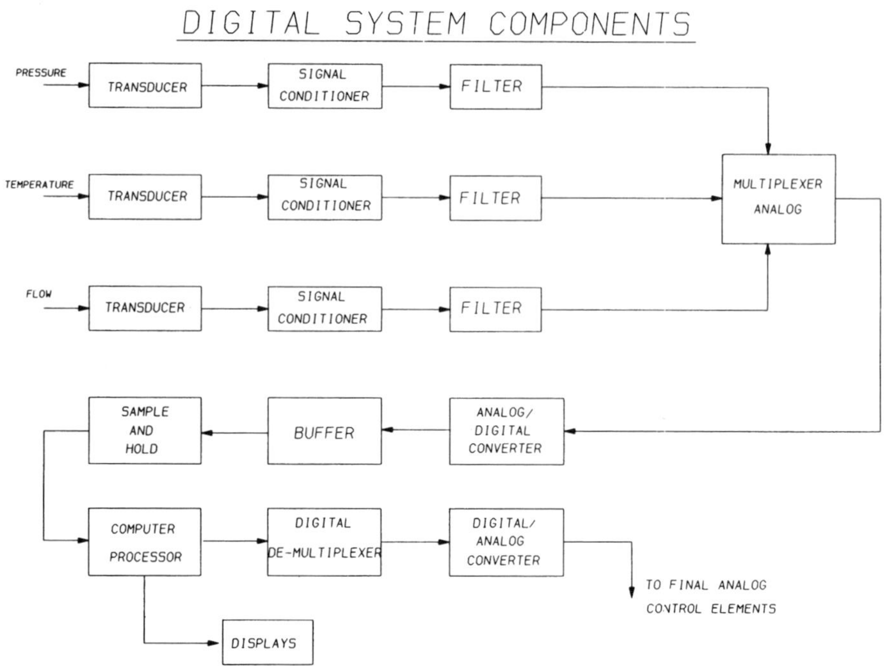

Figure 24-7. Digital system components

some remote location connected to the central computer, each arrangement having advantages depending on application.

Filters. Digital and analog systems, pneumatic or electronic, function as carriers of information signals with respect to the process they monitor. These signals consist of two parts: the actual representation of the variable being monitored and the unwanted noise or information that is totally unrelated to the process. The unwanted noise serves no useful purpose. It masks the true picture of the measured variable and in many instances can cause serious problems in the operation of a system. A filter, specifically designed to attenuate the most common sources of noise, is a necessary component in a digital system. Different filters are selected to remove noise generated from 60-cycle current, momentary high voltage spikes due to ground loops, and radio frequency transmission. Any of these interferences may cause system operational problems if they are not filtered out of the system prior to entering the central operating system computer.

Analog to digital converter. The computer section of the central operating system functions in a digital format only; therefore, all analog signals must be converted into a series of discrete bits of information for processing by the computer by means of an analog to digital (A/D) converter. The speed with which the A/D unit processes the signal makes the information appear continuous to any observer. Figure 24-8 shows how a continuous signal is digitized. After processing by the A/D converter and the computer, the information appears as a number, or a group of numbers, for future use.

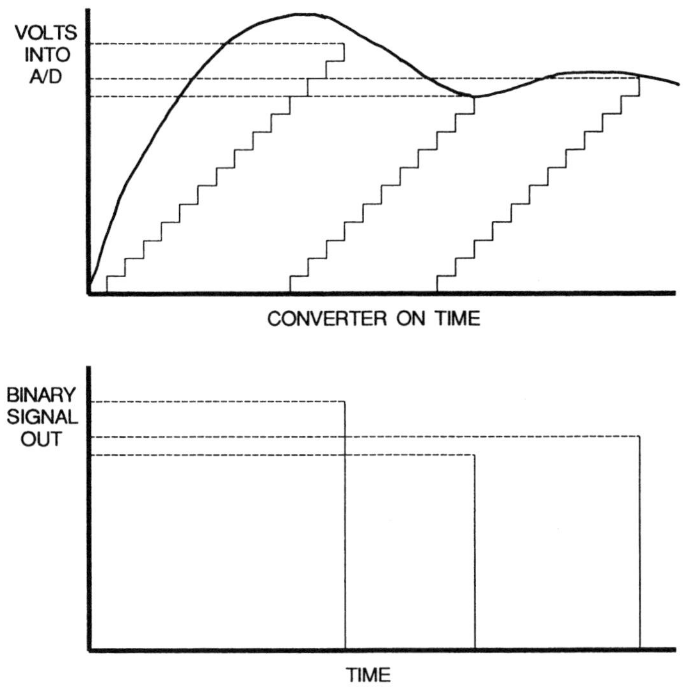

Figure 24-8. A/D conversion process

The computer. A computer is composed of five basic parts:

1. Internal storage—the memory bank used as the residence for instructions, computations, and other functions that determine the characteristics of the system.
2. Internal control—the unit that interprets the instructions in a specified order and acts like a "traffic cop" to insure that the instructions do not conflict.
3. Arithmetic unit—a component that gives the computer the ability to add, subtract, multiply, divide, and do logical comparisons.
4. Input registers—units to provide a path into the internal storage for the instructions or data upon which the computer operates as a system.

5. Output registers—providers of the path for the results of any operation directed by the system to be output to the observer or the central system.

The central operating system requires additional comments:

1. A real-time clock
2. An external interrupt for operator communication with the system
3. A man-machine interface such as a keyboard, keypad, or typewriter
4. Control system input and output so that continuous and digital parameters may enter and leave the system
5. Intra- and intersystem communications
6. Memory protection in event of electrical failure
7. Dual redundancy of critical components

Digital to analog converter. In many central operating systems digital information must be communicated to the outside world in analog form. An electric motorized analog valve may be driven directly from the digital computer. The information represented by these digital bits is converted to an analog format and the motor sees a continuous rather than a discrete signal. Figure 24-9 shows the conversion process.

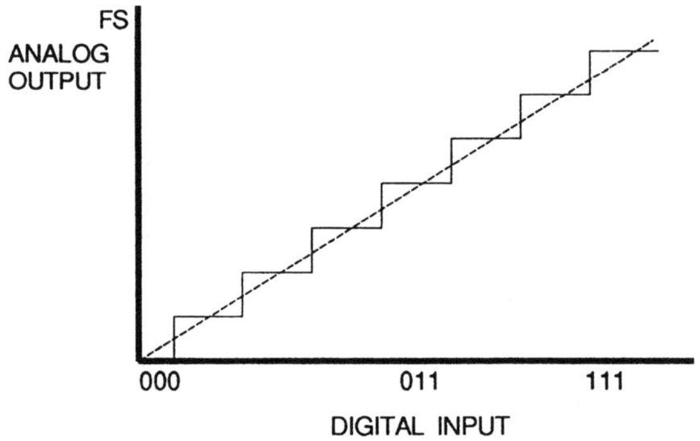

Figure 24-9. D/A conversion process

Man-machine interface. The engineering officer on watch in the central control room is required to perform the following duties:

1. Monitor for malfunction of operating equipment.
2. Monitor key operating variables for deviation from the norm.
3. Detect and interpret alarms.
4. Obtain prompt access, display, and control of systems during malfunctions.

5. Implement proper emergency procedures.
6. Implement proper start-up and shutdown procedures.
7. Check control and monitoring system calibration and performance.

The format in which the information is presented must be carefully engineered to enable the officer to perform his duties. A 30- to 40-foot panelboard is no longer required for operation of engine spaces on new complicated vessels. Microprocessor-based digital control and monitoring systems have reduced these panels to small desk size consoles containing computer-generated CRT displays in detail limited only by system capacity. Figure 24-10(a) is a typical console for a central control system of the 1960s and 1970s; Figure 24-10(b) shows a console for the 1980s. Presentations of the engineering functions can be programmed to allow the operating engineer to select dynamically active overviews of the entire plant or finite components of every subsystem in the operating spaces.

Figure 24-10(a). Typical 1960s control panel. Courtesy General Regulator Division of Forney Engineering, Carrollton, Texas.

Memory. Data, control, and information processing require storage in the computer for further processing. The storage takes place in banks located either in the computer or in an externally controlled memory device accessed by the computer. Memory can be stored on paper tape, magnetic tape, disk, or electronic components. Of the four mentioned, central operat-

Figure 24-10(b). Typical 1980s control panel.
Courtesy Bailey Meter Co., Cleveland, Ohio, a McDermott company.

ing systems generally rely on electronic memory and disk. Here are some definitions:

Disk. A plastic, iron-impregnated wafer that can store information and programs in a specified order, to be later accessed by the computer.

Random access memory (RAM). An electronic method of storing information and retrieving it in any order as directed by the computer.

Read only memory (ROM). An electronic method of retrieving information from the memory, but not a two-way communication device.

Programmable read only memory (PROM). A programmable electronic device in which information can be processed on a one-time basis during an initial phase and permanently stored, to be accessed later by the computer.

Erasable read only memory (EPROM). Same as a PROM except that the storage disk can be erased (by exposing the unit to an ultraviolet light) and then reused.

Electrically erasable programmable read only memory (E^2PROM). Same as EPROM except that the memory can be erased by electrical means and re-programmed.

System components can be configured in several different ways to suit the requirements of central operating and control design objectives.

DIGITAL SYSTEMS

Configurations

Digital centralized control and monitoring systems can be configured as:

1. Supervisory monitoring and control advisory
2. Set point manipulation and monitoring
3. Direct digital monitoring and control
4. Distributed control and monitoring

Supervisory monitoring and control advisory. Supervisory monitoring and control advisory is a digital computer system that accepts key operating parameters of the engine spaces, external parameters that may affect the operation of the vessel, and manual inputs by authorized personnel concerning some other variables. All the inputs are processed by the computer; the results are advisory outputs to the operating engineer and selected other authorized personnel, who determine control set points, speeds, course, draft, etc. Any variation in operation is manually controlled by the engineer. Figure 24-11 shows such a system. The computer can be programmed to calculate best speed for vessel structural loading, best speed for minimum fuel consumption, and offers optimum operating conditions. These computations were impossible to calculate before digital computers.

Set point manipulation and control. The basic objective of a vessel's operation is to optimize the financial return on investment. The economic return on an operation depends upon a number of factors, especially the day-to-day operating strategy. It is frequently not clear to the operating personnel what is optimum. A ship's plant is a complex, interacting entity, and the optimum operating strategy can only be ascertained after considering the combined effects of many different options. The digital computer can be used to perform the analysis, select the optimum, and implement it with analog control systems.

A process model is needed to relate all of the various factors to the economic return on the operation.

Although the computer application software determines the optimum operating strategy, the analog control system still implements the decisions. In many cases the control computer simply provides the set points to the analog control loops. The computer system does not replace

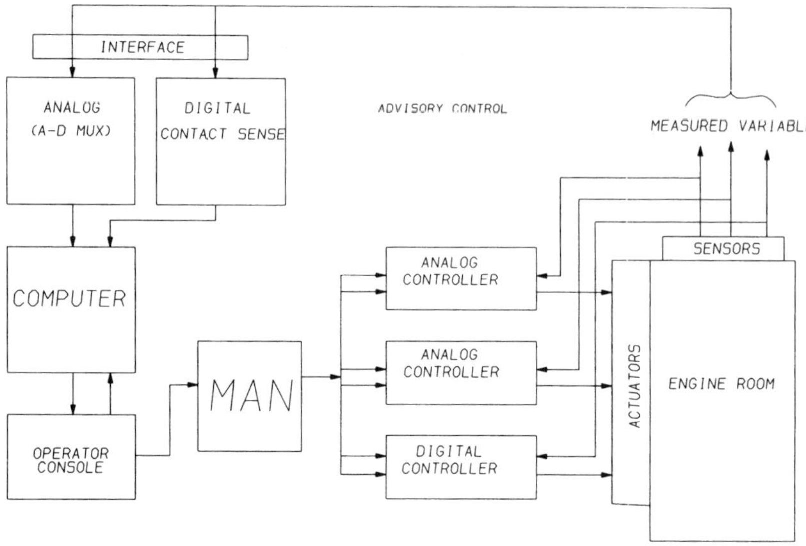

Figure 24-11. Advisory control

any analog hardware. The system backup problem is not critical since, if there is computer failure, the set points simply remain at their last settings. Figure 24-12 shows the component configuration of a set point manipulation system with analog backup, i.e., supervisory control.

Direct digital control and monitoring system. In direct control (DDC) the computer calculates the values of the manipulated variables (e.g., valve positions) directly from the values of the set points, controlled variables, and other measurements on the process. The decisions of the computer are applied directly to the process, hence the name direct digital control. This control arrangement is illustrated in Figure 24-13.

All manipulated variables, control functions, set points, and other conditions necessary for the ship's operation are preprogrammed in the computer. These programs are called algorithms.

An early incentive for using DDC systems was economic savings. The fundamental premise was that, since one computer could provide the same functions as several analog controllers, there must be some economic point at which the cost of these several analog controllers would equal the cost of the digital system. Two problems were encountered:

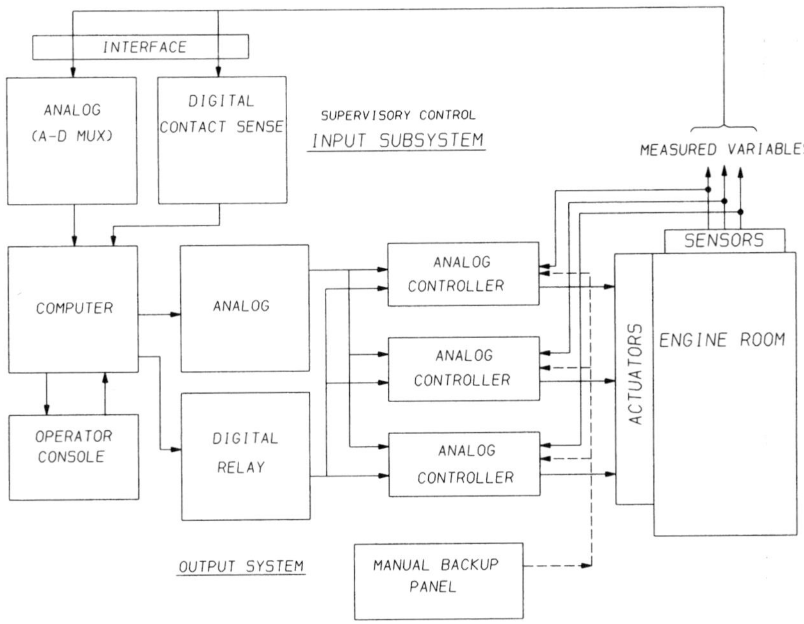

Figure 24-12. Supervisory control

1. Programming costs. With no prior experience and without DDC software packages or proven monitors, the programming effort far exceeded that which was anticipated. This problem no longer exists since most manufacturers now provide the required software packages.
2. Backup hardware. This problem stems from the fact that operating personnel must be able to exercise effective control over the plant in the event of total computer failure. In many cases this backup was a complete analog system, thus eliminating any hardware savings.

With the use of larger integrated circuits (IC) and better system configuration, the present reliability of computer systems is greater than that of the instrumentation system. Most manufacturers also can provide various degrees of backup at a reasonable cost, thereby eliminating the above problem.

An important justification for DDC is the application of control techniques which are either impractical or impossible to realize with analog hardware. New control methods can be easily implemented, tried out, and changed at any time in a digital system.

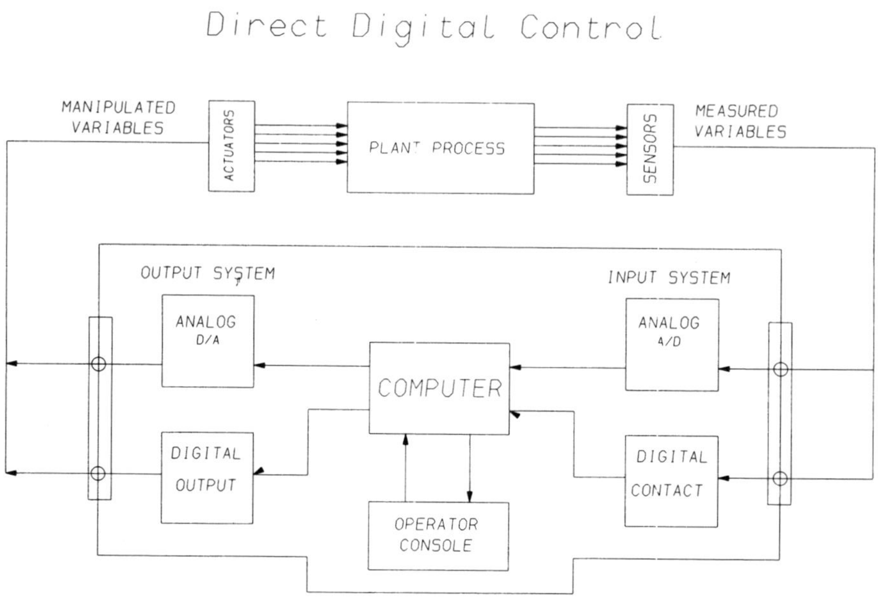

Figure 24-13. Direct digital control

Distributed control. All distributed control systems designs (see Figure 24-14) have the following generic features in common:

The heart of each control and monitoring system is located at the local machinery unit, in the engine space, close to the sensing point of the primary variable being measured.

All of the local controllers' house microprocessor-based hardware is capable of "predigesting" and, thus, converting control signal and set point information into digital pulses acceptable to a communications system.

The information is then placed on a redundant communications link that is run throughout the ship's spaces.

Much of the primary signal and control wiring is multiplexed; that is, it is run through a local switching system prior to placement on this communications link, so that only two wires are run around the plant.

Conventional control board size in the central control room is also drastically reduced. CRT screens are used for flow sheet, process value, set point, and alarm information.

These screens, plus host computer memory, enable storage of current and past ship's operating conditions by the operator or by management.

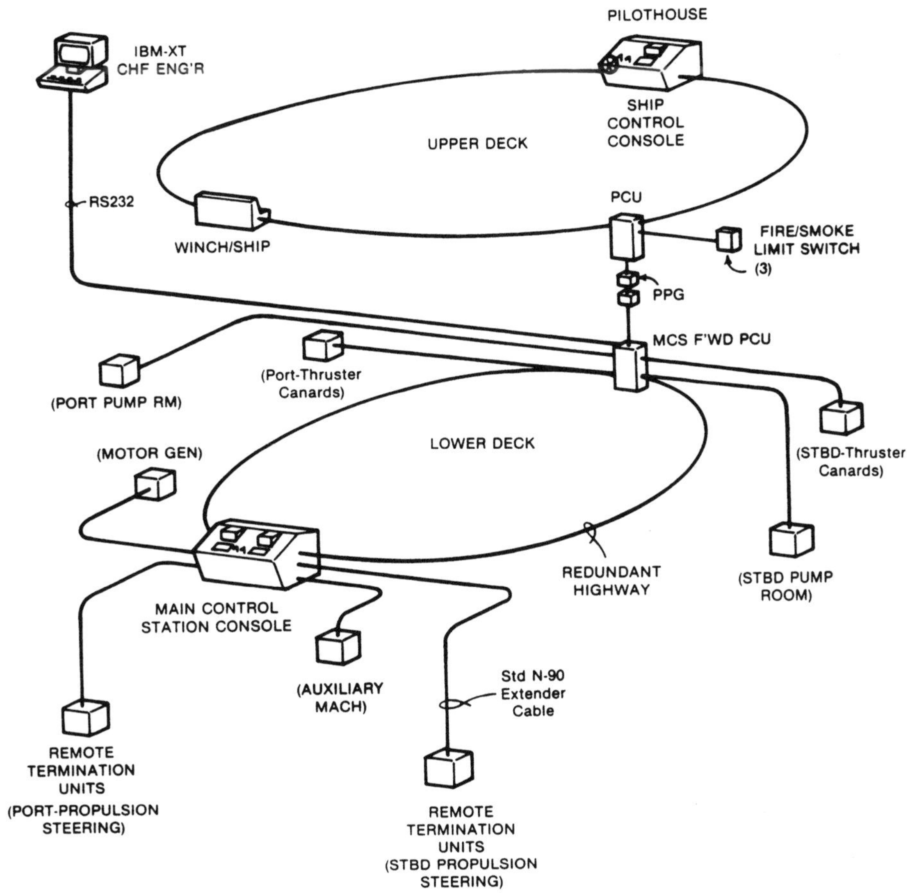

Figure 24-14. Distributed system with peripheral equipment. Courtesy Bailey Controls Co., Cleveland, Ohio, a McDermott company.

In summary, distributed control is appearing in both computer and instrumentation systems designs. The development of the microprocessor has made possible the location of microcomputers closer to the point of final control and the redesign of the conventional analog instrumentation control and monitoring boards. Variations in both are beginning to appear in the marine marketplace. Sophisticated controls, not previously practical, are now possible. Plant optimization, on-line, and managerial decision-making information will become commonplace in the industry if applications are properly implemented.

INSTALLATION AND MAINTENANCE OF DIGITAL SYSTEMS

History

When the application of computer systems began in the 1960s, the interfacing of the sensing devices with the digital system created many problems. These were caused by the use of improper signal cable, improper grounding techniques, poor cable routing, and poor layout of termination cabinets. In general, the methods for interfacing devices with computers were not fully developed. These early failures showed the importance of adhering to proper interfacing techniques. Reliable measurements were essential, especially when the computers were intended for control. Signal noise could not be tolerated and therefore it was imperative to find ways to eliminate this along with other problems.

Today these methods have been developed and are in practice. For example, high and low level signals are not mixed in the same cable tray. Shielded cable is used and the shield is grounded only at one end. In addition to the basic techniques, signal multiplexing is employed to reduce the cost of transmitting signals. New ways of interfacing pneumatic signals with digital devices are being utilized.

The emergence of the microprocessor into the instrument environment has renewed interest in interfacing with digital devices. Initially, most of the sensing devices will send analog or contact signals to a microprocessor-based receiving device. Eventually, the sensors themselves will use microprocessors, thus allowing sensor measurements to be sent over a digital communications channel. The types of transducers discussed herein are shown in Table 24-1.

TABLE 24-1

Summary of Transducers

Output	*Typical Applications*	*Interfacing Method*	*Cable*
Contact	Alarms, status, binary	Digital/multiplexer	Twisted pair, overall shield
High level analog	Flow, level, pressure, etc.	Analog/multiplexer	Twisted pair, overall shield
Low level analog	Temperature, smoke	Analog/multiplexer	Twisted pair, individual shield and overall shield

Output	*Typical Applications*	*Interfacing Method*	*Cable*
Pneumatic	Flow, level, pressure, etc.	P/I converter, P/D converter, pneumatic/multiplexer	Not applicable
Digital	Flow, level, pressure, temperature, RPM, torque	Parallel, serial	As required by communications channel
Discontinuous	Composition analysis (salinity, dissolved oxygen, etc.)	Digital interface, analog/multiplexer	Digital—as required by communications channel; Analog—twisted pair, overall shield

Digital Contacts

Relay contacts must be interfaced with a computer. (Consider a computer to be any digital device, such as a microcomputer, data logger, programmable logic computer, minicomputer, etc., which has an analog or digital multiplexer associated with it.)

Contacts are used to represent binary information, such as the on/off status of a motor or the presence of an alarm. A group of contacts can be used together to represent digital information being sent in parallel. A contact should be isolated from ground and all other signals, except those from the computer. This is sometimes referred to as a dry contact. A typical connection for a contact computer input is shown in Figure 24-15. The contact input, or digital input, is connected to a digital multiplexer within

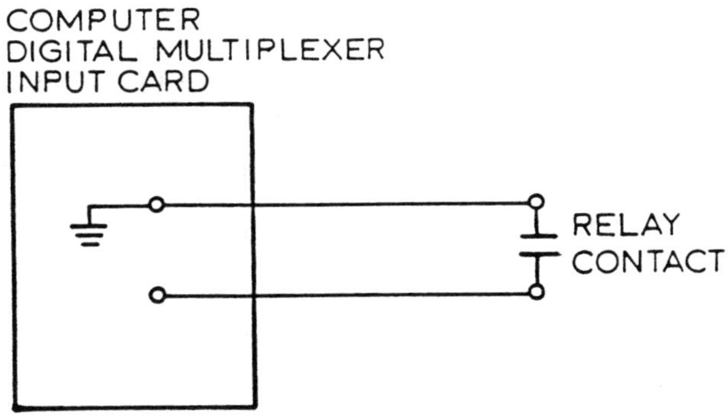

Figure 24-15. Connection for a contact input

the computer system. Note that the computer sends out a signal (a ground in this example) and senses if the signal is returned. If the computer were to be disconnected from the contact, it would be completely isolated. It is imperative that the contact is rated for the voltage and current of the computer sensing signal, although, in general, these voltages and currents are not very high.

Analog Signals

Analog signals are divided into two categories, high level and low level. While there is not a universally agreed-upon break point for these two categories, signals below 200 millivolts are considered to be high level. For example, a 4-20 mA signal working across a 250 ohm impedance has a voltage drop of 1 to 5 volts and is considered to be a high level signal, while thermocouple signals are classified as low level. Analog signals are connected to an analog multiplexer.

High level. A typical hookup of a 4-20 mA sensor in a plant with electronic analog panelboard instrumentation is shown in Figure 24-16. Resistors R1

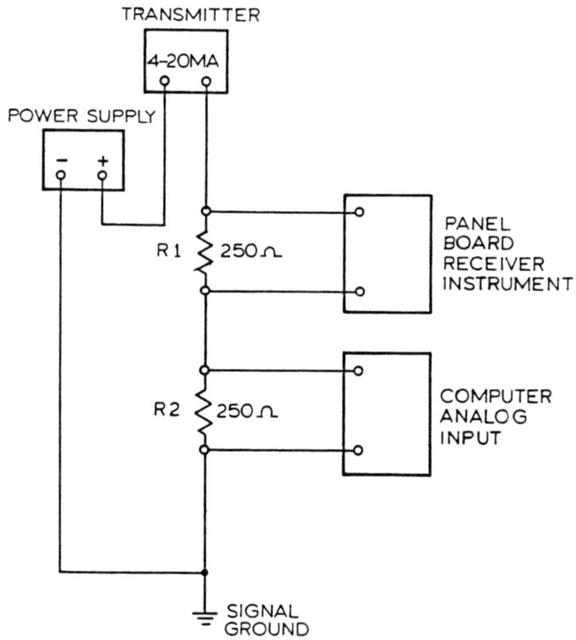

Figure 24-16. Connection for a 4-20 mA input

and R2 should be located in a terminal board cabinet, where they will not be disturbed during equipment maintenance. In this way, the computer analog input card can be pulled from its slot for servicing and the measurement circuit will not open up. Similarly, the panelboard receiver instrument, such as a controller or indicator, can be disconnected without affecting the computer reading. Also, a short across R1 or R2 will not affect the other reading. It is imperative that this circuit be grounded in only one place, i.e., at a connection to signal ground. Signal ground should be a separate ground connection running into the plant grid or several feet into the ground. The wire connecting the signal to earth ground at the equipment should be large enough to present a low impedance. (In applications where intrinsic safety is required, barriers should be employed between the field transmitter and the control room/computer room equipment.)

Low level. Low level analog signals present more problems than high level signals because the low level signals are more susceptible to noise. Noise can be a combination of:

1. Electrostatic coupling
2. Inductive pickup
3. Common mode noise

Electrostatic Noise

This type of noise is the result of coupling the electric field with the signal wires because of the capacitance between the signal conductors and between the conductors and ground. Since the capacitance between the conductors increases as the length of the cable increases, electrostatic coupling is more of a problem in long signal cables. To eliminate this noise (also referred to as capacitive coupling or static noise), shielded cable should be used. One end of the shield must be grounded. The use of aluminum-polyester tape and a drain wire as the shield is becoming popular. To achieve a higher degree of noise rejection, a cable should be used which provides a separate shield for each pair of wires and a shield overall.

Inductive Pickup Noise

Power lines, motors, transformers, etc., generate electromagnetic fields that can be picked up inductively. (This is also called electromagnetic noise.) The use of twisted pairs greatly reduces inductive pickup. The more twists per foot, the greater the noise reduction. Six twists per foot should be sufficient. In situations where a low level signal cable is in close proximity to high voltage cables, the low level cable can be run in conduit.

Common Mode Noise

A common mode voltage is a signal that appears simultaneously at both input terminals with respect to a common reference point. This common reference point is usually the electrical ground. The common voltage contains no useful information. The useful signal is the voltage that is present across the input terminals, i.e., the normal mode voltage. The analog input multiplexers used in computer systems are designed to reject common mode noise. Almost all common mode noise is at power line frequency, 60 Hertz. Typically, these analog front ends provide 120 dB attenuation of 60 Hz common mode voltage.

Common mode noise can be caused by inductive coupling, capacitive coupling, and by improper grounding procedures which result in ground loops.

Ground Loops

A signal circuit should be grounded in one place only. If one signal lead is grounded at the analog input multiplexer and also in the field at the thermocouple junction, and these two grounds are several hundred feet apart, there may be a considerable potential difference between the two grounds. This potential difference causes a current to circulate through both signal conductors. This current combines with the signal current, causing a common mode voltage to be present at the analog input terminals. The correct connection is shown in Figure 24-17. No ground loops are present. Note that if the cable shield were also grounded at the multiplexer end, a ground loop would result. Therefore, the cable shield should be grounded in one place only.

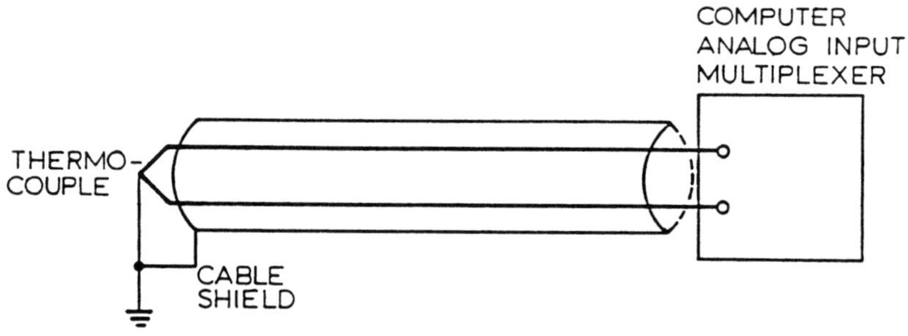

Figure 24-17. Correct connection for a low level signal

Cabling Procedures

As much as is practical, cables carrying different types of signals should be segregated into separate trays. Signal cables must be run in trays that are separate from trays containing power wiring. Where signal cable must cross power wiring, it should be done at right angles, with as much separation as is practical.

Which End to Ground?

We have established that both the signal circuit and the cable shield should be grounded only in one place. However, there is some question whether this grounding should be done at field end or at the computer end of the cable. For example, thermocouples may be either grounded or ungrounded in the field. The grounded variety provides a connection between the tip of the thermocouple and the sheath. This contact with the sheath provides a quicker recognition of temperature changes at the measuring point. However, as shown in Figure 24-18, any accidental grounding of the signal cable will result in common mode noise. Therefore, there is no clear-cut answer to whether grounded or ungrounded thermocouples are preferable. In cases where thermocouples are being used for control, the quicker response time is very important and the grounded type may be preferred. Often millivolt-to-current transmitters are installed in the field for temperature control loops. This allows the temperature measurement to be sent as a 4-20 mA high level signal, thereby reducing the noise considerations.

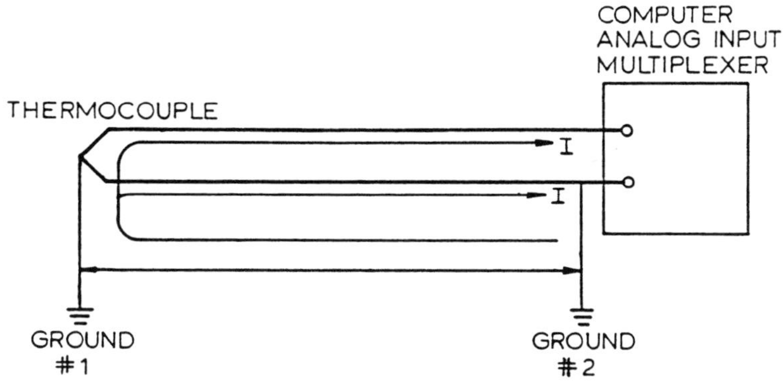

Figure 24-18. Common mode noise caused by ground loop

A convention should be established for an installation as to whether cable shields are to be grounded at the computer terminal boards or at the sensor in the field. This convention must be strictly followed.

Pneumatic Instruments

A generation ago, many plants used pneumatic instrumentation. Today some new ships are still being built which employ pneumatic instruments, but the vast majority are electronic. Many of the older pneumatic-operated ships are candidates for the addition of new computer systems and, therefore, the subject of interfacing pneumatic transducers with digital systems is important today.

The use of a P/I (pneumatic to current) converter is one solution. This device receives a 3 to 15 psig signal and converts it to a 4-20 mA electronic signal. The 4-20 mA signal can then be connected to a digital device in the same manner as any other analog input. This requires one P/I converter for each pneumatic signal. In a situation where there are many pneumatic signals, this may be costly. Also, the maintenance work needed to keep a large number of converters in calibration should be considered. In existing plants, the space to mount racks of P/I converters may be too difficult to obtain. While this solution might not be suitable for many pneumatic signals, it might be the best solution for interfacing a small number of signals with a computer.

Another solution is to use P/D (pneumatic to digital) converters. One of these devices accepts a 3 to 15 psig pneumatic signal and converts it to digital signals which can be sent over a communications channel (see "Digital Transducers" below). This method is particularly applicable to a situation where the digital system does not have the capability of receiving analog signals. As with the use of P/I converters, this solution becomes less practical as the number of signals increases.

A pneumatic multiplexer is often used in situations where there are many pneumatic signals to be interfaced with a computer. This device senses the pneumatic signals one at a time and converts each signal to analog. Some of these devices provide digital output. The switching is slow (only a few points per second), but it is much less expensive than purchasing individual P/I converters, and the maintenance time is less. If needed because of speed requirements, several units can be operated simultaneously.

Digital Transducers

An increasing number of transducers now provide digital outputs. The fact that a device has a digital output does not necessarily make it easier to interface to a computer than an analog output device. Often special programs must be written for the computer and these programs can be expensive. If these development costs can be amortized over several applications, a digital interface will often be less expensive.

Digital interfaces can be divided into two categories: parallel and serial.

Figure 24-19 shows a typical parallel interface for 16 bits of information. The data to be transferred to the computer is held in a register. The 16 bits of data each have dedicated signal paths. During the period when the register is being updated, the data ready signal will be false. When the register contains valid data, the data ready signal will be true. The computer should be programmed to read the data whenever the data ready

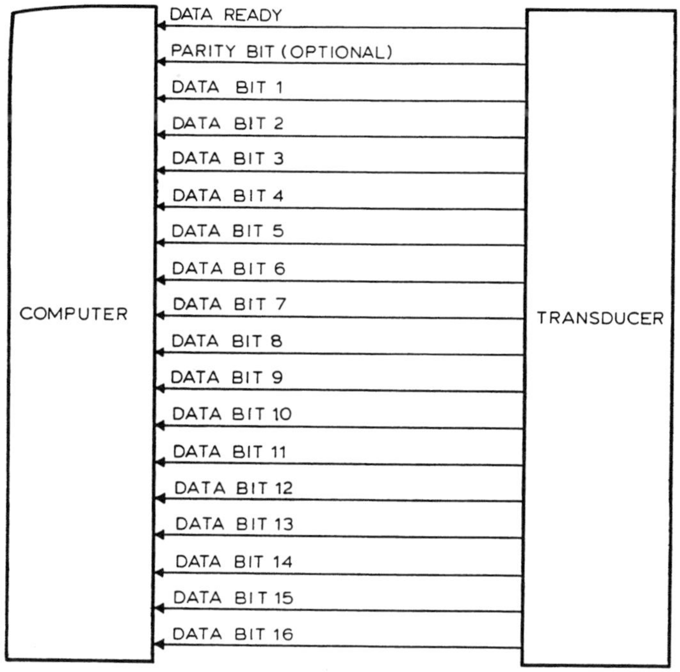

Figure 24-19. Parallel digital interface

signal makes a transition from false to true. This will allow the computer to retrieve information as soon as it is available. The computer should disregard any information read when the data ready signal is false.

Sometimes a parity check bit is included. A parity check is the simplest but lease secure method of error checking. This involves adding an extra bit, called a parity bit, to the message. This bit is set or reset so that the total number of binary ones in the message is always odd. (This is referred to as "odd parity." Similarly, some systems use "even parity," where the number of binary ones in a message is always even.)

A serial interface uses one communication path as opposed to the parallel interface, where a separate path is needed for each bit. In the serial interface, the message is sent a bit at a time, with the bits following in order. The parallel interface is, of course, much faster. An entire block of data (17 bits in this example) can be sent in the same time as one bit can be sent over a serial interface. But the parallel interface requires a separate communications channel for each bit and is therefore more expensive. For communications over long distances, the serial interface is preferred, because transmission time becomes expensive. Parallel interfaces are used for short distances where high speed is essential. Serial interfaces are gaining in popularity even for short distances because data rates of 1 million bits per second can be achieved today over a communications channel. This is more than adequate to meet the interfacing requirements of any transducer. Also, using a serial interface allows the use of the same software for interfacing with a digital transducer, regardless of how far from the computer it may be. For long distances, greater than a few hundred feet, digital data are normally transmitted over the communications channel as modulated signals. At the receiving end, the signals are demodulated back into digital signals. The transformation of digital logic to modulated signals and the transformation from modulated signals to digital logic is performed by a modulator-demodulator, commonly referred to as a modem or data set. A modem is not required for short distances. When the system is selected and the central operating philosophy established, detailed installation engineering of system interfacing should be done to insure a successful project.

USE OF DIGITAL CENTRAL OPERATING, MONITORING, AND CONTROL SYSTEMS

The state-of-the-art digital technology can provide, at present, the following typical functions for use aboard ships:

1. Main engine condition monitoring, comparing with trial results or company standards.
2. Individual cylinder operating conditions, comparing with engine mean values.
3. Hull performance related to main engine performance:
   a. Evaluating hull condition, e.g., detection of hull fouling.
   b. Evaluating and comparing hull coatings.
   c. Determining optimum periods between drydocking, various trade routes.
4. Fuel efficiency (relates to Items 1, 2, and 3 above).
5. Fuel quality evaluation (including chemical treatment application).
6. Lube oil and cylinder oil inventory control. Establish and control planned ordering and storing cycles.
7. Lube oil analysis control.
8. Shipboard spare part management, inventory control, cost evaluation and control.
9. Overtime distribution, i.e., separating operating overtime from M&R overtime.
10. Assistance with crew payroll computations and payoffs.
11. Personnel records.
12. Training and instruction manuals.
13. Inventory control, requisition printing.
14. Routine maintenance planning and record keeping.
15. Machinery running hours record with maintenance projection.

16. Auxiliary machinery performance analysis.
17. Maintenance of shipyard and drydock records, and assistance in planning of future drydock work.
18. Storing and printing drydock specifications.
19. Part incorporation of existing preventive maintenance reportable items.
20. Trend analysis of machinery performance, voyage reports, etc.
21. American Bureau of Shipping survey planning, record keeping, and projection.
22. U.S. Coast Guard inspection/reinspection, work lists/record keeping.
23. Company directives.
24. Training programs.
25. Safety programs.
26. Miscellaneous record keeping.
27. Ballast control and trim.

The engine room data logger, chief engineer's office event recorder, and bridge bell recorder should interface with the shipboard management computer so that information/data in memory of the former can be communicated to the latter and stored in a logical format or in a reliable storage medium. These stored data are then available for local display and printout as well as transmission ashore by satellite and/or modem to company headquarters.

The shipboard management computer should be capable of receiving transmitted data via satellite.

Future applications may include (1) remote operation vessels, totally unmanned and supervised by a host ship, and (2) automatic cargo loading and unloading.

The implementation of a digital central operating system and its uses are limited only by the imagination of the engineer and the desires of the owner. The potential of these systems is virtually untapped!

A COMPREHENSIVE AUTOMATION SYSTEM

Introduction

The advent of sophisticated digital electronic components at reasonable cost has made possible the development of comprehensive automation systems for control, monitoring, communication, record keeping, and fault alarm of shipboard machinery and equipment. An example of such a system is the integrated ship control developed and manufactured by Kongsberg Norcontrol of Norway. This system manages (1) alarm, monitoring, and control of cargo and ballast; (2) alarm, monitoring, and control of machinery; (3) propulsion control; (4) bridge control and navigation; and (4) ship administration and data communication to shore. Standard components are used to design a system that meets the needs of a particular ship.

The System

Figure 24-20 shows a diagram of an automation system suitable for comprehensive automation of a crude oil carrier, product carrier, bulk carrier, OBO carrier, containership, chemical carrier, or LPG carrier. This configuration is suitable for between 1,000 and 2,000 channels of input, output, and control. The practical upper limit for such channels is 20,000. In this example, there are two segregated process buses for port and starboard control and the same arrangement for cargo process buses. Table 24-2 is a tabulation of the major functions of the system.

System Hardware

Figure 24-21 illustrates the arrangement of the major hardware components for a twin-screw diesel propulsion system. An important feature of the arrangement is the redundancy of the computer local area networks and the controller area networks. The physical system is decentralized for safety and protection from flooding, fire, hull damage, etc. but the operation may be centralized from several locations using display workstations, i.e., a remote operator system (ROS) consisting of a display screen (VDU), control panel (OCP) or computer keyboard, and a computer unit (MCU) as shown in Figure 24-22. Local operator system (LOS) panels distributed to appropriate areas of the vessel provide personnel with local control when needed. Figure 24-23 shows a typical modern main propulsion control panel including a ROS workstation.

The key component of this distributed system is a family of intelligent input/output units called distributed processing units (DPU) which communicate among themselves on the redundant high capacity process bus (CAN). The distributed processing units carry out all monitoring and automation functions while the remote operator system terminals provide the interface with the ship's personnel. Each DPU has a capacity of 8 to 32 input/output channels and is configured for specific alarm, control, and/or safety tasks by downloading the appropriate software into the unit. Should any failure to a DPU, cabling, sensor, or the process bus occur, an extensive built-in system test located in each DPU will warn the ship personnel through the remote operator system. The system is designed for replacement of a faulty DPU while the system remains in operation. When a new DPU is installed, the correct software is automatically downloaded from the remote operator system before the DPU starts.

The ROS stations are the redundant central control locations for the entire system. They also serve as the database for the process data, engineering data, and system software that operates in a Windows technology environment. ROS stations are usually located in the engine control room,

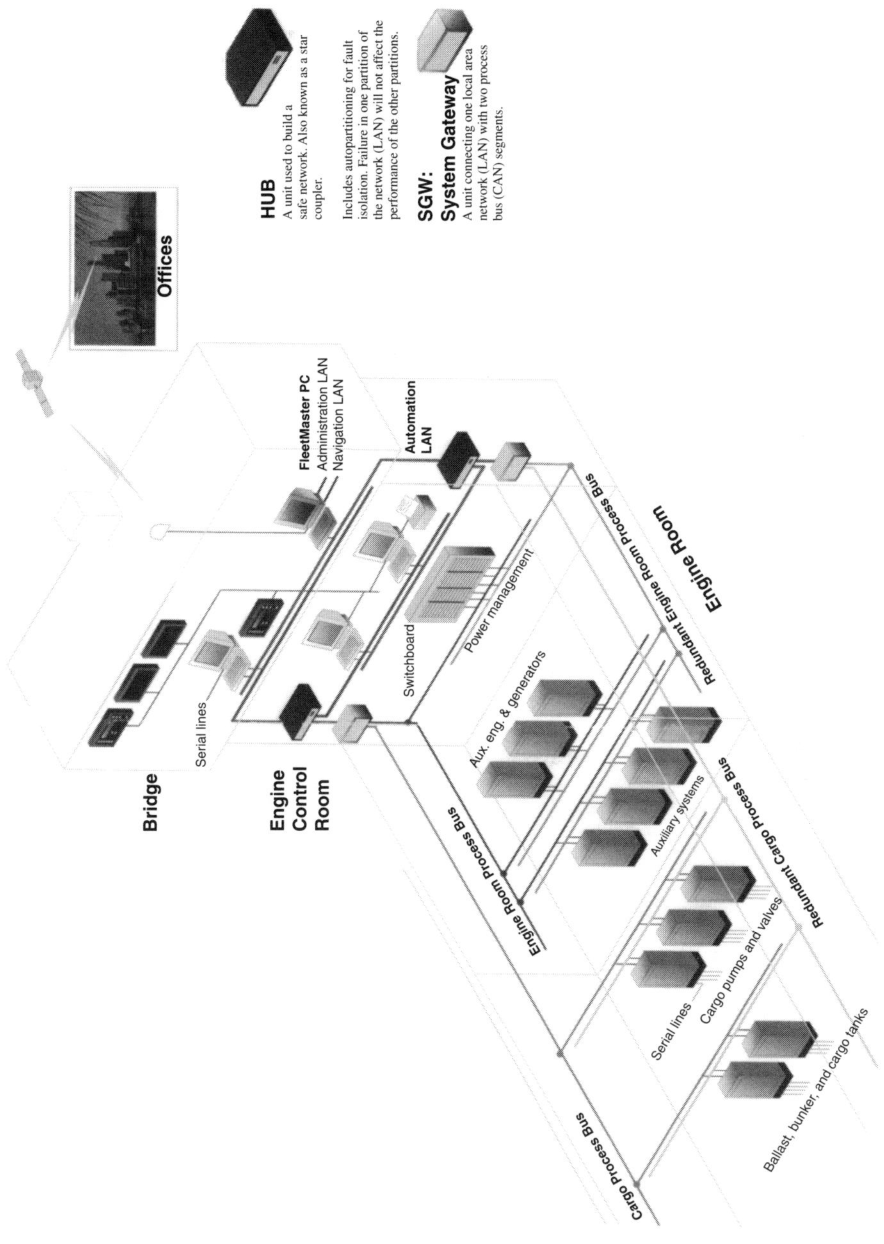

Figure 24-20. Arrangement of distributed automation system with segregation and redundancy of components (Kongsberg Norcontrol)

TABLE 24-2

System Functions

Alarm and Monitoring System Functions
Alarm and monitoring functions
Alarm extension system
Personnel alarm system

Control Functions
Standby pump start
Restart of pumps after loss of power
Sequential start/stop of pumps, fans, compressors, etc.
Main propulsion

Control of Auxiliary Systems
Lubrication oil system
Sea water syst3em
Fuel oil system
Cooling systems
Steam generation system
Purifier plant
Miscellaneous systems

Electric Power Management
Automatic start and stop of generator drive engines
Automation of generators and switchboards
Frequency control
Load control and automatic load shedding

Cargo and Ballast Control
Level gauging
Cargo pump control
Valve control
Trim control
Ballast control

Record Keeping
Machinery monitoring data records
Alarm records
Voyage records
Ship position records
Ship business records
Voyage data recorder

Communications
Company office communications
Regulatory organization communications
Ship data communications
Charterer communications

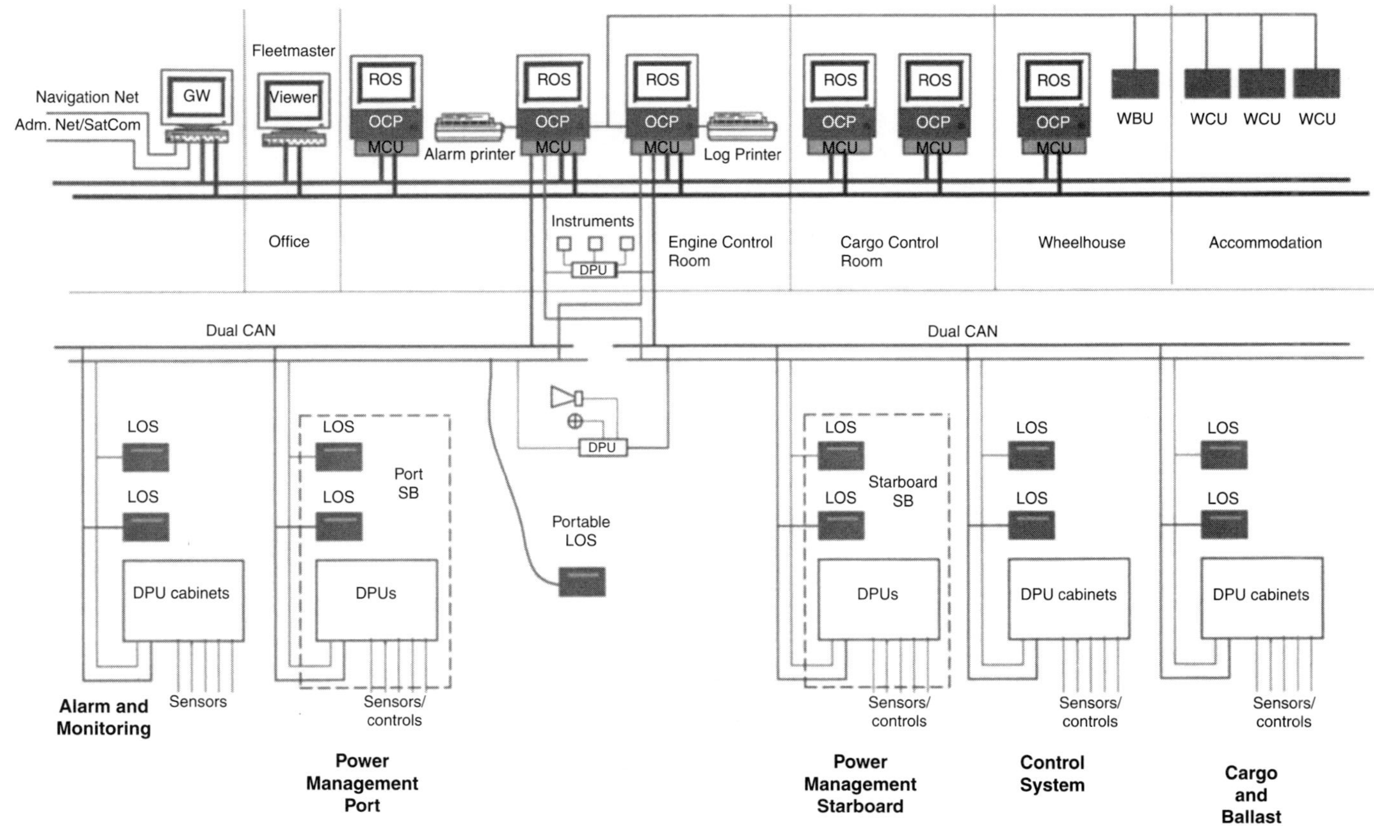

Figure 24-21. Diagram of a distributed automation system for a twin-screw cargo vessel showing connections among the remote operating system workstations and the monitor, alarm, and control devices (Kongsberg Norcontrol)

bridge, cargo control room, damage control room, and the ship office. All the ROS stations work in parallel and none acts as the system master. All ROS stations are identical, but operator access to various control functions differs, depending on the system configuration. ROS stations used only for monitoring are usually located in the ship office or engineer officer cabins. Other ROS stations are used for partial or full control of the automation process depending on the individual's use of password access. Changes of limit parameters in one ROS station will automatically result in the changes in all the other ROS stations. All operator actions are logged by the system.

Figure 24-22. Remote operator workstation

Figure 24-23. Main propulsion control panel

The LOS is a communication panel in the automation system that enables the engineer officer to have local access to the DPUs for inspection of process variables, local operation of equipment, simulation of input/output signals, adjustment of parameters, and inspection of the built-in diagnostics. All DPUs connected to the same process bus as the LOS are accessible via the LOS. The LOS panel has a liquid crystal display (LCD) and push buttons for operator interaction.

A dual redundant local area network (LAN) connects the ROS stations and the system gateways (SGW). The SGW is a DPU programmed for the special purpose of providing a connection between the LAN and the controller area network (CAN). Both paths are active at all times. The different partitions of the LAN are segregated by the use of hubs that are duplicated for redundancy.

Both the SGW and the hub are shown in Figure 24-20. The redundant CAN process bus is employed to connect the various units of the distributed processing system. The dual processing system controller (dPSC) provides a redundant gateway between two CAN segments. The main purpose of both the system gateways and the dPSC gateways is to provide galvanic isolation between the parts of the total network and to segregate process bus traffic.

The extended alarm system incorporates a watch bridge unit (WBU) and several watch cabin units (WCU) interconnected by a CAN process bus. The extended alarms are required to provide information to the bridge and engineer officer cabins when a vessel has an unmanned engine room.

System Software Controlled Functions

The computer database holds several standard pictures for display of common information concerning engine and ancillary equipment. Typical mimic windows are (1) exhaust gas temperatures and average temperature bar graph, (2) stand-by pump status, (3) controllers and valves status, (4) bar graphs, (5) trend curves, (6) power management, and (7) switchboard and generator status.

A variety of logs for alarms and data may be configured and printed on request or at specified intervals. The ROS stations display the built-in self-test of the DPU units. A fire alarm system displays deck plans with fire indication in the ROS mimic windows. A system for personnel tracking displays deck plans with indications of the presence of operators in the ROS windows.

The alarm and monitoring system permits alarm of analog signals and high/low or on/off signals. If an alarm is conditional, then it may be blocked when the specified condition is present. The control and logic functions performed by the process control, generator control, and data acquisition systems generate defined alarm messages that are presented locally and in the central control room. Acknowledgment of separate or group alarms may be accomplished by a panel push button.

Software is included in the system for constant monitoring, calculating, summarizing, and alarming, when necessary, of useful machinery performance data such as engine exhaust gas temperature, fuel economy data, boiler performance parameters, power generation and distribution, component running hours, ship speed, shaft power, voyage duration hours, distance traveled, etc.

All functions for data acquisition signal conditioning, signal scaling, and process control are performed in the DPU located close to the monitored process. This assures that local operational control and oversight is maintained on all units even if the remote centralized system is not operating.

Programs in the local DPU control all pump functions, pump monitoring, and alarms. Remote operation from the ROS station is possible. All valves may be activated for the ROS mimic windows. Depending on emergency needs for start/stop of standby units, software provides the automatic operation of valves.

The basic electric-power-management system consists of one DPU per connected generator unit programmed for the power management function. The system is suitable for generators driven by any prime mover including the propeller shaft. Depending on the type of drive, the DPU must have different configurations and control strategies. Engineer officers may control the generators locally or by an ROS. The power management system design details must be adapted to the specific configuration of power generators and the electrical bus system. All generator functions including synchronizing, frequency control, load sharing, start, and stop are automated within the management system. This management system accommodates different operational modes required of a vessel including harbor, at sea, cargo handling, and maneuver.

Tank level measuring for ballast, bunker, fresh water, waste, cargo, etc. is provided by appropriate sensor signals to local DPU devices. For ballast, bunker, and fresh water, submersible pressure transducers are used. Radar sensors are used for cargo tanks. During cargo operations this system provides all the information necessary for maintaining the vessel safety by calculating the tank levels; correcting levels for trim, list, and density; and presenting the results on the ROS mimic window diagram and printed text.

The comprehensive system includes the information management system illustrated in Figure 24-24. The shipboard functions of this system include logging data from the bridge, alarm monitoring, and control systems, organizing and storing of data onboard, preparation of reports of logged data, and transfer of data to owner office and other contacts. At the owner's office the information management system organizes and stores data received from the ship, and handles presentation of reports received

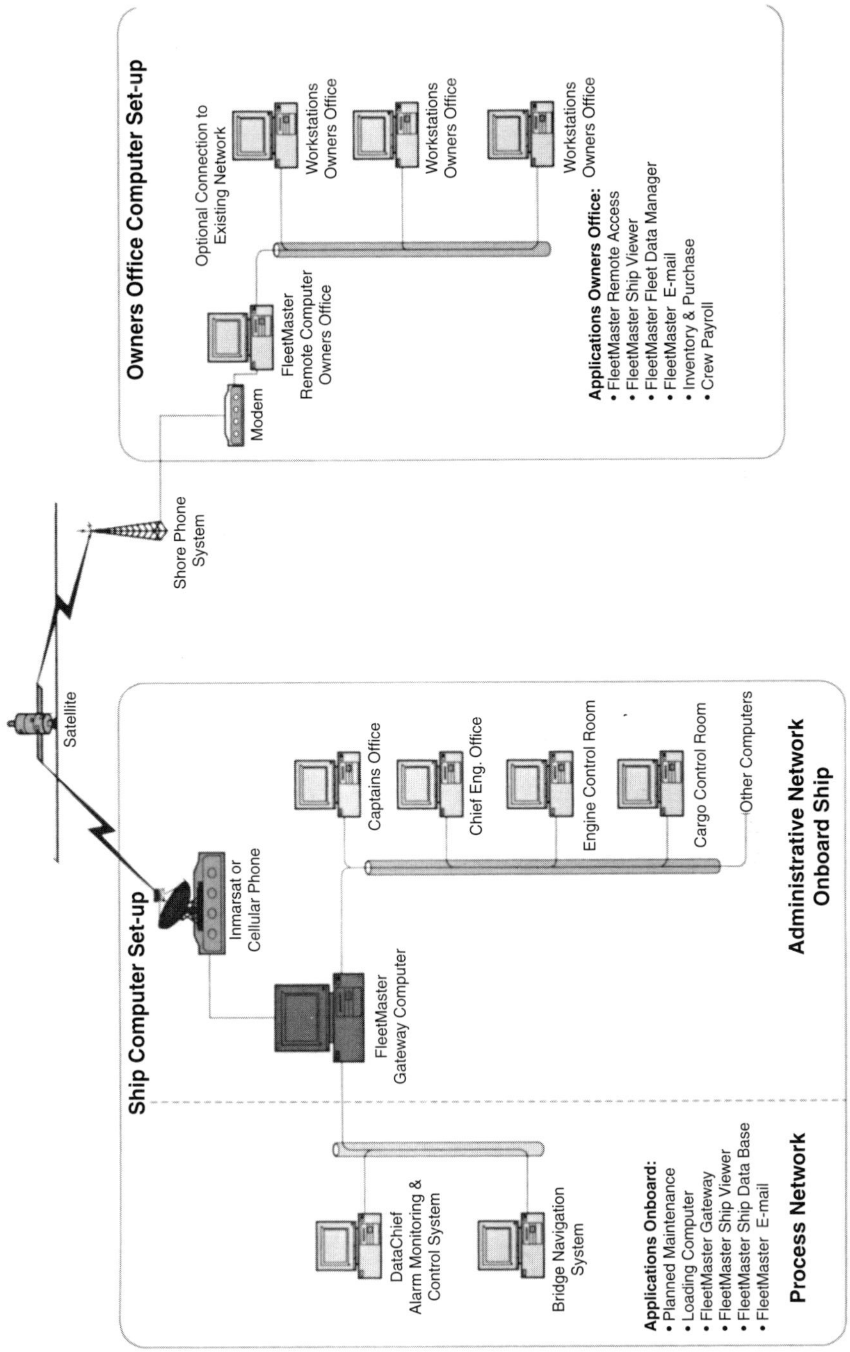

Figure 24-24. Automation system administration network onboard ship and ashore (Kongsberg Norcontrol)

from the ship, trending and evaluation of ship data, and distribution of data to external organization such as charterers, class society, or flag authority.

REVIEW

1. List the principal sensing devices found in the engine room.
2. What are the four means of signal transmission for engine room controls?
3. What are the factors that affect the transducer input/output relationship?
4. What are the general classifications of a central operating system?
5. What is a multiplexer?
6. What is the function of an A/D converter?
7. What are the basic parts of a computer?
8. What is the difference between a set point manipulation and a distributed control digital system?
9. How is electrostatic noise controlled or eliminated?
10. What is a ground loop?
11. What are the major functions of a comprehensive distributed digital automation system?
12. What are the events following the failure of a DPU in an automation system?
13. What is the practical upper limit for input/output units in an automation system?
14. Why is segregation among redundant components important in an automation system?
15. What data would you specify for collection to monitor fuel economy?
16. How is an automation system employed in port?

REFERENCES

Ambrosius, E. E., R. D. Fellows, and A. D. Brickman. 1966. *Mechanical Measurement and Instrumentation.* New York: The Ronald Press Company.

Blanding, H. May 1987. "Automation of Ships and the Human Factor." STAR Symposium. Society of Naval Architects and Marine Engineers.

Ewart, W. D. April 1986. "Reduced Manning in Japan—What Others Could Learn." *Marine Engineers Review.* Institute of Marine Engineers.

Gordon, M. December 1989. "Retrofitting Automation on Steamships for Periodically Unattended Machinery Spaces—A US Flag Operator's Experience." *Transactions 1989.* Society of Naval Architects and Marine Engineers.

Gondovin, D. May 1999. "Automation Systems in Modern Diesel Electric Marine Applications." London: IME Conferences and Symposia. Institute of Marine Engineers.

Harris, R. J. C. September 1979. "Engine Surveillance Today and Tomorrow." *Marine Engineers Review*. Institute of Marine Engineers.

Hansen, L. R. September 1990. "Control of Propulsion System." Newcastle-upon-Tyne: Conference of Maritime Systems Integrity. Institute of Marine Engineers.

Lough, A. November 1999. "Marine Automation—past, present and future." London: IME Conferences and Symposia. Institute of Marine Engineers.

MAN. March 1980. "MAN Turns to Automation To Increase Efficiency." *Marine Engineers Review*. Institute of Marine Engineers.

Marine Engineers Review. January 1984. "Integrated Machinery Monitoring." Institute of Marine Engineers.

Marine Engineers Review. October 1989. "Integrated Remote Control of Propulsion Plant." Institute of Marine Engineers.

Marine Engineers Review. November 1996. "Intelligent Drive Control on *Grand Princess*." Institute of Marine Engineers.

Miller, I. C. February 1984. "Modern Technology and Its Impact on Marine Training in Merchant Navy." London: Conference on Maritime Systems Integrity. Institute of Marine Engineers.

ACKNOWLEDGMENTS

The author acknowledges with gratitude the assistance of Margaret Hoering and Lillian Ferrari of the State University of New York Maritime College Engineering Department for their assistance, his wife Iris who made literary sense out of his scrambled thoughts, and Kongsberg Norcontrol of Norway for permission to publish the material on the modern integrated ship control system.

CHAPTER 25

Shipboard Vibration Analysis

EVERETT C. HUNT

INTRODUCTION

THERE are many ways to determine that operating machinery is in satisfactory condition, including (a) measurement of temperatures, pressures, flows; (b) recording and analysis of noise; (c) analysis of lubricating oil; (d) ferrography; (e) borescope inspections; (f) thermography; and (g) the observation of machinery response to changing demands. Engineers have always included a variety of techniques for the observation of vibration characteristics among the methods of determining the condition of machinery. A crude but useful method used by engineer officers in the early years of steam engine propulsion was to place a coin on the bearing cap of a rotating machine. If the coin remained on the cap, the vibration level was judged to be satisfactorily low. Modern vibration analysis techniques, instruments, and mathematical methods have greatly refined and expanded the engineer's ability to measure the condition of operating machinery. This chapter will introduce basic methods, some underlying theory, and the various equipment associated with vibration analysis.

Vibration

A rotating or reciprocating machine is subject to the influence of oscillatory forces. These oscillatory forces repeat themselves at equal time intervals, hence, they are periodic. For example, a small weight attached to a perfectly balanced rotating cylinder will cause the cylinder to vibrate periodically. That is, the period or time elapsed between maximum vibration forces will be equal to the time for one revolution of the cylinder. The reciprocal of the period is the frequency of the vibration or the number of vibrations which occur per unit time. Frequency is measured in cycles per second, or cycles per minute, where the cycle is the motion completed during the period. Vibrations may be divided into two general types, free and forced.

Free Vibration

If an elastic system, such as a weight connected to a fixed position by a spring, is moved so that the spring is compressed and then released, with the assumption that there is no friction in the subsequent movement of the weights, the weights will adopt an oscillating motion. This free vibration will occur at the natural frequency of the system. The natural frequency is a result of the physical properties of the spring and the weight. Figure 25-1 is a schematic of a simple free vibration system.

Forced Vibration

When an oscillatory external force is applied so that the system vibrates at the frequency of this exciting force, the system is said to have a forced vibration. If the frequency of the external exciting force is the same as one of the natural frequencies of the vibrating system, then a condition of resonance exists. In resonance, the external force reinforces the natural vibration during each cycle so that the amplitude or range of motion increases with the ultimate possibility of machinery failure.

Damped Vibration

Actual vibrating systems, such as rotating machinery, have internal forces due to friction which act to reduce the forces of excitation and prevent the amplitude of motion from becoming excessive during resonance conditions. For example, a steam turbine passing through its first critical speed has exciting forces at the same frequency as the natural rotor frequency. Nevertheless, damage is prevented because the vibration amplitude is damped by physical features of the rotor and the internal friction of the lube oil in

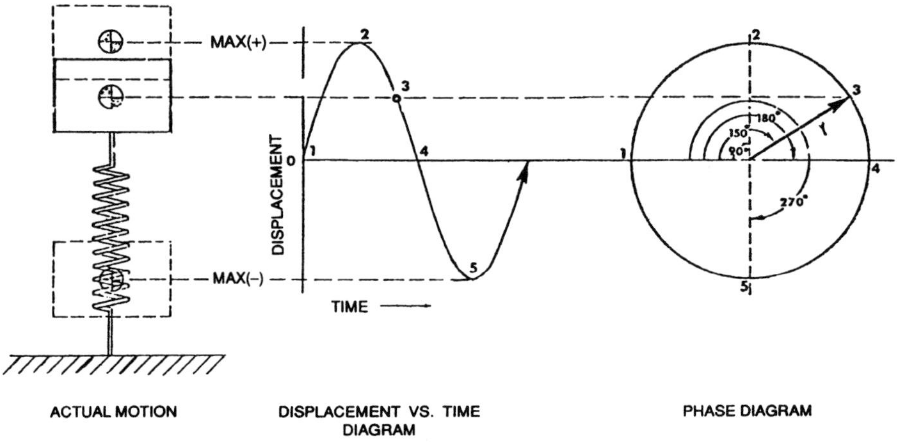

Figure 25-1. A simple free vibration system

the bearings. Figure 25-2 shows a simple damped system with forced vibrations.

Vibration Measurement

There are three common measurements to describe the periodic motion of vibration: displacement, velocity, and acceleration. These three measurements are related to each other by mathematical formulations. For example, the periodic motion may be represented by a unit vector rotating at ω radians per second. A radian per second equals 2π times revolutions per second. In this case, the vibratory motion or amplitude represented by x is

$$x = \sin \omega t$$

and the velocity, v, is the rate of change of x or

$$v = \omega \cos \omega t$$

and the acceleration, a, is the rate of change of the velocity or

$$a = -\omega^2 \sin \omega t$$

Plotting these values on a time scale as shown in Figure 25-3 illustrates the relationship among these vibration measurements. The velocity leads the displacement by 90°. The acceleration leads the displacement by 180°.

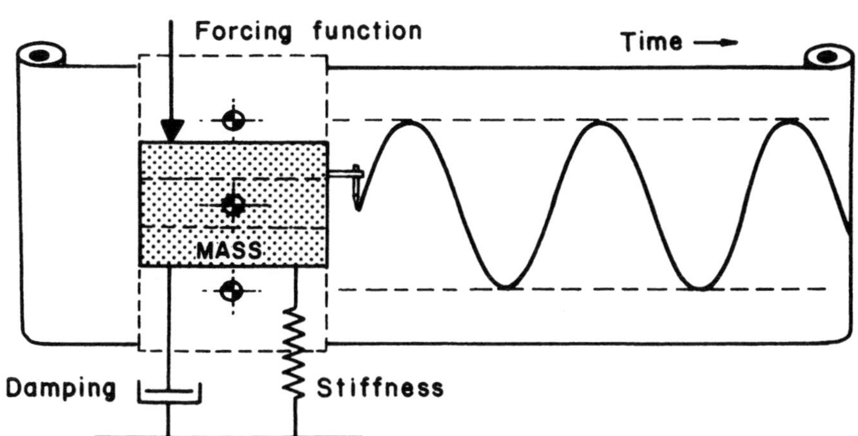

Figure 25-2. A simple damped system with forced vibrations

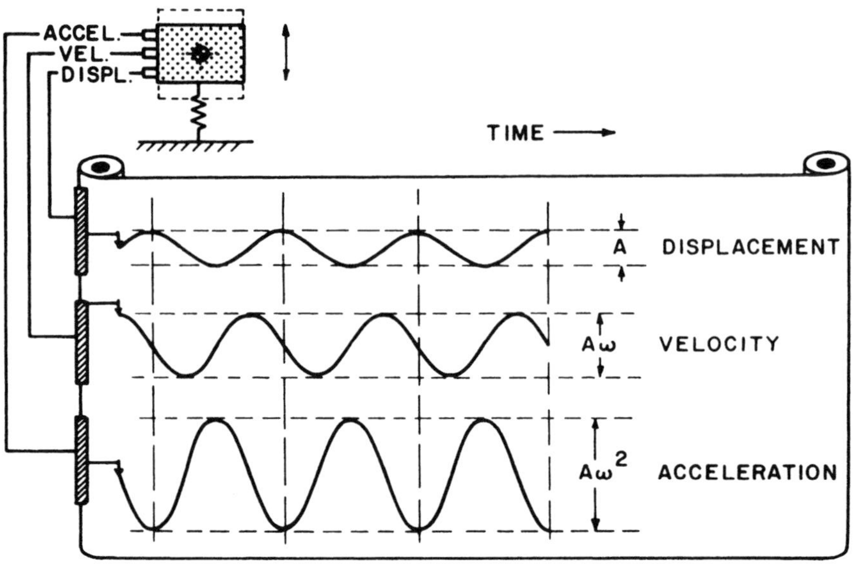

Figure 25-3. Relationship among displacement, velocity, and acceleration in a vibration system

Displacement is the distance the mass travels in a simple vibrating system. In a complex system such as the bearing cap of a boiler feed pump, displacement would be the actual distance the part moves in a selected direction such as vertical or horizontal. Displacement is commonly measured in fractions of an inch, such as the mil or .001 inch. Displacement measurements have many uses but are the least satisfactory for measuring the severity of vibration.

The proximity pickup is used to measure displacement. This pickup is a noncontact device which establishes an electric field in the vicinity of the moving part that generates eddy currents inversely proportional to the distance between the pickup and the moving part. The eddy currents modulate the amplitude of the carrier frequency which is converted to distance or displacement by the circuits of the analyzer. Figure 25-4 illustrates the operations principle of an eddy current pickup.

Velocity is movement per unit time of the vibrating part during its periodic or oscillatory motion. Velocity is the most useful measurement of the severity of vibration. This measurement is normally used in the frequency range of 400 cycles per minute to 60,000 cycles per minute. Below this range displacement is most effective while acceleration is commonly used for higher frequency vibration.

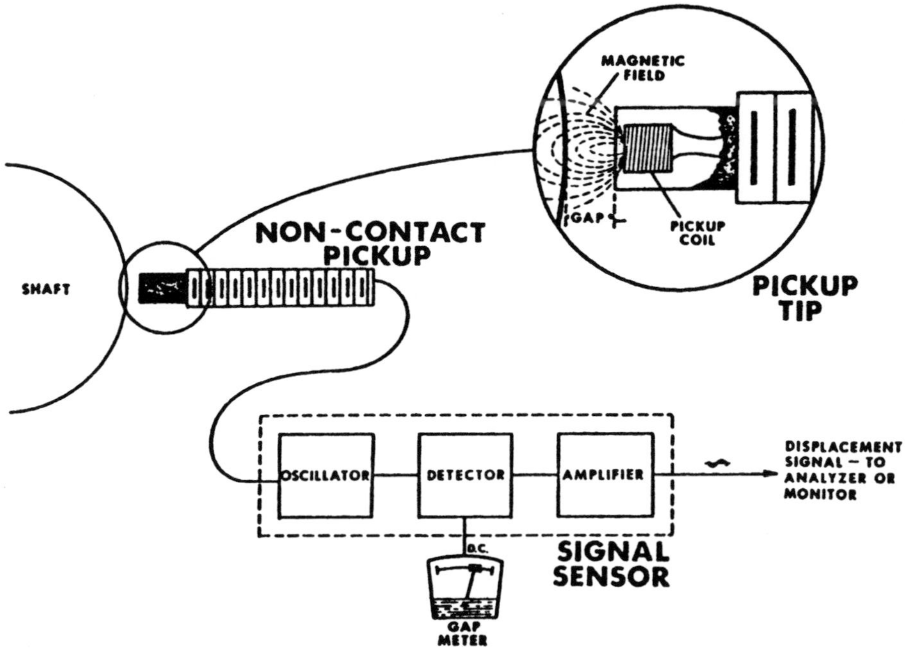

Figure 25-4. Operating principle of a reluctance pickup used to measure displacement

A typical velocity sensor consists of a spring-held coil of wire which is surrounded by a permanent magnet moving at the velocity of the vibrating part. This device produces a voltage which is proportional to inches per unit time movement of the oscillating part. Figure 25-5 illustrates the operating principle of the velocity pickup.

Acceleration is the rate of change of the velocity of movement of the vibrating part. Since the part is oscillating, the acceleration is constantly changing and reversing direction. The part slows down to approach the maximum displacement and reverses direction to accelerate in the opposite direction. Acceleration reaches its maximum when the part has traveled the maximum displacement of its periodic motion.

The acceleration sensor or accelerometer consists of a mass resting on a piezoelectric crystal. The force of the mass acting on the crystal due to change in acceleration produces an electric voltage proportional to the acceleration. This voltage is converted to an acceleration indication by the output device or meter. Figure 25-6 is a cross section illustrating the operating principle of an accelerometer.

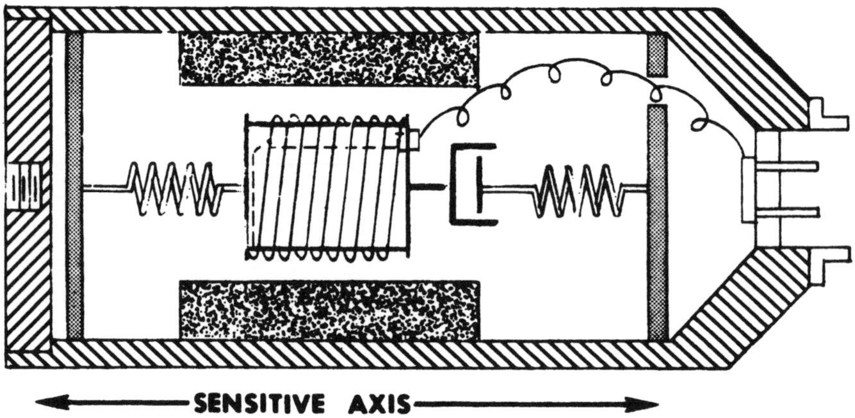

Figure 25-5. Operating principle of velocity pickup

Phase angle is the angle between the instantaneous position of a vibrating part and a reference position which is the fractional part of the vibration cycle through which the part has advanced relative to the reference position. The phase angle between two vibrating parts is the angular difference between the two when both are compared with the same reference position.

Analyzers are fitted with strobe lights to provide a phase angle measurement. By tuning the analyzer filter to a machine's shaft rotational frequency and directing the strobe light at a reference mark placed on the

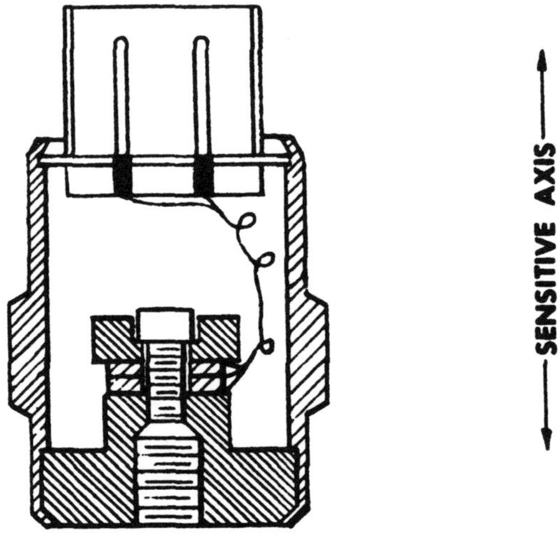

Figure 25-6. Operating principle of an accelerometer

shaft, the mark will appear stationary at a certain angular position. Should the phase of the vibration change, the mark will shift by an angular amount equal to the phase shift of the vibration.

SPECTRAL ANALYSIS

The ability to generate frequency spectrums rapidly, to represent them with a Fourier series, and to quickly calculate a Fourier transform to yield a frequency versus amplitude signal from the amplitude versus time signal has made possible the real-time analysis of vibration for measurement purposes. Furthermore, a single instrument with multiple data channels can simultaneously compute spectrum analysis for several channels. Such data when stored digitally and compared to identical information sources over time are the key to detecting machinery changes. The details concerning the effective use of a modern spectrum analyzer will depend on the particular device selected from the many available. The number of channels, the line resolution, bandwidth, dynamic range, and numerous other capability measurements determine the use of an analyzer in a shipboard environment. The manufacturer's literature is an important source for needed information. Some professional groups such as the Vibration Institute provide formal training in the use of analyzers and the analysis of the data produced. This section on spectral analysis is intended to provide an introduction to the underlying principles of the spectral analyzer.

Fourier Series

A powerful method for frequency analysis is the Fourier series which states that any periodic curve, such as the curve representing vibration amplitude versus time, regardless of its complexity may be represented by a combination of sine and cosine curves. In discrete form, the Fourier series relationship is

$$f(x) = A_o + \sum_{n=1}^{\infty}\left(A_n \cos\frac{2n\pi x}{P} + B_n \sin\frac{2n\pi x}{P}\right)$$

And in the interval (a,a+P) the coefficients are determined by:

$$A_o = \frac{1}{P}\int_a^{a+P} f(x)dx$$

$$A_n = \frac{2}{P}\int_a^{a+P} f(x)\cos\frac{2n\pi x}{P}dx$$

$$B_n = \frac{2}{P}\int_a^{a+P} f(x)\sin\frac{2n\pi x}{P}dx$$

Where P is the period, i.e, P=2L, (–L,L) is the interval, and n is the number of periodic curves needed to fit the complex curve.

In the above equation, a characteristic such as vibration amplitude is represented as a function of time in the interval of the analysis. A computer-based algorithm calculates and displays this time domain analysis.

Fourier Transform

The Fourier transform of the Fourier series f(x) permits the determination and display of a vibration characteristic, such as amplitude, versus the frequency of vibration, i.e., a frequency domain analysis:

$$x(f) = \int_{-\infty}^{\infty} x(t)e^{-j2\pi ft}dt$$

The fast Fourier transform (FFT) is a computer procedure for calculating discrete frequency components from sampled time data. In this special case of the discrete Fourier transform, the number of samples is constrained to a power of 2.

Spectrum Analyzer

Analog vibration spectrum analyzers, once very common, have diminished in popular use. In such analyzers, an analog filter is used to isolate frequencies of interest to the engineer. The signal power which passed through a certain filter was measured to determine its strength in frequency bands of interest. Tuning the filters provided the measurements over a range of frequencies.

In the FFT analyzer, which is now used as both a portable and permanent instrument on shipboard, the input signal is digitized at a high sampling rate. Numerical methods are then employed to generate a Fourier series and to apply the transform to produce the frequency spectrum.

Since the FFT spectrum analyzer measures all frequency components at the same time, it is possible for this method to be hundreds of times faster than the analog method. It is necessary to increase the time record to obtain higher time signal resolution. This is possible with the higher speeds and simultaneous recording of all frequencies.

Modern FFT spectrum analyzers are available in a variety of configurations, physical sizes, and capabilities. Figure 25-7 shows a very capable and small hand-held FFT analyzer.

Dual-channel FFT analyzers are available. A useful application of such analyzers is to measure the transfer function or the ratio of the output spectrum to the input spectrum of a mechanical system.

Shipboard Application

The shipboard application of the modern FFT analyzer becomes increasingly important as shipping economics require smaller crews, increased

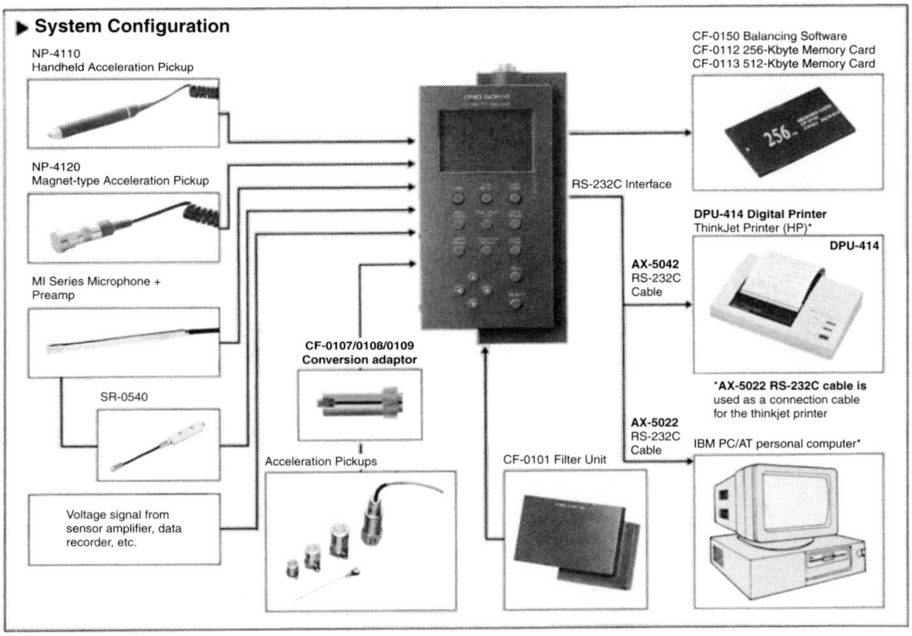

CF-0101 Filter Unit

- Vibration mode display
  Acceleration (HI: 7kHz-40kHz, LOW: 1kHz-10kHz)
  RMS, Peak, Crest factor
  Velocity (10kHz-1kHz)
  P-P
- Filter and envelope functions
  Acceleration (HIGH) : 7kHz-40kHz
  Acceleration (LOW) : 1kHz-10kHz
  Velocity and displacement : 10Hz-1kHz
  Filter characteristics : –18dB/OCT, 3rd order butterworth filter
  Envelope : Absolute value detection method
- Measurement ranges
  Acceleration (HIGH) : 0.1-100G (1-1000m/s2)
  Acceleration (LOW) : 0.1-100G (1-1000m/s2)
  Velocity : 1-100mm/s
  Displacement : 50-5000μm
- General
  Outer dimensions : Approx. 117.5 (W) × 80.6(H) × 46(D)mm
  Weight : Approx. 150g
  Continuous operation time: Approx. 6 hours with alkaline batteries

CF-0150 Balancing Software

- Balancing calculation
  Calculation method : Vector calculation
  Number of correcting plane: 1
  Number of conditions: 1
- Measuring range
  Rotational speed : 15-80000 r/min
  Input voltage ranges : 12 ranges (2mV-10V)
  Frequency range : 8 ranges (100 Hz-20kHz)
  Circumference error : Error in calculated angle depends on the rpm where the measurement is made.

15-5,000 r/min	: Within 1°
5,000-10,000	: Within 2°
10,000-20,000	: Within 3°
20,000-40,000	: Within 5°
40,000-80,000	: Within 10°

- Memory
  512K bytes data memory

Figure 25-7. Modern hand-held FFT spectrum analyzer with accessories. Courtesy Ono Sokki Technology, Inc.

availability and reliability, and improved control of maintenance costs. Vibration measurement is a valuable supplement to traditional preventive maintenance techniques.

Numerous mechanical problems are detectable in their early stages with vibration measuring equipment. Problems such as unbalance, misalignment, defective bearings, loose parts, and gear problems may be easily identified and corrected before serious damage or failure occurs.

Ships have special conditions, such as propeller, hull, wave induced, and other environmental vibrations, that could interfere with accurate machinery vibration measurement and analysis. The FFT analyzer, properly employed, permits the accurate separation of environmental vibrations from those of the machine under study.

DESIGN ENGINEER'S APPROACH

Mathematical Models

During the design phase of machinery or ship systems, the design engineer calculates the natural vibration frequencies of the machine or system. The natural vibration frequencies are compared to the anticipated periodic exciting forces to determine the cases where a resonant condition can exist. For example, a turbine designer must ensure that the turbine blades are not excited by the periodic force of the blades passing the nozzles. The ship designer must assure that structural arrangements do not have natural frequencies which will resonate with the periodic forces generated by the propeller.

When the design engineer determines that an external vibratory force may excite resonant conditions in machinery or structure, he modifies the design to (a) tune the natural frequency of the machine or structure away from the frequency of the external force; (b) change the frequency of the external vibrating force (for example, changing the blades of a propeller will modify the frequency of propeller blade forces acting on the hull); or (c) add damping characteristics to the machine or structure so that it can operate successfully in a resonant condition.

Machinery Balancing

Machinery design engineers also specify the accuracy of rotating balance which a machine must have to operate smoothly without vibration due to unbalanced weights. Manufacturers utilize dynamic balancing machinery in their shops to achieve the specified level of balance before delivery of the machinery. Figure 25-8 shows a typical balancing machine used by a machinery manufacturer.

If repairs, wear, or accidents cause a machinery rotor to become unbalanced, it may be necessary to remove it from the ship for rebalancing in a dynamic balancing machine. It is often possible to correct unbalance aboard ship by using portable equipment with the machinery operating in its own bearings. Machinery unbalance manifests itself as a one-per-revolution vibration.

Figure 25-8. A typical balancing machine

SHIPBOARD APPROACH TO VIBRATION ANALYSIS

During the life of machinery, the engineering officer can use portable instruments and analyzers for monitoring conditions and solving problems which manifest themselves as vibrations. Vibration analysis involves the measurement of the frequency of vibration which in turn permits identification of the source of excitation. Once the source is discovered, shipboard personnel can take corrective action.

Inspection of New Machinery and Preventive Maintenance

Vibration analysis of new machinery installed aboard ship provides evidence that the manufacturer has met the specifications and the normal standards for smooth running machinery. A planned program of vibration analysis can also be a key to preventive maintenance. Levels, or intensities, and frequency of vibration taken on a regular basis reveal the deterioration of antifriction bearings and facilitate replacement prior to failure. Preventive maintenance based on vibration analysis identifies machinery that has become unbalanced so that it can be repaired before the problem leads to complete failure.

Investigation of Suspected Problems

The source of malfunction of any rotating or reciprocating machine may be uncovered as a result of vibration analysis. It is often possible to determine that an external oscillatory force is causing a problem in a machine. A hot bearing may raise questions which can be answered by measuring the severity and frequency of vibration in the machine.

Evaluation of Machinery Repairs

A frequent shipboard maintenance task requires the removal of rotors from pumps, turbines, fans, etc., for rebalance in a shoreside repair facility. Vibration analysis provides the ship's engineering officers with a tool to evaluate the repair and rebalance job which the repair facility completed. Such evaluation may save them from returning to sea with an unsatisfactory piece of machinery.

Baseline Vibration Signature

A very valuable technique of vibration analysis is the vibration signature. This procedure requires that the amplitude of vibration over a range of frequency be recorded when the machinery is placed in service. During the subsequent life of the machine this "signature" is recorded at intervals and compared to past vibration amplitudes and frequencies. Changes may be interpreted to provide an understanding of the current condition of the machinery.

Economic Benefits of Vibration Analysis

Vibration analysis results in reduced machinery downtime. Downtime can cause expensive delays and disruptions in sailing schedules. Vibration analysis helps avoid costly machinery repairs caused by unexpected major breakdowns. Furthermore, early correction of faults means lower repair bills and longer machinery life. Finally, the classification societies, such as the American Bureau of Shipping, have recognized the value of vibration analysis in their rules to permit an extension of the period between required surveys of machinery. This extension reduces the cost of machinery survey during the life of a vessel. In 1971, the American Bureau of Shipping added the following statement to the "Rules for Building and Classing Steel Vessels":

> In addition to the foregoing requirements, turbine blading and rotors, cylinders, pistons, valves, condensors and such other parts of main and auxiliary machinery as may be considered necessary, are to be opened up for examination. At Special Periodical No. 1 only, for vessels having more than one main

propulsion ahead turbine with emergency steam crossover arrangements, the turbine casings need not be opened provided vibration indicators and rotor position indicators are fitted and that the operating records are considered satisfactory by the Surveyor. An operational test of the turbines may be required if considered necessary by the Surveyor.

Shipboard Applications of Vibration Analysis

With some particular types of machines there are specific tasks which are notable because they can be accomplished with vibration analysis equipment. These machines and tasks include the following:

General propulsion systems. For geared-turbine propulsion drives, fore and aft vibrational displacements and frequencies are measured in the machinery during trials to determine whether excessive motions exist in the system. Vibration displacement and frequency are also an important measure of the stiffness of the main thrust bearing, the thrust foundation, and the supporting structure of the ship's inner bottom. Propulsion turbine-gear unit spline and coupling motions may be determined and evaluated by measuring the relative and absolute vibratory motion of casings and shafts of the turbines, reduction gears, and thrust bearing units. Vibration analysis provides an effective measurement of excessive athwartship vibration of diesel engine and reciprocating steam engine propulsion systems. Continuous monitoring of steam or gas turbine propulsion systems permits analysis which provides early identification of impending trouble.

Shafts and propellers. Where outboard shafting is a part of a propulsion system, vibration measurements may be used to analyze structural rigidity and the effect of adding struts or other shaft support members. In the same manner, propellors, main shafting, and other rotating parts of the propulsion train can be checked for balance, indications of misalignment, and excessive clearance in bearings, all of which may contribute to unsatisfactory operations.

Air-handling equipment. Fans and blowers driven by motors or turbines are easily balanced in place after reblading or other repair using vibration analysis equipments. Fan or blower aerodynamic vibrations, misalignment, and wear may be identified with vibration analysis. Pumps and compressors are similar high speed machines for which vibration analysis helps to prolong service life and avoid operating problems.

Motors, generators, and similar rotating machinery. All electric motors and generators aboard ship can be analyzed and compared against established standards for permissible vibrations to obtain an evaluation of the general condition. Unbalance, electrically caused vibrations, bearing wear, and machine misalignment are readily detected. For large critical electrical machines, vibration signature data should be accumulated over time for preventive maintenance programs.

Marine gas and diesel engines. On ships with industrial or aircraft-type gas turbines, vibration measurement determines the presence of excessive vibration levels. Causes of vibration in gas turbines may be determined. For marine diesel engines, dynamic balancing of engine assemblies and components such as flywheels and turbochargers is feasible.

Gears and gear drives. Vibration analysis will reveal the condition of gear tooth contact, tooth geometry, tooth spacing, gear concentricity, and rotor balance. Bent shaft and gear box misalignment are also revealed through vibration analysis. This can be accomplished without the expensive disassembly and physical measurements which are normally required to detect these conditions in gears.

MACHINERY VIBRATION ACCEPTABLE LIMITS

An essential part of any vibration analysis program is the establishment of acceptable and unacceptable limits of vibration for the machines in the program. Such a program is only useful when there is confidence that certain vibration levels accurately indicate when a machine is in good mechanical condition, or conversely that the levels are too high and the machine must be shut down to effect repairs. Establishing such limits cannot be done casually and must draw on a number of sources of information. These include manufacturers' recommendations, vibration severity guideline charts, vibration measurements of machines in the program, and increases in vibration levels of machines in the program.

Manufacturers' recommendations on vibration limits are usually established on the basis of experience. For example, certain limits on design tolerances, balance levels, and general vibration, which insured long machine life, have evolved into recommended limits for specific classes of machines.

Vibration severity guideline charts are indirectly related to the manufacturers' recommendations. Basically, they represent the experience of those who operate, maintain, inspect, and approve machinery. These data have been collected, evaluated, and reduced to guideline format for ease of use. Some of the guideline charts cover specific types of machines. Others are generalized guidelines. Figure 25-9 is an example of a general vibration severity chart prepared by IRD Mechanalysis, Inc.

If vibration measurements of a number of similar machines in a program can be obtained, these, in themselves, form a good representative

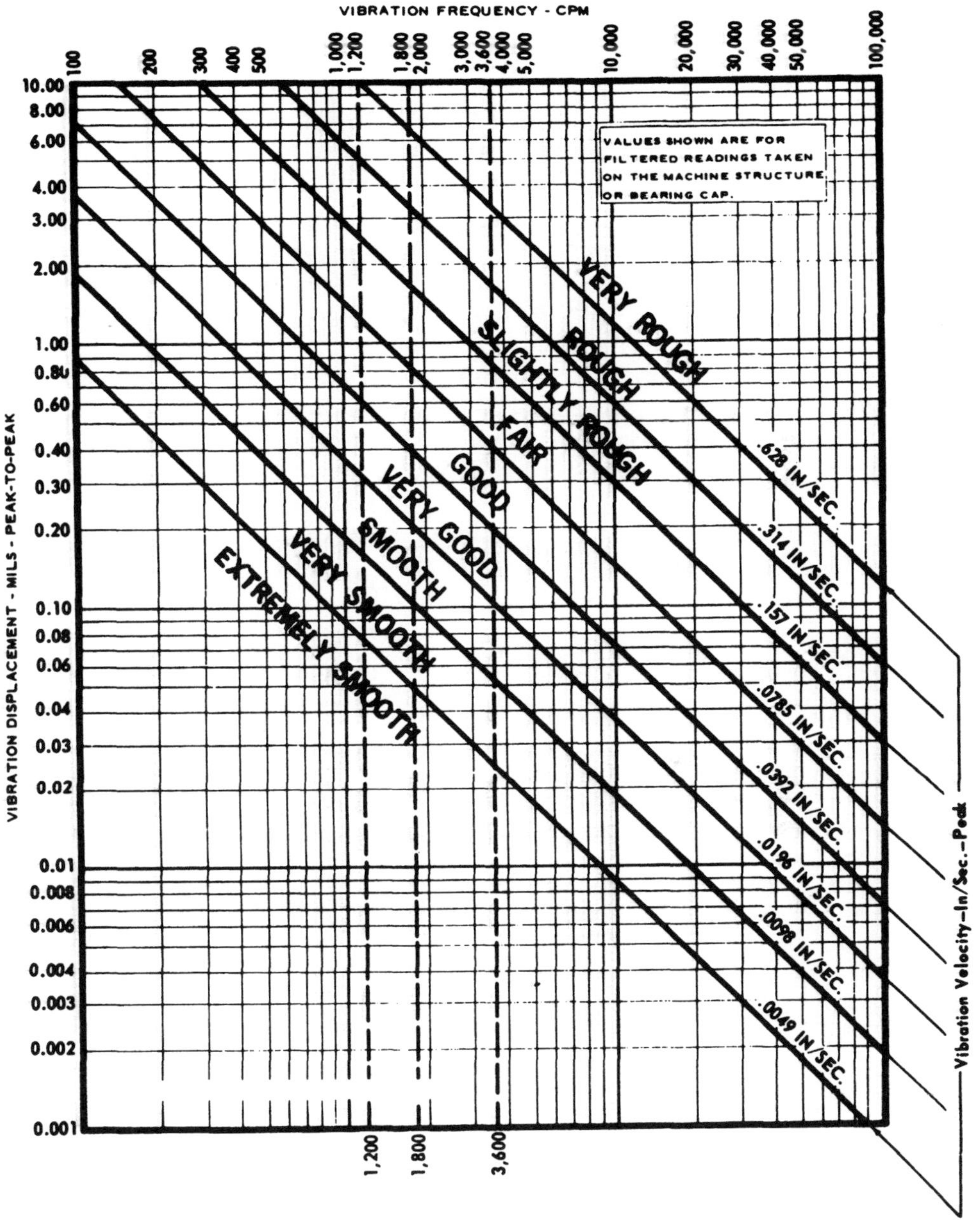

Figure 25-9. General machinery vibration severity chart

vibration sample which can be used to establish levels for machines in good operating condition. Of course, some analysis and judgment are required to insure that the sample includes only machines in good mechanical condition. Those which have defects are usually readily detected because their vibration levels are much higher than the average.

High vibration levels in a machine, which are later found to have been caused by a specific defect, provide useful vibration limit data. Such levels can be used as a vibration limit guideline to warn of a similar defect when seen in other machines.

Still another method which assists in establishing vibration limits makes use of the fact that increases in vibration levels of machines in a program indicate machinery defects and hence a tentative upper limit. When a machine's vibration level remains unchanged over an extended operating period (i.e., on the order of six months), it is reasonable to assume that the machine is in good mechanical condition and its vibration levels indicate acceptable limits. On the other hand, vibration levels which are changing generally indicate a deterioration of machine conditions.

While vibration limits which must be met by some machinery are used as a simple "go" or "no-go" tool, more often the limits are set up as a series of graded levels. One of these graded guidelines, "Vibration Velocity Standards for Shipboard Machinery," is tabulated below:

Peak Velocity inch / second	*Severity Rating*	*Severity Classification*
Above .6	Very rough	High vibration level, considered potentially unsafe. Detailed vibration analysis required to identify trouble. Excessive vibration may cause oil-film breakdown. Consider shutdown.
.4 to .6	Rough	Considered hazardous. Detailed vibration analysis needed to identify trouble. Rapid wear expected. More frequent periodic vibration checks needed to detect further increases. Schedule for repair.
.2 to .4	Slightly rough	Faults likely. Detailed analysis needed. Continue periodic checks. Schedule repair, if needed.
.1 to .2	Fair	Minor faults. Continue periodic checks to detect increase.
0 to .1	Good	Typical of well-balanced, well-aligned equipment.

Acceptable vibration levels for speed reduction gears have been published by the American Gear Manufacturers Association. Figure 25-10 shows these levels in mils peak to peak amplitude versus frequency in hertz or cycles per second.

VIBRATION MEASURING EQUIPMENT

Some of the equipment used for measuring vibrations include the pickup devices, the operation principles which are illustrated in the introduction to this chapter, and electronic devices which display and analyze the pickup signals. These electronic "black boxes" range from simple hand-held vibration meters, which display the displacement and velocity of vibration, to complicated systems which provide an analysis in real time of the complete spectrum of vibration frequencies and the amplitudes in each of the frequency bands. Devices of primary interest to the shipboard marine engineer include the basic vibration meter, the vibration analyzer, and the permanently installed vibration monitors.

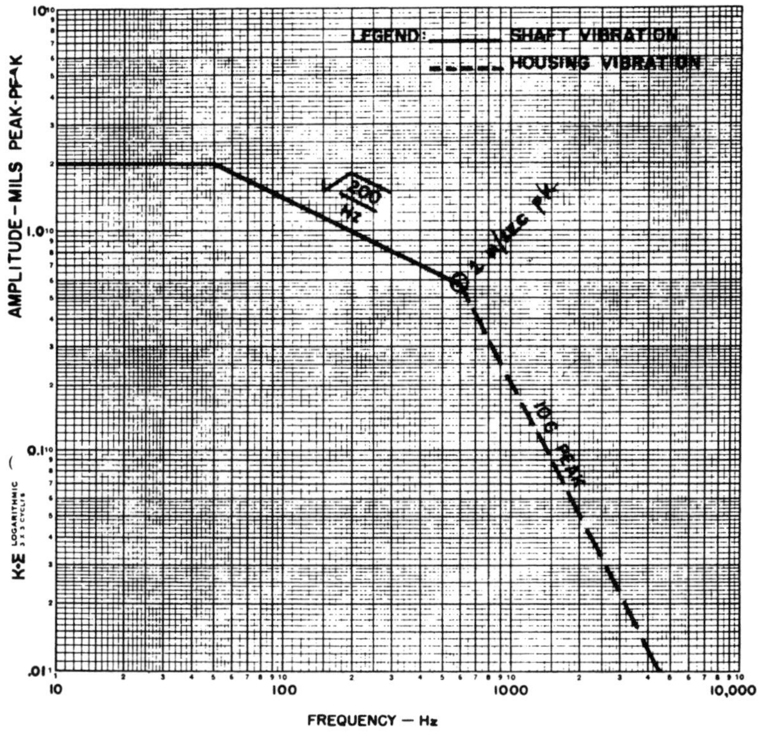

Figure 25-10. Acceptable vibration levels for reduction gears. Extracted from AGMA Standard Specification for Measurement of Lateral Vibration on High Speed Helical & Herringbone Gear Units (AGMA 426-01), with the permission of the publisher, The American Gear Manufacturers Association, 1330 Massachusetts Avenue, N.W., Washington, D.C. 20005.

Vibration Meters

The portable vibration meter and self-generating velocity pickup is a basic vibration measuring device. Such devices typically cover a frequency range of 600 to 600,000 CPM (cycles per minute). Displacement in the range of 0 to 100 mils peak to peak and velocity of vibration in the range of 0 to 100 in/sec (inches per second) peak may be displayed by selecting from the overlapping ranges of the instrument. Little training is required to use such an instrument in a vibration monitoring program. Model 308 Vibration Level Meter shown in Figure 25-11 is typical of portable vibration meters.

Vibration Analyzer

An analyzer permits the trained shipboard engineering officer to measure frequency of vibration in addition to the velocity and displacement. It is the frequency and phase which normally reveal the source of a vibration problem. A tunable filter allows the user to measure the frequency. A typical range of frequency measurement is 50 to 500,000 CPM. Analyzers are normally fitted with stroboscopic lights which permit visual identification of the vibrating part for which a displacement, velocity, or frequency measurement is being made. A plotter may be connected to the analyzer to

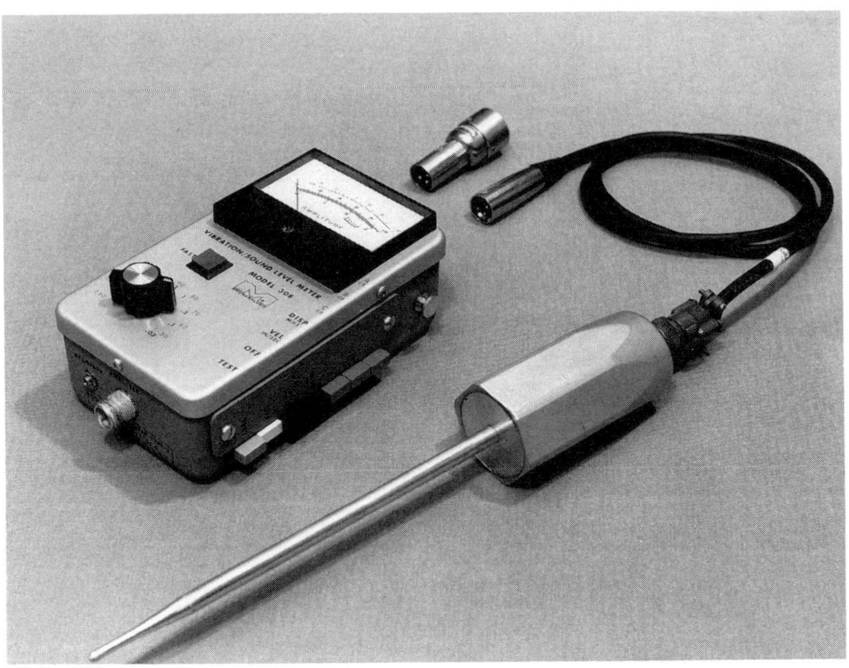

Figure 25-11. Typical portable vibration meter

record vibration amplitude versus frequency. This plot is called a "vibration signature" and it provides the information used for comparison of changes in machinery conditions over a period of time. The Model 350 Vibration Analyzer with a plotter attached, as shown in Figure 25-12, is typical of this type of instrument. This equipment can be used with the stroboscopic light for in-place balancing of rotating machinery.

Monitoring Equipment

Main propulsion units may have permanently installed vibration monitoring equipment as specified by the propulsion plant designer. Acceleration transducers and noncontact transducers, mounted on the turbine bearing housings, sense vibration and axial turbine shaft positions. The monitor receives the signals from the transducers and compares the vibration levels and rotor position to quantities specified by the machinery manufacturer. If the vibration level or axial position exceeds predetermined quantities, the monitor sounds a warning alarm providing the engineer the opportunity to correct the problem before machinery is damaged. The monitor also provides for display of the vibration level and the axial position of the rotor. In some installations, the monitor provides a signal which shuts down the machinery if vibration or axial position varies excessively from acceptable

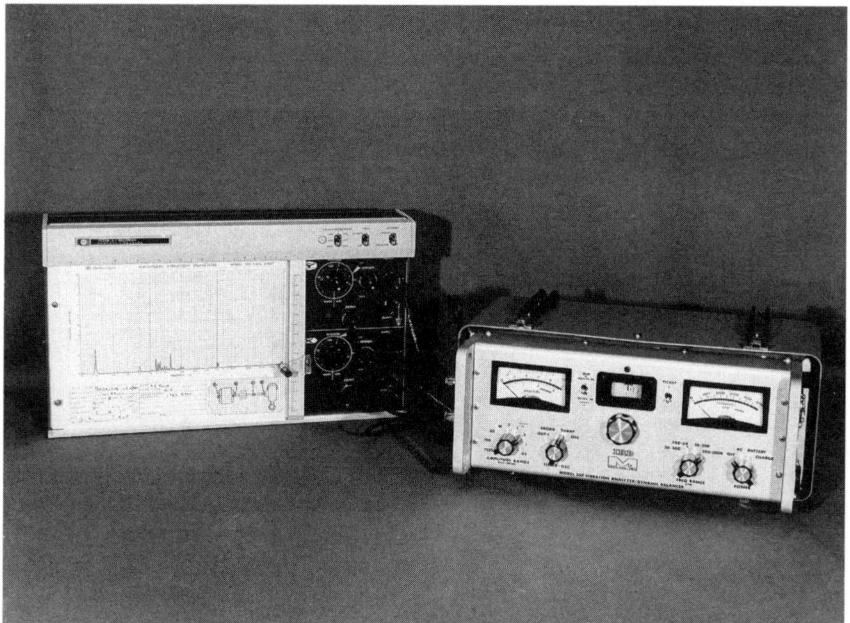

Figure 25-12. Typical vibration analyzer with plotter accessory

limits. Figure 25-13 is a schematic of a typical four-pickup vibration monitor arrangement and Figure 25-14 shows a typical permanently installed vibration monitor.

PROGRAMS FOR PREVENTIVE MAINTENANCE

Simple Program Using Portable Meter

The portable meter which indicates overall vibration level in mils displacement or in in/sec velocity may be employed to establish a simple but useful program of vibration monitoring. Following installation or overhaul of a machine, the overall vibration levels are recorded for selected points on the bearing caps. These points, usually in the vertical, horizontal, and axial directions, are recorded so that identical location readings may be taken on a regular basis, e.g., biweekly. The vibration levels are compared to general severity data to provide an indication of the condition of the machinery. A tabulation of these readings for a period of time provides trend information which is most valuable. A sudden or gradual change in identically recorded vibration levels indicates a change in the internal condition of a machine which should be investigated to correct the cause before damage occurs.

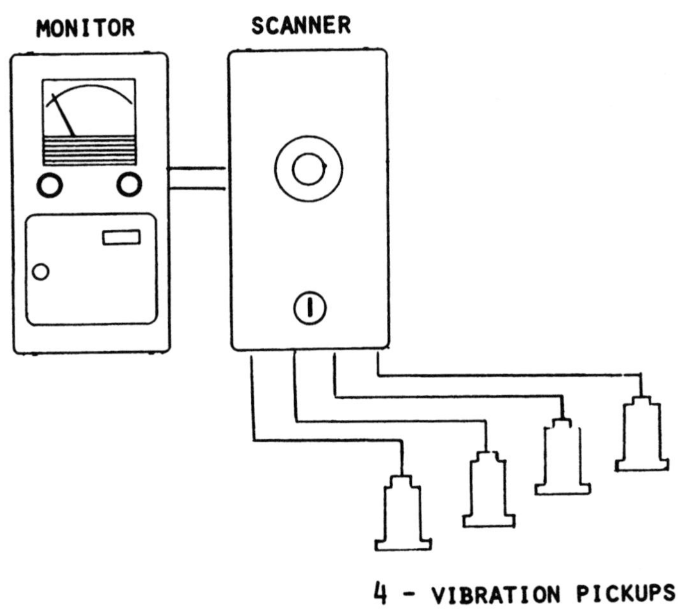

Figure 25-13. Permanently installed vibration monitor arrangement used on main propulsion machinery. IRD vibration monitor with automatic scanner provides intermittent sampling of four vibration pickups.

Figure 25-15 shows a typical set of vibration readings taken in the athwartship direction for a one-year period. Eventual high vibration readings led to an investigation and correction of a problem. Figure 25-16 illustrates the use of the hand-held vibration meter while recording an axial overall vibration level of a refrigeration compressor.

A Maintenance Program Using a Vibration Analyzer

When measurements are limited to overall vibration levels, the ship's engineering officer does not have all the information needed to identify the source of the problem. The ability to indicate vibration levels and the frequency at which they occur is necessary for complete understanding of the machinery condition. A vibration analyzer provides this information. Such information may be recorded in tabular form by measuring the amplitude and frequency of the machine's major vibration peaks. If an X-Y plotter is

Figure 25-14. Typical vibration monitor display and control cabinet for permanently installed device

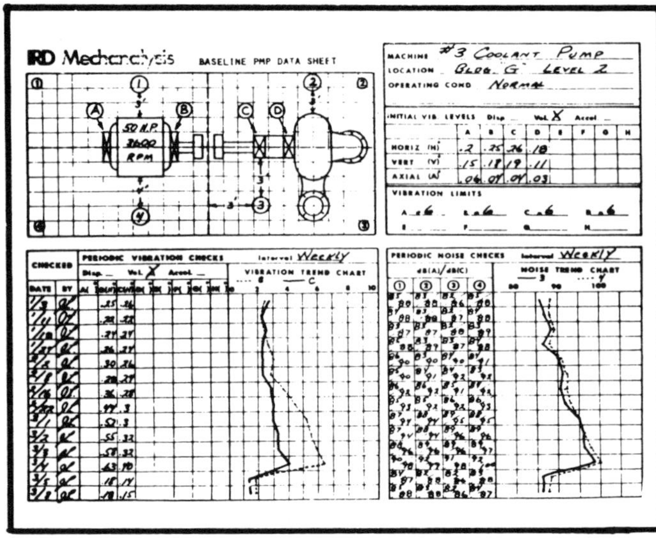

Figure 25-15. Periodic vibration readings using a portable meter provide valuable trend information

Figure 25-16. A hand-held vibration meter employed to determine overall axial vibration level of a refrigeration compressor

available as an accessory to the vibration analyzer, a continuous plot of vibration level versus frequency can be prepared as a vibration signature. Figure 25-17 shows a typical plotted vibration velocity peak measurement versus the frequency.

Establishing a Shipboard Baseline Program

A preventive maintenance baseline program may be established aboard ship utilizing vibration analyzer trend information to determine the condition of machinery. Eight steps are required to set up the program.

1. Select the shipboard machinery to be included in the program. This selection must be made on the engineering officer's judgment on the importance of the machine to the operation of the vessel and the avoidance of costly repairs. Certainly all the turbines, pumps, fans, and compressors on the main and auxiliary machinery lists would be candidates for a baseline program.
2. Establish and record on appropriate forms the individual measurement points on each machine. Typically, the points will include horizontal, vertical, and axial measurement locations on each of the machine's bearings.

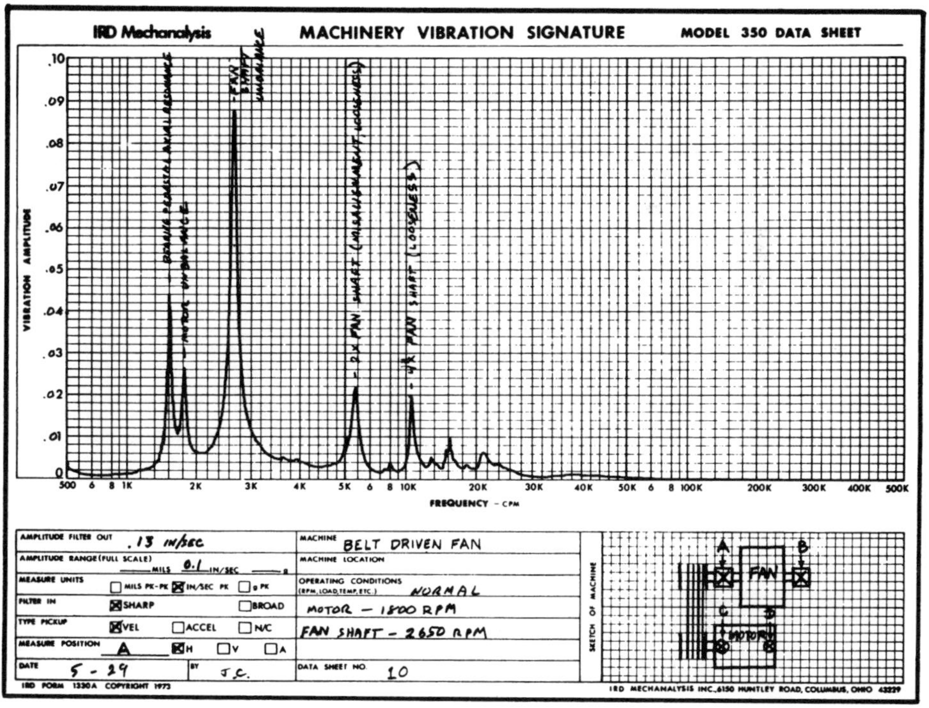

Figure 25-17. A vibration signature plotted by an X-Y plotter accessory to the vibration analyzer

3. Select the type of measurements to be taken at each of the points. For typical machines, the measurements should include amplitude (velocity in/sec) versus frequency in a tabulation form or a plot if the plotter accessory is available. It is also desirable to record the overall vibration level by switching off the analyzer's tunable filter. Figure 25-18 is a comparison of vibration signatures for the same pump bearing using three different measurements—displacement (mils), velocity (mils/sec), and acceleration (in/sec/sec)—at identical operating conditions.
4. Select the interval at which measurements will be recorded. Aboard ship, a biweekly schedule of overall vibration measurement will be adequate. The vibration signature should be recorded at quarterly intervals or when an overall noise measurement suggests a developing problem.
5. Establish and record the standard operating conditions at which vibration levels will be measured. It is essential that the machine be in the same generating mode for each measurement. Depending on the machine, speed, pressure, temperature, flow, and load should be considered.
6. Make sure the machine's foundation, piping, and associated piping are not changed during the intervals between measurements. Changes in these structural elements may introduce variations in the vibration characteristics. If repairs or modifications result in structural changes, it is necessary to establish new baseline data.
7. Compare the original baseline data for each machine with the manufacturer's vibration data or the industry standards for vibration severity to ensure that the machine is in good condition at the start of the monitoring program.
8. Maintain an effective record keeping system. This is essential to the success of a baseline monitoring program. With forms and plotter sheets available from instrument manufacturers, individual files can be kept for each machine.

INTERPRETING RESULTS OF VIBRATION MEASUREMENTS

One of the reasons that vibration analysis has proved to be such a powerful tool is the fact that small changes in a machine's mechanical condition produce significant changes in the machine's vibration level. Overall vibration level increases on the order of a factor of ten are not uncommon as a machine changes from smooth to very rough running. Even the smoothest running machine has characteristic vibrations which can be measured. These are caused by minute variations in the dimensions that are inherent in the machine's manufacture. In fact, the machine's design, and particularly its tolerances (i.e., the allowable variations in its dimensions), determine the absolute at which the machine will characteristically vibrate. When the amplitude of these vibrations increases, it is an indication that the machine has developed a defect.

Overall vibration is particularly valuable in vibration analysis for a number of reasons. First, its measurement produces a single number that

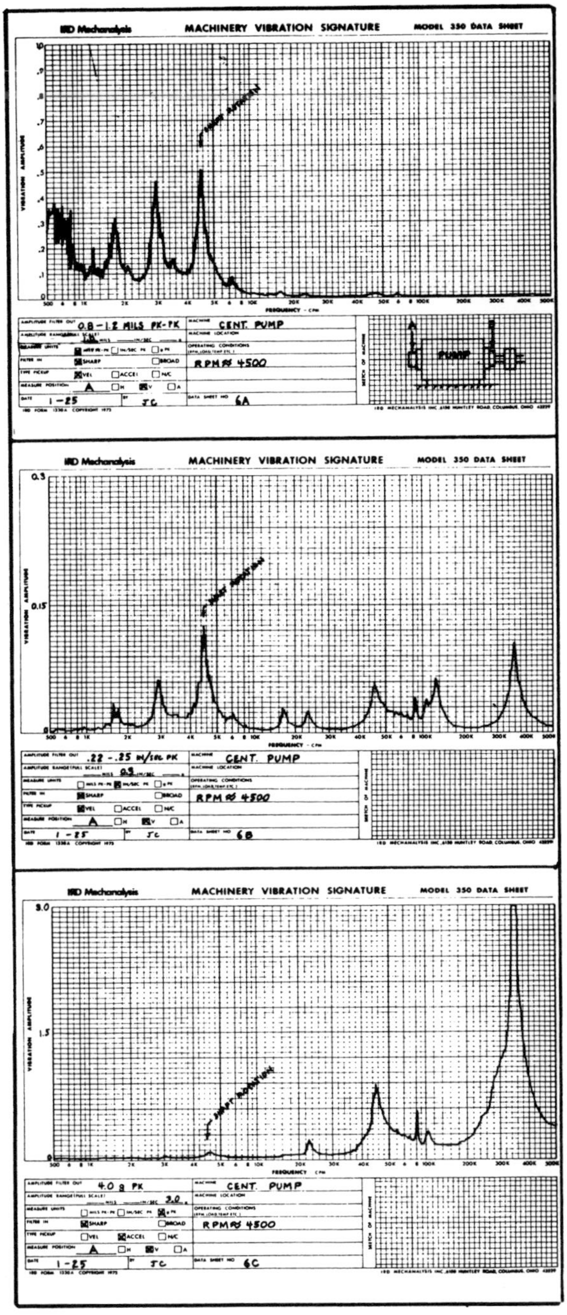

Figure 25-18. Comparison of amplitude, velocity, and acceleration vibration measurement for the same pump

includes all the different defects which may be present in a machine (e.g., unbalance, misalignment, bad bearings, etc.), thereby providing a total assessment of the general mechanical condition of the machine. Second, for record keeping purposes overall amplitude (vibration) provides a very compact data record. Third, because it requires only a few seconds to make a measurement, it is a very rapid method of analyzing a machine. Fourth, when the overall amplitude is compared with the sum of the individual frequency peaks which have been analyzed, it provides a check as to whether or not all the individual frequency peaks have been measured. It is also a remarkable fact that the vast majority of rotating machines generally have similar overall vibration levels which indicate their mechanical condition. For example, an overall level of .1 in/sec-pk (peak) generally indicates a smooth operating machine with no defects, while a level of .6 in/sec-pk indicates a machine with severe defects. Of course, there are many exceptions, but these levels do provide a valuable guideline.

No less important than a specific measurement of the overall vibration level is the pattern—or trend—indicated by a series of such measurements over a period of time. It is almost invariably true that if a vibration level of a machine remains essentially constant over a period of time (e.g., six months or one year), the machine is in good mechanical condition. In contrast, a machine with developing defects will show a rising trend in overall vibration. The use of the trend is valuable because it lessens the dependence on the absolute vibration level as an indication of mechanical condition.

Frequency Analysis

Frequency analysis is the method utilized to pinpoint the source of a vibration and thereby detect the defective component within a machine. This is accomplished by relating the measured discrete frequencies to the characteristic frequencies of the machine.

It is possible to calculate most of the characteristic frequencies. For example, a 5-vane impeller centrifugal pump which is rotating at 1,800 RPM (revolutions per minute) will have a vibration at 1,800 CPM that is indicative of imbalance, misalignment, or a bent shaft. A vibration at twice rotational speed, 3,600 CPM, is characteristic of looseness in the pump, while vibration at five times rotational speed, 9,000 CPM, is caused by the vanes of the pump impeller passing the pump cutwater and creating a hydraulic pulsation. In similar manner gear tooth vibrations can be calculated by counting the number of teeth and multiplying by the rotational speed.

Some frequencies which appear in a vibration signature cannot be readily calculated and are identifiable on the basis of experience. Typically, ball bearings have vibrations in the 10,000 to 60,000 CPM range and also at very high ultrasonic frequencies. When in its incipient stage, vibration caused by cavitation can appear at high frequencies (i.e., above 60,000 CPM) and will tend to produce a broad peak of frequencies, rather than a single peak. As vibration caused by cavitation becomes more severe, it tends to shift down to lower frequencies. Natural frequencies of the machine and its supporting structure can also appear in a machine's vibration signature. These can sometimes be identified by striking the structure while the machine is shut down. For this to be successful the structure must be relatively undamped. If the natural frequency is pronounced, it will be momentarily registered on the analyzer frequency meter. Machine critical speeds can be identified in many cases by a similar technique. As the machine coasts down, the amplitude of vibration will momentarily increase (when it passes through the critical speed) and the meter will register the frequency of the critical speed. An excellent way of separating and identifying natural frequencies, criticals, and various types of shaft whirling vibrations is by changing machine speed. Those vibrations such as unbalance, etc., which are related to machine speed will change in frequency when the machine speed changes. However, natural frequencies, criticals, and shaft whirl will tend to remain at the same frequencies, although their amplitudes may change as the forces which are exciting them change. Of course, there are a few exceptions to this rule. In certain types of aerodynamic and hydrodynamic pulsations which involve resonance, there may be a small change in frequency.

Most frequencies observed tend to have individual, discrete peaks. These are the vibrations caused by unbalance, looseness, misalignment, etc. Some vibrations, however, cover a broad range of frequencies. Cavitation-caused vibration, which was previously mentioned, is one of these. Steam, liquid, and gas flow through pipes, valves, and ducts also produce a broad band of frequencies. In fact, most of the broad band vibrations measured can be related to some type of flow mechanism.

On occasion, these broad band vibrations will excite the natural frequencies of the piping and other structures with which they come in contact. When this occurs these vibrations tend to take on a discrete frequency appearance, rather than a broad band. If the amplitudes of these peaks are observed, you will note that they tend to vary in a random manner. This provides a clue to the source of the vibration since vibrations caused by unbalance, misalignment, impeller vane passing, etc., tend to have fairly steady amplitudes.

Impulse-type vibrations occur in reciprocating machines, or when one component in a machine strikes another. These vibrations are a harmonic series of discrete frequencies which occur as multiples of the impulse repetition frequency. Generally, the fundamental impulse repetition frequency has the highest amplitude, with the higher harmonics decreasing in amplitude. In some cases, however, one of the higher harmonics may coincide with a natural frequency of the structure and consequently have a larger amplitude than adjacent peaks. As many as seven or more harmonics can sometimes be measured.

Beat frequency vibration occurs when two vibrations are present in a machine at almost the same frequency. In most cases the analyzer filter bandwidth is not sufficiently narrow to separate these two vibrations into their individual peaks. Instead, the vibration signal appears to be a single vibration whose amplitude is changing in a periodic or random manner. Two machines running at almost the same speed on a common foundation are often the cause of such a beat frequency. A beat frequency in itself is not usually an indication of any mechanical defects, but it may be mistaken for an amplitude modulation vibration which can be an indicator of mechanical problems.

Amplitude modulation vibration occurs when the amplitude of a relatively high frequency vibration is caused to vary in a periodic manner. For example, when a 24-tooth gear turning at 1,800 RPM has some eccentricity in the gear wheel, the amplitude of the tooth contact vibration (43,200 CPM) will vary at the rotational frequency (1,800 CPM) of the machine.

Phase Angle Analysis

Phase can be very useful in diagnosing specific machinery defects. For example, the vibration measurement of a motor-driven overhung fan may reveal a relatively high vibration amplitude at rotational frequency in the axial direction. This indicates that a mechanical defect is present which could be any one of the following: unbalance of the fan rotor, coupling misalignment, bearing misalignment, or a bent shaft. It is phase analysis which aids in determining which of these is the actual problem.

To diagnose the problem, phase measurements are made in the axial direction at four or more equally spaced positions around the shaft on each bearing housing, as shown in Figure 25-19. If large phase differences (i.e., on the order of 180°) are seen for different positions on the bearing housing, it indicates that a severely bent shaft (small shaft bow may not cause a phase difference), a cocked ball, or roller bearing is the cause. Should the phase, however, be approximately the same (i.e., on the order of 0°) at each position around the shaft at all bearing housings, a further phase comparison must be made among all the bearing housings to pinpoint the problem.

If the phase at all the bearing housings is the same (i.e., approximately 0°), it is an indication of unbalance or possibly an axial resonance of the foundation. The latter can be ascertained by tests for resonance. If, however, the two motor bearings have a large phase difference between them, it indicates a bent shaft or severely misaligned bearings in the motor. Furthermore, if the two motor bearings are in phase with each other and the two fan bearings are in phase with each other, but there is a large phase difference between the motor and the fan, then the coupling is most probably misaligned.

Note: In recording phase measurements, care must be exercised to subtract 180° from the measured phase at each position where the transducer direction for measurement is opposite from the first position measured.

CHARACTERISTICS OF SPECIFIC MACHINERY DEFECTS

Some of the more common machinery problems and their associated characteristics of amplitude, frequency, and phase are described in the following paragraphs:

Mechanical Unbalance

This is the most common cause of vibration. Mechanical unbalance can be recognized by the fact that the vibration occurs at rotational frequency and the amplitude in the radial direction is more than twice the axial amplitude. It can often be separated from electrical defects by cutting the power

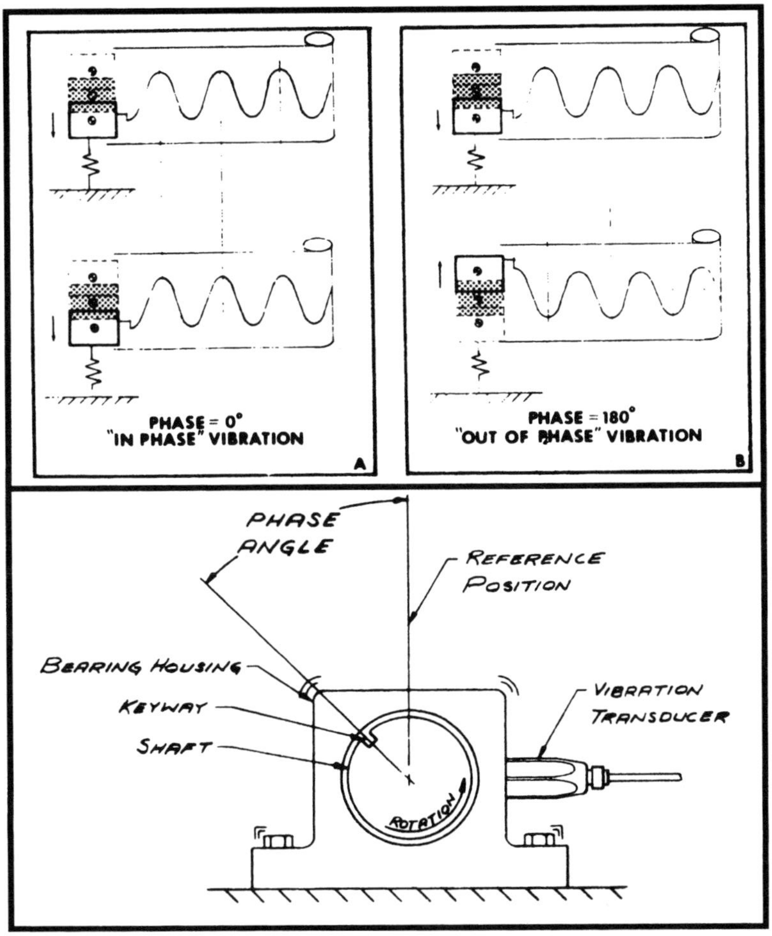

Figure 25-19. Measurement locations for vibration phase on bearing housing using strobe light and vibration analyzer

and noting whether the amplitude drops off gradually or abruptly. A gradual drop indicates mechanical unbalance; an abrupt drop indicates electrical unbalance. The strobe phase measurement shows a single steady reference mark.

Misalignment (Shafts, Couplings, Bearings)

This is the second largest cause of vibration even when self-aligning bearings and flexible couplings are used. Vibration occurs at shaft rotational frequency and can also occur at two and three times the rotational frequency. Amplitude in the axial direction is 50 percent or more of the radial direction. Phase shows a single reference mark. Phase readings in the axial direction taken around the shaft on the bearing housing are approximately constant. Misaligned bearings, as well as shafts misaligned at couplings, can cause these vibration characteristics.

Bent Shaft

This problem may show the same vibration characteristics as misalignment. The phase, however, when measured axially around the shaft will vary through approximately 180°, particularly when the bend is close to the bearing. Misaligned ball and roller bearings also show these characteristics. A further indication of a bent shaft is an approximate 180° phase difference between axial vibration at opposite ends of the machine.

Defective Ball and Roller Bearings

This vibration is characterized by a number of high frequency vibration peaks, plus frequency regions of fluctuating vibration levels. The lower frequencies are usually not multiples of shaft rotation frequency but can be roughly calculated from bearing dimensions and shaft speed. The high frequencies emitted by the bearings cannot be readily calculated. Phase is unsteady.

Defective Gear Teeth

Vibration occurs at shaft rotational frequency times the number of gear teeth. Side band frequencies may also occur at frequencies equal to the tooth contact frequency plus and minus the gear rotational speed. Phase shows multiple reference marks and is unsteady.

Mechanical Looseness (Sleeve Bearings, Bearing Pedestals, Foundations, Etc.)

Vibration occurs at twice shaft rotational frequency. Phase shows two reference marks and is usually somewhat unsteady. Change in balance and alignment generally affects this vibration.

Bad Drive Belts

Vibration occurs at three and four times, as well as twice, the rotational frequency of the belts. Frequencies and amplitudes are somewhat unsteady because of belt slippage. Phase shows one or two reference marks which are also unsteady.

Aerodynamic/Hydraulic Pulsation

Vibration occurs at shaft rotational frequency times the number of blades or impeller vanes. Amplitude is generally low, except when amplified by resonance.

Rubbing

Vibration can occur at many frequencies, each with fluctuating amplitudes (i.e., random vibration) simultaneously, when rub is continuous. It may also occur at one and two times shaft rotational frequency when rub is intermittent. Phase is unsteady.

Resonance

Vibration occurs at a single "natural" frequency of the part which is vibrating. This is a frequency at which the part "likes" to vibrate. For example, a resonance occurs when the shaft RPM reaches the shaft critical frequency. To check for resonance, let the machine coast down from operating speed. Watch the analyzer amplitude and frequency meter. If the amplitude drops off at first, then rises, and then drops off again, the frequency at which the rise occurs is a natural frequency. A resonance will occur if the machine is operated at a speed equal to the natural frequency. Phase will shift by a significant amount, as much as 180°, as machine speed moves through a resonant frequency.

Summary of Measurement Interpretation

Figure 25-20 is a vibration identification chart which shows the amplitude, frequency, and phase characteristics of the most common machinery defects.

MARINE VIBRATION CASE HISTORIES

Some case histories of shipboard vibration analysis will suggest the range of application and the value of vibration monitoring.

Severe vibration of a fan was noted aboard a ship. The engineering officers believed that severe imbalance was the result of bad bearings and recommended removal of the fan for bearing replacement. However, vibration analysis indicated that it was not bad bearings but vibration caused by the fan running in a partially stalled condition due to a dirty intake filter. Cleaning of the intake filter eliminated the vibration problem.

In another case, apparent severe vibrations of a fan were assumed to be caused by unbalance in the fan rotating parts. The shipboard engineers recommended removal of the fan for balancing. Vibration analysis showed

VIBRATION IDENTIFICATION

CAUSE	AMPLITUDE	FREQUENCY	PHASE	REMARKS
Unbalance	Proportional to unbalance. Largest in radial direction.	1 x RPM	Single reference mark.	Most common cause of vibration.
Misalignment couplings or bearings and bent shaft	Large in axial direction 50% or more of radial vibration	1 x RPM usual 2 & 3 x RPM sometimes	Single double or triple	Best found by appearance of large axial vibration. Use dial indicators or other method for positive diagnosis. If sleeve bearing machine and no coupling misalignment balance the rotor.
Bad bearings anti-friction type	Unsteady - use velocity measurement if possible	Very high several times RPM	Erratic	Bearing responsible most likely the one nearest point of largest high-frequency vibration.
Eccentric journals	Usually not large	1 x RPM	Single mark	If on gears largest vibration in line with gear centers. If on motor or generator vibration disappears when power is turned off. If on pump or blower attempt to balance.
Bad gears or gear noise	Low - use velocity measure if possible	Very high gear teeth times RPM	Erratic	
Mechanical looseness		2 x RPM	Two reference marks. Slightly erratic.	Usually accompanied by unbalance and/or misalignment.
Bad drive belts	Erratic or pulsing	1, 2, 3 & 4 x RPM of belts	One or two depending on frequency. Usually unsteady.	Strob light best tool to freeze faulty belt.
Electrical	Disappears when power is turned off.	1 x RPM or 1 or 2 x synchronous frequency.	Single or rotating double mark.	If vibration amplitude drops off instantly when power is turned off cause is electrical.
Aerodynamic hydraulic forces		1 x RPM or number of blades on fan or impeller x RPM		Rare as a cause of trouble except in cases of resonance.
Reciprocating forces		1, 2 & higher orders x RPM		Inherent in reciprocating machines can only be reduced by design changes or isolation.

IRD # 303

Figure 25-20. Vibration identification chart

that it was not unbalance but a resonance in the fan mounts. The vibration was eliminated by adjustment of the mounts.

Following the installation of a new bearing in a feed pump motor, ship's engineers conducted a vibration analysis of the pump to determine that the repair was satisfactory. The analysis showed that the repair was not satisfactory since severe vibrations existed. The new bearing was removed and a metallic chip was found under the bearing, causing improper seating of the bearing. Vibration was eliminated with the correct seating of the bearing.

Vibration analysis was conducted on two diesel generators. One of the generators had significantly increased vibration levels over a period of several months. The engine was secured and dismantled for inspection. The flexible coupling between the engine and the generator was worn and a complete overhaul of the coupling was required.

A water pump had a defective upper antifriction bearing which caused high vibration levels in the frequency range of 60,000 CPM. An adjacent pump was found to have an overall vibration level higher than the pump with the defective bearing. The adjacent pump was found to have a locked-up flexible coupling which caused the high axial vibration at twice the pump fundamental frequency in the 63 CPS (cycles per second) band. Correction of vibration level was achieved by properly fitting the flexible coupling.

Figure 25-21 shows vibration signature analysis comparison of a machine with and without defective drive belts. This signature, taken at the horizontal position of the motor bearing at the belt end, reveals a dramatic change in the signature when comparing the baseline to the condition with a defective drive belt.

GLOSSARY

Alignment. A condition in which the axes of machine components are coincident, parallel, or perpendicular according to design requirements.

Amplitude. The magnitude of vibration expressed as peak-to-peak, zero-to-peak, or rms. Dynamic signal analyzers generally read rms for spectral components and peak for time components.

Averaging. In a dynamic signal analyzer, digitally averaging several measurements to improve accuracy or to reduce the level of asynchronous components.

Balancing. Procedure for adjusting the radial mass distribution of a rotor to move the mass centerline close to the geometric centerline.

Bandwidth analysis. Analysis over a frequency range from DC to specified frequency.

Baseline spectrum. Vibration spectrum taken when a machine is in correct operating condition for comparison use during monitoring.

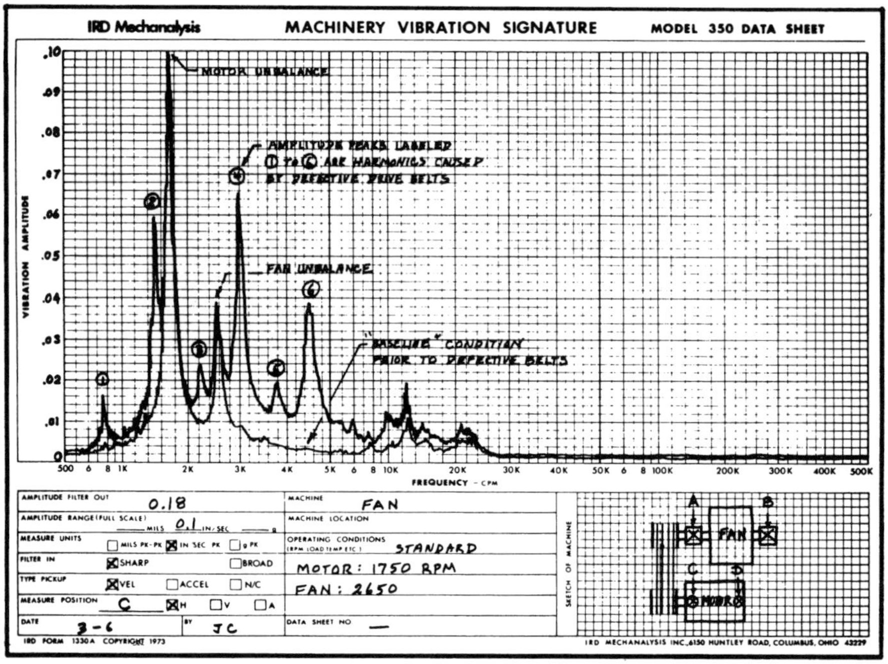

Figure 25-21. Vibration signature analysis comparison of a machine with and without defective drive belts

Blade passing frequency. Vibration frequency on a bladed machine such as a compressor or turbine that equals the number of blades times the rotational frequency.

Block size. Number of samples used by a dynamic signal analyzer to compute the fast Fourier transform or the number of samples in a DSA time display.

Critical speed. The rotor speed which corresponds to the natural frequency of the system usually resulting in high vibration levels.

Damping. The characteristic of a mechanical system that restrains the amplitude of motion with each successive cycle. For example, dampening motion is provided by lubricants in bearings.

Discrete Fourier transform. A method for calculating discrete frequency components from sampled time data. The number of points is equal to one-half the number of samples. The dynamic signal analyzer uses this method to display the frequency components or the maximum amplitude of a vibration at discrete frequencies.

Dynamic signal analyzer. Vibration analyzer that employs digital signal processing and the fast Fourier transform to determine and display vibration frequency components such as maximum amplitude.

External sampling. In a dynamic signal analyzer, external sampling is control of data sampling by a multiplied tachometer signal.

Fast Fourier transform. A computer method for calculating discrete frequency components , such as vibration amplitude, from sampled time data.

Frequency. The rate of a periodic event, expressed in cycles per second or revolutions per minute.

Gear mesh frequency. Potential vibration frequency of a gear in mesh equal to the number of teeth times the rotational frequency.

Hertz. Frequency unit in cycles per second.

MFPT. Society for Machinery Failure Prevention Technology, Haymarket, Virginia, 20169-2420. A source for information and training on vibration analysis.

Overlap processing. Concept of performing a new analysis on a segment of data in which only a portion of the signal has been updated, i.e., new data combined with old data.

Processing time. FFT computing device processing time is the total amount of time required to calculate a desired parameter such as signal amplitude.

Period. The reciprocal of frequency or the time of an event.

Sampling rate. The rate at which analog signals are sampled and then digitized. Expressed as samples per second, it is the inverse of sampling interval.

Signature. Vibration frequency spectrum which is distinctive to a tested machine at a specific point in time under specific operating conditions. The signature is used for comparison of machine vibration condition over time for maintenance purposes.

Spectrum. Amplitude of the components of a time domain signal as a function of frequency, for example, vibration amplitude versus frequency.

Time record. The sampled time data converted to frequency domain by a fast Fourier transform in a dynamic signal analyzer.

REVIEW

1. What is vibration?
2. Define the units used to measure vibration.
3. What is the relationship among vibration velocity, displacement, and acceleration?
4. Describe the operation of a reluctance pickup.
5. What are the design engineer's concerns with vibration?
6. What is a baseline vibration signature?
7. How would you employ vibration analysis to balance a forced draft fan in place?
8. A machine operating at 1,000 RPM has a once-per-revolution vibration displacement of .20 mil peak-to-peak. Is this acceptable operation?

9. What is a vibration meter?
10. What is a vibration analyzer? What is the difference between an analog and digital analyzer?
11. How would you employ a vibration meter to set up a maintenance program for a machine?
12. What are the steps to establish a shipboard baseline program?
13. How is frequency analysis used to find the source of vibration in a machine?
14. How does phase angle measurement help with the analysis of a machine?
15. Describe the typical vibration characteristic of a defective anti-friction bearing?
16. What is a Fourier series? What is a Fourier transform?
17. What is a fast Fourier transform?
18. What information can be obtained from a spectral analyzer?

REFERENCES

Broch, J. T. October 1980. "Mechanical Vibration and Shock Measurement." Denmark: Bruel & Kjaer.

Catlin, J. B. 1984. "Micro-manual Vibration Monitor Predicts Maintenance Needs." *Marine Engineers Review Marine.* Institute of Marine Engineers.

Cavaliere, L. A. April 1984. "Condition Monitoring of Shipboard Machinery Using Vibration Analysis: An Operator's Experience." New York. Symposium. Institute of Marine Engineers and SUNY Maritime College.

Church, A. H. 1963. *Mechanical Vibrations,* 2nd ed. New York: John Wiley & Sons.

Collacott, R. A. 1979. *Vibration Monitoring and Diagnosis: Techniques for Cost Effective Plant Maintenance.* London: George Godwin Press.

DeBond, Frank, William Hennessy, and Joseph McDonald. January 1998. "Measurement and Analysis of Shipboard Vibrations." *Marine Technology,* Volume 35, Number 1. Society of Naval Architects and Marine Engineers.

Den Hartog, J. P. 1956. *Mechanical Vibrations,* 4th ed. New York: McGraw-Hill.

Elsevier. 1982. *Handbook of Noise and Vibration Control,* 6th ed. Oxford, U.K.: Elsevier Science Publishers.

Harris, C. M. and C. E. Crede. 1976. *Shock and Vibration Handbook.* New York: McGraw Hill.

Hinson, A. R. 1961. *Some Notes on Vibration Problems.* London: Lloyd's Register.

Hunt, Everett C., editor. April 1984. "Proceedings: A Symposium on Shipboard Vibrations, Measurements, Analysis and Correction." New York: Institute of Marine Engineers and SUNY Maritime College.

Institute of Marine Engineers. 1987. "Vibration Analysis in the Marine Engineering Environment." *Marine Engineers Review.*

International Standard ISO 10816-3. 1998. "Measurement of Vibration Acceleration, Velocity and Displacement." Geneva. International Organization for Standardization.

Lalanne, M., P. Berthier and Johan Der Hagopian. 1983. *Mechanical Vibrations for Engineers.* New York: John Wiley and Sons.

Leach, R. M. April 1978. "Rotating Machinery Survey by Vibration Monitoring." *Marine Engineers Review.* Institute of Marine Engineers.

Lundgaad, Bertel. 1982. "Machinery Condition Analysis." Pacific Northwest Section, Society of Naval Architects and Marine Engineers.

Newland, D. E. 1989. *Machinery Vibration Analysis and Computation.* Harlow, U.K.: Longman Scientific Press.

Pusey, H. C. May 1999. "Turbomachinery Condition Monitoring and Failure Prognosis Technology." *Sound and Vibration.*

Reeves, Charles W. 1998. *The Vibration Monitoring Handbook.* Oxford, U.K.: Coxmoor Publishing Company.

Smith, J. D. 1989. *Vibration Measurement and Analysis.* London: Butterworths.

Society of Naval Architects and Marine Engineers. October 1978. "Ship Vibration Symposium." Arlington, Virginia. SNAME.

Society of Naval Architects and Marine Engineers. May 1987. "Guidelines for the Use of Vibration Monitoring for Preventive Maintenance." New York. Technical and Research Bulletin 3-42. SNAME.

Taylor, James I. 2000. "The Gear Analysis Handbook." Willowbrook, Il. Vibration Institute.

Thomas, B. January 1990. "Vibration—Identifying the Source, Solving the Problem." *Marine Engineers Review.* Institute of Marine Engineers.

Timoshenko, S., D. H. Young, and W. Weaver. 1974. *Vibration Problems in Engineering.* New York: John Wiley & Sons.

Tse, F. S., I. E. Morse, and R. T. Hinkle. 1963. *Mechanical Vibrations.* Boston: Allyn Bacon.

ACKNOWLEDGMENTS

The editors are pleased to acknowledge and thank IRD Mechanalysis, Inc. for much of the material included in this chapter.

CHAPTER 26

Inert Gas Systems and Crude Oil Washing Machinery

EVERETT C. HUNT
AND JAMES MERCANTI

INTRODUCTION AND BACKGROUND

INERT gas is a gas, or a mixture of gases, such as flue gas, containing insufficient oxygen to support the combustion of hydrocarbons. It can be obtained from a ship's main or auxiliary boiler, from the exhaust of an internal combustion engine, from a separately fired generator, or from independent storage tanks.

An inert gas system consists of the means of distributing inert gas from a specific source to cargo tanks. It includes all the equipment specially fitted to supply, cool, clean, pressurize, monitor, and control delivery of inert gas to the cargo-tank system. It also includes the necessary piping, valves, and associated fittings to properly distribute the inert gas to the cargo tanks. The primary purpose of an inert gas system is to provide protection against a tank explosion, achieved by introducing inert gas into the tank to bring the oxygen content far below the flammable range. Furthermore, by allowing the atmosphere in the tank to be expelled and replaced by inert gas, the hydrocarbon gases as well as the oxygen are reduced to safe levels. In addition to its primary purpose, the inert gas system is often used as a common vent for venting gases from the cargo tanks to the atmosphere, and also as a common manifold for protection of cargo tanks against excessive pressure or vacuum.

Inert gas systems have been used on oil tankers for over fifty-five years. Between 1925 and 1927, the Standard Oil Company of California installed a combination flue gas, tank-cleaning, and venting system on twelve of their seagoing tankers. An application for a patent covering the system was filed in 1928, and a patent was granted in 1932. In 1932, after a serious casualty during gas-freeing operations, Sun Oil Company installed an inert gas system on all of its oceangoing tankers. Although over thirty companies installed the system during that period, the use of flue gas to protect cargo tanks never became widespread until the advent of the very large crude carriers (VLCCs).

In 1969, a series of explosions aboard the VLCCs *Mactra, Marpessa,* and *King Haakon VII* caused grave concern about the safety of such ships. The common factor in all of the casualties was that at the time the explosions occurred, tank-cleaning operations were being carried out.

A formal board of inquiry investigating the *Mactra* explosion recommended that VLCCs, including ships with cargo tanks exceeding 10,000 cubic meters in capacity, be fitted with an inert gas system. In 1972, the Intergovernmental Maritime Consultative Organization (IMCO, now IMO) advocated the use of inert gas systems on all new tankers, and in 1973, IMCO's resolution was adopted by the international marine community, requiring inert gas systems on all new tankers of 100,000 DWT and over, and on combination carriers of 50,000 DWT and over. During the winter of 1976-77, several tanker casualties occurred in or near United States waters which caused great public concern about the risks associated with the transportation of oil by tankers, and resulted in demands for the United States government to take steps toward the improvement of tanker safety and the prevention of pollution. In 1978, the Port and Tanker Safety Act of 1978 (PTSA) became law, and mandated that inert gas systems be included on certain new vessels and retrofitted on certain existing tankers. In 1979, the U.S. Coast Guard issued its final rules concerning the requirements for installation of inert gas systems. Existing crude carriers over 20,000 DWT must be fitted with inert gas systems by June 1, 1983, in order to enter a United States port. Crude carriers and product carriers between 20,000 DWT and 40,000 DWT not fitted with high capacity tank-washing machines may be granted an exemption by the office of merchant marine safety, if the owner can show that compliance would be unreasonable and impractical within the vessel's design characteristics. Existing crude carriers and product carriers above 70,000 DWT were required to have inert gas systems installed by June 1, 1981. The law provides that any vessel in noncompliance can be barred from United States waters. Furthermore, civil and criminal penalties can be imposed by the Coast Guard against the owner, operator, or any individual crew member for violating certain aspects of this law.

In the early 1980s, there has been a recurrence of major explosions aboard tankers. Of particular concern is the fact that a number of these casualties has occurred on ships fitted with inert gas systems. One extensive investigation by an independent regulatory association revealed that an alarming percentage of inert gas system equipped ships had systems inoperative because of various design and material problems. Simultaneously, the improper use of the system may jeopardize its effectiveness, while at the same time bring a false sense of security to those involved. Should the inert gas system fail, for whatever reason, or be used improperly, problems could arise that would not occur in its absence. Therefore, it is imperative that the system be designed, installed, and operated properly.

PRINCIPLES OF INERT GAS SYSTEMS

An explosion within a cargo tank can only occur when these three components are all present at the same time: hydrocarbon gas, oxygen, and a source of ignition. A mixture of hydrocarbon gas and oxygen can only ignite if its composition falls within a certain range, known as the "flammable range." The hydrocarbon gas and the oxygen contents are expressed in percentages by volume. Although flammable limits vary for different types of hydrocarbon gases, they are generally assumed to fall within a range of 1 to 10 percent for oil cargoes carried in tankers. The flammable limits for oxygen fall within approximately 11 and 21 percent (the 21 percent representing fresh air). Possible sources of ignition are smoking, welding, metal-to-metal sparking, spontaneous combustion, static electricity, and lightning.

As shown in Figure 26-1, any point on the flammability triangle diagram represents a hydrocarbon gas-air-inert gas mixture, specified in terms of its hyrocarbon and oxygen content. Hydrocarbon-air mixtures

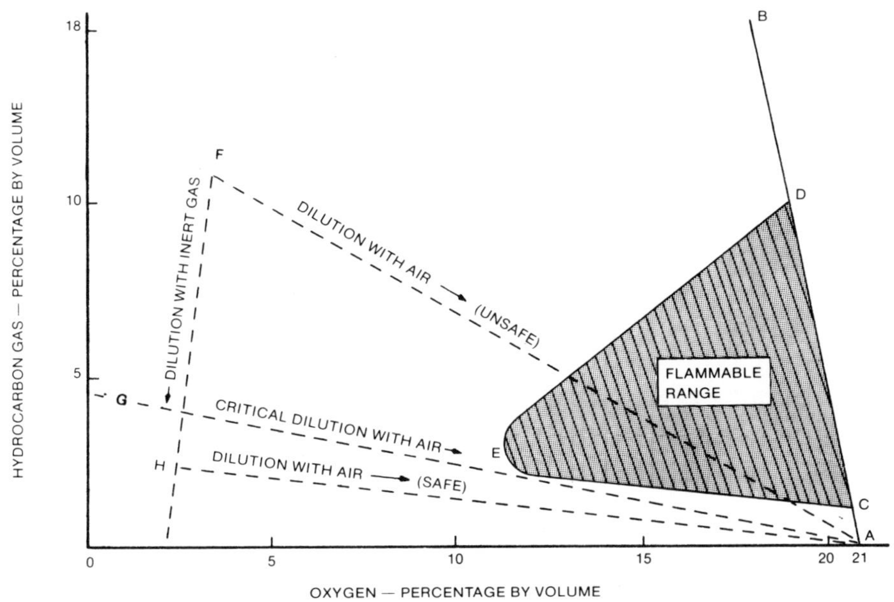

Figure 26-1. Flammability triangle

without inert gas are on the line AB, the slope of which shows the reduction in oxygen content as the hydrocarbon content increases. Points to the left of line AB represent mixtures with their oxygen content further reduced by the addition of inert gas. It is evident from the diagram that as inert gas is added to hydrocarbon-air mixtures, the flammable range progressively decreases until the oxygen content reaches a level generally taken to be about 11 percent by volume, at which level no mixture can burn. Therefore, an inert gas system, with a design point of 5 percent and maximum allowable operating point of 8 percent oxygen, allows a good margin of safety.

The lower and upper flammable limit mixtures for hydrocarbon gas in air are represented by points C and D in the diagram. As the inert gas content increases, the flammable-limit mixtures change. This is indicated by lines CE and DE, which converge at point E. Only those mixtures represented by points in the shaded area within the loop CED are capable of burning. Changes of composition, due to the addition of either air or inert gas, are represented by movements along straight lines. These lines are directed either towards point A, representing pure air, or towards a point on the oxygen content axis corresponding to the composition of the added inert gas. Such lines are shown for the gas mixture represented by point F.

When an inert mixture, such as that represented by point F, is diluted by air, its composition moves along line FA and therefore enters the shaded area of flammable mixtures. This means that all inert mixtures in the region above line GA, the "critical dilution" line, pass through a flammable condition as they are mixed with air, for example, during a gas-freeing operation. Those below line GA, represented by point H, do not become flammable upon dilution. It is possible to move from a mixture, as represented by point F, to one represented by point H, by dilution with additional inert gas, i.e., purging.

Methods of Gas Replacement

In order to properly understand the inerting process, the various operations which involve replacement of gas in cargo tanks—namely, *inerting, purging,* and *gas-freeing*—need to be discussed. In each of these replacement operations, one of two processes can predominate: *dilution*, which is a mixing process; and *displacement,* which is a layering process.

These two processes have a marked effect on the method of monitoring the tank atmosphere and the interpretation of the results. Figures 26-2 and 26-3 show that an understanding of the nature of the gas replacement process actually taking place within the tank is necessary for the correct interpretation of the reading shown on the appropriate gas-sampling instrument.

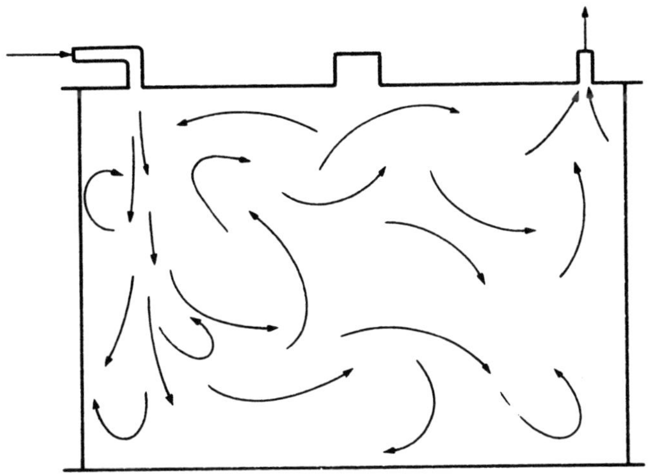

Figure 26-2. Dilution concept

Dilution. The dilution theory illustrated in Figure 26-2 assumes that the incoming gas mixes with the original gases to form a homogeneous mixture throughout the tank. The result is that the concentration of the original gas decreases exponentially. In practice the actual rate of gas replacement depends upon the volume flow of the incoming gas, its entry velocity, and the dimensions of the tank. For complete gas replacement it is important that the entry velocity of the incoming gas be high enough for the jet to reach the bottom of the tank. It is therefore important to confirm the ability of every installation using this principle to achieve the required degree of gas replacement throughout the tank.

Displacement. The displacement theory, illustrated in Figure 26-3, is based on expelling the heavier gases only. Ideal replacement requires a stable horizontal interface between the lighter gas entering at the top of the tank and the heavier gas being displaced from the bottom of the tank through some suitable piping arrangement. This method requires a relatively low entry velocity of gas and in practice more than one volume change is necessary. It is therefore important to confirm the ability of every installation using this principle to achieve the required degree of gas replacement throughout the tank. Figure 26-3 shows an inlet and outlet configuration for the displacement process, and indicates the interface between the incoming and outgoing gases. Also shown are typical curves of gas concentration against time for three different sampling levels within the cargo tank.

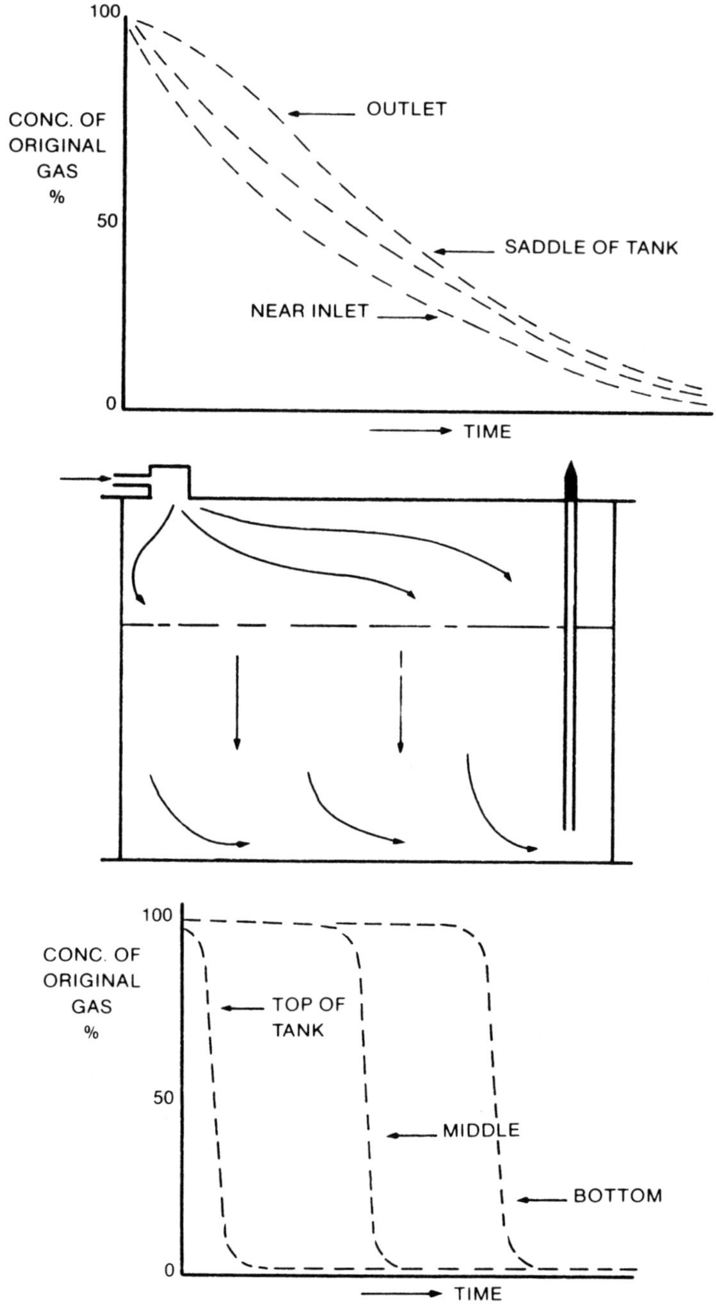

Figure 26-3. Displacement concept

General Policy of Cargo Tank Atmosphere Control

Tankers fitted with an inert gas system must have their cargo tanks kept in a nonflammable condition at all times. Therefore, tanks should be kept in the inert condition whenever they contain cargo residues or ballast. The oxygen content should be kept at 8 percent or less by volume with a positive gas pressure in all the cargo tanks. The atmosphere within the tank should make the transition from the inert condition to the gas-free condition without passing through the flammable condition. In practice this means that before any tank is gas freed, it should be purged with inert gas until the hydrocarbon content of the tank atmosphere is below the critical dilution line as shown in Figure 26-1.

When a ship is in a gas-free condition before arrival at a loading port, tanks should be inerted prior to loading. In order to maintain cargo tanks in a nonflammable condition, the inert gas plant will be required to inert empty cargo tanks; to be operated during cargo discharge, deballasting, and necessary in-tank operations; to purge tanks prior to gas freeing; and to top up pressure in the cargo tanks when necessary, during other stages of the voyage.

TYPES OF INERT GAS SYSTEMS

Flue Gas System

Most marine inert gas systems utilize flue gases from the main or auxiliary boilers. A typical composition of such flue gas, based on burning bunker C fuel, is (by volume):

Oxygen (O2)	4.2%
Carbon dioxide (CO2)	13.5%
Sulfur dioxide (SO2)	0.25%
Nitrogen (N2)	77.0%
Water vapor (H2O)	5.0%
Solids (soot)	250 mg/m^3

Before being distributed to the cargo tanks, this gas must first be cooled and cleaned. This is normally accomplished by the scrubber which, if properly designed and operated, will produce a gas free of corrosive components and solid particles.

The cooling effect, using seawater for scrubbing, should yield a gas temperature that is within 5°C above the seawater temperature.

Independent Generator

If there are no main or auxiliary boilers, or if there is a need for special quality inert gas, such as is required on LNG tankers, a separately fired inert gas generator can be used. This system generates its own gas through independent combustion. The fuel most often used is low sulfur diesel oil in order to obtain cleaner combustion and less corrosion. Some independent generator systems are able to burn heavier fuels such as bunker C, but with sacrifice to gas quality.

Exhaust Gas Generator

Although little has been done to develop the potential of this concept, there are a few installations aboard tankers utilizing a gas turbine driving an electric generator. Because the exhaust gases are normally too high in oxygen content, an afterburner is used to reduce the oxygen level to 5 percent by volume. Diesel engines have been used in similar fashion, using a separate afterburner section to burn off excess oxygen. A serious problem is the difficulty in maintaining constant load on the prime mover to ensure stable combustion. Because of its questionable suitability for this application, no further reference is made to this system.

Independent Storage Tanks

Where small quantities of inert gas are required, such as in small size LPG tankers, storage tanks of an inert gas, such as nitrogen, are installed aboard ship, and used to fill the void areas surrounding the tanks with inert gas. Although extremely simple in concept, the major disadvantage of this system is limited capacity. Furthermore, refilling facilities are not easily available on a worldwide basis.

DESIGN OF INERT GAS SYSTEMS

This section deals generally with inert flue gas systems. The design of systems other than inert flue gas systems should take into account, whenever applicable, the general design principles outlined in this section.

General Description

A typical arrangement for an inert flue gas system is shown in Figure 26-4. It consists of flue gas isolating valves located at the boiler uptake through which pass hot, dirty gases to the scrubber. Here the gas is cooled and cleaned before being piped to the blowers which deliver the gas through the deck water seal, the nonreturn valve, and the deck-isolating valve, thence to the cargo tanks. A gas pressure regulating valve is fitted downstream of the blowers to regulate the flow of gases to the cargo tank. A liquid-filled pressure-vacuum breaker is fitted to prevent excessive pres-

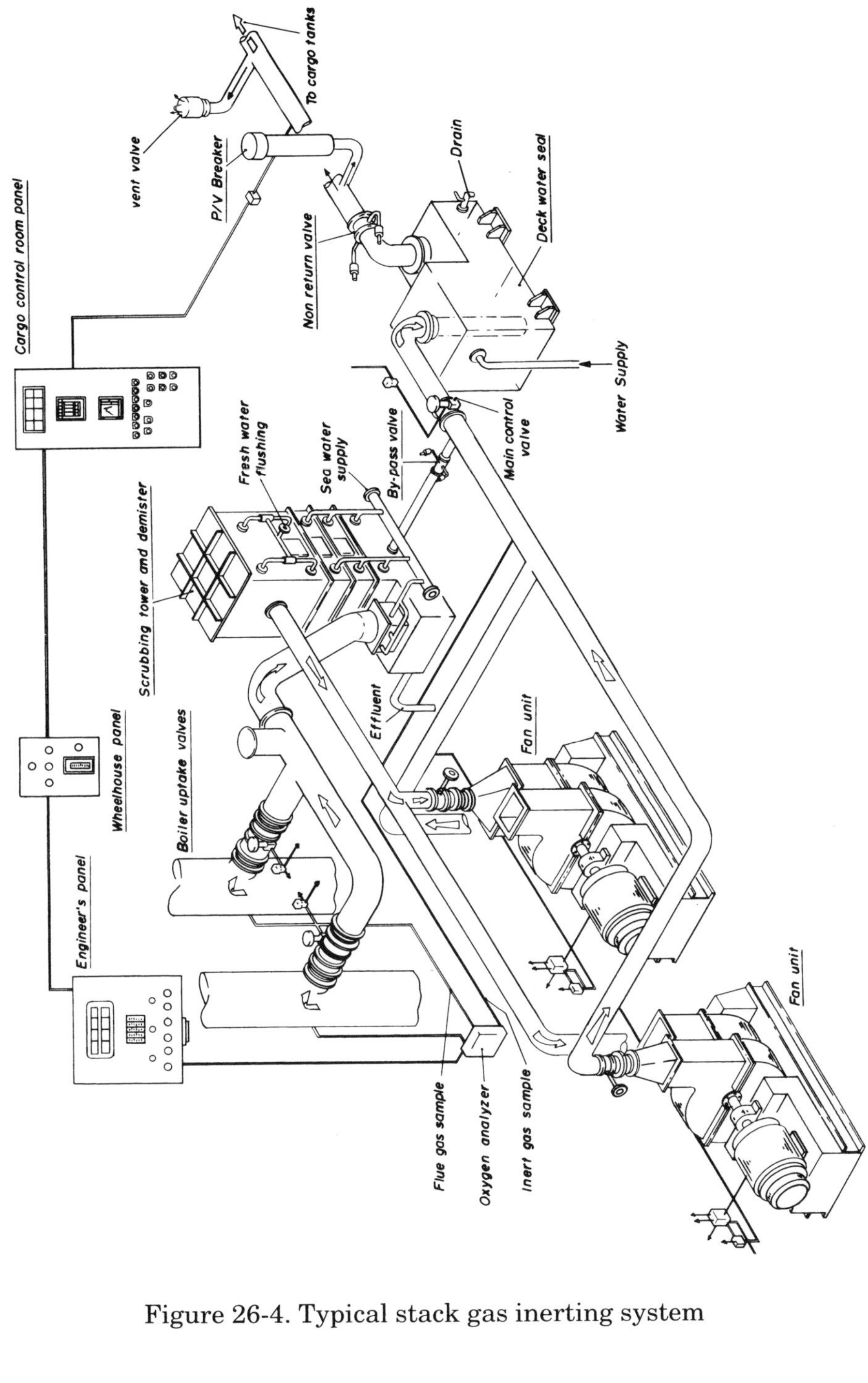

Figure 26-4. Typical stack gas inerting system

sure or vacuum from causing structural damage to the cargo tanks. A vent is sometimes fitted between the deck water seal and the gas pressure regulating valve to vent any leakage when the plant is shut down.

For delivering inert gas to the cargo tanks during cargo discharge, deballasting, tank cleaning, and for topping up the pressure of gas in the tanks during other phases of the voyage, an inert gas deck main runs forward from the deck-isolating valve for the length of the cargo deck. From this inert gas main, inert gas branch lines lead to the top of each cargo tank, or to each cargo tank hatch.

Scrubber

Function. The purpose of the scrubber is to cool the flue gas and remove most of the sulphur dioxide and particulate soot. This is achieved by direct contact between the flue gas and large quantities of seawater. Before entering the bottom of the scrubbing tower, the gas is cooled by being passed through a water spray. Within the scrubbing tower, the gas moves upwards through downward-flowing water. For maximum contact between gas and water, several layers made up of one or more of the following arrangements may be fitted: spray nozzles; trays of "packed" stones or plastic shapes; perforated "impingement" plates; or venturi nozzles and slots.

At the top, or downstream, of the scrubbing tower, water droplets are removed by a demister which may be a polypropylene pad or a cyclone separator. Figure 26-5 illustrates the principles of inert gas scrubber operation.

Design considerations. The scrubber should be capable of cooling the flue gas to within 9°F (5°C) of the cooling water. The performance of the scrubber at all gas flows should be such as to remove at least 90 percent of sulphur dioxide and to remove solids effectively at the designed pressure differential of the system. In product carriers more stringent requirements may be needed for product quality. The internal parts of the scrubber should be constructed in corrosion-resistant materials because of the corrosive effect of the gas. Alternatively, the internal parts may be lined with rubber, fiberglass epoxy resin, or other equivalent material, in which case the flue gases may require cooling before they are introduced into the lined sections of the scrubber. Adequate openings and sight glasses should be provided in the scrubber shell for inspection and cleaning. The sight glasses should be reinforced to withstand impact and be suitable for the temperatures encountered.

The design should be such that under normal conditions of trim and list the scrubber efficiency will not drop by more than 3 percent; and the temperature rise at the gas outlet not exceed the designed gas outlet

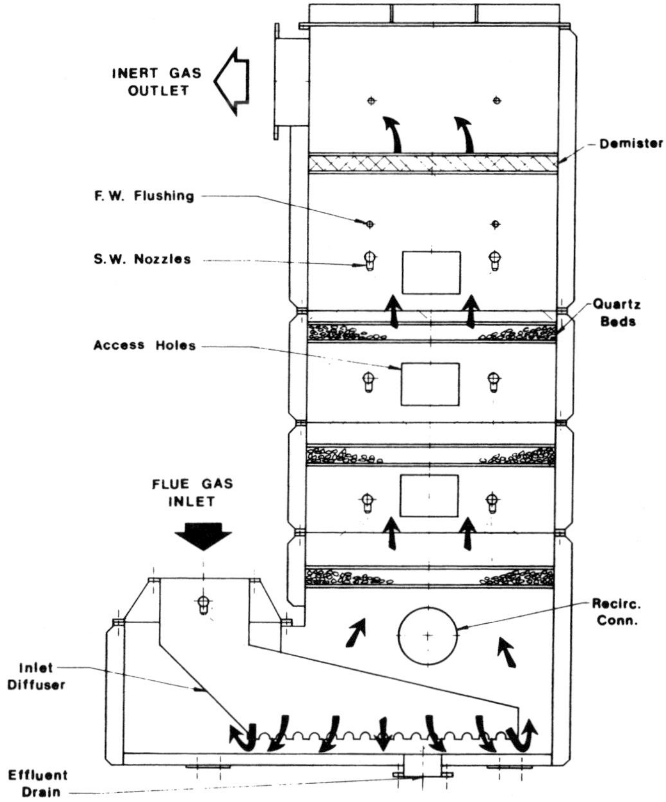

Figure 26-5. Gas scrubber section drawing

temperature by more than 5°F (3°C). The location of the scrubber above the load waterline should be such that the drainage of the effluent is not impaired when the ship is in the fully loaded condition.

Inert Gas Blowers

Function blowers are used to deliver the scrubbed flue gas to the cargo tanks. Regulations require that at least two blowers be provided which together will be capable of delivering inert gas to the cargo tanks at a rate of at least 125 percent of the maximum rate of cargo discharge capacity of the ship expressed as a volume.

In practice, installations vary from those which have one large blower and one small blower, whose combined total capacity meets the regulations, to those in which each blower can meet this requirement. The advantage claimed for the former is that it is convenient to use a small

Figure 26-6. Typical gas scrubber

capacity blower when topping up the gas pressure in the cargo tanks at sea. The advantage claimed for the latter is that if either blower is defective, the other one is capable of maintaining a positive gas pressure in the cargo tanks without extending the duration of the cargo discharge.

Design considerations. The blower casing should be constructed of corrosion resistant material or alternatively of mild steel lined with rubber, fiberglass epoxy resin, or other equivalent material to protect it from the corrosive effect of the gas. The impellers should be manufactured in a corrosion-resistant material having high tensile strength. Inconel impellers are excellent for this service. Aluminum-bronze impellers should be tested by overspeeding to 20 percent above the design running speed of the electric motor or 10 percent above the speed at which the overspeed trip of the turbine would operate, whichever is applicable. Large-sized drains, fitted with adequate water seals, should be provided for each casing to prevent damage by an accumulation of water and soot. Means should be provided for freshwater washing of the impellers to remove buildup of deposits which would cause vibration during blower operation.

The casing should be adequately ribbed to prevent panting and should be so designed and arranged as to facilitate the removal of the rotor without disturbing major parts of the inlet and outlet gas piping. Sufficient openings in the casing should be provided to permit inspection. Where separate shafts are provided for the prime mover and the blower, a flexible coupling with protective cover between the shafts should be installed. When roller or ball bearings are used, proper attention should be paid to the problem of brinelling and the method of lubrication. The type of lubrication chosen, i.e., oil or grease, should take into account the diameter and rotational speed of the shaft. If sleeve bearings are fitted, resilient mountings are not recommended.

The blower performance characteristics should be matched to the maximum system requirements. The characteristics should be such that in the event of the discharge of any combination of cargo tanks at the maximum discharge rate, a minimum pressure of 200 millimeters water gauge is maintained in any cargo tank after allowing for pressure losses due to any of the following: the scrubber tower and demister; the hot gas piping from the stack to the scrubbing tower; the cold gas distribution piping downstream of the scrubber; the deck water seal; all valves and fittings; and the length and diameter of the inert gas distribution system.

When the blowers are not of equal capacity, the pressure-volume characteristics and inlet and outlet piping should be so matched that if both blowers can be run in parallel, they are able to develop their designed outputs. The arrangements should be such as to prevent the blower on load from motorizing the blower that is stopped or tripped out. Figure 26-7 is a sectional view of the blower. Figure 26-8 is a photograph of a typical blower.

The prime mover should be of sufficient power to ensure that it will not be overloaded under all possible operating conditions of the blower. The overload power requirement should be based on the blower inlet conditions of 23°F (–5°C) at –400 millimeters water gauge, outlet conditions of 32°F (0°C), and atmospheric pressure.

Nonreturn Devices

Function. The deck water seal and mechanical nonreturn valve together form the means of automatically preventing the backflow of cargo gases from the cargo tanks to the machinery space or other safe area in which the inert plant is located.

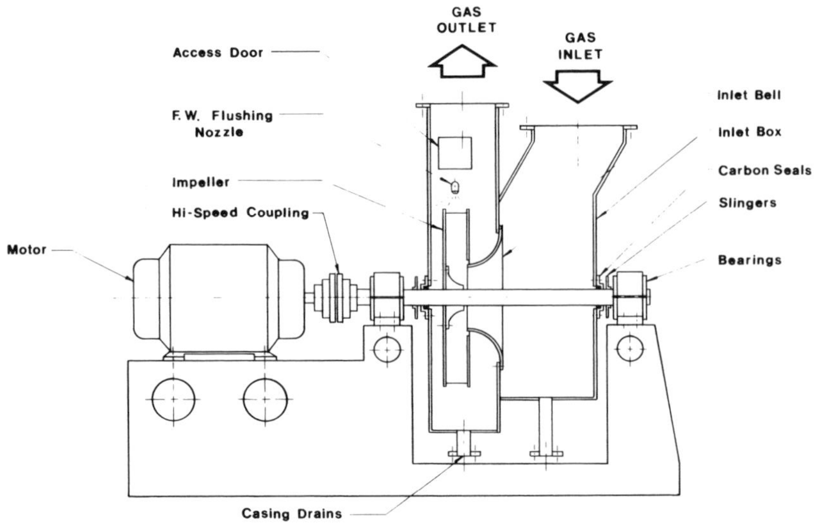

Figure 26-7. Fan section drawing

Figure 26-8. Typical gas fan assembly

Deck water seal. This is the primary barrier between the safe area and the hazardous area. For safety, it is the most important device in the inert gas system. A water seal is fitted which permits inert gas to be delivered to the deck main but prevents any backflow of cargo gas even when the inert gas plant is shut down. It is vital that a supply of water is maintained at the seal at all times, particularly when the inert gas plant is shut down. In addition, drains should be led through a vented loop-seal directly overboard, and should not pass through the machinery spaces. There are different designs but one of three principal types of deck water seals may be adopted. The three types are the following:

1. Wet type, which is the simplest type of water seal. When the inert gas plant is operating, the gas bubbles go through the water from the submerged inert gas inlet pipe, but if the tank pressure exceeds the pressure in the inert gas inlet line, the water is pressed up into this inlet pipe and thus prevents backflow. The drawback of this type of water seal is that water droplets may be carried over with the inert gas which, although it does not impair the quality of the inert gas, could increase corrosion. Therefore, a demister is usually fitted in the gas outlet from the water seal to reduce carryover. Figure 26-9 shows an example of this type.

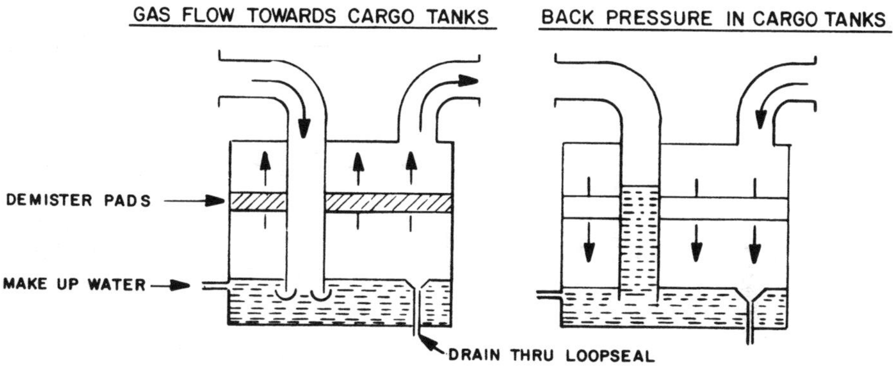

Figure 26-9. Wet type of deck water seal

2. Semidry type, in which instead of bubbling through the water trap, the inert gas flow draws the sealing water into a separate holding chamber by venturi action, thus avoiding or at least reducing the amount of water droplets being carried over. Otherwise, it is functionally the same as the wet type. Figure 26-10 shows an example of this type.
3. The dry type, in which the water is drained when the inert gas plant is in operation (gas flowing to the tanks) and filled with water when the inert gas plant is either shut down or the tank pressure exceeds the inert gas blower discharge pressure. Filling and drainage are performed by automatically operated pneumatic valves controlled by the levels in the water seal and drop tank, and by the operating state of the blowers. The advantage of this type is that water carryover is prevented. The drawback could be the risk of failure

of the automatically controlled valves which may render the water seal ineffective. Figure 26-11 shows an example of this type.

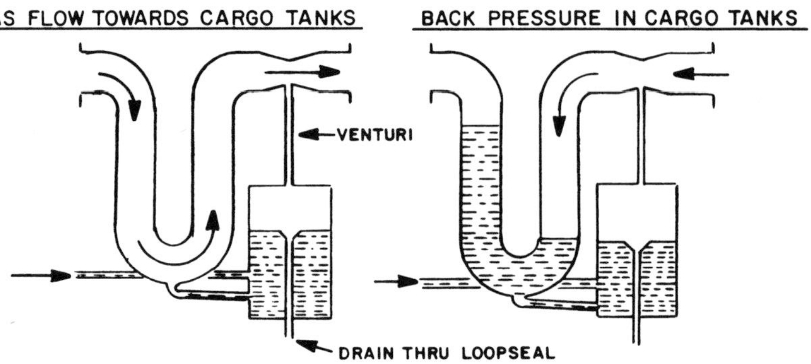

Figure 26-10. Semidry type of deck water seal

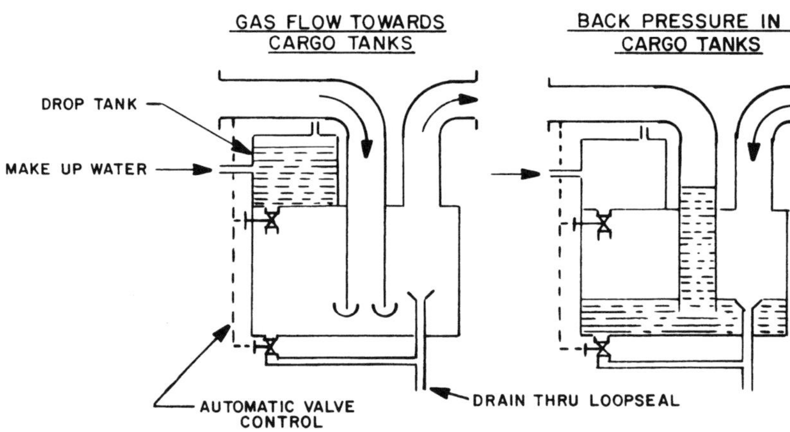

Figure 26-11. Dry type of deck water seal

Nonreturn and isolation valves. As an additional precaution to avoid any backflow of gas from the cargo tanks, and to prevent any backflow of liquid which may enter the inert gas main if the cargo tanks are overfilled, regulations require a mechanical nonreturn valve, or equivalent, which should operate automatically at all times.

This valve should be provided with a positive means of closure or, alternatively, a separate deck isolation valve fitted forward of the nonreturn valve, so that the inert gas deck main may be isolated from the nonreturn devices. The separate deck isolation valve has the advantage of enabling maintenance work to be carried out on the nonreturn valve.

Design considerations. The material used in the construction of the nonreturn devices should be resistant to fire and to the corrosive attack from acids formed by the gas. Alternatively, low carbon steel protected by a rubber lining, fiberglass epoxy resin, or equivalent material may be used. Particular attention should be paid to the gas inlet pipe to the water seal, which is most susceptible to corrosion, because it is constantly immersed in salt water.

The deck water seal should present a resistance to backflow of not less than the pressure setting of the pressure-vacuum breaker on the inert gas distribution system and should be so designed as to prevent the backflow of gases under any foreseeable operating conditions. The water in the deck seal should be maintained by a regulating flow of clean seawater through the deck seal reservoir. Sight glasses and inspection openings should be provided on the deck seal to permit satisfactory observation of the water level. The sight glasses should be reinforced to withstand impact.

Piping and Valves

Function. The inert gas distribution system, together with the cargo tank venting system, where applicable, must provide the means of delivering inert gas to the cargo tanks during discharge, deballasting, and tank-cleaning operations, and for topping up the pressure of gas in the tanks; venting tank gases to atmosphere during cargo loading and ballasting, including an additional inlet or outlet points for inerting, purging, and gas freeing; isolating individual tanks from the inert gas main for gas freeing; and protecting tanks from excessive pressure and vacuum.

Design considerations. The flue gas uptake point should be selected so that the gas is not too hot for the scrubber, or that it causes hard deposits on the flue gas isolating valves. It should not be too close to the uptake outlet so that air is drawn into the system. When boilers are fitted with rotary air heaters, the offtake point should be before the air heater inlet.

The materials used for flue gas isolating valves should take into account the temperature of gas at the offtake. Cast iron is questionable even for temperatures below 428°F (220°C). Valves exposed to a temperature exceeding 220°C should be made from a material not only compatible with the temperature but also resistant to the corrosive effect of stagnant flue gases laden with sulfuric and sulfurous acids.

Flue gas isolating valves should be provided with facilities to keep the seatings clear of soot unless the valve is designed to close with a seat-cleaning action. Flue gas isolating valves should also be provided with air-sealing arrangements, usually fed from the forced draft blowers. If expansion bellows are considered necessary, they should have a smooth internal sleeve and preferably be mounted so that the gas flow through them is vertical. They should be constructed of material resistant to sulfuric and sulfurous acids. The pipework between the flue gas isolating valve and the scrubber should be made from Corten steel or heavy gauge steel resistant to corrosion and so arranged to prevent the accumulation of damp acidic soot by the avoidance of unnecessary bends and branches. The inlet piping to the scrubber should be arranged to permit positive isolation from the flue gases prior to gas freeing the scrubber for entry for maintenance purposes. This may be accomplished by the removal of a suitable length of pipe section and blanking, or a spectacle flange. The gas outlet piping from the scrubber to the blowers and recirculating lines should be made from steel suitably coated internally. Suitable isolating arrangements should be incorporated in the inlet and outlet of each blower to permit overhaul and maintenance of one blower safely while the other blower is in operation. The gas pressure regulating valve should be provided with means to indicate whether the valve is open or shut. Where the valve is used to regulate the flow of inert gas, it should be controlled by the inert gas pressure sensed between the deck-isolating valve and the cargo tanks.

Inert gas deck lines should be of steel and arranged to be self-draining. They should be firmly attached to the ship's structure with suitable arrangements to take into account movement due to heavy weather, thermal expansion, and flexing of the ship.

The diameter of the inert gas main, valves, and branch pipes should be sized to avoid excessive pressure drop. The inert gas velocity should not exceed 40 meters per second in any section of the distribution system when the inert gas system is operating at its maximum capacity. If the inert gas main is used for venting during loading, other factors may need to be taken into consideration.

All pressure and vacuum relief openings should be fitted with U.S. Coast Guard approved flame screens, with easy access for cleaning and renewal. The flame screens should be at the inlets and the exits of any relief device and be of robust construction sufficient to withstand the pressure of gas generated at maximum loading and ballasting operations while presenting minimum resistance.

Gas Pressure Regulating Valve

Pressure control arrangements should be fitted, fulfilling two functions: (1) to prevent automatically any backflow of gas in the event of either a failure of the inert gas blower, scrubber pump, etc., or when the inert gas plant is operating correctly but the deck water seal and mechanical nonreturn valve have failed and the pressure of gas in the tank exceeds the blower discharge pressure, e.g., during simultaneous stripping and ballasting operations; and (2) to regulate the flow of inert gas to the inert gas deck main.

A typical arrangement by which gas regulation can be achieved is when a system with automatic pressure control and a gas recirculating line permits control of inert gas pressure in the deck main without requiring adjustment of the inert gas blower speed. Gas not required in the cargo tanks is recirculated to the scrubber or vented to the atmosphere. Gas pressure regulating valves are fitted in both the main and recirculating lines. One is controlled by a gas pressure transmitter and regulator, while the other may be controlled either in a similar manner or by a weight-operated valve. The pressure sensor is located downstream of the deck isolation valve which enables a positive pressure to be maintained in the cargo tanks during discharge. (Figure 26-12 shows a typical automatic pressure control system.)

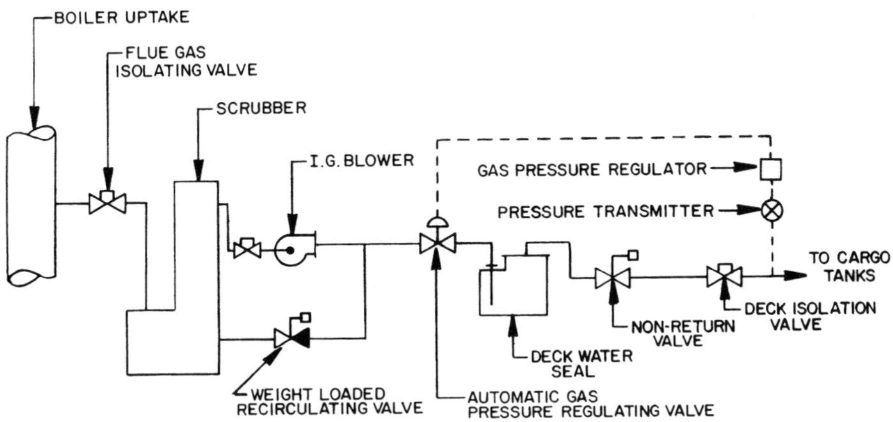

Figure 26-12. Automatic pressure control system

P-V (Pressure-Vacuum) Breaker

One or more liquid-filled pressure-vacuum breakers should be fitted, unless pressure-vacuum valves have the capacity to prevent excessive pressure or vacuum. These devices required little maintenance. However, they will only operate at the required pressure if they are filled to the correct level with liquid of the correct density. Either a suitable oil or a freshwater-glycol mixture should be used to avoid freezing in cold weather. Evaporation, ingress of seawater, condensation, and corrosion should be taken into consideration and adequately compensated for. In heavy weather, the pressure surge caused by the motion of liquid in the cargo tanks may cause the liquid of the pressure-vacuum breaker to be blown out. (Figure 26-13 illustrates the principles of a liquid-filled pressure-vacuum breaker.)

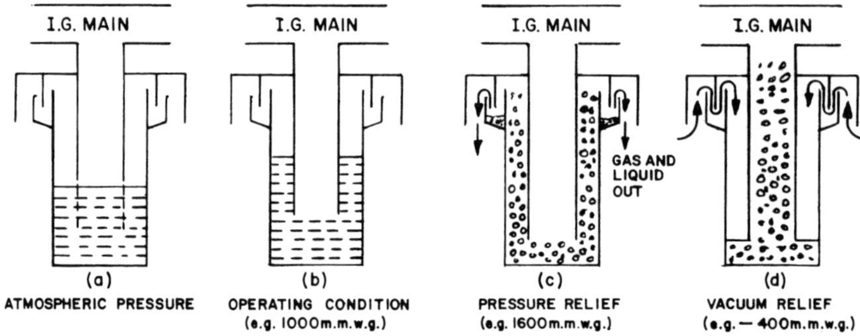

Figure 26-13. Liquid-filled pressure-vacuum breaker

The characteristics of the deck water seal, pressure-vacuum breakers, and pressure-vacuum valve, and the pressure settings of the high and low inert gas deck-pressure alarms must be compatible. It is also desirable to check that all pressure-vacuum devices are operating at their designed pressure settings.

High Velocity Vent Valve

Also referred to as the mast riser (vent) valve, this assembly consists of a hydraulically actuated or manual butterfly valve, a high velocity cone and mixing chamber, and a flame safety screen. The assembly is normally connected to the forward section of the inert gas main, and is used to discharge the inert gas at high velocity into the atmosphere to reduce its toxic effect on personnel on deck. According to regulatory requirements, the outlet of the high velocity vent valve should be at least 2 meters above deck level. Numerous installations have the gas outlet located on top of the mast (thereby being previously referred to as a "mast riser" vent), but this arrangement has been under question following an accident aboard an American flag tanker.

Instrumentation and Alarms

Certain fixed and portable instruments are required for the safe and effective operation of an inert gas system. It is desirable that all instruments be graduated to a consistent system of units. Clear instructions provide for operating, calibrating, and testing all instruments and alarms, including suitable calibration facilities. All instrumentation and alarm equipment required should be suitably designed to withstand supply voltage variation, ambient temperature changes, vibration, humidity, shock, impact, and corrosion normally encountered on board ships.

To fulfill regulation requirements the arrangement of scrubber instrumentation and alarm equipment should be set up as follows: With the water flow to the scrubber monitored either by a flowmeter or by pressure gauges, an alarm will be initiated when the water flow drops below the designed flow requirements by a predetermined amount, and the inert gas blowers are stopped automatically in the event of a further reduction in the flow. The precise setting of the alarm and shutdown limits are related to individual scrubber designs and materials. The water level in the scrubber is monitored by a high water level alarm. This alarm will be activated when predetermined limits are reached, and the scrubber pump will be shut down when the level rises above set limits. These limits will be set with regard to the scrubber design as well as with regard to the flooding of the scrubber inlet piping from the boiler uptakes.

The inert gas temperature at the discharge side of the gas blowers is monitored and an alarm given when the temperature reaches 149°F (65°C). Automatic shutdown of the inert gas blowers will occur if the temperature reaches 167°F (75°C).

If a precooler is necessary at the scrubber inlet to protect coating materials in the scrubber, the arrangements for giving an alarm will also apply to the outlet temperature from the precooler. To monitor the scrubber efficiency, it is desirable that the cooler water inlet and outlet temperatures and the scrubber differential pressures be indicated. All sensing probes, floats, and sensors required to be in contact with the water and gas in the scrubber are made from materials resistant to acidic attack.

For the deck water seal, an alarm should be given when the water level falls to a predetermined level, but before the seal is rendered ineffective. For certain types of deck water seals, such as the dry type, the water level alarm may need to be suppressed when inert gas is being supplied to the inert gas distribution system. The pressure of the inert gas in the inert gas main should be monitored, and an alarm given when the pressure reaches the set limit. The pressure limit should be set with regard to the design of cargo tanks, the mechanical nonreturn valve, and the deck water seal.

The arrangement of the oxygen analyzer, recorder, and indicating instruments should meet the following specifications. The gas-sampling connection must be located in the pipe at a point where turbulent flow conditions prevail for all possible blower operating conditions. The sampling point, located after the blower and before the gas regulating valve, must be easily accessible and have air- or steam-cleaning connections. The probe used for sampling must be fitted with a dust filter. Both the probe and the filter are designed to be removed for cleaning or replacement. The sensing pipe from the probe to the analyzer is designed to drain condensate to preclude interruption of the gas flow to the analyzer. Ingress of air is minimized by limiting the number of joints in the sampling pipe.

The position of the analyzer is chosen so that it is protected from heat and adverse ambient conditions since it is placed as close as practicable to the sampling point in order to reduce to a minimum the time between the extraction of a sample and its analysis.

The oxygen analyzer should have an accuracy of ±1 percent of the full-scale deflection of the indicator.

Between the automatic gas pressure regulating valve and the deck water seal a sampling point is provided for use with portable instruments.

The inert gas pressure sensor and recorder obtains the signal from a point in the inert gas main between the deck isolation nonreturn valve and the cargo tanks. When the pressure in the inert gas main forward of the nonreturn devices falls below 50 millimeters on the water gauge, a means is provided to shut down the main cargo pumps automatically, or to give an audible alarm on the navigating bridge and in the machinery space.

Portable instruments provide for measuring oxygen and flammable concentration. Hydrocarbon vapor meters working on the catalytic filament principle are unsuitable for measuring hydrocarbons concentration in oxygen deficient atmospheres. Furthermore, meters using this principle cannot measure concentrations of hydrocarbon vapors above the lower flammable limit. It is, therefore, advisable to use meters utilizing a principle which is not affected by oxygen deficiency, and is capable of measuring hydrocarbon concentration in and above the flammable range. For measuring below the lower flammable limit, provided sufficient oxygen is present, the catalytic filament meter is suitable.

All metal parts of portable instruments and sampling tubes requiring to be introduced into tanks should be securely grounded to the ship structure while the instruments and sampling tubes are being used. These portable instruments should be of an intrinsically safe type approved by the U.S. Coast Guard.

Sufficient tubing in cargo tanks should be provided to enable a fully representative sampling of a cargo tank atmosphere to be obtained as well as suitable openings to enable fully representative samples to be taken from each tank. Where tanks are subdivided by complete or partial wash bulkheads, additional openings will be needed for each subdivision.

Effluent and Drainpiping

The effluent piping from scrubbers and deck water seal drainpipes, where fitted, are of corrosion-resistant material, or of carbon steel suitably protected internally against the corrosive nature of the fluid. The scrubber effluent pipe and deck water seal drainpipe, where fitted, are not to be led to a common drainpipe, and the deck seal drain is to be led clear of the engine room and any other gas-safe space.

Piping made in glass-reinforced plastic of acceptable manufacture, substantial thickness, pressure tested, and adequately supported may be acceptable for effluent piping from scrubbers when effluent lines are, as far as possible, led through cofferdams or ballast tanks and are in accordance with the load line regulations in force. Glass-reinforced plastic is acceptable also where effluent lines are led through machinery spaces in which the arrangements include the following:

1. A valve fitted to a stub piece at the shell, actuated both from inside and outside the machinery space by pneumatic or hydraulic means led through steep piping, which will close automatically in the event of failure of the operating media. The valve should have a position indicator and is to be closed at all times when the plant is not in operation as well as in the event of a fire in the machinery space. Suitable instructions to this effect are to be given to the master.
2. A flap-type nonreturn valve.
3. A short length of steel pipe, or spool piece, lined internally and fitted between the valve referred to in 1 above and the nonreturn valve referred to in 2 above fitted with a 12.5-millimeter diameter-flanged drain branch pipe and valve.
4. A further spool piece fitted inboard of, and adjacent to, the nonreturn valve referred to in 2 above, similarly fitted with a drain. (Note: The purpose of this arrangement is to enable the tightness of the valves and nonreturn valves referred to in 1 and 2 above to be checked and to facilitate the removal of the nonreturn valve for examination and replacement.
5. A means outside the machinery space for stopping the scrubber pump. (A suitable arrangement is illustrated in Figure 26-14.)

A water seal in the shape of a U-bend at least 2 meters (6.5 feet) in depth is fitted at least 2 meters below the equipment to be drained and a means of draining the lowest point of the bend must be provided. In addition the seal should be adequately vented to a point above the water level in the scrubber or deck water seal.

The diameter of the effluent and drainpipes should be adequate for the duties intended, and the pipe run should be self-draining from the water seal.

Seawater Service

It is advisable that the main supply of water to the inert gas scrubber be from an independent pump. The alternative source of supply of water may be from another pump, such as the sanitary, fire, bilge, and ballast pumps, provided that the quantity of water required by the inert gas scrubber is readily available, and the requirements of other essential services are not thereby impaired. The requirement for two separate pumps to be capable

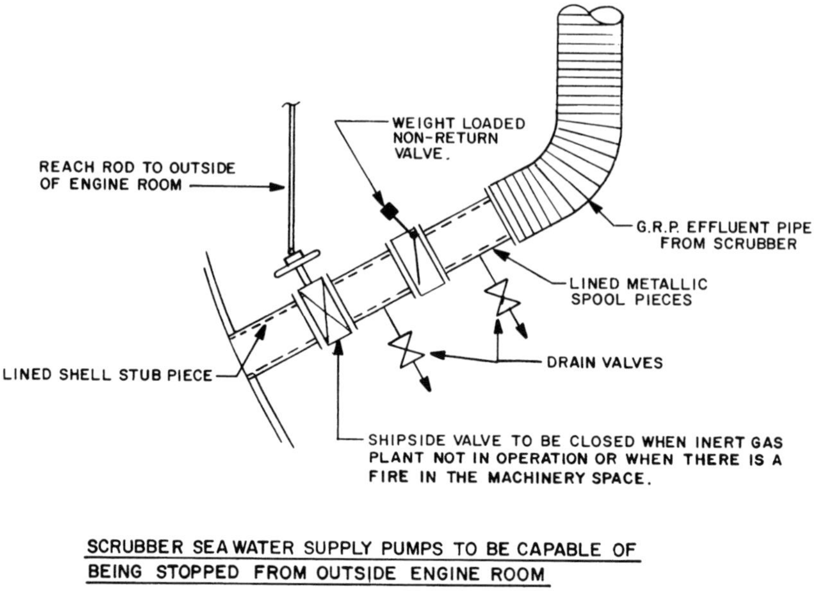

Figure 26-14. Scrubber effluent arrangement

of supplying water to the deck water seal can be met by any of the pumps referred to above under alternative source of supply.

The pumps supplying water to the scrubber and the deck water seal should be such as to provide the required throughput of water at light draft conditions. The quantity of water at all other draft conditions should not flood the scrubber or increase the gas flow resistance excessively. A loop seal with vent should be provided in the supply water piping to the deck seal to prevent the backflow of hydrocarbon vapor or inert gas and should be positioned outside the machinery space, suitably protected against freezing. With reference to the deck water seal arrangement, provisions should be made to prevent any pneumatically controlled system from freezing.

OPERATION OF THE INERT GAS SYSTEM

Though inert gas systems differ in detail, certain fundamentals remain the same: (1) starting up the inert gas plant; (2) shutting down the inert gas plant; and (3) utilizing safety checks when the inert gas plant is shut down. In all cases manufacturer's detailed instructions should be followed.

Start-Up

Start-up procedures involve the following steps:

1. Ensure boiler is producing flue gas with an oxygen content of 5 percent by volume or less (for existing ships, a maximum of 8 percent by volume and, whenever practicable, less).
2. Ensure that power is available for all control, alarm, and automatic shutdown operations.
3. Ensure that the quantity of water required by the scrubber and deck water seal is being maintained satisfactorily by the pump(s) selected for this duty.
4. Test operation of the alarm and shutdown features of the system dependent upon the throughput of water in the scrubber and deck seal.
5. Check that the gas-freeing fresh air inlet valve where fitted is shut and the blank is securely in position.
6. Open the flue gas isolating valve. Ensure that the air for sealing is shut off when the valve is open.
7. Open the selected blower suction valve. Ensure that the other blower suction and discharge valves are shut unless it is intended to use both blowers simultaneously.
8. Start the blower.
9. Test blower “failure” alarm.
10. Open the blower discharge valve (this should be done without delay to prevent overheating within the blower casing).
11. Ensure that the blower recirculating valve opens automatically to enable plant to stabilize.
12. Open the inert gas regulating valve.
13. Check that oxygen content in the cold inert gas is 5 percent by volume or less (for existing ships 8 percent by volume or, wherever practicable, less). (Note: Some oxygen analyzers require as much as two hours to stabilize before accurate readings can be obtained.)

The inert gas system is now ready to deliver gas to the cargo tanks. Continue the start-up steps as follows:

14. Open main deck isolation valve.
15. Open individual cargo tank branch valves if fitted.

Shutdown

Shutdown procedures should be as follows: When all tank atmospheres have been checked for an oxygen level of not more than 8 percent and the required in-tank pressure has been obtained,

1. Shut the deck isolation valve and the gas pressure regulating valve.
2. Shut down the inert gas blower.

3. Close the blower suction and discharge valve; check that the drains are clear; open the water-washing system on the blower while it is still rotating with the power off the driving motor, unless otherwise recommended by the manufacturer; and shut down the water-washing system after a suitable period.
4. Close the flue gas isolating valve and ensure the air sealing system functions.
5. Keep the full water supply on the scrubber tower in accordance with the manufacturer's recommendation; then flush with fresh water.
6. Ensure that the water supply to the deck water seal is running satisfactorily, that an adequate water seal is retained, and that the alarm arrangements for it are in order.

Safety Checks

When an inert gas plant is shut down, safety checks are in operation as follows:

1. The water supply and water level in the deck seal are ascertained at regular intervals of at least once per day depending on weather conditions.
2. The water level in loop seals installed in pipework for gas, water, or pressure transducers is checked to prevent the backflow of hydrocarbon gases into gas-safe areas.
3. In cold weather arrangements are ensured to prevent the freezing of water in the deck seal, pressure vacuum breaker, etc.
4. Inerted cargo tanks are repressurized with inert gas before their pressure drops to 100 millimeters.

Failure Symptoms

Among the possible causes for failure of an inert gas system is high oxygen content, which may be caused or indicated by the following conditions:

1. Poor combustion control at the boiler, especially under low load conditions;
2. Air drawn down the boiler uptake when gas output is less than the inert gas blower demand, especially under low load conditions;
3. Air leaks between the inert gas blower and the boiler uptake as well as faulty operation or calibration of the oxygen analyzer;
4. Operating in the recirculation mode; and/or
5. Entry of air into the inert gas main through the pressure-vacuum valve, pressure-vacuum breaker, or mast riser valve, due to misoperation. If the inert gas plant is delivering inert gas at an oxygen content of more than 5 percent, the fault is to be traced and repaired. All cargo tank operations shall be suspended if the oxygen content exceeds 8 percent unless the quality of the gas is improved.

Another cause for failure is the inability of the system to maintain positive pressure during cargo discharge or deballasting operations. This may be caused by:

1. Inadvertent closure of the inert gas valves;
2. Faulty operation of the automatic pressure control system;
3. Improper opening of the recirculating valve;
4. Inadequate blower pressure; or
5. A cargo discharge rate in excess of the blower output.

The cargo discharge or deballasting should be stopped or reduced depending on whether or not the positive pressure in the tanks can be maintained while the fault is rectified.

Standard Procedures

The inert gas system should be used during the full cycle of tanker operation as described in this section.

Inerting of tanks. Tanks that have been cleaned and gas freed should be reinerted preferably during the ballast voyage to allow the inert gas system to be tested fully prior to cargo handling. Purge pipes/vents should be opened to atmosphere. When the oxygen concentration of the atmosphere in the tank has fallen below 8 percent, the purge pipes/vents should be closed and the tank pressurized with inert gas.

During the reinerting of a tank following a breakdown and repair of the inert gas system, nongas-free and noninerted tanks should be reinerted. During inerting, no ullaging, dipping, sampling, or other equipment should be inserted unless it has been established that the tank is inert. This should be done by monitoring the efflux gas from the tank being inerted until the oxygen content is less than 8 percent by volume and for such a period of time as determined by previous test records when inerting gas-free tanks to ensure that the efflux gas is fully representative of the atmosphere within the tank.

When all tanks have been inerted, they should be kept common with the inert gas main and maintained at a positive pressure in excess of 100 millimeters water gauge during the rest of the cycle of operation.

Discharging water ballast. Before the discharge of cargo tank ballast is undertaken, the following conditions are checked as indicated:

1. All cargo tanks are connected up to the inert gas system and all branch isolation valves (if fitted) in the deck pipework are locked open.
2. All other cargo tank openings are shut.
3. The valve isolating the high velocity vent from the inert gas system is shut.
4. The inert gas plant is producing gas of an acceptable quality.
5. The deck isolation valve is open.

During the deballasting operation, the oxygen content of the gas and its pressure in the inert gas main should be continuously recorded.

Loading cargo. When loading cargo, the deck isolation valve is closed. The inert gas plant may be shut down unless other cargo tanks are being deballasted simultaneously. All openings to the cargo tanks except the connections to the high velocity vent arrangement are to be kept closed to minimize flammable vapor on deck. Before loading commences the flame screens in the high velocity venting arrangement should be inspected, and any stop valves isolating the cargo tanks from the inert gas main locked in the open position.

Loaded passage. During the loaded passage a positive pressure of inert gas of at least 100 millimeters water gauge should be maintained in the cargo tanks and topping up of the pressure may be necessary. When topping up the inert gas of pressure in the cargo tanks, particular attention should be paid to obtaining an oxygen concentration of 5 percent or less in the inert gas supply before introducing the gas into the cargo tanks.

On motor tankers, the auxiliary boiler loading may have to be increased in order that the low oxygen concentration in the inert gas supply can be achieved. It may also be necessary to restrict the output of the inert gas blowers to prevent air being drawn down the uptake during the topping up operation. If by these means inert gas of sufficient quality cannot be achieved, then inert gas from an alternative source of supply, such as an inert gas generator, might have to be used.

Discharging cargo. It may be necessary to relieve the inert gas pressure in the cargo tanks on arrival to permit manual measurement before cargo is discharged. If this is done, no cargo or ballasting operation is to be undertaken and a minimum number of small tank openings are to be uncovered for as short a time as necessary to enable these measurements to be completed. The tanks should then be repressurized before cargo discharge commences. Cargo discharge should not be commenced until all the conditions have been checked and are in order. During discharge the oxygen content and pressure of the inert gas in the inert gas main should be continuously recorded.

Crude oil washing. Before each tank is crude oil washed, the oxygen level shall be determined at a point 1 meter below the deck and at the middle region of the ullage space, and neither of these determinations should exceed 8 percent by volume. Where tanks have complete or partial wash bulkhead, the determination should be taken from similar levels in each section of the tank. The oxygen content and pressure of the inert gas being delivered during the washing process should be continuously recorded.

Descriptions of crude oil washing and regulations governing it will be presented at the end of this chapter.

Ballasting cargo tanks. The conditions for ballasting of cargo tanks are the same as those for loading. When, however, simultaneous discharge and ballasting is adopted, a close watch must be kept on the inert gas main pressure.

Ballast voyage. During a ballast voyage, tanks other than those required to be gas free for necessary tank entry should be kept inerted with the cargo tank atmosphere at a positive pressure of not less than 100 millimeters water gauge, and an oxygen level not exceeding 8 percent by volume especially during tank cleaning. Before any inert gas is introduced into cargo tanks to maintain a positive pressure, it should be established that the inert gas contains not more than 5 percent by volume of oxygen.

Tank cleaning. Cargo tanks should be washed in the inert condition and under a positive pressure. The procedures adopted for tank cleaning with water should follow those for crude oil washing.

Purging prior to gas freeing. When it is desired to gas free a tank after washing, reduce the concentration of hydrocarbon vapor by purging the inerted cargo tank with inert gas. Open purge pipes/vents to atmosphere, introduce inert gas into the tank until the hydrocarbon vapor concentration measured in the efflux gas has been reduced to 2 percent by volume and until such time as determined by previous tests on cargo tanks has elapsed to be confident that readings have stabilized and the efflux gas is representative of the atmosphere within the tank.

Gas freeing. Gas freeing of cargo tanks is only to be carried out when tank entry is necessary (e.g., for essential repairs). It should not be started until it is established that a flammable atmosphere in the tank will not be created as a result. Hydrocarbon gases are to be purged from the tank.

Gas freeing may be effected by pneumatically, hydraulically, or steam-driven portable blowers, or by fixed equipment. In either case it is necessary to isolate the appropriate tanks to avoid contamination from the inert gas main. Gas freeing should continue until the entire tank has an oxygen content of 21 percent by volume and until a reading of less than 1 percent of the lower flammable limit is obtained on a combustible gas indicator. Care must be taken to avoid the leakage of air into inerted tanks, or inert gas into tanks which are being gas freed.

Preparation for tank entry. The entry of personnel into the cargo tank should only be carried out under the close supervision of a responsible ship's officer and in accordance with all applicable safety rules. Some safety precautions to observe are listed as follows:

1. Secure inert gas branch line gas valves and blank(s) in position or, if gas freeing with the inert gas blower, isolate the scrubber from the flue gases.
2. Close all drain lines entering the tank from the inert gas main and secure relevant cargo line valves or controls in the closed position.
3. Keep the inert gas deck pressure on the remainder of the cargo tank system at a low positive pressure such as 200 millimeters water gauge. (This minimizes the possible leakage of inert hydrocarbon gas from other tanks through possible bulkhead cracks, cargo lines, valves, etc.) Clean sample lines well into the lower regions of the tank in at least two locations. Keep these locations away from both the inlet and outlet openings used for gas freeing. After it has been ascertained that a true bottom sample is being obtained, require the following readings: 21 percent on a portable oxygen analyzer; and less than 1 percent lower flammable limit on a combustible gas indicator.
4. Observe use of breathing apparatus whenever there is any doubt about the tank being gas free, e.g., in tanks where it is not possible to sample remote locations. (This practice should be continued until all areas, including the bottom structure, have been thoroughly checked.)
5. Ventilate continuously and sample regularly the tank atmosphere whenever personnel are in the tank.
6. Observe carefully normal regulations for tank entry.

Reinerting after Tank Entry

When all personnel have left the tank and the equipment has been removed, the inert gas branch line blank, if fitted, should be removed, the hatch closed, and the main deck isolation valve reopened. As soon as a gas-free tank is reconnected to the inert gas main, it should be reinerted to prevent transfer of air to other tanks.

Special Precautions for Product Carriers

The basic principles of inerting are exactly the same on a product carrier as on a crude oil tanker. However, there are differences in operation of these vessels as outlined below. Product carriers transport products having a flash point exceeding 140°F (60°C) (closed cup test) as determined by an approved flash point apparatus. Product carriers may carry bitumens, lubricating oils, heavy fuel oils, high flash point jet fuels, some diesel fuels, gas oils, and special boiling point liquids without inert gas systems having to be fitted, or, if fitted, without tanks containing such cargoes having to be kept in the inert condition. If cargoes with a flash point over 60°C are carried heated at temperatures near to or above their flash point (some bitumen cut-backs and fuel oils), a flammable atmosphere can occur. When cargoes with a flash point above 140°F (60°C) are carried at a temperature higher than 9°F (5°C) below their flash point, they should be carried in an inerted condition. When a nonvolatile cargo is carried in a tank that has not been previously gas freed, then that tank shall be maintained in an inert condition.

Product contamination. Contamination of a product may affect its odor, acidity, or flash point specifications, and may occur in several ways. Those ways which are relevant to ships with an inert gas main connecting all cargo tanks are contamination by other cargo, liquid contamination due to overfilling a tank, and vapor contamination through the inert gas main. The latter is largely a problem of preventing vapor of low flash cargoes (typically gasolines) from contaminating the various high flash cargoes, aviation gasolines, and most hydrocarbon solvents. This problem can be overcome by removing vapors of low flash point cargoes prior to loading and by preventing ingress of vapors of low flash point cargoes during loading and during the loaded voyage. When carrying hydrocarbon solvents where quality specifications are stringent and where it is necessary to keep individual tanks positively isolated from the inert gas main after a cargo has been loaded, pressure sensors should be fitted so that the pressure in each such tank can be monitored. When it is necessary to top up the relevant tanks, the inert gas main should first be purged of cargo vapor.

For a well-designed and -operated flue gas system, experience suggests that petroleum cargoes traditionally carried on product tankers do not suffer contamination from the flue gas itself, as opposed to contamination from other cargoes. However, unacceptable contamination from the flue gas may be encountered if proper control is not exercised over fuel quality, efficiency of combustion, scrubbing, and filtering. The more critical petrochemical cargoes which may be carried by product carriers can be contaminated by flue gas.

All lubricating oils and jet fuels are acutely water critical. Current practice requires full line draining and mopping up of any water in tanks before loading. Water contamination may occur on inerted ships because of water carryover from the scrubber or deck water seals due to inadequacies in design or maintenance of the various drying arrangements and from condensation of water from warm, fully saturated flue gas delivered to the tanks.

Requirements for additional purging and gas freeing. Gas freeing is required on noninerted product carriers more frequently than on crude carriers because of the greater need both for tank entry and inspection, especially in port, and for venting vapors of previous cargoes. On inerted product carriers any gas-freeing operation has to be preceded by a purging operation, but gas freeing for purely quality reasons may be replaced by purging only. It should be recognized that there are increased risks of air leaking into inert tanks and of inert gas leaking into a tank being entered; that purging is not a prerequisite of gas freeing when the hydrocarbon gas content of a tank is below 2 percent by volume; and that the operation of gas freeing for product purity and where tank entry is not contemplated does not require the atmosphere to contain an oxygen content of 21 percent by volume.

Special Precautions for Combination Carriers

The basic principles of inerting are exactly the same on a combination carrier as on a tanker. However, there are differences in the design and operation of these vessels, and in this respect, considerations are outlined below.

Static electricity in slack holds. It is particularly important for combination carriers to have their holds inerted because whenever a hold in an OBO (oil-bulk-ore) carrier (which could extend the full breadth of the ship) is partially filled with clean or oily ballast, water agitation of this ballast can occur at small angles of roll and this can result in the generation of static electricity. The agitation is sometimes referred to as sloshing, and it can happen whenever the ullage of the liquid content of the hold is more than 10 percent of the depth of the hold, measured from the underside of the deck. (See Figure 26-15 for remedy condition.)

Figure 26-15. Cargo hold of OBO carrier

Leakage through hatches. To ensure that leakage of tank gas, particularly through the hatch centerline joints, is eliminated or minimized, it is essential that the hatch covers are inspected frequently to determine the state of their seals, their alignment, etc. When the hatch covers have been opened, particularly after the ship has been carrying a dry bulk cargo, the seals and trackways should be inspected and cleaned of any foreign matter.

Ballast and void spaces. The cargo holds of combination carriers are adjacent to ballast and void spaces. Leakages may occur in pipelines or ducts in these spaces, or by a fracture in the boundary plating. In this event there is a possibility that oil, inert gas, and hydrocarbon gas may leak into the ballast and void spaces. As a consequence gas pockets may form, and difficulty with gas freeing could be anticipated due to the considerable steelwork, in the form of stiffening, which is characteristic of these spaces. Therefore, personnel need be alerted to the hazard.

Inert gas distribution system. Due to the special construction of combination carriers, the vent line from the cargo hatchway coaming is situated very close to the level of the cargo surface. In many cases, the inert gas main line passing along the main deck may be below the oil level in the hold. During rough weather, oil or water may enter these lines and completely block the opening, preventing an adequate supply of inert gas during either tank cleaning or discharge. Vent lines should therefore have drains fitted at their lowest point, and these should always be checked before any operation takes place within the cargo hold.

Applications carrying oil and nonoil cargo. On combination carriers the inert gas system should be utilized whenever the ship is engaged exclusively in the carriage of oil. When a combination carrier is carrying a cargo other than oil, it should be considered as a tanker. When cargoes other than oil are intended to be carried it is fundamental that all holds of cargo tanks other than slop tanks be emptied of oil and oil residues, and cleaned and ventilated to such a degree that the tanks are completely gas free and internally inspected. The pump room, cargo pumps, pipelines, duct keel, and other void spaces are to be checked to ensure that they are free of oil and hydrocarbon gas. Where holds are required to carry cargo other than oil, they should be isolated from the inert gas main and oil cargo pipeline by means of blanks which should remain in position at all times cargoes other than oil are being handled or carried. During the loading and discharging of solid cargoes and throughout the intervening periods all holds of cargo tanks other than slop tanks, cargo pump rooms, cofferdams, duct keels, and other adjacent void spaces should be kept in a gas-free condition and checked periodically at intervals of not more than two days to ensure that:

1. There has been no generation of hydrocarbon gas or leakage of hydrocarbon gas from the slop tanks. If concentrations of more than 20 percent of lower flammable limit are detected, the compartment(s) should be ventilated.

2. There is no deficiency of oxygen which could be attributable to leakage of inert gas from another compartment.

As an alternative to this procedure, those cargo tanks which are empty of cargo may be reinerted provided they are subsequently maintained in the inert condition and at a minimum pressure of 100 millimeters water gauge at all times, and provided that they are checked at intervals of not more than two days to ensure that any generation of hydrocarbon gas does not exceed one percent by volume. If such a concentration is detected the compartment(s) should be purged with inert gas.

Slops should be contained in a properly constituted slop tank and should be:

1. Discharged ashore, and the slop tanks cleaned and ventilated to such a degree that the tanks are completely gas free and then inerted; or
2. Retained on board for not more than one voyage when, unless the vessel reverts to carrying oil, the slop tank should be inerted; or
3. If retained on board for more than one voyage because reception facilities for oily residues are not available, the slop tank should be inerted.

When slop tanks have not been discharged, it is required that they be isolated from other tanks by blank flanges which will remain in position at all times when cargoes other than oil are being carried. On combination carriers where there are also empty cargo tanks which are not required to be isolated from the inert gas main, the arrangement for isolating the slop tanks from these tanks should be such as to:

1. Prevent the passage of hydrocarbon gas from the slop tanks to the empty tanks; and
2. Facilitate monitoring of and, if necessary, topping up of the pressure in slop tanks and any empty cargo tanks if the latter are being kept in the inert condition. (A suggested arrangement is shown in Figure 26-16.) In addition, all cargo pipelines to or from the slop tanks should be blanked off.

The above requirements need not apply and instead a combination carrier may be operated as a bulk carrier without having to use its inert gas system:

1. If it has never carried a cargo of oil;
2. If, after its last cargo of oil, all of its cargo tanks, including slop tanks, the pump room, cargo pumps, pipelines, cofferdams, duct keel, and other void spaces are emptied of oil and oil residues, cleaned, and made completely gas free and the tanks and void spaces internally inspected to that effect; or
3. If, in addition, the monitoring for hydrocarbon gas is continued until it has been established that generation of hydrocarbon gas has ceased.

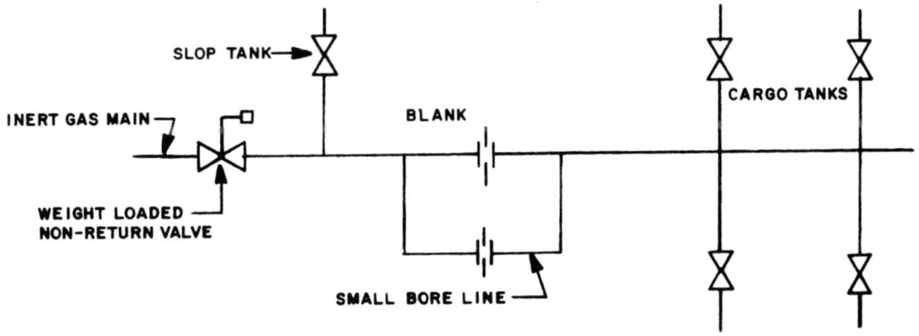

Figure 26-16. Bypass arrangement for topping cargo tanks

Emergency Procedures

In the event of total failure of the inert gas system to deliver the quality and quantity of inert gas to maintain a positive pressure in the cargo tanks and slop tanks, action should be taken immediately to prevent any air being drawn into the tank. All cargo tank operations should be stopped, and the deck isolating valve should be closed.

If it is assessed to be totally impractical to effect a repair to enable the inert gas system to deliver the quality of gas and maintain a positive pressure in the cargo tanks, cargo discharge and deballasting should only be resumed provided that the following precautions are taken:

1. The flame screens are checked to ensure that they are in a satisfactory condition.
2. The valve on the high velocity vent is opened.
3. No free-fall of water or slops is permitted.
4. No dipping, ullaging, sampling, or other equipment should be introduced into the tank unless essential for the safety of the operation. If such equipment must be introduced into the tank, it should only be done after at least 30 minutes have elapsed since the injection of inert gas has ceased. All metal components of equipment to be introduced into the tank should be securely grounded. This restriction should not be applied until a period of five hours has elapsed since injection of inert gas had ceased.

If it is essential to clean tanks following the failure of the inert gas system, and satisfactory inerted conditions cannot be maintained, the following precautions are necessary:

1. Tank washing should only be carried out one tank at a time.
2. The tank should be isolated from other tanks and from any common venting system or the inert gas main and maximum ventilation output should be

concentrated on that tank both before and during the washing process. Ventilation should provide as far as possible a free flow of air from one end of the tank to the other.

3. The tank bottom should be flushed with water and stripped. The piping system including cargo pumps, crossovers, and discharge lines should also be flushed with water.
4. Washing should not commence until tests have been made at various levels to establish that the vapor content in any part of the tank is below 10 percent of the lower flammable limit.
5. Testing the tank atmosphere should continue during the washing process.
6. If the vapor level rises to within 50 percent of the lower flammable limit, washing should be discontinued until vapor level has fallen to 20 percent of the lower flammable limit or less.
7. If washing machines with individual capacities exceeding 60 cu. meters/hr are to be used, only one such machine shall be used at any one time on the ship. If portable machines are used, all hose connections should be made up and bonding cables tested for continuity before the machines are introduced into the tank, and should not be broken until after the machines have been removed from the tank.
8. The tank should be kept drained during washing. If buildup of wash water occurs, washing should be stopped until the water has been cleared.
9. Only clean, cold seawater should be used. Recirculated water should not be used.
10. Chemical additives should not be used.
11. All deck openings, except those necessary for washing and designed venting arrangements, should be kept closed during the washing process.
12. The regulations of the local port authorities shall take precedence over any of the foregoing emergency procedures during cargo operations in port.

Safety Considerations

Backflow of cargo gases. To prevent the return of cargo gases or cargo from the tanks to the machinery spaces and boiler uptake, it is essential that an effective barrier is always present between these two areas. In addition to a nonreturn valve, a water seal and deck isolation valve should be fitted on the deck main. It is of prime importance that these devices are properly maintained and correctly operated at all times.

Health hazards

Oxygen deficiency. Exposure to an atmosphere with a low concentration of oxygen does not necessarily produce any recognizable symptom before unconsciousness occurs (at which time the onset of brain damage and risk of death can follow within a few minutes). If the oxygen deficiency is not sufficient to cause unconsciousness, the mind is liable to become apathetic and complacent, and even if these symptoms are noticed, and escape is attempted, physical exertion will aggravate the weakness of both mind and body. It is therefore necessary to ventilate thoroughly so no pockets of oxygen-deficient atmosphere remain. When testing for man entry, a steady reading of 21 percent oxygen by volume is required.

Toxicity of hydrocarbon vapors. Inert gas does not affect the toxicity of hydrocarbon gases, and the problem is no different from that of ships without an inert gas system. Because of possible gas pockets, regeneration, etc., gas freeing must continue until the entire compartment shows a zero reading with a reliable combustible gas indicator or equivalent, or a 1 percent lower flammable limit reading should the instrument have a sensitivity scale on which a zero reading is impractical.

Toxicity of flue gas. The presence of toxic gases such as sulphur dioxide, carbon monoxide, and oxides of nitrogen can only be ascertained by measurement. However, provided that the hydrocarbon gas content of an inerted tank exceeds about 2 percent by volume before gas freeing is started, the dilution of the toxic components of flue gas during the subsequent gas freeing can be correlated with the readings of an approved combustible gas indicator or equivalent. If, by ventilating the compartment, a reading of 1 percent lower flammable limit or less is obtained in conjunction with an oxygen reading of 21 percent by volume, the toxic trace gases will be diluted to concentrations at which it will be safe to enter. Alternatively, and irrespective of initial hydrocarbon gas content, ventilation should be continued until a steady oxygen reading of 21 percent by volume is obtained. Inert gas is asphyxiating. Great care must be exercised when work on the plant is undertaken. Although the worker may be in the fresh air, inert gas leaking from the plant could render him unconscious very quickly. Before opening up any equipment, therefore, it is recommended that the inert gas plant is completely gas freed. If any unit (e.g., the inert gas scrubber) is to be examined internally, the standard recommendations for entering enclosed spaces must be followed. Blind flanges should be fitted where applicable, or the plant should be completely isolated.

Tank pressure. When an inerted cargo tank is maintained at a positive pressure, personnel should be advised of the practical hazards of this pressure. Such pressure must be adequately reduced before opening any tank lids, ullage plugs, or tank washing openings.

Electrostatic hazards. Small particulate matter carried in flue gas can be electrostatically charged. The level of charge is usually small, but levels have been observed well above those encountered with water mists formed during tank washing. Because cargo tanks are normally in an inerted condition, the possibility of electrostatic ignition has to be considered only if the oxygen content of the tank atmosphere rises as a result of an ingress of air or if it is necessary to inert a tank which already has a flammable atmosphere.

MAINTENANCE AND TESTING

The safety arrangements are an integral part of the inert gas system and it is important for the ship's staff to give special attention to them during any inspection.

Inspection routines for some of the main components are dealt with in this section.

Inert Gas Scrubber

Inspection may be made through the manholes. Checks should be made for corrosion attacks, fouling, and damage to:

1. Scrubber shell and bottom.
2. Cooling water pipes and spray nozzles (fouling).
3. Float switches and temperature sensors.
4. Other internals such as trays, plates, and demister filters.

Checks should be made for damage to nonmetallic parts such as internal lines, demisters, and packed beds.

Inert Gas Blowers

To a limited degree, internal visual inspection will reveal damage at an early stage. Diagnostic monitoring systems should be used as they greatly assist in maintaining the effectiveness of the equipment. By fitting two equal-sized blowers or, alternatively, supplying and retaining on board a spare impeller with a shaft for each blower, an acceptable level of availability is ensured. Visual inspection through the available openings in the blower casing are adequate for this purpose.

An inspection of inert gas blowers should include:

1. Internal inspection of the blower casing for soot deposits or signs of corrosive attack.
2. Examination of fixed or portable washing system.
3. Inspection of the functioning of the freshwater flushing arrangements, where fitted.
4. Inspection of the drain lines from the blower casing to ensure that they are clear and operative.
5. Observation of the blower under running conditions for signs of excessive vibration, indicating too large an imbalance.

Deck Water Seal

This unit performs an important function and must be maintained in good condition. Corroded inlet pipes and damage to flat control valves and level indicators are not uncommon. The overboard drain line and connection are also possible sources of trouble. An inspection of the deck water seal should include opening for internal inspection to check for:

1. Blockage of the venturi lines in semidry-type water seals.
2. Corrosion of inlet pipes and housing.
3. Corrosion of heating coils.
4. Corroded or sticking floats for water drain and supply valves and level monitoring.

Appropriate testing for the function of the deck water seal should include automatic filling and draining (check with a local level gauge if possible) and presence of water carryover (open drain cocks on inert gas main line) during operation.

Nonreturn Valve

The nonreturn valve should be opened for inspection to check for corrosion and also to check the condition of the valve seat. The functioning of the valve should be tested in operation.

Scrubber Effluent Line

The scrubber effluent line cannot normally be inspected internally except when the ship is in dry dock. The ship side stub piece and the overboard discharge valve should be inspected at each dry-docking period.

Testing of Other Units and Alarms

A method should be devised to test the correct functioning of all units and alarms, and it may be necessary to simulate certain conditions to carry out an effective testing program. An effective testing program should include checking for:

1. All alarm and safety functions.
2. The functioning of the flue gas isolating valves.
3. The operation of all remotely or automatically controlled valves.
4. The functioning of the water seal and nonreturn valve (with a backflow pressure test).
5. The vibration level of the inert gas blowers.
6. Leakages (in systems four years old or more, deck lines should be examined for gas leakage).
7. The interlocking of the soot blowers.

8. Oxygen-measuring equipment, both portable and fixed, for accuracy by means of both air and a suitable calibration gas.

Component	*Preventive Maintenance*	*Maintenance Interval*
Flue gas isolating valves	Operate the valve.	Before start-up and one week
	Clean with compressed air or steam.	Before operating valve
	Dismantle for inspection and cleaning.	Boiler shutdown
Flue gas scrubber	Water flush.	After each use
	Clean demister.	Three months
	Dismantle level regulators and temperature probes for inspection.	Six months
	Open for full internal inspection.	Dry-docking
Overboard pipes and valve from flue gas scrubber	Flush with scrubber water and pump for about one hour.	After each use
	Dismantle the valve for overhaul; inspect pipeline and overboard end.	Dry-docking repair period
Blowers	Check vibration.	During operation
	Flush.	After each use
	Inspect internally through hatches.	After flushing and six months
	Dismantle for full overhaul of bearings, shaft tightenings, and other necessary work.	Two years or more frequently if required or at dry-docking
Deck water seal	Dismantle level regulators/float valves for inspection.	Six months
	Open for total internal inspection.	One year
	Overhaul automatic valves.	One year
Deck mechanical nonreturn valve	Move and lubricate the valve if necessary.	One week and before start
	Open for internal inspection.	One year to 18 months
Pressure-vacuum valves	Operate and lubricate the valves.	Six months
	Open for full overhaul and inspection.	One year
Deck isolating valve	Open for overhaul.	One year
Gas pressure regulating system	Remove condensation in instrument, air supply.	Before start
	Open gas pressure regulating valves for overhaul.	As appropriate
Liquid filled pressure-vacuum breaker	Check liquid level when system is at atmospheric pressure.	When opportunity permits and every six months

INSTRUCTION MANUALS

Instruction manuals are required to be provided on board by regulations, and should contain the following information and operational instructions:

1. A line drawing of the inert gas system showing the positions of the inert gas pipework from the boiler or gas generator uptakes to each cargo tank and

slop tank; gas scrubber; scrubber cooling water pump and pipework up to the effluent discharge overboard; blowers including the suction and discharge valves; recirculation or other arrangements to stabilize the inert gas plant operation; fresh air inlets; automatic gas pressure regulating stop valve; deck water seal and water supply; heating and overflow arrangements; deck nonreturn stop valve; water traps in any supply, vent, drain, and sensing pipework; cargo tank isolation arrangement; purge pipes/vents; pressure-vacuum valves on tanks; pressure-vacuum breakers on the inert gas main; permanent recorders and instruments and the takeoff points for their use; arrangements for using portable instruments, complete and partial wash bulkheads, mast risers, mast riser isolating valves; high velocity vents; and manual and remote controls.

2. A description of the system and a listing of procedures for checking that each item of the equipment is working properly during the full cycle of tanker operation. This includes a listing of the parameters to be monitored such as inert gas main pressure, oxygen concentration in the delivery main, oxygen concentration in the cargo tanks, temperature at the scrubber outlet and blower outlet, blower running current or power, scrubber pump running current or power, deck seal level during inert gas discharge to cargo tanks at maximum rate, deck seal level at nil discharge, etc. Established values for these parameters during acceptance trials should be included, where relevant.
3. Detailed requirements for conducting all operations particular to the installation of the ship such as times to inert, purge, and gas free each tank; sequence and number of tanks to be inerted, purged, and gas freed; sequence and number of purge pipes/vents to be opened or closed during such operations, etc.
4. Precautions to be taken (relating to the particular construction and equipment on board) to prevent leakage of inert gas and hydrocarbon vapors.
5. Precautions to be taken to prevent dangers arising from cargo over- and underpressure during various stages in the cycle of tanker operation.
6. Precautions to be taken to prevent dangers relating to the particular construction of the vessel or equipment on board.

CRUDE OIL WASHING

The pumping out operation for crude oil tank ships has been historically complicated by asphaltic compounds, waxes, and petroleum sands which form a difficult-to-remove sediment in the tanks. The buildup of this sediment over a number of voyages reduced the cargo-carrying capacity of the vessel, and proved to be expensive and difficult to remove when it was necessary to gas free and clean the tanks for shipyard entry. The International Convention for the Prevention of Pollution from Ships, 1973 (MARPOL), provided further incentive for crude oil washing by requiring that crude oil carriers over 40,000 DWT be equipped with segregated ballast tanks or be fitted with crude oil washing systems to be used instead of the traditional water washing system which contributes to pollution.

Since crude oil washing techniques require inert gas systems, it is appropriate to consider the important design and operating characteristics of crude oil washing machinery in this chapter on inert gas systems.

Techniques of Crude Oil Washing

For reliable details on crude oil washing, refer to "Revised Specifications for the Design, Operation and Control of Crude Oil Washing Systems" published by IMO. The appendix to this chapter includes extracts of text from the "Annex to Tanker Safety And Pollution Prevention 1978 Protocol."

Since crude oil is a good solvent and carrier of sediment, it is an effective medium for the removal of the voyage accumulation of sediment from the cargo tanks. With an inert gas system and fixed-in-place tank-cleaning machines similar to those described in the following paragraphs, it is possible to use the crude oil being discharged to clean the cargo tanks. As the cargo tank is discharged, crude oil is pumped under pressure to the tank-cleaning machines where it is directed from the nozzle at high velocity to the tank walls. The jet of crude oil removes sediments clinging to the tank's top, walls, and bottom. The washing machines are programmed to effectively cover the entire tank area. The sediments are mixed with the cargo oil and discharged to the shoreside storage tanks. Crude oil washing may only be undertaken in tanks fitted with an inert gas system which is operational and capable of maintaining the tank atmosphere at not more than 8 percent oxygen content and a positive pressure. Because the shipboard engineer officer is not usually involved in crude oil washing operations, it is not necessary to provide a complete description of the complicated procedure required for safe operation of the crude oil washing system.

The Crude Oil Washing System

Crude oil washing systems include permanently fixed-in-place tank-washing machines with fluid supplied from the main cargo pumps through permanent pipework. A typical modern installation will have single nozzle, high capacity, programmable, self-powered washing machines which are deck mounted. For difficult-to-clean areas, submersible machines supplied from permanent pipework are located within the tank. Tank-cleaning machine nozzles normally have nozzle diameters from 1⅝ to 1¼ inch which establish the flow of the cleaning jet. The submersible machines located in the bottom of the tank are inaccessible except during yard periods. Therefore, they must be highly reliable and rugged. The submersible machines are nonprogrammable.

A Typical Washing Machine

Typical of the deck-mounted machines in use on modern crude oil carriers is the Lavomatic SA tank washing machine. This equipment includes three main components: the gearbox unit, the drop pipe, and the wash head/nozzle assembly. The general arrangement of this machine is shown in Figure 26-17. Figure 26-18 is a photograph of the machine.

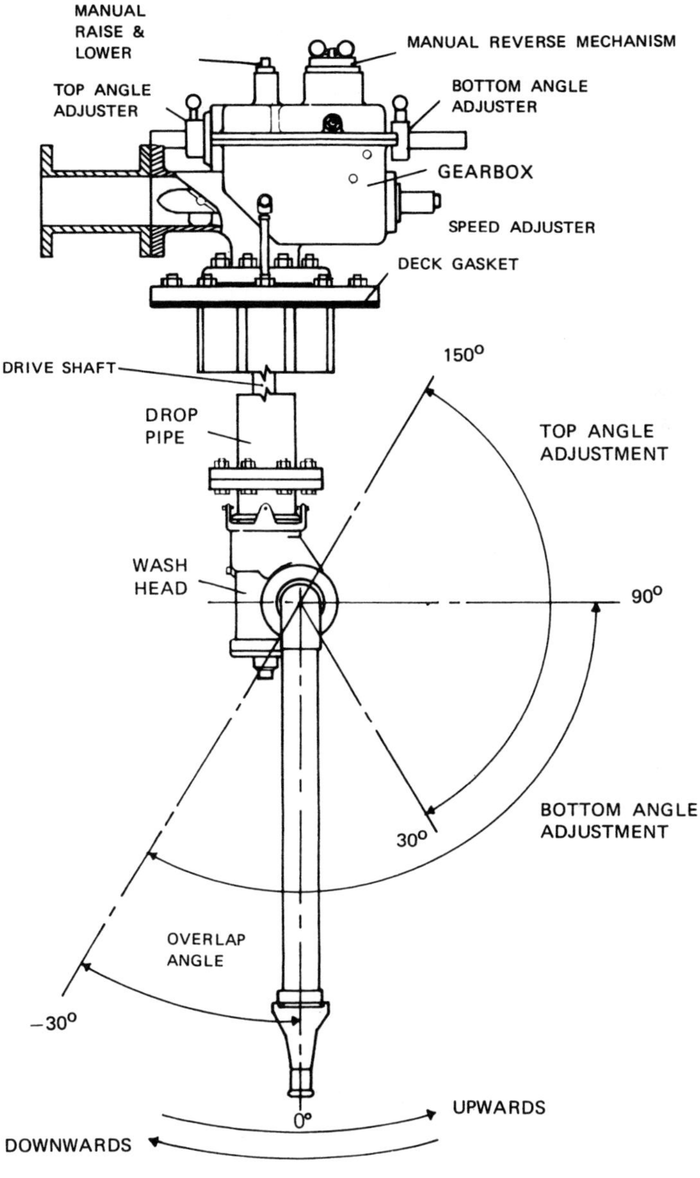

Figure 26-17. General arrangement of Lavomatic SA tank washing machine

Figure 26-18. Lavomatic SA tank washing machine

The gearbox unit shown in Figure 26-19 houses the impeller assembly which converts energy in the fluid stream to rotary power to drive the machine. The drive end of the impeller assembly is supported in the gearbox casing by a static bearing sleeve and seal housing. The non-drive end of the impeller shaft is supported in a bushing located in an end cap fitted on the gearbox casing. The cross shaft is located above the idler shaft, from which the drive to the cross shaft is transmitted via a spur gear on

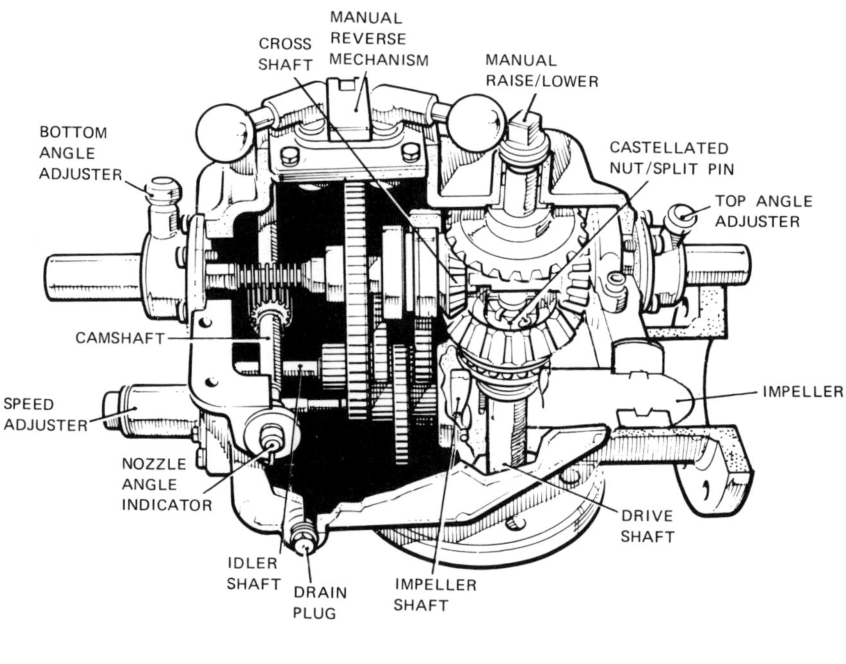

Figure 26-19. Gearbox unit

the cross shaft. The cross shaft assembly, together with the top and bottom nozzle angle adjusters, controls the total angular movement of the nozzle, the selected arc angles, and the automatic reversing of nozzle movement as programmed. A preselect facility is built into the assembly to enable the next new top angle setting to be selected during a wash cycle. The cross shaft assembly, through a dog clutch mechanism, drives a bevel gear keyed to the top of the drive shaft that turns the wash head assembly mounted on the bottom of the drop pipe. Where the drive shaft and the impeller penetrate the gearbox, double seals are provided. Any washing fluid seeping past the impeller seal or drive shaft O-rings drains into the tank and will not contaminate the gearbox. The rack end of the cross shaft is fitted with a shaft steady which is clamped to the gearbox casing by four screws. The two bevel pinion gears on the cross shaft are fitted with ball bearings. The drive for the main shaft is transmitted from the rotating bevel pinions on the cross shaft assembly to the bevel pinion keyed to the top of the main shaft assembly. In addition to the key, a castellated nut and split pin are also fitted for security. A robust thrust bearing is fitted below the bevel pinion to take the weight of the revolving wash head/nozzle assembly. Where the main shaft penetrates the gearbox, double O-ring seals are provided in a flanged plain bearing bush fitted to the top of the main shaft. The extension of the separate shaft protrudes through the top of the gearbox lid casing, and has a squared end to enable the machine to be wound manually. This shaft is supported in two self-lubricating bushes. Shaft sealing is provided by an O-ring fitted to a groove machined on the shaft. Manual drive is transmitted to the main drive via a bevel pinion keyed to the upper main shaft extension.

As shown in Figure 26-20 the drop pipe assembly basically comprises a long tube and a solid drive shaft. The outer tube (i.e., the drop pipe) extends from the main deck level into the cargo tank. Inside the drop pipe is the drive shaft (which provides the rotary wash head movement) and the vertical nozzle movement by means of spur and worm gearing fitted on the shaft. The gearbox unit is bolted directly to the drop pipe. The drive shaft provides the nozzle vertical movement and the rotary movement of the wash head assembly. Gear teeth cut on the bottom of the main inlet casting teeth mesh with a double pinion gear mounted on an end cap. The end cap is keyed to the drive shaft. The upper end of the shaft passes up through the gearbox; the bottom end passes through the wash head, both secured by a castellated nut and split pin.

The wash head/nozzle assembly is situated at the lower end of the drop pipe assembly inside the cargo tank. This assembly is illustrated in Figure 26-21. The assembly comprises three main parts: the main housing of gun metal, the flanged inlet housing of aluminum bronze, and the nozzle tube assembly. The main housing incorporates the worm and gear drive together with the nozzle tube assembly. The nozzle tube assembly primary

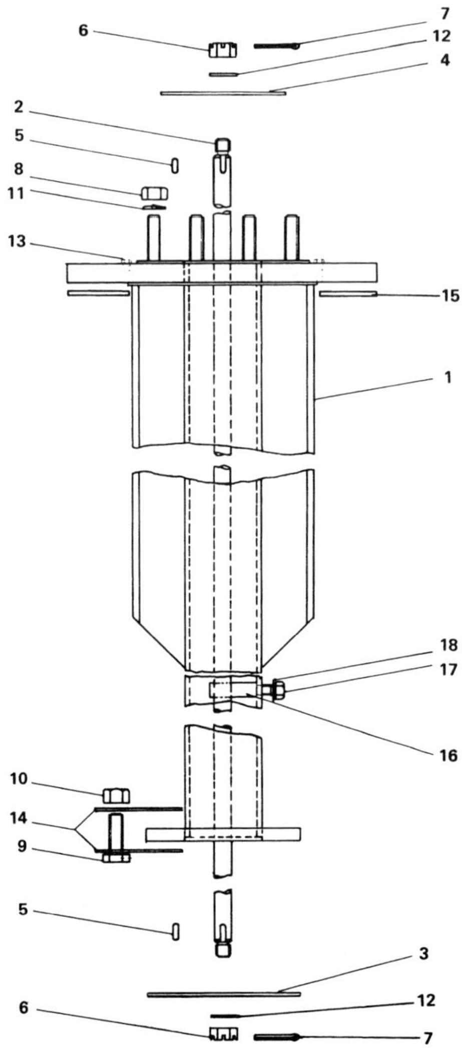

1	Flange & Pipe Assembly (ASA)	10	Hexagon Nut
2	Drive Shaft	11	Lock Washer
3	Flange Gasket	12	Flat Washer
4	Inlet Gasket	13	Plug
5	Key	14	Tab Washer
6	Hexagon Slotted Nut	15	Deck Gasket
7	Split Pin	16	Steady
8	Cap Nut	17	Nut
9	Bolt	18	Washer

Figure 26-20. Drop pipe assembly

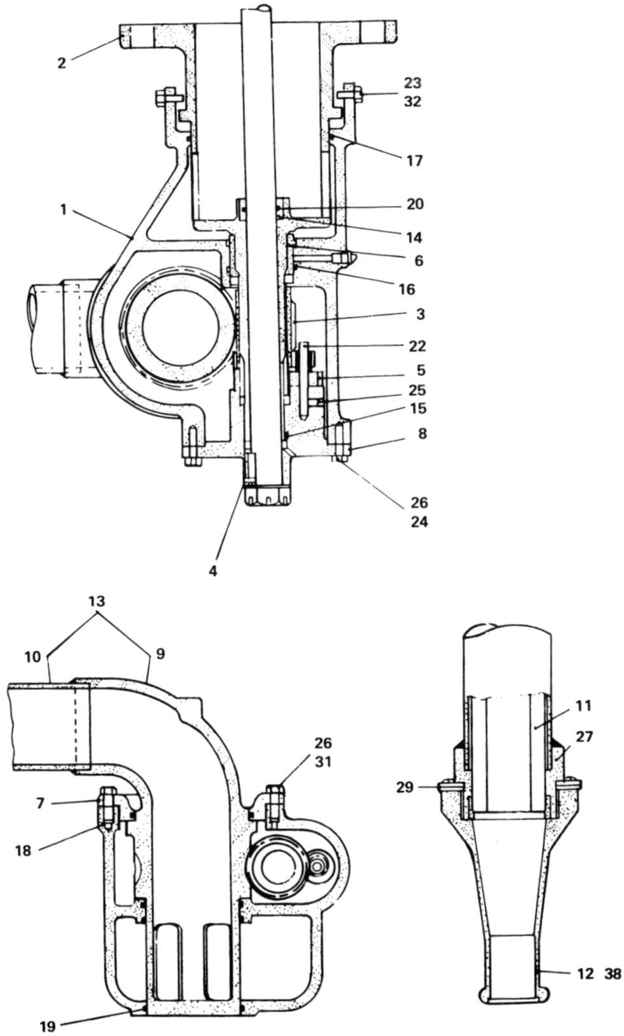

1 Main Housing Casting
2 Flanged Inlet Casting
3 Worm & Pinion
4 Groove Pin
5 26T & 14T Pinions
6 Sleeve
7 Retaining Plate
8 End Cap
9 Pipe Bend & Worm Wheel
10 Nozzle Tube
11 Stream Straightener
12 Nozzle Tip (38 mm)
13 Nozzle Tube Assy.
14 Seal Sleeve
15 O-Ring
16 O-Ring
17 O-Ring
18 O-Ring
19 O-Ring
20 O-Ring
22 Dowel Pin
23 Set Screw
24 Bolt
25 Grub Screw
26 Spring Washer
27 Adaptor
29 Grub Screw
31 Bolt
32 Spring Washer
38 Nozzle Tip (29 mm)

Figure 26-21. Wash head assembly

bend is secured to the main housing by means of a retaining plate fitted over the tube flange and secured in position by eight hexagonal headed bolts. An O-ring seal fitted between the flange and the main housing prevents leakage. An extension of the primary bend carries the worm wheel. This extension has four cast ports which allow the washing fluid to enter the nozzle tube assembly. Two O-ring seals are fitted to the extension to prevent leakage.

Washing machine maintenance considerations. At three-month intervals, check the washing machine for signs of leakage and ensure that the gearbox lubricating oil is at the proper level. Check all nuts for tightness. Periodically manually rotate the cross shaft assembly to recoat gearbox internals with protective lubricating oil.

At six-month intervals the gearbox should be checked for signs of water or crude oil, an indication of leaking O-ring seals. Fill lubricating oil to the proper level during this check.

APPENDIX: EXTRACTS FROM IMO REGULATIONS CONCERNING CRUDE OIL WASHING

Piping

1. The COW (crude oil washing) system shall consist of permanent pipework and be independent of the fire mains. (Flexible hoses can be used to connect the COW system to tank cleaning machines in cargo tank hatch covers.)
2. Pipes and valves shall be of steel or equivalent.
   The supply piping must have a pressure release device.
3. Hydrant valves must be blanked off.
4. The piping system shall be of such a diameter that the greatest number of tank-cleaning machines required can work simultaneously at the designed pressure and throughput.
5. The piping shall be tested to 1½ times normal working pressure once installed.
6. No part of the COW system shall enter the machinery space.

Tank-Cleaning Machines

1. Machines must be permanently mounted and must have their own valve to allow isolation.
2. Where separate drive units are used, there must be enough units on board to ensure no unit has to be moved more than twice from its original position to enable the washing cycle to be completed.
3. The number and location of machines in each cargo tank shall be such that all horizontal and vertical areas are washed by direct impingement or effectively by splash back or deflection of the impinging jet. In assessing an

acceptable degree of jet deflection and splashing, particular attention shall be paid to the washing of upward-facing horizontal areas and the following parameters shall be used:

(a) For *horizontal areas* of a tank bottom and the upper surfaces of a tank's stringers and other large primary structural members, the total area shielded from direct impingement by deck or bottom transverses, main girders, stringers or similar large primary structural members *shall not exceed* 10 percent of the total horizontal area of the tank bottom, the upper surfaces of stringers and other large primary structural members.

(b) For *vertical areas* of the sides of tanks, the total area of the tanks' sides shielded from direct impingement by deck or bottom transverses, main girders, stringers or similar large primary structural members *shall not exceed* 15 percent of the total area of the tanks' sides.

4. To confirm the cleanliness of the tank, a visual inspection will be made after COW and prior to water rinsing.
5. To verify the effectiveness of the stripping and drainage arrangements, a measurement will be made of the amount of free oil floating on top of the *departure ballast*. The ratio of the volume of oil (on top of the *departure ballast*) to the volume of tanks that contain this water, shall not exceed .00085. Note: This is 8.5 tons oil per 10000 tons water. So, for a 250,000- DWT tanker, this represents 212 tons of oil.
6. To verify the design, installation and operation of the system, the *arrival ballast* will be discharged through an oil-monitoring system and the oil content of the effluent *must not exceed 15 parts per million.*
7. The *deck-mounted tank-cleaning machines* must have an external indicator to indicate the rotation and arc of movement of the machine.
8. The *submerged machines* must be nonprogrammable to verify their operation.
   (a) They can have an external indicator.
   (b) Their characteristic sound pattern can be checked during operation (i.e., noise of the jet on deck structure). If more than one machine is mounted on the same supply line, they must have separate valves to verify operation of all the machines.
   (c) Their operation can be visually checked during the ballast passage. This check must be carried out after a maximum of six usages or at least once a year.

Pumps

1. The capacity of the pumps must be sufficient to provide the required throughput and the pressure for the maximum number of machines that will run simultaneously. *This must be met with one pump inoperative.*
2. Where the back pressure presented by the shore terminal is below the required pressure for COW, provision must be made to ensure the pressure can be maintained to run all the machines (again with *one pump inoperative*). If minimum supply pressure condition (usually 8kg/cm^2) cannot be met, *crude oil washing operations shall not be carried out.*

Stripping System

1. The stripping system must be able to remove oil at a rate of 1.25 times the total throughput of all the machines to be operated simultaneously when washing the bottom of the tank.
2. Verification of the stripping system will be by level gauges and "hand-dipping." Suitable arrangements for "hand-dipping" will be in the aftermost position of the tank as well as three other positions.
3. The line-and-pump draining must be discharged ashore through a special small diameter line connected outboard of the ship's manifold valve. This line shall not exceed 10 percent of the main cargo line for new vessels, or 25 percent for existing vessels already having fitted the line.

Operation

1. The number of tanks that have to be crude oil washed will be checked:
   (a) to ensure the vessel can comply with minimum draught and trim requirements;
   (b) to ensure that any additional ballast water required during the voyage is not put into tanks that have not been crude oil washed;
   (c) in addition, approximately one quarter of the remaining tanks must be crude oil washed for sludge control. No tank for sludge control need be washed more than once every four months.
2. Crude oil washing *cannot* be done during the ballast voyage.
3. Crude oil washing can *only* be done at sea when the vessel is between multiple discharge ports.
4. The *inert gas system* must be in proper use and prior to crude oil washing the oxygen level at a point *one metre* from the deck and at the middle of ullage space *must not exceed* 8 percent by volume.
5. Crude oil washing *must be stopped* if:
   (a) Oxygen level of inert gas being delivered exceeds 8 percent by volume, or
   (b) The pressure of the atmosphere in the tank is no longer positive.
   (c) To avoid excessive electrostatic generation in washing due to the presence of water, the content of any tank to be used as a source of crude oil washing fluid must be discharged by at least one metre.

Qualification of Personnel

Where a person, such as the Master, Chief Officer or Cargo Control Officer, assumes overall command of the COW operation, he must:

1. Have at least one year's experience on an oil tanker where his duties have included the discharge of cargo and associated crude oil washing;
2. Have participated in crude oil washing programmes *twice—once on that particular ship*.

Other nominated personnel must have had *six months'* experience on oil tankers, where they had been involved in cargo discharge.

REVIEW

1. What is the dilution concept?
2. What is the displacement concept?
3. What are the four basic types of inert gas systems?
4. What is the function of the deck water seal?
5. Describe the operation of a liquid-filled pressure-vacuum breaker.
6. Describe a typical stack gas inert system.
7. Describe the start-up procedure for an inert gas system.
8. Describe the condition of an inert gas system during loaded passage.
9. List the safety considerations of an inert gas system.
10. List the routine inspection checks for an inert gas scrubber.
11. What are the elements of an effective testing program for an inert gas system?
12. Describe a typical modern tank washing machine.
13. What are the maintenance considerations for a crude oil washing machine?

REFERENCES

American Bureau of Shipping. 1980. *Guidance Manual for Material Selection and Inspection of Inert Gas Systems.* New York: American Bureau of Shipping.

Atkinson, F.H. and L.J. Crighton. 1978. "Design Considerations of New Oil Tankers," London, IMCO Tanker Safety and Pollution Conference.

Baker, W. D. and T. W. Allsop. 1981. "Crude Oil Washing." *Transactions.* Institute of Marine Engineers.

Bell, A. June 1980, "IGS Hardware to Satisfy IMCO Requirements." *Marine Engineers Review,* Institute of Marine Engineers

Boroe, B. June 1983. "Safe Operating of Inert Gas Systems." *Marine Engineers Review.* Institute of Marine Engineers.

Chand, Dumbleton and Reiter. February 1982. "A Classification Society's Experience with IMCO Requirements for Crude Oil Washing and Gas Systems." Texas Section, Society of Naval Architects and Marine Engineers.

Cowles, W. C. March 1980. "A Practical Guide to Design and Installation of Crude Oil Washing Systems." New York Section, Society of Naval Architects and Marine Engineers.

Crowley, J. 1978. *Design, Construction and Operation of Oil Tankers—Crude Oil Washing.* London: International Maritime Organization.

Crowley, J. March 1979. "Crude Oil Washing of Oil Tankers." *Marine Engineers Review,* Institute of Marine Engineers.

Institute of Marine Engineers. November 1976. "Inert Gas Systems." *Marine Engineers Review.* Institute of Marine Engineers.

Institute of Marine Engineers. August 1978. "Crude Oil Washing, A Giant Step for Pollution Control." *Marine Engineers Review.* Institute of Marine Engineers.

Institute of Marine Engineers. June 1980. "Crude Oil Washing Can Save Money." *Marine Engineers Review.* Institute of Marine Engineers.

Institute of Marine Engineers.October 1982. "Auto-control of Crude Oil Washing." *Marine Engineers Review.* Institute of Marine Engineers.

Institute of Marine Engineers. July 1995. "New Cargo Pump Designs." *Marine Engineers Review.* Institute of Marine Engineers.

Institute of Marine Engineers. January 1997. "Alternate Crude Oil Washing Program Said to Offer Improved Sludge Removal." *Marine Engineers Review.* Institute of Marine Engineers.

Institute of Marine Engineers. January 1998. "How Additives Can Make Tank Cleaning More Efficient." *Marine Engineers Review.* Institute of Marine Engineers.

International Maritime Organization. Publication Number IMO-617E. *Crude Oil Washing Systems.* London: International Maritime Organization.

Norledge, T. K. June 1980. "Marine Inert Gas Systems." *Marine Engineers Review.* Institute of Marine Engineers.

ACKNOWLEDGMENTS

The editors are pleased to acknowledge and thank the Camar Corporation, Worcester, Massachusetts, and Butterworth Systems, Inc., Floram Park, New Jersey, for the material presented in this chapter.

CHAPTER 27

Coal Burning Technology

EVERETT C. HUNT

INTRODUCTION

Economic Factors

CONDITIONS in the marketplace periodically cause steep increases in petroleum fuel prices leading some shipowners to the consideration of coal as the fuel for ship propulsion. Worldwide, there are few modern coal fueled vessels. Nevertheless, even without unusually high petroleum fuel costs that make alternative fuels attractive, there are some unique trades in the world were coal fuel has significant economic advantage when it is readily available at low cost. Such trades are usually coastwise and involve coal or other bulk transportation.

Special Considerations in the Operation and Design of Coal Burning Propulsion Plants

The use of coal is, of course, conditioned by regulations both local and international, and by ship construction rules. Codes governing the building and operation of coal burning propulsion plants come from such sources as classification societies, the International Maritime Organization (IMO), the U.S. Coast Guard, and the U.S. Occupational Safety and Health Administration (OSHA).

Coal availability is an important consideration in the selection of coal fired boilers. Ships using coal will undoubtedly be those that follow trade routes with easy access to this fuel.

It will require an analysis to determine the types of coals available. This in turn will be a factor in the selection of particular coals and in the choice of firing methods. Due to the variety of coals which a ship may be required to use, the overfed stoker with traveling grate which can handle a wide range of coal characteristics will be the most common marine coal firing system selected. Other coal firing processes such as pulverized coal and fluid bed systems have serious disadvantages for shipboard application.

The means of handling coal and ashes influences the operation and design of a coal burning vessel. In loading coal aboard the vessel, methods for controlling noise and dust must be provided while assuring rapid movement of the fuel. A choice must be made between mechanical and pneumatic systems of conveying coal to the day bunker and to the boilers. Ash transfer, storage, and disposal methods must be selected to meet economic and regulatory requirements.

The owner and designer must give careful attention to the general arrangement of the ship for the economic utilization of coal. Considerations such as loss of cargo space, round trip bunker capacity, bunker geometry, need for cofferdams, inert gas systems, corrosion of bunkers, and watertight integrity are all important in providing a suitable general arrangement.

Combustion control systems for induced draft and forced draft combinations bring about an additional complication for coal fired vessels. Due to the slower response rate inherent in stoker fired systems, it is necessary to have a steam dump connected to the main condenser to avoid lifting safety valves during routine maneuvers. The main condenser must be specially designed to accept the steam dump load.

The steam power plant cycle is influenced by the selection of coal fuel. Because of the increasing cost of oil fuels, the steam plant design trend has been toward higher efficiency with a consequent increase in complexity. Coal fired steam plants have lower initial steam conditions and fewer feedwater heaters since these plant features cannot be economically justified with the lower cost coal. Figure 27-1 shows a typical proposed steam plant cycle for a 20,000 SHP coal fired vessel.

COAL FUEL

Physical and Chemical Characteristics

In order to use coal effectively its physical and chemical characteristics must be known and the relationship of these characteristics to boiler operation and design understood. Marine engineers who specify and operate coal fired marine boilers must understand the methods a boiler designer uses to relate coal analysis to boiler design.

Moisture Content and Heating Value

Coals are identified or ranked by their moisture content and heating value (mineral matter free) as shown in Figure 27-2. This method is based on age and degree of coalification. Bituminous coal is relatively low in moisture and high in heating value. Subbituminous and lignite are progressively higher in moisture content and lower in heating value. Although the classification system provides some information relative to a coal's application for steam generation, it does not provide sufficient information for the boiler designer or operator who needs a complete analysis to design and operate correctly all of the boiler components and to be able to predict boiler performance when coals of different characteristics are burned.

Coal Analysis

For marine applications, coal will often be acquired from more than one source and will vary in characteristics. This will require consideration of each of the several different coals to be used in order to design a boiler which will perform satisfactorily with each. It is important that analysis provided to the boiler designer describe the specific coals selected for use and that it not represent average figures derived from the analysis of a large number of coals. Using average results could lead to the adoption of a boiler larger than required or one which will not operate satisfactorily with the available coals.

There are a large number of analyses available to the designer, some routine and others very specialized, all of which may be important in the design and operation of a boiler. Figure 27-3 shows an example of an analysis for normal design requirements.

The American Society for Testing Materials Standards (Part 26) describes the procedures used for measuring the physical and chemical properties of coal. The significant properties are (a) total moisture content; (b) volatile matter, fixed carbon, and ash determined by proximate analysis; (c) carbon, hydrogen, nitrogen, sulfur, and oxygen determined by ultimate analysis; (d) BTU per pound measured by heating value; and (e) ash fusion temperatures.

Effect of Coal Properties on Design and Operation

Coal moisture can effect boiler design and performance in many ways. Moisture tends to reduce flame temperature and increases the specific heat of the flue gases. Large changes in the moisture content can alter the heat transfer characteristics in the furnace and thus the steam temperature. Moisture will increase the volume of flue gas which in turn will affect induced draft fan size, boiler draft loss, particulate collection equipment size, and boiler efficiency.

The extent of volatile matter in coal is an indication of the ease of its ignition, the higher the volatile matter in coal, the easier the coal is to ignite. Various types of firing equipment will require different limits of volatile matter for proper operation.

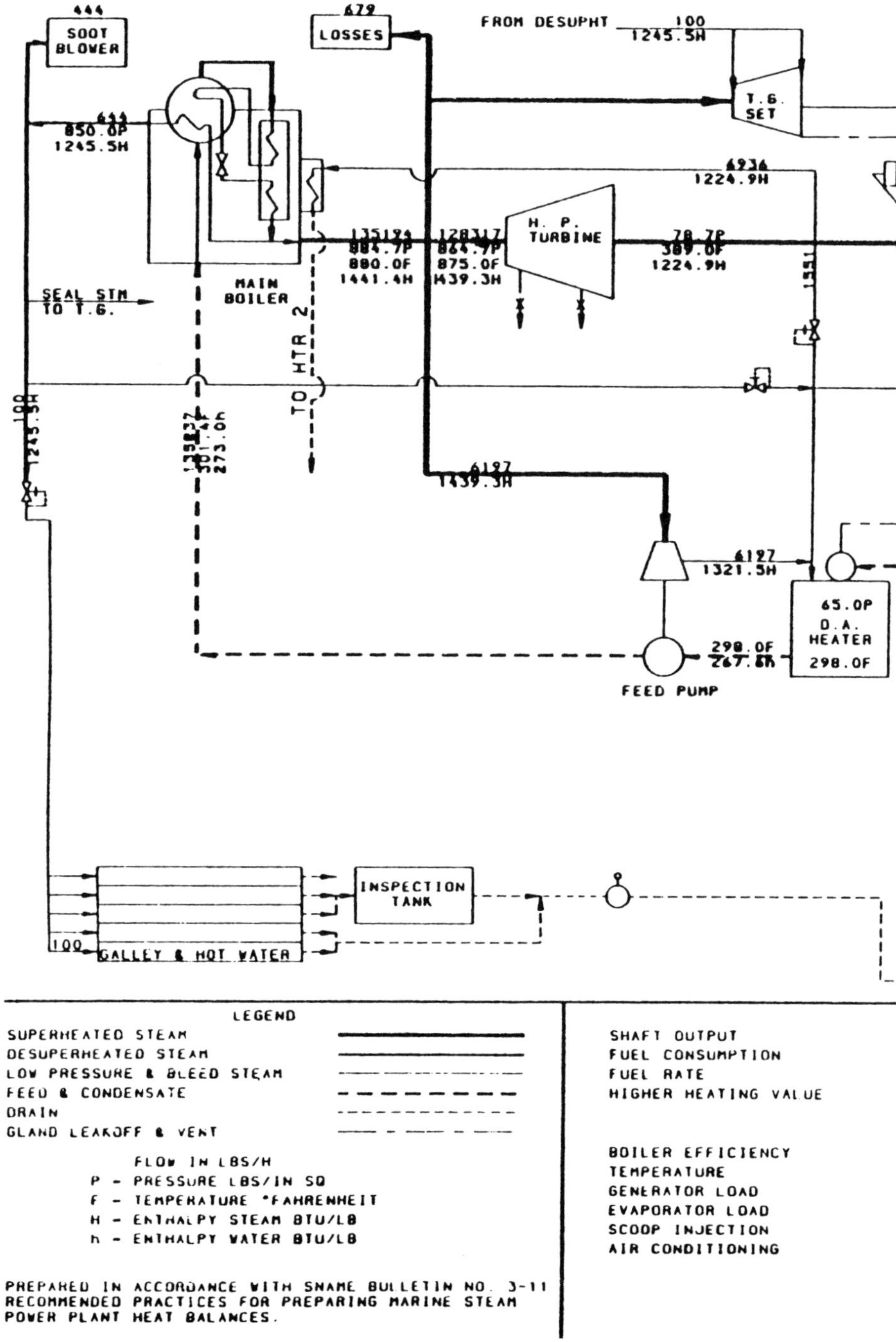

Figure 27-1. Typical 20,000 SHP coal fired steam

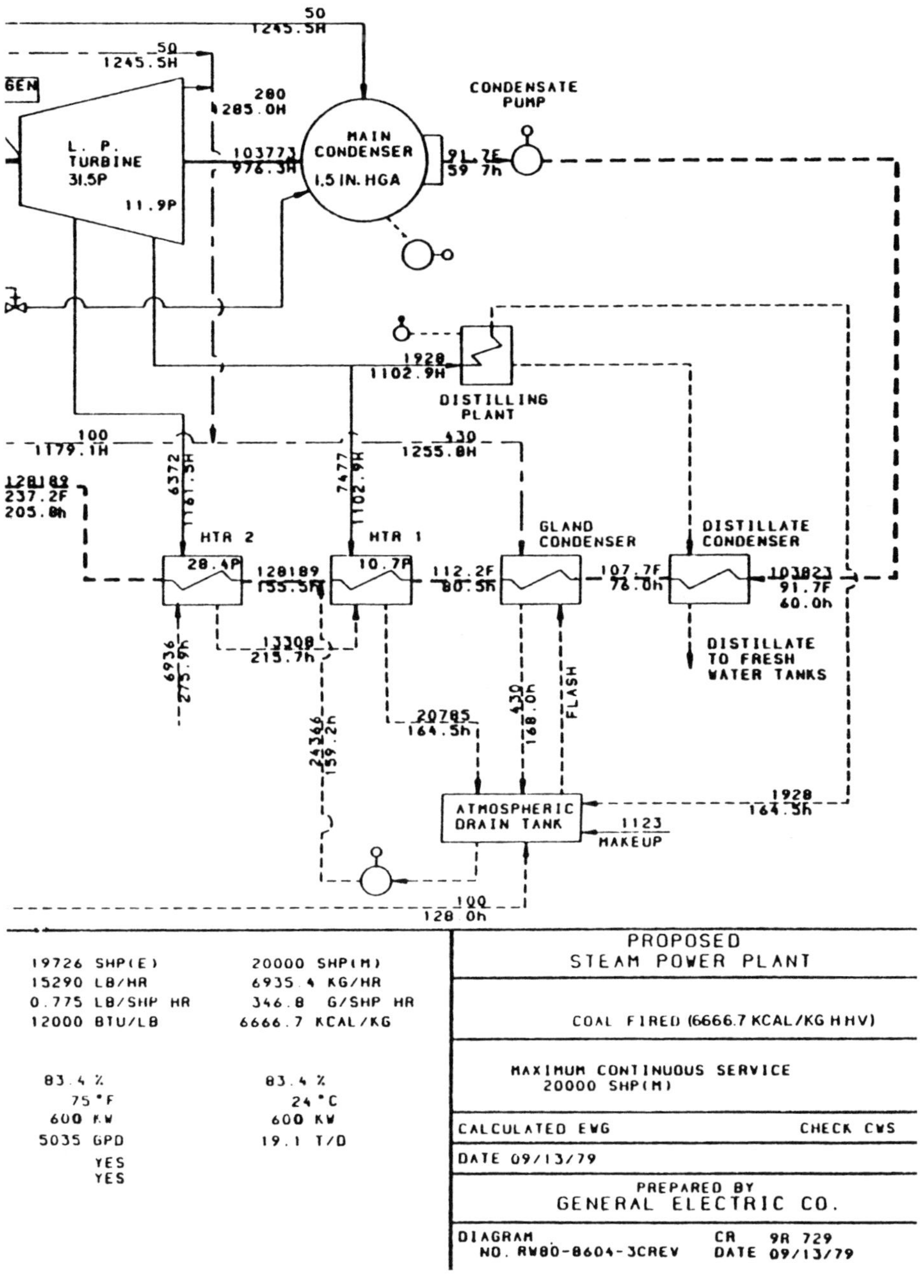

plant cycle. Courtesy General Electric Co.

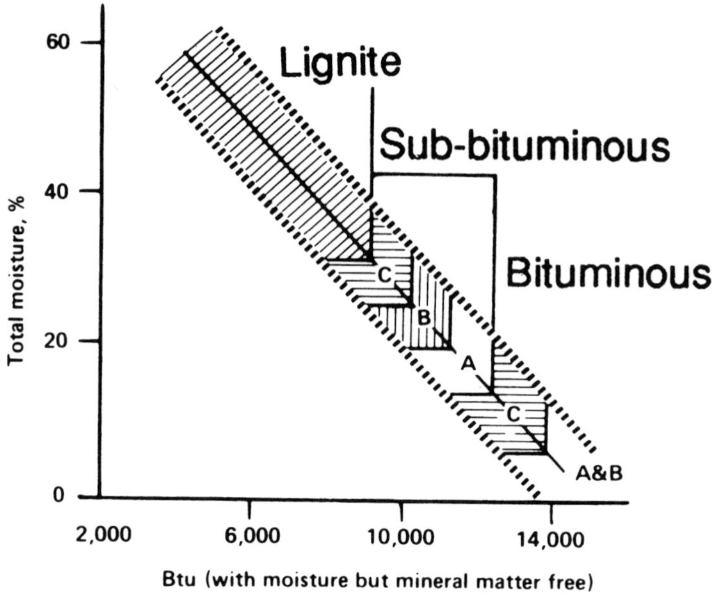

Figure 27-2. Coal classification by moisture content and heating value. Courtesy Babcock & Wilcox.

Fixed carbon is the combustible residue remaining after volatile matter has been driven off and represents the portion of the fuel which is burned in a solid form. Again, some types of firing equipment will require limitations of the amount of fixed carbon in the coal.

The ash in coal has to be paid for, transported, conveyed, heated, collected, and disposed of, but adds nothing useful to the operation of the boiler. The amount of ash is used to determine the dust loading to particulate control equipment and to determine the capacity of the ash removal and storage systems, and it may influence the design of the soot blowing system. The chemical characteristics of the ash are important and will be described herein.

The heating value of the coal will determine the amount to be stored, handled, and burned to provide the steam output required. It will directly affect the size of the burning and handling equipment.

The ultimate analysis breaks down the coal composition into its elemental components in addition to moisture and ash. Carbon, hydrogen, and oxygen are used, together with BTU per pound and moisture, to calculate the combustion air requirements, flue gas weight and composition, and boiler efficiency.

Sulfur content is important because the acid dew point of the flue gas is related to the percent of sulfur in the coal. Reduction of the flue gas temperature below the dew point can cause condensation on air heater or

Proximate Analysis %

Moisture	3.4
Volatile Matter	39.1
Fixed Carbon	48.9
Ash	8.6

Gross Heating Value

BTU per lb.	12940

Ultimate Analysis %

Moisture	3.4
Carbon	71.7
Hydrogen	5.1
Nitrogen	1.38
Sulfur	2.2
Ash	8.6
Oxygen (difference)	7.62

Ash Fusion Temperatures - F

	Reducing	**Oxidizing**
Initial Deformation	1940	2310
Softening	2010	2470
Softening	2140	2490
Fluid	2450	2660
Fluid	2730	2749

Ash Analysis (Spectographic):

Silicon as SiO_2	42.77
Aluminum as Al_2O_3	30.37
Iron as Fe_2O_3	17.50
Titanium as TiO_2	1.10
Calcium as CaO	1.95
Magnesium as MgO	0.60
Sodium as Na_2O	0.46
Potassium as K_2O	1.74
Sulfur as SO_3	1.87
Phosphorus as P_2O_5	0.15

Figure 27-3. Typical coal analysis. Courtesy Babcock & Wilcox.

economizer surfaces which will result in tube corrosion and excessive ash deposition. Figure 27-4 shows the tube metal temperatures required for different fuels to minimize corrosion.

Nitrogen content is also important because it will affect the amount of fuel NOx generated in the boiler. NOx is also generated from the reaction of nitrogen and oxygen in the combustion air, but fuel NOx can be a major contributor to the total amount leaving the stack.

Ash Deposition

Ash fusion temperatures provide information relative to the melting characteristics of the ash. These temperatures are determined in both

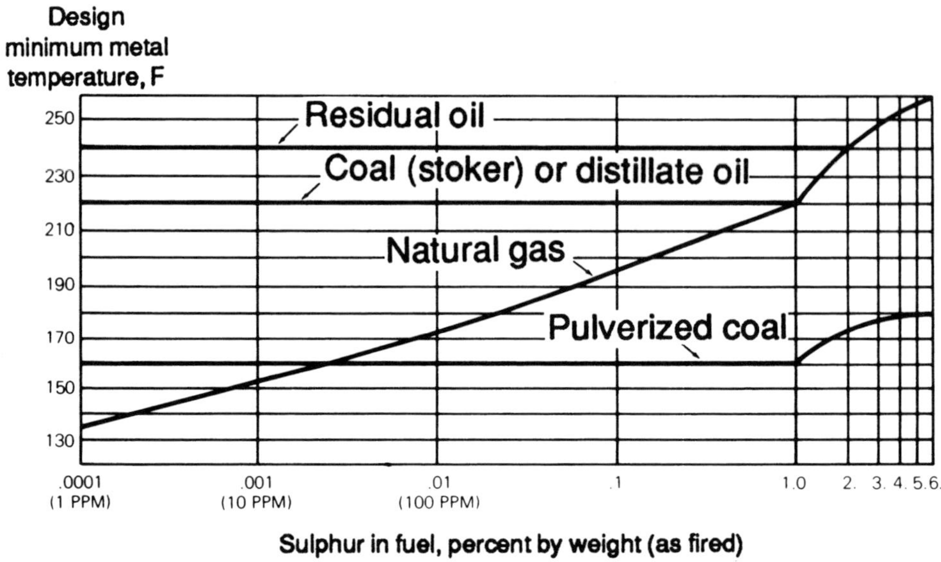

Figure 27-4. Tube metal temperatures to minimize corrosion. Courtesy Babcock & Wilcox.

reducing conditions in a fuel bed and oxidizing atmospheres at the top of the furnace since both conditions may exist in a boiler. These temperatures are used to determine the applicability of a coal to a specific type of firing, the size of the furnace, and the slagging tendency of the ash. The softening temperature is the temperature used by the designer to set the furnace exit temperature.

The ash spectrographic analysis provides an elemental analysis of the ash reported as the oxide. This information is the basic data needed to characterize the ash relative to its tendency to deposit on boiler heat transfer surfaces. Deposition is a major concern to the boiler designer and operator. Proper application of the data derived from this analysis is necessary to insure reliable operation of the boiler.

There are two major types of ash deposition in addition to the low temperature deposits associated with flue gas dew point temperature previously described. The first of these, *slagging,* is caused primarily by the physical transport of molten or partially fused ash particles entrained in the flue gas stream. When these particles strike a furnace wall or tube surface, they are chilled and solidify on the surface. The strength of their attachment is influenced by the temperature and physical contour of the surface, the direction and force of impact of the ash, and its melting characteristics. As a general rule, coals with low ash fusion temperatures have a high slagging potential.

The second type of deposition is called *fouling*. These high temperature bonded deposits are caused by constituents of the ash which are volatilized during combustion and condense on fly ash particles, tube surfaces, and existing deposits where the temperatures are such that the constituents remain liquid. These constituents react chemically with fly ash, other deposits, and flue gas to form the bonded deposit. The severity of fouling is related to sintering strength of the ash as determined in the laboratory.

Slagging and Fouling

Figure 27-5 shows the areas of a boiler which may be subject to slagging or fouling. In order to design a boiler for reliable operation it is necessary to minimize the tendency for slagging and fouling. By using a large number of coal analyses and comparing ash properties in operating boilers with those produced in laboratory combustors it has been possible to develop slagging and fouling indices. These indices reflect the severity of each type of deposition and are used by the boiler designer in a number of ways.

Characterization of Slagging

The first step in characterizing the slagging and fouling potentials of any coal ash is to determine whether the coal is a *bituminous* or *lignite* type since different tests for slagging and fouling potential for each type of ash have been found necessary and have been developed. A lignite ash is defined as one in which the total of CaO and MgO in the ash is greater than the amount of Fe_2O_3. When the amount of Fe_2O_3 is greater than the total of CaO and MgO, the ash is considered a bituminous type. This is shown in Figure 27-6.

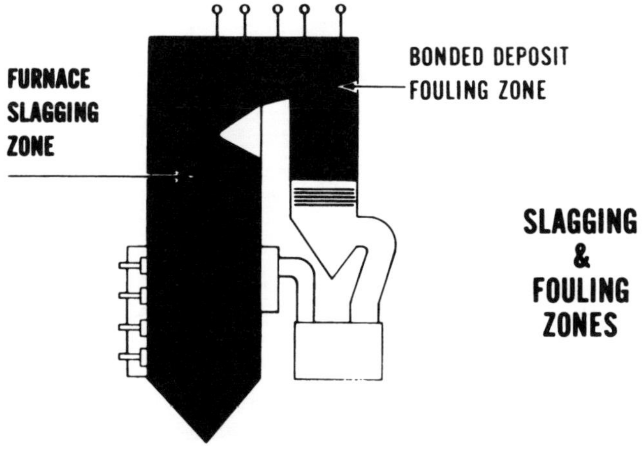

Figure 27-5. Boiler areas subject to slagging and fouling. Courtesy Babcock & Wilcox.

- Defined by ratio of iron (Fe_2O_3) to sum of calcium (CaO) and magnesium (MgO) in ash
- Bituminous type when $Fe_2O_3 > CaO + MgO$
- Lignitic type when $Fe_2O_3 < CaO + MgO$

Figure 27-6. Bituminous and lignitic ash. Courtesy Babcock & Wilcox.

There are three different methods for the determination of a slagging index described as low, medium, high, or severe. The first method is used for bituminous ash only and relates the base (B) to the acid (A) ratio of the ash components. The formula for the slagging index is:

$$R_S = \frac{B}{A} \times S$$

where $B = CaO + MgO + Fe_2O_3 + Na_2O + K_2O$
$A = SiO_2 + Al_2O_3 + TiO_2$
$S =$ weight % S in coal, dry basis

The classification of slagging potential using R_s as a function of base-acid ratio is shown in Figure 27-7.

$$R_S = \frac{\text{base}}{\text{acid}} \times S$$

Slagging Type	Index R_S
Low	< 0.6
Medium	0.6–2.0
High	2.0–2.6
Severe	> 2.6

Figure 27-7. Slagging potential classification by base/acid ratio. Courtesy Babcock & Wilcox.

The second method is used for lignite ash only and is based on ash fusion temperature. The formula for this slagging index is:

$$R_S = \frac{(\text{Max HT}) + 4\,(\text{Min IT})}{5}$$

where Max HT = higher of the reducing or oxidizing "hemispherical softening temperatures," °F

Min IT = lower of the reducing or oxidizing "initial deformation temperatures," °F

The classification of slagging potential using R_s as a function of fusion temperature is shown in Figure 27-8.

$$Rs = \frac{(\text{max. HT}) + 4(\text{min. IT})}{5}$$

Slagging factor Rs	Slagging classification
> 2250 F	Medium
2250 - 2100 F	High
< 2100 F	Severe

Figure 27-8. Slagging potential classification by fusion temperature. Courtesy Babcock & Wilcox.

The third method can be used with either type ash and is considered to provide the best indication of slagging potential. This method requires a determination of slag viscosity either by calculation or measurement. For a lignitic-type ash a direct determination using a high temperature rotating bob viscometer is recommended. The formula for this slagging index is:

$$R_{vs} = \frac{(T\ 250\ ox) - (T\ 10{,}000\ red)}{97.5\ fs}$$

where T 250 ox = temperature, °F, corresponding to a viscosity of 250 poise in an oxidizing atmosphere

T 10,000 red = temperature, °F, corresponding to a viscosity of 10,000 poise in a reducing atmosphere

fs = temperature correlation factor

The classification of slagging potential using R_{vs} together with detailed information relative to fs is shown in Figure 27-9.

The value of 250 poise is a viscosity at which slag will flow on a horizontal surface. The 10,000 poise value indicates a solid slag. Between the liquid and solid viscosity levels the slag is plastic or sticky, has a tendency to build up on furnace surfaces, and is difficult to remove. Figure

Slagging index

as a function of viscosity

$$Rvs = \frac{T_{250\ poise\ (Ox)} - T_{10,000\ poise\ (Red)}}{97.5\ fs}$$

Slagging index Rvs	Slagging classification
0.5-0.99	Medium
1.0-1.99	High
> 2.00	Severe

fs (severity factor as a function of mid-point temperature at 2000 poise)

F	fs
1900	1.00
2000	1.25
2100	1.60
2200	2.00
2300	2.60
2400	3.25
2500	4.10
2600	5.20
2700	6.55
2800	8.30
2900	11.00

Figure 27-9. Slagging potential classification by viscosity. Courtesy Babcock & Wilcox.

27-10 shows the viscosity-temperature relationship of two different coals. It can be seen that the Kentucky coal is plastic only in a very narrow temperature zone and therefore would probably not cause a slagging problem. The Illinois coal is plastic over a wider temperature range and consequently over a greater part of the furnace surface. The possibility of troublesome slag is much greater for the Illinois coal.

The boiler designer will use the appropriate slagging potential in several ways. A coal ash having a severe slagging potential will require a boiler with more conservative design features than one with a low potential. These design features would include (a) increased furnace plan area; (b) lower furnace exit gas temperature; and (c) furnace wall soot blowers.

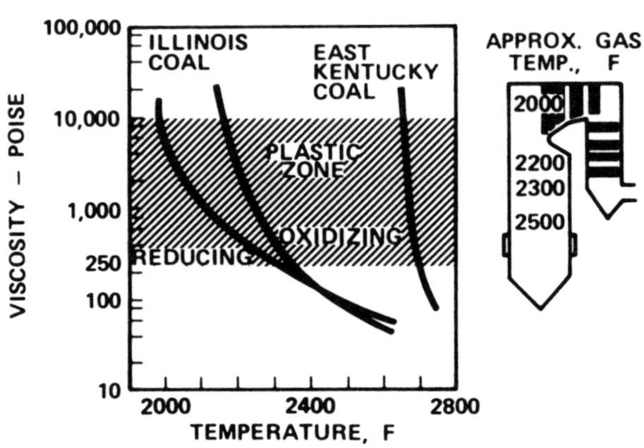

Figure 27-10. Viscosity-temperature relation. Courtesy Babcock & Wilcox.

Figure 27-11 illustrates one application of the slagging index. This curve is used to determine the spacing of furnace wall soot blowers for different degrees of slagging potential.

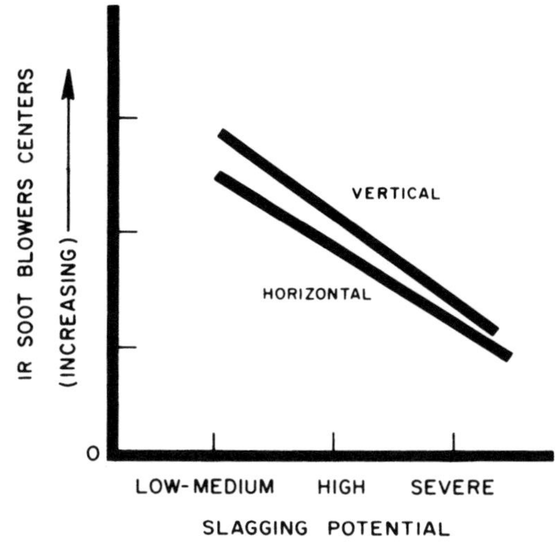

Figure 27-11. Soot blower spacing versus slagging potential. Courtesy Babcock & Wilcox.

Characterization of Fouling

There are two methods for determination of fouling potential, described as low, medium, high, or severe. The first method is applicable to bituminous ash only and is calculated using the base (B) to acid (A) ratio of the ash components. The formula for the fouling index is:

$$R_f = \frac{B}{A} \times Na_2O$$

where B and A are as defined in equations in the section on slagging and Na_2O is weight percent from ash analysis. Figure 27-12 shows the classification of fouling potential using R_f.

The second method is based on the sodium content of the ash expressed as the weight percent of Na_2O in the ash. Classification of fouling potential using Na_2O for lignitic ash only is as follows:

$Na_2O < 3 =$ low to medium
$3 < Na_2O < 6 =$ high
$6 < Na_2O =$ severe

$$R_F = \frac{\text{base}}{\text{acid}} \times Na_2O$$

Fouling

Type	Index R_F
Low	< 0.2
Medium	0.2-0.5
High	0.5-1.0
Severe	> 1.0

Figure 27-12. Fouling potential classification by base/acid ratio. Courtesy Babcock & Wilcox.

The potential for fouling increases with the presence of alkaline metals like sodium because these metals tend to vaporize during combustion and condense on the convection heating surface. The combination of high Na_2O, high base to acid ratio, and high gas temperature can create a severe fouling potential. Figure 27-13 shows the effect of Na_2O content of a lignitic type ash on sintering strength. While the strength of the lignitic ash is lower than that of a bituminous ash, the amount of sodium increases the fouling potential.

Another variable which influences the tendency for fouling is time. The sintering strength at a given temperature increases sharply with time

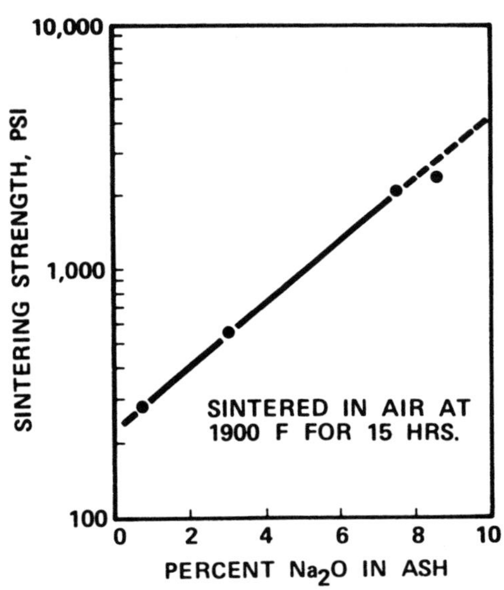

Figure 27-13. Sintering strength versus Na_2O. Courtesy Babcock & Wilcox.

particularly for bituminous-type ashes, as shown in Figure 27-14. From these data it is obvious that frequent soot blowing would be required for ashes having high fouling tendencies.

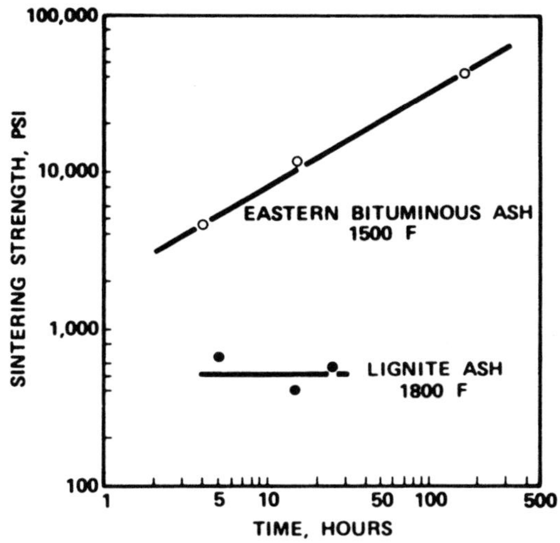

Figure 27-14. Sintering strength versus time.
Courtesy Babcock & Wilcox.

The boiler designer will relate the appropriate fouling potential data to boiler design in several ways. A coal producing an ash with a severe fouling potential as compared to one with low fouling potential will require: (a) lower furnace exit gas temperature; (b) fewer tubes in the direction of flow; (c) increased open area between tubes; and (d) additional soot blowers.

Figures 27-15 and 27-16 illustrate two applications of the fouling index. Figure 27-15 shows the relationship of superheater tube spacing to fouling potentials for two different gas temperatures. Figure 27-16 indicates the need for low soot blower cleaning radii for high fouling ashes, particularly at high gas temperatures.

The goal of the boiler designer is to prevent slagging and fouling from occurring. Therefore, the operator must be sure that he uses coals for which the boiler was designed.

Special Coal Analysis

There are several special analyses which are of interest particularly when the coal in question is unknown to the designer or for which there is no burning experience in the type of installation under consideration. The first of these is the *burning profile.* This test employs a derivative thermogravimetric technique in which a sample of the fuel is oxidized under controlled conditions. The burning profile, shown in Figure 27-17, is a plot

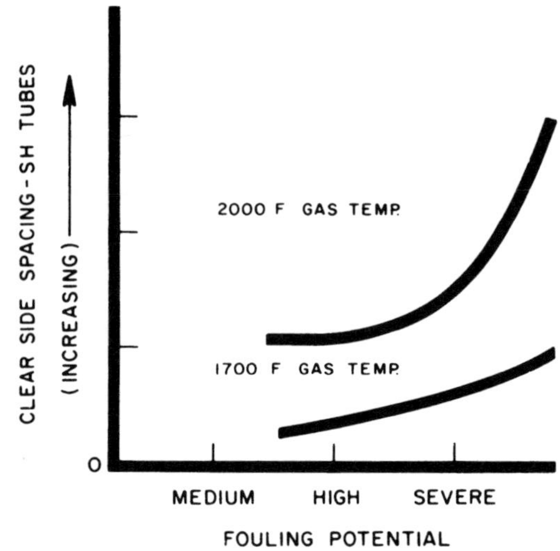

Figure 27-15. Fouling potential versus superheated tube spacing. Courtesy Babcock & Wilcox.

of the rate of weight loss of a coal sample versus the furnace temperature. The boiler designer compares the profile of the coal in question with those of coals having known combustion characteristics in similar furnaces. Using this comparison, the ignitability, residence time required for complete combustion, and furnace arrangement can be predicted with a high degree of success.

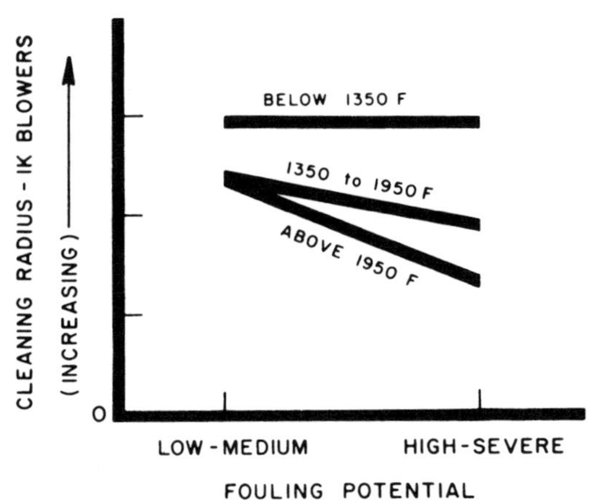

Figure 27-16. Soot blower cleaning radius versus fouling potential. Courtesy Babcock & Wilcox.

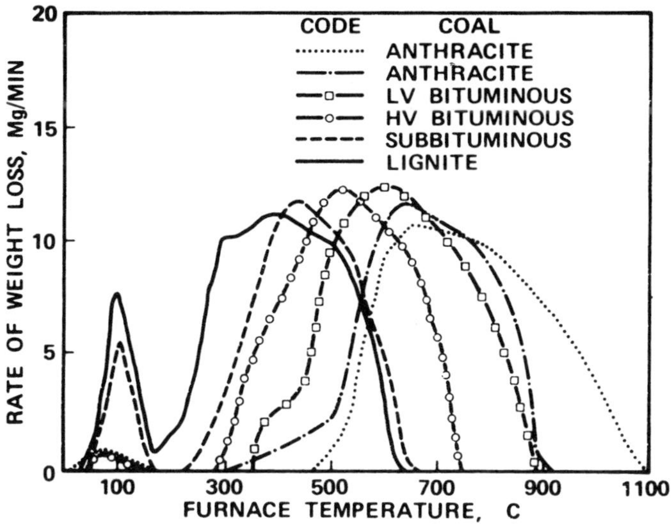

Figure 27-17. Burning profile for typical coals.
Courtesy Babcock & Wilcox.

The second test is the *volatile release profile* which is also a plot of the rate of weight loss of a coal sample versus the furnace temperature. However, in this case the sample is heated in an inert atmosphere rather than oxidized. The principle function of this test is to indicate the ignition characteristics of the fuel. It has been found that fuels having the same amount of volatile matter do not always release it at the same temperature and therefore have different ignition characteristics. By comparing the profile of an unknown fuel with those of fuels with known ignition characteristics the boiler designer can gain valuable information.

One property of coal which does not usually appear on a standard analysis, but which can be very important in shipboard applications of coal firing, is bulk density. This is the density of coal as handled naturally or stored with no compaction. Bulk density depends on a number of factors, such as ash content, moisture content, and sizing, and is used to size bunkering and handling equipment.

Importance of Correct Fuel Analysis

From the preceding discussion it is apparent that the boiler designer has a number of tools at his disposal which make it possible for him to design an economical and reliable boiler. However, the design will only be as good as the fuel analysis he has to work from.

Unlike an oil fired boiler where the changes in the characteristics of that fuel rarely require major changes in boiler configuration, major changes in boiler design can be brought about from utilization of coal. It is obvious then that the boiler designer needs the best information about the coal or coals to be used to design the best boiler for all specified conditions. Similarly, the boiler operator must have the analysis of all coals loaded in the ship's bunkers and the analysis must comply with the range of acceptable coal for which the boiler has been designed and which the manufacturer specifies in the operating instructions. Otherwise, the boiler operator may unwittingly defeat the purpose of the designer.

COMBUSTION ENGINEERING (C-E) COAL FIRED MARINE BOILERS

General

The following sections describe modern marine coal fired boilers. For information on marine boilers in general including description, operation, and maintenance, see *Modern Marine Engineer's Manual,* Volume 1.

C-E V2M9S Boiler

The C-E V2M9S boiler is similar to an oil fired unit except that it has been suitably modified to fire coal and is physically larger for any given horsepower than its oil fired counterpart. The cross section of the V2M9S boiler is shown in Figure 27-18.

The V2M9S boiler design has a dropped furnace and is supported at the midpoint by the sidewall header and the water drum saddles. As compared to a bottom or top supported unit this midsupport feature reduces the effects of the forces imposed on the boiler as a result of the vessel's motion. It also minimizes the effects of thermal expansion. The furnace consists of welded wall panels in way of the roof, front, rear, and sidewalls, and this construction continues on the sidewalls to the beginning of the generating bank.

A single row of 2-inch OD screen tubes on $4\frac{1}{2}$-inch centers originating from the screen header and terminating in the steam drum is utilized to reduce the radiation heat input entering the superheater. One row of screen tubes is satisfactory for these units as compared to oil fired units because the gas temperature leaving the furnace is considerably lower due to the significantly lower furnace release rates utilized when firing coal.

The superheater has a primary and a secondary stage, and each is of the vertical in-line, multipass type. The primary superheater consists of a series of 4-loop elements; each element is 2-inch OD tubing, on $4\frac{1}{2}$-inch centers. This spacing is the same as that utilized for the screen thereby providing clear gas paths through the screen and first stage of the superheater. Tube spacing is wider than normally utilized for an oil fired unit to preclude ash buildup.

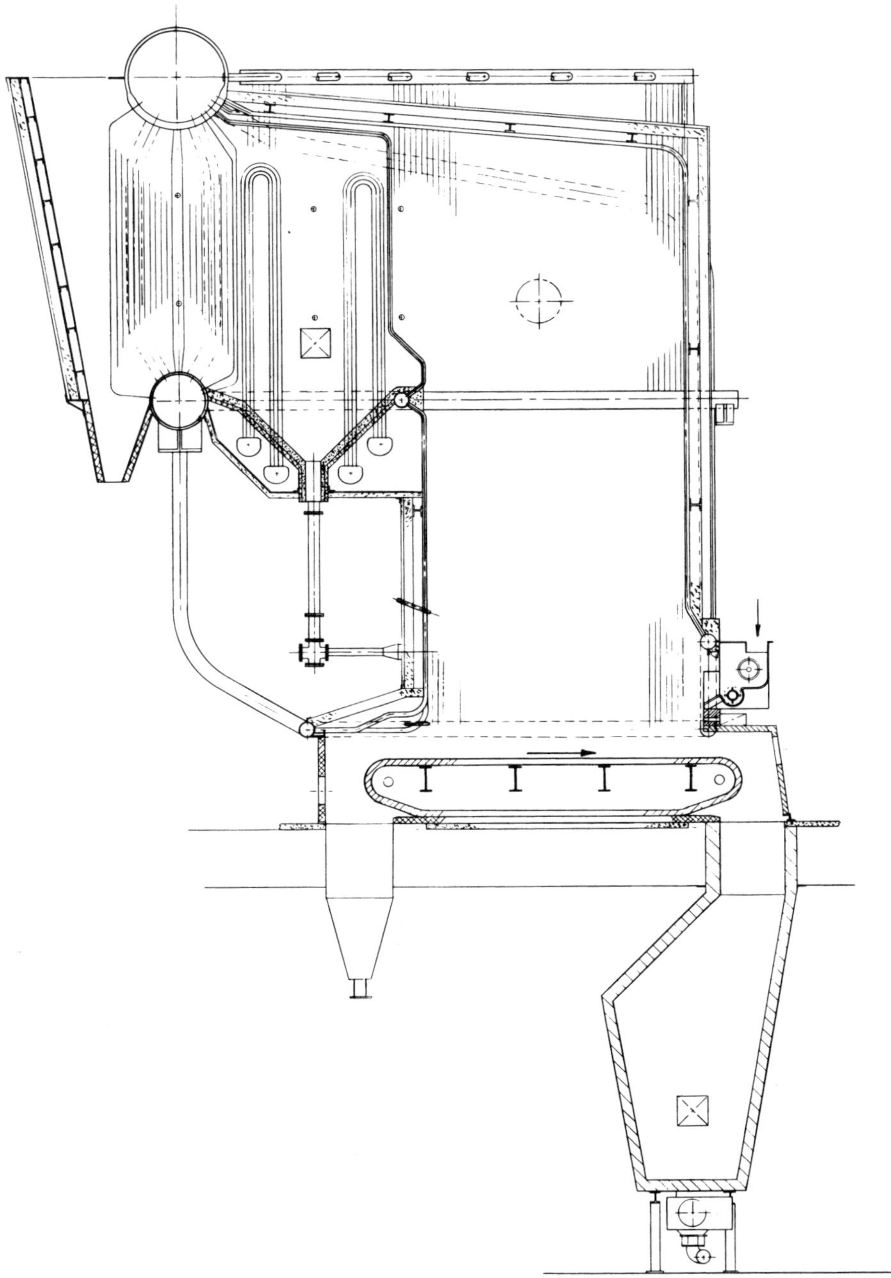

Figure 27-18. V2M9S boiler internal arrangement.
Courtesy Combustion Engineering.

The secondary superheater consists of a series of 4-loop assemblies, each element manufactured from 2-inch OD tubing on 3½-inch centers. Because some of the heavier cinders will drop out into the reinjection hopper between the stages due to the reduction in velocity as the gases pass through the primary superheater, assemblies are placed somewhat closer together in the secondary superheater than they are in the primary superheater. The superheater floor is sloped to direct the cinders which collect in this area into the hopper for reinjection.

The steam generating bank consists of in-line tubes of 1½-inch OD on 2½-inch centers.

An economizer is installed either at the boiler outlet or is separately supported by the ship's structure. It is of the conventional counterflow multibank type utilizing extended surface elements of spiral finned design. Elements are installed in-line, and fin spacing is considerably greater than for an oil fired unit. Each element is manufactured from 2-inch OD tubing utilizing a spiral steel fin which is ¾-inch high by .105-inch thick with 1½ fins per inch.

A balanced draft system is utilized for each boiler which requires fans for forced draft (FD) and for induced draft (ID). Because this system results in a slightly negative pressure in the furnace and eliminates gas and soot leakage, a single casing construction is permitted for the boiler. In the furnace area, the welded wall panels act as the casing and in all other areas a single layer of reinforced sheet steel forms the casing. In addition to the FD and ID fans, a small overfire fan is required. In providing air which is injected above the fuel bed through overfire air nozzles located in the rear wall of the furnace and below the distributors, additional turbulence in the furnace is afforded.

A suitable number of soot blowers are installed to ensure adequate cleaning of all external surfaces from the superheater through the economizer. As shown in Figure 27-18 two retractable soot blowers are installed before the screen to clean the leading edge of the primary superheater. Two additional ones are also installed in the access space between the primary and secondary superheater stages to ensure adequate cleaning of the rear portion of the primary stage and the leading edge of the secondary stage.

Rotary blowers are installed in the generating bank and in the economizer. From the screen tubes through the generating bank and in the economizer, the in-line tube configuration, as well as the wide tube spacing, assists the soot blowers to ensure adequate cleaning of all external surfaces.

C-E Continuous Discharge Spreader Stoker

Coal is fired on a C-E continuous discharge spreader stoker. A cross section of this equipment is shown in Figure 27-19. The traveling grate portion

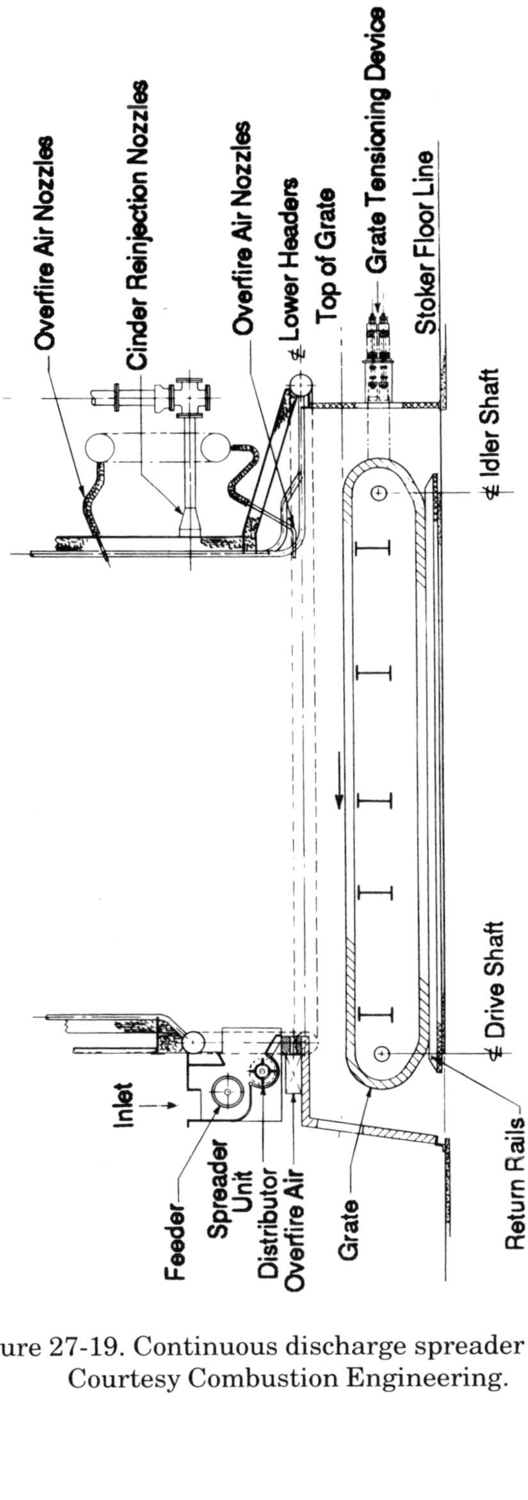

Figure 27-19. Continuous discharge spreader stoker.
Courtesy Combustion Engineering.

comprises carrier bars and keys and, depending upon the size and application of the equipment, is driven from the drive shaft through either a hydraulic cylinder activated ratchet device or an electric motor powered gear. The idler shaft end is fitted with Belleville spring devices which are used to maintain optimum chain tension and to permit compensation for chain elongation during operation. Return rails are located at the underside of the grate and serve to support and guide the grate during operation. The combination of the tensioning device, the return rails, and the use of thrust collars on the drive and idler shafts provides for excellent resistance to the forces caused by ship motion during operation. The speed of the grate is adjusted to maintain a constant thickness of ash at the discharge end, usually 2 to 3 inches. The maximum grate speed is about 15 to 20 feet per hour with a 6 to 1 reduction to provide for reduced ratings. The standard direction of grate travel is toward the distributors with the ash discharging at the stoker front into a hopper located below. A thermocouple monitoring system is provided at the top of the grate supports just beneath the chain to warn of abnormally high grate metal temperatures.

The coal spreading units are located at the front end of the grate directly above the drive shaft. Two or more of these units are fitted depending upon boiler capacity and the grate dimensions. A constant head of coal is maintained at the spreader inlet by the coal forwarding equipment. The spreader contains a feeder and a distributor. The coal feeder is a variable speed device regulated by the combustion control system to admit coal to the distributor rotor located directly below it, in response to boiler pressure. This device performs the same function as a fuel oil control valve. The function of the distributor is to properly distribute the coal supplied by the feeder onto the grate. The larger coal particles are thrown to the rear of the grate where they will have the maximum time to burn and the smaller particles which require less burning time are thrown to the front of the grate. Depending upon the type of coal and the mixture of particle sizes, the speed of the distributor is varied manually to obtain an optimal coal spread. This adjustment is necessary when new bunkers are received.

Even distribution of coal on the grate is important in obtaining smokeless combustion at reasonable excess air levels. In order to obtain an even distribution, a good mixture of fines and coarse particles must be present in the coal fired. For optimum results, the following coal sizing criteria are recommended: (a) top size 1½ inches; (b) through a ¼-inch round hole screen 95 percent; and (c) through a ¼-inch round hole screen no more than 50 percent. In most cases this consistency can be achieved by use of a coal crusher without the necessity of screening out excessive fines. Actually a certain percentage of fines is beneficial in fully utilizing the furnace of a spreader stoker fired boiler. As the coal is thrown out from the distributor, the fines ignite and burn in suspension while the larger particles fall and are burned on the grate.

The volatile matter, which in most stoker grade coal ranges from 25 percent to 35 percent, is distilled off from the larger particles and is burned above the grate as a gas. Therefore, one can readily observe that an excessive trend toward coarse or fine particles could upset one or more of the simultaneous combustion processes, resulting in a loss of combustion efficiency.

Approximately 80 percent of the combustion air is admitted to the interior section of the grate by means of wind boxes located on each side of the grate. Perforations and channels in the grate keys allow the air to pass through the grate thereby cooling it as well as providing air for the combustion process. Zone air control dampers are located in the grate interior to allow air flow adjustments to be made in order to accommodate wind box design variances as well as to allow for adjustments to obtain good combustion at reduced ratings. The air to this section is provided by a forced draft blower(s) and is proportioned by the combustion control system to the amount of coal fired. This air may be preheated to a maximum temperature of 400°F, depending upon the cycle selected.

C-E Overfire Air System

Approximately 20 percent of the combustion air is supplied by overfire air nozzles which are located in the furnace waterwall above the grate. These nozzles are supplied by a separate fan which is not controlled by the combustion control system. This air is supplied at approximately 25 inches water gauge static pressure and is not preheated. The purpose of this air is to provide agitation above the grate to assure good mixing of the fines and gases which are burning above the grate with the combustion air.

Cinder Reinjection System

The overfire air system additionally provides an air source which is used to reinject a portion of the coal ash into the furnace for reburning. This arrangement is shown in Figure 27-20. The ash hopper located below the superheater cavity collects the larger unburned coal particles which exit the furnace. This area is a logical place for large particle dropout due to the slowing of the gas velocity as it exits the front superheater stage prior to entering the second superheater stage. At the base of the cinder reinjection ash hopper, a rotary air lock valve is fitted to prevent combustion gas from being circulated in this system.

The rotary valve feeds the cinders into the reinjection piping where a portion of the overfire air then propels them into the furnace for further burning. Boiler efficiency is improved approximately 4 percent by use of this system.

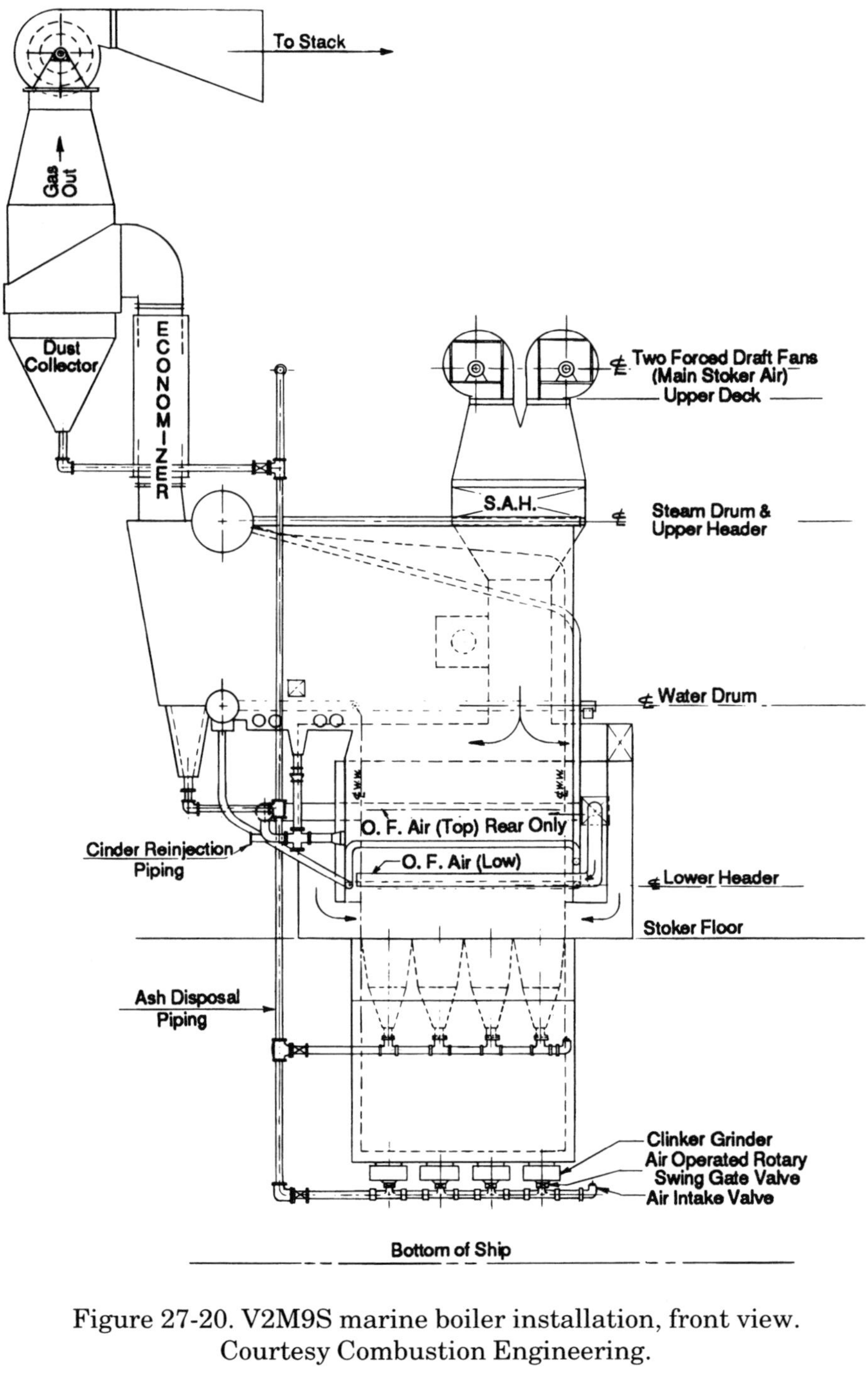

Figure 27-20. V2M9S marine boiler installation, front view. Courtesy Combustion Engineering.

Ash Handling

The coal used for stoker firing, depending upon origin, may have an ash content as high as 20 or 25 percent. The boiler must be designed with provisions to remove and store the ash for its eventual disposal. There are four locations for the removal of ash from the boiler system. Most of the ash (approximately 70 percent) falls off the end of the traveling grate at a point just below the spreader unit, as shown in Figure 27-21. This ash will be hot and may still contain some glowing embers. Therefore, the ash collection hopper must be refractory lined. A small amount of ash sifts through the stoker keys and falls into the interior of the grate. This ash falls out of the grate as it turns on the idler shaft and drops into the siftings hoppers. A third ash collection point is located below the exit of the boiler main bank which catches particles which drop out of the gas stream as it turns to exit the boiler. The final ash collection point is located at the base of the mechanical dust collector.

The ash collection system is designed for a 24-hour capacity assuming full power operation with coal containing approximately 15 percent ash. In most cases, this would require ash disposal once per day and would allow for several days in port without the necessity of handling ash. The ash handling system selected is a pneumatic system. A steam eductor utilizing 150 psi auxiliary steam creates the vacuum in the ash handling system piping. Valves or gates are then opened sequentially at the base of each ash collection hopper. A vacuum switch located at each hopper signals the ash handling system that a particular hopper is empty before the system empties the next hopper. Clinker grinders are fitted at the base of the main grate discharge hoppers to insure that large ash particles do not enter and jam the system. The other hoppers will contain only fine ash which does not require further processing prior to removal.

Depending upon trade route and possible regulatory body or governmental regulations, the ash removed from the boiler ash hoppers may be discharged directly overboard or stored onboard the vessel for shore-based disposal. Some combination of the two may be used.

Air and Gas Systems

Combustion air is supplied by two systems: the forced draft system, which supplies 80 percent of the air required, and an overfire air fan, which supplies 20 percent of the total. The overfire air is supplied at ambient temperature and is not preheated. The combustion air does not require preheating. However, where cycle economics dictate, preheating up to a limit of about 400°F is possible depending upon the steam cycle utilized.

The combustion air is supplied at a positive pressure up to the point of its introduction into the furnace. The furnace itself is maintained at a negative pressure of approximately 0.1 inch or 0.2 inch water gauge by use

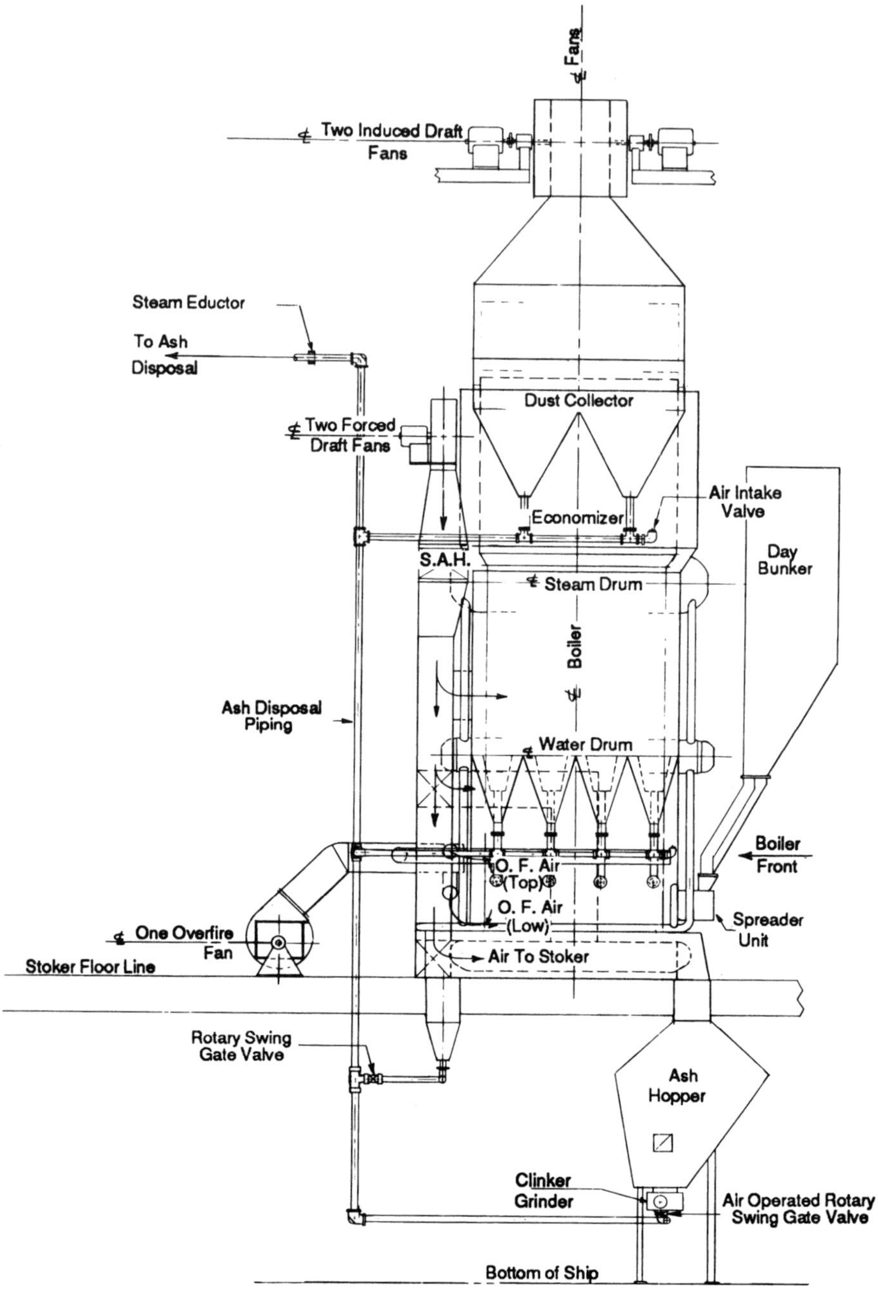

Figure 27-21. V2M9S marine boiler installation, side view. Courtesy Combustion Engineering.

of an induced draft fan which is located at the base of the ship's stack. The entire boiler with its attendant heat recovery equipment and dust collector is therefore protected against gas and soot leakage under normal circumstances. The forced draft and overfire air fans are typical marine fan designs and require no special consideration. The forced draft fan contains a two-speed motor and is inlet vane controlled. The overfire air fan has a single speed motor with inlet vane control.

Due to its elevated operating temperature and the erosive and corrosive nature of the flue gas which it handles, the induced draft fan requires special consideration. Typically, a mechanical dust collector would be fitted prior to the induced draft fan inlet. Depending upon the efficiency of the dust collector, 95 percent or more of the dust present in the flue gas may be removed. If a low efficiency collector is fitted (collection efficiency 85 percent) the induced draft fan is limited to 900 RPM maximum to avoid excessive blade erosion. If a high efficiency collector is fitted (collection efficiency 95 percent or greater) then the maximum fan speed may be raised to 1,200 RPM. Three-speed fans are utilized. Maximum fan RPM is only necessary at boiler overload. The medium speed is selected for boiler full power or normal load, whichever is the prevalent operating rate. Low speed is for use in port only. By selecting these speed ranges, fan blade life may be increased significantly. The induced draft fans contain replaceable fan blade wear surfaces to avoid the entire fan rotor having to be pulled for worn blades. The induced draft fan flow is controlled by inlet vane dampers which are actuated by the combustion control system in response to furnace pressure and rate changes.

Dust Collectors

Due to the environmental restrictions on stack opacity which are in effect for potential ports of call, coal fired vessels are fitted with some form of mechanical or other type of dust collector. Depending upon collector efficiency, most visible flue gas particles will be removed. The dust collector is located after the economizer and immediately prior to the induced draft fan. Depending upon the type of coal fired and the economizer surface selected, it may be desirable in some cases to locate the dust collector immediately after the boiler bank rather than after the economizer. Because of the increased gas temperature, this alternate location will significantly increase the size and cost of the collector.

A centrifugal-type dust collector is shown in Figure 27-22. This device removes dust by creating a cyclone-type motion as the dust-laden gas enters the annular space around the collector tubes. The dust then drops into the hopper below and the cleaned gas moves up the center of the cyclone vortex through the center of the collector tube. This device has no moving parts and, other than infrequent water washing, requires little or no maintenance. Depending upon the diameter and type of precipitator

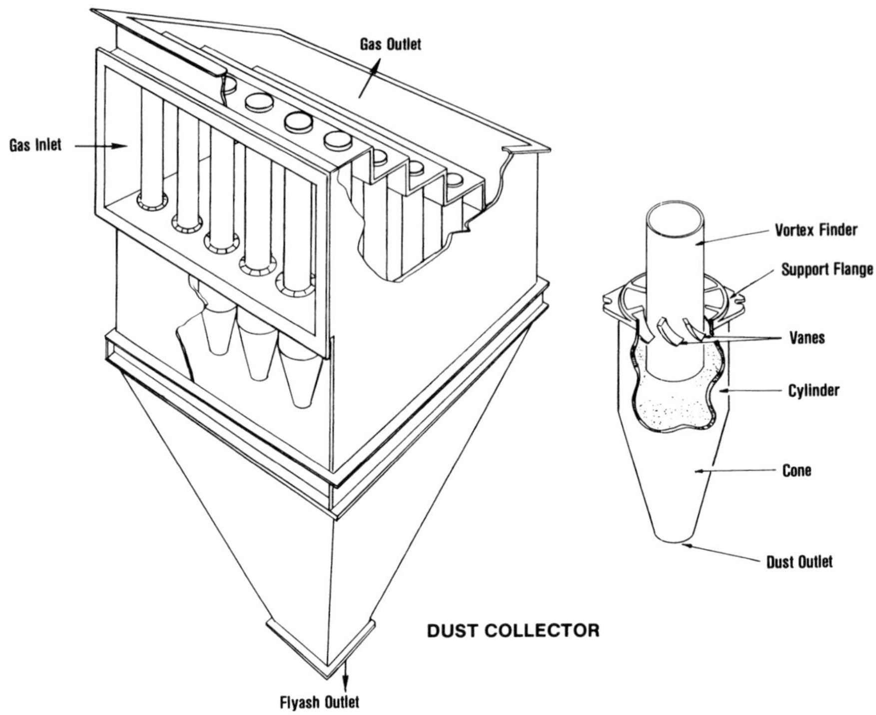

Figure 27-22. Mechanical dust collector.

tubes used in the mechanical dust collector, collection efficiency of 95 percent may be achieved. If a higher collection efficiency is desired, two cyclone separators located in series may be utilized. Collection efficiencies in excess of 98 percent are then possible.

Coal Bed Ignition

A coal fired boiler is initially started by building a wood fire on top of the grate and then spreading coal which quickly ignites. For a unit which must be restarted at infrequent intervals this method is quite simple and adequate. However, if frequent boiler shutdown is required, a light oil ignitor in the lower furnace section should be installed and sized to fire number 2 oil at a rate of 200 to 250 pounds of oil per hour. The light-off sequence involves lighting off the ignitor and preheating the grate to approximately 700°F (measured by the grate thermocouple monitoring system) then starting the distributors and depositing a layer of coal on the grate. After the deposited coal ignites, the ignitor is secured, and the boiler is then brought up to pressure by feeding coal to the grate. As required by various regulatory agencies, the ignitor has a fan and flame scanner. The fan also supplies cooling air to the ignitor after it is secured.

Supplementary Oil Firing

Depending upon the operating requirements of the vessel and projected trade route, it may be desirable to have a supplementary oil firing capability. This can be readily accommodated by installing a conventional oil fired burner register(s) in the upper furnace section. As long as a 2- or 3-inch protective layer of ash remains on the grate, and a small amount of cooling air is supplied to the underside of the grate (supplied by the forced draft blower), oil firing may be continued indefinitely. The combustion air is supplied by the same forced draft blower that supplies air to the grate for coal firing. The induced draft fan is also utilized during oil firing; however, the overfire air system is secured.

Supplementary oil firing capability substantially increases the complexity of the boiler control system by requiring the installation of a burner management system as well as increasing the complexity of the combustion air ductwork.

C-E Boiler Performance

Table 27-1 shows typical performance for a single coal fired C-E boiler applied to a 24,000 SHP plant.

TABLE 27-1

Typical Performance of a C-E V2M9S Coal Fired Boiler

Number of boilers	1
Normal power	24,000 SHP
Operating condition	Normal
Rating	100%
HHV fuel	9,387 BTU/lb
Superheater outlet pressure	870 psig
Feedwater teperature to economizer	302°F
Steam pressure to air heater	55 psig
Superheater steam flow	165,600 lb/hr
Steam temperature	920°F
Efficiency	82.7%
Fuel firing rate	24,575 lb/hr
Air temperature leaving air heater	278°F
Air flow at 100°F	55,397 CFM
Gas flow	253,943 lb/hr
Total draft loss	7 inchWG
Excess air	30%
Furnace release rate	25,009 BTU/hr-ft^3

TABLE 27-1 (continued)

Absorption rate/RHAS	7,038 BTU/hr-ft^2
Firing rate/RHAS	10.86 lb coal/hr-ft^2
Furnace exit temperature	1,900°F

Coal fired boilers have lower efficiency than comparable capacity oil fired boilers. The difference in efficiency is due to carbon loss which is unburned in the ash and the loss of moisture inherent in the content of coal. These losses will vary depending on the analysis of the purchased coal.

Operating Procedure during Light-Off with Oil Burner and Coal Ignitor

1. Open superheater drains and vents.
2. Purge furnace (FD and ID fans running). Set fans to go to low fire after purge. Establish fire with coal ignitor.
3. Spread layer of coal on furnace when grate reaches 700°F.
4. Allow enough air under stoker to burn coal.
5. Establish 50-75 mm (2-3 inches) of ash on stoker grate. Monitor grate temperature and keep below 1,000°F by adjusting air flow under grate.
6. Bring boiler up to pressure at low oil fire. When pressure reaches 3.5 kg/cm^2 (50 psi), close superheater drains.
7. Establish steam flow when boiler is up to pressure. Put boiler on line, and close superheater vents.
8. Spread coal on grate, and establish stoker fire with coal ignitor on leaving port.
9. Shut down coal ignitor once fire is established. Keep ignitor fan on in order to cool ignitor throat.
10. Start grate and overfire air fan. Bring coal rate up to hold boiler load.
11. Shut down oil burner (some leakage will be required to cool oil burner throat).

Operating Procedure Coming into Port

1. Reduce coal rate.
2. Establish oil fire.
3. Shut off coal. Keep a layer of ash on grate.
4. Take load on oil. Shut down grate.
5. Monitor grate temperature to hold under 1,000°F by adjusting stoker air damper.
6. Shut down overfire air fan and coal ignitor fan.

Operating Procedure during Light-Off with Coal Ignitor

1. Open all superheater vents and drains.

2. Purge furnace (FD and ID fans running). Set fans to go to low fire after purge.
3. Light coal ignitor, monitor grate temperature. Spread layer of coal on grate when grate temperature reaches 700°F (371°C).
4. Start grate once fire is established, and secure coal ignitor (ignitor fan will remain on to provide cooling for ignitor throat).
5. Start overfire air fan.
6. Bring boiler up to pressure at low firing rate.
7. Close superheater drains upon reaching 3.5 kg/cm^2 (50 psi).
8. Put boiler on line and close superheater vents when boiler is up to pressure and steam flow can be established.
9. Firing rate minimum will be approximately 25 percent of maximum load. Steam dump will be in operation below this point.

FOSTER WHEELER (F-W) COAL FIRED BOILER

General Description

The Foster Wheeler (F-W) stoker coal fired boiler is a marine "D"-type that is supported at the lower drum and from the furnace walls just above the stoker grate. Figures 27-23 (a) and (b) are sectional views of a F-W stoker coal fired marine boiler. The pressure vessel sections of this boiler, including steam drum, water drum, headers, tubes, superheater, and economizers, are all very similar to the oil fired boiler. Arrangements to accommodate and remove ash deposits on the heating surfaces are the only significant differences from the oil fired boiler design in practice. This is also designed for oil firing for in-port or maneuvering operations of the ship.

Air and Flue Gas Flow

Air for combustion is delivered by a forced draft fan and heated by a steam coil air heater. The warmed air then enters the plenum chamber under the continuous ash discharge grate. The air is evenly distributed through the grate via orifices and grate clearances. The air mixes with the coal particles on the grate and in suspension to complete the combustion process. The flue gas then passes through the screen tubes and superheater banks and continues through the generating tube section. The flue gas enters a mechanical dust collector to remove the fly ash particles before entering the economizer. The cleaned flue gas then enters the induced draft fan and flows out through the stack.

The coal fired boilers are designed for balanced draft operation, achieved by maintaining a slightly negative pressure (.1 inch water gauge) in the furnace. The induced draft fan controls the furnace pressure, and the forced draft fan controls the air flow to insure proper combustion in the furnace.

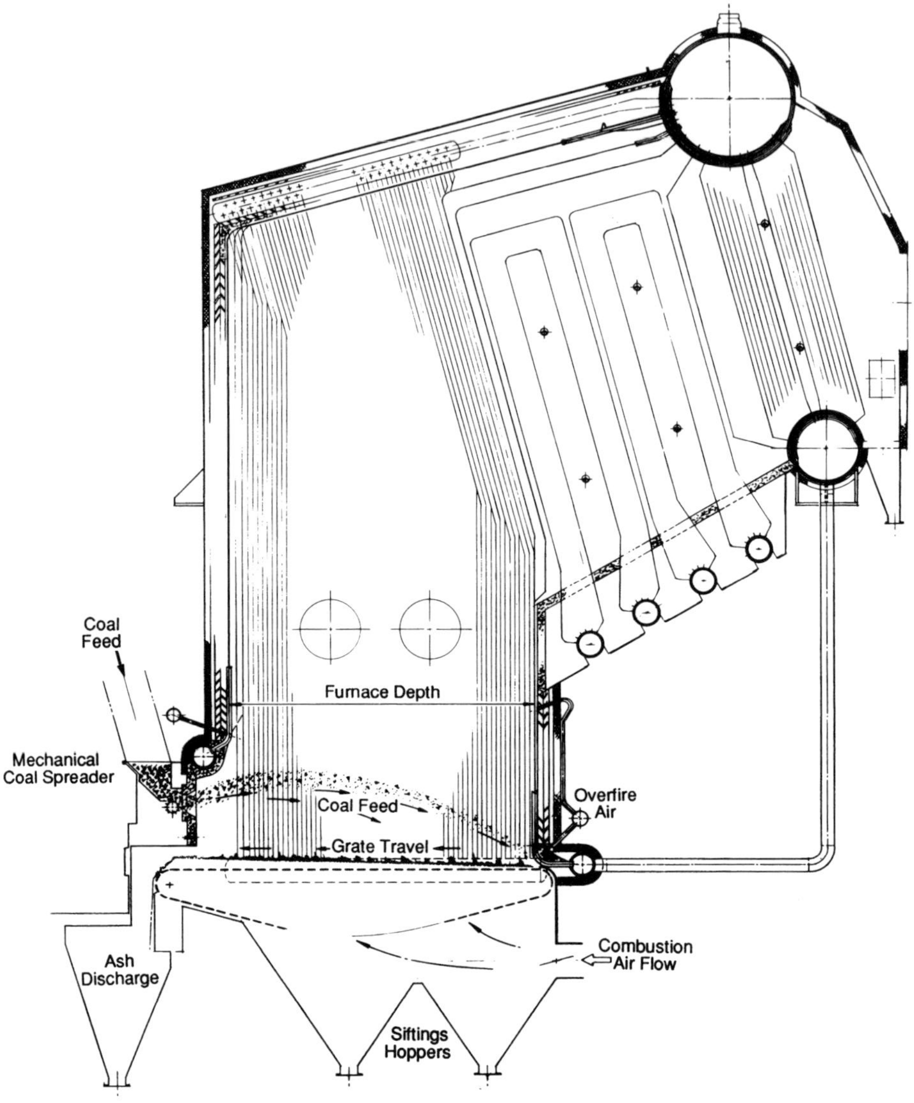

Figure 27-23 (a). F-W stoker fired boiler.
Courtesy Foster Wheeler Boiler Corp.

A higher pressure overfire air fan is used to blow air through nozzles over the coal bed on the front and rear walls to optimize combustion condition in the freeboard volume above the grate.

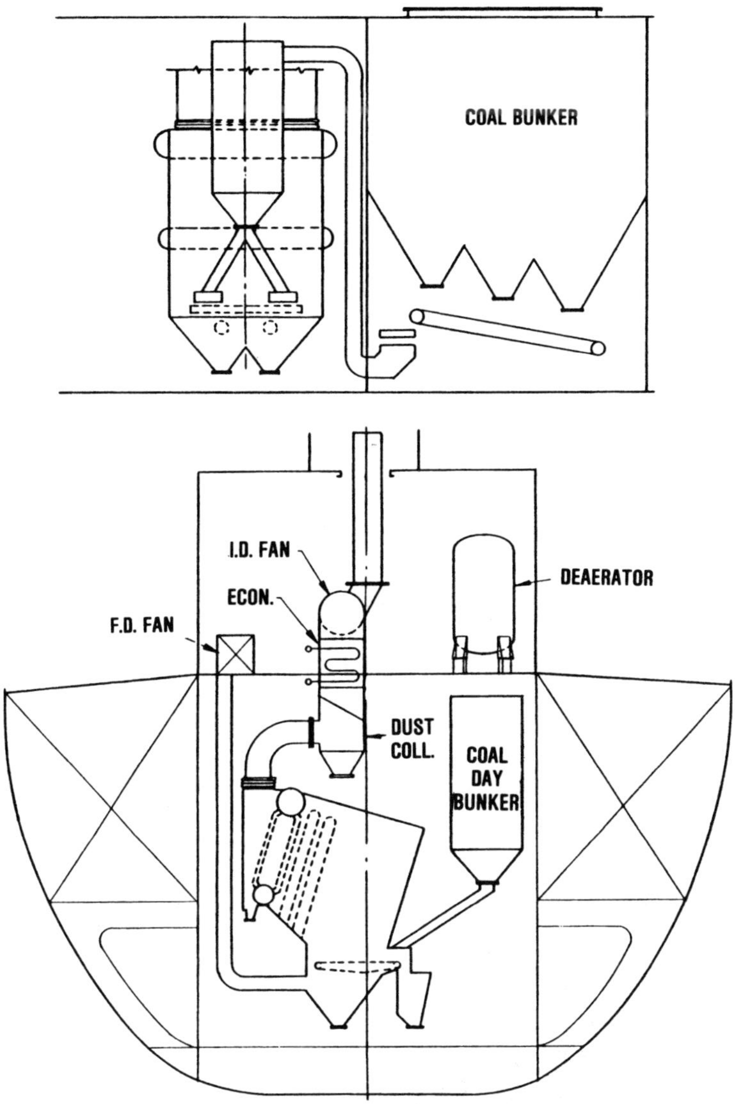

Figure 27-23 (b). Typical arrangement of single boiler.
Courtesy Foster Wheeler Boiler Corp.

Coal Feed

The stoker size coal (1¼ inch) with fines not more than 40 percent is supplied to the boiler coal day bunkers from the main bunkers by a mechanical or pneumatic transport system.

The coal day bunkers have multiple outlets to feed each mechanical spreader stoker for the unit. The number of spreaders depends on the size of the boiler, but will range from two to six per boiler. A variable speed mechanism feeds the coal to the rotating paddle wheel spreader. The spreader paddle wheels evenly distribute the heavier coal particles on the continuous ash discharge stoker grate. A small portion of the coal fines are burned in suspension before they reach the grate.

The stoker grate is moving forward continuously to dump the ash formed on the grate during the burning of the coal. A minimum depth of ash is maintained on the stoker grate to protect the moving parts from overheating.

The unburned carbon and fly ash are carried in the flue gas and collected in the boiler outlet hopper and a mechanical dust collector. This mixture of fly ash and unburned carbon is reinjected into the furnace just above the stoker grate to improve the boiler combustion efficiency.

Fly Ash System

The ash from the various hoppers is pneumatically transported overboard or to ash holding bins. The holding bins can be unloaded by the same methods upon reaching port with a facility for receiving the fly ash. This system normally operates under vacuum produced by a steam exhauster or mechanical exhauster. The ash from the various collection points is pneumatically conveyed to the ash holding bin. The system cycles through each ash collection point on a timed basis to keep the hoppers from getting overfilled. The system can be designed to allow ash transfer to the land side equipment or possibly pumped overboard when at sea. Figure 27-24 is a schematic diagram of the ash system.

Operational Considerations

Some units are furnished with load carrying oil burners for low load and in-port operation. These burners are located far enough above the stoker grate to prevent overheating of the grate. It is normally recommended that the stoker grate remain covered with ash during oil firing periods. This provides additional protection to the stoker grate system.

The coal firing can be established manually by igniting some easily combustible material such as wood or rags. Newer units may be equipped with oil ignitors close to the stoker grate to establish the coal bed burning. Once the coal ignition is stabilized, the ignitors are removed from service.

Close operator observation of the furnace combustion conditions is recommended to maintain satisfactory combustion over the load range. The supervision of the operating equipment for coal and fly ash handling is very important to maintain continuous, reliable operation of the boilers.

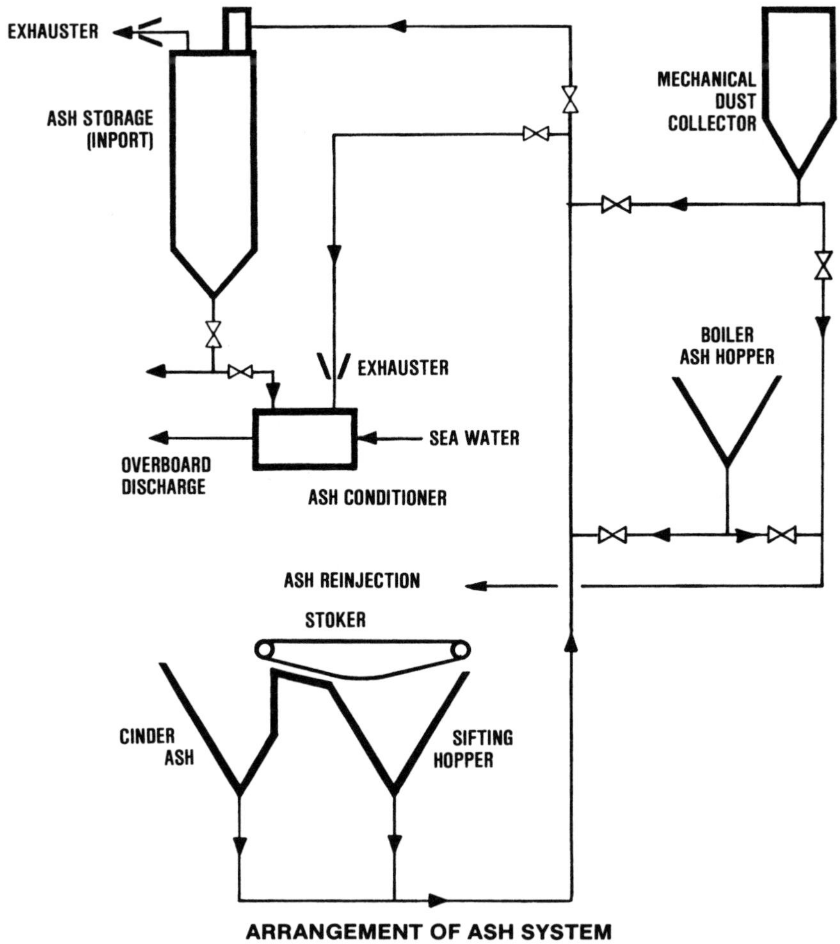

Figure 27-24. Arrangement of ash system.
Courtesy Foster Wheeler Boiler Corp.

Performance

Tables 27-2 and 27-3 show performance comparisons between coal fired and oil fired boilers for typical single boiler and two boiler shipboard arrangements.

TABLE 27-2

Comparison of Coal Fired Boiler with Full Load Oil Firing
(Single Boiler Arrangement)

	Oil Fired	*Coal Fired*
Boiler capacity, lb/hr	162,800	162,800
Design pressure, psi	1,100	1,100
Superheater outlet pressure, psi	870	870
Superheater outlet temperature, °F	950	935
Feedwater temperature, °F	284	284
Stack temperature, °F	330	330
Boiler efficiency, %	87.8	83.9
Fuel rate, lb/hr*	12,319	17,959
Furnace volume, cu ft	10,790	10,790
Furnace ratings:		
Liberation, BTU/cu ft	21,770	21,927
Firing rate, lb/hr sq ft RHAS	4.87	7.11
Weight/ship (based on single boiler arrangement)		
Boiler dry, lb	618,000	618,000
Stoker, lb	—	190,000
Total operating, lb	618,000	808,000
Basic dimensions, per boiler		
Furnace width	18′- 3½″	18′- 3½″
Overall boiler width (fore and aft)	26′- 0″	26′- 0″
Furnace depth	18′- 0″	18′- 0″
Overall boiler depth	41′- 0″	41′- 0″
Height between drums at centerline	14′- 3″	14′- 3″
Height, furnace floor of stoker to top of drum	43′- 2″	43′- 2″

TABLE 27-3

Comparison of Coal Fired Boiler with Full Load Oil Firing
(Two Boiler Arrangement)

	Oil Fired	*Coal Fired*
Boiler capacity, lb/hr	47,925	47,925
Design pressure, psi	1,100	1,100
Superheater outlet pressure, psi	870	870
Superheater outlet temperature, °F	900	900
Feedwater temperature, °F	297	297
Stack temperature, °F	330	330
Boiler efficiency, %	87.8	83.9
Fuel rate, lb/hr*	3,362	5,406
Furnace volume, cu ft	2,413	2,413
Furnace ratings:		

*Fuel oil: 18,500 BTU/lb; stoker coal: 12,500 BTU/lb.

	Oil Fired	*Coal Fired*
Liberation, BTU/cu ft	26,809	28,056
Firing rate, lb/hr sq ft RHAS	3.32	5.36
Weight/ship (based on two boiler arrangement)		
Boiler dry, lb	650,000	650,000
Stoker, lb	—	100,000
Total operating, lb	650,000	750,000
Basic dimensions, per boiler		
Furnace width	8′- 0″	8′- 0″
Overall boiler width (fore and aft)	16′- 0″	16′- 0″
Furnace depth	12′- 0″	12′- 0″
Overall boiler depth	32′- 6″	32′- 6″
Height between drums at centerline	14′- 3″	14′- 3″
Height, furnace floor of stoker to top of drum	35′- 4″	35′- 4″

MARINE COAL HANDLING SYSTEM

Shore Based Coal Preparation

Proper preparation of coal prior to loading aboard ship will minimize most coal handling problems. The land based coaling station must perform a series of functions to prepare coal for shipboard use including the following:

1. Crush coal to optimal size.
2. Wash coal to remove clay.
3. Screen coal to remove oversize pieces, noncombustibles, and trash.
4. Dry coal to reduce surface moisture to a point where it does not cake and will not be dusty (5 to 8 percent surface moisture for most coals).
5. Remove iron from coal by passing it over a magnetized head.

Shipboard Coal Handling System (Stock)

The Stock Equipment Company has proposed the typical coal handling system for shipboard use shown in Figure 27-25. The coaling station (or lighter) should be equipped with a hammerhead crane mounting 1 or 2 belt conveyors which carry the coal to a hatch on the fore and aft centerline of the vessel. The belt conveyors place the coal on 1 or 2 shuttle conveyors. These are belt conveyors mounted on carriages which allow them to be moved back and forth over the bunker for layer loading of the coal. The shuttle conveyors are reversible so that port and starboard sides will be filled alike. A movable inlet chute is required for this mode of operation.

Operation of the shuttle conveyors is manually controlled to provide the highest reliability. A gallery at the upper part of the boiler room allows the operator to control the operation. Bull's-eyes along the bunker provide for visual inspection. These are hinged-type portholes for cleaning of the bunker side when necessary. Bull's-eyes must be in the closed position at all times when underway in order to maintain watertightness of the bulkhead.

The coal bunker runs the full usable width of the ship. It is as high as possible consistent with ship stability considerations. The bunker length varies according to the BTU of the coal anticipated and the cruising range desired. The greater the number of bunker outlets provided, the more coal can be stored in a given bunker volume. The fewer the number of bunker outlets, the lower the investment cost for coal handling machinery. These factors are carefully weighed in each design. There are 16 outlets at the bottom of the bunker shown in Figure 27-25. Each outlet is 18 inches square. The bunker pockets are designed with a slope angle of not less than 50°. The lower half of each bunker outlet pocket is made of type 304 stainless steel. No roughness or ledges are permitted in this area since these may initiate stoppage of coal flow. Even a horizontal welding bead may cause coal flow stoppage in a hopper.

The coal valves are equipped with motor operators to make the system automatic. The outlet is somewhat larger to provide a minimum face to face dimension without creating space under the gate for tramp coal to collect. With the arrangement shown, tramp coal falls directly down into the drag conveyor.

A drag conveyor is a heavy malleable iron chain with stainless steel pins and cotters, running in a trough. When the chain moves, coal moves with it. In this design, two 80-foot drag conveyors operate simultaneously in parallel with one another. One conveyor runs from starboard to port and the second from port to starboard. This arrangement loads the gravity discharge conveyor elevator without permitting coal segregation. Both drag conveyors are always run at the same time. There is only one coal valve open on the port side and one coal valve open on the starboard side when the conveyors are operating.

The gravity discharge conveyor elevator consists of two chains, one on each side connected by a V-shaped bucket. This bucket acts as a flight to push the coal along horizontal runs and to carry the coal in the bucket vertically in the vertical runs. At the end of a horizontal run there is a chain sprocket that allows the chain and bucket to go upward. The flight which has pushed the coal horizontally now carries the coal vertically. In this design, there is a length of the gravity discharge conveyor elevator operating horizontally underneath the drag conveyors. This turns upward to carry coal the full height of the bulkhead before ending in another short length of the horizontal conveyor that moves the coal to the day bunker.

As shown, the forward and aft sides of the bunker continue down to the bottom of the vessel and make a watertight bulkhead. In such an arrangement, watertight access doors are required for access to the conveying machinery under the bunker.

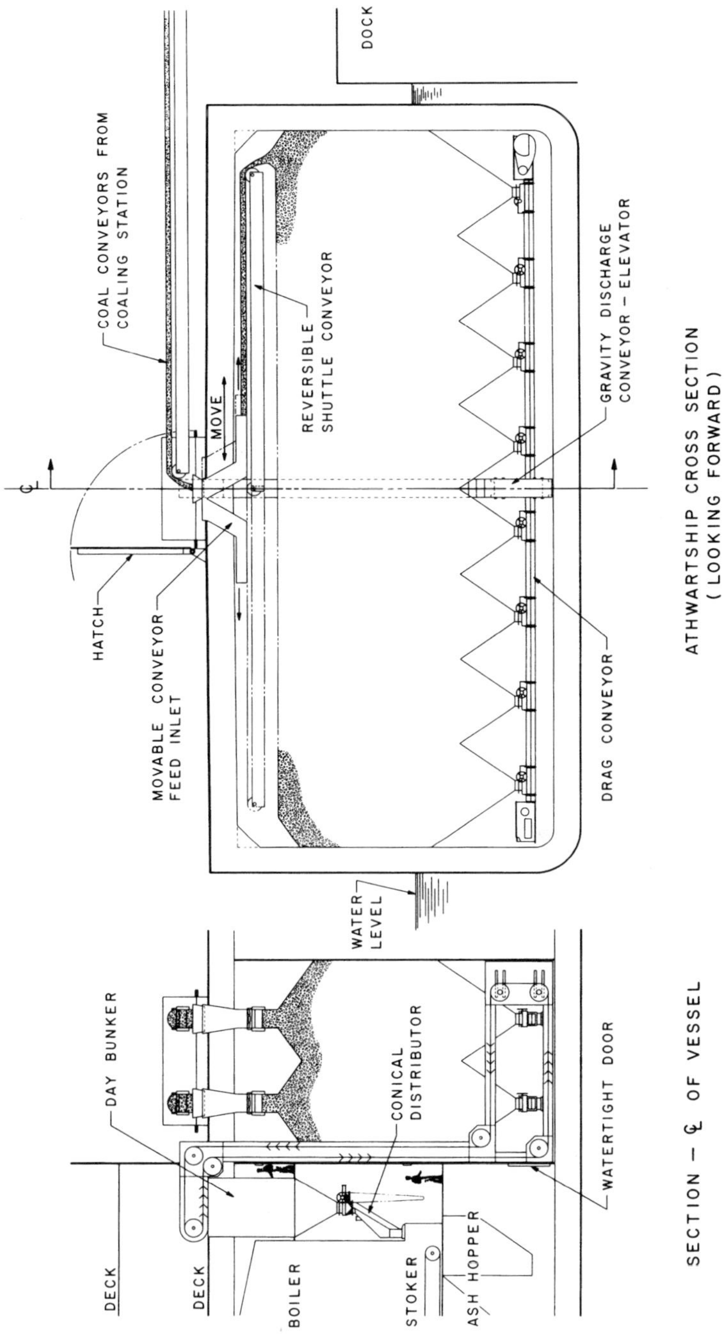

Figure 27-25. Coal handling system. Courtesy Stock Equipment Co.

The day bunker receives coal from the gravity discharge conveyor elevator. This bunker will have sufficient capacity for between 30 minutes and 2 hours of full power steaming. This should be adequate if fed by a reliable conveying system under automatic control.

To achieve automatic operation, a paddle-type stop switch in the upper part of the day bunker stops the gravity discharge conveyor elevator. A timer on a 20-minute or half-hour cycle restarts the coal conveyors. The automatic controls will open and close the coal valves so that coal is withdrawn from each bunker outlet equally and therefore all bunkers will become empty at approximately the same time.

At the bottom of the day bunker, there is a manually operated coal valve. Normally, this valve is used to shut off the flow of coal when the ship is reaching port, allowing the conical nonsegregating coal distributor and the stoker hopper to be emptied if desired.

The conical nonsegregating coal distributor shown has a conical bottom and top plate. Therefore, coal is placed into the stoker hopper without segregation of the coarse and fine pieces. The tops of the stoker hopper are closed with dust-tight slip joints, making a final dust-tight enclosure for the entire system. In front of the slip joints is an access door for inspection of the coal flow if desired and a drive gate to shut off the flow of coal to any particular stoker hopper.

Operational Considerations

Bunker fires are not a problem with the system described unless the type of coal used is subject to spontaneous combustion. The small day bunker permits the bunker to be emptied at frequent intervals, reducing the bunker coal temperature and the chance of fire.

The entire coal handling system below the bunkers is dust-tight, eliminating the air ingress and the consequent risk of bunker fires. Provision for the introduction of CO_2 into the bunkers is another means of controlling bunker fires.

Coal flow problems will be minimized if proper preparation of the coal ashore reduces the surface moisture to acceptable levels. The shore removal of oversized coal, foreign material, tramp iron, and clay are important to the free flow of coal in the shipboard system.

The traveling shuttle conveyors provide for layer loading of the bunker to reduce bunker coal segregation. Dual drag conveyors loading the gravity discharge conveyor elevator from opposite directions should eliminate segregation at that point. The gravity discharge conveyor elevator loads the day bunker along a line perpendicular to the boiler front. If loaded parallel to the boiler front, coal segregation in the day bunker would go through the conical distributor, and uniform firing would be difficult to maintain.

Since the coal conveying system is designed to be dust-tight, it is possible to introduce a slight vacuum in the bunker during coaling operations. The coal-dusty air is discharged through filter bags to control dust during loading.

This typical automatic mechanical coal handling method is the system used in many shore-based power plants. Other coal handling systems such as a pneumatic system are available and may be applied to shipboard use.

STOKER SYSTEM (DETROIT ROTOGRATE)

Introduction

The typical automatic stoker system (the Detroit RotoGrate) is illustrated in Figures 27-26 and 27-27. The system includes fuel receiving hoppers from which the fuel is picked up by revolving rotors and distributed into the furnace. The rotor blades are curved in a manner which causes fuel to be distributed uniformly on the grate. The fine particles of fuel are burned rapidly before they fall to the traveling gate. High pressure overfire air jets are located around the furnace wall above the grate to provide mixing of the fuel and air needed for complete combustion. The heavier fuel particles are spread evenly on the grate to form a thin, fast-burning fuel bed. The traveling grate discharges ash continuously over the front end of the grate into the ash receiver.

The grate is a hinged bar design which permits the individual grate bars to open at the lower portion of the catenary during the return to the rear of the furnace as illustrated in Figure 27-27. This facilitates air admission to the fuel bed and permits discharge of any ash which drops through the grate. Figures 27-27(a) and 27-27(b) show the self-adjusting air seal arrangements at the front and rear ends of the traveling grate. The bearings for the coal feeder drums are water cooled to prevent failure from overheating. Shearing devices are installed in the grate drive assemblies to prevent damage to the system from foreign material in the coal.

Lighting-Off Procedure

In a marine boiler, the normal operating procedure will be to raise steam and maneuver the ship from port using the oil fired operating mode. During this time the grate must be covered with about 2 inches of ash to protect it from overheating.

When ready for coal firing, the coal hoppers are filled and the stoker drive motor is started. The stoker is operated manually by occasionally engaging the feed latches (4 in Figures 27-28 and 27-29) while adding coal a little at a time until all the coal on the grates is ignited and burning freely. The oil burners must be turned down as necessary to maintain drum pressure. The feed latches may be engaged and coal fed continuously at

Figure 27-26. Detroit stoker system. Courtesy Detroit Stoker Co.

this time. The arms positioning crosshead (5) should now be raised to the top notch of the link for feed control (7). The handwheel (6) is backed off to prevent overfeeding coal. Once the fire has been started and the fuel feed and distribution adjusted, the grate drive can be started.

In adjusting the speed of the grate drive, the governing factors are the load being carried and the thickness of the ash coming off the front end of the grate. Normally the ash should be 3 to 4 inches deep where it comes off the grate. If the ash is too thin, the grate is moving too fast. The grate speed may be reduced by turning the handwheel on the safety release link to variable speed reducing. (The variable speed reducing unit must always be running when making this speed change adjustment.) If the ash is too thick, it will tend to form clinkers and carry unburned coke forward on top of the clinkers. If this condition develops, the grate speed should be

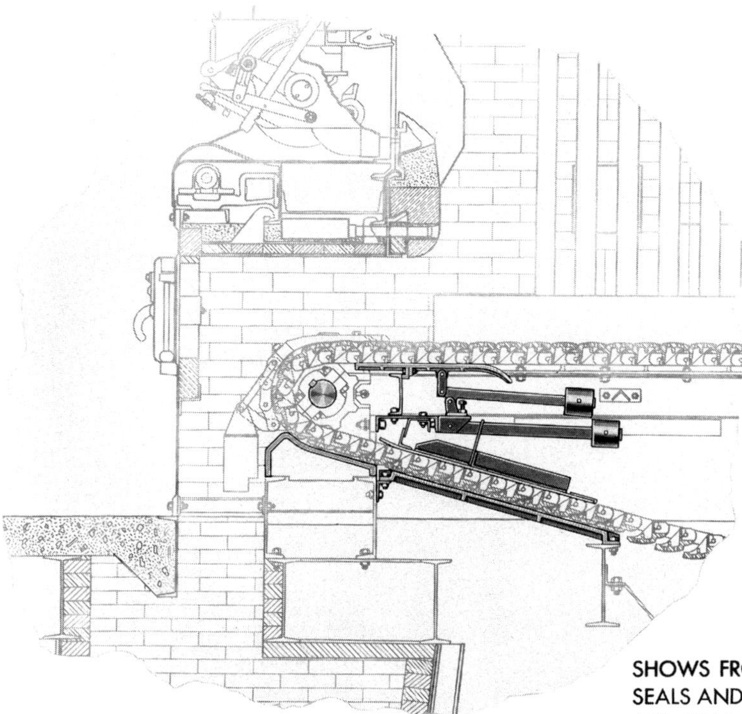

Figure 27-27(a). Front end of chain grate assembly.
Courtesy Detroit Stoker Co.

increased by turning the handwheel on the safety release link in a clockwise direction.

If the ash bed is too thin, as little as two inches or less, and forms clinkers with unburned coke riding forward on top of the clinkers, it indicates that the fuel bed is too heavy and the CO_2 is too high. This condition will often cause smoke to be emitted from the stack. To correct this condition the coal feed is reduced by turning the handwheels (shown as 6 in Figures 27-28 and 27-29) counterclockwise one-quarter turn a time until the desired results are obtained.

Combustion Control Adjustments

The most satisfactory method of adjusting the combustion control is to start at the minimum or closed position, take gas samples to determine the percentage of CO_2, adjust the coal feed and fan damper to produce the correct coal-to-air ratio, then move the power jack or piston about 2 inches,

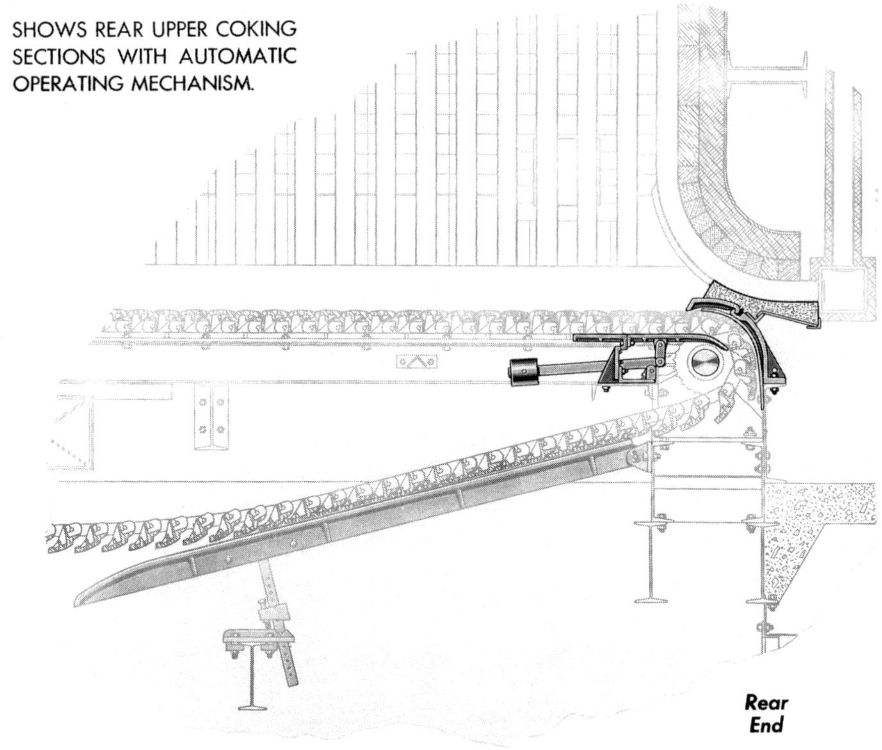

Courtesy Detroit Stoker Co.

and again check the CO_2 at the furnace outlet. If satisfactory CO_2 readings are obtained at the low position and low CO_2 readings at the open position, it will be necessary to increase the arc of travel of the coal feed control shaft. When making this adjustment the connecting arms to notched links (12) on the coal feed control shaft (13) must be reset so that the crossheads will be at the top ends of the slotted arms (2) when the master regulator is at the closed or minimum position.

Low CO_2 can also be corrected by reducing the air flow. This can best be done by moving the pin in the fan damper arm to a longer radius, thereby reducing the travel as well as the maximum opening of the fan damper. In making this change the closed position of the fan damper will have to be reset when the master regulator power unit is in the closed or minimum position. The link to the fan damper must be lengthened to bring the damper back to the original setting at the closed position. The air pressure under the grates should increase about in proportion to the square of the load on the unit.

1	Feed plate	10	Inner arm operating feed plate
2	Slotted arm	11	Rotor housing, lower
3	Link locating crosshead	12	Connecting arm to notched link
4	Latch for coal feed	13	Shaft for control
5	Arm locating crosshead	14	Crosshead
6	Adjusting handwheel	15	Outer arm operating feed plate
7	Notched link for coal feed control	16	Shoulder screw for arm locating crosshead
8	Spilling plate	18	Bearing for control shaft
9	Screw for spilling plate	24	Sifting pan under feeder

Figure 27-28. Coal feed assembly. Courtesy Detroit Stoker Co.

By continuing this process of combustion control adjustment, it is possible to establish the proper relation between the coal feed and the air supply so that a given setting of the coal feed adjusting handwheels (6) and the crossheads (14) will give a rising characteristic of CO_2 leaving the furnace as the control opens. This total rise should be approximately 3 or 4 percent to change CO_2 in the furnace from about 9 or 10 percent to about

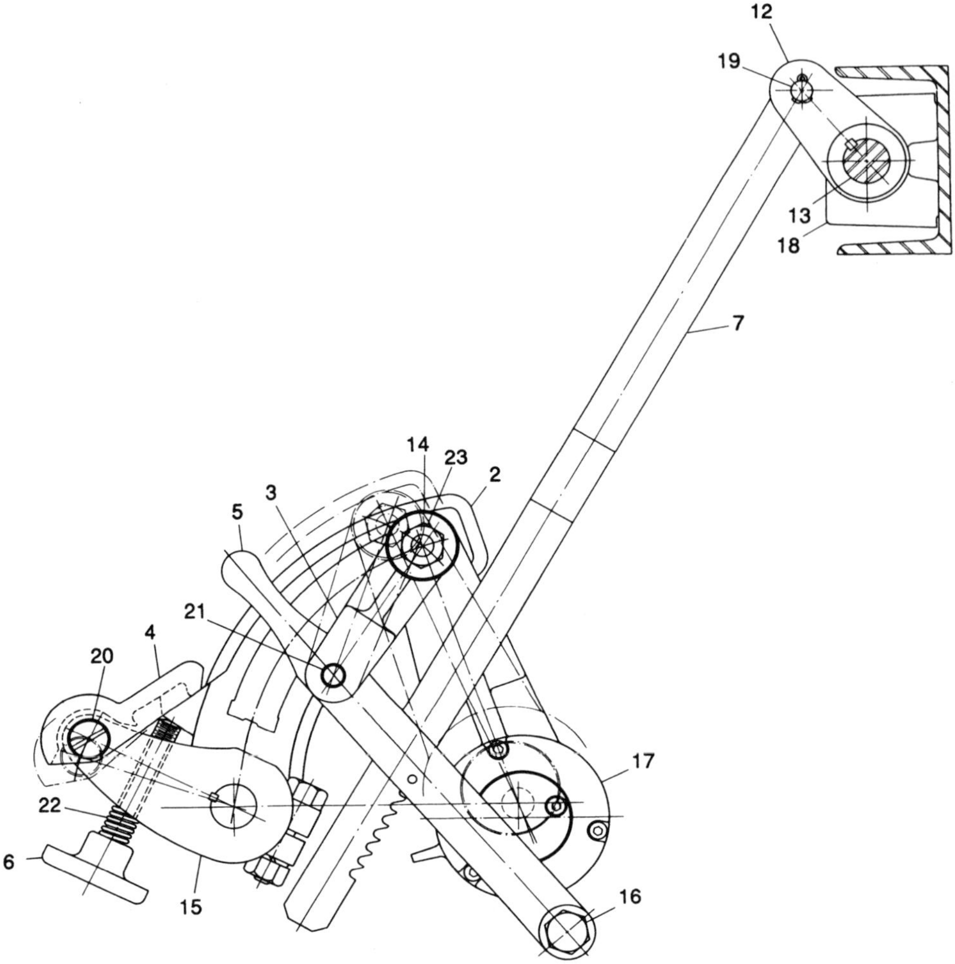

2 Slotted arm
3 Link locating crosshead
4 Latch for coal feed
5 Arm locating crosshead
6 Adjusting handwheel
7 Notched link for coal feed control
12 Connecting arm to notched link
13 Shaft for control
14 Crosshead
15 Outer arm operating feed plate
16 Shoulder screw for arm locating crosshead
17 Eccentric link assembly
18 Bearing for control shaft
19 Pin for notched link
20 Screw for feed latch
21 Pin for arm and link
22 Screw & spring for handwheel
23 Pin for crosshead

Figure 27-29. Coal feed control. Courtesy Detroit Stoker Co.

13 to 14 percent CO_2 at the open or maximum position of the master regulator.

Coal Feed Control

The coal feed is automatically controlled by the master regulator of the combustion control through a system of arms or links or a special power jack connected to the stoker control shaft and from there to the crossheads (14) located in the slotted arms (2). Raising the crossheads in the slotted arms decreases the coal feed and lowering the crossheads in the slotted arms increases the coal feed.

The crossheads are positioned by using one of the several notches in the links (7). The proper notch to be used depends on the amount of coal required, the size of the coal, and the amount of moisture in the coal. The crosshead position should be lower for coarse or wet coal than for fine or dry coal.

If it is found that proper coal feed cannot be obtained with the adjustments outlined above, the travel of the feed plates (1) can be increased or decreased by changing the location of the pin, bushing, and block assembly in arms (10) thereby increasing or decreasing the coal feed. Placing the pins and blocks farther from the shaft increases the travel of the feed plate thereby increasing the coal feed and vice versa. (There are three locations for the pin and block, and feeders are shipped with the pins in the center holes. Therefore, if making a change in block location, all feeders of the stoker must be changed to correspond.)

In case too much coal is being fed to the grates when the crosshead (14) is carried in the upper notch of the links (7), the adjusting handwheels (6) should be turned counterclockwise one-quarter or one-half turn at a time until the desired coal feed is obtained. Where not enough coal is being fed when the crossheads are being carried in the upper notch of links (7) and the handwheels are turned up tight, the crosshead should be lowered to the second notch. If this setting feeds too much coal the handwheel can be turned one-quarter turn at a time counterclockwise until the desired amount of coal is obtained. The crossheads should always be carried as high as possible. One-half turn lost motion of the handwheels (6) is normally equivalent to one notch in links (7). In other words, if the crossheads are carried in the second notch with the handwheels backed off one-half turn or more, the crosshead should be raised to the top notch and the handwheel readjusted for the desired condition.

The notches in the links are the coarse adjustment and the handwheels are the fine adjustment of the coal feed. The adjustments should be kept the same on all feeders of the stoker provided the coal in all hoppers is of uniform size and moisture content. After the coal feed has been properly adjusted, the master regulator will raise and lower the crossheads in the slotted arms in proportion to the load requirements, as indicated by changes in steam pressure. It will likewise increase or decrease the air supply and pressure under the grates and change stoker grate speed.

Furnace Draft Regulator Adjustment

The furnace draft regulator of the combustion control system is provided to maintain the furnace draft, or draft over the fire, at a predetermined point. It is desirable to hold the furnace draft somewhere between negative 0.05 and 0.10 inches water column. The exact setting will have to be determined to suit the shipboard installation. Best results can be obtained by holding the furnace draft as near zero as possible without either causing gas to escape from around the doors or leak through the setting or overheating of the stoker parts exposed to the furnace. The furnace walls and the ash must also be considered in respect to amount of draft, for if held too low, the walls may slag and the ash may fuse into clinkers.

Coal Distribution

Strictly speaking, there is no fuel bed with the Detroit RotoGrate Stoker shown in Figure 27-30. The depth of the burning coal on the grates or on top of the ashes will vary from practically nothing to about 1.0 or 1.5 inches, depending on the size and characteristics of the fuel. It is of importance to prevent the burning coal from reaching a greater thickness than this because of the tendency to smoke and the possibility of forming clinkers.

The fuel bed is kept level by maintaining the proper adjustments of the rotor speed and the proper position of the spilling plate. Normally it is recommended that the spilling plate (8) be moved forward until the center rib bears against the end of the adjusting screws (9) by turning a screw (9) clockwise, and then controlling the fuel distribution by adjusting the rotor speed.

To feed the coal towards the rear end of the grate, speed up the rotors by turning the handwheel on variable speed drive adjusting base to decrease centers between the motor and the driven pulley. To feed the fuel forward on the grate or towards the front end of the stoker, turn the handwheel on variable speed drive adjustable base to increase distance between centers of the motor and the driven pulley. For normal operation turn the handwheel about one revolution at a time and wait for the fuel bed to readjust itself.

If mechanical coal handling equipment is used to fill the stoker hopper, and one rotor section of a hopper consistently receives either coarser or finer coal than another section, this inequality can be compensated for in fuel size by adjusting the spilling plate (8) using the adjusting screw (9). To feed the coal toward the rear of the stoker, move the spilling plate (8) forward by turning the adjusting screws (9) clockwise. To feed the coal

Figure 27-30. Coal feed assembly installation.
Courtesy Detroit Stoker Co.

toward the front end of the grate, move the spilling plate rearward by turning the adjusting screws (9) counterclockwise. Coarse, nut, and dry coal will feed farther back for the same setting than will fine or wet coal. For coarse or dry coal, screws (9) should be turned counterclockwise. For fine or wet coal, screws (9) should be turned clockwise.

The fuel distribution should be adjusted so that the coal will build up on the grate just at the forward end of the rear tuyeres. If the coal piles up on top of the rear tuyeres, the coal is being fed back too far; if the coal does not feed back to the rear tuyeres, it is not being fed back far enough.

Grate Speed Adjustment

The grate speed is determined to a large extent by the load being carried, the amount of coal being burned, and the percentage and characteristics of the ash in the fuel. Normally the ash should be maintained at a thickness of 3 to 4 inches where it comes off the front end of the grate. With some fuels this may be too thick due to the clinkering characteristics of the ash. When the ash fuses into sheet clinkers with unburned coke riding forward on top of the clinker, it indicates that the grate speed is too slow. The grate speed should be increased by turning the handwheel on the safety release link to variable speed reducing unit to lengthen the safety release link.

The above is based on the assumption that the coal feed control is properly adjusted and that too much fuel is not being fed. Coal feed control can be determined by observing the fire, the gases leaving the stack, and the carbon dioxide (CO_2). If the fuel feed is too heavy, the furnace will be smokey, the carbon dioxide (CO_2) will be too high, and the stack will be smoking. A change in the grate speed will also have some effect on the distribution adjustment. If the grate speed is increased, it may be necessary to increase the rotor speed in order to keep the fuel back to the forward edge of the rear tuyeres. And, if the grate speed is reduced, the rotor speed will need to be reduced to keep the coal from piling on top of the rear tuyeres.

The safety release link for the variable speed reducing unit disengages if the control shaft is rotated while the grate drive variable speed reducing unit is idle or stopped. In order to synchronize the relation between the grate speed and the coal feed, a handwheel is provided on the end of the safety release link, to increase or decrease its length. Increasing the length of this safety link assembly proportionally increases the speed of the grate for any given coal feed adjustment and vice versa. To increase the length of the assembly the locknut is backed off, the handwheel is turned clockwise, and the locknut is retightened. Should the plunger assembly disengage when the grate drive variable speed reducing unit is operating, the plunger is reengaged. To prevent repetition of this difficulty the tension on the spring is increased by tightening the pipe cap.

Safety Devices

Safety shear pins are provided on the drive shaft for protection of each of the feeder units. If foreign materials such as tramp iron, wood, etc., lodge in the feeder mechanism, the small pin in the safety device will shear and stop the feeder before breakage occurs. Very often when the feeder is stopped, the foreign material will dislodge itself, but if this does not happen and the cause of stoppage is not readily removed, it may be necessary to remove the lower part of the rotor housing (11). This is accomplished by taking out 2 or 3 bolts at the upper front edge of the lower section of housing allowing this section to be dropped and slipped forward for inspection of the rotor drum and blades. It is suggested that when the feeder is open, a check be made to insure no accumulation of grease and coal dust around the ends of the rotor drum. After the feeder is cleaned, the cover should be replaced and bolted securely.

Before replacing the shear pin the clutch must be disengaged. The coal feed latch (4) should also be disengaged, and after the pin is in place, the shear pin assembly is rotated by hand to insure that the feeder is entirely free. If the shear pin assembly and rotor shaft turn freely, the clutch may be reengaged, putting the feeder back in service.

Safety devices or shear pins are provided for the protection of the grate driving mechanism. Two shear pins are located in the large sprocket mounted on the end of the front grate drive shaft. When these pins shear, it is very important that the grates and grate supporting chains and mechanisms be inspected. This is done by entering the wind box or air chamber through the access door in the bottom of the air chamber. It is also necessary to inspect the underside of the grates by entering the air chamber through the access door in the side support. When these pins shear, some foreign material has probably been wedged at some point. After the obstruction has been removed, new pins can be inserted in the shear pin bushings of the large sprocket. It will be necessary to start the grate drive and rotate the large sprocket until the holes of the shear pin assembly are in line.

Overfire Air and Cinder Return System

The high pressure fan of the system serves to return the cinders and fly ash from the hoppers under the boiler mud drum, economizer, air heater, dust collector, or other traps to the furnace in the case of pneumatic systems and to the intake manifold of cyclone assembly in the case of pneumatic gravity flow systems. The same fan also serves in supplying air to create furnace turbulence. The cinder return fan should be started when the fire is started and remain in operation as long as there is fire in the furnace.

All settling hoppers should be thoroughly cleaned out after erection as any foreign matter such as pieces of brick, mortar, etc., will obstruct the openings of the cinder return pickup assemblies. Hoppers should be kept empty down to the level of the pickup assemblies and should also be inspected periodically to make sure that the cinders are being picked up. For pneumatic gravity flow systems the cyclone vanes and tubes should be inspected whenever there is a boiler outage to make certain there are no obstructions. To check the operation of each cinder carrier jet, look through the fly ash observer to see whether cinders are passing through the carrier pipes.

Required air pressures will have to be determined in the field as each installation has its own peculiarities of layout and operating conditions which make adjustment necessary. The character of fuel burned must also be taken into consideration.

In the case of the pneumatic gravity systems the setting of the slide gate in the exhaust duct from the gravity cinder return system manifold should be determined by taking pressure readings in the fly ash settling hopper of the system. This is done by setting the slide gate for a pressure in the settling hopper which will always be on the plus side of the negative pressure, or draft, found at the rear of the furnace and at the stoker grate line to eliminate any possibility of flue gases short-circuiting up through the discharge pipes of the fly ash settling hopper of the gravity flow system.

It is very important that the entire cinder return system be maintained in good operating condition and kept in operation at all times when the stoker is carrying a fire. If the high pressure fan is shut down while there is a fire in the furnace, the hot gases and flame may short-circuit through the cinder return piping and cause serious damage to the system. If the straight gravity flow cinder return system is used, the high pressure fan will serve only for overfire air and turbulence purposes.

Inspection for Clinkers

Periodic inspection should be made of the fuel bed by observing the fire through the front fire and access doors and the side wall observation doors or ports. Ash or small clinkers may have a tendency to build up at the front end of the grates at either side next to the sidewalls. If this condition develops, these clinkers should be removed by using a hook through the front door. If they are not removed, they will probably grow in size and impede the normal progress of the ash to the front end discharge point.

Apron Tuyeres

The apron tuyeres are designed to admit air into the furnace over the fire and under the coal stream leaving the feeders. These tuyeres are connected to the overfire air duct mounted below the feeders at the stoker front. The dampers should be set below each feeder so that the air pressure at the apron tuyeres is just sufficient to keep them cool and free from buildup.

The apron tuyeres should be examined every time the unit is shut down for inspection. If the tuyeres should become plugged and the flow of air shut off, they will heat up and, in time, burn out if this condition is not corrected.

When the apron tuyeres become overheated, the coal sticks to the top surface and obstructs the rotor opening. This affects the fuel distribution. Fuel piles up near the front end of the grate which overheats the front wall and causes considerable damage.

Coal Feeder or Rotor Operation

The stoker should not be operated with one feeder or rotor shut down while the grate drive is in operation because the grate section back of the idle feeder would soon become bare and be exposed to high furnace temperature. This would probably result in grate bar warpage or other damage. If for any reason it is found necessary to stop one feeder for a short period, the grates should also be stopped, so that they remain covered with ash and thus are protected from the radiant heat of the fire.

Ash Pit Cleaning

The ash pit should be cleaned periodically long before it becomes nearly filled. If the fire should get out of control for a short time and it is found necessary to run hot clinkers, ash, and coke over into the ash pit, the ash pit should be cleaned promptly. If this refuse is allowed to remain there, it may fuse into a solid mass of clinkers.

Stoker Out of Service

Whenever the boiler and stoker are taken out of service for cleaning and inspection or repairs, the coal hoppers should be emptied before letting the fire go out. If the stoker is stopped with coal in the hoppers, the coal may ignite and cause serious damage to the rotor or feeder assemblies. The grate drive should be stopped as soon as the hoppers are empty so that the ash will remain on the grates and serve as an insulator against the high furnace temperature.

The feeder rotors should be kept rotating, with feed latches disengaged so that rotor drum temperatures will be kept uniform on all sides and prevent warping. Three to four hours after the fire has gone out, the furnace temperature should be sufficiently lowered to permit starting the grate drive and discharging the ashes into the ash pit leaving the grates uncovered for inspection.

Important Operating Considerations

Keep reasonably clean all settling chambers in the boiler, breeching, and heat traps where cinders and ash can accumulate. Keep the ash pits clean.

Keep the combustion control system in proper adjustment and working freely.

Always keep cooling water flowing through water cooled bearing casings when stoker is in operation.

Always keep the cinder return fan running when stoker is in operation.

Grease all bearings at least every eight hours. Check oil level of mechanical drive gear case daily.

Check oil level of enclosed variable speed reducing units daily and check chain tension after every 1,000 hours of operation.

Clean and oil all silent chain drives at regular intervals.

Clean fans and fan blades and repair any blades that may have become loose or damaged.

Clean and check motors and starters periodically.

Examine grates frequently to be certain that all air holes are open.

Keep coal and air properly proportioned so that fire burns clean and without smoke.

Do not allow coal to accumulate on top of the deflector tuyeres, back of the coal hoppers, and above each feeder opening.

Check all moving parts frequently and replace worn or burned parts.

COMBUSTION CONTROL SYSTEM

Introduction

Figure 27-31 is the logic diagram for a combustion control system developed by the Forney Engineering Company. A combustion control system regulates the amount of air and fuel to maintain a preset steam pressure for any boiler load and for any rate of change in boiler load. In doing so, a predetermined ratio is maintained between the fuel and the air. A pneumatic two-element combustion control system is selected for this application, whereby the complete range from maximum overload to minimum load can be obtained by firing oil only, coal only, or a coal and oil combination. The system has 3 modes of operation:

1. Automatic control (through pneumatic control functions and solid-state burner management interface protection);
2. Remote manual control (operated from the manual/auto stations in the control room console); and
3. Local manual control (operated at the boiler, by means of mechanical devices on the final control elements).

As shown in Figure 27-31 the combustion control system consists of seven subloops (plant master controller, steam flow feed forward, FD fan controller, ID controller, coal fire control, oil fire control, and burner management interface), each of which is briefly described in the following text.

Plant Master Controller

A steam pressure transmitter senses the pressure in the common superheated steam header. The output signal from the transmitter is compared with the set point and through proportional and integral action of the master controller, a load demand signal is forwarded to both boilers. High limit and low limit relays are included in the controller output line, with

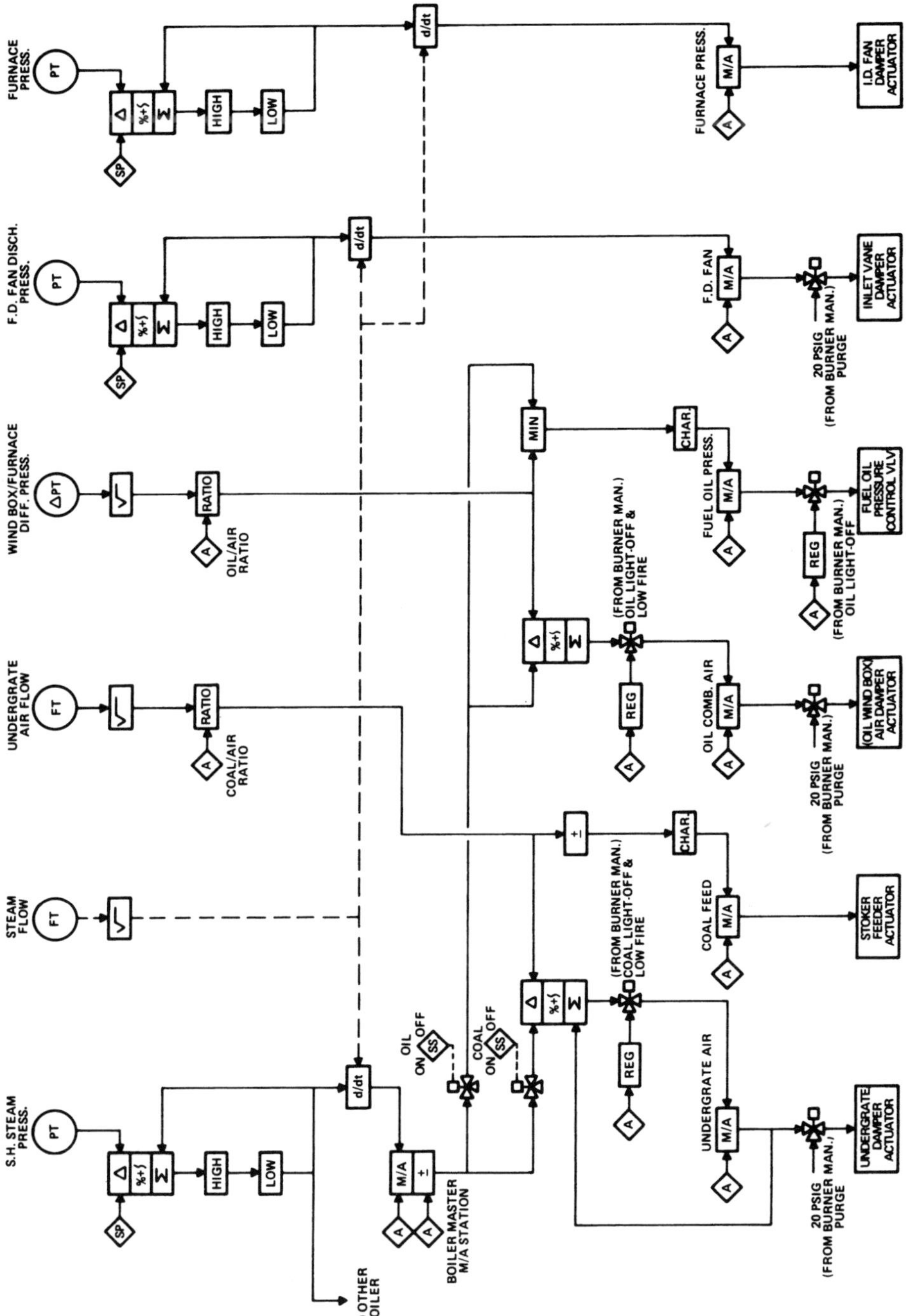

Figure 27-31. Combustion control system. Courtesy Forney Engineering Co.

a feedback to prevent the controller from winding out of control range. A master manual/auto station for each boiler with a built-in adjustable bias function provides the capability to divide the load share between the boilers.

The plant master controller is common to the two-boiler plant. All other subloops are provided in duplicate, one for each boiler.

Steam Flow Feed Forward

The steam flow is measured by means of a flow nozzle in the saturated steam line. The output signal of the steam flow transmitter is introduced as a rate of change action signal in the impulse relay (d/dt). As a result, whenever the steam flow changes, the control signal receives a temporary impulse to commence a firing rate change which is proportional to the rate of load change. This control action anticipates the change in steam pressure that will result in a firing rate change of the master controller. The result is a faster response from the combustion control system. The same impulse signal from the steam flow is used in the FD fan and the ID fan control loop.

FD (Forced Draft) Fan Controller

The function of the FD fan control loop is to provide a constant combustion air pressure in the FD fan discharge ducting. An air pressure transmitter senses the pressure on the discharge side of the FD fan. The output signal from the transmitter is compared with the set point and, through proportional and integral action of the FD fan controller, a demand signal is forwarded to the inlet vane damper actuator. High and low limit relays with controller feedback are included to prevent the controller from winding out of control range. A manual/auto station is included for remote manual operation of the inlet vane damper actuator.

ID (Induced Draft) Fan Controller

The function of the ID fan control loop is to provide a constant furnace pressure. A furnace pressure transmitter senses the pressure in the furnace. The output signal from the transmitter is compared with the set point and, through proportional and integral action of the ID fan controller, a demand signal is forwarded to the ID damper actuator. High and low limit relays with controller feedback are included to prevent the controller from winding out of control range. A manual/auto station is included for remote manual operation of the ID damper actuator.

Coal Fire Control

The coal fire control loop controls the amount of coal and corresponding combustion air to satisfy the plant master demand. When the coal firing mode is selected, the load demand signal from the plant master controller is introduced into a proportional plus reset airflow controller which directs the undergrate damper actuator to the desired position. An undergrate airflow transmitter is installed, measuring the differential pressure across an air duct resistance as a function of underfire airflow. Via a square root extractor and a coal/air ratio relay, the transmitter output signal becomes the feedback signal to the airflow controller, thus closing the airflow control loop. The output signal from the coal/air ratio relay is also introduced into a characterizing relay which is preprogrammed to correct the signal for any nonlinearity of the stoker feeder. The programing also includes an additional amount of predetermined excess air at lower firing conditions. The output signal from the characterizing relay directs the stoker feeder actuator to the desired position. The grate speed is linked directly to the stoker feeder so that the traveling speed of the grate is directly proportional to the amount of coal fed onto the bed.

When firing coal, an overfire air fan provides a fixed amount of combustion air which is blown into the furnace across the surface of the coal bed. Approximately 15 percent of the full load airflow is thus introduced and compensated for by the coal characterizing relay.

Manual/auto stations are provided for remote manual operation of both the undergrate damper actuator and the stoker feeder actuator.

Oil Fire Control

The oil fire control loop controls the amount of fuel oil and corresponding combustion air to satisfy the plant master demand. When the oil firing mode is selected, the load demand signal from the plant master controller is introduced into a proportional plus reset airflow controller which moves the air damper actuator on the entry of the oil firing wind box to the desired position. An airflow transmitter measures the differential between wind box pressure and furnace pressure. Via a square root extraction and an oil/air ratio relay, the transmitter output signal becomes the feedback signal to the airflow controller, thus closing the airflow control loop.

The boiler load demand signal is introduced into a minimum selector relay. The output signal from the oil/air ratio relay is also introduced into this minimum selector relay, where the lower of the two input signals (actual airflow and boiler demand) is passed on as a fuel demand signal. The oil/air ratio relay is adjustable and actually subtracts the desired percentage of excess combustion air from the total amount of air available in the wind box.

Thus the relay output signal becomes the stoichiometric combustion air signal and the corresponding fuel oil signal. The fuel oil demand signal is entered into a characterizing relay which is preprogrammed to correct the signal for any nonlinearity of the fuel oil control valve. The programming also includes an additional amount of predetermined excess air for low fire conditions. The output signal from the characterizing relay moves the fuel oil pressure control valve to the desired position. Manual/auto stations are provided for remote manual operation of both the air damper actuator and the fuel oil control valve.

Burner Management Interface

The burner management interface instrumentation has the function of assuring that protective conditions are satisfied prior to lighting the oil burners or the coal bed. The burner assures management interface that the fuel supply to the boiler is limited or shut off whenever a condition exceeds preset limits. The burner management interface accomplishes these functions by overriding the regular combustion control signals.

The burner management interface consists of a set of solenoids and pressure regulators which can be divided into the following functions.

Boiler prepurge. The prepurge section has the function of assuring that the furnace is thoroughly purged with combustion air prior to allowing the initial fire. Via solenoids, a maximum air signal is introduced to the undergrate damper actuator, the wind box air damper actuator, and the inlet vane damper actuator which forces these dampers to the full open position. Upon completion of the purge function, the solenoids are released, and the dampers return to a position which corresponds to the combustion control demand signals.

Oil light-off and low fire. Prior to lighting off the first oil burner, low fire control signals are introduced to the air damper actuator and to the fuel oil control valve. These signals force the air damper and the fuel oil control valve to a position corresponding with the required amount of air and oil for lighting off a burner. As soon as the oil flame is established (proven by the flame scanner), the oil solenoid is released back to the combustion control. The airflow solenoid is released when the flame is established, the drum pressure is satisfactory, and the coal/oil selector switches are set properly.

Coal light-off and low fire. Prior to introducing coal to the stoker, a low fire control signal is introduced to the undergrate damper actuator. This signal forces the undergrate damper to a position corresponding to the minimum airflow through the grate. As a result of the combustion control system, the stoker feeder will be in the minimum feed position. Positioning the coal/oil selector switches in the desired mode will release the coal light-off solenoid and will return the operation back to the combustion control.

Coal/oil select. A two-position coal selector switch introduces the master demand signal to the coal firing control loop. In the same way, a two-position oil selector switch introduces the master demand signal to the oil firing control loop.

Before the first fire is introduced into the furnace, the operator turns the selected fuel switch to the "on" position. This will open the corresponding solenoid, and the control signal can enter the subloop. If oil firing is selected and the oil firing switch is therefore in the "on" mode, the coal switch must be in the "off" mode. When the flame is established and the drum pressure is satisfactory, the oil light-off solenoid will be released, and the combustion control will command the oil and the corresponding amount of combustion air.

To initiate coal firing, the coal selector switch must be turned to the "on" position. The coal fire will then be initiated, but will remain in the low fire position as long as the oil burner is operated and controlled by the combustion control. To increase the base coal fire load, the manual/auto station for undergrate air and/or the manual/auto station for the stoker feeder can be switched to manual. The fuel oil demand will automatically decrease proportionally as the coal firing is manually increased.

When the coal fire is established, the oil firing can be stopped by switching the oil selector switch to the "off" position. This results in shutting down the oil flame while at the same time it releases the coal firing from the light-off mode, so that the combustion control takes over control of the coal and undergrate air, proportional to the steam pressure demand.

The next time oil firing is desired while coal fire is on, the oil selector switch must be switched to the "on" position. A fifteen-second wind box purge will be initiated to remove any coal dust from the wind box. Then the oil burner will be ignited, but the oil will be held in low fire position as long as the coal firing is modulated by the combustion control. To increase the base oil fire load, the manual/auto station for the air damper and/or the manual/auto station for the fuel oil valve can be switched to manual. The coal demand will automatically decrease proportionally. To summarize, in the automatic mode, the first fuel (coal or oil) switched in the "on" position will modulate automatically in response to the combustion control demand. The other fuel can be either shut down (switch in "off" position) or in base load condition (switch in "on" position). This base load can be increased by switching the corresponding manual/auto station into manual mode and the other fuel demand will decrease automatically.

REVIEW

1. Name some trades where you might expect to find modern coal burning vessels.
2. How do coal properties affect the design and operation of a propulsion system?
3. What is the significance of sulfur content in coal fuel?
4. What is slagging?
5. What features are designed in a boiler when severe slagging potential is anticipated?
6. What is fouling on a steam generator fire side?
7. What is the burning profile of coal?
8. What features are designed into a steam generator with severe fouling potential?
9. What is the burning profile of coal? How is it used by boiler designers?
10. What is a balanced draft system?
11. Describe the operation of a continuous discharge spreader stoker.
12. What is a cinder reinjection system?
13. Why does a coal fired boiler have both forced and induced draft fans?
14. How is a coal fired boiler started?
15. Compare the typical efficiencies of coal and oil fired boilers.
16. Describe a stoker-fired boiler.
17. Describe a typical ash handling system.
18. How is coal delivered from the bunker to the stoker?
19. What are apron tuyeres?
20. Describe the elements of a coal fired combustion control system.
21. How is furnace pressure controlled?

REFERENCES

American Bureau of Shipping. 1986. *Guide For Ships Burning Coal.* New York: American Bureau of Shipping.

Babcock & Wilcox. 1978. *Steam/Its Generation and Use,* 39th ed. New York: Babcock & Wilcox.

Beggs G. C. 1982. "Coal Burning Bulk Carriers for an Australian Coastal Trade." *Transactions.* Institute of Marine Engineers.

Combustion Engineering, Inc. 1981. *Combustion/Fossil Power Systems,* 3rd ed. Windsor, Ct.: Combustion Engineering, Inc.

Institute of Marine Engineers. September 1983. "Quincy Completes New US Coal-fired Ship." *Marine Engineers Review.* Institute of Marine Engineers.

Institute of Marine Engineers. January 1981. "British Design for Coal Burning Ships." *Marine Engineers Review.* Institute of Marine Engineers.

Joshi, P. B. and A. D. Taylor. January 1982. "Fuel Handling Aboard Coal-fired Ships," *Marine Engineers Review,* Institute of Marine Engineers.

ACKNOWLEDGMENTS

The editors are indebted to the following companies for permission to publish the material included in this chapter:

Babcock & Wilcox, New York, New York
Combustion Engineering, Inc., Windsor, Connecticut
Detroit Stoker Company, Monroe, Michigan
Forney Engineering Company, Addison, Texas
Foster Wheeler Boiler Corporation, Livingston, New Jersey
Stock Equipment Company, Cleveland, Ohio

CHAPTER 28

Waste Disposal Systems

EVERETT C. HUNT

INTRODUCTION

SHIP designers and ship operators have always found it necessary to deal with trash, garbage, and sewage resulting from the ship's operation. In the past when environmental concerns had not yet reached a high priority, waste materials were usually dumped in the oceans. Untreated sewage was often flushed in the world's harbors while the vessels were in port.

Since the early 1970s, a complicated set of national and international regulations, laws, protocols, and agreements have placed restrictions on the discharge of waste into the sea, and have required ship designers to incorporate new systems to handle the waste. As a result, the shipboard engineer officer has additional systems to operate and maintain.

General Restrictions on Discharge into the Sea

Garbage and trash. The Convention on the Prevention of Marine Pollution by Dumping of Wastes and Other Matter at Sea (LDC 1972) prohibits discharge of all plastics. Dunnage and similar floating materials may be discharged within twenty-five nautical miles of land. Food waste and other garbage may be discharged within 12 nautical miles of land. Current trends in international agreements suggest that complete prohibition of the discharge of garbage and trash into the oceans will be required in the future.

Sewage and wastewater. No ship is permitted to discharge untreated sewage in any U.S. controlled waters. There is no restriction, however, on the discharge of gray waste, i.e., sink and shower drains.

Oil and oily mixtures. A series of international agreements place complicated restrictions on the discharge of oil or oily mixtures from any vessel. Some of these restrictions are discussed in Chapter 26.

SEWAGE TREATMENT

The methods now in use on ships for the treatment of sewage include the following: collection and retention; maceration and chlorination; physical separation; physical/chemical treatment; and biological treatment.

Collection and Retention

Collection and retention systems involve, as the name implies, no more than simple holding tanks. These tanks must be capable of retaining the collected waste until it can be discharged ashore, or in unrestricted waters that are acceptable. Where there are no shore receiving facilities, it may be necessary to retain the waste for a considerable period. Under such circumstances measures must be taken to prevent septic conditions occurring which generate toxic, corrosive, and inflammable gases.

Maceration and Chlorination

Macerator/chlorinators are simple and relatively inexpensive systems. They include a macerator which discharges the sewage into a small holding tank in which the waste is treated with a metered dose of disinfectant (usually a sodium hypochlorite) and held for approximately 30 minutes before it is discharged overboard. However, this type of unit is not usually capable of discharging an effluent product acceptable to the legislating authorities.

Physical Separation

A number of systems utilize physical separation of sewage. This generally involves the use of some form of filter, and in some instances includes a settling tank. The separated liquid effluent is disinfected before being discharged overboard. Sludge accumulated by filtration and settling must be stored or incinerated. If stored, measures have to be taken to prevent the onset of septic conditions.

Physical/Chemical Treatment

Physical/chemical treatment systems use flocculating agents to aid the physical separation and precipitation of solids. These systems may be started in a very short period of time and can discharge good effluents almost immediately. Like physical separation units they produce sludge that must be continuously removed and eliminated.

Biological Method

There are a number of different processes in the biological treatment of sewage universally used for the management of municipal wastes. Several of these processes have been applied to the treatment onboard ships, and have become, for the most part, the preferred systems. The basis of all these treatment variations is nature's own purification process in which microorganisms use the waste as food material. Two variations of the biological method found in modern marine sewage treatment units are extended aeration and the trickling filter.

In the extended aeration process the sewage is held in an aeration tank for approximately 24 hours, and air is continuously bubbled through the liquid (mixed liquor, as it is called). Mixed liquor is then displaced by incoming sewage into a settling tank where the biological floc is formed and allowed to settle under quiescent conditions. The settled material (activated sludge) is continuously withdrawn from the bottom of the settling chamber and recycled to mix with the waste material entering the aeration tank. Clear effluent from the top of the settling tank is discharged after disinfection.

The trickling filter process utilizes the ability of certain aggregate materials to hold large communities of microorganisms, allowing the waste to trickle over the material and the organisms to remove the substrate. The effluent draining from the bottom of the filter then settles and any entrained floc is removed before it is disinfected and discharged overboard. Variations of this process utilize submerged filters with air bubbling through the material to provide the oxygen required by the microorganisms.

Assessment of the advantages and disadvantages of these processes must include such factors as ease of maintenance, degree of supervision required, running costs, capital costs, and, where installation in existing vessels is being considered, the cost of installation. The designer of shipboard equipment must take into consideration all of the advantages and disadvantages of these processes. In addition, he must examine his design with respect to manufacturing techniques, because the manufacturing of the unit has a significant bearing on its final cost. Finally, the equipment must, of course, comply with the requirements established by the legislating authorities and must be approved by the relevant local authority.

DESIGN OF SEWAGE TREATMENT SYSTEMS

Almost all of the early sewage treatment plant designs were of the biological type, and many of these used the anaerobic process similar to septic tanks found ashore. Although they required the absolute minimum of supervision and had only one moving component—a discharge pump—these anaerobic plants were not capable of producing effluents of the quality proposed by IMO and other legislating bodies. Consequently, anaerobic plants were superseded by equipment using other biological processes that were capable of meeting the proposed standards for discharges.

Aerobic biological plants, which superseded the earlier anaerobic units, are based on the extended aeration process of waste treatment. As may be expected the designers of early units drew on experience gained from the operation of industrial equipment, and while the process remained the same as in the earlier anaerobic plants, the practical problems associated with shipboard use were not insignificant. The evolution of the present-day unit—the extended aeration method—has been along a path of trial and error where the need to make it as simple as possible has been a prime objective.

Since the plants had to be small, compact, and suitable for installation in the limited spaces available onboard ship, the biological oxygen demand per volume of aeration space (BOD/vol.) imposed on them was considerably higher than those accepted ashore and in some cases was more than double. This deviation from land practice means that there is a need for fairly frequent, periodic desludging. For example, a shipboard unit needs to be desludged at intervals of 2 to 3 months, whereas an industrial unit for an equivalent population on shore would be expected to operate for 10 to 12 months without desludging. Sludge built up in an extended aeration unit consists of biological solids together with any nondegradable material that may be present in the waste. The nondegradable material increases the solid content of the liquid in the aeration chamber until a point is reached where settling in the final sedimentation chamber is impaired, and a reduction in the solids level becomes necessary. The reduction in the solids content of the mixed liquor is achieved by desludging, an operation consisting of pumping out 90 percent of the contents of the aeration tank and refilling with clear water. Although the sludge, 99 percent water and handled entirely by the discharge pump, may not be dumped into waters covered by regulations, it may be discharged at sea well clear of the protected areas. If the vessel is in port when desludging becomes necessary, the sludge will have to be retained on board or discharged to a shore receiving facility. Where ships are fitted with incinerators, small quantities of sludge may be incinerated daily with other wastes having higher caloric value.

The size of a waste treatment plant is determined by the biological load, or material to be digested per unit time, and the hydraulic load, or liquids to be physically separated per unit time. While it is possible to increase biological load, the limitation is established by the hydraulic load on the final settling chamber where the minimum retention time of fluids is 4 to 5 hours to achieve the effluent condition required. The size of the extended aeration sewage treatment unit for a given population equivalent will depend on a number of parameters. For example, the separation of black and gray wastes in the shipboard systems is of prime importance because if both wastes are introduced into the sewage treatment plant, the size of the plant would be unacceptably large. Units using flocculating agents to aid settling can usually accept higher hydraulic loads or be reduced in size. The sludge storage or disposal unit is independent of the sewage unit and therefore is sized for the material it must handle for the assumed time period.

Aeration of the mixed liquor is a fundamental requirement of aerobic treatment systems. The air supplied is necessary as a source for the oxygen required by the bacteria for metabolism. It also provides the means of creating the turbulence needed to promote mixing of the bacterial population with the waste material and to prevent settling in the aeration tank. A number of different aerating methods have been used on marine installations, including (a) coarse bubble, (b) venturi tube, and (c) fine bubble aeration system. Each of these methods has unique advantages and disadvantages.

Coarse Bubble System

The coarse bubble system has a simple perforated tube with relatively large holes, not easily blocked, and not requiring particularly clean air. Since the oxygen transfer efficiency is not high because of large bubble size, air requirement is greater than in other methods. However, the large bubble size creates good mixing.

Venturi Injection System

The venturi injection system has high oxygen transfer. No air compressor is necessary because the power source for the injector can be a simple centrifugal pump used to pump the recirculated, activated sludge. The venturi nozzle is prone to blockage, and the minute air bubbles coming out of solution in the aeration tank adhere to sludge floc, causing the sludge floc to have a tendency to rise.

Fine Bubble Aeration System

The fine bubble aeration system, provided by aerator diffusers, has a relatively high oxygen transfer efficiency, and consequently the quantity of air required for bacterial metabolism is low (the quantity of air approximates 100 m^3/kg of BOD). The aerator diffusers are invariably of a porous material and relatively clean air is necessary to prevent diffuser blockages. This system, in most common use at the present time, requires additional air to be supplied to power air lifts for recycling sludge and for skimming settling tanks. The air is supplied in most instances by small compressors of either the rotary vane or lobe type. Since the air quantity is not large, the pressure necessary is that which is needed to overcome the depth of submersion of the aerator diffuser.

Figures 28-1 through 28-3 show the stages of development of the marine extended aeration sewage treatment unit. Figure 28-1 depicts an early design utilizing an aeration system of the coarse bubble type. Figure 28-2 shows a similar tank design with an aeration system of the venturi type. The unit illustrated in Figure 28-3 is of a design in which the arrange-

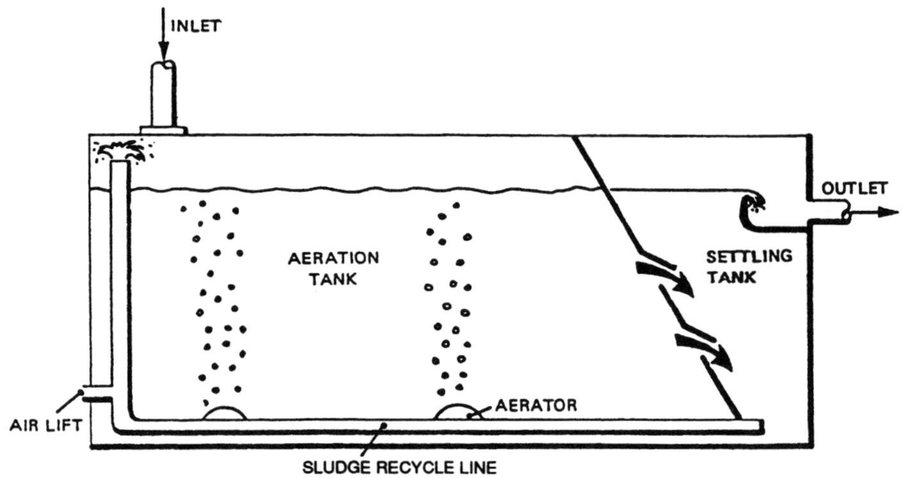

Figure 28-1. Early model of an extended aeration sewage system

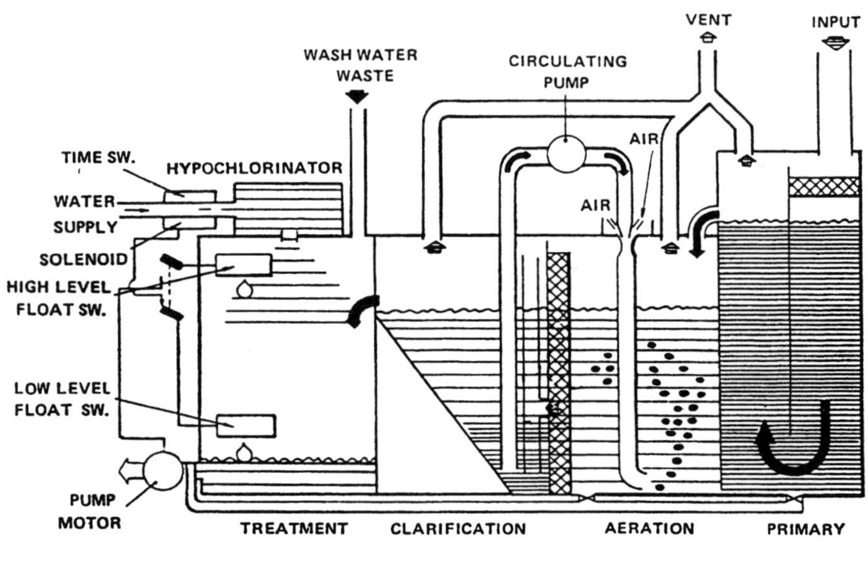

Figure 28-2. Improved model of an extended aeration sewage system

ment of process compartments provides more compact configuration and permits the discharge pump to be built onto the unit, and is a good example of the fine bubble aeration type. Design for the settling tank of the fine bubble aeration system has required more effort than any other section of any treatment plant. This is because the floc formed in the biological plants is much lighter than that formed in physical/chemical plants where the flocculating agents are added. For this reason, the settling time is longer with the tank requirement being proportionately larger. The retention time desirable in the settling tank of a biological unit is on the order of 4 to 5 hours with, ideally, a mean upward velocity of the effluent of approximately 1 meter per hour (3.28 ft./hr.). Unfortunately, the restrictions imposed by the need to build marine units as compact as possible have made this objective difficult to achieve. As a result the final design has invariably been a compromise. To reduce the effect of ship motion on the settling process, the position of the settling tank has been changed from time to time and later designs have incorporated the tank in the center of the complete plant with the aeration tank wrapped around it in a U form. Several designs have used settling tanks which are slightly pressurized and which have no free surface.

Disinfection of the final effluent before discharge overboard is necessary to reduce coliform to a level acceptable to legislating authorities. In most cases the reduction is brought about by treating the effluent with sodium or calcium hypochlorite. The Saint Lawrence Seaway and Great Lakes authorities impose severe limitations on the quantities of residual disinfectants that may be discharged in an effluent. These limitations have

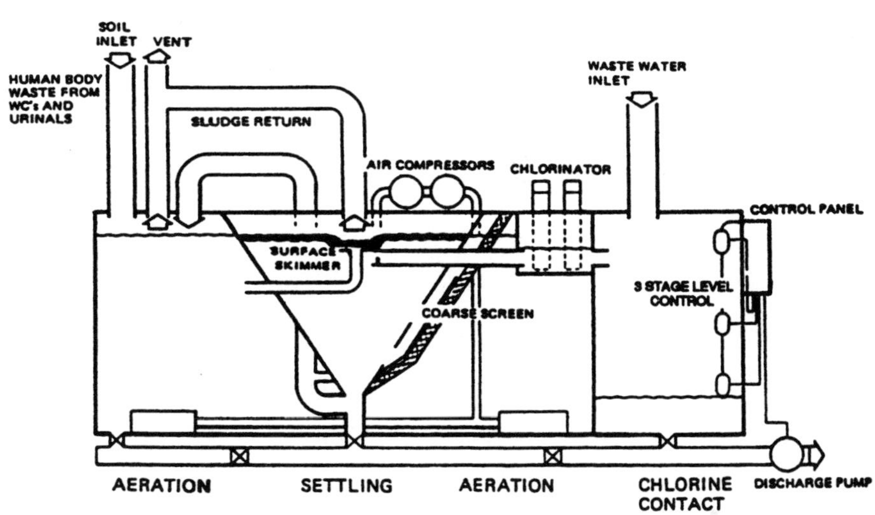

Figure 28-3(a). Advanced model of an extended aeration sewage system

1 Raw sewage inlets
2 Visual indication pipe for activated sludge return
3 Screen
4 Vent
5 Skimmer to remove floating debris
6 Control panel
7 Chlorinator for continuous chlorination of effluent
8 Wastewater inlet
9 Emergency overflow
10 Air diffuser assemblies
11 Aeration tank
12 Settling tank
13 Discharge pump (2 pumps fitted as standard to units ST15-ST30)
14 Filling connection
15 Aeration compressors
16 Float switches

Figure 28-3(b). Cutaway view of an advanced extended aeration sewage system

stimulated interest in disinfection by means of ultraviolet light. This method is now being widely used and disinfection is achieved by passing the clean effluent through tubes fitted with ultraviolet lamps. The effect of the ultraviolet radiation at a wavelength of 254 n.m. is that it breaks down the outer membrane of the microorganism and destroys the nucleus. Ultraviolet disinfection has no residual effect and is therefore suitable for treating effluents being discharged into confined receiving waters. A typical ultraviolet module is shown in Figure 28-4.

Most manufacturers use steel tanks in their sewage treatment plants and these must be suitably protected from the corrosive effects of their

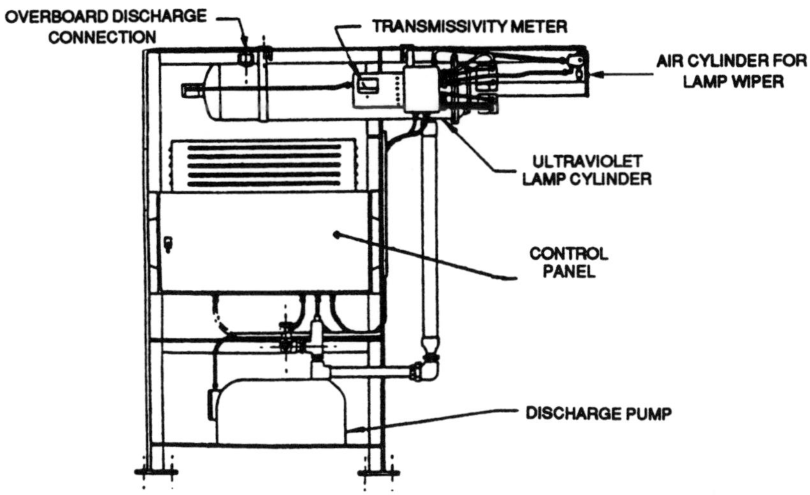

Figure 28-4. Arrangement of ultraviolet disinfection unit

contents, although it should be noted that where aerobic conditions prevail, these effects are likely to be no worse than those encountered in seawater ballast tanks. Where holding tanks are in use and anaerobic conditions can occur, it is then probable that the space above the liquid level would be subjected to considerable corrosive conditions created by the conversion of hydrogen sulphide to sulphuric acid. Therefore, it is essential that care is taken to ensure that adequate plate protection is provided. From experience with sewage plants already in service and from tests carried out on development units, it has been found that the coal tar epoxy compounds provide one of the most effective coatings available, and in terms of cost about the most economical. Some manufacturers use fiberglass for tank materials, but it seems that cost and legislation will limit further use of this material.

Figure 28-5 shows the arrangement, connections, and dimensions of a typical biological sewage treatment plan rated for 5 cubic meters per hour 1,321 gal/hr) of sewage. Figure 28-6 provides similar information for a physical/chemical sewage treatment plant rated for 36,000 liters per day (9,511) gal/day) of sewage.

OPERATION OF A SEWAGE TREATMENT PLANT

Figure 28-3(b) is a cutaway view of a typical modern sewage treatment plant for shipboard use. This type of unit is manufactured in sizes for daily sewage flow range of 250 to 3,830 gallons suitable to support 9 to 127 persons per day at the anticipated rate of 30 gallons per day per person. The unit is divided into 3 compartments: the aeration section (11), settling

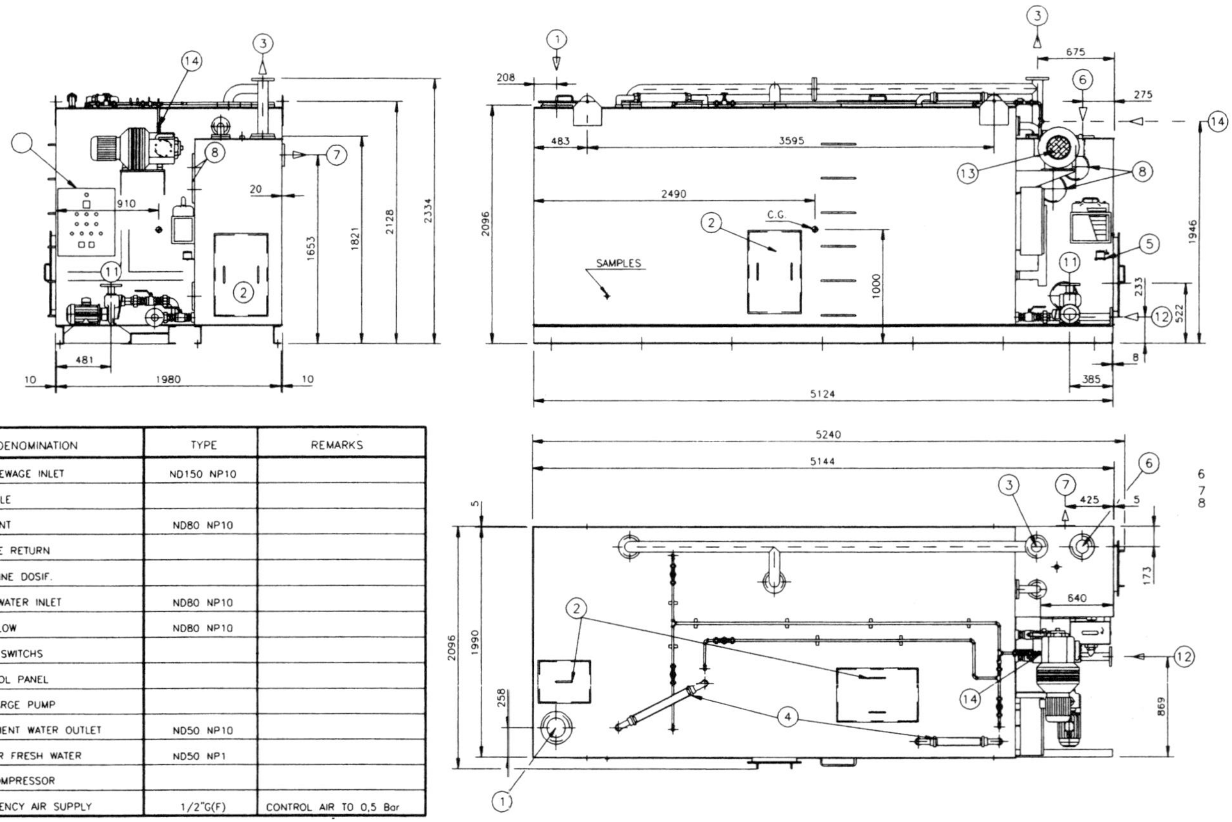

°	DENOMINATION	TYPE	REMARKS
1	RAW SEWAGE INLET	ND150 NP10	
2	MANHOLE		
3	AIR VENT	ND80 NP10	
4	SLUDGE RETURN		
5	CHLORINE DOSIF.		
6	GRAY WATER INLET	ND80 NP10	
7	OVERFLOW	ND80 NP10	
8	LEVEL SWITCHS		
9	CONTROL PANEL		
10	DISCHARGE PUMP		
11	TREATMENT WATER OUTLET	ND50 NP10	
12	SEA OR FRESH WATER	ND50 NP1	
13	AIR COMPRESSOR		
14	EMERGENCY AIR SUPPLY	1/2"G(F)	CONTROL AIR TO 0,5 Bar

Figure 28-5. Arrangement of typical biological sewage treatment plant, capacity 120,000 liters per day (31,704 gallons per day). Dimensions are millimeters. Courtesy DETEGASA.

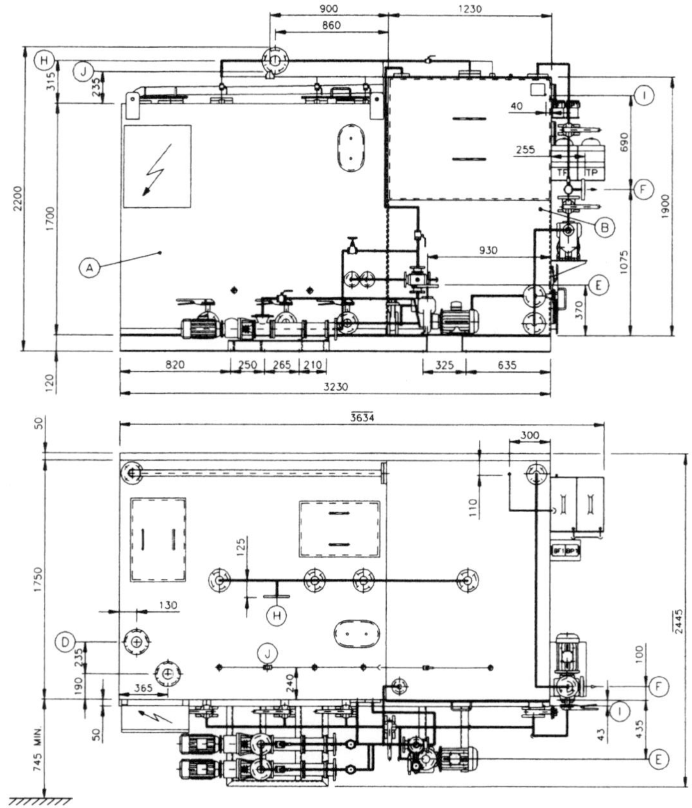

A	COLLECTION TANK
B	DISINFECTION TANK

N°	DENOMINATION	TYPE
D	RAW SEWAGE INLET	ND65 NP10
E	SLUDGE OUTLET	ND50 NP10
F	EFLUENT OUTLET	- -
H	AIR VENT	ND80 NP10
I	AIR INLET	1/4"G
J	FLUSHING WATER INLET	3/4"G

TOLERANCE ±2,5

Figure 28-6. Arrangement of typical physical/chemical sewage treatment plant, capacity 36,000 liters per day (9,511 gallons per day). Dimensions are millimeters. Courtesy DETEGASA.

chamber (12), and the chlorine contact tank adjacent to the pumps and compressors. Sewage enters the aeration compartment (11), via the soil inlet (1), and is retained for approximately 24 hours. During this period, it is mixed and aerated by the compressed air delivered to the bottom of the chamber by the aerator (10). The aerobic bacteria and microorganisms break down the organic waste material into carbon dioxide, water, and inert organic material. The mixture passes through a coarse screen into the settling compartment (12).

The settling compartment is designed to precipitate all solid material to the bottom of the hopper as sludge. The sludge is returned by pneumatic lift (2) to the aeration compartment, where it mixes with the incoming raw sewage. The clear liquid in the settling compartment is displaced into the chlorination compartment via a flow-through type chlorinator. A combined surface skimmer and outlet weir (5), located in the center of the settling compartment surface, removes any floating debris and controls the flow of clear liquid into the chlorinator.

The chlorinator compartment provides residence time for the chlorine to kill any remaining bacteria. The discharge of the chlorinated effluent is controlled by a float-switch operated pump.

DISCHARGE OF OILY WATER

In recent years the oil pollution of the seas has probably been discussed more often and at greater length than almost any other maritime topic. In addition to the many attempts over the years to analyze the sources of this pollution, there have been an equal number of efforts made to tighten up regulations controlling the discharge of oil-contaminated wastes. Table 28-1 gives the reader an indication of the varying quantities of oil that are being deposited in the world's oceans, and although the actual total figure may be subject to some disagreement among authorities, very few will question the validity of the relative percentages. As indicated, the marine transportation industry is the greatest offender, and tankers contribute the greatest quantity.

As discussed in Chapter 26, there has been a gradual strengthening of regulations controlling the discharge of oil-polluted water. An IMO convention sets out the "requirements for the control of pollution." The convention covered not only the problem of tanker discharges, but also discharges from general cargo vessels.

To achieve the quality of overboard discharge required by the regulations, it has now become essential to install an efficient oily water separator as a standard item of ship's equipment. In the past the regulations only required the maintenance of a logbook detailing the position of the vessel when discharges took place. The new requirements calling for the oily water separator are in addition to the inclusion of a monitoring system to upgrade the quality of overboard discharge.

TABLE 28-1

Sources of Oil Input to the Oceans

Source	*Annual Input, 1973 (million tons)*
Marine transportation	
L.O.T. tankers	0.31
Non-L.O.T. tankers	0.77
Dry docking	0.25
Terminal operations	0.003
Bilges bunkering	0.50
Tanker accidents	0.2
Nontanker accidents	0.1
Subtotal	(2.133)
Offshore oil production	0.08
Coastal oil refineries	0.2
Industrial waste	0.3
Municipal waste	0.3
Urban runoff	0.3
River runoff (including input from recreational boating)	1.6
Natural seeps	0.6
Atmospheric rainout	0.6
Total	6.133

In order to meet these new requirements, most oily water separators consist of three stages. The first, a gravimetric stage, involves a simple chamber where the velocity of the mixture is reduced to permit large oil globules to rise to the surface and heavier debris to settle. The second stage is usually an inclined plate coalescer in which a series of plates, either corrugated or flat, are set at an angle of approximately 40/50 slope in the oily water stream. The plates can be either in line with the flow or across the flow and collect the small oil droplets on the underside of one plate, allowing the droplets to coalesce with the larger globules and then escape at the top of the plate pack, while the solid debris particles fall and collect on the upper surface of the lower plates, gradually sliding to the bottom of the separator. Dependent upon flow rate, pump type, etc., the plate type separator can produce very good effluents, certainly below 100 ppm. However, very few are capable of performing to the 15 ppm standard and to meet this higher quality an additional polisher coalescer is fitted. Figure 28-7 shows a typical separator/coalescer.

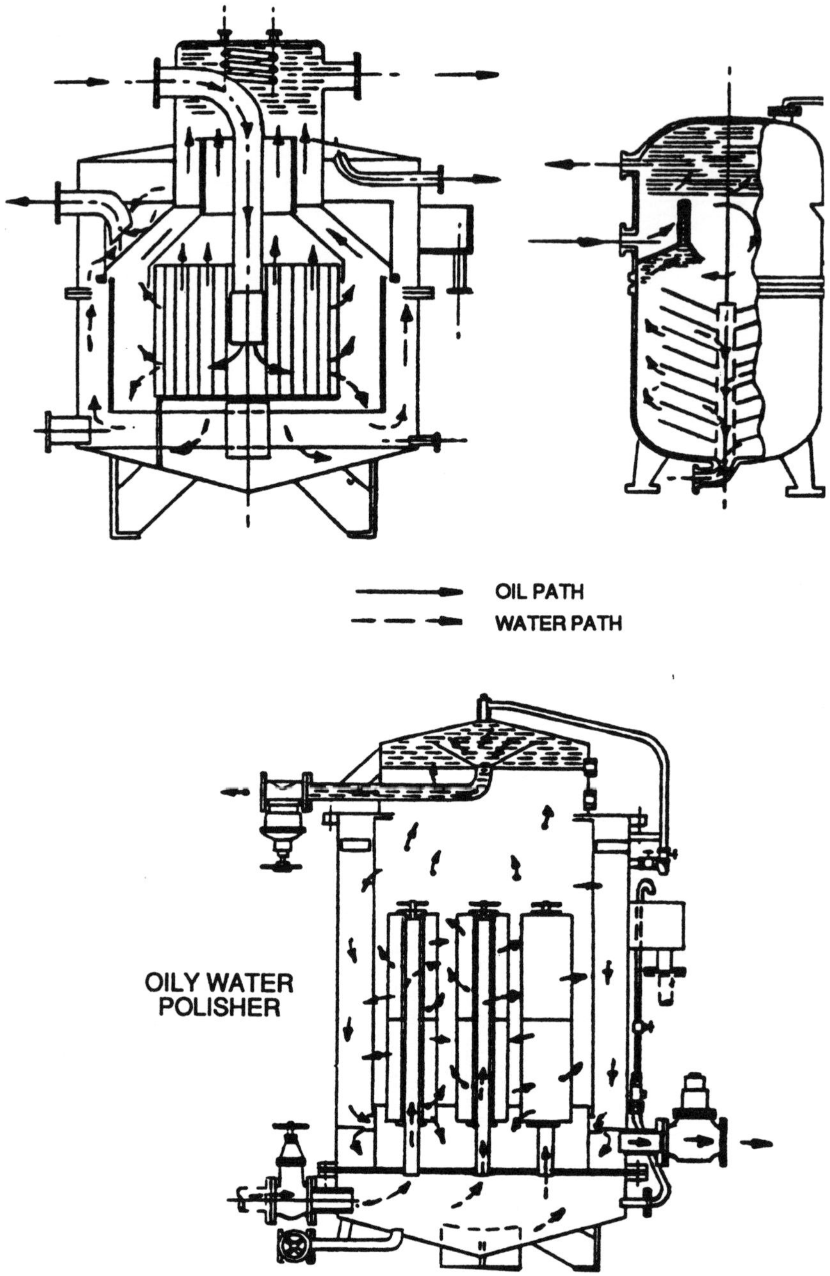

Figure 28-7. Typical oily water separator/coalescers

The third stage of an oily water separator consists of a polisher coalescer usually made of a fibrous or knitted material. This material, formed in either blocks, cartridges, or socks, has the properties to attract oil (oleophilic) or to attract water (hydrophilic). In the oleophilic coalescer the minute particles of oil remaining after the plate separator stage collect on the fibers of the coalescer material where they form larger droplets which rise relatively easily to the surface of the water and are removed. Unfortunately, the very nature of the material of the polisher coalescer makes it an extremely good filter for any other debris which may come through a plate separator and will in time cause the final coalescer to block, necessitating replacement of the coalescing material. In this respect the essential requirements of a streamlined flow with settling velocities in excess of the hydraulic velocity will be seen to have a tremendous influence on the life of the final coalescer.

Coupled with the improved design of the separator, there has been a greater realization of the effect that the pump supplying the separator has on the performance of the separator. For instance, a centrifigal pump is a very good mixer, and to use a unit of this type with an oily water separator is unwise. Similarly, a reciprocating pump with plate-type valves produces a very homogeneous mixture which is difficult to separate. Ideally, a diaphragm pump is the most suitable for use with an oily water separator. However, such a pump is not particularly suited to shipboard application and as a reasonably good alternative a screw pump can be used. Figure 28-8 shows curves depicting the effect of pump type on oil droplet size.

It is also important that the separator's proportions be such that the settling velocity of both oil droplets and/or debris particles not be exceeded by the hydraulic velocities in critical areas of the separator. Stokes' law is applicable in this instance as it applies at low Reynolds numbers. To ensure that streamline flow is achieved, a Reynolds number of 2,000 should not be exceeded. The combined effect of a poorly selected feed pump and

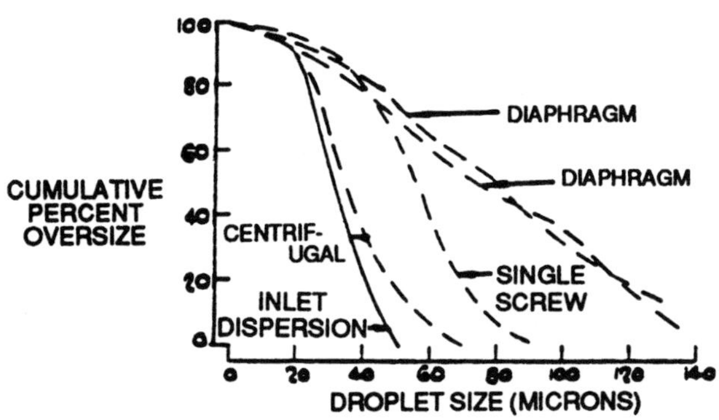

Figure 28-8. Oily water separator pump droplet size

excessive velocity will create small particles with low separating velocities which are therefore able to escape overboard. Figure 28-9 shows how the particle size influences the performance of a separator.

Figure 28-10 shows the arrangement, connections, and dimensions of a typical oily water separator rated for 5,000 liters per hour (1,321 gal/hr).

INCINERATION OF OIL WASTE AND GARBAGE

The disposal of solid and liquid wastes other than sewage has become the subject of investigation following the publication of the 1973 IMCO proposal. Garbage can be contained and compacted, but storage is necessary and not always available. Packaging will inevitably include a large quantity of plastics which must be retained on board since discharging overboard at sea will not be permitted regardless of distance offshore. Liquid wastes, mainly oil sludges, must be retained on board and, along with garbage, will necessitate the provision of storage space. Incineration of these materials is without doubt the most effective means of dealing with the storage problem because the resultant residues are extremely small in volume and easily disposed of.

Before deciding on the type of incinerator to be adopted, an assessment of storage requirements are necessary. Equally important in this assessment is the categorization of the waste to be consumed. These categories are: (a) galley waste including food scraps, bones, cans, bottles, plastics, etc.; (b) accommodation wastes such as paper, cardboard, cans, plastics, textiles; (c) oils/sludge including bilge, oil purifiers, lubricating oils, etc.; and (d) sewage sludges.

Once the various materials are identified, it is necessary to obtain estimates of the quantities of such materials. An accepted estimate is 2 to 2.5

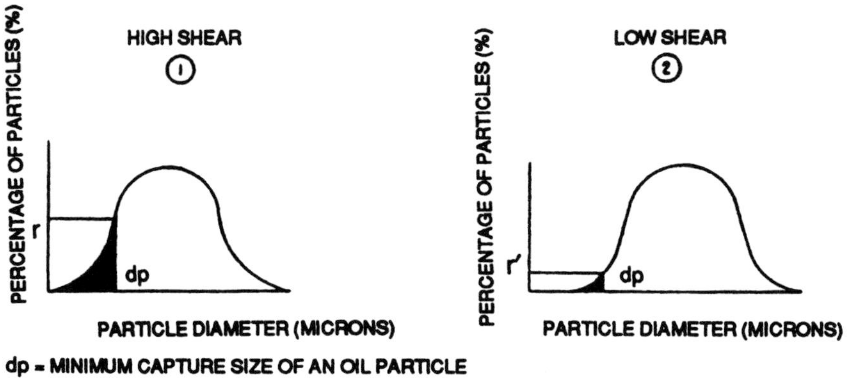

Figure 28-9. Particle size in oily water mixture and influence on particle capture

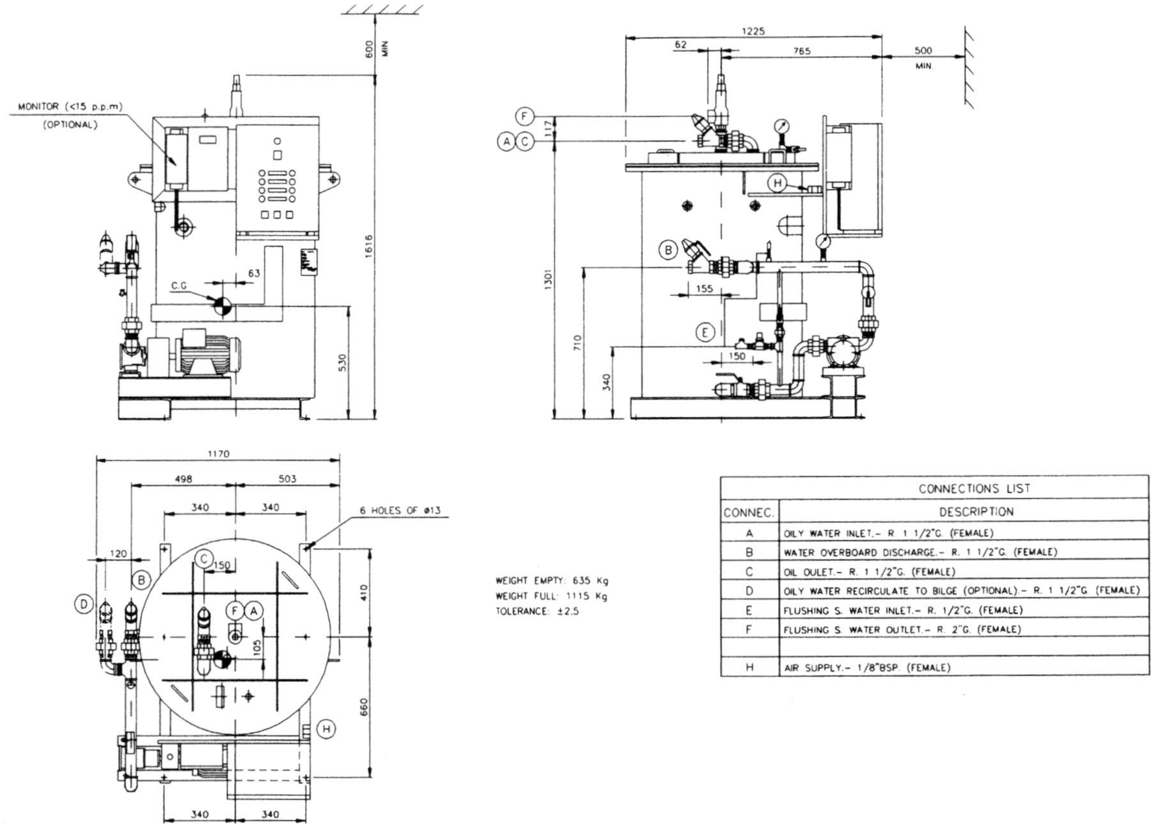

Figure 28-10. Arrangement of typical oily water separator, capacity 5,000 liters per hour (1,321 gallons per hour). Dimensions are millimeters. Courtesy DETEGASA.

kg (4.4 to 5.5 pounds) per man per day. The quantity of solid garbage covers both categories (a) and (b) above. In steamships the accumulation of oily wastes is almost negligible, whereas the amounts obtained for motor ships vary from ½ to 2 percent of the daily fuel consumption, and include lubricating and fuel oil sludges.

Sewage sludge quantities vary with the type of treatment process. For example, a biological treatment plant may be desludged at sea before entering waters subject to legislation and will then not require desludging for another 2 or 3 months. On the other hand, if daily desludging is preferred, the quantity to be incinerated will be approximately equivalent to 2.5 liters (.6 gallons) per man per day. A physical or physical/chemical system will produce approximately 5 liters (1.3 gallons) of sludge per man per day. The heating values for these materials are:

Garbage	2,750kcal/kg	(4,959BTU/lb)
Oily water/sludge 50% water	5,000kcal/kg	(8,998BTU/lb)
Sewage sludge 98-99% water	—	—

For a vessel of 20,000 BHP with a crew of 40, this gives an approximate heat release of:

Garbage 100 kg / day (220 lbs / day)	275,000 kcal (1.091 × 10^6 BTU)
Oily sludge 775 kg / day (1,708 lbs / day)	3,875,000 kcal (15.37 × 10^6 BTU)
	4,150,000 kcal (16.47 × 10^6 BTU)

From this estimated heat release rate, the size of a suitable incinerator can be calculated on the basis of normal daily usage of 8 hours. The hourly heat release rate will then be approximately 518,000 kcal (2.055) 10^6 BTU).

DESIGN FEATURES OF INCINERATION SYSTEMS

The requirements for a successful incinerator are (a) The operating temperature must be high enough to sterilize all residues and ensure that all discharges are innocuous; (b) gas residence time must be as long as possible; (c) the temperature must not be high enough to melt glass or metal; (d) auxiliary fuel usage must be at a minimum; (e) simplicity of operation consistent with safety must be achieved; (f) capital cost must be as low as possible; (g) the number of man-hours required in daily attendance must be kept to a minimum; (h) ash removal must be as easy as possible.

Various designs of incinerators are available to the shipowner. These are mostly of the single chamber, multiple chamber, or vortex types. All the vortex types are of the vertical form while the single and multiple chamber units can be either vertical or horizontal. Figures 28-11, 28-12, and 28-13 show three typical units.

While single and multiple chamber units can be used for incinerating liquid sludges, the vortex types available at present are capable of higher throughputs and, with the tightening of rules covering the discharge of oil, this aspect of incineration is very important.

The physical dimensions of any incinerator are also important because when available space is allocated invariably the incinerator comes close to the bottom in the list of priorities. As a general rule, the single and multichamber incinerators will occupy more deck area than the vortex types, but the latter are invariably somewhat higher, a factor which may prove to be a disadvantage where installation between decks is necessary.

Figure 28-14 shows arrangement, dimensions, and connections for a typical solid waste incinerator rated for 50 kilograms per hour (110.2

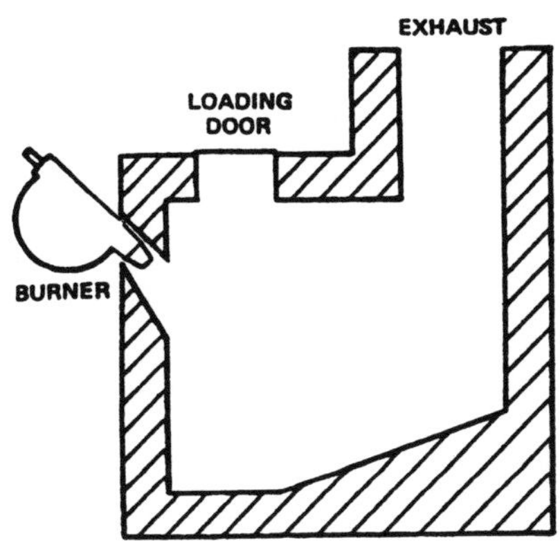

Figure 28-11. Single chamber incinerator

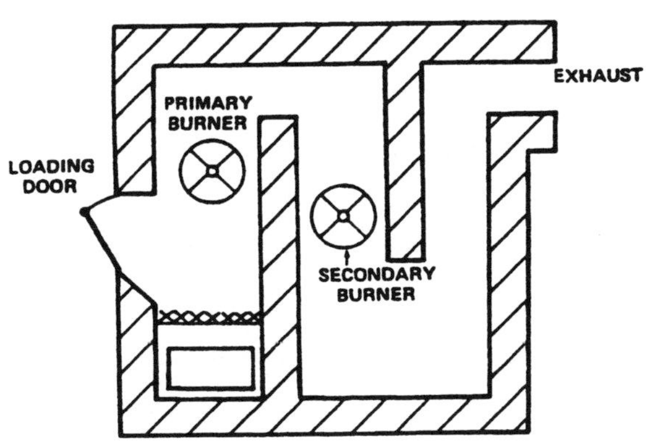

Figure 28-12. Multiple chamber incinerator

lbs/hr) of solid waste. This type of incinerator is capable of burning oil sludge.

FEEDING AND CONTROL OF INCINERATORS

Various methods of automatic garbage feeding have been tried on marine incinerators, including chute, screw feed, and ram feed, and the problems that have been encountered with all of these have been significant. If the limitation of time spent on maintenance is of great importance, then there is no doubt that hand loading utilizing paper or plastic sacks is the most reliable. Segregation of waste is not practical, and the incinerator must be able to accept anything that is put into a sack. Accordingly, the loading door must be of dimensions that are big enough to accept a fully loaded sack

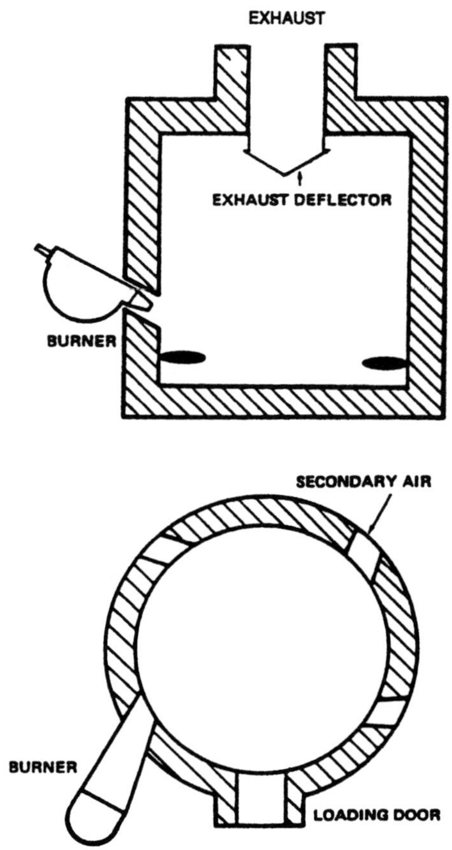

Figure 28-13. Typical marine incinerator arrangement

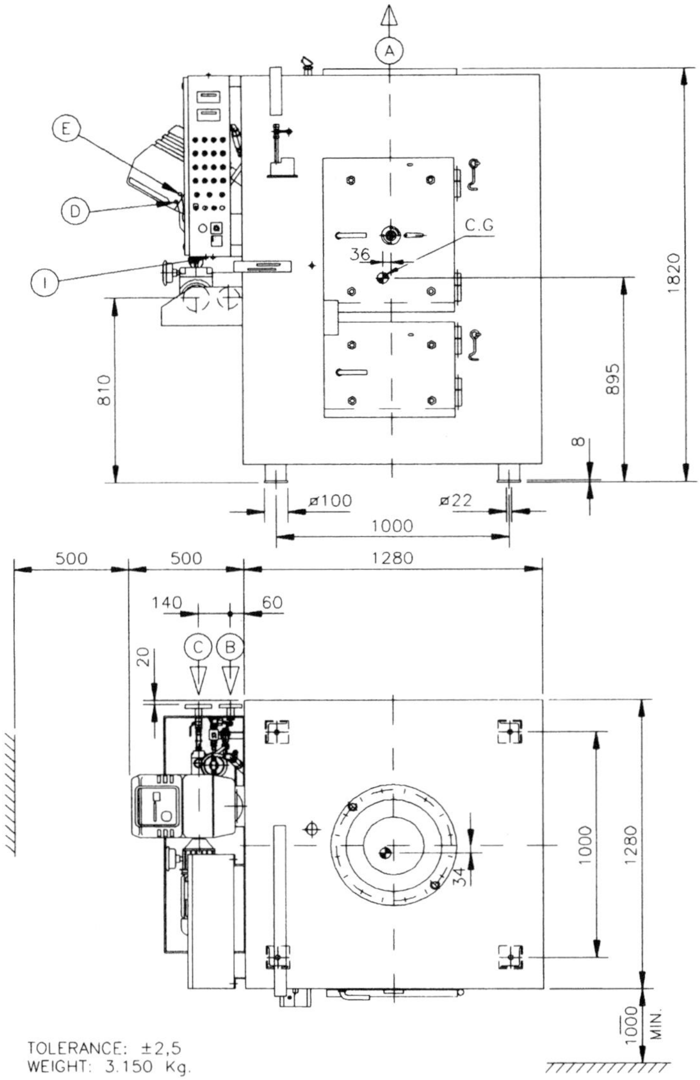

CONN.	DESCRIPTION	TYPE	REMARKS
A	SMOKE OUTLET	FLANGE ND400 NP6	16 HOLES OF M20
B	AIR OR STEAM INLET	FLANGE ND15 NP16	
C	SLUDGE INLET	FLANGE ND25 NP16	
D–E	D.O. SUCTION & RETURN	3/8"BSP. (M)	
I	ELECTRIC CONNECTIONS		

Figure 28-14. Arrangement of solid waste incinerator, capacity 50 kilograms per hour (110 pounds per hour). Dimensions are millimeters. Courtesy DETEGASA.

of approximately 450 mm (17.7 inches) diameter by 750 mm (29.5 inches) long. The average contents of one of the sacks will weigh approximately 6 kg (13 lb) and on the basis of the combustion rate mentioned earlier, it will be noted that a crew of 40 producing approximately 100 kg (220 lb) of garbage per day will require a daily attendance of 20 to 30 minutes for manual loading for the incinerator.

Liquid wastes and sludges are easier to handle and store although it must be borne in mind that sludges can contain solids of 6 to 8 mm (.2 to .3 inches) diameter as well as abrasive materials from fuel oil separators. Consequently, the pump or pumps designated for handling these media should be properly selected. Because of the possibility of encountering relatively large solids, restrictions in transfer pipework should be avoided if blockages are to be prevented.

The previously mentioned points must also be considered when the type of burner used is being appraised. Sludge and liquid waste burners at present are of the rotary cup type, low pressure weir type, or combustion chamber gutter type. Jet burners should be avoided unless there is no chance of the entrainment of solids.

The preparation of liquid wastes is important, and provision must be made for heating and mixing these wastes before incineration. A good homogeneous mixture is necessary where oil and wastewater sludges are being burned because separation of the constituents will result in combustion at a steady rate, becoming impossible to contain. Evaporation of water in wastes is most economically achieved by using the available heat in waste oils. If this cannot be done, it is necessary to use a supporting oil burner with its attendant cost. In practice, it has been found that with efficient mixing, a combination of oil and water sludges containing 50 to 55 percent water can be consumed without the aid of a support burner.

The control of an incinerator must be such that all aspects of safety are covered. Exhaust gas temperatures must be controlled to ensure that they are high enough to destroy odors, but not so high as to melt glass. Temperature of exhaust gases can also be used to control the operation of the auxiliary or support burner so that, whenever possible, the use of additional fuel is avoided. All the usual interlocks and lockouts associated with an oil-fired device must be fitted, and include flame failure, forced draft fan failure, and high exhaust gas temperature.

Since incinerator exhaust gas temperatures are considerably higher than those from heat exchange devices such as boilers, it is necessary to reduce them to more acceptable levels. This can best be achieved by introducing cold-diluted air into the exhaust stream at a point as close to the incinerator discharge as possible, and before the exhaust trunking passes through spaces adjacent to spaces where people are accommodated. The temperature of the exhaust after dilution should be no more than 662°F (350°C). Some classification societies require this exhaust gas temperature to be monitored and excessively high temperatures to instigate shutdown procedures.

OPERATION OF A MARINE INCINERATOR

A cutaway view of a typical modern incinerator is shown in Figure 28-15. This incinerator is a vertical cyclone type with a rotating arm device (6) to improve combustion and remove ash and noncombustibles from the furnace. A sludge burner (4) is incorporated to dispose of sewage sludge and waste oil. The auxiliary oil burner (12) is fitted to ignite the refuse. Automatic controls secure the ignitor when the refuse burns without it. Combustion air is furnished by a forced draft fan (8). The loading door (11) for the refuse is pneumatically operated and is interlocked with the burner and forced draft fan.

Solid waste is fed into the incinerator in standard plastic bags, about three at one time. A single "close door" button initiates automatic functioning of the incinerator. A charge is incinerated in 5 to 10 minutes.

Liquid waste may be introduced through the liquid waste burner when the refractory is hot. The sludge-burning cycle is fully automatic. Following use, the incinerator is allowed to cool, and ash and noncombustibles are removed by pulling out the ash slide door in the hearth. The rotating arm will scrape all the solid residue into the ash box from which it may be disposed.

SEWAGE REGULATIONS

Annex IV of MARPOL 73/78, titled "Regulations for the Prevention of Sewage from Ships" sets forth the international agreements concerning sewage. Engineers involved in the design, operation, and maintenance of ships must be familiar with these regulations. In brief, the regulations deal with the following topics concerning sewage: (1) definitions, (2) types of ships to which regulations apply, (3) the survey process for sewage systems, (4) conditions for issue by the Flag state and duration of an International Sewage Pollution Certificate, (5) conditions for discharge of sewage into the sea, and (6) requirements for shore reception facilities and the international standard for dimensions of piping flange used by both the ship and shore reception facility.

GARBAGE REGULATIONS

Annex V of MARPOL 73/78, titled "Regulations for the Prevention of Pollution by Garbage from Ships" sets forth the international agreements concerning garbage. These regulations require ships to have a management

1 Air diluter
2 Exhaust gas deflector
3 Rotating rabble arm shaft
4 Liquid waste burner
5 Combustion air inlet
6 Rabble blades
7 Rabble shaft drive
8 Forced draft fan
9 Control panel
10 Sight glass
11 Pneumatically operated garbage door
12 Auxiliary burner
13 Ash slide
14 Ash hopper

Figure 28-15. Cutaway view of marine incinerator

system that specifies policy, procedures, and instructions for handling ship-generated garbage. Each ship must have a "Garbage Management Manual" describing the management system that usually assigns specific responsibilities and duties to certain engineering officers and others.

The management system deals with the following topics concerning garbage: (1) definitions of garbage; (2) assignments of responsibilities; (3) procedures to minimize the garbage generated; (4) procedures for collection, processing, marine disposer use, incinerator use, storage, and disposal; (5) training; (6) noncompliance criminal penalties; and (7) record keeping.

Table 28-2 is a summary of the international regulations for disposal of garbage at sea. Figure 28-16 shows the options available in the garbage management system for the collection, processing, storage, and disposal of garbage.

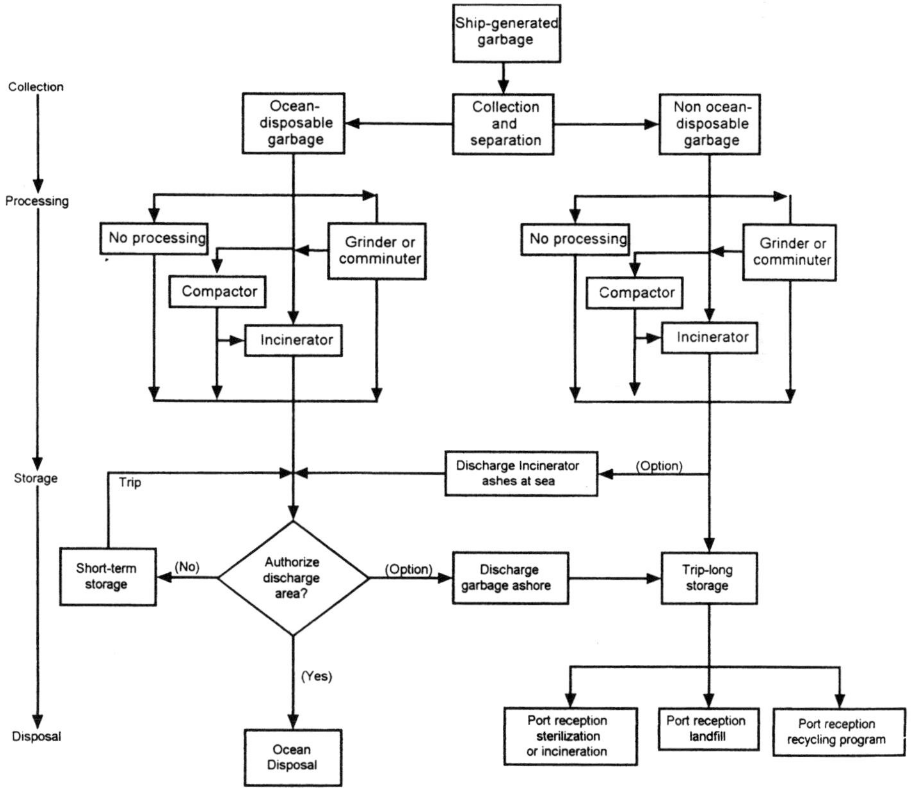

Figure 28-16. Options available under international regulations for shipboard collection, processing, storage, and disposal of garbage

TABLE 28-2

Summary of International Regulations for Disposal of Garbage at Sea

Garbage Type	*Outside Special Areas*	***In special areas*
Plastics—includes synthetic ropes and fishing nets and plastic garbage bags	Disposal prohibited	Disposal prohibited
Floating dunnage, lining and packing materials	>25 miles offshore	Disposal prohibited
Paper, rags, glass, metal, bottles, crockery, and similar refuse	>12 miles	Disposal prohibited
All other garbage including paper, rags, glass, etc. comminuted or ground	>3 miles	Disposal prohibited
Food waste not comminuted or ground	>12 miles	>12 miles
*Food waste comminuted or ground	>3 miles	>12 miles
Mixed refuse types	***	***

*Comminuted or ground garbage must be able to pass through a screen with mesh size no larger than 25 mm.

**Garbage disposal regulations for special areas shall take effect in accordance with regulation 5(4)(b) of Annex V.

***When garbage is mixed with other harmful substances having different disposal or discharge requirements, the more stringent disposal requirements shall apply.

BILGE WATER DISCHARGE

Bilge water results from a variety of spills, leaks, and washing operations in the engine room space between the tank tops and the floor plates. Various oils are always found in bilge water. Safety and good housekeeping require that bilge water accumulations be pumped overboard at regular intervals. Current international regulations require that bilge water discharge to the ocean contain less than 15 ppm oil. A high capacity centrifugal separator provides one of the acceptable solutions to meeting the requirements for oil content in overboard discharge. Figure 28-17 shows a popular high capacity separator used for processing bilge water. This separating plant consists of a compact separator, pump, pre-heater, filter, and controls. The feed pump permits the throughput to be controlled. Typical bilge water feed with the following can be accommodated without absorption filters or emulsion-breaking chemicals: solids less than .1%, oil less than 5%, chlorides less than 30,000 ppm, and a pH of 6 to 9.

MEMBRANE TECHNOLOGY

In anticipation of the IMO Marine Environmental Protection Committee adoption of new regulations concerning all ship discharges including treated water, manufacturers of sewage treatment equipment have developed proposals for plants based on membrane technology that are capable of treating gray and black water and bilge water to produce an acceptable overboard discharge free of chlorine. At present, there are only a few experimental membrane based systems installed in ships. However, the technology is used ashore with both advantages and disadvantages.

Rochem has developed one of the current seagoing pilot plants. In this plant, gray water is treated by low-pressure reverse osmosis that separates and rejects the bacteria and suspended solids. Ninety percent of the

Figure 28-17. Oil water centrifugal separator plan suitable for bilge water and sludge separation. Bilge water capacity is 6,000 liters per hour (1,585 gallons per hour). Courtesy Westfalia Separator Mineraloil Systems GmbH.

gray water processed is suitable for discharge overboard or a suitable shipboard use. The remaining ten percent is discharged to the continuing plant process.

In the next stage, the remaining gray water is joined with galley water and black water and treated in a biological reactor. The treated mixture is then passed through an ultra-filtration membrane barrier following which the fluid is suitable for overboard discharge in any waters. Expectations for the discharge are 5 mg/liter (i.e., parts/million) of total suspended solids, 20 mg oxygen per liter of biochemical oxygen demand, and a most probable number of 10 living organisms per 100ml of effluent containing fecal coliform. Ultra-filtration depends on porous membranes that reject particles larger than the menbrane's pore size. The ultra-filtration membrane is a reliable barrier for bacteria and viruses including colloids and macromolecules. Some plants under development are designed to effectively handle gray water, black water, and oil.

There may be some disadvantages to the application of membrane technology to shipboard sewage treatment. For example, maintainability may be difficult due to problems in handling the large membranes, and repair and operation of membrane tanks may be beyond the capability of ship personnel or require special training including understanding of the chemical reactions in the process.

GLOSSARY

Aeration. The process or method of bringing about intimate contact between air and liquid to dissolve oxygen into the liquid.

Aerobic. In the presence of air.

Aerobic digestion. A waste degradation process utilizing microorganisms requiring oxygen for their metabolism and growth.

Anaerobic. In the absence of air.

Anaerobic digestion. A waste degradation process utilizing microorganisms which do not require oxygen for their metabolism and growth.

Biochemical oxygen demand (BOD). The amount of oxygen utilized by microorganisms in the stabilization of organic matter.

Chlorine, residual. The total amount of chlorine (combined and free available chlorine) remaining in water, sewage, or industrial waste at the end of a specified contact period following chlorination.

Colloidal solids. Extremely finely divided suspended materials which will not settle.

Comminutor. Sewage grinder or shredder.

Denitrification. The reduction of nitrates in solution by biochemical action.

Diffuser. A device through which air is forced and divided into small bubbles for diffusion in liquids.

Disinfection. The killing of the larger portion (but not necessarily all) of harmful and objectionable microorganisms.

Effluent. Sewage, water, or other liquid, partially or completely treated or in its natural state, flowing out of a reservoir, basin, or treatment plant.

Floc. Small gelatinous masses, formed in a liquid by the addition of coagulants, through biochemical processes or by agglomeration.

Flocculation. The process of raising suspended matter to the surface of the liquid and the subsequent removal of the matter by skimming.

Hypochlorite. Compound of chlorine in which the radical (OCl) is present.

Hydrogen concentration. A measure of the acidity or alkalinity of a solution (See pH).

Imhoff cone. A conically shaped graduated glass vessel used to measure the volume of settled solids.

Influent. Sewage, water, or other liquid, partially or completely treated, or in its natural state, flowing into a reservoir, basin, or treatment plant.

Microorganism. A minute organism either plant or animal, invisible or barely visible to the naked eye.

Mixed liquor. A mixture of activated sludge and sewage in the aeration tank.

Nitrification. The conversion of nitrogenous matter into nitrates through biochemical action.

Oxygen, available. The quantity of uncombined or free oxygen dissolved in a liquid.

Oxygen, dissolved. Usually designated as DO. The oxygen dissolved in sewage, water, or other liquid usually expressed in parts per million.

Oxygen deficiency. The additional quantity of oxygen required to satisfy the biochemical oxygen demand in a given liquid. Usually expressed in parts per million.

Oxygen demand. Oxygen required for oxidation of unstable matter in sewage.

Parts per million (ppm). A ratio of pounds per million pounds, grams per million grams (approximately equal to milligrams per liter), expressing the concentration of a specified component.

Pathogens. Disease-causing organisms.

pH value. The logarithm to the base 10 of the reciprocal of the concentration of hydrogen ions in a solution. If the pH is below 7.0 , the solution is acidic, above 7.0, the solution is alkaline.

Process, activated sludge. A biological sewage treatment process in which a mixture of sewage and activated sludge is agitated and aerated. The activated sludge is subsequently separated from the treated sewage (mixed liquor) by sedimentation, and wasted or returned to the process as needed. The treated sewage overflows the weir of the settling tank in which separation from the sludge takes place.

Process, biological. The process by which the life activities of bacteria and other microorganisms break down complex organic materials into simple, more stable substances. Self-purification of the sewage-polluted streams, sludge digestion, and all so-called secondary sewage treatments result from this process. Also called biochemical process.

Putrefaction. Biological decomposition of organic matter with the production of ill-smelling products, associated with anaerobic conditions.

Sedimentation. The process by gravity of settling and deposition of suspended matter carried by water, sewage, or other liquids. It is usually accomplished by reducing the velocity of the liquid below the point where it can transport the suspended material. Also called settling.

Septic sewage. Sewage undergoing putrefaction in the absence of oxygen.

Skimming. The process of removing floating grease or scum from the surface of a tank.

Sludge, activated. Sludge produced in raw settled sewage by the growth of Zoogloea (bacteria) and other organisms in the presence of dissolved oxygen and accumulated in sufficient concentration by returning floc previously formed.

Sludge bulking. A phenomenon that occurs in activated sludge plants in which the sludge occupies excessive volumes and will not concentrate readily.

Sludge digestion. The process by which organic or volatile matter in sludge is gasified, liquefied, mineralized, or converted into more stable organic matter, through the activities of living organisms.

Solids, dissolved. Solids which are in solution.

Solids, settleable. Suspended solids which will subside in quiescent water, sewage, or other liquid within a reasonable period. Such period is commonly, though arbitrarily, taken as one hour. Also called settling solids.

Solids, suspended. Solids physically suspended in sewage which can be removed by proper laboratory filtering.

Solids, total. The total amount of solids in solution and suspension.

Solids, volatile. The quantity of solids, in water, sewage, or other liquid, lost on ignition of the total solids.

Sterilization. The destruction of all living organisms, ordinarily through the agency of heat or some chemical.

Treatment, primary. The first major (sometimes the only) treatment, usually sedimentation, in a sewage works. The removal of a high percentage of suspended matter but little or no colloidal and dissolved matter.

Treatment, secondary. The treatment of sewage by biological methods.

Water, potable. Water which does not contain objectionable pollution, contamination, minerals, or infection, and is considered satisfactory for domestic consumption.

REVIEW

1. What are the principal methods available for shipboard handling of sewage?
2. What is the difference between a coarse bubble and a fine bubble system?
3. Describe the design and operation of an oily water separator.
4. Why are incinerators needed aboard ship?
5. Describe the operation of a modern incinerator.
6. What materials are incinerated aboard ship?
7. What is effluent?
8. What are pathogens?
9. What is mixed liquor?
10. What is the function of an ultraviolet disinfection unit?
11. Why is it desirable to separate gray and black water aboard ship?
12. What materials are fed to an incinerator?
13. What solid wastes may be dumped into the ocean? Where?
14. What is the maximum ppm of oil which may be pumped overboard with bilge water?
15. What are the options for garbage management aboard ship?

REFERENCES

Champ, Michael A. 1989. *Marine Waste Management: Science and Policy.* Malabar, Fla.: R. E. Krieger Publishing Company.

Institute of Marine Engineers. November 1997. "Waste Management." *Marine Engineers Review.* Institute of Marine Engineers.

Institute of Marine Engineers. November 1998. "Sewage Treatment." *Marine Engineers Review.* Institute of Marine Engineers.

International Maritime Organization. 1997. *Guidelines for the Implementation of Annex V of MARPOL 75/78.* London: International Maritime Organization.

International Maritime Organization. Publication Number IMO-592E. *Guidelines for Performance Tests of Sewage Treatment Plants.* London: International Maritime Organization.

International Maritime Organization. Publication Number IMO-608E. *Oily-water Separators and Monitoring Equipment.* London: International Maritime Organization.

Lloyd's of London. 1993. *Marine Environment Law.* London, Lloyd's of London.

Vie, R. H., 1990. "Shipboard Generated Waste-treatment and Disposal—Operational Experience on Passenger Cruise Vessels." Paper #22, IMarE Symposium. London. Institute of Marine Engineers.

Wang, L. K. 1980. *Handbook of Environmental Engineering,* Vol. II. Clifton, N.J.: Humana Press.

ACKNOWLEDGMENTS

The editors are indebted to Hamworthy Engineering Limited, Dorset, England for most of the material in this chapter. The arrangement drawings were provided by Desarrollos Technicas Industriales De Galicia (DETEGASA), La Coruna, Spain. The separator photograph was provided by Westfalia, Oelde, Germany.

CHAPTER 29

Propellers and Propulsion

WILLIAM B. MORGAN

INTRODUCTION

A PROPELLER is a device which transforms energy from a rotating shaft into kinetic energy in the slipstream. Thus, the propeller adds momentum to the fluid passing through it, and an axial force is generated on the shaft which pushes the ship. This energy is transferred through the blades which act physically like wings. However, to cope with the combination of rotational velocity and axial velocity, the blades are manufactured in a general helical shape. Because of this general helical shape of each blade, similar to a screw thread, the name "screw" is often used interchangeably with "propeller." The analogy is often made that each blade is like a part of a screw thread and that the pitch of the propeller blade is like the pitch of a screw. The propeller advances a distance through the water in one revolution equal to its pitch minus a slippage, called slip, which is equivalent to the axial velocity induced by the propeller. This analogy is crude and will not be carried further since physically the blades are analogous to airplane wings and not to screw threads.

There are many definitions associated with propellers, and these tend to change somewhat over the years as new terms are added and old terms are dropped. More recently, an effort has been made to standardize terminology internationally. In 1975, the International Towing Tank Conference adopted a set of terms for propulsion and propellers, and some of these definitions are reproduced in the glossary. There are also a number of propulsor types included in the definitions. In general, these propulsor types operate along the same principle as the screw propeller, and they will also be discussed in this chapter.

SHIP RESISTANCE AND PROPULSION

Resistance

When a ship moves through the water, the viscosity of the water gives rise to shear forces on the surface of a ship and her appendages. These shear forces cause the hull to have a *drag,* called frictional resistance or skin friction drag. In addition to the shear forces, the boundary layer modifies the pressure distribution around the ship giving rise to a pressure or form drag. If this pressure distribution is adverse enough, the boundary layer will completely separate from the ship's stern. When this occurs, the pressure drag greatly increases. These two components, frictional resistance and pressure or form drag, compose the total *viscous drag* of the ship. Since a ship operates in a free surface, there is another component of drag which arises, and that is the *wave drag* caused by the disturbance of the surface (wave formation) by the ship. There may be other components of drag caused by operating near other boundaries, shallow water, etc. In any case, the propeller must overcome all these drag components.

Another effect of the viscosity of the water is that there is a wake formed behind the ship. (The wave component may also affect the flow behind the ship.) The ship, as it moves through the water, actually drags some of the water with it, the velocity of which depends on how close a particular particle of water is to the hull. In fact, a method used for determining the ship's viscous resistance is to survey the wake downstream of the ship and to determine the velocity of the water being dragged along. The energy in the water behind the ship is representative of the viscous resistance.

Ship resistance is normally predicted from model tests where the model is towed through the water and the drag measured. The model must be towed through the water at a speed corresponding to the square root of the linear ratio of the ship to model length. This is called the Froude-scaled speed. At this speed, the wave formation is the same for the model as for the ship. An example is that for a 500-foot ship with a 20-foot model, the model speed would be one-fifth of the ship speed.

The viscous part of the model drag is calculated by assuming that the viscous drag corresponds to the drag of a flat plate the same length as the model and with the same wetted surface. In actuality, therefore, only the frictional resistance part of the viscous drag is considered by the flat plate drag. This frictional resistance is subtracted from the total drag of the model, and the resulting drag, called *residuary resistance,* is essentially that of the wave drag and pressure or form drag. The residuary resistance of the model is nondimensionalized by the dynamic head ($\frac{1}{2}\rho V^2$) times the area of the model-wetted surface. It is assumed that the ship coefficient of residuary resistance has the same value as the model coefficient, and hence the ship residuary resistance is calculated from this coefficient using the ship speed and the area of the ship's wetted surface. As for the model, the viscous drag (actually frictional resistance) of the ship is calculated by assuming that it is equal to the drag of a flat plate of the same length as the ship and with the same wetted surface. Curves have been developed for the drag coefficient of a flat plate. This drag coefficient is a function of a dimensionless parameter defined as the velocity times the length divided by the kinematic viscosity:

$$\frac{V \times L}{\nu}$$

This parameter is called the Reynolds number. The flat plate drag coefficient decreases as the Reynolds number increases. In addition to these two drag components, an additional resistance must be added to the smooth ship prediction to complete the model-ship balance. This additional resistance has been called the roughness allowance, supposedly to account for the additional drag of a rough ship versus a smooth model. Actually the roughness allowance covers more factors such as differences in flow separation between model and ship, air resistance, etc. Therefore, the term *roughness allowance* has been changed to *correlation allowance.* Drag due to wind and waves in rough weather can also be added to the list of drag components. Other techniques for scaling model resistance to ship resistance have been proposed or are in use. For instance, the use of the drag of an equivalent body of revolution the same length as the model or ship and with the same wetted surface distribution has been proposed. And form factors to account for the ship form in scaling the various drag components are in use. Series data from model tests have been developed so that the resistance can be predicted for many ship types. Two series of note are the Taylor Standard Series (see Taylor, 1933) and David Taylor Model Basin Series 60 (see Todd, 1963). There are also series for high speed craft (see Yeh, 1965).

It should be noted that the ship resistance increases significantly as the hull becomes fouled. Hull fouling increases shaft torque for a given ship speed and decreases ship speed for a given shaft torque. The net effect is an increase in fuel cost for a given ship speed. The increase in drag over the time a vessel is out of dry dock is a function of the amount of time the ship is at sea and the time in port and the location of the port since ships in warm waters foul much faster than ships in cold waters. The resistance can increase as much as 0.25 percent per day in temperate waters and 0.50 percent per day in tropical waters and will cause significant increases in fuel costs. Special paints to inhibit fouling and periodic hull cleaning should be considered.

Fitting the Propeller

A propulsor must be added to the ship to overcome the total resistance of the ship. Only propellers will be considered and not sails or other propulsion devices.

The propeller is usually located at the stern of the ship although most positions have been investigated. The reason for this location is that when the propeller operates in the wake of the ship, there is improved performance since the propeller "fills in the wake defect" to some extent. Also, if the propeller is in front of the hull, the induced velocity from the propeller increases the drag of the ship. Furthermore, there is improved protection of the propeller with a stern location, as well as some payoff with regard to control by locating the propeller at the stern with a rudder behind. For whatever reason, then, the propeller is located at the stern of the ship and normally within the viscous wake of the ship. The propeller accelerates the flow in the wake relative to the ship's motion in such a way that the wake behind the ship becomes essentially momentumless on an average.

Standard practice has been for propellers on single-screw ships to be installed on a shaft with no supporting system outside the stern tube bearing and for propellers on twin-screw ships to be installed on a shaft-strut system. In some cases, however, skegs have been used for both. Recently, propellers have been located on streamlined pods at the ship's stern. In this configuration, the propellers are driven by electric motors inside the pods and the propellers may be located on the forward or after ends of the pods. The pods can either be stationary or rotatable. If the pods are rotatable, then there may be no need for a rudder. In any case, the use of pods does not change the principle of the propeller action.

Wake

The wake is the term used to describe the motion imparted to the water by the passage of the ship. On the other hand, the velocity field of the water is usually measured with the measuring apparatus moving with the ship (or model). Thus, the velocity field measured is the ship speed minus the wake. There is a tendency to use these terms interchangeably in discussion of the interaction between a ship and a propeller, but when actual data are presented, it is made clear which is being used.

The propeller operates behind the ship in the wake caused by the viscosity of the water and the wave action. If we make a survey of the velocity field behind a single-screw ship, a pattern something like Figure 29-1 is obtained.

The two components of velocity shown are the axial, V_x, and tangential, V_t. There is also the radial component of velocity, V_r, which must be considered. As it rotates, a blade section operating in this velocity field encounters a velocity which fluctuates both in magnitude and direction. Thus, the propeller operates in an unsteady flow field, and the propeller blades produce an unsteady force. For design purposes, the velocity is analyzed by a harmonic analysis; i.e., at each radius r a harmonic analysis is carried out with respect to the angular position θ on the wake components:

$$\frac{V_x(r,\theta)}{V} = \frac{V_{x\,aver}(r)}{V} + \sum_{n=1}^{\infty} V_{nx}(r)\cos n\theta$$

$$\frac{V_t(r,\theta)}{V} = \frac{V_{t\,aver}(r)}{V} + \sum_{n=1}^{\infty} V_{nt}(r)\sin n\theta$$

The V_{aver} terms which are the circumferential averages of the velocity components give rise to the steady forces on the propeller, and the V_n terms which are the *n*th harmonic amplitudes of the velocity components give rise to the unsteady forces on the propeller. It is assumed in the design process of the propeller that the circumferential average velocity and the harmonic amplitudes of the velocity can be considered separately; i.e., the propeller can be designed to meet the conditions of power, speed, and RPM independent of the harmonic content of the velocity. It will be shown later that the harmonic amplitudes of the velocity V_n must be considered when designing the propeller for good cavitation performance and minimum vibration.

Hull-Propeller Interaction

You should not infer that the only interaction between the ship and its propeller is in the requirement that the propeller operate in the wake of the

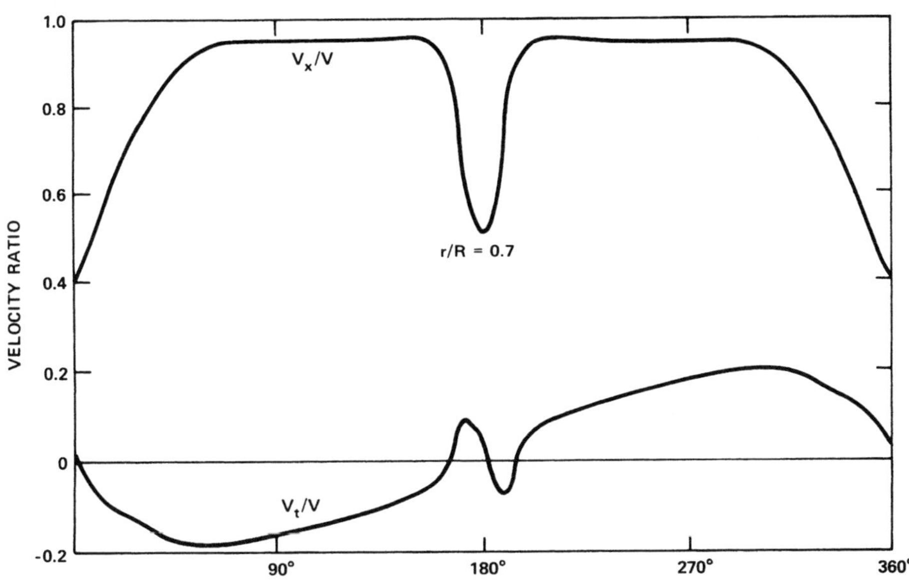

Figure 29-1. Representative velocity distribution in place of the propeller for a single-screw ship

ship. Since it must accelerate a mass of fluid in order to produce thrust, the propeller accelerates part of the fluid over the stern of the ship as well as over the rudder. This acceleration of the fluid changes the pressure distribution over the stern and has two effects: (1) there is a change in the drag, and (2) the boundary layer is affected.

This change in resistance or drag is called the thrust deduction, t, and is defined as

$$t = \frac{T - R}{T}$$

where T is the propeller thrust and R is the total ship drag. The effect of the acceleration of fluid is normally to increase the drag. The propeller causes a favorable pressure gradient. In a poorly designed stern, it is possible that the propeller so alters the flow that separation may be prevented and the viscous drag reduced. There are theoretical methods for calculating the thrust deduction, but for conventional hulls it is a complicated procedure and none too satisfactory. Nevertheless, the thrust needs to be known, and it is not sufficient only to know the resistance of the hull.

To obtain the thrust, a self-propulsion experiment is conducted on a model with a stock propeller, i.e., a propeller which closely approximates the design propeller. Thus, it is assumed that the thrust of the stock propeller can be used to determine the thrust required of the design propeller.

In addition to the effect of the propeller action on the resistance of the hull, the pressure defect caused by the propeller on the hull may also affect the boundary layer formation and, hence, the wake. Since this effect of the propeller acts on each particle of fluid, it affects all the velocity components at the stern. There are not yet techniques sophisticated enough to determine how the propeller affects all these velocity components. It is standard practice in model testing to measure the velocity field at the propeller plane without the propeller in place. This velocity is called the nominal velocity or in terms of wake, the nominal wake. If the interaction of the propeller on the boundary layer is considered, the velocity is called the effective velocity or, in terms of wake, the effective wake. These definitions seem simple enough but one should recognize that effective velocity is not a measurable quantity since it does not include the velocities induced by the propeller. The definition for effective velocity is derived from an attempt to determine the average speed of advance of the propeller in the ship's wake.

The speed of advance of the propeller through the water has been defined in terms of the effective wake as

$$\frac{V_A}{V} = (1 - w)$$

where w is the Taylor wake fraction. The Taylor wake fraction is a measure of how much the mean velocity at the propeller deviates from the ship speed. For the propeller design, the wake is estimated from self-propulsion tests on a ship model using a stock propeller. However, prior to preliminary design or selection of a stock propeller, estimates of the mean wake must be made. Data are available from series of model tests to estimate the Taylor wake fraction, e.g., Taylor Standard Series and DTMB Series 60. These data show that, for single-screw ships, the Taylor wake fraction increases with increasing ship block coefficient. (Note that increase in the wake fraction implies that velocity at the propeller decreases.) For low block coefficient single-screw ships ($C_B<0.5$), the wake fraction can be as low as 0.2 and for high block coefficient single-screw ships ($C_B>0.85$), the wake fraction can be as high as 0.5. For twin-screw ships, especially with propellers supported by struts, the Taylor wake fraction is much lower and can approach 0.1 or less. For high speed military-type ships with struts, the wake fraction usually lies between −0.02 and +0.02.

The distribution of the velocity at the propeller is important for cavitation and vibration considerations. As mentioned above, the harmonics in the wake are obtained from a velocity survey in the propeller plane. Hadler and Cheng (1965) have given data for a large number of twin-screw and single-screw ships. Their data consist of the mean velocity, the distribution of the circumferential average velocity, and the wake harmonics. Thus, one can obtain a reasonable estimate of the nominal wake field behind the ship. However, one should be aware that the data are for a model and that the wake behind a full-sized ship is somewhat thinner.

Thrust deduction can also be estimated from model series data. For single-screw ships, the thrust deduction is expressed as a percentage of the Taylor wake fraction. For instance, for ships equipped with streamlined rudders or contra-rudders, the thrust deduction varies between 0.5 and 0.7 of the Taylor wake fraction.

For reasons of economy, it is said to be desirable to propel the ship with the minimum possible power. Ideally, what is desired is the minimum fuel consumption, and not minimum power. However, this requires knowledge of engine fuel rate, and so, when considering the propeller independent of the main engine, it is sufficient to consider minimum power. Thus, when designing the propeller without knowing the main engine fuel rate, a propeller should be selected which uses the minimum power. Because of the hull-propeller interaction, it may be that the propeller with the maximum efficiency may not use the minimum power. For a ship with a given speed, the power required to overcome the resistance is the resistance times the ship speed. This power is called the effective power P_E. The delivered power at the propeller, P_D, is 2π times the shaft torque times the RPM. For efficient propulsion, it is desirable to maximize the ratio of the delivered power to the effective power. This ratio is known as the propulsive efficiency, η_D:

$$\eta_D = \frac{P_E}{P_D} = \frac{R \times V}{2\pi Qn} = \frac{(1-t)\, T \times V_A}{(1-w)\, 2\pi Qn}$$

If η_H is defined as the hull efficiency and is

$$\eta_H = \frac{(1-t)}{(1-w)}$$

and the behind propeller efficiency is

$$\eta_B = \frac{T \times V_A}{2\pi Qn}$$

then the propulsive coefficient can be defined in terms of a hull efficiency and a behind propeller efficiency:

$$\eta_D = \eta_H \eta_B$$

It is a common practice to break down the behind propeller efficiency, η_B, into two components, the open water propeller efficiency, η_O, and a term called the relative rotative efficiency, η_R.

$$\eta_B = \eta_O \eta_R$$

The relative rotative efficiency is the ratio of the propeller efficiency behind the hull and in open water. The difference in the behind and open water efficiency is that in the behind condition the velocity to the propeller varies in both the radial and circumferential directions, and in open water the inflow is uniform. The relative rotative efficiency usually lies between 1.0 and 1.05 for most merchant ship forms. This means that the propeller is slightly more efficient in the behind condition than in open water.

PROPELLER DESIGN

Theory

For many years, propellers were designed using design charts developed from model tests with a series of propellers. A number of series have been developed, the most extensive of which is the Wageningen B-Series, or Troost Series, published by the Netherlands Ship Model Basin (see Wageningen Series, 1969). The series extends from 2 through 7 blades with a variation of a number of other parameters. Table 29-1 lists the blade area ratios and range of pitch ratios tested for each number of blades. Starting in the early 1960s, significant advances were made in the design of marine propellers. These advances were related to the practical application of theory and the use of the computer in propeller design. Even where propeller series are used, e.g., for preliminary calculations, they have been computerized. Today it is practical to design propellers theoretically, and the use of charts is becoming a thing of the past. Since the theories are very complicated and computers are necessary for execution of the calculations, a description of the present-day design theories will not be given. It suffices to say only that propeller theory stems from wing and airfoil theory but instead of the blade being in a plane as a wing, the blade lies along a helix. This geometry greatly complicates the theory and the interference between the blades. The mathematical representation of the blades in the design theory is normally in two parts. One part, called lifting line theory, is where each blade is represented by a single line with the slipstream shed behind each line. Lifting line theory is used to predict the performance of a propeller and to select the pitch of the shed helical vortices. The other part, called lifting surface theory, is where each blade is represented mathematically both in blade camber and in blade thickness. Lifting surface theory is used to give the blade distortion in both camber and pitch to achieve the performance predicted from lifting line theory. Although the complete design can be carried out with lifting surface theory, for practical computational reasons, and with no loss in generality, lifting line theory is used. A further discussion of the theoretical design of propellers would include a rather complete discussion of the vortex theories entailed in aerodynamics and hydrodynamics. However, a good understanding of propeller action can be obtained from momentum theory.

TABLE 29-1

Summary of the Wageningen B-Series

Blade Number	*Blade Area Ratio*												*Pitch Ratio*
2	0.30		0.38										0.6-1.4
3		0.35		0.50			0.65			0.80			0.5-1.4
4			0.40		0.55			0.70			0.85	1.00	0.5-1.4
5				0.45		0.60			0.75			1.00	0.5-1.4
6				0.50			0.65			0.80			0.6-1.4
7					0.55			0.70			0.85		0.6-1.4

In momentum theory, the propeller is replaced by a hypothetical disk called the actuator disk. In practice, this is equivalent to a propeller with an infinite number of blades and no blade area and no thickness. Despite being far away geometrically from an actual propeller, the theory does give a good idea of the velocity induced by the propeller and the shape of the slipstream. The propeller will be assumed to be at rest with a steady flow of water past the propeller.

To calculate the propeller action, three laws of hydrodynamics are required:

1. Law of Momentum. The rate of change of momentum in a fluid domain bounded by a fluid surface is equal to the resultant of the external forces. From the law of momentum it follows that the forces acting on the fluid inside the surface are determined by the forces on the surface itself in steady flow; i.e., forces on the propeller can be derived without knowledge of velocities and pressures at the propeller itself.

2. Bernoulli's Equation. In a flow of incompressible, nonviscous fluid, the total energy per unit volume of fluid is constant along a streamline. If energy is added or taken away at any one position, the total energy will be changed downstream by the amount of energy added or subtracted.

3. Equation of Continuity. From the law of conservation of mass, fluid matter can neither be created nor destroyed, and therefore the mass of fluid passing one cross section per unit time must be equal to the mass of fluid passing a downstream cross section per unit time, where the cross section lies between streamlines. For an incompressible flow, where the fluid density is a constant, the cross-sectional area times the velocity through the area is a constant.

If the propeller disk in Figure 29-2 is considered, the equation of continuity gives:

$$A_O V_A = A_1(V_A + U_{A_1}) = A_2 U_2 = A_2(V_A + U_{A_2})$$

The momentum equation states that:

$$T_i = \rho A_1(V_A + U_{A_1})(U_2 - V_A)$$

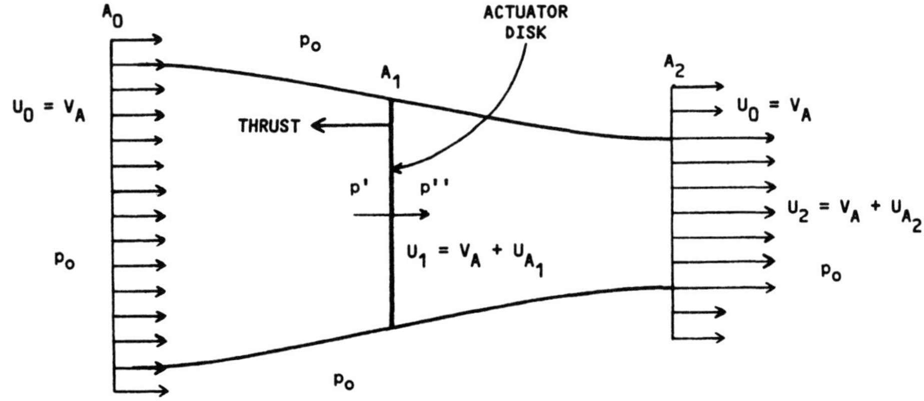

Figure 29-2. Momentum diagram

The subscript "i" denotes that the force is for inviscid flow. The work done on the actuator disk is the thrust times the velocity at the disk, and it is equal to the kinetic energy imparted to the slipstream, i.e.:

$$T_i(V_A + U_{A_1}) = \frac{\rho}{2} A_1(V_A + U_{A_1})(U_2{}^2 - V_A{}^2)$$

The combination of this equation and the one previous states that:

$$U_{A_1} = \tfrac{1}{2} U_{A_2}$$

Therefore, the velocity induced at the actuator disk is one-half the total velocity induced far downstream.

The work done on the actuator disk is equal to the power absorbed, while the useful work is equal to the thrust times the speed of advance. The efficiency of the propeller is defined as the power output over the power input, which for the actuator disk is

$$\eta_i = \frac{T_i V_A}{T_i(V_A + U_{A_1})} = \frac{1}{1 + \dfrac{U_{A_1}}{V_A}}$$

or in terms of thrust loading on the disk:

$$\eta_i = \frac{2}{1 + \sqrt{1 + C_{T_{h_i}}}}$$

where

$$C_{T_{h_i}} = \frac{T_i}{\dfrac{\rho}{2} A_1 V_A{}^2}$$

This equation gives an upper bound to the propeller efficiency since it does not include the rotational energy losses in the slipstream, the frictional losses on the blades, or the effect of a finite number of blades, etc. The equation does show that as the load coefficient, $C_{T_{h_i}}$ increases, the efficiency decreases.

A general momentum theory can be derived which takes into account the rotational losses in the slipstream and the actual radial load distribution. To take the viscous losses on the blades into account, the blade area and pitch angle, besides the viscosity of the fluid, must be known. Approximations for this relationship have been derived.

From momentum theory, the following conclusions can be drawn:

1. The axial velocity induced by the propeller increases from zero forward of the propeller to a maximum constant value downstream. The induced velocities at the propeller are half their value far downstream (axial, tangential, and radial components).
2. The increase in axial velocity through the propeller causes the slipstream diameter to reduce and is a function of the thrust loading.
3. The propeller efficiency decreases with increasing loading.

Series Data

Propeller series data are very useful for checking operation of propellers and in determining optimum RPM and diameter. Originally, all propeller series were based on experiments as the theory was not adequate for predicting propeller performance. With the advent of high-speed computers and the development of theoretical design techniques, theoretical propeller series of performance can be derived. The series data are plotted in a number of various forms. Figures 29-3 to 29-7 show data from the Wageningen 4-bladed series with a blade area ratio of 0.55. Figure 29-3 shows data in the standard K_T-K_Q-J system (J is defined as the advance

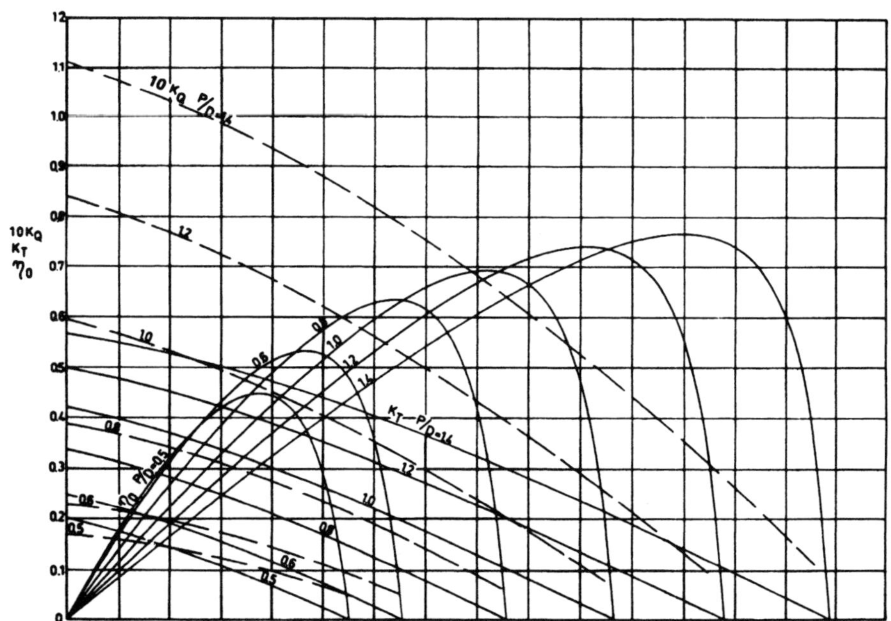

Figure 29-3. Open water test results of B 4-55 screw series in standard coefficient form

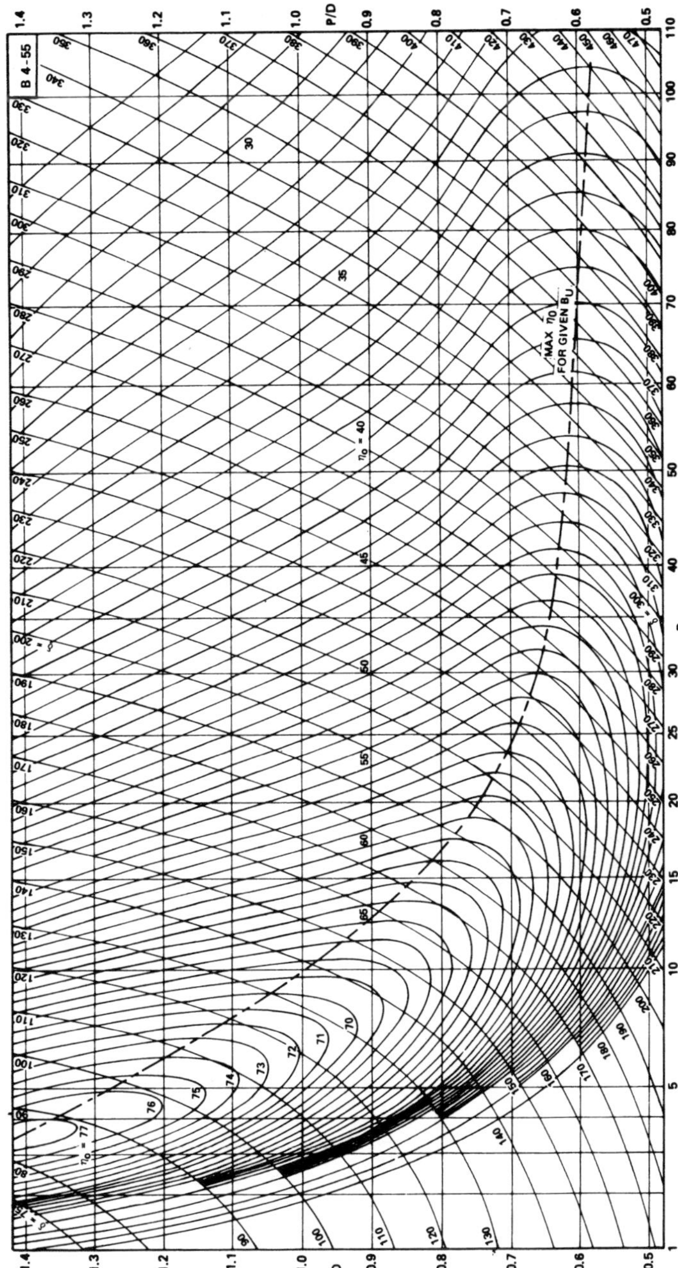

Figure 29-4. Open water test results of B 4-55 screw series in B_U-δ coefficient form

coefficient). This system presents the complete performance of the propeller when the propeller is moving ahead through the water and rotating in the direction for which it is designed. Obviously, there are many other modes of operation—crash back, crash ahead, and steady backing—for which data are available, but the ahead operation is of interest for propeller design. In Figure 29-3, each set of curves for a given pitch-to-diameter ratio represents the data from one propeller test. For ease in design, these data, cross-plotted in various other coefficient forms, are much more useful for design purposes. One of the oldest forms is that used by Taylor, the B_U-B_P-δ system, Figures 29-4 and 29-5. The B_U coefficient is a thrust coefficient and the B_P coefficient is a power coefficient, where

$$B_U = \frac{N\sqrt{T}}{V_A^{\,2}}(0.05541)$$

$$\delta = \frac{ND}{V_A}$$

$$B_P = \frac{N\sqrt{P}}{V_A^{5/2}}$$

In these equations, the symbols are not in consistent units, e.g.,

D = diameter in feet
N = RPM
P = delivered horsepower in fresh water
T = thrust in fresh water in pounds
V_A = speed of advance in knots

Therefore, the equations are in dimensional form for fresh water. This causes some difficulties when, for instance, the propellers operate in salt water and corrections must be made to power and thrust. With the charts it is easy to obtain the optimum diameter since the diameter does not appear in B_U or B_P. It is only necessary to calculate B_U or B_P and read the δ value from the max η_O line. Calculate D from δ. For optimum RPM, values of the coefficient must be assumed and cross curves plotted on the diameter. For instance:

$$B_U = \frac{\delta\sqrt{T}}{DV_A}(0.05541)$$

The coefficient B_U is calculated for various δ values and the results plotted on Figure 29-5. Calculation of RPM from the δ, where B_U has the largest efficiency, gives the optimum RPM.

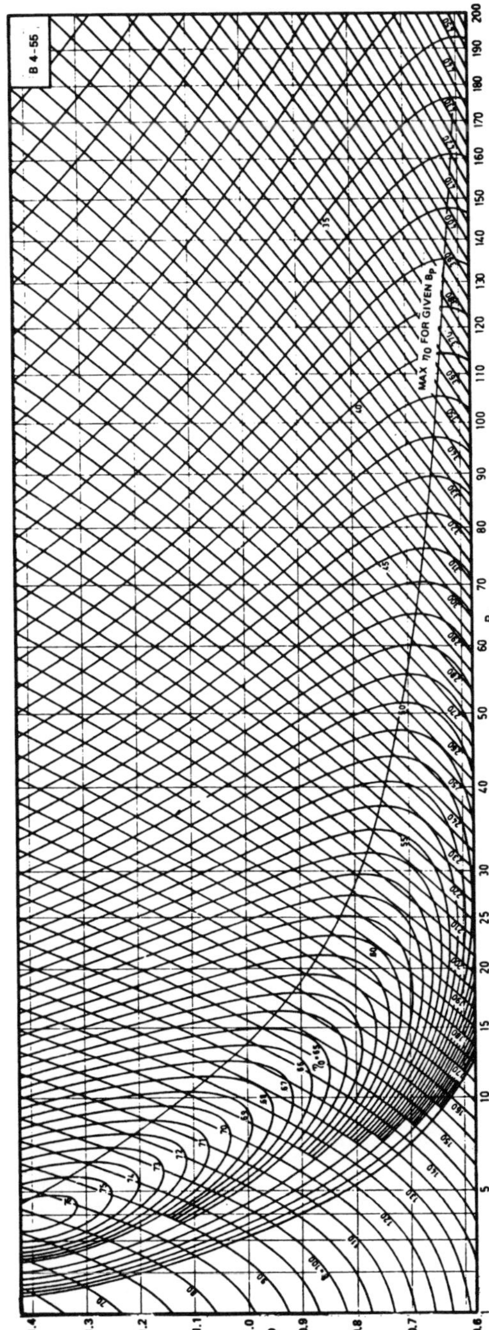

Figure 29-5. Open water test results of B 4-55 screw series in BP-δ coefficient form

Another common set of coefficients derived from cross-plotting the experimental data is the C_{Th}-C_P-J system. This set is shown in Figures 29-6 and 29-7, where the coefficients are defined as follows:

$$C_{T_h} = \frac{T}{\frac{\rho}{8}\pi D^2 V_A^2} = \text{thrust loading coefficient}$$

$$J = \frac{V_A}{ND} = \text{advance coefficient}$$

$$C_P = \frac{2\pi Q N}{\frac{\rho}{8}\pi D^2 V_A^3} = \frac{\text{power}}{\frac{\rho}{8}\pi D^2 V_A^3} = \text{power loading coefficient}$$

All the symbols are in consistent units so that the coefficients are nondimensional. To obtain the optimum RPM, calculate C_{Th} or C_P and read the J value from the optimum RPM line (line A-B). Calculate RPM from J. For the optimum diameter, assume a value of D and calculate C_{Th} or C_P and J. Plot this point on the appropriate curve and draw a straight line through the origin. Where the efficiency is maximum on this line, read the

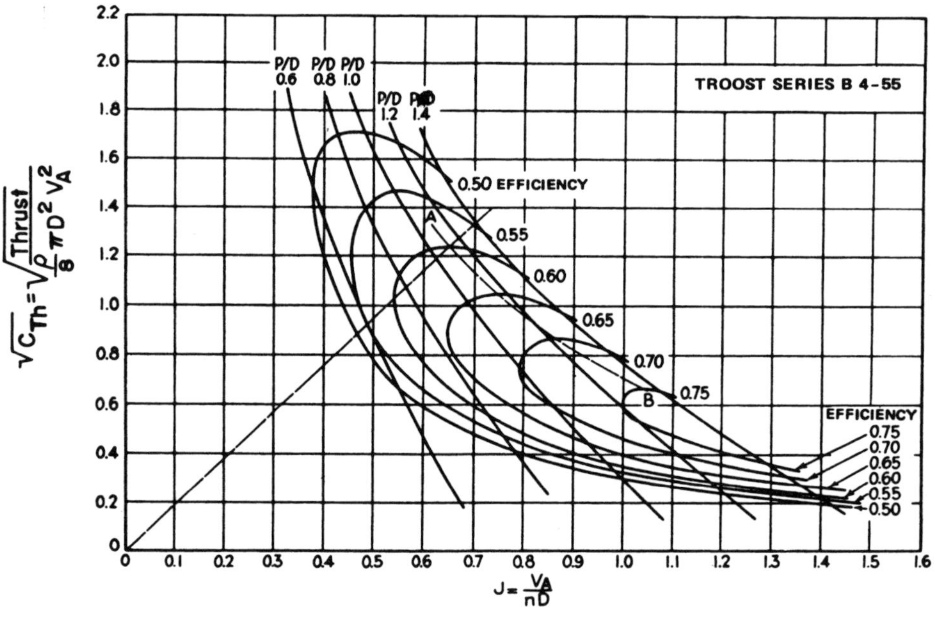

Figure 29-6. Open water test results of B 4-55 screw series in C_{Th}-J coefficient form

J and then calculate the diameter. This is the optimum diameter for this design condition. The C_{T_h}-C_P-J system has advantage over the B_U-B_P-δ system in that it is based on a consistent set of units, and the curves are independent of the fluid in which the propeller operates, i.e., salt water, fresh water, or even air. The only requirement is that the correct density of the fluid be used. Two examples of the use of the C_{T_h}-C_P-J curves in salt water are as follows:

First example:

V_A = 20 knots = 33.76 fps
Thrust = 915,000 lb
RPM = 120
ρ = 1.9905 lb = sec^2/ft^4, fluid density

What is the optimum diameter? Assume:

D = 20 feet

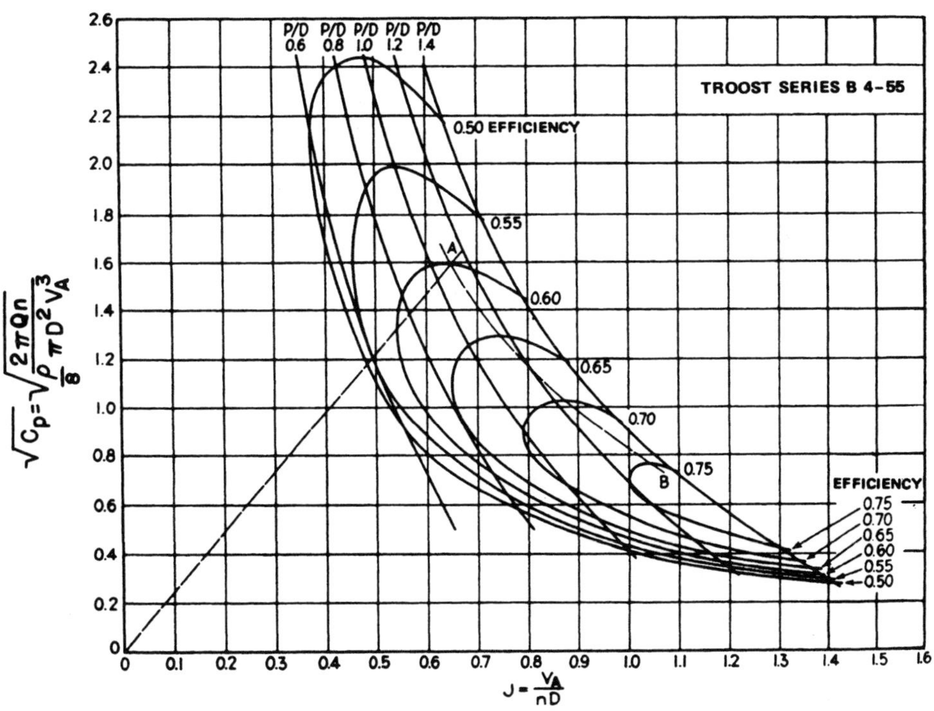

Figure 29-7. Open water test results of B 4-55 screw series in C_P-J coefficient form

$$J = \frac{20 \times 1.688 \times 60}{120 \times 20} = 0.844$$

$$\sqrt{C_{T_h}} = \left[\frac{915{,}000}{\frac{\rho}{8} \pi 20^2 (20 \times 1.688)^2} \right]^{1/2} = 1.6$$

Plot this point on $\sqrt{C_{T_h}}$ versus J curve, draw a straight line through origin, and read off J at maximum efficiency.

$$J = 0.6 \text{ and optimum diameter} = 28.1 \text{ feet, } \eta = 0.5$$

Second example:

$V_A = 15$ knots $= 25.32$ fps
Power $= 20{,}000$ hp
$D = 25$ ft
$\rho = 1.9905$ lb $= \text{sec}^2/\text{ft}^4$

Find optimum RPM.

$$C_P = \frac{20{,}000 \times 550}{\frac{1.99}{8} \pi \, 25^2 \, (15 \times 1.688)^3} = 1.387$$

$$\sqrt{C_P} = 1.1777$$

Read J at this value of $\sqrt{C_P}$ for optimum efficiency

$J = 0.8$
$RPM = 76$
$\eta O = 0.67$

The series are all for propellers in the unfouled condition. Fouling can change the propeller performance significantly. Efficiency can be reduced by 25 to 50 percent, torque is increased significantly, and thrust is reduced somewhat. The overall effect on a ship is a significant increase in power required for a given speed and an increase in RPM. This means that the shaft torque goes up significantly. Propeller cleaning should be done often since it is a small area to clean relative to the hull area, and the power reduction can be significant. One-fourth of the power increase for a fouled ship can be due to the propeller fouling.

Cavitation

Cavitation is the process in which the formation of the vapor phase of the liquid takes place as the liquid is subjected to reduced pressure at a constant ambient temperature. Cavitation occurs in the low pressure region on the propeller blade surfaces and in the low pressure regions of the hub and tip vortices. Surface cavitation, when it covers a significant part of the blade, can cause thrust breakdown. This results in the requirement for increased power and RPM to maintain a given speed. In other words, the performance of the propeller can significantly decrease when compared to a noncavitating condition. The majority of problems with cavitation are with blade erosion and increased vibration. Performance breakdown occurs for high speed ships, and cavitation is a product of high speed. The following formula indicates this. The cavitation number:

$$\sigma = \frac{2gH}{V^2}$$

where g = acceleration due to gravity

H = absolute static head at the place cavitation occurs minus the head due to the vapor pressure of the liquid. For a propeller this is the head due to the atmospheric pressure plus depth of submergence minus the head due to vapor pressure. Oftentimes the depth of submergence is taken to be the shaft centerline in order to give an overall cavitation number for a propeller.

V = local velocity at the point on the propeller where the cavitation number is calculated. For convenience, the velocity is often taken as the speed of advance of the propeller or the ship speed to give a cavitation number that is representative of the whole propeller.

Propeller series have been developed where the propellers have varying amounts of cavitation (Gawn-Burrill Series, 1957, and for the Wageningen B-Series, 1969) so that the cavitating performance of propellers can be predicted. Theoretical series for supercavitating or ventilated propellers have been developed by Caster, 1963. These propellers operate with a complete cavity on the blades extending beyond the trailing edge of the blades. These propellers are for very high speed craft. They are not subject to large decreases in efficiency as conventional propellers under severe cavitation, and the blades are not subject to erosion since the cavities collapse beyond the blade trailing edge. Figure 29-8 gives an indication of when supercavitating propellers should be used. The ordinate of this figure is the absolute pressure (includes atmospheric pressure) in feet of water at the propeller divided by the speed of advance in knots, and the abscissa is the propeller advance coefficient. Region I is practical for supercavitating propellers, and Region III is best for subcavitating propellers. Region II between Regions I and III indicates where partial cavitation is likely to occur on either type of propeller. Region IV for J less than 0.4 indicates a region of low efficiency. The regions on this figure should be used only as a guide to what type of propeller to design.

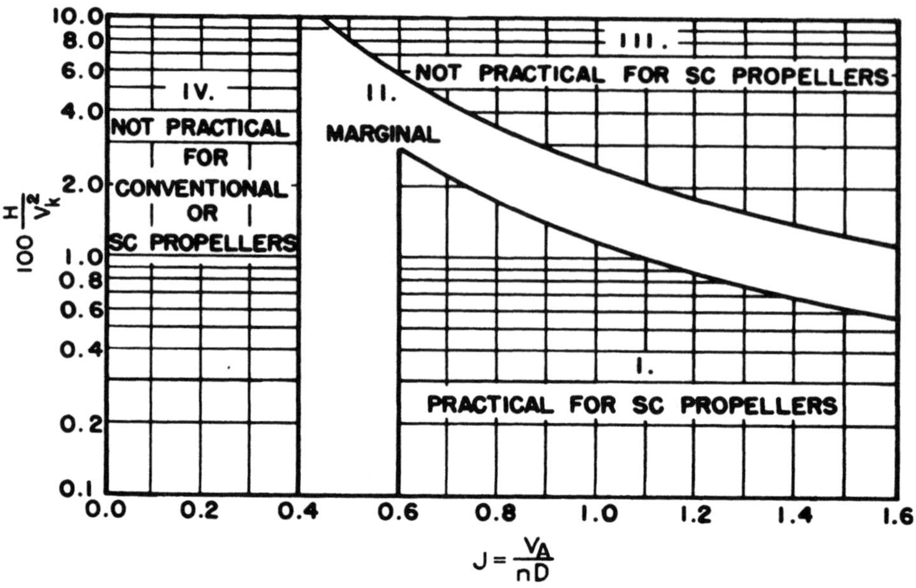

Figure 29-8. Practicability of supercavitating propellers

Blade surface cavitation on conventional propellers is controlled by the blade area. It is desirable to have as low a blade area as possible, within the strength constraints, for efficiency reasons, i.e., low viscous drag of the blades. However, the blade must have enough blade area to prevent thrust breakdown. Propellers that operate at high speeds and higher blade loading require higher blade areas. Various criteria have been derived to estimate the cavitation performance of propellers, and an example from the work of Burrill is given in Figure 29-9. The curves give suggested criteria for various ships and the extent of back cavitation as a function of a thrust loading coefficient and cavitation number. The ordinate is the thrust loading per square foot of projected blade area divided by the relative velocity of the flow to the blade section at the 0.7 radius of the propeller. The abscissa is the local section cavitation number of the propeller at the 0.7 radius.

$$\sigma = \frac{64.32H}{V_R^{\ 2}}$$

$$V_R^{\ 2} = V_A^{\ 2} + (0.7\pi ND)^2,\ V_A \text{ in fps, N in rps}$$

$$A_P = A_O\,(1.067 - 0.229\,{}^{P}\!/_{D})$$

If cavitation is suspected to be a problem, cavitation tests are conducted on a model propeller, and the ship propeller performance under cavitating conditions is predicted from these tests. The cavitation tests are normally conducted in a cavitation tunnel which is a large circulating flow facility where the velocity is controlled by a pump in the circuit. It is necessary to control the pressure so that tests can be carried out at the full-scale cavitation number. Many tunnels do not have sufficient velocity to test the propeller at full-scale speeds. Also, many model propellers are made of material not as strong as the material used for full-scale propellers and since full-scale speeds on a model would imply, approximately, full-scale stresses, it may be necessary to test the models at less than full-scale speed. Some tunnels are large enough to take 7-meter ship models so that the wake at the propeller can be reproduced, while in others the wake is simulated by partial hulls or screens. There are a few cavitation facilities which are towing tanks where the pressure can be controlled. In these facilities, the model is tested as in a standard towing tank but at reduced pressure.

Cavitation can be caused by fouling on the propeller and by nicks on the blades near the leading edge and the tip. The propeller should be inspected for these defects and cleaning and repairs made.

Strength

A propeller blade must contain enough material to keep the stresses within the blade below a certain level. This level depends both on the steady state and fatigue strength of the material and on both the mean and unsteady blade loadings. The material selected controls the allowable stress level, and the blade chord and thickness are the main parameters which control the blade stress for a given blade loading. For highly skewed propellers, the skew is also an important propeller geometric parameter influencing the stress level.

The maximum stress for a propeller normally occurs near the point of maximum blade thickness unless the blade has an unusual configuration, such as high skew. Without consideration of the centrifugal forces and blade rake and skew, the required maximum blade thickness/chord, t/c, at a given propeller radius can be approximated by

$$t/c = \left[\frac{M_{x_0}}{C_s C_c c^3 \sigma_a} \right]^{1/2}$$

where M_{x_0} is the bending movement at maximum power about the blade chord, σ_a is the allowable blade stress, C_s is a coefficient dependent on the chordwise thickness distribution (for standard sections, varies between 0.8

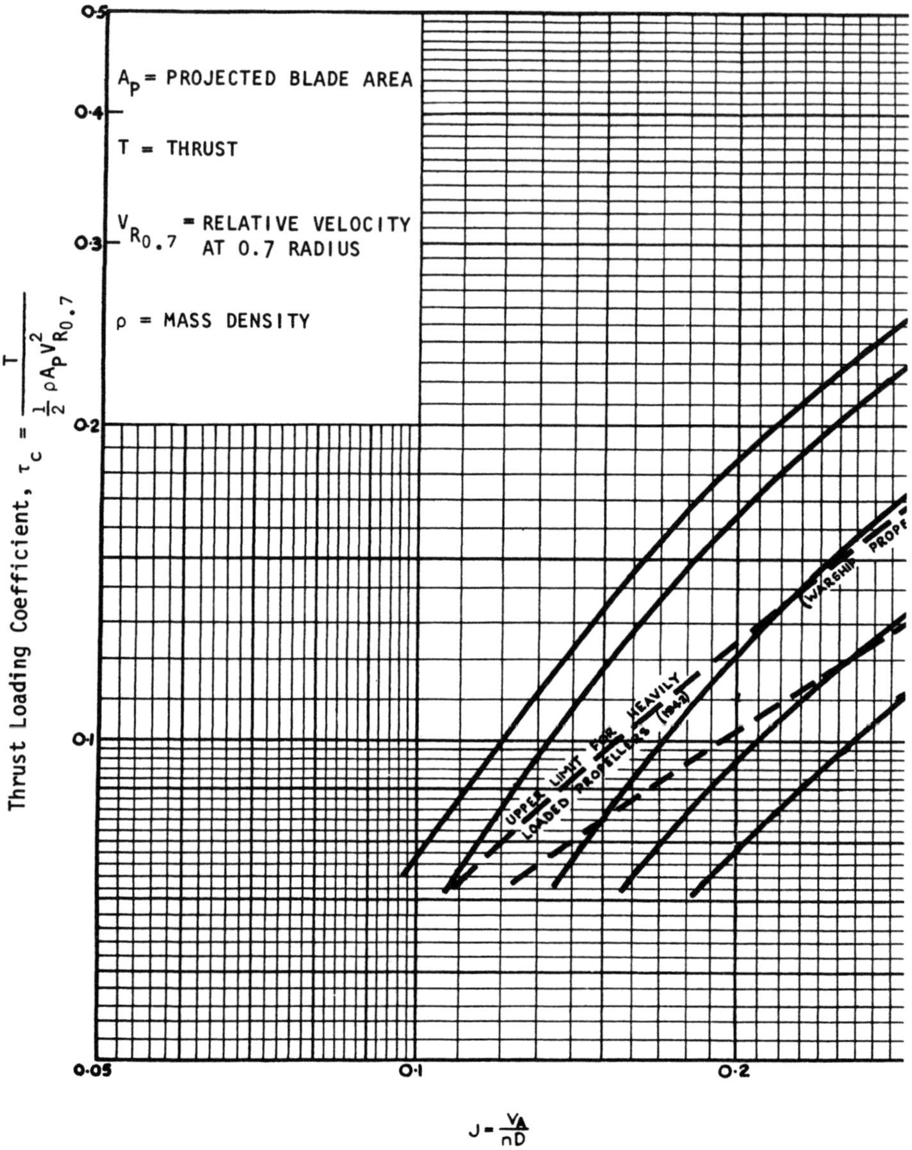

Figure 29-9. Simple pro-

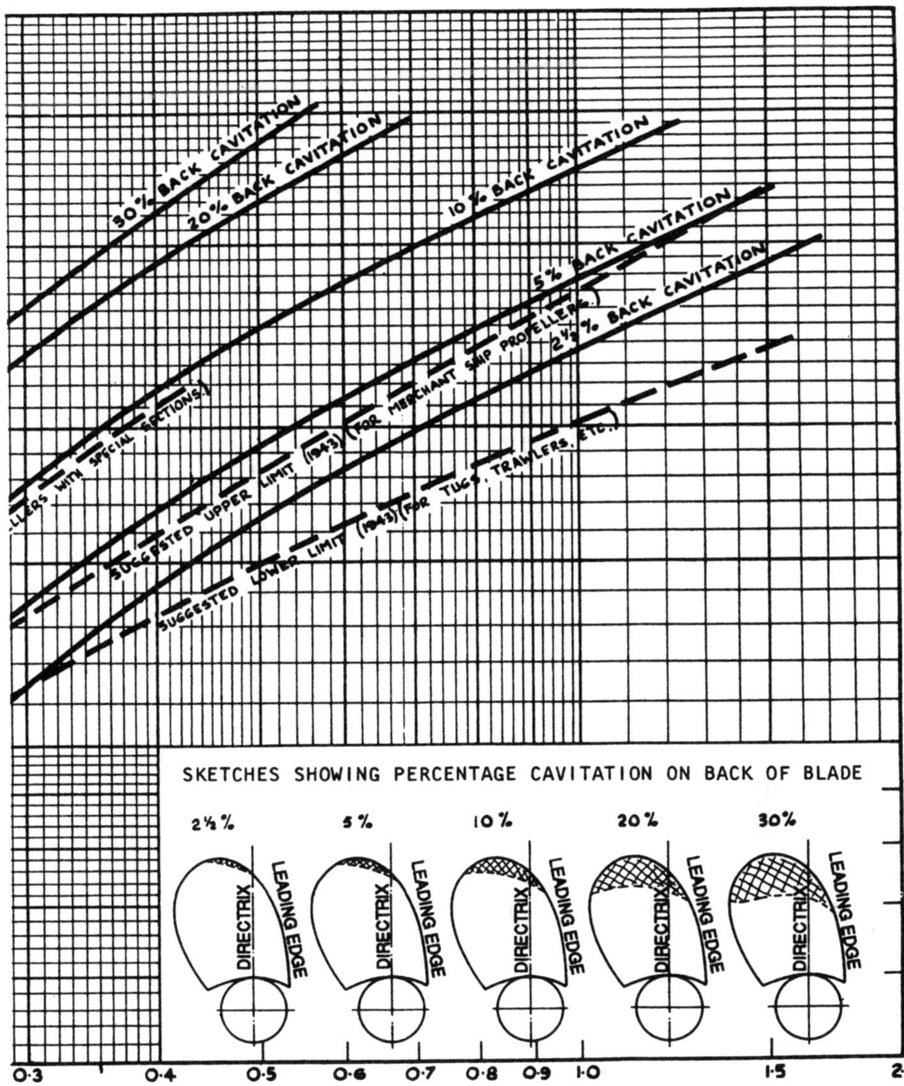

peller cavitation diagram

and 0.9), C_c is a coefficient dependent on the chordwise camber distribution (for standard sections, assume it is equal to one), and c is the section chord.

All propellers should be designed considering both mean and unsteady loads acting on the blade. These must be considered together since the allowable unsteady stress decreases as mean stress increases. A simple equation for allowable stress can be derived from the Goodman diagram for both mean and unsteady loads. This equation is known as the equation for the Goodman safe region.

$$\sigma_f = \frac{\sigma_{max}\,(2 - R_s)}{3}$$

where R_S is the ratio of minimum stress to maximum stress during a cycle, σ_f is the fatigue strength for a given number of cycles, single amplitude, and σ_{max} is the maximum stress for a given number of cycles. The maximum value of unsteady stress for single screw cargo ships is considerably greater than the mean stress. The difficulty in making strength predictions of propeller blades is the lack of knowledge of the unsteady bending movements on the propeller blade. A few measurements have been made which show that the unsteady load for a single screw cargo ship can be as much as 50 percent larger, and even higher, than the mean value. For design purposes, unsteady propeller theory must be relied upon for predicting the unsteady blade loads. In practice, the American Bureau of Shipping publishes criteria for blade thickness which standard propellers must meet before they can be certified for use on a particular ship. These criteria depend on the propeller geometry, the horsepower, and the RPM. All criteria are for the steady state condition, but it is obvious that unsteady loading is accounted for in the factor of safety used.

Many factors determine the choice of the structural material for marine propellers. These are weight, material and fabrication costs, yield strength, fatigue strength, castability, machinability, weldability for repair, corrosion resistance, impact resistance, and erosion resistance. The propeller operates in the hostile environment of salt water. Because of the corrosive nature of salt water on most materials and the substantial decrease in fatigue strength of most materials in salt water, and considering the other factors for selecting propeller materials, only a few materials are suitable for use in propellers. Most common materials are some of the bronzes and some types of stainless steels. Cast iron was used some years ago but is not common today. Also, propellers have been built of Inconel, titanium, and glass reinforced plastics, but these materials are not yet used for large size propellers. It is quite probable that composite materials such as glass reinforced plastics will find more common usage. The more exotic materials such as Inconel and titanium have been used for special application to high speed craft where cavitation erosion and strength are a particularly difficult problem. Table 29-2 shows the fatigue strength in salt water of some of the more common propeller materials. Which material to select depends not only on the fatigue strength, but its other properties such as cost and ease to manufacture and repair.

TABLE 29-2

Fatigue Strenghts (psi) in Seawater, 10^8 cycles

Material	*ABS** *Type*	σa
Manganese bronze	2	6,000
Nickel manganese bronze	3	6,000
Nickel aluminum bronze	4	12,500
CF-4 stainless steel	—	6,500 (much higher with cathodic protection)

Vibration and Noise

The propeller can induce ship vibration in two ways: (1) unsteady forces transmitted through the shaft bearings and (2) pressure forces which act on the hull plating and appendages in the vicinity of the propeller. The unsteady forces come from the blades operating in the spatially varying wake field with unsteady forces being introduced through the shaft. Bearing forces are little affected by cavitation. They can be minimized by selecting the blade number to avoid strong wake harmonics at blade number and blade number ±1. The wake harmonics at blade number give rise to fluctuating axial force and torque on the shaft while the wake harmonics at blade number ±1 give rise to side forces and movements on the shaft. Harmonic analysis of the wake at the propeller will provide excellent guidance to the selection of number of blades. Also, in the selection of the number of blades, it is essential to avoid exciting hull, superstructure, and machinery resonances at normal operating speeds. By changing blade number, it is often possible to alter the blade frequency so that hull and machinery resonances are avoided.

The propeller pressure forces are dependent on the clearances between the hull and the propeller, the blade thickness, and the blade load distribution. The clearance should be selected with care. Blade cavitation can greatly enhance the propeller pressure forces especially if the unsteadiness in the cavitation coincides with the blade frequency which, of course, is the usual case.

Blade rake and skew can be used to minimize vibration. Blade skew, i.e., differential displacement of the blade chords along the helical reference lines, permits a more gradual entry of the blade into a wake region. The idea is to mismatch the wake with the blade shape. Highly skewed

* American Bureau of Shipping

blades can reduce the bearing forces to one-fourth the value for unskewed blades, and the pressure forces can be reduced by one-half. Figure 29-10 shows a drawing of a highly skewed propeller and Figure 29-11 is a picture of a highly skewed propeller.

Recommendations have been developed by most all of the classification societies for propeller clearances, and theoretical and experimental techniques have been developed for predicting unsteady bearing forces. However, the main source of propeller induced vibration, especially on high speed ships, is the unsteady pressure force caused by a cavitating propeller. This problem is being widely investigated, and no clear criteria have yet been established.

The cavitation problem is very severe if the cavitation is particularly unsteady. The cavitation can also cause noise problems as well as vibration problems within the hull and, if the noise is severe enough, can cause hearing problems and be extremely uncomfortable. When this happens, the problem is usually associated with a hull form which gives rise to an extremely sharp wake peak at the upper part of the propeller aperture. Oftentimes, it is necessary to modify the hull form, such as by putting fins on the hull, to alleviate the noise and vibration problems.

Another propeller noise is singing. This is intense discrete frequency sound radiating from the propeller and is much like a steady bell tone. This noise is generally associated with vortices shedding from the blade trailing edges in such a way that the whole trailing edge is at resonant frequency. Sometimes singing is associated with a damaged blade, excitation from the shaft, or cavitation. However, the self-excitation from vortex shedding is the most common cause of singing. It can generally be controlled by sharpening the blade trailing edge so that the included angle between the blade face and back is 25 degrees or less.

TYPES OF PROPELLERS

There are a number of propulsor types used for propelling ships and many of these are defined in the glossary. Many more tend to be one of a kind, have limited usage, or are inventor's names for variations in otherwise common propulsors. The following discussion includes the most common types, except for oars and sails, but is not meant to be all-inclusive. Where the description is adequate under the definition, in the glossary, only the advantages and disadvantages will be given here. In all cases, these propellers are more expensive than the standard screw propeller.

Controllable Pitch Propeller

Advantages: Capable of giving reverse thrust without reversing shaft and allows full power to be absorbed for all loading conditions.

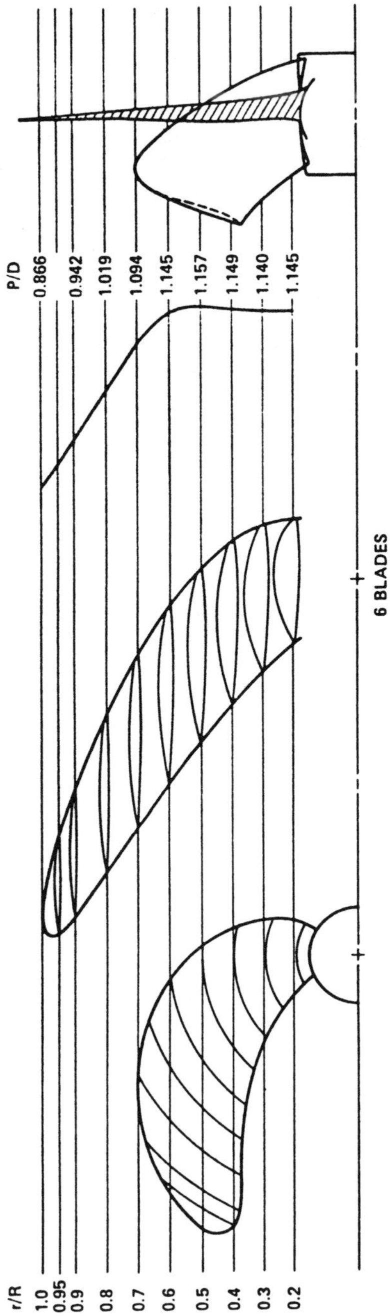

Figure 29-10. Schematic drawing of a highly skewed propeller

Figure 29-11. Photograph of a highly skewed propeller. Courtesy Ferguson Propeller and Reconditioning, Ltd.

Disadvantages: More complicated mechanically and subject to failure. Less efficient at design condition (for standard single-screw hull forms, the efficiency loss is on the order of 2 percent and for shaft-strut systems where the shaft is large and struts must be large to support the extra weight of the propeller and shaft, the net efficiency loss can be on the order of 6 to 8 percent).

(See Figure 29-12 for hub details.)

Cycloidal Propeller

Advantages: Thrust is produced in any direction normal to axis of rotation. Axis of rotation would usually be vertical so that thrust is produced in any horizontal direction. This propeller is used where excellent maneuverability, as for ferries and tugs, is required.

Disadvantages: Very complicated mechanism and more subject to failure. Efficiency is about 25 percent less than standard propeller.

Ducted Propeller

Advantages: More efficient when propeller is heavily loaded, more bollard pull, better protection of the propeller.

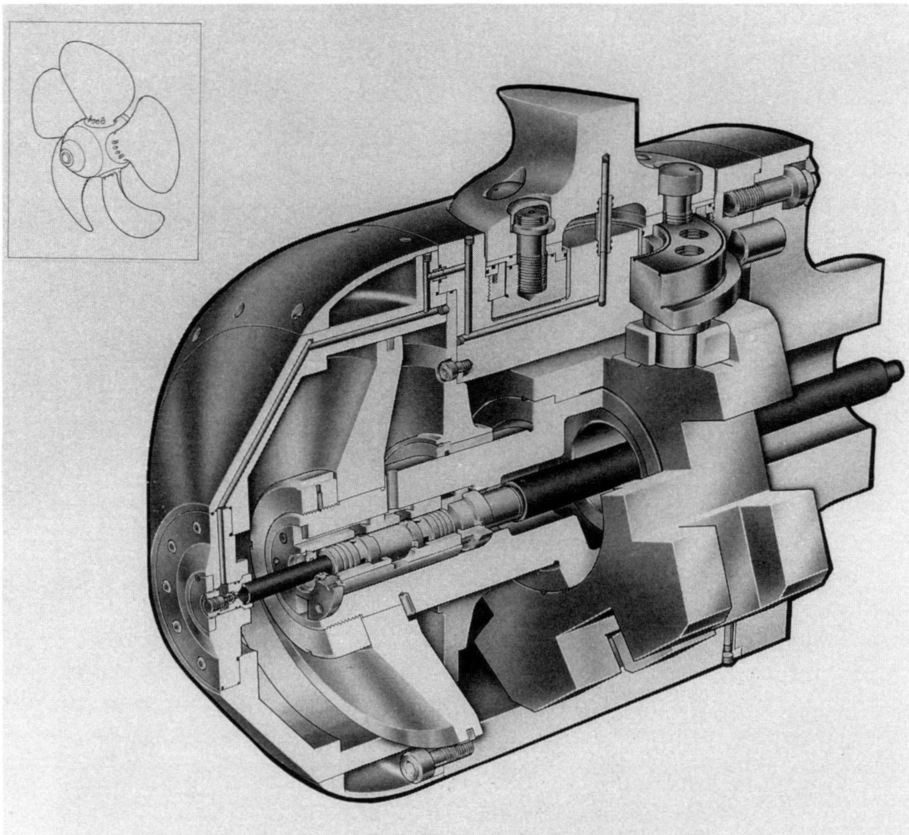

Figure 29-12. Hub details of a controllable pitch propeller. Courtesy Bird-Johnson Co.

Disadvantages: Less efficient when the propeller is lightly loaded. Actual efficiency difference (can be more or less) depends on loading and can be several percent. Types of ships which use Kort nozzles, or some form thereof, are ships which typically have high thrust and low speed, such as tugboats, trawlers, and full form tankers. Gains in efficiency can be on the order of 10 percent. On the other hand, for fast cargo ships, the loss in efficiency can be on the order of 10 percent. The decelerating pumpjet types are not commonly used except for torpedoes and some military ships.

Fully Cavitating Propeller

Advantages: More efficient under fully cavitating conditions. Applicable only to high speeds (40 knots or above), same for Propeller Types, Interface Propellers (see the glossary), except speed can be somewhat lower.

Water Jet

Advantages: No exposed propulsor parts, good for shallow water and high-speed operation. By rotating, exit nozzle can control thrust in any direction. Used for high-speed ships such as fast ferries and often used for small, high-speed craft.

Disadvantages: More complicated, heavier, and less efficient for ships of conventional speeds. For these ships, efficiency can be on the order of 25 percent less than a standard propeller.

MAINTENANCE AND REPAIR OF PROPELLERS

The construction of propellers must be carried out according to the rules of the classification society which certifies a propeller for a particular ship. These rules consist mainly of the material specifications and specification of the blade thickness. Also, the protection of the steel shaft from corrosion due to action of the water is specified.

The International Organization for Standardization (ISO) has recommendations for the manufacturing tolerances for casting and finishing of ship propellers. These recommendations discuss methods of measuring propellers and tolerances on pitch, radius, thickness, blade section lengths, location of design median lines, longitudinal position, and surface finish of the propeller blade. The tolerances are broken into four classes. For instance, Class I for the majority of merchant vessels gives tolerances on pitch for total pitch of the blades of ±0.75 percent with any local pitch on the blade being ±2 percent. Tolerances on thickness are an upper deviation of +3 percent of the maximum thickness and a lower deviation of −1.5 percent of the maximum thickness.

The classification societies require the stamping of certain data on the propeller hub in a conspicuous place. The following data are usually stamped on the propeller hub:

Name of Ship
Builder
Owner's approval
Plan number
Hull number
Diameter
Pitch (at 0.7 radius)
Classification society approval
Manufacturer's Heat Number
Finished Weight

Inspection of the Propeller

The propeller should be inspected whenever the opportunity arises for bent blades, erosion on the blades, nicks and gouges to the blades, and fouling. Nicks and gouges, particularly near the leading edges and blade tip, should be particularly noted since they can give rise to cavitation as well as increase the blade drag. Damages should be repaired by qualified personnel. Repair of propellers will be discussed in a subsequent section.

Marine fouling should be removed from the blades and hub in all cases. As explained in the previous section, marine fouling can greatly increase the blade frictional resistance resulting in a significant decrease in propeller efficiency. Also, it can give rise to cavitation. Marine fouling can usually be removed with a wooden scraper or a wire brush. When a wire brush or a more abrasive tool is used, great care must be taken to insure that the propeller surface is not damaged. Underwater inspection for fouling should be considered before ship operation when there is a possibility that the propeller is fouled, such as when the ship has been inactive for a long time or has been in warm waters. The propeller can be cleaned underwater by divers by the methods just described. Commercial firms also are available for this.

Propeller Measurement

Laser and other methods are available to accurately measure propellers on or off a ship. However, in some cases, it is desirable to know about certain aspects of the propeller without use of the more elaborate methods, for example, the shape of the leading and trailing edges and the blade pitch. Some idea of the local leading and trailing edge shapes, as compared to design, can be obtained by using leading and trailing edge templates. Female templates can be cut from stock material using the propeller drawing, but care must be used in applying them at the correct propeller radius. A crude estimate of the blade pitch at a particular radius can be made by the following technique.

Turn the propeller blade so that the radius section R at which the pitch is to be measured is horizontal. Drop plumb bobs from the leading edge and trailing edge of the propeller blade at this section to a level surface, making certain that the distances from the shaft centerline to both plumb bobs are the same. Measure the distance between the two plumb bobs, A, between the level surface and the blade leading edge, B, and between the level surface and the blade trailing edge, C. The tangent of the pitch angle is given by the formula:

$$\tan\phi = \frac{|B - C|}{A}$$

and the blade pitch is

$$2\pi \tan\phi = 2\pi R \frac{|B - C|}{A}$$

The absolute sign about B – C means only that the distance should be taken as positive whether the leading edge is up or down. Errors in this procedure occur in making the linear measurements, locating the propeller section horizontally, determining how level the surface is, finding the radius of the leading and trailing edges (the radius affects where the cord on the plumb bob leaves the blade, which may not be precisely the location of the pitch line), and locating the leading and trailing edges at the same propeller radius. It is estimated that the accuracy of the pitch measured in this way is no better than ±2 percent.

Repair of Propellers

In this section, the repair of bronze propellers only will be discussed since this is the material used for most propellers. The repair of stainless steel propellers or propellers of other material should be carried out according to the special considerations for those propellers. Repair of bronze propellers should be carried out only by experienced personnel with the proper equipment. Some of the bronzes, e.g., manganese (Mn) bronze, nickel manganese (Ni Mn) bronze, and manganese nickel aluminum (Mn Ni Al) bronze, are subject to cracking due to stress corrosion, and any welding or bending must be followed by stress relieving to assure that the residual stresses have been reduced to safe levels. There have been numerous instances where blades and tips of blades have broken due to fatigue failure where the original crack arose from stress corrosion. Nickel aluminum (Ni Al) bronze is not subject to stress corrosion cracking and stress relief is not required.

For actual repair of bronze propellers, one should refer to the publication by the American Bureau of Shipping entitled *Guidance Manual for Making Bronze Propeller Repairs*. The section below will give a brief summary of important points in the manual.

There are major and minor repairs. Minor repairs are defined as those repairs to the edges on the outer one-third of the propeller blade and limited to the repair of sections under one and one-fourth inches thick. Propellers of Mn or Ni Mn bronze must be stress relieved for both minor and major repairs; Mn Ni Al bronze needs to be stress relieved for only major repairs; and Ni Al bronze need not be stress relieved for either minor or major repairs. In general, welding for cosmetic purposes should not be done on the blade. A critical area of the blade is defined as the area on the pressure side of the blade between the fillet and 0.4 radius and starting at the leading edge and extending to 80 percent of the chord length. No repairs in the critical area should be undertaken without prior approval of the American Bureau of Shipping surveyor.

Special preheat procedures are recommended before undertaking welding repairs. These procedures depend on the welding process and the welding material used. Welding should be done only by a qualified welding operator.

Straightening can be carried out while the propeller is cold by means of dynamic loads, but this procedure should be restricted to minor repairs. Hot straightening by means of dynamic loads, or slowly applied loads, can be carried out for major repairs. The area of the propeller being repaired should be kept within the recommended temperature range during the whole straightening process, e.g., the temperature range for major straightening for Ni Al bronze is 760-954 degrees centigrade.

As previously stated, Mn bronze, Ni Mn bronze, and Mn Ni Al bronze should all be stress relieved after all welding or straightening repair to prevent stress corrosion cracking. Furnace stress relief is recommended wherever possible. However, local stress relief can be carried out if necessary.

GLOSSARY

Except where stated, the entries refer generally to screw propellers.

Advance coefficient (J). A parameter relating the speed of advance of the propeller, V_A, to the rate of rotation, n, given by $J = V_A/nD$, where D is the propeller diameter. The advance coefficient may also be defined in terms of ship speed, V, in which case it is given by:

$$J_V = V/nD$$

Advance coefficient, Taylor's (δ). A parameter defined as:

$$\delta = nD/V_A = 101.27 / J$$

where n is the rate of propeller rotation in revolutions per minute, D is the propeller diameter in feet, and V_A is the speed of advance in knots.

Angle, advance (or a propeller blade section) (β). The inflow angle to a propeller blade section determined by the rotative speed, ωr, the axial velocity of the fluid, V_x, and the tangential velocity of the fluid, V_θ, according to the equation:

$$\beta = \tan^{-1} \{ V_x (r,\theta) / [\omega r - V_\theta (r,\theta)] \}$$

The induced velocities are not included in the determination of the advance angle (see Figure 29-13), r is the radius of the blade section, ω the angular rate of rotation and θ the angular position of the blade section.

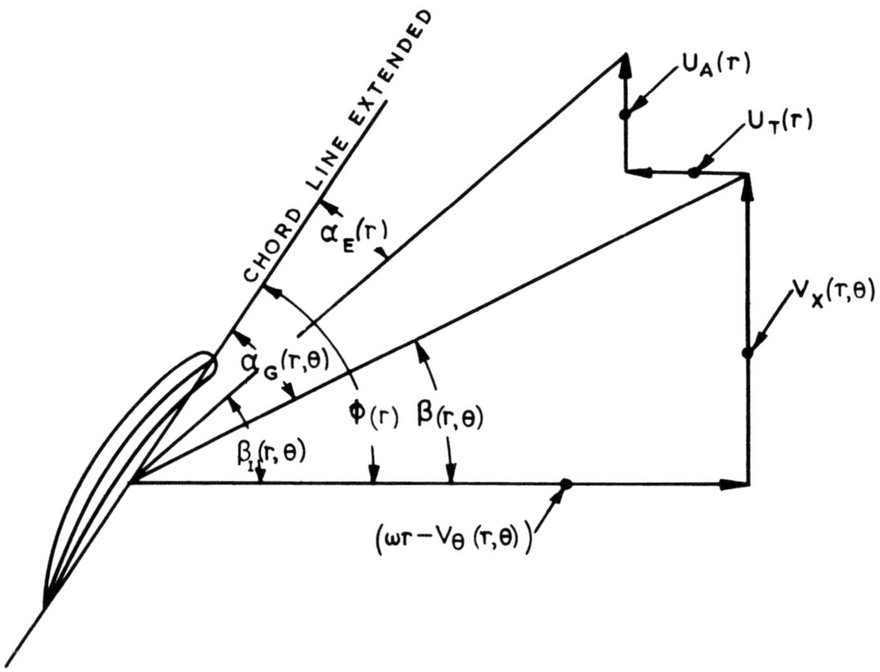

Figure 29-13. Typical velocity diagram for a propeller blade section at radius r

Angle, hydrodynamic flow (β_I). The inflow angle to a propeller blade section including the axial and tangential induced velocities given by the equation:

$$\beta_I = \tan^{-1} \{ [V_x (r, \theta) + U_A (r)] / [\omega r - V_\theta (r, \theta) - U_T(r)] \}$$

U_A and U_T are induced axial and tangential velocities respectively as indicated. For other items see *Angle, advance*. (See also Figure 29-13.)

Angle, shaft. The angle or angles made by a shaft axis with the center plane and/or the base plane of a ship. If a craft significantly changes attitude at speed, the shaft angle may, if so indicated, be measured between the shaft axis and the direction of motion.

Angle of attack (α). The angle, measured in the plane containing the lift vector and the inflow velocity vector, between the velocity vector representing the relative motion between a body and a fluid and a characteristic line or plane of the body such as the chord line of an airfoil or hydrofoil.

Angle of attack, effective (α_E). The angle of attack relative to the chord line including the induced velocities. (See Figure 29-13.)

Angle of attack, geometric (α_G). The angle of attack relative to the chord line of a section neglecting the induced velocities. (See Figure 29-13.)

Angle of zero lift (α_0). The angle of attack relative to the chord line for which the lift is zero.

Area, developed (A_D). An approximation to the surface area of the propeller equal to the area enclosed by an outline of a blade times the number of blades. The outline of a blade is constructed by laying off, at each radius r, the chord length along an arc whose radius of curvature, r_1, is equal to the radius of curvature of the pitch helix given by $r_1 = r/\cos\phi$, where ϕ is the pitch angle at that radius. The outline is formed by the locus of the end points of the chord lines laid out in the above manner.

Area, disc (A_o). The area of the circle swept out by the tips of the blades of a propeller of diameter D:

$$A_O = \pi D^2/4.$$

Area, expanded (A_E). An approximation to the surface area of the propeller equal to the area enclosed by an outline of a blade times the number of blades. The outline of a blade is constructed by laying off at each radius r the chord length along a straight line. The outline is formed by the locus of the end points of the chord lines laid out in the above manner.

Area, projected (A_p). The area enclosed by the outline of the propeller blades outside the hub projected onto a plane normal to the shaft axis. The outline is constructed by laying off, along each radius r, the extremities of each section as determined in a view along the shaft axis. The locus of the end points of the chord lines laid out in the above manner is the required outline.

Back (of blade). The side of a propeller blade which faces generally in the direction of ahead motion. This side of the blade is also known as the suction side of the blade because the average pressure there is lower than the pressure on the face of the blade during normal ahead operation. This side of the blade corresponds to the upper surface of an airfoil or wing.

Blade area ratio. A term used to denote the ratio of either the developed or expanded area of the blades to the disc area. The terms *expanded area ratio* or *developed area ratio* are recommended in order to avoid ambiguity.

Blade section. Most commonly taken to mean the shape of a propeller blade at any radius when cut by a circular cylinder whose axis coincides with the shaft axis.

Blade thickness fraction. If the maximum thickness of the propeller blades varies linearly with radius, this variation of thickness of the blades may be imagined to extend to the axis of rotation. The hypothetical thickness at the axis of rotation, t_o, divided by the diameter is known as the blade thickness fraction or blade thickness ratio. If the thickness does not vary linearly with radius, then the blade thickness fraction is not uniquely defined.

Body of revolution. A symmetrical body having the form described by rotating a plane curve about an axis in its plane.

Bollard pull. The pull force exerted by a ship at zero ship speed. It is the sum of the propeller thrust and the interaction force on the hull.

Boundary layer. The region of fluid close to a solid body where, due to viscosity, transverse gradients of velocity are large as compared with longitudinal variations, and shear stress is significant. The boundary layer may be laminar, turbulent, or transitional.

Camber. The maximum separation of the mean line and the nose-tail line.

Camber ratio. The camber divided by the chord length, f/c.

Cap, propeller. See *Cone, propeller*.

Cavitation. In most engineering contexts, cavitation is defined as the process of formation of the vapor phase of a liquid when it is subjected to reduced pressure at constant ambient temperature. In general, a liquid is said to cavitate when vapor bubbles are observed to form and grow as a consequence of pressure reduction. The following definitions refer to cavitation.

Back cavitation. Cavitation occurring on the suction side (back) of a propeller blade.

Cavitating flow. A two-phase flow composed of a liquid and its vapor is called a cavitating flow when the phase transition is a result of hydrodynamic pressure change.

Cavitation damage. Deformation and/or erosion of materials in cavitated regions, associated primarily with the high pressures developed during cavity collapse.

Cavitation number (σ). The ratio of the difference between absolute ambient pressure p and cavity pressure p_c to the free stream dynamic pressure q:

$$\sigma = \frac{p - p_c}{q}$$

When the cavity pressure is assumed to be the vapor pressure p_v the term is generally called *vapor cavitation number*.

Face cavitation. Cavitation occurring on the pressure side (face) of a propeller blade. It is generally a result of operation in which the local blade angle of attack is excessively negative.

Fully developed cavity. A cavity formed on a body which terminates sufficiently far downstream so that the flow at the downstream region does not influence the body itself. For example, the cavity is fully developed when the re-entrant jet formed at the downstream end of the cavity is dissipated without impinging on the body.

Hub vortex cavitation. Cavitation in the vortex produced by the blades of a propeller at the hub.

Orange peel surface appearance. Description of a surface moderately damaged by cavitation in which the appearance is that of the surface of the Jaffa or California orange.

Pitted surface appearance. Description of a surface damaged by cavitation in which pits are formed either by craterlike deformation without loss of material or by actual loss of material following work hardening or fatigue.

Propeller-hull vortex cavitation. Propeller tip vortex cavitation that extends intermittently to the surface of the hull.

Root cavitation. Cavitation in the low pressure region of the blade roots on a marine propeller.

Spongy surface appearance. Description of a surface badly damaged by cavitation in which erosion has taken place to a considerable depth and has the appearance of a sponge. This description is particularly characteristic of brittle materials and other materials after long exposure.

Supercavitating flows. Cavity flows in which attached, fully developed cavities extend beyond the trailing edge of the body about which the cavity is formed.

Tip cavitation. Surface cavitation which occurs near the tip of a propeller blade.

Tip vortex cavitation. Cavitation occurring in the low pressure core of the tip vortex of a propeller.

Chord line. The straight line connecting the extremities of the mean line. The length of this line is called the chord length or simply the chord. It passes through, or nearly through, the fore and aft extremities of the section. Synonymous with nose-tail line.

Cone, propeller. The conical cover placed over the after end of the propeller shaft for the purpose of protecting the nut and forming a hydrodynamic fairing for the hub. Also known as a propeller fairwater or a propeller cap.

Correlation allowance, model ship (R_A). This is the addition which has to be made to the resistance of the "smooth" ship, as predicted from the model results, to bring it into agreement with the actual ship performance determined from full-scale trial or service results. The correlation allowance depends upon the method used to extrapolate the model results to the "smooth" ship, the ship length and type, the basic shell roughness of the newly painted ship, fouling, weather conditions at the time the ship measurements were taken, scale effects on the factors making up the model, and ship propulsive coefficients.

Developed area ratio. The ratio of the developed area of the propeller blades to the disc area.

Drag (D). The fluid force acting on a moving body in such a way as to oppose its motion; the component of the fluid forces parallel to the axis of motion of a body. *Drag* is the preferred term in aerodynamics and for submerged hydrodynamic bodies, while *resistance* is generally used in ship hydrodynamics. The various forms of drag are defined in relation to resistance. See also *Resistance.*

Drag coefficient (C_D). The nondimensional ratio of the drag per unit of a representative area of a body to the dynamic pressure far ahead of the body.

Efficiency, gearing (η_G). The ratio of the power output to the power input of a set of reduction or multiplying gears between an engine and a propulsion device.

Efficiency, hull (η_H). The ratio between the useful work done on the ship and the work done by the propeller or other propulsion devices in a given time that is effective power P_E and thrust power P_T respectively.

$$\eta_H = \frac{P_E}{P_T} = \frac{R_T V}{T V_A} = \frac{1 - t}{1 - w} \text{ in Taylor notation}$$

or

$$\eta_H = (1 + w_F)(1 - t) \text{ in Froude notation}$$

where R_T is the total resistance, V the ship speed, T the propeller thrust, and V_A the speed of advance; t is the thrust deduction fraction; w and w_F are the wake fractions according to Taylor and Froude respectively.

Efficiency, propeller, behind hull (η_B). The ratio between the power P_T developed by the thrust of the propeller and the power P_D absorbed by the propeller when operating behind a model or ship:

$$\eta_B = \frac{P_T}{P_D} = \frac{T V_A}{2\pi Q_o n} = \eta_O \times \eta_H \times \eta_R$$

where T is thrust, V_A speed of advance, Q_o shaft torque, and n rate of propeller rotation; η_O and η_R are the open water propeller and relative rotative efficiencies respectively.

Efficiency, propeller, open water (η_O). The ratio between the power developed by the thrust of the propeller P_T, and the power absorbed by the propeller P_D, when operating in open water with uniform inflow velocity V_A:

$$\eta_O = \frac{P_T}{P_D} = \frac{T V_A}{2\pi Q_o n}$$

where T is the thrust, Q_o the torque in open water, and n the rate of propeller rotation.

Efficiency, propulsive, or quasi-propulsive (η_D) The ratio between the useful or effective power P_E and the power delivered to the propeller or the propulsion device P_D.

$$\eta_D = \frac{P_E}{P_D} = \eta_O \times \eta_H \times \eta_R$$

where η_O, η_H, and η_R are the open water propeller, hull, and relative rotative efficiencies respectively.

Efficiency, relative rotative (η_R). The relative rotative efficiency is the ratio of the propeller efficiencies behind the hull and in open water, as already defined.

$$\eta_R = \frac{\eta_B}{\eta_O}$$

Emergence, tip. The vertical distance from the top of the propeller tip circle to the at-rest water surface when the tips are exposed.

Expanded area ratio. The ratio of the expanded area of the blades to the disc area.

Face (of blade). The side of the propeller blade which faces downstream during ahead motion. This side of the blade is also known as the pressure side because the average pressure on the face of the blade is higher than the average pressure on the back of the blade during normal operation. The face corresponds to the lower surface of an airfoil or wing.

Fillet. The transition region (fairing) between the propeller hub and the blades at the blade root.

Flow, separated. The detachment of the main fluid flow from a solid surface due to an adverse longitudinal pressure gradient sometimes caused by a sudden change of the direction or the curvature of the surface. The fluid in the separated flow contains eddies, and may be nearly static or may contain a region of reversed flow.

Flow, turbulent. A flow in which there are rapid and apparently random fluctuations both in the magnitude and in the direction of velocity. The velocity fluctuations may also be described by a random spectrum of vortices of varying size and strength. Turbulent resistance is higher than that in laminar flow at the same Reynolds number because of the high momentum exchange by transverse fluctuations.

Flow, viscous. The flow of a fluid where the flow characteristics include the effects of the shear forces acting on the fluid, and within it.

Froude number (F_n). A dimensionless parameter expressing the condition of dynamical similarity for flow systems influenced by gravity and inertia alone. In particular the Froude number defines the speed at which geometrically similar models and a ship will develop wave systems which are geometrically similar. It is given by:

$$F_n = V / \sqrt{gL}$$

The length term L is usually the length of the ship. Other forms of the Froude number use some other characteristic dimension, such as the cube root of the volume of displacement, the submergence depth, or the depth of water in restricted waterways.

Gap (G). The distance between the chord lines of two adjacent propeller blade sections measured normal (or perpendicular) to the chord. This distance is given by the formula:

$$G = (2\pi r \sin\phi) / Z$$

where r is the radius in question, ϕ is the pitch angle of the chord line at radius r (geometric pitch), and Z is the number of blades.

Generator line. The line formed by the intersection of the pitch helices and the plane containing the shaft axis and the propeller reference line. The distance from the propeller plane to the generator line in the direction of the shaft axis is called the rake. The generator line, the blade reference line, and the propeller reference line each intersects the shaft axis at the same point when extended thereto. Because of ambiguities which can arise in so extending the generator line and blade reference line when nonlinear distributions of rake and skew angle are used, it is recommended that each of these lines be defined as originating at the reference point of the root section (see Figure 29-14). The rake and skew angle of the root section will thus be defined as zero, and the propeller plane will pass through the reference point of the root section.

Hub. The central portion of a screw propeller to which the blades are attached and through which the driving shaft is fitted. Also known as the boss.

Hub diameter (d). The diameter of the hub where it intersects the generator line.

Hub ratio. The ratio of the diameter of the hub to the maximum diameter of the propeller, d/D.

Hydrofoil. A structure externally similar to an airplane wing designed to produce lift and which operates in water.

Hydrofoil section. The cross-sectional shape of a hydrofoil.

Immersion (h). The depth of submergence of the propeller measured vertically from the shaft axis to the free surface.

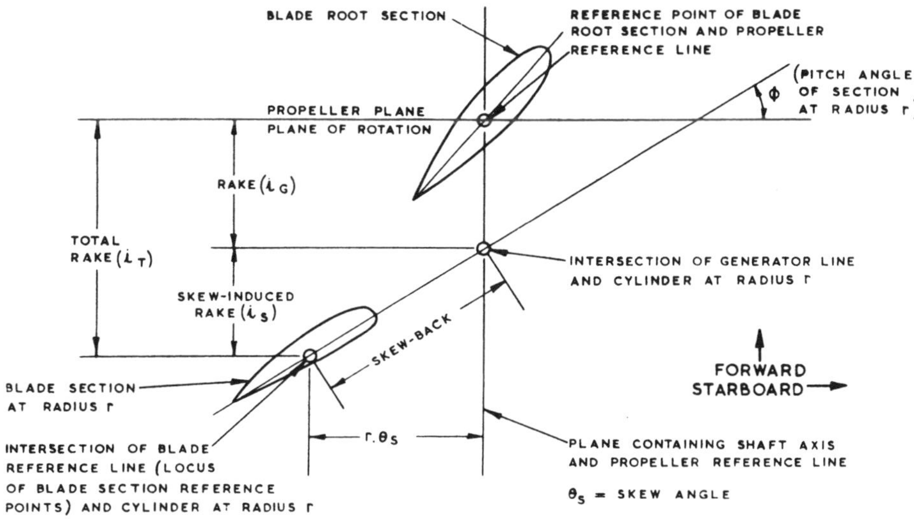

Figure 29-14(a). View of unrolled cylindrical sections at blade root and at any radius r of a right-handed propeller (looking down) showing recommended location of propeller plane

Inboard rotation. A propeller which is not located on the centerline of the ship is said to have inboard rotation if the blades move toward the centerline as they pass the upper vertical position. The opposite direction of rotation is called outboard rotation. Also called inward and outward rotation respectively.

Induced velocity, axial (U_A). The change in the velocity component in the direction parallel to the propeller axis due to the presence of the propeller but not including any change in the wake field due to propeller-hull interactions. Positive upstream.

Induced velocity, radial (U_R). The change in the velocity component in the radial direction due to the presence of the propeller but not including any change in the wake field due to propeller-hull interactions. Positive outward.

Induced velocity, tangential (U_T). The change in the velocity component in the tangential direction due to the presence of the propeller but not including any change in the wake field due to propeller-hull interactions. Positive clockwise looking forward.

Left-handed propeller. A propeller which rotates in the counterclockwise direction when viewed from astern.

Lift (L). The fluid force acting on a body in a direction perpendicular to the motion of the body relative to the fluid.

Mean line. The mean line is the locus of the midpoint between the upper and lower surface of an airfoil or hydrofoil section. The thickness is

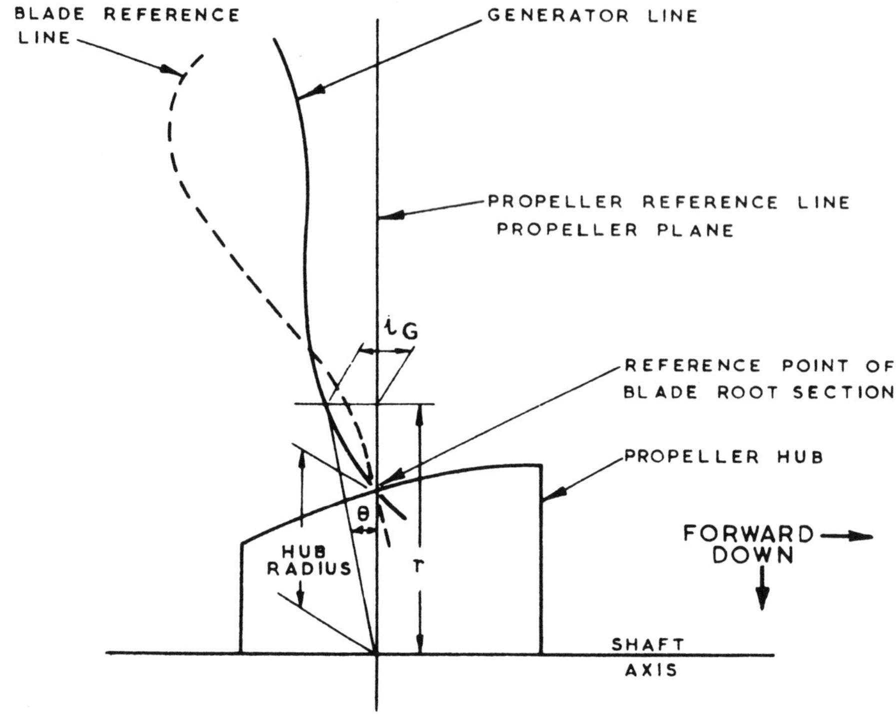

Figure 29-14(b). Diagram showing recommended reference lines (looking to port)

generally measured in the direction normal to the chord rather than normal to the mean line. The maximum distance between the mean line and the chord line, measured normal to the chord line, is called the camber. The term camber line is often used synonymously with mean line.

Mean width ratio. Mean expanded or developed chord of one blade divided by the propeller diameter. Equal to the inverse of one-half the aspect ratio for a wing.

Nose-tail line. Synonymous with chord line.

Nozzle. The duct portion of a ducted propeller. Synonymous with duct or shroud.

Ogival section. A type of an airfoil or hydrofoil section having a straight face, a circular arc or parabolic back, maximum thickness at the midchord, and relatively sharp leading and trailing edges.

Outboard rotation. A propeller not located on the centerline of the ship is said to have outboard rotation if the blades move away from the centerline as they pass the upper vertical position. The opposite direc-

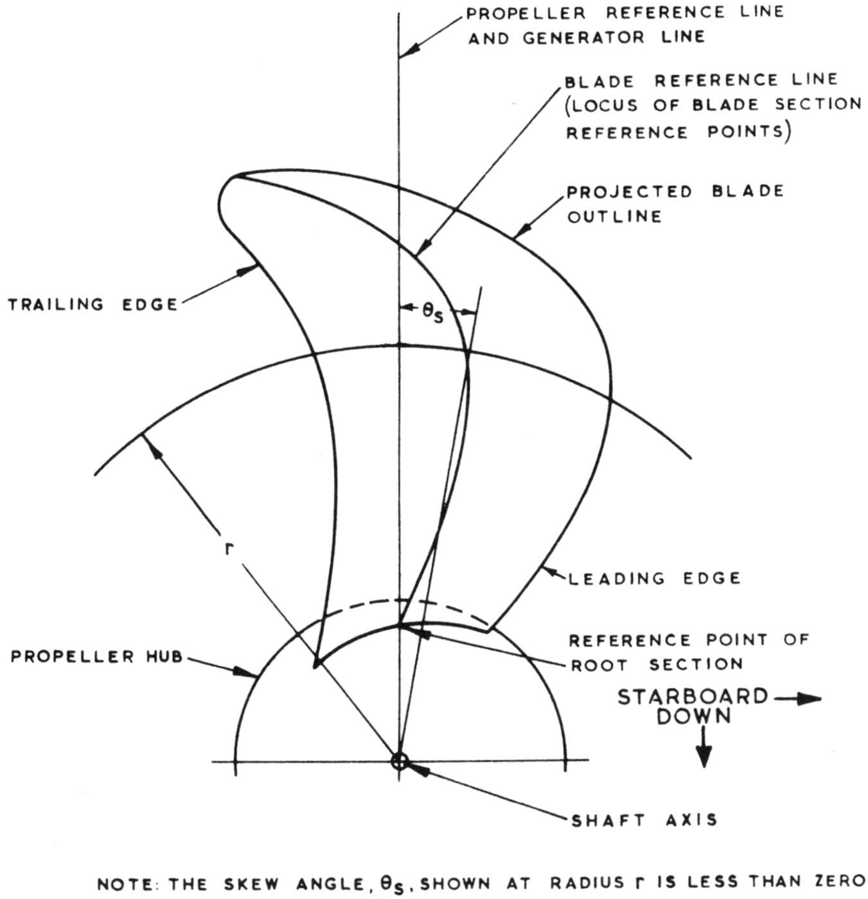

Figure 29-14(c). Diagram showing recommended reference lines (looking forward)

tion of rotation is called inboard rotation. Also called outward and inward rotation respectively.

Pitch (P). The pitch of a propeller blade section at radius r is given by

$$P = 2\pi r \tan\phi$$

where ϕ is the angle between the intersection of the chord line of the section and a plane normal to the propeller axis. This angle is called the pitch angle. Also called geometric pitch. (See *Pitch, geometric.*)

Effective pitch. Weighted value of geometric pitch when pitch is not constant. Both the radius and the thrust distribution (if known) have been used as weighting factors.

Face pitch. The pitch of a line parallel to the face of the blade section. Used only for flat-faced sections where offsets are defined from a face reference line. Also called nominal pitch.

Geometric pitch. The pitch of the nose-tail line (chord line). It is equal to the face pitch if the setback of the leading and trailing edges of the section are equal.

Hydrodynamic pitch. The pitch of the streamlines passing the propeller including the velocities induced by the propeller at a radial line passing through the midchord of the root section. See *Angle, hydrodynamic flow*.

Mean pitch.

(i) Generally synonymous with effective pitch.

(ii) The pitch of a constant pitch propeller which would produce the same thrust as a propeller with radially varying pitch when placed in the same flow.

Pitch analysis. Advance per revolution at zero thrust as determined experimentally.

Pitch angle (ϕ). See *Pitch*.

Pitch ratio. The ratio of the pitch to the diameter of the propeller. Generally, the face pitch or geometric pitch at the 70 percent radius is used to compute the pitch ratio. Any measure of pitch can be used with the diameter to form a pitch ratio.

Variable (or varied) pitch. A propeller blade for which the pitch is not the same at all radii. A propeller which has the same pitch at all radii is said to be a constant pitch propeller.

Power, effective (P_E). The power required to tow a ship, usually without its propulsive device, at constant speed V in unlimited undisturbed water:

$$P_E = R_T V$$

The power may be for the ship either with or without appendages. If the latter, it is usually known as the naked or bare hull, effective power.

Power, shaft (P_S). The power delivered to the shafting system by the propelling machinery.

Power, thrust (P_T). The power developed by the propeller thrust T, at the speed of advance V_A.

Power coefficient, Taylor's (B_p). The horsepower absorbed by the propeller, P, expressed in coefficient form:

$$B_P = n\,(P_D)^{1/2} / (V_A)^{5/2}$$

where n is revolutions per minute and V_A is the speed of advance in knots.

Power delivered (P_D). The power delivered to the propeller:

$$P_D = 2\pi Q n$$

Power loading coefficient (C_P). The power absorbed by the propeller, P_D, expressed in coefficient form:

$$C_P = P_D / [½\rho V_A^3 (\pi D^2 / 4)] = (K_Q / J^3)(8/\pi)$$

where ρ is the fluid density, V_A is the speed of advance, and D is the propeller diameter. This coefficient may be defined in terms of ship speed V and is then denoted by the symbol C_{PS}. K_Q and J_Q are the torque and advance coefficients respectively *(qq.v.)*.

Pressure side. The side of the propeller blade having the greater mean pressure during normal ahead operation. Synonymous with the face of the blade. Analogous to the lower surface of a wing.

Projected area ratio. The ratio of the projected area to the disc area.

Propeller. Most generally, any device which will produce thrust to propel a vehicle. The most common form is the screw propeller, which basically consists of a central hub and a number of fixed blades extending out radially from the hub. Lift is generated by the blades when the propeller is rotated. One component of the lift force produces the desired thrust and the other component creates torque which must be overcome by the engine to sustain rotation.

Propeller plane. The plane normal to the shaft axis and passing through the intersection of the generator line and the shaft axis when the generator line is extended thereto. Also called the plane of rotation (see Figure 29-14). It is recommended that the plane be defined instead to contain the propeller reference line, i.e., contain the reference point of the root section, in order to avoid the ambiguities which can arise when nonlinear distributions of rake and skew are used.

Propeller Types. The basic screw propeller may be described as fixed pitch, subcavitating, open (unducted), and fully submerged. Variations on this basic type are listed below.

Adjustable pitch propeller. A propeller whose blades can be adjusted to different pitch settings when the propeller is stopped.

Contrarotating propeller. Two propellers rotating in opposite directions on coaxial shafts.

Controllable pitch propeller. A propeller having blades which can be rotated about a radial axis so as to change the effective pitch of the blades while the propeller is operating. This allows full power to be absorbed for all loading conditions. If the pitch can be adjusted to the extent that reverse thrust can be achieved without reversing the direction of rotation of the shaft, then the propeller is sometimes called a controllable reversible pitch propeller.

Cycloidal propeller. A propeller consisting of a flat disc set flush with the under surface of the vessel with a number of vertical, rudderlike blades projecting from it. The disc revolves about a central axis, and each of the blades rotates about its own vertical axis. The axis of each blade traces a cycloidal path. The blade motion can be varied so as to produce a net thrust in any desired direction in a plane normal to the axis of rotation. It is used where excellent maneuverability is required.

Ducted propeller. A propeller with a short duct mounted concentrically with the shaft. The duct, or nozzle, is shaped so as to control the expansion or contraction of the slipstream in the immediate vicinity of the propeller. In one form (the Kort nozzle) the flow is accelerated, whereas in the other form (pump jet) the flow is decelerated. A pump jet is sometimes also defined as a ducted propeller with stator vanes regardless of whether the flow is accelerated or decelerated.

Fully cavitating propeller. A propeller designed to operate efficiently at very low cavitation numbers where a fully developed cavity extends at least to the trailing edge of the blade. The blade sections of such propellers have relatively sharp leading edges for more efficient supercavitating operation and thick trailing edges for strength. Also known as a supercavitating propeller.

Interface propeller. A propeller of the fully cavitating ventilated type designed to operate with only a portion of the full disc area immersed. These propellers are considered for high speed applications to vehicles such as surface effect ships where the appendage drag associated with the shafts and struts of a fully submerged propeller would result in a considerable increase in resistance. Also known as partially submerged or surface propellers.

Ring propeller. A propeller with a very short duct attached to the tips of the blades and rotating with the propeller. Also called a banded propeller.

Steerable ducted propeller. A ducted propeller in which the duct can be pivoted about a vertical axis so as to obtain a steering effect.

Supercavitating propeller. See *Fully cavitating propeller.*

Tandem propeller. Two propellers fitted to the same shaft, one behind the other, and rotating as one.

Ventilated propeller. A propeller of the fully cavitating type, but with provision to introduce air into the cavities in order to achieve fully developed, stable cavities at lower speeds than would otherwise be possible.

Vertical axis propeller. Synonymous with cycloidal propeller.

Pump jet. See *Propeller Types, Ducted propeller.*

Race, propeller. The accelerated, turbulent column of water forming the outflow from a screw propeller.

Radius (r). Radius of any point on a propeller.

Rake (i_G). The displacement, i_G, from the propeller plane to the generator line in the direction of the shaft axis. Aft displacement is considered positive rake (see Figure 29-14). The rake at the blade tip or the rake angle is generally used as a measure of the rake. The rake angle is defined as:

$$\theta = \text{rake angle} = \tan^{-1} [i_G(r) / r]$$

where r is the radius.

Rake, skew-induced (i_S). The amount of axial displacement (rake) of a blade section which results when skewback is used (see Figure 29-14). It is the distance, measured in the direction of the shaft axis, between the generator line and the blade reference line and is given by $r\theta_S \tan\phi$, where r is the local radius, θ_S is the local skew angle, and ϕ is the local pitch angle. It is positive when the generator line is forward of the blade reference line.

Rake, total (i_T). The sum of the rake and the skew-induced rake. (See Figure 29-14.)

Reference line, blade. The locus of the reference points of the blade sections (see Figure 29-14). Sometimes used synonymously with generator line.

Reference line, propeller. The straight line, normal to the shaft axis, which passes through the reference point of the root section (see Figure 29-14). It lies in the plane containing the shaft axis and the generator line.

Resistance (R). The fluid force acting on a moving body in such a way as to oppose its motion; the component of the fluid forces acting parallel to the axis of motion of a body. Resistance is the preferred term in ship hydrodynamics, while drag is generally used in aerodynamics and for submerged bodies. Total resistance is denoted by R_T and various (not mutually exclusive) components of resistance are defined below. See also *Drag*.

Appendage resistance (R_{AP}). The increase in resistance relative to that of the naked, or bare, hull resistance, caused by appendages such as bilge keels, rudders, bossings, struts, etc.

Frictional resistance (R_F). The component of resistance obtained by integrating the tangential stresses over the surface of a body in the direction of motion.

Pressure resistance (R_P). The component of resistance obtained by integrating the normal stresses over the surface of a body in the direction of motion.

Resistance coefficient (C_F, C_R, C_S, C_T, C_V, C_W, etc.). The nondimensional ratio of any specific component of resistance per unit area to the dynamic pressure far ahead of the body.

Resistance in waves, mean increase in (R_{AW}). The mean increase in resistance in wind and waves as compared with the still water resistance at the same mean speed.

Residuary resistance (R_R). A quantity obtained by subtracting from the total resistance of a hull a calculated friction resistance obtained by any specific formulation.

Roughness resistance (R_{AR}). The increase in resistance relative to the resistance of a hydraulically smooth hull due to the effect of roughness. The hull roughness may be of different types, such as structural roughness caused by the method of hull construction, waviness of plating, scoops, valve openings, etc.

Spray resistance (R_S). The component of resistance associated with the expenditure of energy in generating spray.

Viscous resistance (R_V). The component of resistance associated with the expenditure of energy in viscous effects.

Wavebreaking resistance (R_{WB}). A resistance component associated with the breakdown of the ship bow wave.

Wavemaking resistance (R_W). The component of resistance associated with the expenditure of energy in generating gravity waves.

Wind resistance (R_{AA}). The fore and aft component of the resistance of the above-water form of a ship due to its motion relative to still air or wind. When there is no natural wind, this is called the still air resistance.

Reynolds number (R_n). A dimensionless parameter expressing the condition of dynamical similarity for flow systems influenced by viscosity and inertia alone. For equal values of Reynolds number and the same orientation to the flow, the specific resistance coefficients of all geometrically similar smooth surfaces are identical as long as the uninfluenced speed fields are similar, and the flow is influenced by viscosity and inertia alone. It is given by:

$$R_n = \frac{VL\rho}{\mu} = \frac{VL}{\nu}$$

The length term L is usually the length of the surface, but the distance from the leading edge of the surface to a specific point, the diameter of a body, or the thickness of the boundary layer are sometimes used as length terms.

Right-handed propeller. A propeller which rotates in the clockwise direction when viewed from astern.

Root. The part of the propeller blade adjacent to the propeller hub.

Rudder, active. A propulsion device installed in the rudder for ship maneuvering at low or zero speed.

Scale effect. The change in any force, moment or pressure coefficients, flow pattern, or the like, due to a change in absolute size between geometrically similar models, bodies, or ships. These variations in performance due to differences in absolute size arise from the inability to satisfy simultaneously all the relevant laws of dynamical similarity (e.g., gravitational, viscous, and surface tension).

Setback. The displacement of the leading edge or trailing edge of a propeller blade section from the face pitch datum line when the section shape is referenced to that line. Also called wash-back or wash-up. It is called wash-down if negative. The setback ratio is the setback divided by the chord length.

Shroud. The duct portion of a ducted propeller concentric with the axis of rotation of the propeller blades. In some cases the duct may be rotated about a vertical axis to provide steering forces. Synonyms: duct, nozzle.

Singing. Intense discrete frequency sound radiated from the propeller due to resonant vibrations of the blades. Generally thought to be due to the shedding of Karman vortices from the trailing edges of the blades at a resonant frequency of the blade vibration.

Skew. Synonymous with skewback but sometimes used (incorrectly) to denote the skew angle.

Skew angle (θ_S). The angular displacement about the shaft axis of the reference point of any blade section relative to the generator line measured in the plane of rotation (see Figure 29-14). It is positive when opposite to the direction of ahead rotation. This angle is the same as the warp. The skew angle at the tip is often used as a measure of the skewback of a propeller.

Skewback. The displacement of any blade section along the pitch helix measured from the generator line to the reference point of the section (see Figure 29-14). Positive skewback is opposite to the direction of ahead motion of the blade section. Also called skew.

Slip ratio, real (S_R). This is defined by the ratio:

$$S_R = \frac{Pn - V_A}{Pn} = 1 - \frac{V_A}{Pn}$$

where P is the nominal, geometrical pitch, or the effective pitch of the propeller (i.e., advance per revolution at zero thrust), V_A is the speed of advance, and n the rate of propeller rotation.

Slipstream. See *Race, propeller*

Span. The distance from tip to tip of a hydrofoil. The distance from root to tip is the semispan.

Speed of advance of a propeller (V_A). Speed of advance of a propeller in open water. When a propeller behind a ship or model is producing the same thrust at the same rate of rotation as in open water, the corresponding speed V_A determined from the open water propeller characteristics is termed the speed of advance of the propeller. This is usually less than the ship speed V. (See also *Wake fraction, effective.*) This is based on thrust identity. There is another corresponding speed based on torque identity.

Spindle axis. The axis about which a controllable pitch propeller blade is rotated to achieve a change in pitch.

Spindle torque (Q_S). The torque acting about the spindle axis of a controllable pitch propeller blade resulting from the hydrodynamic and centrifugal forces exerted on the blade. This torque is positive if it tends to rotate the blade toward a higher positive pitch.

Suction side. The low pressure side of a propeller blade. Synonymous with the back of the propeller blade. Analogous to the upper surface of a wing.

Thickness, maximum (t). The maximum thickness of a propeller blade section, generally measured normally to the chord line.

Thickness ratio. The ratio of the maximum thickness, t, of a hydrofoil section to the chord length, c, of that section.

Thrust (T). The force developed by a screw propeller in the direction of the shaft.

Thrust breakdown. The phenomenon of loss of thrust due to excessive cavitation on a subcavitating-type propeller. The torque absorbed by the propeller is affected similarly and is called torque breakdown. Both the thrust and torque coefficients may increase slightly above non-cavitating values near the initial inception of cavitation. In general, the changes in thrust and torque are such that propeller efficiency is reduced.

Thrust coefficient (K_T). The thrust, T, produced by a propeller expressed in coefficient form:

$$K_T = T/\rho n^2 D^4$$

where ρ is the mass density of the fluid, n is the rate of propeller rotation, and D is the propeller diameter.

Thrust deduction fraction (t). It is logical to view the effect of the propeller behind the hull as causing an increase in resistance. However, it is also common practice to look upon this increase in R_T as a deduction from the thrust T available at the propeller, i.e., to assume that of the total thrust T only R_T is available to overcome resistance. This "loss of thrust," $T - R_T$, expressed as a fraction of the thrust T, is called the thrust deduction fraction, t, where

$$t = \frac{T - R_T}{T}$$

$$\text{or } R = (1 - t)\, T$$

Thruster. A propulsion device for zero or low speed maneuvering of vessels.

Thrust loading coefficient (C_{Th}). The thrust, T, produced by the propeller expressed in coefficient form:

$$C_{Th} = T / [(\rho/2)\, V_A^2\, (\pi D^2/4)] = (K_T/J^2)\,(8/\pi)$$

where ρ is the density of the fluid, V_A is the speed of advance, D is the propeller diameter, and K_T and J are the thrust and advance coefficients respectively *(qq.v.)*. The symbol C_{TS} is used when this coefficient is based on ship speed instead of speed of advance.

Torque (Q). The torque delivered to the propeller aft of all bearings.

Torque breakdown. See *Thrust breakdown.*

Torque coefficient (K_Q). The torque, Q, delivered to the propeller expressed in coefficient form:

$$K_Q = Q / (\rho n^2 D^5)$$

where ρ is the density of the fluid, n is the rate of propeller rotation, and D is the propeller diameter.

Wake. A term used to describe the motion imparted to the water by the passage of the ship's hull. It is considered to be positive if its direction is the same as that of the ship.

Wake, frictional. The component of the wake which results from the frictional action of the water when moving along the solid surface of a body or ship.

Wake, potential. The component of the wake due to the potential flow around a body or ship, with velocity and pressure relationship in accordance with Bernoulli's theorem.

Wake, wave or orbital. The component of the wake set up by the orbital motion in the waves created by a body or ship.

Wake fraction (w, w_F). The difference between the ship speed V and the speed of advance V_A is called the *wake speed* ($V - V_A$). Froude expressed the wake speed at the position of the propeller as a fraction of the speed of advance, calling this ratio the wake fraction w_F, so that

$$w_F = \frac{V - V_A}{V_A} \text{ and } V_A = \frac{V}{1 + w_F}$$

Taylor expressed the wake speed at the position of the propeller as a fraction of the ship speed, so that

$$w = \frac{V - V_A}{V} \text{ and } V_A = V\,(1 - w).$$

Wake fraction, effective (w_Q, w_T). A propeller will develop the same thrust T at the same revolutions per unit time, n, when working behind a hull advancing at speed V and in open water at a speed of advance V_A. The effective wake fraction will then be

$$w_T = \frac{V - V_A}{V}$$

This depends on identity of thrust. A similar wake fraction W_Q can be found for identity of torque Q and revolutions n.

Wake fraction, nominal. Wake fraction calculated from speeds measured at the propeller position by pitot tubes, vane wheels, etc., in the absence of the propeller are called nominal wakes.

Water, standard fresh. Water having zero salinity and a temperature of 15°C (59°F) with

$$\text{density } \rho = 999.00 \text{ kg/m}^3 \text{ (or } 1.9384 \text{ lb s}^2/\text{ft}^4)$$

$$\text{kinematic viscosity } \nu = 1.13902 \times 10^{-6} \text{ m}^2/\text{s (or } 1.22603 \times 10^{-5} \text{ ft}^2/\text{s)}$$

Water, standard salt. Water having 3.5 percent salinity and a temperature of 15°C (59°F) with

$$\text{density } \rho = 1{,}025.87 \text{ kg/m}^3 \text{ (or } 1.9905 \text{ lb s}^2/\text{ft}^4)$$

$$\text{kinematic viscosity } \nu = 1.18831 \times 10^{-6} \text{ m}^2/\text{s (or } 1.27908 \times 10^{-5} \text{ ft}^2/\text{s)}$$

Water jet. A form of propulsion in which water is taken into the hull of the ship by means of ducting, and energy is imparted to the water with a pump. The water is then ejected astern through a nozzle.

Windmilling. The rotation of a propeller caused by flow past the propeller without power being applied to the propeller shaft. This action may take place while the ship is moving under its own momentum, while it is being towed, or while it is being propelled by other means of propulsion.

REVIEW

1. Describe the various components of ship resistance.
2. Define Froude's law for calculating ship resistance from model tests.
3. How does the propeller affect the wake of a ship?
4. Define propulsive efficiency. What part can be measured full scale and what part must be determined from model tests?
5. Define Taylor wake fraction. How is it used in propeller design?
6. What is the maximum efficiency that a propeller can obtain for a thrust loading coefficient of 1.0?
7. What is the optimum diameter for a propeller of 30,000 horsepower operating at 120 RPM with a speed of 15 knots in salt water? Assume a 4-bladed propeller with a blade area ratio of 0.55.
8. Define cavitation. What can be the effects of extensive cavitation?
9. For an RPM of 360, a diameter of 10 feet, a speed of 35 knots, and shaft centerline submergence of 10 feet, is a design of a supercavitating propeller practical?

10. What blade area would be required for the previous propeller to have no thrust breakdown?
11. What is the fatigue strength of practical propeller matierals?
12. What propeller design features are used to minimize vibration?
13. What is propeller singing?
14. How much can marine fouling of a propeller reduce its efficiency?
15. Describe a method for measuring the pitch of a propeller.
16. What are some of the problems associated with repairing propellers made out of certain materials?

REFERENCES

Caster, E. B. 1963. "TMB 2, 3 and 4-Bladed Supercavitating Propeller Series." *David Taylor Model Basin Report 1637,* January.

[Gawn-Burrill Series] Gawn, R. W. L., and L. C. Burrill. 1957. "Effect of Cavitation on the Performance of a Series of 16-inch Model Propellers." *Transactions,* Royal Institution of Naval Architecture, Vol. 99, 690-728.

Hadler, J. B., and H. M. Cheng. 1965. "Analysis of Experimental Wake Data in Way of Propeller Plane of Single and Twin-Screw Ship Models." *Transaction of the Society of Naval Architects and Marine Engineers,* Vol. 73, 287-414.

Taylor, D. W. 1933. "Speed and Power of Ships." Published by the Department of Commerce, Washington, D. C. See also Gertler, M., 1954. "A Reanalysis of the Original Test Data for the Taylor Standard Series." *David Taylor Model Basin Report 806,* March. (Both "Speed and Power of Ships" and "DTMB Report 806" were reprinted by the Society of Naval Architects and Marine Engineers in 1998 as part of the celebration of the Centennial of the David Taylor Model Basin.)

Todd, F. H. 1963. "Series 60, Methodical Experiments with Models of Single-Screw Ships." *David Taylor Model Basin Report 1712,* July.

[Wageningen B-Series] Van Lammeren, S. P. S., J. D. Van Manen, and M. W. C. Oosterveld. 1969. "The Wageningen B-Screw Series." *Transaction of the Society of Naval Architects and Marine Engineers,* Vol. 77, 269-317.

Yeh, H.Y.H.1965."Series 64—Resistance Experiments on High-Speed Displacement Forms." *Marine Technology,* Vol. 2, No. 3.

CHAPTER 30

Machinery Tests and Trials

ROBERT S. BEHR AND EVERETT C. HUNT

INTRODUCTION

A MODERN ship consists of many complex mechanical, electrical, and electronic components and systems. Before delivery to the owner, tests and trials are required to demonstrate the operation and performance of these components and systems. Regulatory agencies and statutory law also require certain tests and trials to assure the safety of personnel and the seaworthiness of the ship. Tests may be performed at the equipment manufacturer's facilities, in the shipyard during construction, or on sea trials. This chapter provides an outline of the types of tests and trials with which the engineering officer should be familiar.

TESTS AT MANUFACTURER'S PLANT

Classification societies, governmental regulations, and contractual requirements require tests of some machinery components and systems at the manufacturer's plant. These tests include material and metallurgical tests, hydrostatic and strength tests, performance tests, equipment inspections, and operational demonstrations. Component tests may be required by the regulations, the shipbuilding contract, or the machinery purchase contract.

Equipment for which manufacturer's operational tests are normally required include

Air compressors
Refrigeration and air-conditioning units
Major control systems
Deck machinery

Diesel engines
Large pumps and fans
Packaged boilers
Distillers
Generators
Oil purifiers
Large motors
Turbines and gears
Inert gas generators

SHIPYARD INSTALLATION TESTING

As equipment is installed and systems are completed, shipboard installation tests are performed to confirm the adequacy of the selected equipment, the system design, and the installation. Specific tests may be required by the regulations or by the shipbuilding contract. Detailed written test procedures are normally prepared by the shipbuilder and reviewed or approved by the owner before testing. Normal practice is for shipyard personnel to perform a pre-test before notifying regulatory and owner's representatives of the official test.

DOCK TRIALS

Trials at dockside are conducted to verify that the newly installed machinery and systems are operable and ready for the more rigorous sea trials. During dock trials, it is desirable to test equipment and controls at as close to rated operating conditions as possible. Testing of the ship's propulsion machinery is, of course, limited by the ability of the mooring arrangement to withstand the thrust developed by the propeller. The ship's lines may be doubled up and tugboats may be utilized during a part load test of the main propulsion equipment. The high minimum operational speed of direct connected diesel engines can cause significant thrust with fixed pitch propellers. During a spin test on steam plants, the reduction gear is uncoupled from the fixed pitch propeller to check high speed operation and overspeed settings.

The dock trial also provides the builder an opportunity to test controls and instrumentation that are essential for a successful sea trial. Again, every effort should be made to simulate full load operating conditions including artificial load generation or simulation. Final adjustments are made to many machinery components during the dock trial. Equipment manufacturer's representatives are often in attendance to complete adjustment and confirm proper operation of their equipment.

SEA TRIALS

Preliminary Builder's Sea Trial

A builder's sea trial is sometimes conducted to provide an opportunity to make final adjustments to machinery and systems to ensure the success of the contract sea trial which must demonstrate conformance to contractual requirements. The builder's trial normally follows a procedure similar to that of the contract sea trial. Because of the high cost of maintaining a ship at sea during trials, some shipbuilders have abandoned the practice of builder's trials for commercial vessels.

Contract Sea Trial

The contractual sea trial is a formal and highly organized series of events. Typical trials last two to four days depending on the extent of testing and demonstrations required. There are many reasons for conducting the sea trial, and there are many parties interested in this trial and its results. The trial can extend to every area of the ship, including hull, machinery, communications, cargo gear, and navigational equipment. The extent and complexity of the trial varies depending on the desires of the owner as expressed in the contract, the shipyard's normal practices, regulatory requirements, the complexity of the vessel, and whether the ship is the first of a class. The first ship of a class normally requires more extensive testing to confirm performance in accordance with the design specifications.

Performance tests are scheduled to demonstrate that the vessel meets the design or contractual speed, power, and propeller RPM. Other tests demonstrate safety equipment and systems for classification societies and regulatory organizations. An airborne noise survey is usually performed to demonstrate contract and regulatory conformance. A vibration survey is sometimes required to confirm equipment operation and to establish a baseline that can be used to identify changes in machinery condition.

The endurance portion of the trial demonstrates the ability of all of the ship's systems to function for a prescribed period of time at maximum speed and power without the failure of the components. Economy tests demonstrate that the design or contractual fuel consumption is achieved. A deep-sea environment is usually required for sea trials. Water depth should exceed five times the mean draft, with even greater depth desirable during speed and power trials.

The sea trial provides operating data for the ship in an as-built condition. These data can be very valuable to the owner for future trend analysis and other baseline comparisons. The use of modern data collection systems makes trend analysis a useful tool for the ship's engineering officer and the owner's technical staff ashore. The sea trial also provides the designer with data to confirm model test and calculation methods and to improve future designs.

Trial Specifications

Requirements for trials are defined in the contract specification. Trials are normally conducted in accordance with the Society of Naval Architects and Marine Engineers (SNAME) T&R Bulletin 3-47, *Guide for Sea Trials.* If conflicts or contradictions exist between the SNAME codes and the contract specifications, the specifications take precedence.

Trial Schedule

The trial schedule lists each event to be completed on the trial, the scheduled time, and often, the individual responsible for conducting the event. The schedule is usually updated periodically during the trial as events are rescheduled. The following is a typical schedule for the contract trials of a slow speed diesel ship. Note that this schedule includes a diesel engine break-in period during which the main propulsion engine speed and power are gradually increased per the requirements of the engine manufacturer for new engines. This break-in period provides an excellent opportunity to adjust and tune the power plant, particularly the control systems.

Time	*Event*
First Day	
0000	Energize gyro and electronic equipment at least eight hours prior to departure.
0400	Sound all tanks. Make preparations for getting underway. (The sea trial director usually has prepared a detailed check-off list for these preparations.)
0600	Test navigation gear.
0800	Depart shipyard and proceed to test area.
0900	Adjust magnetic compass.
1000	Begin adjustment of main propulsion plant and control system. Begin main engine break-in. Increase speed to 108 RPM.
1300	Increase speed to 113 RPM.
1600	Main engine examination.
1700	Continue main engine break-in. Increase speed to 116 RPM.
2000	Increase speed to 119 RPM.
2300	Increase speed to 122 RPM (maximum rated speed).
Second Day	
0200	Main engine examination.
0300	Main engine examination and main engine break-in complete, vessel underway.
0400	Begin unattended engine room demonstration. Conduct distiller test on steam.
1000	Complete control system demonstration.
1030	Begin horsepower meter turning gear zero check. Begin anchor windlass test.

1200	Complete anchor windlass test and turning gear zero check. Increase speed for economy power trials.
1300	Begin economy (80%) power trials and distiller test on jacket water.
1700	Complete economy (80%) power trials. Build up to maximum power.
1730	Begin maximum (100%) power trial.
1900	Begin ahead steering test.
1930	Complete ahead steering test and maximum (100%) power trial.
2000	Quick reversal from full power ahead to full power astern.
2015	Begin astern endurance run.
2030	Begin astern steering test.
2045	Complete astern steering test and astern endurance run.
2115	Begin final horsepower meter turning gear zero test.
2145	Vessel underway.
2400	Moored at shipyard.

Trial Report

The results of trials are documented in a comprehensive report issued by the builder. A typical formal trial report will include the following information:

Hull description
Machinery description
Log of trial events
List of personnel in attendance
Draft, displacement, and ballast data
Anchor windlass test results
Steering gear test results
Machinery test results
Endurance trial results
Economy trial results
Control system tests
Distiller test results
Fuel rate test results
Model basin curves
Standardization data and curves
"Z" maneuvering test results
Crash stop test results
Turning circle results
Supporting data, such as design correction factors

PROPULSION PLANT AND MACHINERY TESTING

Ahead Endurance Test

The ahead endurance test is also referred to as the full or maximum power run. It is normally a four-hour test of the main propulsion and auxiliary machinery at maximum rated power. The ahead endurance trial demonstrates the ability of each power plant component to reach a level of performance consistent with the maximum power of the system and to sustain acceptable operation at that level for a time period of sufficient duration to suggest the capability of the plant to operate indefinitely at the maximum power level. All important ship and power plant parameters are recorded including ship speed, shaft speed, power, fuel flow, temperatures, pressures, and operating conditions. As in all power trials, every effort must be made to prevent fluctuation of the power plant parameters during the test period. Steering is kept to a minimum except that the ahead steering test is sometimes performed at the end of this test. On ships with diesel engines, which are torque critical, care must be taken to avoid overtorque while steering at high power. This situation may require reducing power once the ship has entered the turn.

Astern Endurance Test

The astern endurance test is also referred to as the astern full or maximum power run. The astern endurance is normally a one-half to one hour test of the main propulsion machinery at maximum design astern power. As with the ahead endurance test, the objective is to prove the ability to reach and sustain maximum power. The critical measurement during this period is the torque and propeller speed (PRPM) developed. All significant ship and power plant parameters are recorded. The astern steering test is often conducted during this test. On steam turbine ships, the temperature rise in the ahead turbine elements in the low pressure turbine casing and the differential growth of the turbine casings and rotors are carefully monitored to demonstrate the adequacy of design and installation.

Economy Test

The economy test is also referred to as the normal power run. The economy test demonstrates that the power plant overall fuel rate, corrected for differences between the design and the actual sea trial conditions meets the contractual requirements. The economy test is usually a four- to six-hour run at service power conditions. Critical data measured during this test are the power and the fuel consumption. All important ship and power plant parameters are recorded, including ship speed, shaft speed, power, fuel flow, temperatures, pressures, and operating conditions. Useful data on fuel consumption versus shaft horsepower are generated during this test.

Control System Test

This operational test demonstrates the ability of the central control system to function properly under steady state and transient conditions in all operating modes including maneuvering and rapid emergency load changes.

Manual local, manual remote, and automatic control from all control stations is demonstrated. Transfer between modes and control locations is also demonstrated. A control system test is also required to obtain reduced manning certification by classification societies and often requires extended demonstrations of reduced or unmanned operation.

Diesel Propulsion Trials

Fuel rate guarantees on main propulsion and electrical generator diesel engines are normally based on the engine factory test. If fuel guarantees are based on factory tests, overall fuel rates are still normally measured on sea trials. Since there are multiple fuel consumers on diesel vessels, it is important to carefully define the intent of the trial. Fuel flow to main propulsion, auxiliary generator engines, and oil fired boilers may need to be separately measured. Separate calculations of fuel rates for main engine (lb_m/SHP-hr) and electrical generators (lb_m/KW-hr) can be performed. Overall fuel rate should always be calculated based on the fuel flow to all consumers and the propulsion shaft horsepower generated. Where shaft horsepower meter is not installed, power may be estimated from fuel rack position, indicator cards, or engine analyzer data.

For fuel systems that return fuel upstream of the supply meter, a return meter is required. To properly evaluate the performance of diesel engines, measured fuel rates must be corrected for ambient conditions in accordance with data provided by the engine manufacturer or SNAME publication T&R Bulletin 3-27, *Marine Diesel Power Plant Performance Practices,* or T&R Bulletin 3-49, *Marine Diesel Power Plant Practices.* When comparing the engine fuel rate of different designs, the presence of attached auxiliaries should be taken into account.

An engine break-in period may be required by the manufacturer to ensure seating of new rings. This normally requires operation at increasing levels of power for specific time periods often interrupted for engine inspections between load increases.

Diesel engines on vessels with fixed pitch propellers are normally selected to produce only 80–90% power at full rated shaft speed with normal sea conditions and a clean bottom. This is done to avoid overtorquing the engine as the hull fouls or in adverse sea/wind conditions. Since sea trials are normally conducted with a clean bottom in reasonable sea/wind conditions, it may not be possible to reach maximum rated engine power at maximum rated shaft speed. To resolve this, engine manufacturers normally grant special permission to exceed maximum rated engine RPM for a limited time on sea trials only. Although increasing speed allows the engine to reach maximum rated power, maximum torque and MEP are not normally obtained.

Where torsional vibration calculations indicate that a torsional vibration critical is within the engine operating speed range, torsiograph tests are performed to confirm the location of the barred speed range.

Gas Turbine Propulsion Trials

Because there have been few commercial gas turbine vessels, criteria for sea trials are not well established. Although gas turbine fuel rates are measured during the engine factory test, the performance of combined plants should be proven on the vessel on sea trials.

Since there are multiple fuel consumers on gas turbine vessels, it is important to carefully define the intent of the trial. Fuel flow to main propulsion, auxiliary generator engines, and oil-fired boilers may need to be separately measured. Separate calculations of fuel rates for main engine (lb_m/SHP-hr) and electrical generators (lb_m/KW-hr) can be performed. Overall fuel rate should always be calculated based on the fuel flow to all consumers and the propulsion shaft horsepower generated.

To properly evaluate the performance of gas turbines, measured fuel rates must be corrected for ambient conditions in accordance with data provided by the engine manufacturer or SNAME publication T&R Bulletin 3-28, *Marine Gas Turbine Power Plant Performance Practices.*

Where a shaft horsepower meter is not installed, power may be estimated from engine RPM, internal gas temperatures and pressures, and/or fuel flow.

Steam Propulsion Trials

Since the parts of a steam propulsion plant are only assembled into a propulsion system on board the vessel, guaranteed fuel rate must be proven on the sea trials. Only overall fuel rate is calculated. Special tests sometimes conducted during steam plant trials include a steam rate test to demonstrate the nonextracting guaranteed steam rate for the propulsion turbine and a boiler overload test conducted at the maximum design fuel flow to the boiler to demonstrate the overload capacity of the boiler and its associated auxiliaries. Special instrumentation sometimes included on steam system sea trials includes steam flow and/or condensate flowmeters.

Where a shaft horsepower meter is not installed, power may be estimated from turbine RPM, steam temperature, first stage steam pressure, condenser vacuum, and turbine manufacturer's data.

Auxiliary Machinery Tests

Auxiliary equipment tested on sea trials includes main, auxiliary, and backup steering systems; anchor systems; distilling plant; and all auxiliary machinery supporting the propulsion and electrical generating plants. Where multiple units are installed, all units should be demonstrated during the sea trial.

MANEUVERING, STANDARDIZATION AND MISCELLANEOUS TESTS

In addition to machinery system testing, numerous navigation, communications, maneuvering, and standardization tests; trials; and demonstrations are conducted on sea trials. These include events required by regulatory agencies as well as tests required by the shipbuilding contract.

Navigation and Communications Tests

Ship's navigation systems are used throughout the trial, and formal tests and demonstrations are performed as required by regulations and contract.

Maneuvering Tests

Maneuvering tests always include steering gear tests, quick reversal from ahead to astern and from astern to ahead, and thruster operational tests. They may also include turning circles, "Z" maneuver, initial turning, pull-out, direct or reverse spiral maneuver, thruster maneuvering, and low speed controllability test. These optional tests are often performed only on the first ship of a class.

The shipbuilder must prepare and install maneuvering information required by the U.S. Coast Guard in the pilothouse. This information can be based on the maneuvering trials performed during the sea trial or on information gathered from a similar or earlier vessel in the class.

Standardization Trials

Standardization trials are performed to establish a relationship between ship speed, shaft horsepower, and shaft RPM at trial drafts. This relationship is established on all trials at normal and maximum power. On standardization trials, the relationship is also established at a number of other speeds. Standardization is normally only performed on the first ship of a class.

SPECIAL SHIPBOARD INSTRUMENTATION

Sea trials require accurate instrumentation for the measurement of vital performance characteristics such as fuel flow, shaft horsepower, ship position, and ship speed. On many vessels accurate fuel flow and horsepower instrumentation is installed as permanent ship equipment. Whether to use installed instrumentation (specially calibrated or off the shelf) or to install specially calibrated trial instrumentation should be established in the contract based on the criticality of the trial data and the quality of the installed instruments. The need for radiometric ranges to establish ship position for standardization or maneuvering tests should also be established.

Modern navigational equipment can often provide ship position information that is sufficiently accurate for trial purposes.

Permanently installed instrumentation, if maintained in calibration, can be used during informal post-delivery tests to provide comparisons of current performance to original sea trial performance throughout the life of the ship. Such information is vital for economic decisions on machinery overhaul, hull and propeller cleaning, and power plant optimization. The engineering officer on a properly instrumented ship has the opportunity to play a key role in the control of operating costs for fuel and maintenance.

Shaft Horsepower Measurement

Although horsepower can be estimated from engine and/or power plant data, direct measurement of power is required for all but the most basic sea trials. In addition, permanently installed power meters provide the owner and operator with critical economic and mechanical information for the operation of the vessel.

The measurement of horsepower requires the measurement of both shaft torque and shaft RPM. Torque is obtained by measuring twist of the shaft either by strain gauges or by measuring circumferential displacement over a fixed length of shaft (LVDT meters). Shaft RPM is measured by standard, noncontact shaft speed sensors.

Modern shaft horsepower meters are brushless, requiring no connection between the on-shaft sensors and the off-shaft instrumentation. FM radio transmitters transmit data from the rotating shaft mounted equipment to receivers adjacent to the shaft. Power is supplied to the on-shaft sensors and transmitter by a transformer with a stationary primary external to the shaft and a rotating secondary on the shaft.

A typical shaft horsepower meter is illustrated in Figure 30-1. Meters can normally display shaft torque, shaft RPM, and shaft horsepower. Shaft horsepower is calculated by multiplying RPM by torque. Torque is a function of the twist in the shaft, the length over which twist is measured, the geometry of the shaft, and the shear modulus of elasticity. At one time, shear modulus of elasticity was measured in elaborate test stands for each meter shaft section. It is now generally accepted to be 11.9×10^6 psi for Grade 2 ABS steel shafting.

The shaft horsepower is calculated as follows:

(1) Torque

$$T = \frac{\theta JG}{L}$$

T = line shaft torque, lb-inch

θ = shaft twist, radians

L = length of shaft section over which the torque meter measures the twist, inches

J = polar moment of shaft $(\text{inches})^4$
G = shear modulus of elasticity, 11.9×10^6 psi

(2) Polar moment of inertia $J = \frac{\pi}{32}\left(D_O{}^4 - D_I{}^4\right)$

D_O = outside diameter of shaft, inches
D_I = inside diameter of shaft, inches
π = 3.14159
D_I = 0 for solid shaft

(3) Shaft horsepower $SHP = \frac{Tn}{63{,}000}$

SHP = shaft horsepower
n = shaft speed, RPM

Example:

Consider a 22-inch-diameter shaft section with a torsion meter that measures twist in radians, over a 6-inch section of solid shaft calibrated

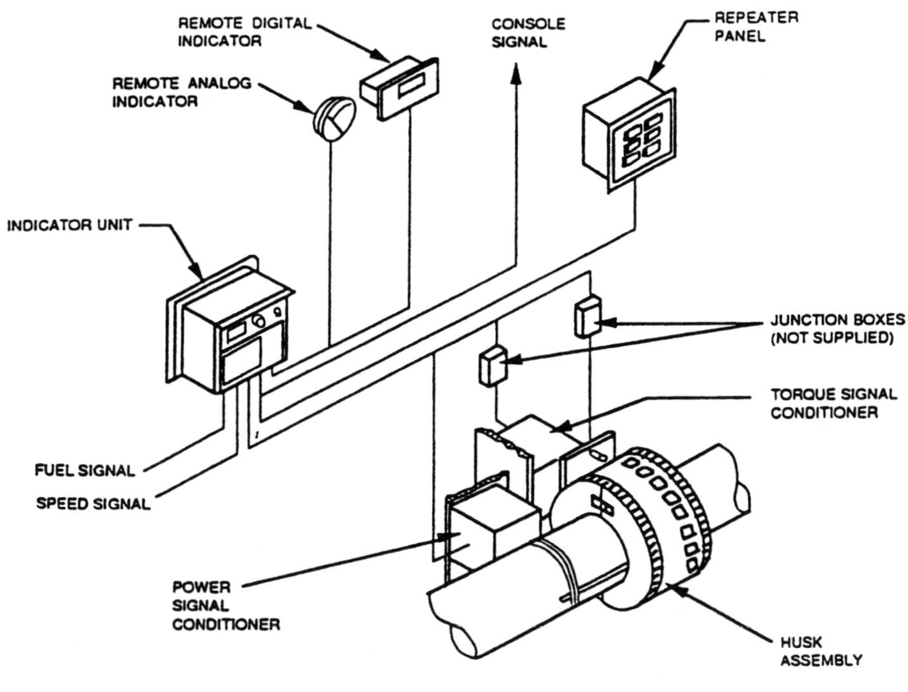

Figure 30-1. McNab shaft horsepower meter

for a modulus of 11.9×10^6 lb/sq inch. At 80 PRPM, the twist in the shaft is measured at 0.00052 radians. What is the horsepower delivered to the propeller?

$$J = \frac{\pi}{32}(22)^4 = 23{,}000\ (\text{inches})^4$$

$$T = \frac{(0.00052)(23{,}000)\left(11.9 \times 10^6\right)}{6}$$

$$T = 23{,}700{,}000 \text{ lb-inches}$$

$$\text{SHP} = \frac{23{,}700{,}000(80)}{63{,}000}$$

$$\text{SHP} = 30{,}100 \text{ SHP}$$

Horsepower Meter Calibration

Since the shaft horsepower meter will be used to verify contractual requirements and, if permanently installed, will be used to monitor and correct power plant performance, it is essential that the metering system be carefully calibrated and that the results of the calibration be recorded. Details of calibration procedures will be found in the manufacturer's instruction manual. In general, calibration of modern meters is limited to electronic checks and a turning gear zero which establishes a zero torque point by rotating the shaft in both directions with the turning gear. Calibration is normally performed at the pier before the trial, immediately before the power runs, and immediately after the power runs.

Fuel Consumption Measurement

Since a fuel rate calculation is a critical outcome of the sea trial, specially calibrated trial fuel meters are commonly used. Since fuel oil is a major factor in the cost of ship operations, it is also desirable to have an accurate and reliable system to determine fuel consumption rates in service. Modern fuel meters offer high accuracy and ruggedness at reasonable costs. Among the multitude of meter types available, the most commonly used include positive displacement, turbine, and mass flowmeters. Positive displacement meters are normally of the nutating disc-type used for many years for both day-to-day and trial service. Turbine flowmeters are used primarily for sea trials or performance tests. The newest type commonly used is the coriolis effect mass flowmeter which can be employed both as a normal installed meter and for sea trials and performance tests.

Coriolis Effect Mass Flowmeters

In recent years, coriolis effect mass flowmeters have gained popularity in the marine field. Mass flowmeters are simple and accurate. Unlike volumetric devices, mass flowmeters do not require correction for fluid density or temperature because they read directly in mass.

The meter has no moving parts in contact with the fluid. The process fluid entering the sensor is split, half passing through each of two U-shaped flow tubes. A drive coil causes the tubes to oscillate up and down in opposition to one another. While one tube moves downward, the other tube moves upward and then vice versa. When fluid is moving through the sensor's tubes, coriolis forces are induced. These forces cause the flow tubes to twist in opposition to each other. See Figure 30-2. When the tube is moving upward during half of its vibration cycle, the fluid flowing into the sensor resists moving upward, by pushing down on the tube. Since the fluid has the tube's upward momentum as it travels around the bend, the fluid flowing out of the sensor resists having its vertical motion decreased by pushing up on the tube. The opposing forces at the inlet and outlet to the tube cause the tube to twist.

Magnet and coil assemblies, called pick-offs, are mounted on the flow tubes. Wire coils are mounted on the side legs of one flow tube, and magnets are mounted on the side legs of the opposing flow tube. Each coil

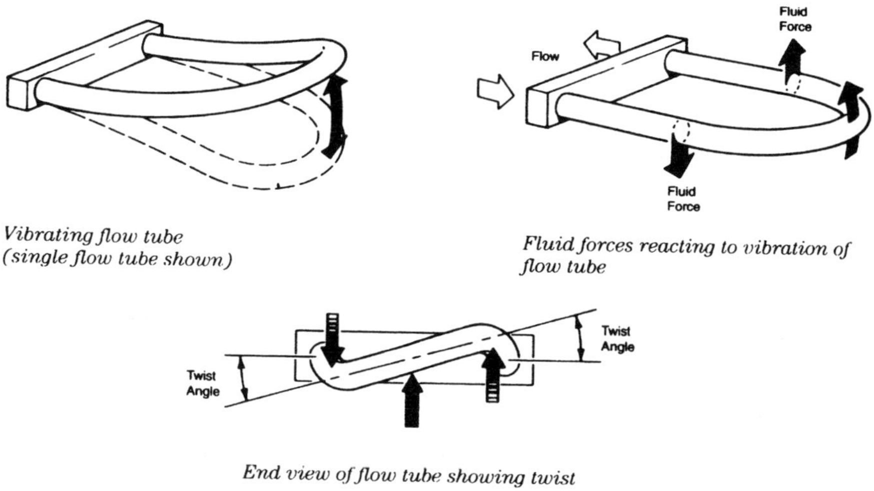

Vibrating flow tube (single flow tube shown)

Fluid forces reacting to vibration of flow tube

End view of flow tube showing twist

Figure 30-2. Micro-Motion coriolis effect mass flowmeter—principle of operation

moves through the uniform magnetic field of the adjacent magnet. The voltage generated from each pick-off coil creates a sine wave. Because the magnets are mounted on one tube, and the coils on the opposing tube, the sine waves generated represent the motion of one tube relative to the other.

Both pick-offs, the one on the inlet side and the one on the outlet side, generate sine wave current continuously when the tubes are oscillating. See Figure 30-3 for an illustration of the Micro-Motion design. When there is no flow, the sine waves are in phase. When there is flow in the tubes, the tubes twist as described above. As a result of the twist in the flow tubes, the sine waves generated by the pick-offs are now out of phase because the inlet side is lagging behind the outlet side. The amount of time difference between the sine waves is measured in microseconds. The time difference is directly proportional to the mass flow rate. The greater the time difference, the greater the flow.

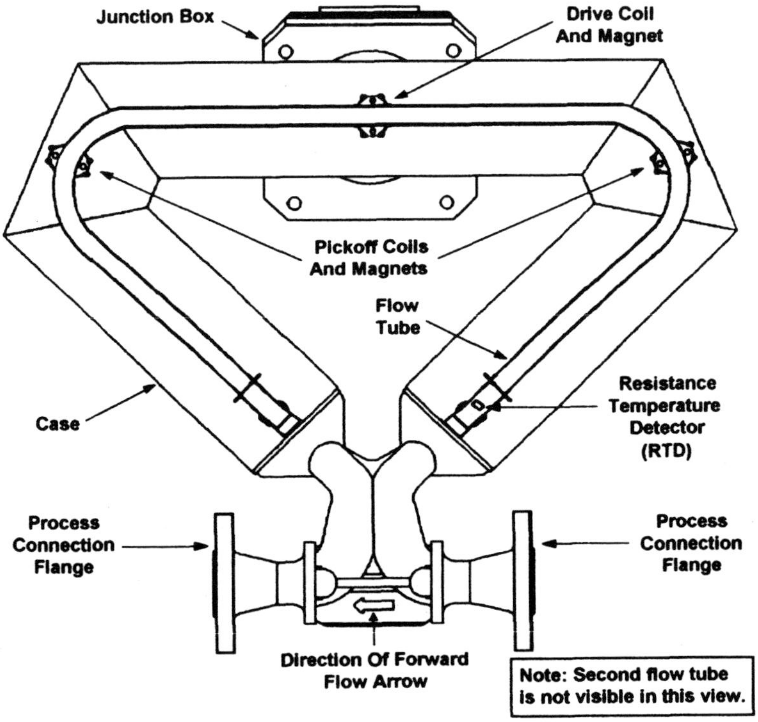

Figure 30-3. Micro-Motion coriolis effect mass flowmeter—arrangement

Turbine Flowmeter

This very accurate meter is simply a miniature turbine wheel suspended in a length of pipe. The flow of fluid through the meter rotates the bearing-supported axial turbine at a speed that is directly proportional to the velocity of the fluid. Since the diameter of the pipe is fixed, the turbine rotational speed is proportional to the volumetric flow rate of the fluid. External electronic devices count the rotations of the turbine rotor and convert the pulses to gallons per minute or any other desired volumetric flow rate. Figure 30-4 illustrates a typical turbine flowmeter manufactured by Flow Technology, Inc. When used with a liquid it is possible to achieve a calibration accuracy of ±0.05 percent.

There is no direct physical connection between the rotor and the housing other than the rotor bearings. Rotor rotation is sensed through the flowmeter by an external pick-off mounted on the surface directly above the flowmeter rotor. The rotation produces a train of electrical pulses in the pick-off. The frequency of these pulses is directly proportional to the volumetric flow rate. In fuel oil measurement, it is essential that the temperature to the meter be monitored to provide an indication of change in specific gravity needed to calculate mass flow of fuel from the meter volumetric flow. The head loss of turbine meters is moderate compared to other types of flowmeters.

Since fuel oils are likely to contain a high proportion of suspended solids, it is desirable to use this meter only for periodic special tests to avoid the loss of calibration because of bearing problems and wear or fouling of surfaces.

Fuel Rate Calculation

Fuel rate is defined as the quantity of fuel consumed to produce a unit of power for a specific length of time. The measured fuel rate is corrected based on a fuel meter correction curve that plots correction factor against flow rate. The basic fuel rate equation requires information on the mass fuel rate and power developed.

$$\text{Fuel rate} = \frac{\text{lb}_m/\text{hr}}{\text{SHP}} = \frac{\text{lb}_m}{\text{SHP-hr}}$$

Example:

Given that a ship operating at 29,925 SHP consumes 18,163 gallons/hr of fuel with a specific gravity of 0.973 at metered temperature and a meter calibration factor of 0.991 (see Figure 30-5), determine the fuel rate.

In this case, volumetric fuel flow rate is provided. Volumetric flow rate must be corrected for specific gravity to determine the mass fuel flow.

Figure 30-4. Flow Technology Turbine flowmeters

$$\text{Fuel flow}\left(\text{gal/hr}\right) = 18{,}163\ \text{gal/hr} \times\ 0.991$$
$$= 18{,}000\ \text{gal/hr}$$
$$\text{Fuel flow}\left(\text{lb}_\text{m}/\text{hr}\right) = \left(18{,}000\ \text{gal/hr}\right)\left(8.337\ \text{lb}_\text{m}/\text{gal}\right)\left(0.973\right)$$
$$\text{Fuel flow}\left(\text{lb}_\text{m}/\text{hr}\right) = 146{,}000\ \text{lb}_\text{m}/\text{hr}$$

where 0.973 is the specific gravity of the fuel.

With fuel flow in terms of mass, fuel rate may be calculated:

$$\text{Fuel rate}\left(\text{lb}_\text{m}/\text{SHP - hr}\right) = \frac{146{,}000\ \text{lb}_\text{m}/\text{hr}}{29{,}925\ \text{SHP}} = 4.88\ \text{lb}_\text{m}/\text{SHP - hr}$$

FUEL OIL METER CALIBRATION CURVE

Meter Serial No. 83027 Meter Model No. 602 Meter Type: Turbine Meter Mfr: Ajax Meter Co.	Calibration Fluid: Diesel Fuel Marine (DFM) Calibration Fluid Temperature: 40 C Calibration Fluid Viscosity: 35.9 SSU Calibration Fluid Density: 6.967 lb/gallon

Remarks:

1. The meter was calibrated gravimetrically in the horizontal position
2. Actual flow rate equals indicated flow rate multiplied by meter faftor

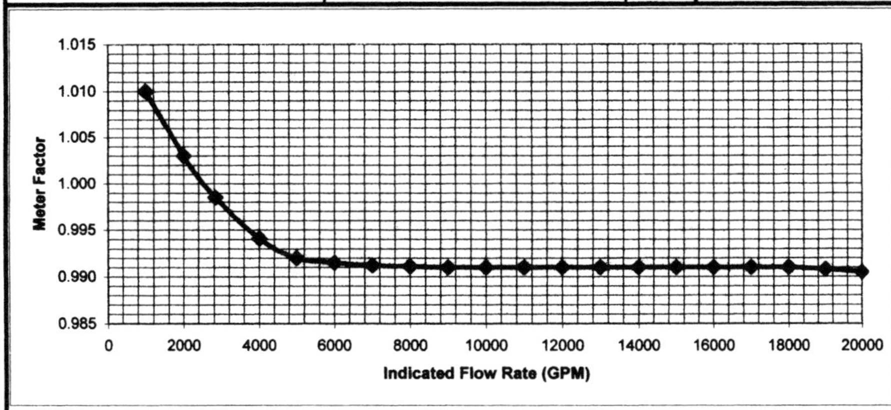

Figure 30-5. Fuel oil meter calibration curve

USE OF SEA TRIAL DATA

Sea trial data are used as a baseline for comparison and troubleshooting throughout the life of the ship. Machinery data provides a record of power plant performance with new equipment presumably tuned for optimum performance. Speed and power data are used for ongoing analysis of the combined performance of the hull, propeller, and propulsion machinery. The corrected sea trial curves of ship speed versus shaft horsepower and propeller RPM are compared to model predictions. Figure 30-6 shows the plots of typical data used in trial analysis.

The data in Figure 30-6 are of interest to the engineer because they provide a performance benchmark. By gathering performance data at drafts and sea conditions similar to the sea trial, he can compare performance with that of the ship in its original condition. Since the new data are not likely to be as accurate as those gathered on trials, multiple points should be plotted on the original trial curves. This comparison will provide a good indication of the current hull and propeller condition.

A more flexible method of evaluating performance data taken after the sea trial utilizes propeller characteristic curves. In analyzing ship trial performance, it is possible to calculate and plot the propeller characteristic curves. One useful technique is to plot the speed coefficient versus the torque coefficient as shown in Figure 30-7.

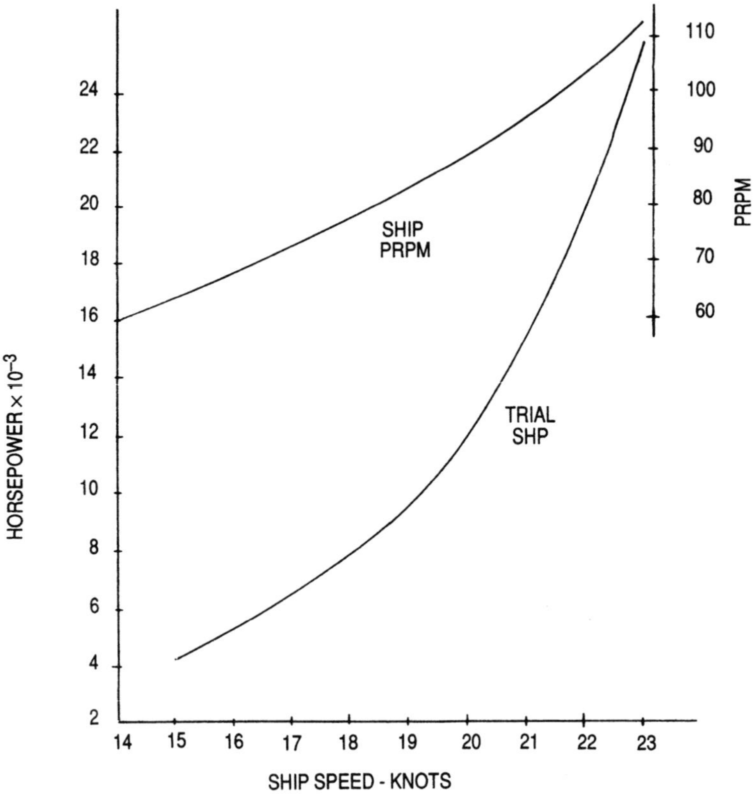

Figure 30-6. Typical trial performance data

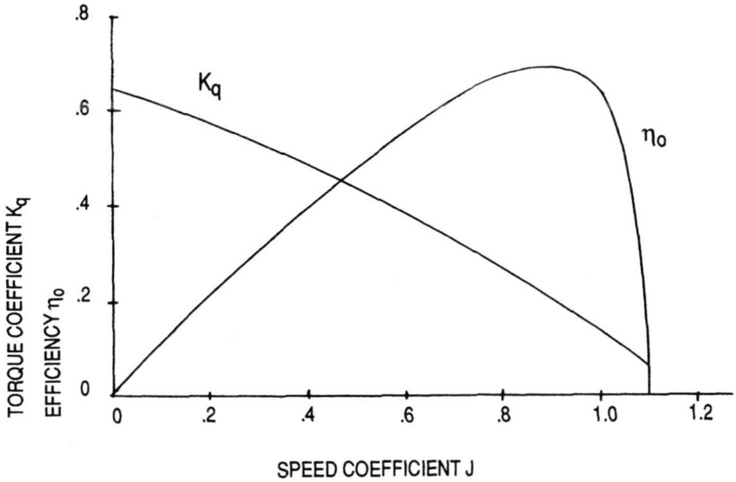

Figure 30-7. Typical propeller coefficient curves

The torque coefficient K_Q is defined as follows:

$$K_Q = \frac{3{,}600\,Q}{\rho n^2 D^5}$$

where: Q = torque in lb-ft
ρ = water density
n = PRPM
D = propeller diameter (ft)

The advance ratio, J, is defined as follows:

$$J = \frac{101.33\,V}{n\,D}$$

where: V = ship speed (knots)

It will be noted in Figure 30-7 that the plot of K_Q versus J closely approaches a straight line for a range of speed coefficient from .5 to 1.0. This fact can be employed to develop a useful benchmark for comparison of the combined hull, propeller, and propulsion machinery with original trial performance. If we assume that ρ and D are constant for a given vessel and that Q times n is proportional to fuel consumption, then:

$$K_Q \approx g\left(f/n^3\right)$$

where f is fuel rate in any consistent units of measurement and n is propeller speed (PRPM).

By similar reasoning:

$$J \approx f\left(V/n\right)$$

where V is ship speed in knots and n is PRPM.

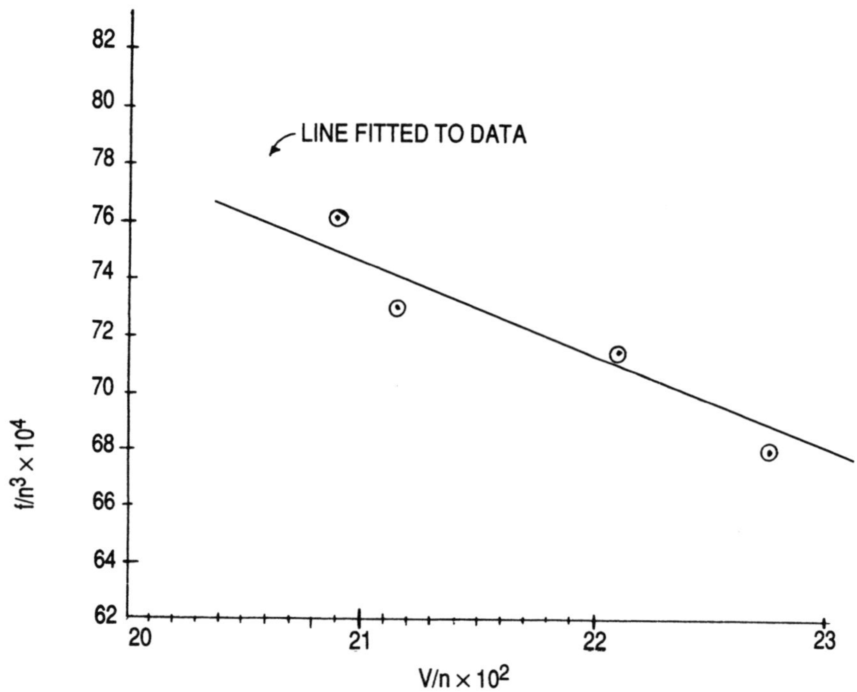

SEA TRIAL DATA

Speed (knots)	PRPM	SHP	Fuel (lbs/hr)	f/n^3	V/n
20	88	11,500	4,634	.0068	.2273
21	95	15,000	6,045	.0071	.2211
22	104	20,000	8,211	.0073	.2115
23	110	25,000	10,075	.0076	.2091

Figure 30-8. V/n versus f/n^3 curve

Using trial data, the ship's officer can fit a straight line to V/n versus f/n as shown in Figure 30-8. With this curve, derived from the original combination of hull, propeller, and propulsion system performance, future data under similar conditions can be plotted to provide a useful measure of deterioration of the ship's performance. Any performance data taken which are significantly different from the data plotted from trial conditions require an investigation to determine a cause for such a deviation.

REVIEW

1. Discuss the purpose of dock trials.
2. At what power level should dock trials be performed?
3. Discuss the purpose of a builder's sea trial.
4. What determines the tests that will be performed on a contract sea trial?
5. Compare the purposes of the ahead endurance test and the economy test.
6. Discuss the measurement of fuel flow on diesel propelled ships. How may the requirements differ from those on a steam propelled ship?
7. On a diesel propelled ship, explain why maximum rated engine speed may have to be exceeded to achieve maximum rated power.
8. What is the purpose of standardization trials?
9. Discuss the advantages of permanently installed, accurate fuel flow and horsepower instruments on a vessel.
10. Describe the principle of operation of a SHP meter.
11. What is the difference between a torque meter and a horsepower meter?
12. Describe two methods used in modern torque and horsepower meters to measure the twist of the shaft.
13. Consider a 20-inch-diameter shaft section with a torsion meter that measures twist in radians over a 4-inch section of solid shaft with a modulus of 11.9×10^6 lb/sq inch. At 80 PRPM, the twist in the shaft is measured at 0.0003 radians. What is the horsepower delivered to the propeller?
14. Consider a hollow shaft, 22-inch outside diameter, and 14-inch inside diameter with a torsion meter that measures twist in radians over a 4-inch section of solid shaft with a modulus of 11.9×10^6 lb/sq inch. At 100 PRPM, the twist in the shaft is measured at 0.0005 radians. What is the horsepower delivered to the propeller?
15. How is the calibration of a horsepower meter checked on sea trials?
16. Describe the principle of operation of a coriolis effect mass flow meter.
17. Describe the principle of operation of a turbine flow meter.
18. Given that a ship operating at 20,450 SHP consumes 13,500 gallons/hr of fuel with a specific gravity of 0.976 at metered temperature

and a meter correction factor of 0.993, determine the fuel rate in lb_m/SHP-hr.

19. Discuss how the engineering officer can use original sea trial data during normal ship operation.
20. Discuss the use of propeller characteristic curves in evaluating ship performance data after sea trial.

REFERENCES

Society of Naval Architects and Marine Engineers. 1989. *Guide for Sea Trials,* T&R Bulletin 3-47. New Jersey: Society of Naval Architects and Marine Engineers,

Society of Naval Architects and Marine Engineers. 1985. *Guide for Shop and Installation Tests,* T&R Bulletin 3-39. New York: Society of Naval Architects and Marine Engineers.

Society of Naval Architects and Marine Engineers. 1976. *Marine Gas Turbine Power Plant Performance Practices,* T&R Bulletin 3-28. New York: Society of Naval Architects and Marine Engineers.

Society of Naval Architects and Marine Engineers. 1975. *Marine Diesel Power Plant Performance Practices,* T&R Bulletin 3-27. New York: Society of Naval Architects and Marine Engineers.

Society of Naval Architects and Marine Engineers. 1990. *Marine Diesel Power Plant Practices,* T&R Bulletin 3-49. New Jersey: Society of Naval Architects and Marine Engineers.

Society of Naval Architects and Marine Engineers. 1973. *Marine Power Plant Heat Balance Practices,* T&R Bulletin 3-11. New York: Society of Naval Architects and Marine Engineers.

Society of Naval Architects and Marine Engineers. 1987. *Guidelines for the Use of Vibration Monitoring for Preventative Maintenance,* T&R Bulletin 3-42. New York: Society of Naval Architects and Marine Engineers.

ACKNOWLEDGMENTS

The editors acknowledge the assistance of McNab, Inc, Mount Vernon, New York, for providing material and information on the shaft horsepower meter, of Flow Technology, Inc., Phoenix, Arizona, for providing material and information on the turbine flowmeter, and of Micro Motion, Inc., Boulder, Colorado, for providing material and information on the coriolis effect flow meter.

APPENDIX

Useful Conversion Factors

To convert from	*To*	*Multiply by*
atm	bar	1.012325
atmospheres	feet of water @ 4°C	3.39E+1
atmospheres	inches Hg @ 0°C	2.992E+1
atmospheres	meters Hg @ 0°C	7.6E–1
atmospheres	Kg/sq cm	1.0333E+00
atmospheres	pounds f/sq in	1.47E+1
bar	atm	0.98692
bar	newton/square meter	1.00E+5
bar	pascal	1.00E+5
barrels (U.S. liquid)	gallons	3.15E+1
barrels (oil)	gallons (oil)	4.2E+1
Btu	foot-pounds	7.7816E+2
Btu	horsepower-hours	3.927E–4
Btu	joules	1.055E+3
Btu	kilowatt-hours	2.928E–4
Btu/hour	horsepower	3.929E–4
Btu/hour	watts	2.931E–1
Btu/minute	horsepower	2.356E–2
Celsius degree	Fahrenheit degree	C × 9/5 + 32
Celsius degree	Kelvin degree	C + 273.18
centimeters	feet	3.281E–2
centimeters	inches	3.937E–1
centimeters	mils	3.937E+2

To convert from	*To*	*Multiply by*
centimeters/second	knots	1.943E–2
centimeters/second	miles/hour	2.237E–2
centimeters/sec/sec	feet/sec/sec	3.281E–2
centipoise	gram/cm-sec	1.0E–2
centipoise	pound/ft-sec	6.72E–4
cubic centimeters	cubic feet	3.531E–5
cubic centimeters	cubic inches	6.102E–2
cubic centimeters	gallons U.S.	2.642E–4
cubic feet	cubic inches	1.728E+3
cubic feet	cubic meters	2.832E–2
cubic feet	cubic yards	3.704E–2
cubic feet	gallons U.S.	7.48052E+0
cubic feet	liters	2.832E+1
cubic inches	cubic feet	5.787E–4
cubic inches	gallons U.S.	4.329E–3
cubic meters	cubic feet	3.531E+1
cubic meters	gallons U.S.	2.642E+2
days	seconds	8.64E+4
days	minutes	1.44E+3
degrees angle	radians	1.745E–2
dynes/square centimeter	inches of Hg	2.953E–5
dynes/square centimeter	inches of H_2O @ 4°C	4.015E–4
dynes	poundals	7.233E–5
ergs	Btu	9.486E–11
ergs	foot-pounds	7.376E–8
ergs	horsepower-hrs	3.725E–14
ergs/second	kilowatts	1.0E–10
faradays	ampere-hour	2.68E+1
faradays	coulombs	9.649E+4
fathoms	meters	1.8288E+0
fathoms	feet	6.0E+0
feet	kilometers	3.048E–4
feet	meters	3.048E–1
feet	nautical miles	1.645E–4
feet	statute miles	1.894E–4
feet of water	inches of Hg	8.826E–1

To convert from	*To*	*Multiply by*
feet of water	pounds/sq inch	4.335E–1
feet/second	knots	5.921E–1
foot-candle	lumens/sq meters	1.0764E+1
foot-pounds	Btu	1.286E–3
foot-pounds	horsepower-hours	5.05E–7
foot-pounds	joules	1.356E+0
foot-pounds	kilowatt-hours	3.766E–7
foot-pounds/minute	horsepower	3.03E–5
foot-pounds/minute	kilowatts	2.26E–5
gallons	cubic feet	1.337E–1
gallons	cubic meters	3.785E–3
gallons of water, fresh	pounds of water	8.337E+0
gallons/minute	cubic feet/second	2.228E–3
gausses	lines/sq inch	6.452E+0
gilberts	ampere-turns	7.958E–1
grams	avdp ounces	3.527E–2
grams	troy ounces	3.215E–2
horsepower	Btu/minute	4.244E+1
horsepower	foot-pounds/minute	3.3E+4
horsepower metric	horsepower	9.863E–1
horsepower	kilowatts	7.457E–1
horsepower boiler	Btu/hour	3.352E+4
horsepower-hours	Btu	2.547E+3
horsepower-hours	foot-pounds	1.98E+6
horsepower-hours	joules	2.684E+6
horsepower-hours	kilowatt-hours	7.457E–1
inches	centimeters	2.54E+0
inches of Hg	feet of water	1.133E+0
inches of Hg	pounds/sq inch	4.912E–1
inches of H_2O @ 4°C	pounds/sq inch	3.613E–2
joules	Btu	9.486E–4
joules	foot-pounds	7.736E–1
joules	watt-hours	2.778E–4
kilograms	pounds	2.2046E+0
kilograms	long tons	9.842E–4
kilograms	short tons	1.102E–3

To convert from	*To*	*Multiply by*
kilograms	avdp ounces	3.5274E+1
kilo-calories	Btu	3.968E+0
kilo-calories	foot-pounds	3.086E+3
kilo-calories	horsepower-hours	1.558E–3
kilo-calories	joules	4.183E+3
kilo-calories	kilowatt-hours	1.163E–3
kilometers	feet	3.281E+3
kilometers	statute mile	6.214E–1
kilometers	nautical mile	5.396E–1
kilowatts	Btu/minute	5.692E+1
kilowatts	horsepower	1.341E+0
kilowatt-hours	Btu	3.413E+3
kilowatt-hours	foot-pounds	2.655E+6
kilowatt-hours	joules	3.6E+6
knots	feet/hour	6.076E+3
knots	kilometer/hour	1.852E+0
knots	statute miles/hour	1.151E+0
knots	feet/second	1.688E+0
liters	cubic feet	3.531E–2
meters	fathoms	5.4681E–1
meters	feet	3.281E+0
meters	inches	3.937E+1
meters	nautical miles	5.4E–4
metric horsepower	kilowatts	7.355E–1
miles nautical	feet	6.076E+3
miles nautical	meters	1.852E+3
miles/hour	feet/minute	8.8E+1
miles/hour	feet/second	1.467E+0
miles/hour	kilometers/hour	1.6093E+0
miles/hour	knots	8.684E–1
newtons	dynes	1.0E+5
newtons	pounds-force	2.248E–1
pascals	newtons/square meters	1.000
poise	gram/centimeter-second	1.0E+0
poundals	dynes	1.3826E+4
poundals	joules/centimeter	1.383E–3

To convert from	*To*	*Multiply by*
pounds	kilograms	4.536E–1
pounds troy	long tons	3.6735E–4
pounds troy	metric tons	3.7324E–4
pounds troy	short tons	4.1143E–4
radians	degrees	5.7296E+1
slugs	kilograms	1.459E+1
slugs	pounds	3.217E+1
square centimeters	square inches	1.55E–1
square feet	square meters	9.29E–2
square meters	square feet	1.076E+1
tons long	kilograms	1.016E+3
tons long	pounds	2.24E+3
tons long	tons short	1.12E+0
tons metric	kilograms	1.0E+3
tons metric	pounds	2.205E+3
tons short	kilograms	9.0718E+2
tons short	pounds	2.0E+3
tons short	tons long	8.9287E–1
tons short	tons metric	9.078E–1
watts	Btu/hour	3.4129E+0
watts	foot-pounds/minute	4.427E+1
watts	horsepower	1.341E–3
watts	metric horsepower	1.36E–3

Index

US $85.00

ISBN: 978-0-87033-537-2